Contributions to Statistics

AF786118

The series **Contributions to Statistics** features edited and conference volumes in theoretical and applied statistics. Composed of refereed selected contributions, the volumes present the latest developments in all the exciting areas of contemporary statistical research.

Germán Aneiros • Enea G. Bongiorno
Aldo Goia • Marie Hušková
Editors

New Trends in Functional Statistics and Related Fields

Springer

Editors
Germán Aneiros
Department of Mathematics
University of A Coruña
A Coruña, Spain

Aldo Goia
Department of Economics
and Business Studies
Università del Piemonte Orientale
"Amedeo Avogadro"
Novara, Italy

Enea G. Bongiorno
Department of Economics
and Business Studies
Università del Piemonte Orientale
"Amedeo Avogadro"
Novara, Italy

Marie Hušková
Department of Probability
and Mathematical Statistics
Charles University in Prague
Prague, Czech Republic

ISSN 1431-1968 ISSN 2628-8966 (electronic)
Contributions to Statistics
ISBN 978-3-031-92385-2 ISBN 978-3-031-92383-8 (eBook)
https://doi.org/10.1007/978-3-031-92383-8

Mathematics Subject Classification (2020): 60B12, 62G05, 60G15, 62H25, 60G25, 62J05, 62Pxx, 62G10

© The Editor(s) (if applicable) and The Author(s), under exclusive license to Springer Nature Switzerland AG 2025, corrected publication 2025

Chapter 42 is licensed under the terms of the Creative Commons Attribution 4.0 International License (http://creativecommons.org/licenses/by/4.0/). For further details see license information in the chapter.
This work is subject to copyright. All rights are solely and exclusively licensed by the Publisher, whether the whole or part of the material is concerned, specifically the rights of translation, reprinting, reuse of illustrations, recitation, broadcasting, reproduction on microfilms or in any other physical way, and transmission or information storage and retrieval, electronic adaptation, computer software, or by similar or dissimilar methodology now known or hereafter developed.
The use of general descriptive names, registered names, trademarks, service marks, etc. in this publication does not imply, even in the absence of a specific statement, that such names are exempt from the relevant protective laws and regulations and therefore free for general use.
The publisher, the authors and the editors are safe to assume that the advice and information in this book are believed to be true and accurate at the date of publication. Neither the publisher nor the authors or the editors give a warranty, expressed or implied, with respect to the material contained herein or for any errors or omissions that may have been made. The publisher remains neutral with regard to jurisdictional claims in published maps and institutional affiliations.

This Springer imprint is published by the registered company Springer Nature Switzerland AG
The registered company address is: Gewerbestrasse 11, 6330 Cham, Switzerland

If disposing of this product, please recycle the paper.

Preface

Over the years, the International Workshop on Functional and Operatorial Statistics (IWFOS) has evolved alongside the rapid advancements in Functional Data Analysis and related fields. Since its inaugural meeting in 2008 in Toulouse, IWFOS has been held every three years, bringing together researchers from around the world. However, due to the COVID-19 pandemic and the retirement of Professor Philippe Vieu – one of the founders and the main driving force behind the conference – the regular schedule was disrupted. As a result, the most recent editions took place in 2021 (online) and now in 2025.

This volume gathers the works selected for presentation at the 6th edition of IWFOS, held in Novara, Italy, from June 25 to 27, 2025. As in previous editions, the workshop aimed to foster discussions on theoretical advancements, methodological innovations, and applications in functional and operatorial statistics. Continuing the tradition of past meetings, this edition also welcomed contributions from related fields such as high-dimensional statistics, machine learning, and data science.

Diversity has always been a core value of IWFOS, and this is reflected in various aspects of this volume. The contributions cover a broad spectrum of topics, tackling both fundamental theoretical challenges and practical applications. Moreover, this diversity extends to the contributors themselves, encompassing both renowned senior statisticians and promising young researchers from across the globe.

We would like to express our deepest gratitude to the invited speakers who kindly accepted our invitation and significantly enriched the scientific program of IWFOS 2025:

Anne van Delft (Columbia University, NY, USA)
Holger Dette (Ruhr-University Bochum, Germany)
Wenceslao González Manteiga (University of Santiago de Compostela, Spain)
Karel Hron (Palacký University Olomouc, Czech Republic)
Ioannis Kalogridis (University of Glasgow, Belgium)
Alessandra Menafoglio (Politecnico di Milano, Italy)
Pavlo Mozharovskyi (Télécom Paris, France)
Anuj Srivastava (Florida State University, USA)

Shahin Tavakoli (University of Geneva, Switzerland)
Simone Vantini (Politecnico di Milano, Italy)

Their insightful contributions have been invaluable to the success of this event.
Our sincere thanks also go to the members of the Scientific Committee for their expertise and dedication in selecting the invited speakers, reviewing submissions and ensuring the high quality of the workshop's program:

John Aston (University of Cambridge, UK)
Graciela Boente (Universidad de Buenos Aires, Argentina)
Hervé Cardot (Université de Bourgogne, Dijon, France)
Manuel Febrero Bande (University of Santiago de Compostela, Spain)
Sonja Greven (Humboldt-Universität zu Berlin, Germany)
Eduardo García Portugués (Carlos III University of Madrid, Spain)
Daniel Hlubinka (Charles University, Prague, Czech Republic)
Siegfried Hörmann (Graz University of Technology, Austria)
David Kraus (Masaryk University, Brno, Czech Republic)
Dominik Liebl (University of Bonn, Germany)
Sara Lopez-Pintado (Northeastern University, Boston, USA)
Stanislav Nagy (Charles University, Prague, Czech Republic)
Juhyun Park (ENSIIE and LaMME, Évry, France)
Matthew Reimherr (Pennsylvania State University, State College, USA)
Laura Sangalli (Politecnico di Milano, Italy)
Piercesare Secchi (Politecnico di Milano, Italy)
Hanlin Shang (Macquarie University, Australia)
Rosanna Verde (Università della Campania, Caserta, Italy)
Jane-Ling Wang (University of California, Davis, USA)
Yoav Zemel (University of Cambridge, UK)

Nothing would have been possible without the hard work of many people. Our thanks also go to the members of the Organizing Committee, to the secretaries of Università del Piemonte Orientale, and to Veronika Rosteck, Gerlinde Schuster and Francesca Ferrari from the Springer's team.

We also would like to thank all the institutions that have financially supported IWFOS 2025: Università del Piemonte Orientale (UPO) and the International Association for Statistical Computing (IASC). A special thank for the support also goes to University of A Coruña (UDC), the Charles University, the Politecnico di Milano, Université Perpignan, Università degli Studi di Torino, the Italian Statistical Society (SIS) and the Université Franco-Italienne / Università Italo-Francese. Moreover, Springer International Publishing deserves a particular thank for granting the use of EquinOCS platform and publishing of this book.

Our gratitude is also extended to the current Editor-in-Chief Professor Dietrich von Rosen of the Journal of Multivariate Analysis, who kindly accepted our proposal for a special issue on Functional Data Analysis and related topics.

We extend our heartfelt thanks to Professor Philippe Vieu, whose dedication and professionalism in overseeing the scientific organization of all previous IWFOS editions have been instrumental to the workshop's success. His contributions have left a lasting mark on our community, and we are truly grateful for his years of commitment.

Last but not least, we would like to thank all the authors who contributed to this volume, as well as all participants who attended the workshop and engaged in fruitful discussions. Their enthusiasm and research efforts continue to drive the progress of functional and operatorial statistics.

We hope that this collection of papers will serve as a valuable reference for researchers and practitioners, further advancing the field and inspiring future developments.

Novara
June 2025

Germán Aneiros
Enea G. Bongiorno
Aldo Goia
Marie Hušková

Contents

List of Contributors

Stéphane Abide
LJAD, Université Côte d'Azur, France
e-mail: Stephane.ABIDE@univ-cotedazur.fr

Nihan Acar-Denizli
Universitat Politècnica de Catalunya, Barcelona, Spain
e-mail: nihan.acar.denizli@upc.edu

Mohamed Alahiane
National School of Business and Management, Chouaib Doukkali University, LERSEM Laboratory, El Jadida, Morocco
e-mail: alahiane.mohamed@ucd.ac.ma

Germán Aneiros
Department of Mathematics, Research Group MODES, CITIC, University of A Coruña, A Coruña, Spain
e-mail: ganeiros@udc.es

Raffaele Argiento
Università degli Studi di Bergamo, Dipartimento di Scienze Economiche, Bergamo, Italy
e-mail: raffaele.argiento@unibg.it

Eleonora Arnone
Università degli Studi di Torino, Italy
e-mail: eleonora.arnone@unito.it

Natalya Pya Arnqvist
Department of Mathematics and Mathematical Statistics, Umeå University, Umeå, Sweden
e-mail: natalya.pya@umu.se

John Aston
Statistical Laboratory, University of Cambridge, UK
e-mail: jada2@cam.ac.uk

Alexander Aue
Department of Statistics, University of California, Davis, USA
e-mail: aaue@ucdavis.edu

Luigi Augugliaro
Department of Economics, Business and Statistics, University of Palermo, Italy
e-mail: luigi.augugliaro@unipa.it

Patrick Bastian
Ruhr Universität Bochum, Bochum, Germany
e-mail: patrick.bastian@rub.de

Rupsa Basu
Universität zu Köln, Köln, Germany
e-mail: rbasu@uni-koeln.de

Xavier Bay
Mines Saint-Étienne, Univ Clermont Auvergne, CNRS, and Institut Henri Fayol, Saint-Étienne, France
e-mail: bay@emse.fr

Mauro Bernardi
Dipartimento di Scienze Statistiche, Università degli Studi di Padova, Italy
e-mail: mauro.bernardi@unipd.it

Sarah Beutler-Traktovenko
Department of Psychology, Friedrich Schiller University of Jena, Jena, Germany
e-mail: sarah.beutler@uni-jena.de

Sayan Bhadra
Dept. of Statistics, Florida State University, USA
e-mail: sbhadra@fsu.edu

Filip Bočinec
Faculty of Mathematics and Physics, Charles University, Department of Probability and Mathematical Statistics, Prague, Czech Republic
e-mail: bocinec@karlin.mff.cuni.cz

Enea G. Bongiorno
Dipartimento di Studi per l'Economia e l'Impresa, Università del Piemonte Orientale, Italy
e-mail: enea.bongiorno@uniupo.it

Teresa Bortolotti,
MOX - Department of Mathematics, Politecnico di Milano, Milano, Italy
e-mail: teresa.bortolotti@polimi.it

Alain Boudou
IMT, Université Paul Sabatier, Toulouse, France
e-mail: palainboudou@gmail.com

Antonio Canale
University of Padova, Padova, Italy
e-mail: antonio.canale@unipd.it

Robert Cantwell
Statistical Laboratory, University of Cambridge, UK
e-mail: rc936@cam.ac.uk

Christian Capezza
University of Naples Federico II, Naples, Italy
e-mail: christian.capezza@unina.it

Hervé Cardot
Université Bourgogne Europe, Institut de Mathématiques de Bourgogne, UMR UBE-CNRS 5584, Dijon, France
e-mail: herve.cardot@ube.fr

Roberto Casarin
Ca' Foscari University of Venice, Italy
e-mail: r.casarin@unive.it

Cristian Castiglione
Bocconi Institute for Data Science and Analytics, Bocconi University, Italy
e-mail: cristian.castiglione@unibocconi.it

Michele Cavazzutti
Politecnico di Milano, Milan, Italy
e-mail: michele.cavazzutti@polimi.it

Lax Chan
School of Mathematics, University of Birmingham, UK
e-mail: l.chan.1@bham.ac.uk

Yiya Chen
Centre for Linguistics and Leiden Institute for Brain and Cognition, Universiteit Leiden, Netherlands
e-mail: yiya.chen@hum.leidenuniv.nl

Francesca Chiaromonte
Dept. of Statistics, Pennsylvania State University, PA, USA and Institute of Economics and L'EMbeDS, Sant'Anna School of Advanced Studies, Pisa, Italy
e-mail: fxc11@psu.edu

Gerda Claeskens
ORSTAT - Faculty of Economics and Business, KU Leuven, Belgium
e-mail: gerda.claeskens@kuleuven.be

Aldo Clemente
Politecnico di Milano, Italy
e-mail: aldo.clemente@polimi.it

Letizia Clementi
Fondazione IRCCS Istituto Neurologico Carlo Besta, SC Neurologia, Salute Pubblica, Disabilità, Italy
e-mail: letizia.clementi@istituto-besta.it

Bianca Maria Colosimo
Politecnico di Milano, Milano, Italy
e-mail: biancamaria.colosimo@polimi.it

Radu Craiu
University of Toronto, Canada
e-mail: radu.craiu@utoronto.ca

Marzia A. Cremona
Dept. of Operations and Decision System, Université Laval, Québec, Canada and CHU de Québec – Université Laval Research Center, Québec, Canada
e-mail: marzia.cremona@fsa.ulaval.ca

Rosa M. Crujeiras
CITMAga, Universidade de Santiago de Compostela, Spain
e-mail: rosa.crujeiras@usc.es

Adéla Czolková
Department of Mathematical Analysis and Applications of Mathematics, Faculty of Science, Palacký University, Olomouc, Czech Republic
e-mail: adela.czolkova01@upol.cz

Xuanye Dai
University of Florence, Florence, Italy
e-mail: xuanye.dai@unifi.it

Florence d'Alché-Buc
LTCI, Telecom Paris, Institut Polytechnique de Paris, France
e-mail: florence.dalche@telecom-paris.fr

Marco F. De Sanctis
MOX, Dipartimento di Matematica, Politecnico di Milano, Milano, Italy
e-mail: marcofrancesco.desanctis@polimi.i

Luca Dede'
Politecnico di Milano, Milano, Italy
e-mail: luca.dede@polimi.it

Pedro Delicado
Universitat Politècnica de Catalunya, Barcelona, Spain
e-mail: pedro.delicado@upc.edu

Laurent Delsol
Institut Denis Poisson, Université d'Orléans, Orleans, France
e-mail: Laurent.Delsol@univ-orleans.fr

Holger Dette
Ruhr Universität Bochum, Bochum, Germany
e-mail: holger.dette@rub.de

Emanuele Di Angelantonio
Health Data Science Research Centre, Human Technopole, Milan, Italy
e-mail: emanuele.diangelantonio@fht.org

Ilenia Di Battista
MOX, Dipartimento di Matematica, Politecnico di Milano, Milano, Italy
e-mail: ilenia.dibattista@polimi.it

Tonio Di Battista
Dipartimento di Studi Socio-Economici, Gestionali e Statistici, Università G. d'Annunzio Chieti Pescara, Italia
e-mail: tonio.dibattista@unich.it

Luigi Di Carlo
Department of Economics, University G. d'Annunzio of Chieti Pescara, Italy
e-mail: elledicarlo@gmail.com

Simone Di Gregorio
Department of Computer, Control and Management Engineering Antonio Ruberti, Sapienza University of Rome, Italy
e-mail: simone.digregorio@uniroma1.it

Jacopo Di Iorio
Dept. of Quantitative Theory & Methods, Emory University, Atlanta, GA, USA
e-mail: jacopo.di.iorio@emory.edu

Chiara Di Maria
Department of Economics, Business and Statistics, University of Palermo, Italy
e-mail: chiara.dimaria@unipa.it

Daniel Diz-Castro
Department of Statistics, Mathematical Analysis and Optimization, University of Santiago de Compostela, Spain
e-mail: danieldiz.castro@usc.es

Patric Dolmeta
Bocconi University, Department of Decision Sciences, Italy
e-mail: patric.dolmeta@unibocconi.it

Mélanie Dreina
LAMPS, Université de Perpignan, France
e-mail: melanie.dreina@univ-perp.fr

Matteo Farnè
University of Bologna, Bologna, Italy
e-mail: matteo.farne@unibo.it

Manuel Febrero-Bande
Department of Statistics, Mathematical Analysis and Optimization, University of Santiago de Compostela, Spain
e-mail: manuel.febrero@usc.es

Stefania Fensore
Dipartimento di Studi Socio-Economici, Gestionali e Statistici, Università G. d'Annunzio Chieti Pescara, Italia
e-mail: stefania.fensore@unich.it

Nicole Fontana
MOX, Department of Mathematics, Politecnico di Milano and Health Data Science Research Centre, Human Technopole, Milan, Italy
e-mail: nicole.fontana@polimi.it

Davide Forcina
University of Naples Federico II, Naples, Italy
e-mail: davide.forcina@unina.it

Alexander Fottner
Augsburg University, Augsburg, Germany
e-mail: alexander.fottner@uni-a.de

Pedro Galeano
Departamento de Estadística, Universidad Carlos III de Madrid, Spain
e-mail: pedro.galeano@uc3m.es

Antonino Gagliano
Department of Economics, Business and Statistics, University of Palermo, Italy
e-mail: antonino.gagliano02@unipa.it

Eduardo García-Portugués
Department of Statistics, Universidad Carlos III de Madrid, Spain
e-mail: edgarcia@est-econ.uc3m.es

Günter Gauglitz
Eberhard Karls Universität Tübingen, Germany
e-mail: guenter.gauglitz@uni-tuebingen.de

Marc G. Genton
King Abdullah University of Science and Technology, Thuwal, KSA
e-mail: marc.genton@kaust.edu.sa

Matteo Giordano
University of Turin, Department of Economics, Social Studies, Applied Mathematics and Statistics, Italy
e-mail: matteo.giordano@unito.it

Aldo Goia
Dipartimento di Studi per l'Economia e l'Impresa, Università del Piemonte Orientale, Italy
e-mail: aldo.goia@uniupo.it

Wenceslao González-Manteiga
Department of Statistics, Mathematical Analysis and Optimization, University of Santiago de Compostela, Spain
e-mail: wenceslao.gonzalez@usc.es

Nouhaila Goujili
L@bISEN, Vision-AD, ISEN Yncréa Ouest and Kardham Group, Paris, France
e-mail: nouhaila.goujili@isen-ouest.yncrea.fr, ngoujili@kardham.com

Sonja Greven
Humboldt-Universität zu Berlin, Berlin, Germany
e-mail: sonja.greven@hu-berlin.de

Marc Hallin
ECARES and Département de Mathématique, Université libre de Bruxelles, Belgium
e-mail: mhallin@ulb.ac.be

Jana Heckenbergerová
University of Pardubice, Faculty of Economics and Administration, Czech Republic
e-mail: jana.heckenbergerova@upce.cz

Nicolás Hernández
Queen Mary University of London, School of Mathematical Sciences, UK
e-mail: n.hernandez@qmul.ac.uk

Zdeněk Hlávka
Faculty of Mathematics and Physics, Charles University, Department of Probability and Mathematical Statistics, Prague, Czech Republic
e-mail: hlavka@karlin.mff.cuni.cz

Daniel Hlubinka
Faculty of Mathematics and Physics, Charles University, Department of Probability and Mathematical Statistics, Prague, Czech Republic
e-mail: hlubinka@karlin.mff.cuni.cz

Karel Hron
Department of Mathematical Analysis and Applications of Mathematics, Faculty of Science, Palacký University, Olomouc, Czech Republic
e-mail: karel.hron@upol.cz

Šárka Hudecová
Faculty of Mathematics and Physics, Charles University, Department of Probability and Mathematical Statistics, Prague, Czech Republic
e-mail: hudecova@karlin.mff.cuni.cz

Marie Hušková
Faculty of Mathematics and Physics, Charles University, Department of Probability and Mathematical Statistics, Prague, Czech Republic
e-mail: huskova@karlin.mff.cuni.cz

Francesco Iafrate
Department of Mathematics, University of Hamburg, Hamburg, Germany
e-mail: francesco.iafrate@uni-hamburg.de

Francesca Ieva
MOX, Department of Mathematics, Politecnico di Milano and Health Data Science Research Centre, Human Technopole, Milan, Italy
e-mail: francesca.ieva@polimi.it

Rosaria Ignaccolo
Dipartimento di Economia e Statistica "Cognetti de Martiis", Università degli Studi di Torino, Italy
e-mail: rosaria.ignaccolo@unito.it

Luigi Ippoliti
Department of Economics, University G. d'Annunzio of Chieti Pescara, Italy
e-mail: luigi.ippoliti@unich.it

Yuwei Jiang
Department of Mathematics and Mathematical Statistics, Umeå University, Umeå, Sweden
e-mail: yuwei.jiang@umu.se

Maher Jridi
L@bISEN, Vision-AD, ISEN Yncréa Ouest, Carquefou, France
e-mail: maher.jridi@isen-ouest.yncrea.fr

Ioannis Kalogridis
University of Glasgow, UK
e-mail: ioannis.kalogridis@glasgow.ac.uk

Alexander Krauth
Eberhard Karls Universität Tübingen, Germany
e-mail: alexander.krauth@uni-tuebingen.de

Sebastian Kühnert
Fakultät für Mathematik, Ruhr-Universität Bochum, Germany
e-mail: sebastian.kuehnert@ruhr-uni-bochum.de

Theodoros Ladas
King's College London, UK
e-mail: theodoros.ladas@kcl.ac.uk

Alessandro Lanteri
Università degli Studi di Torino, Dipartimento di Economia e Statistica "Cognetti De Martiis", Torino, Italy
e-mail: alessandro.lanteri@unito.it

Salvatore Latora
Department of Economics, Business and Statistics, University of Palermo, Palermo, Italy
e-mail: salvatore.latora@unipa.it

Antonio Lepore
University of Naples Federico II, Naples, Italy
e-mail: antonio.lepore@unina.it

Niels Lundtorp Olsen
Technical University of Denmark, Copenhagen, Denmark
e-mail: nalo@dtu.dk

Hassan Maatouk
LAMPS, Université de Perpignan, Perpignan, France
e-mail: hassan.maatouk@univ-perp.fr

Jitka Machalová
Department of Mathematical Analysis and Applications of Mathematics, Faculty of Science, Palacký University Olomouc, Czech Republic
e-mail: jitka.machalova@upol.cz

Eva-Maria Maier
Humboldt-Universität zu Berlin, Berlin, Germany
e-mail: eva-maria.maier@hu-berlin.de

Terence Kevin Manfoumbi Djonguet
Laboratoire de Probabilités, Statistique et Informatique, Unité de Recherche en Mathématiques et Informatique, Université des Sciences et Techniques de Masuku, Gabon
e-mail: tkmpro95@gmail.com

Valentina Masarotto
Mathematical Institute, Universiteit Leiden, Netherlands
e-mail: v.masarotto@math.leidenuniv.nl

Caterina May
King's College London, UK and Università degli Studi del Piemonte Orientale, Italy
e-mail: caterina.may@kcl.ac.uk, caterina.may@uniupo.it

Alessandra Menafoglio
MOX, Department of Mathematics, Politecnico di Milano, Italy
e-mail: alessandra.menafoglio@polimi.it

Erik Mendroš
Faculty of Mathematics and Physics, Charles University, Department of Probability and Mathematical Statistics, Prague, Czech Republic
e-mail: erikmendros@gmail.com

Alejandra Mercedes Martínez
Departamento de Ciencias Básicas, Universidad Nacional de Luján, and CONICET, Argentina
e-mail: ale_m_martinez@hotmail.com

Silvia Montagna
Università degli Studi di Modena e Reggio Emilia, Dipartimento di Economia "Marco Biagi", Italy
e-mail: silvia.montagna@unimore.it

Pavlo Mozharovskyi
LTCI, Telecom Paris, Institut Polytechnique de Paris, France
e-mail: pavlo.mozharovskyi@telecom-paris.fr

Tomáš Mrkvička
University of South Bohemia, Faculty of Agriculture and Technology, Czech Republic
e-mail: mrkvicka.toma@gmail.com

Kalliopi Mylona
King's College London, UK
e-mail: kalliopi.mylona@kcl.ac.uk

Stanislav Nagy
Faculty of Mathematics and Physics, Charles University, Department of Probability and Mathematical Statistics, Prague, Czech Republic
e-mail: nagy@karlin.mff.cuni.cz

Federica Nicolussi
Politecnico di Milano, Milano, Italy
e-mail: federica.nicolussi@polimi.it

Gilles Nisol
ECARES and Département de Mathématique, Université libre de Bruxelles, Belgium
e-mail: gnisol@gmail.com

Eugenia Nissi
Department of Economics, University G. d'Annunzio of Chieti Pescara, Italy
e-mail: eugenia.nissi@unich.it

Guy Martial Nkiet
Laboratoire de Probabilités, Statistique et Informatique, Unité de Recherche en Mathématiques et Informatique, Université des Sciences et Techniques de Masuku, Gabon
e-mail: guymartial.nkiet@univ-masuku.org

Silvia Novo
Department of Statistics, Instituto Flores de Lemus, UC3M-Santander Big-Data Institute, Universidad Carlos III de Madrid, Spain
e-mail: snovo@est-econ.uc3m.es

Idir Ouassou
National School of Applied Sciences, Cadi Ayyad University, Marrakech, Morocco
e-mail: i.ouassou@uca.ma

Biagio Palumbo
University of Naples Federico II, Naples, Italy
e-mail: biagio.palumbo@unina.it

Alessandro Palummo
MOX, Department of Mathematics, Politecnico di Milano, Milan, Italy
e-mail: alessandro.palummo@polimi.it

Todd C. Pataky
Department of Human Health Sciences, Kyoto University Graduate School of Medicine, Kyoto, Japan
e-mail: pataky.todd.2m@kyoto-u.ac.jp

Giulia Patanè
Politecnico di Milano, Milano, Italy
e-mail: giulia.patane@polimi.it

Caroline Peltier
Inrae, Centre des Sciences du Goût et de l'Alimentation, UMR CNRS-INRAE-Institut Agro, Dijon, France and Probe Research Infrastructure, Chemosens facility, CNRS-INRAE, Dijon, France
e-mail: caroline.peltier@inrae.fr

Davide Pigoli
King's College London, UK
e-mail: davide.pigoli@kcl.ac.uk

Alessia Pini
Department of Statistical Sciences, Università Cattolica Del Sacro Cuore, Milano, Italy
e-mail: alessia.pini@unicatt.it

Carlo Andrea Pivato
Department of Biomedical Sciences, Humanitas University, Pieve Emanuele, Milan, Italy
e-mail: carlo.pivato@humanitas.it

Charl Pretorius
Centre for Business Mathematics and Informatics, North-West University, Potchefstroom, South Africa
e-mail: charl.pretorius@nwu.ac.za

Nicola Pronello
Dipartimento di Economia e Statistica “Cognetti de Martiis", Università degli Studi di Torino, Italy
e-mail: nicola.pronello@unito.it

Mustapha Rachdi
Grenoble Alpes University (UGA), AGEIS laboratory, Grenoble, France
e-mail: mustapha.rachdi@univ.grenoble-alpes.fr

Hedvika Ranošová
Faculty of Mathematics and Physics, Charles University, Department of Probability and Mathematical Statistics, Prague, Czech Republic
e-mail: ranosova@karlin.mff.cuni.cz

Mohammad Reza Seydi
Department of Statistics, Umeå School of Business and Economics, Umeå University, Sweden
e-mail: mohammad.seydi@umu.se

Gregory Rice
Department of Statistics and Actuarial Science, University of Waterloo, Canada
e-mail: grice@uwaterloo.ca

María D. Ruiz–Medina
Universidad de Granada, Spain
e-mail: mruiz@ugr.es

Didier Rullière
Mines Saint-Étienne, Univ Clermont Auvergne, CNRS, and Institut Henri Fayol, Saint-Étienne, France
e-mail: drulliere@emse.fr

Laura M. Sangalli
Politecnico di Milano, Milan, Italy
e-mail: laura.sangalli@polimi.it

Matthieu Saumard
L@bISEN, Vision-AD, ISEN Yncréa Ouest, Brest, France
e-mail: matthieu.saumard@isen-ouest.yncrea.fr

Lina Schelin
Department of Statistics, Umeå School of Business and Economics, Umeå University, Sweden
e-mail: lina.schelin@umu.se

Piercesare Secchi
MOX, Department of Mathematics, Politecnico di Milano, Milan, Italy
e-mail: piercesare.secchi@polimi.it

Diego Serrano
Department of Statistics, Universidad Carlos III de Madrid, Spain
e-mail: dieserra@est-econ.uc3m.es

Han Lin Shang
Department of Actuarial Studies and Business Analytics, Macquarie University, Australia e-mail: hanlin.shang@mq.edu.au

Stanilav Škorňa
Department of Mathematical Analysis and Applications of Mathematics, Faculty of Science, Olomouc, Czech Republic
e-mail: stanislav.skorna01@upol.cz

Veronika Šmajserová
Department of Mathematical Analysis and Applications of Mathematics, Faculty of Science, Olomouc, Czech Republic
e-mail: veronika.smajserova@upol.cz

Gianluca Sottile
Department of Economics, Business and Statistics, University of Palermo, Italy
e-mail: gianluca.sottile@unipa.it

Anuj Srivastava
Dept. of Statistics, Florida State University, USA
e-mail: anuj@stat.fsu.edu

Marco Stefanucci
University of Rome Tor Vergata, Rome, Italy
e-mail: marco.stefanucci@uniroma2.it

Almond Stöcker
EPFL Lausanne, Lausanne, Switzerland
e-mail: almond.stoecker@epfl.ch

Johan Strandberg
Department of Statistics, Umeå School of Business and Economics, Umeå University, Sweden
e-mail: johan.strandberg@umu.se

Ying Sun
King Abdullah University of Science and Technology, Thuwal, KSA
e-mail: ying.sun@kaust.edu.sa

Shahin Tavakoli
Research Institute for Statistics and Information Science, University of Geneva, Switzerland
e-mail: shahin.tavakoli@unige.ch

Roberta Troilo,
MOX - Department of Mathematics, Politecnico di Milano, Milano, Italy
e-mail: roberta.troilo@polimi.it

Pasquale Valentini
Department of Economics, University G. d'Annunzio of Chieti Pescara, Italy
e-mail: pasquale.valentini@unich.it

Romain Valla
LTCI, Telecom Paris, Institut Polytechnique de Paris, France
e-mail: romain.valla@telecom-paris.fr

Stefan Van Aelst
KU Leuven, Belgium
e-mail: stefan.vanaelst@kuleuven.be

Simone Vantini
Department of Mathematics, Politecnico di Milano, Milano, Italy
e-mail: simone.vantini@polimi.it

Marc Vidal
Department of Data Analysis, Ghent University, Belgium; Max-Planck-Instituts für Kognitions-und Neurowissenschaften
e-mail: marc.vidalbadia@ugent.be

Philippe Vieu
Institut de Mathématiques de Toulouse, Université Paul Sabatier, Toulouse, France
e-mail: vieu@math.univ-toulouse.fr

Sylvie Viguier-Pla
LAMPS, Université de Perpignan and IMT, Université Paul Sabatier, Toulouse, France
e-mail: viguier@univ-perp.fr

Valeria Vitelli
Oslo Centre for Biostatistics and Epidemiology, Department of Biostatistics, University of Oslo, Norway
e-mail: valeria.vitelli@medisin.uio.no

Qing Wang
Ca' Foscari University of Venice, Italy
e-mail: qing.wang@unive.it

Carlo Zaccardi
Department of Economics, University G. d'Annunzio of Chieti Pescara, Italy
e-mail: carlo.zaccardi@unich.it

Chapter 1
An Introduction to the 6th Edition of the International Workshop on Functional and Operatorial Statistics

Germán Aneiros, Enea G. Bongiorno, Aldo Goia and Marie Hušková

Abstract The aim of this introductory chapter is to present the various contributions to the sixth edition of the International Workshop on Functional and Operatorial Statistics (IWFOS 2025) held in June 2025 in Novara, Italy. These contributions are framed into the recent developments on Functional Data Analysis and related fields.

1.1 IWFOS Contribution to Functional Data Analysis

The International Workshop on Functional and Operatorial Statistics (IWFOS) is a European international meeting whose first edition was organized in Toulouse in June 2008 by a group of highly active mathematical statisticians in the city. At that time, the field of statistics dealing with infinite-dimensional problems was still in its early stages of development, and its popularity was beginning to grow, mainly due to the influence of the books [1, 3, 7].

Today, nearly two decades later, the landscape has changed substantially: the number of researchers interested in these topics has increased significantly, as evidenced by the growing number of publications, conferences, dedicated sessions, and special

Germán Aneiros
Department of Mathematics, Research Group MODES, CITIC, University of A Coruña, A Coruña, Spain, e-mail: ganeiros@udc.es

Enea G. Bongiorno (✉)
Dipartimento di Studi per l'Economia e l'Impresa, Università del Piemonte Orientale, Italy
e-mail: enea.bongiorno@uniupo.it

Aldo Goia
Dipartimento di Studi per l'Economia e l'Impresa, Università del Piemonte Orientale, Italy
e-mail: aldo.goia@uniupo.it

Marie Hušková
Faculty of Mathematics and Physics, Charles University, Department of Probability and Mathematical Statistics, Prague, Czech Republic, e-mail: huskova@karlin.mff.cuni.cz

© The Author(s), under exclusive license to Springer Nature Switzerland AG 2025
G. Aneiros et al. (eds.), *New Trends in Functional Statistics and Related Fields*,
Contributions to Statistics, https://doi.org/10.1007/978-3-031-92383-8_1

issues in leading statistical journals. Moreover, a wide range of general books on Functional Data Analysis (FDA) is now available, demonstrating how many statistical concepts traditionally applied in the multivariate context can be effectively adapted to this new type of data. Specifically, the books [4, 6, 9, 10, 11] present various methodological advances in FDA, while [2, 8] focus more on applied aspects, and [5] delves into the mathematical foundations necessary for studying infinite-dimensional statistical problems.

Over the same period, five editions of IWFOS have been held. Each edition has played a crucial role in the growth of Functional Statistics, bringing together statisticians from around the world. These meetings have provided key opportunities to share experiences in FDA and foster collaborations among researchers. While the first two editions (Toulouse, France, 2008 and Santander, Spain, 2011) were focused solely on Infinite-Dimensional Statistics, the third edition (Stresa, Italy, 2014) aimed to promote collaboration with experts in High-Dimensional Statistics, recognizing the clear connections between the two fields. Since the fourth and fifth editions (A Coruña, Spain, 2017 and Brno –online due to the pandemic–, Czech Republic, 2021), the scope has further expanded, encouraging contributions from the field of Machine Learning, with the goal of strengthening links between these diverse areas of modern statistics and driving future developments. This sixth edition continues along this path. Of course, contributions on FDA take a major part in this volume. In the next sections we want to provide a summary through a classification of topics. Although this classification is inherently reductive, it offers a useful overview of the diverse range of issues that the scientific community is currently addressing; in particular, Sect. 1.2 collects contributions devoted to new methodological advances on FDA, while Sect. 1.3 is devoted to applied issues. Contributions concerning related fields are gathered all together in Sect. 1.4.

1.2 Methodological Contribution for Statistics with Functional Data

This volume offers methodological contributions on a variety of features of Statistics for Functional Data, including:

i) Estimation of functional features, such as:

- Functional mean (Ch. **5** and **33**), Covariance Operator (Ch. **9**), Spectral density (Ch. **27**) and Functional quantiles (Ch. **19** and **29**);
- Complexity of functional data (Ch. **8**);
- Time-dependent parameters of distributions (Ch. **2**).

ii) Exploration and pre–processing of functional data, and in particular:

- Anomalies detection (Ch. **64**);
- Depth measures (Ch. **7**, **15**, **48** and **50**);
- Dimension reduction (Ch. **13**, **18**, **24** and **53**);

- Factor model (Ch. **25** and **63**);
- Object-oriented data analysis (Ch. **47**);
- Smoothing (Ch. **17** and **52**).

iii) Methodologies for functional regression and forecasting problems:

- Linear models (Ch. **34**, **54**, **61** and **62**), Bayesian estimation (Ch. **40**), Partial Least Square (Ch. **11**), Tensor regression (Ch. **14**) and Optimal design (Ch. **46**);
- Non-parametric regression (Ch.**16**) and Random forests (Ch. **57**);
- Semi-parametric regression (Ch. **3**, **6** and **51**) and Additive models (Ch. **42** and **44**);
- Graphical models (Ch. **38**);
- Forecasting problems (Ch. **59**) and Motif discovery (Ch. **21**).

iv) Testing procedures, with special attention to:

- Causal inference (Ch. **35**);
- Independence of functional variables (Ch. **43**);
- Local testing (Ch. **39**, **49** and **65**);
- Monitoring: control charts (Ch. **12**) and detecting breaks in panel data (Ch. **32**);
- Sample size estimation (Ch. **58**);
- Testing for comparing different samples (Ch. **31** and **45**);
- Testing regression models: significance of covariates (Ch. **22**), linearity and independence (Ch. **26**);
- Testing time reversibility (Ch. **55**) and Long Range Dependence (Ch. **56**).

v) Unsupervised and Supervised classification (Ch. **37** and **66** respectively).

Some of the chapters investigate some situations which go beyond the pattern of samples of independent variables:

- Functional Time series (Ch. **4**, **10**, **27**, **59** and **63**);
- Point processes analysis (Ch. **23**);
- Spatial statistics (Ch. **47**).

Moreover, nonstandard functional data have been analyzed; in particular:

- Compositional data (Ch. **18**, **30**, **41**, **42**, **59** and **60**);
- Categorical and Ordinal variables (Ch. **13** and **53**);
- Manifolds (Ch. **39** and **56**).

1.3 Applied Functional Data Analysis Contributions

Many contributions propose interesting real data applications in the following areas:

- Building Occupancy (Ch. **28**);
- Climatology (Ch. **41** and **50**);
- Economics (Ch. **8**, **14** and **42**);

- Epidemiology (Ch. **6**);
- Energy (Ch. **48**);
- Environmetrics (Ch. **19**, **31** and **36**);
- Flight delays (Ch. **18**);
- Fluvial networks (Ch. **54**);
- Forensic entomology (Ch. **46**);
- Interferometric data (Ch. **9**);
- Geochemical and Geothermical data (Ch. **15**, **24** and **60**);
- Magnetic resonance (Ch. **64**);
- Medicine (Ch. **2**, **11**, **17**, **27**, **29**, **35**, **51** and **52**);
- Mortality curves (Ch. **59** and **63**);
- Physical activity (Ch. **5**);
- Rosting process (Ch. **12**);
- Sensory perception (Ch. **13**);
- Spectrometric data (Ch. **7**, **61** and **62**);
- Tonal coarticulation (Ch. **45**).

1.4 Contributions on AI

In this edition of IWFOS some contributions present interesting links to the artificial intelligence (AI), in particular with:

- Deep Learning (Ch. **28**);
- Machine Learning (Ch. **20** and **36**).

1.5 Concluding Comments

Before concluding this introductory presentation, it is important to highlight the ongoing support we have received from the Journal of Multivariate Analysis. Once again, we are pleased to align this edition of IWFOS with the launch of a new special issue of the journal. We strongly encourage all participants whose contributions demonstrate significant methodological innovation, as well as any researchers (whether or not they are part of IWFOS 2025) who have introduced new advancements in FDA or related fields, to submit their work to this special issue. It reflects the continuous evolution of the IWFOS series.

References

1. Bosq, D.: Linear processes in function spaces: theory and applications. Springer (2000)

2. Crainiceanu, C. M., Goldsmith, J., Leroux, A., Cui, E.: Functional data analysis with R. CRC Press (2024)
3. Ferraty, F., Vieu, P.: Nonparametric Functional Data Analysis. Springer-Verlag, New York (2006)
4. Horváth, L., Kokoszka, P.: Inference for Functional Data with Applications. Springer, New York (2012)
5. Hsing, T., Eubank, R.: Theoretical Foundations of Functional Data Analysis, with an Introduction to Linear Operators. John Wiley & Sons, Chichester (2015)
6. Kokoszka, P., Reimherr, M.: Introduction to functional data analysis. Chapman and Hall/CRC Press (2017)
7. Ramsay, J., Silverman, B.: Functional Data Analysis. Springer, New York (1997)
8. Ramsay, J., Silverman, B.: Applied functional Data Analysis. Springer, New York (1997)
9. Ramsay, J., Silverman, B.: Functional Data Analysis. (Second edition) Springer, New York (2005)
10. Shi, J.Q., Choi, T.: Gaussian process regression analysis for functional data. CRC Press, Boca Raton, FL (2011)
11. Zhang, J.T.: Analysis of variance for functional data. Monographs on Statistics and Applied Probability, 127. CRC Press, Boca Raton, FL (2014)

Chapter 2
Local Constant Likelihood Estimation for Beta Distribution with Time Varying Parameters

Nihan Acar-Denizli and Pedro Delicado

Abstract The analysis of functional data following a distributional model with time-dependent parameters is getting more common in recent studies. This allows to estimate any time-dependent characteristic, in addition to merely estimating a time-dependent mean. The most standard approach is to fit local exponential family one-parameter models by spline smoothing. In this work, we present an alternative non-parametric smoothing procedure based on local constant likelihood approach which is valid for distributions not in the exponential family, or having more than one parameter depending on time. We apply our proposals to model continuous monitoring glucose curves. First, glucose values are re-scaled to the interval $[0, 1]$, considering the historical minimum and maximum observed values. Then Beta distributions with parameters smoothly depending on t are estimated.

2.1 Introduction

The use of wearable sensors in health sciences is getting more common everyday. One of these wearable sensors are glucometers which give allowance to follow glucose concentrations of patients over a continuous time interval and help to recognize diagnosis of diabetes disease. Recently an increasing number of studies use functional data analysis (fda) methods to analyze continuous glucose monitoring data. Main descriptive fda tools used in the initial studies helped to understand temporal glucose variation and to compare glucose profiles of different groups of patients [6, 7, 10]. Functional principal component analysis (FPCA) is used to analyze important variations of time dependent glucose functions and to cluster individuals [5].

Nihan Acar-Denizli (✉)
Universitat Politècnica de Catalunya, Barcelona, Spain,
e-mail: nihan.acar.denizli@upc.edu

Pedro Delicado
Universitat Politècnica de Catalunya, Barcelona, Spain, e-mail: pedro.delicado@upc.edu

© The Author(s), under exclusive license to Springer Nature Switzerland AG 2025
G. Aneiros et al. (eds.), *New Trends in Functional Statistics and Related Fields*,
Contributions to Statistics, https://doi.org/10.1007/978-3-031-92383-8_2

Functional regression models are used to analyze the effect of different factors, such as sleep quality [4] or nutritional composition of the diet [3], on the blood glucose levels (BGL) of patients, or to understand how the glucose concentrations affect different health related biomarkers [9].

In recent years, the estimation of distributional models with time-dependent parameters has gained relevance due to its capacity to estimate any time-dependent characteristic, in addition to the conventional practice of merely estimating a time-dependent mean. [4] proposed a Beta distribution model with parameters changing over time for each patient, considering their subject-specific minimum and maximum BGL to rescale data. This way, the subject-specific mean and variance functions are obtained from following equations:

$$\mu_i(t) = \mathbb{E}(Y_i(t)) = \frac{\alpha_i(t)}{\alpha_i(t)+\beta_i(t)}, \quad \sigma_i^2(t) = \mathrm{Var}(Y_i(t)) = \frac{\mu_i(t)(1-\mu_i(t))}{\alpha_i(t)+\beta_i(t)+1}.$$

In [4], the method of moments is used to estimate parameter curves and truncated functional principal components basis development is used to smooth the estimated parameter curves.

In the present work, we propose an alternative non-parametric procedure based on local constant likelihood which is valid for distributions having two parameters depending on time (as the Beta distribution). We apply our proposals to model continuous monitoring glucose curves.

2.2 Local Constant Log-likelihood in Two Parameter Models

Without loss of generality, we consider that the time interval is $[0, 1]$. Let $Y_1, \ldots, Y_m$ be independent random variables with probability density (or mass) function $f(y; \delta(t_j), \eta(t_j))$, $j = 1, \ldots, m$, where $0 < t_1 < \cdots < t_m < 1$ are known constants, and $\delta(t)$ and $\eta(t)$ are assumed to be *smooth* functions from $[0, 1]$ to $\mathbb{R}$. Let $y_1, \ldots, y_m$ be the observed values of $Y_1, \ldots, Y_m$ at m times for which a given explanatory variable is assumed to have taken the values $t_1, \ldots, t_m$.

The objective of local likelihood estimation is to estimate the functions $\delta(t)$ and $\eta(t)$, $t \in [0, 1]$, in a non-parametric way. We follow [8], where maximum local polynomial likelihood for the one-dimensional parameter case is developed in detail. We focus on estimating $\delta(t)$ and $\eta(t)$ by maximum local constant likelihood.

Let t_0 be the point in $[0, 1]$ at which we want to estimate the values $\delta(t_0)$ and $\eta(t_0)$. For t in a neighborhood of t_0, consider the continuity approximations

$$\delta(t) \approx \delta(t_0) \equiv a, \quad \eta(t) \approx \eta(t_0) \equiv b.$$

The local constant log-likelihood function around t_0 is defined as

$$\mathcal{L}_{t_0}(a, b) = \sum_{j=1}^{m} w_j(t_0)\ell(y_j; a, b), \tag{2.1}$$

where $\ell = \log f$, and the weights $w_j(t_0)$ are given by a kernel function K and a bandwidth parameter h: $w_j(t_0) = K((t_j - t_0)/h)$.

Let $(\hat{a}, \hat{b})$ be the maximum local constant likelihood estimators of (a, b):

$$(\hat{a}, \hat{b}) = \arg\max_{(a,b)} \mathcal{L}_{t_0}(a, b).$$

Observe that $(\hat{a}, \hat{b})$ depends on t_0. The local constant likelihood estimators of $\delta(t_0)$ and $\eta(t_0)$ are defined as

$$\hat{\delta}(t_0) = \hat{a}, \;\; \hat{\eta}(t_0) = \hat{b}.$$

The local polynomial likelihood estimation was covered in detail in the Chapter 4 of [8] for the case of distribution families with one parameter depending on t in a non-parametric way. In [1] the local polynomial fitting was extended to multi-parameter likelihood models. In particular, [1] proved consistency and asymptotic normality of the local polynomial likelihood estimators under typical likelihood regularity conditions of the multiparameter models (fulfilled by Beta distribution with parameters $\alpha(t)$ and $\beta(t)$, defining $\delta(t) = \log \alpha(t)$ and $\eta(t) = \log \beta(t)$), and smoothness conditions on the parameters as functions of t.

We propose to use the leave-one-out cross-validation (LOO-CV) version of the log-likelihood function for bandwidth choice. For a particular value of the bandwidth h, this function is

$$\mathrm{CV}(h) = \sum_{j=1}^{m} \ell(y_j; \hat{\delta}_{(j)}(t_j), \hat{\eta}_{(j)}(t_j)), \tag{2.2}$$

where $\hat{\delta}_{(j)}(t)$ and $\hat{\eta}_{(j)}(t)$ are the estimated parameter functions when the j-th observation has been left out of the sample and a local constant maximum likelihood estimator has been used with bandwidth h. Then, h is chosen to minimize $\mathrm{CV}(h)$:

$$h_{\mathrm{CV}} = \arg\max_{h} \mathrm{CV}(h).$$

2.3 Case Study: Effect of Quinoa Diet on Glucose Concentrations

In the application we use the follow-up data of 9 patients with prediabetes which is represented in [3]. The data set consists of continuously monitored blood glucose values at 15-minute intervals for 28 days. The patients followed a regular diet for the first 14 days and then followed a quinoa based diet for the next 14 days. The main objective of the analysis is to analyze whether or not the quinoa base diet has a lowering effect on blood glucose levels. A total of 252 curves with measurements at discrete time points were recorded, of which 27 were incomplete. Thus, 225 curves were used in the study. We considered specifically the glucose concentrations $G(t)$ in the morning around breakfast. The selected time period included glucose concentrations from one hour before breakfast to three hours after breakfast $t \in [-60, 180]$ where $t = 0$ indicates the time of the breakfast.

We first transform the glucose concentration curves belonging to each patient to the interval [0, 1] using the global minimum (m) and global maximum (M) glucose levels such that

$$Y_i(t) = \frac{G_i(t) - m}{M - m}.$$

Transformed values of the raw glucose concentrations of three patients for the second day are visualized in Fig. 2.1.

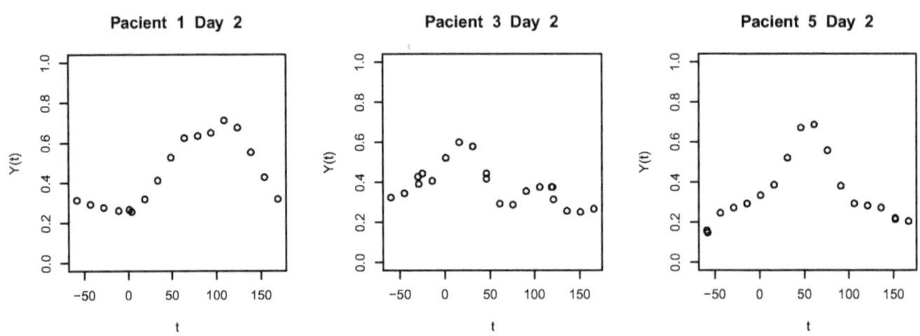

Fig. 2.1: Transformed glucose concentrations of patients one, tree and five from the second day.

These transformed values were then smoothed by using the local constant log-likelihood method where the optimal bandwidth parameter was chosen by LOO-CV for each curve distinctly as seen in Fig. 2.2.

The Beta parameter functions $\hat{\alpha}(t)$, $\hat{\beta}(t)$, $\hat{\delta}(t)$ and $\hat{\eta}(t)$, and finally the mean and variance functions of the Beta distribution were obtained as given in Fig. 2.3. As it is seen, we observed an alignment problem and a high variation in $\hat{\alpha}(t)$ and $\hat{\beta}(t)$ functions due to the differences in observed values.

To solve the alignment problem of the curves, first we registered mean glucose functions by using `fdasrv` function [11] which is based on time-warping functions obtained from square root velocity framework. The results were quite satisfying as shown in Fig. 2.4.

With the aim of understanding the effect of quinoa based diet on BGL of patients, we modeled mean glucose curves on two explanatory variables (diet type and patient effect) by the following function on scalar regression (FoSR) model,

$$Y_{ij}(t) = \beta_0(t) + I_Q^j \beta_Q(t) + \alpha_i(t) + \epsilon_{ij}(t), \tag{2.3}$$

where I_Q^j is the indicator of quinoa diet (0 for regular diet and 1 for quinoa diet) for the j-th day ($j = 1, 2, \ldots, 28$), α_i is the patient effect (for $i = 1, 2, \ldots, 9$) and ϵ_{ij} is a random noise function with expected value of 0 for all t and independent from noise function corresponding to other days or patients. We fitted the model using penalized flexible functional regression (pffr) [2] approach.

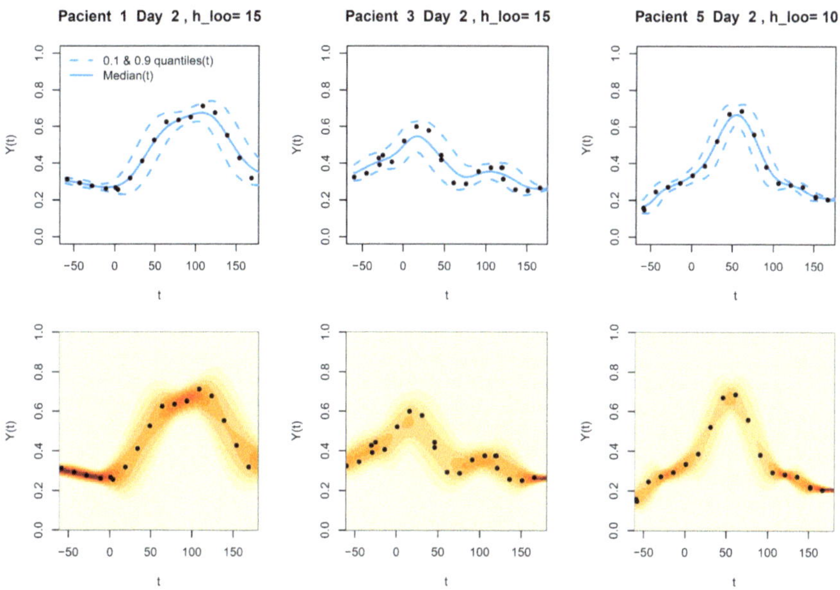

Fig. 2.2: Smoothed glucose curves of patients one, tree and five from the second day. *Up:* Three quantile functions. *Down:* Heatmap plot representing time-dependent Beta distributions.

The results (Fig. 2.5) revealed that the quinoa diet significantly decreases the BGLs. The patients' mean glucose concentration patterns vary from one to another. Some patients have a lower mean glucose level than the global mean at certain time intervals (e.g., see patients 2, 3, 6, and 9 about one hour after starting breakfast), while others have the opposite behavior (e.g., patient 1). Additionally, patients 5 and 8 show a mean glucose pattern similar to the global mean function.

Acknowledgements This research was supported by projects PID2020-116294GB-I00 and PID2023-148158OB-I00 (founded by MICIU/AEI/10.13039/501100011033 and European Union NextGenerationEU/PRTR), by AGAUR under grant 2021 SGR 00613, and by UPC under AGRUPS-2022 and AGRUPS-2023. Nihan Acar-Denizli is a Serra-Húnter fellow.

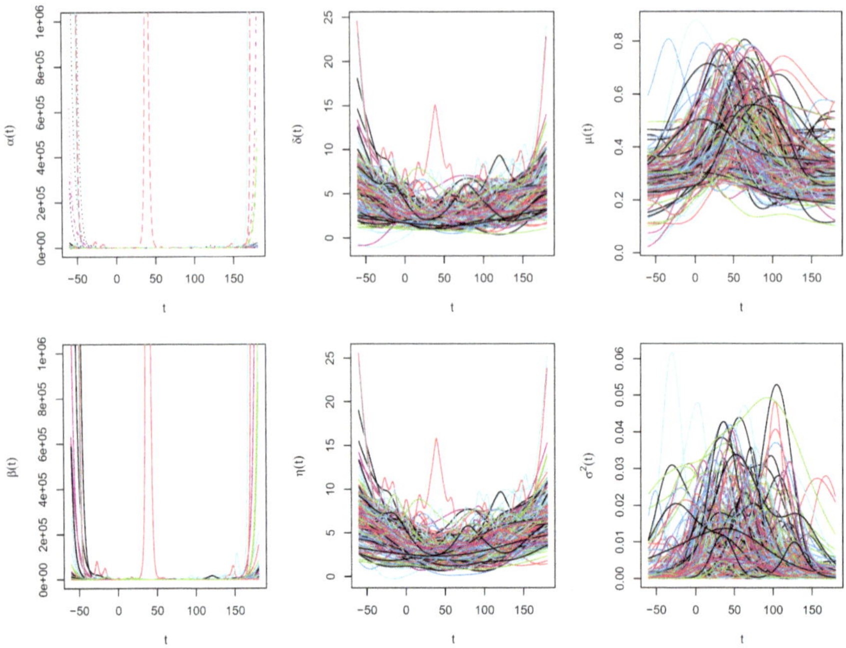

Fig. 2.3: The parameter functions of the Beta distribution.

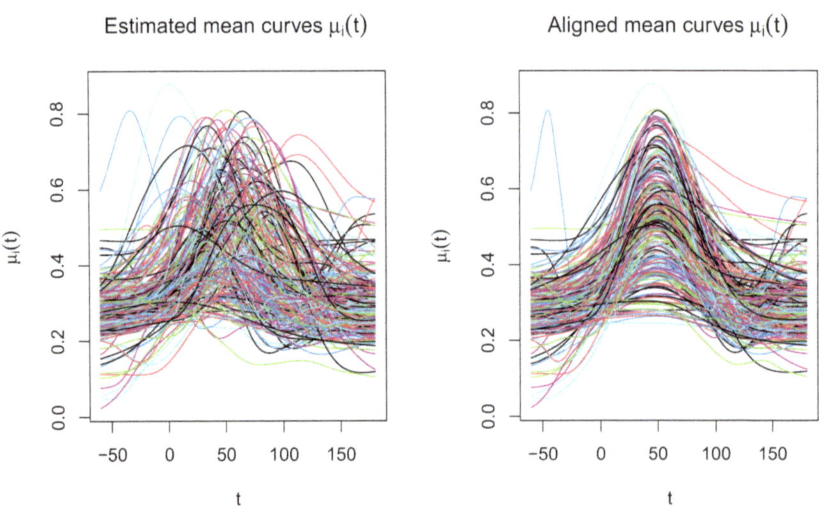

Fig. 2.4: Estimated and aligned mean glucose concentrations.

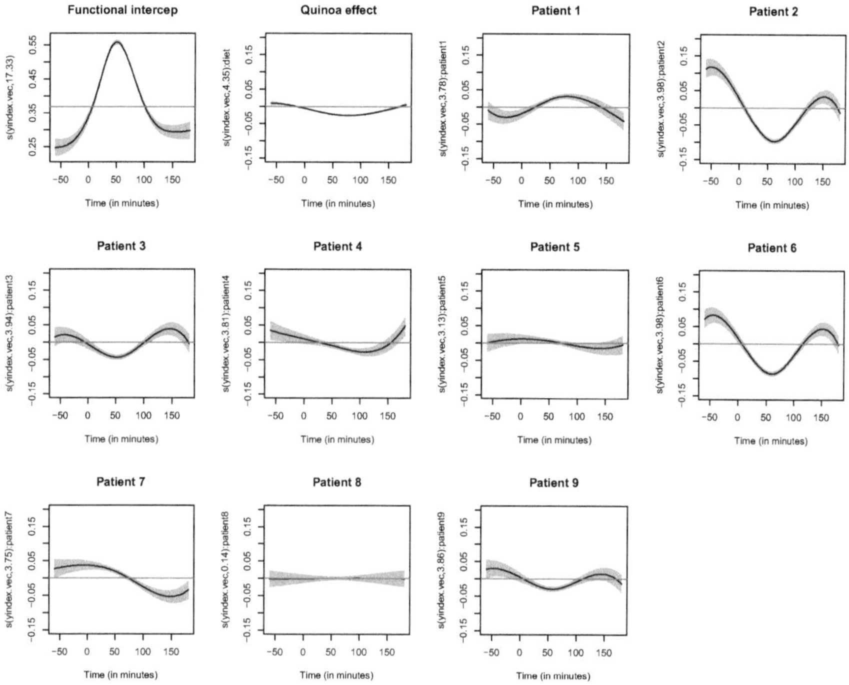

Fig. 2.5: The effects of diet and patient on mean glucose concentrations.

References

1. Aerts, M. and G. Claeskens. Local polynomial estimation in multiparameter likelihood models. Journal of the American Statistical Association. **92**, 1536–1545 (1997).
2. Crainiceanu, C., Reiss, P., Goldsmith, J., Huang, L., Huo, L., and Scheipl, F.: refund: Regression with functional data. r package version 3.0. 1 (2014)
3. Díaz-Rizzolo, D.A., Acar-Denizli, N., Kostov, B., Roura, E., Sisó-Almirall, A., Delicado, P., and Gomis, R.: Glycaemia fluctuations improvement in old-age prediabetic subjects consuming a quinoa-based diet: A pilot study. Nutrients (2022)
4. Gaynanova, I., Punjabi, N., and Crainiceanu, C.: Modeling continuous glucose monitoring (cgm) data during sleep. Biostatistics (Oxford, England), **23**, 223–239 (2022)
5. Gecili, E., Huang, R., Khoury, J. C., King, E., Altaye, M., Bowers, K., and Szczesniak, R. D.: Functional data analysis and prediction tools for continuous glucose-monitoring studies. Journal of clinical and translational science, **5**(1), (2021)
6. Law, G., Alnaji, A., Alrefaii, L., Endersby, D., Cartland, S., Gilbey, S., Jennings, P., Murphy, H., and Scott, E.: Suboptimal nocturnal glucose control is associated with large for gestational age in treated gestational diabetes mellitus. Diabetes Care, **42**(5):810–815, (2019)
7. Law, G., Ellison, G., Secher, A., Damm, P., Mathiesen, E., Temple, R., Murphy, H., and Scott, E.: Analysis of continuous glucose monitoring in pregnant women with diabetes: Distinct temporal patterns of glucose associated with large-for-gestational-age infants. Diabetes Care, **38**(7):1319–1325, (2015)
8. Loader, C.: Local regression and likelihood. Springer, New York (1999)
9. Matabuena, M., Petersen, A., Vidal, J. C., and Gude, F.: Glucodensities: a new representation of glucose profiles using distributional data analysis. Statistical methods in medical research, **30**(6):1445–1464, (2021)
10. Scott, E., Feig, D., Murphy, H., Law, G., and the CONCEPTT Collaborative Group: Continuous glucose monitoring in pregnancy: Importance of analyzing temporal profiles to understand clinical outcomes. Diabetes Care, **43**(6):1178–1184, (2020).
11. Tucker, J. D.: fdasrvf: Elastic Functional Data Analysis. R package version 2.0.3 (2023)

Chapter 3
An Expansion of the Functional Projection Pursuit Regression to Generalized Partially Linear Single Index Models

Mohamed Alahiane, Mustapha Rachdi, Idir Ouassou and Philippe Vieu

Abstract In this manuscript, we introduce a flexible methodology for approximating the systematic component, the single-index vector, and the regression operator in the context of a functional predictor and a scalar response. Following the Fisher-scoring algorithm and the principle of projection pursuit regression, we derive an additive decomposition that exploits the most predictive direction, the most predictive additive component of the functional predictor variable and the single-index component to explain the scalar response. On the one hand, this approach allows us to avoid the well-known problem of the curse of dimensionality in the nonparametric case with the notion of single-index and the projection pursuit regression in the functional case. On the other hand, it can be used as an exploratory tool for the analysis of a multivariate and functional random variable belonging to a separable Hilbert space $\mathcal{H}$. The terms of this decomposition are estimated with an iterative Fisher scoring procedure that uses the Quasi-Likelihood function and an approximation of the nonparametric function by normalized B-splines. The good behavior of our procedure is illustrated from a theoretical and practical points of view. Asymptotic results indicate that the nonparametric function, the single index coefficient and the

Mohamed Alahiane (✉)
National School of Business and Management, Chouaib Doukkali University, LERSEM Laboratory, Corner Avenue Ahmed Chaouki and Rue de Fès, BP. 122 – 24000, El Jadida, Morocco,
e-mail: alahiane.mohamed@ucd.ac.ma

Mustapha Rachdi
Grenoble Alpes University (UGA), AGEIS laboratory, UFR SHS, BP.47, 38040 Grenoble Cedex 09, France, e-mail: mustapha.rachdi@univ.grenoble-alpes.fr

Idir Ouassou
National School of Applied Sciences, Cadi Ayyad University, Marrakech, Morocco,
e-mail: i.ouassou@uca.ma

Philippe Vieu
Institut de Mathématiques de Toulouse, Université Paul Sabatier, 31062 Toulouse Cedex 9, France,
e-mail: vieu@math.univ-toulouse.fr

© The Author(s), under exclusive license to Springer Nature Switzerland AG 2025

G. Aneiros et al. (eds.), *New Trends in Functional Statistics and Related Fields*,
Contributions to Statistics, https://doi.org/10.1007/978-3-031-92383-8_3

terms of the additive decomposition can be estimated without suffering from the curse of dimensionality.

3.1 Introduction

Let $\mathcal{H}$ be the separable Hilbert space endowed with the inner product $\langle f, g\rangle = \int_I f(t)g(t)dt$, for $I \subset \mathbb{R}$, and the induced norm $\|f\|^2 = \langle f, f\rangle$. Let Y be a scalar response variable and $(X, Z) \in \mathbb{R}^d \times \mathcal{H}$ be the predictor vector where $X = (X_1, X_2, \dots, X_d)$ and Z be a functional random variable which is valued in $\mathcal{H}$. For a fixed $(x, z) \in \mathbb{R}^d \times \mathcal{H}$, we assume that the conditional density function of the response Y given $(X, Z) = (x, z)$ belongs to the following canonical exponential family:

$$f_{Y|X=x,Z=z}(y) = \exp(y\xi(x, z) - B(\xi(x, z)) + C(y)),$$

where B and C are two known functions which are defined from $\mathbb{R}$ into $\mathbb{R}$, and $\xi : \mathbb{R}^d \times H \to \mathbb{R}$ is the parameter in the generalized parametric linear model which is linked to the dependent variable through the equality:

$$\mu(x, z) = \mathbb{E}[Y|X = x, Z = z] = B'(\xi(x, z)),$$

where B' denotes the first derivative of the function B.
In what follows we modelize the scalar response Y as a Generalized Functional Projection Pursuit Regression for Partially Linear Single-Index Model by

$$g(\mu(X, Z)) = \eta(\alpha^\top X) + R(Z), \tag{3.1}$$

where $x^\top$ denotes the transpose vector of x, $\alpha = (\alpha_1, \dots, \alpha_d)^\top \in \mathbb{R}^d$ is the $d-$dimensional single-index coefficient vector, η is the unknown single-index link function or the systematic non-linear component which will be assumed to be sufficiently smooth, and R is the regression operator which has to be estimated by:

$$\widehat{R}_n(Z) = \sum_{j=1}^{J+1} g_j(\langle\beta_j, Z\rangle),$$

where $\beta_j \in \mathcal{H}$ such that $\|\beta_j\|^2 = 1$ is the j-th most predictive direction and g_j is the j-th most predictive additive component.

We propose to estimate the unknown single-index vector α, the unknown systematic component $\eta(\cdot)$, the unknown most predictive direction β_j, and the unknown most predictive additive component g_j. Furthermore, we establish their asymptotic distributions based on manuscripts [1], [2], [3], [4], [5], [6],[7], [8], [9], [10], [11] and [12].

3.2 Estimation Methodology

We assume that the function η is supported within the interval $[a, b]$ where $a = \inf(\alpha^\top X)$ and $b = \sup(\alpha^\top X)$.
Let (X_i, Y_i, Z_i) for $i = 1, \dots, n$, be an independent and identically distributed (i.i.d.) n-sample of (X, Y, Z) such that $g(\mu(X_i, Z_i)) = \eta(\alpha^\top X_i) + R(Z_i)$.
Let $\mathcal{S}_n$ be the space of polynomial splines on $[a, b]$ of order $r \geq 1$. For $v \in \mathbb{N}^*$ and $e \in (0, 1]$ such that $p = v + e > 1.5$, we denote by $\mathcal{H}(p)$ the collection of functions g which are defined on $[a, b]$ whose v-th order derivative, $g^{(v)}$, exists and satisfies the following e-th order Lipschitz condition:

$$\left|g^{(v)}(m') - g^{(v)}(m)\right| \leq C\,|m' - m|^e, \quad \text{for all } \; a \leq m, m' \leq b.$$

Let $B_j(u)$ for $j = 1, \dots, N_n$ be the B-spline basis functions of order r, and h the distance between the neighbors knots.
According to [8], we can approximate η, assumed in $\mathcal{H}(p)$, by $\tilde{\eta}(u) = \tilde{\gamma}^\top B(u) \in \mathcal{S}_n$, where $B(u)$ is the spline basis and $\tilde{\gamma} \in \mathbb{R}^{N_n}$ is the spline coefficients vector.
Let $\varepsilon_{1,\beta_1} = Y - g_1(\langle\beta_1, Z\rangle)$. Remark that ε_{1,β_1} and $\langle\beta_1, Z\rangle$ are uncorrelated. So, in an iterative way, we can define $\varepsilon_{j,\beta_j} = Y - \sum_{k=1}^{j} g_k(\langle\beta_k, Z\rangle)$ for $j = 1, \dots, J-1$, and $\varepsilon_{J,\beta_J} = g(\mu(X, Z)) - \eta(\alpha^\top X) - \sum_{k=1}^{J} g_k(\langle\beta_k, Z\rangle)$, with $\mathbb{E}[\varepsilon_{j,\beta_j}|\langle\beta_j, Z\rangle] = 0$ at each stage $j = 1, \dots, J$. For $j = 1, \dots, J$, the j-th direction β_j is obtained by solving the minimum problem

$$\min_{\|\beta_j\|^2=1} \mathbb{E}\left[\left(\varepsilon_{j-1,\beta_{j-1}} - \mathbb{E}\left[\varepsilon_{j-1,\beta_{j-1}}|\langle\beta_j, Z\rangle\right]\right)^2\right], \tag{3.2}$$

and, the j-th component $g_j(u) = \mathbb{E}\left[\varepsilon_{j-1,\beta_{j-1}}|\langle\beta_j, Z\rangle = u\right]$ is estimated by:

$$\widehat{g}_{j,\beta_j}(u) = \frac{\sum_{i=1}^{n} \widehat{\varepsilon}_{j-1,\beta_{j-1},i} K_j\left(\frac{u - \langle\beta_j, Z_i\rangle}{h_j}\right)}{\sum_{i=1}^{n} K_j\left(\frac{u - \langle\beta_j, Z_i\rangle}{h_j}\right)}, \tag{3.3}$$

where, for all $i = 1, \dots, n$ and $j = 1, \dots, J$, $\widehat{\varepsilon}_{0,\beta_0,i} = Y_i$, taking $\varepsilon_{0,\beta_0} = Y$ and $\widehat{\varepsilon}_{j,\beta_j,i} = Y_i - \sum_{k=1}^{j} \widehat{g}_{k,\beta_k}(\langle\beta_s, Z_i\rangle)$, the parameter h_j is the bandwidth and K_j is a kernel.
By setting $W_i = \langle\beta_{J+1}, Z_i\rangle$, the mean function estimator $\widehat{m}(x, z)$ is given by the evaluation of the parameter $\theta = (\alpha^\top, \gamma^\top, \delta^\top)^\top$ and by inverting the following equation

$$g(\mu(X_i, Z_i)) = \sum_{j=1}^{J} \widehat{g}_j(\langle\widehat{\beta_j}, Z_i\rangle) + \widehat{\gamma}^\top B\left(\widehat{\alpha}^\top x\right) + \widehat{\delta}^\top B^{(J)}(W_i).$$

Notice that the parameter $\theta = (\alpha^\top, \gamma^\top, \delta^\top)^\top$ is determined by maximizing the following Quasi-Likelihood rule:

$$\widehat{\theta} = \left(\widehat{\alpha}^\top, \widehat{\gamma}^\top, \widehat{\delta}^\top\right)^\top = \underset{\theta=(\alpha,\gamma,\delta)\in\mathbb{R}^d\times\mathbb{R}^{N_n}\times\mathbb{R}^{lJ}}{\arg\max} L(\theta), \quad (3.4)$$

where $$L(\theta) := L(\alpha, \gamma, \delta) = \frac{1}{n}\sum_{i=1}^{n} Q\left(g^{-1}(m_i), Y_i\right),$$

with $$m_i := \sum_{j=1}^{J} \widehat{g_j}(\langle\widehat{\beta_j}, Z_i\rangle) + \gamma^\top B\left(\alpha^\top X_i\right) + \delta^\top B^{(J)}(W_i).$$

Then, the algorithm is as follows:

The first step: Let $Y_i = R^-(Z_i) + \varepsilon_i$ for $1 \le i \le n$, where $R^-(Z) = \sum_{i=1}^{J} g_j(\langle\beta_j, Z\rangle)$. We consider the nonparametric estimation of the predictive components in some fixed direction. In fact, for each $j = 1, \dots, J$, given $(\beta_1, \dots, \beta_J)$, we estimate the functions $g_{j,\beta_j} = \mathbb{E}\left[\varepsilon_{j-1,\beta_{j-1}} | \langle\beta_j, Z\rangle = u\right]$ by $\widehat{g}_{j,\beta_j}(u)$ given by (3.3).
For a fixed integer J, the predictive directions $\beta_1, \dots, \beta_{J+1}$ are unknown. We therefore use the cross-validation principle based on the leave-one-out statistical sample $\{(X_j, Z_j), j = 1, \dots, n, j \neq i\}$ to detect them recursively, and, at each step, the functional direction is searched over some subset $\Xi = \Xi_n$ with $\Xi \subset \mathcal{H} \bigcap \{\beta : \|\beta\|^2 = 1\}$ as follows:
The most predictive direction $\widehat{\beta}_1$ is estimated by solving the following optimization problem:

$$\widehat{\beta}_1 = \underset{\beta_1\in\Xi}{\arg\min}\; CV_1(\beta_1)$$

with

$$CV_1(\beta_1) = \frac{1}{n}\sum_{i=1}^{n}\left(Y_i - \widetilde{g}^{\,i}_{1,\beta_1}(\langle\beta_1, Z_i\rangle)\right)^2 I_{Z_i\in\mathcal{G}},$$

where $\mathcal{G}$ is a subset of $\mathcal{H}$.When most predictive directions $\widehat{\beta}_1, \dots, \widehat{\beta}_{j-1}$ are estimated, we estimate the next most predictive direction by solving the following optimization problem: $\widehat{\beta}_j = \underset{\beta_j\in\Xi}{\arg\min}\; CV_j(\beta_j)$ with

$$CV_j(\beta_j) = \frac{1}{n}\sum_{i=1}^{n}\left(Y_i - \sum_{k=1}^{j-1}\widehat{g}_{k,\widehat{\beta}_k}(\langle\beta_k, Z_i\rangle) - \widetilde{g}^{\,i}_{j,\beta_j}(\langle\beta_j, Z_i\rangle)\right)^2 I_{Z_i\in\mathcal{G}}$$

Finally, we estimate the last most predictive direction β_{J+1} by solving the following optimization problem: $\widehat{\beta}_{J+1} = \underset{\beta_{J+1}\in\Xi}{\arg\min}\, CV_{J+1}(\beta_{J+1})$ where

$$CV_{J+1}(\beta_{J+1}) = \frac{1}{n}\sum_{i=1}^{n}\left(Y_i - g^{-1}(\widehat{m}_i^{(-i)})\right)^2 I_{Z_i\in\mathcal{G}}$$

and $$\widehat{m}_i^{(-i)} = \sum_{j=1}^{J}\widehat{g}_j(\langle\widehat{\beta_j}, Z_i\rangle) + \widehat{\gamma}^\top B\left(\alpha^\top(\widehat{\tau}_i)X_i\right) + \widetilde{g}^{\,-i}_{J+1,\widehat{\beta}_{J+1}}.$$

The second step: By plug-in the non-parametric estimator in the first step, we consider the model

$$g\left(\mu\left(X_i, Z_i\right)\right) = \widehat{R^-}(Z_i) + \eta\left(\alpha^\top X_i\right) + g_{J+1}(\langle\beta_{J+1}, Z_i\rangle) \text{ for } i = 1, \ldots, n, \quad (3.5)$$

where $\widehat{R^-}(Z_i) = \sum_{j=1}^{J} \widehat{g}_j(\langle\widehat{\beta}_j, Z_i\rangle), \quad g_{J+1}(\langle\beta_{J+1}, Z_i\rangle) \approx \delta^\top B^{(J)}(W_i)$

and $W_i = \langle B^{(J)}, Z_i\rangle$ for $i = 1, \cdots, n$. We seek a function $\eta \in \mathcal{S}_n$ along with a value of α that maximizes the following Quasi-Likelihood function:

$$L(\eta, \alpha) = \frac{1}{n}\sum_{i=1}^{n} Q\left(g^{-1}\left\{\eta\left(\alpha^\top X_i\right) + \widehat{R^-}\left(Z_i\right) + \delta^\top B^{(J)}(W_i)\right\}, Y_i\right), \ \eta \in \mathcal{S}_n. \quad (3.6)$$

The mean function estimator $\widehat{\mu}$ is given by the evaluation of the parameters $\widehat{\theta} = \left(\widehat{\alpha}^\top, \widehat{\gamma}^\top, \widehat{\delta}^\top\right)^\top$ and inverting the equation (3.5). That is determined by maximizing the following Quasi-Likelihood:

$$\widehat{\theta} = \left(\widehat{\alpha}^\top, \widehat{\gamma}^\top, \widehat{\delta}^\top\right)^\top = \underset{\theta=(\alpha^\top,\gamma^\top,\delta^\top)^\top \in \mathbb{R}^d \times \mathbb{R}^{N_n} \times \mathbb{R}^{l_J}}{\arg\max} l(\theta) \quad (3.7)$$

where
$$l(\theta) := l(\alpha, \gamma, \delta) = \frac{1}{n}\sum_{i=1}^{n} Q\left(g^{-1}\left\{m_i\right\}, Y_i\right),$$

with $m_i = \gamma^\top B\left(U_i\right) + \widehat{R^-}\left(Z_i\right) + \delta^\top B^{(J)}(W_i)$, where $U_i = \alpha^\top X_i$.

So, the spline estimator of $\eta(.)$ is $\widehat{\eta}(.) = \widehat{\gamma}^\top B(.)$.

For $l = 1, 2$, we denote

$$\rho_l(m) = \frac{1}{\sigma^2 V\left(g^{-1}(m)\right)}\left[\frac{d}{dm}\left(g^{-1}(m)\right)\right]^l \text{ and } q_l(m, y) = \frac{\partial^l}{\partial m^l} Q\left(g^{-1}(m), y\right).$$

Then, the score vector is

$$S\left(\theta_\tau\right) = \frac{\partial l}{\partial \theta_\tau}(\theta)\Big|_{\theta=\theta_\tau} = \frac{1}{n}\sum_{i=1}^{n} q_1\left(m_i, Y_i\right) \xi_i(\tau, \gamma).$$

Then, for any $1 \leq i \leq n$, we get, after some calculations, the estimators:

$$\widehat{m}_i = \widehat{\gamma}^\top B\left(\alpha^\top(\widehat{\tau})X_i\right) + \widehat{R^-}\left(Z_i\right) + \delta^{(k)\top} B^{(J)}(W_i) \approx \sum_{j=1}^{N_n} \gamma_j^{(k)} B_j\left(\alpha^\top\left(\tau^k\right) X_i\right) + \widehat{R}\left(Z_i\right).$$

Then $\widehat{\mu}_i = g^{-1}\left(\widehat{m}_i\right)$ and $\widehat{\alpha} = \alpha\left(\tau^{(k)}\right)$ is the estimator of the single index coefficient vector of the Generalized Functional Partially Linear Single Index-Model.

3.3 Assumptions

Let $\varepsilon = Y - g^{-1}\left(m_0(T)\right)$ where $T = \left(X^\top, B^{(J)}(W)^\top\right)^\top$. We will need the following assumptions.

(C1) The single-index link function $\eta(\cdot) \in \mathcal{H}(p)$.

(C2) For all $m \in \mathbb{R}$ and for all y in the range of the response variable Y, we have, for $k = 1, 2$ that $q_2(m, y) < 0$ and $c_q < \left|q_2^k(m, y)\right| < C_q$, for some positive constants c_q and C_q.

(C3) The ν-th order partial derivative of the joint density function of X satisfies the Lipschitz' condition of order κ ($\kappa \in (0, 1]$). Moreover, the marginal density function of $\alpha^\top X$ is continuous and bounded away from zero and supported within $[a, b]$.

(C4) For any vector τ, there exists positive constants c_τ and C_τ such that

$$c_\tau I_{t\times t} \leq \mathbb{E}\left[\begin{pmatrix}1\\T\end{pmatrix}\begin{pmatrix}1\\T\end{pmatrix}^\top \left|\alpha^\top(\tau)X = \alpha^\top(\tau)x\right.\right] \leq C_\tau I_{t\times t},$$

where $t = 1 + d + N_n + l_J$ and $T = (X^\top, \bar{W}^\top)^\top$.

(C5) The number N_n of knots satisfies $n^{\frac{1}{2(p+1)}} \ll N_n \ll n^{\frac{1}{8}}$ $(p > 3)$.

(C6) There exists a positive constant $C > 0$ such that the fourth order moment of Z is finite i.e., $\mathbb{E}\|Z(.)\|^4 \leq C$.

(C7) The covariance function $K(t, s) = \mathrm{Cov}(Z(t), Z(s))$ is positive definite.

(C8) For some finite positive constants C_ρ, C_ρ^* and M_0, $|\rho_1(m_0)| \leq C_\rho$ and $|\rho_1(m) - \rho_1(m_0)| \leq C_\rho^* |m - m_0|$ for all $|m - m_0| \leq M_0$.

(C9) For some finite positive constants C_g, C_g^* and M_1, the link function g, in the model (3.1), satisfies:

$$\left|\frac{d}{dm}g(m)\Big|_{m=m_0}\right| \leq C_g \text{ and } \left|\frac{d}{dm}g^{-1}(m) - \frac{d}{dm}g^{-1}(m)\Big|_{m=m_0}\right| \leq C_g^* |m - m_0|$$

for all $|m - m_0| \leq M_1$.

(C10) There exists a positive constant C_0, such that $\mathbb{E}(\epsilon^2 | U_{\tau,0}) \leq C_0$, where $\epsilon = Y - g^{-1}\left(m_0(T)\right)$.

(C11) For all $\beta \in \mathcal{H}$, the random variable $\langle\beta, Z\rangle$ is valued in a compact subset C of $\mathbb{R}$.

(C12) For some $\bar{\delta} > 0$ and some $C = C(\beta_1, \beta_2, \ldots, \beta_J) \in \mathbb{R}^+$, we have, for all $j = 1, \ldots, J$, and for all $(u, v) \in C^2$, $|g_{j,\beta_j}(u) - g_{j,\beta_j}(v)| \leq C|u - v|^{\bar{\delta}}$.

(C13) $\forall r \geq 1$, $\exists C < \infty$, $\mathbb{E}\left[|Y|^r = \sigma_r(Z)\right] < Cr!$, where $\sigma_r(.)$ is continuous and bounded, and we assume that for all $\beta \in \mathcal{H}$, the distribution of the real random variable $\langle\beta, Z\rangle$ is absolutely continuous with respect to the Lebesgue measur, with density f_β satisfying
$0 < \inf_{u\in C} f_\beta(u) \leq \sup_{u\in C} f_\beta(u) < \infty$.

(C14) For any $j = 1, \ldots, J$, the kenel K_j is integrable and bounded within $(-1, 1)$, and there exists $\tau_j > 0$ and $\alpha_j \geq 0$ such that $h_j \sim C^{st}\left(\frac{(\log n)^{\alpha_j}}{n}\right)^{1/(2\tau_j+1)}$.

(C15) For all $k = 1, \ldots, k_j - 1$,

$$\int u^k K_j(u)du = 0 \text{ and } \int u^{k_j} K_j(u)du > 0$$

with $k_j \geq q_j$ and $k_j < k_{j-1}$.

3.4 Some Asymptotics

Convergence of the estimated univariate components
The following theorem states the convergence, with rates, of the estimator $\left(\widehat{g}_j\right)_{j=1,\cdots,J}$.

Theorem 3.1.

(i) Under the conditions (C9)-(C14), with $\tau_j = \bar{\delta}$ and $\alpha_j = 1$ in (C14), for $j = 1, \ldots, J$, we have

$$\sup_{u \in C} |\widehat{g}_{j,\beta_j}(u) - g_{j,\beta_j}(u)| = O\left(\left(\frac{\log n}{n}\right)^{\frac{\bar{\delta}}{2\bar{\delta}+1}}\right), \text{ almost-surely.}$$

(i) Under the conditions (C9)-(C15), with $\tau_j = k_j$ and $\alpha_j = 0$ in (C14), for $j = 1, \cdots, J$, we have

$$\mathbb{E}\left[\int_C \left(\widehat{g}_{j,\beta_j}(u) - g_{j,\beta_j}(u)\right)^2 du\right] \sim C^{st}\left(\frac{1}{n}\right)^{2k_j/(2k_j+1)}.$$

So, for the $(J+1)$-th component, we have the following theorem

Theorem 3.2. *Under assumptions (C1)-(C8), we have*

$$\|\widehat{g}_{J+1} - g_{J+1}\|_2 = O_{\mathbb{P}}\left\{\sqrt{N_n}\left(\frac{1}{\sqrt{nh}} + h^p\right)\right\}.$$

Estimation of the systematic component function

Theorem 3.3. *Under assumptions (C1)-(C7), and the assumptions of theorem (3.1) we have*

$$\|\widehat{\eta} - \eta_0\|_2 = O_{\mathbb{P}}\left\{\sqrt{N_n}\left(\frac{1}{\sqrt{nh}} + h^p\right)\right\} \text{ and } \|\widehat{\eta} - \eta_0\|_n = O_{\mathbb{P}}\left\{\sqrt{N_n}\left(\frac{1}{\sqrt{nh}} + h^p\right)\right\}.$$

Estimation of the parametric components

The next theorem shows that the maximum Quasi-Likelihood estimator is root-n consistent and is asymptotically normal, although the convergence rate of the nonparametric component $\widehat{\eta}$ is slower than root-n. Before enouncing the theorem, let us denote

$$\Upsilon\left(u_{\tau,0}\right) = \frac{\mathbb{E}\left[X\rho_2\left(m_0(T)\right)\left|U_{\tau,0} = u_{\tau,0}\right.\right]}{\mathbb{E}\left[\rho_2\left(m_0(T)\right)\left|U_{\tau,0} = u_{\tau,0}\right.\right]},$$

$$\Gamma\left(u_{\tau,0}\right) = \frac{\mathbb{E}\left[W\rho_2\left(m_0(T)\right)\left|U_{\tau,0} = u_{\tau,0}\right.\right]}{\mathbb{E}\left[\rho_2\left(m_0(T)\right)\left|U_{\tau,0} = u_{\tau,0}\right.\right]},$$

$$\Phi(x) = \Phi\left(U_{\tau,0}, x\right) = x - \Upsilon\left(u_{\tau,0}\right).$$

Theorem 3.4. *Under assumptions $(C_1) - (C_{10})$, the constrained Quasi-Likelihood estimators $\widehat{\alpha}$ and $\widehat{\delta}$ with $\|\widehat{\alpha}\|_d = 1$ are jointly asymptotically normally distributed, i.e.,*

$$\sqrt{n}\begin{pmatrix}\widehat{\alpha} - \alpha_0 \\ \widehat{\delta} - \delta_0\end{pmatrix} \xrightarrow{\mathcal{D}} \mathcal{N}\left(0, R\left(\tau_0\right) D^{-1} R^{\top}\left(\tau_0\right)\right),$$

with $D = \mathbb{E}\left[\rho_2\left(m_0(T)\right)\begin{pmatrix}\eta_0'\left(U_{\tau,0}\right) J^{\top}\left(\tau_0\right)\Phi(X) \\ \Psi(\bar{W})\end{pmatrix}\begin{pmatrix}\eta_0'\left(U_{\tau,0}\right) J^{\top}\left(\tau_0\right)\Phi(X) \\ \Psi(\bar{W})\end{pmatrix}^{\top}\right],$

where $\alpha_0, \delta_0, \tau_0$ and η_0 are the true values of α, δ, τ and η.

References

1. Ait-Saïdi, A., F. Ferraty, R. Kassa, and P. Vieu: Cross-validated estimations in the single-functional index model. Statistics: A Journal of Theoretical and Applied Statistic **42**(6), 475–494 (2008)
2. Alahiane, M., I. Ouassou, M. Rachdi, and P. Vieu: Stats: Functional Data Analysis (FDA) **4**(4), 793–813 (2021)
3. Alahiane, M., I. Ouassou, M. Rachdi, and P. Vieu: High-dimensional statistics: Non-parametric generalized functional partially linear single-index model. Mathematics: Advances in Statistics: Theory, Methodology, Applications and Data Analaysis **10**(15), 2704 (2022)
4. Alahiane, M., M. Rachdi, I. Ouassou, Alahiane A., and Hobbad L: Cross-ValidatedFunctional Generalized Partially Linear Single-Functional Index Model, Mathematics **12**(17) (2024)
5. Aneiros, G. and P. Vieu: Partial linear modelling with multi-functional covariates. Computational Statistics **30**(3), 647-671 (2015)
6. Aneiros-Pérez, G. and P. Vieu: Semi-functional partial linear regression. Statistics & Probability Letters 76(11), 1102–1110 (2006)
7. Carroll, R. J., J. Fan, I. Gijbels, and M. P. Wand: Generalized partially linear single-index models. Journal of the American Statistical Association **92**(438), 477– 489 (1997)
8. De Boor, Carl: A practical guide to splines, **27**, springer-verlag New York (2001)
9. Ferraty, F., A. Goia, E. Salinelli, and P. Vieu: Functional projection pursuit regression. Test **22**(22), 293–320 (2013)

10. Ferraty, F. and Vieu, P: Nonparametric functional data analysis: theory and practice, Volume 76. Springer (2006)
11. M. Rachdi, Alahiane, M., Alahiane, A., I. Ouassou, and L. Hobbad: Cross-Validated Functional Generalized Partially Linear Single-Functional Index Model (2024)
12. Yu, Yan and Ruppert, David: Penalized spline estimation for partially linear single-index models, Journal of the American Statistical Association, **97**(460), 1042-1054 (2002)

Chapter 4
On the Estimation of Invertible Functional Time Series

Alexander Aue, Sebastian Kühnert and Gregory Rice

Abstract This contribution addresses the estimation of invertible processes in a functional time series setting. While asymptotic upper bounds for the estimation error of such operators for specific ARMA processes in the Hilbert space of square-integrable functions have been considered in recent years, this paper adds to the existing research by designing consistent estimates for the operators defining an invertible representation of a stationary process in a general separable Hilbert space under mild conditions that hold for many classes of functional time series. These results imply in turn consistency results with explicit rates for related operator estimates for Hilbert space-valued causal linear processes, as well as functional MA, AR and ARMA processes.

4.1 Introduction

This paper is set in the framework of functional time series analysis; see [4] and [12]. A typical and general setting for functional time series analysis (fTSA) is to assume that the data attain their values in a separable, infinite dimensional Hilbert space $\mathcal{H}$, which is the setting considered throughout. Literature related to the results presented in the following includes [1], [2], [6], [7], [10], [13], [17].

When it comes to estimating the infinite-dimensional parameters defining such processes, early references such as [4] put forward consistent estimators of the

Alexander Aue (✉)
Department of Statistics, University of California, Davis, USA, e-mail: aaue@ucdavis.edu

Sebastian Kühnert
Fakultät für Mathematik, Ruhr-Universität Bochum, Germany,
e-mail: sebastian.kuehnert@ruhr-uni-bochum.de

Gregory Rice
Department of Statistics and Actuarial Science, University of Waterloo, Canada,
e-mail: grice@uwaterloo.ca

© The Author(s), under exclusive license to Springer Nature Switzerland AG 2025

G. Aneiros et al. (eds.), *New Trends in Functional Statistics and Related Fields*,
Contributions to Statistics, https://doi.org/10.1007/978-3-031-92383-8_4

operators defining fAR processes. Limit theorems for $\mathcal{H}$-valued linear processes can for instance be found in [8], [19], [20]. Further, [3] and [15] approached the estimation of the operators in the linear and inverted representation pointwise, without stating explicit estimation rates. In [17], the complete sequences of operators in the linear and inverted representation of invertible linear processes originating from fGARCH models with values in the specific Hilbert space $L^2[0,1]$. Other related work is in [5], [16], [21].

This article formulates consistency results with quantifiable rates for the operators defining a general, $\mathcal{H}$-valued linear and invertible process based on functional Yule-Walker equations and a Tychonoff-regularized estimator, without assuming that the processes stem from specific models. These results are established under a Sobolev–type condition and the well-known and practical weak dependence concept for fTSA called L^p-m-approximibility introduced in [11]. These results can further downstream be applied to obtain consistency results for the operators defining fAR, fMA and fARMA processes for arbitrary AR and MA orders.

4.2 Estimation in the inverted representation

We call a centered process $\boldsymbol{X} = (X_k)_{k\in\mathbb{Z}} \subset \mathcal{H}$ *invertible* if it is stationary and satisfies

$$X_k = \sum_{j=1}^{\infty} \psi_j(X_{k-j}) + \varepsilon_k, \quad k \in \mathbb{Z}, \tag{4.1}$$

where $(\psi_j)_{j\in\mathbb{N}} \subset \mathcal{L}_{\mathcal{H}}$ is a sequence of linear operators with $\sum_{j=1}^{\infty} \|\psi_j\|_{\mathcal{L}} < \infty$, and where the series convergences almost surely and in $L^2_{\mathcal{H}}$.

Assumption 1 *$\boldsymbol{X} \subset L^2_{\mathcal{H}}$ is centered, invertible with representation* (4.1), *and $\boldsymbol{\varepsilon} \subset L^4_{\mathcal{H}}$.*

Assumption 2 *The covariance operator $\mathscr{C}_{\boldsymbol{\varepsilon}}\colon \mathcal{H} \to \mathcal{H}$ is injective.*

Remark 4.1. Injectivity of the covariance operator $\mathscr{C}_{\boldsymbol{\varepsilon}}$ holds if and only if $\overline{\mathrm{im}(\mathscr{C}_{\boldsymbol{\varepsilon}})} = \mathcal{H}$, see [14]. It is also equivalent to the condition that there is no affine, closed, and proper subspace $U \subsetneq \mathcal{H}$ with $\mathbb{P}(\varepsilon_0 \in U) = 1$.

The goal of this section is to define, based on a sample $X_1, \ldots, X_N$ from $\boldsymbol{X}$, consistent estimators of the operators $(\psi_j)_{j\in\mathbb{N}}$ in (4.1). The proposed estimators are based on estimates for lagged (cross-)covariance operators of processes

$$X_k^{[L]} := (X_k, X_{k-1}, \ldots, X_{k-L+1})^\top, \quad k \in \mathbb{Z}, \tag{4.2}$$

taking values in the Cartesian product Hilbert space $\mathcal{H}^L$. We assume that the lag parameter $L = L_N \in \mathbb{N}$ grows as a function of the sample size. We define the empirical version $\hat{\mathscr{C}}_{\boldsymbol{X}^{[L]}}$ of the covariance operator $\mathscr{C}_{\boldsymbol{X}^{[L]}}$ through

$$\hat{\mathscr{C}}_{\boldsymbol{X}^{[L]}} := \frac{1}{N-L+1}\sum_{k=L}^{N} X_k^{[L]} \otimes X_k^{[L]}, \quad L \le N. \tag{4.3}$$

We also define $\hat{\mathscr{D}}_{\boldsymbol{X}^{[L]},\boldsymbol{X}}$, the empirical version of $\mathscr{D}_{\boldsymbol{X}^{[L]},\boldsymbol{X}} = \mathscr{C}^1_{\boldsymbol{X}^{[L]},\boldsymbol{X}}$ by

$$\hat{\mathscr{D}}_{\boldsymbol{X}^{[L]},\boldsymbol{X}} := \frac{1}{N-L}\sum_{k=L}^{N-1} X_k^{[L]} \otimes X_{k+1}, \quad L < N. \tag{4.4}$$

For all L, $\hat{\mathscr{C}}_{\boldsymbol{X}^{[L]}}$ and $\hat{\mathscr{D}}_{\boldsymbol{X}^{[L]},\boldsymbol{X}}$ are unbiased estimators for $\mathscr{C}_{\boldsymbol{X}^{[L]}}$ and $\mathscr{D}_{\boldsymbol{X}^{[L]},\boldsymbol{X}}$, respectively. To derive consistency results for estimates of $(\psi_j)_{j\in\mathbb{N}}$ in (4.1), we assume that these operator estimators have consistency rates for a growing $L = L_N$ as follows.

Assumption 3 *For some sequence $L = L_N \to \infty$ with $L = o(N^{1/3})$,*

$$\|\hat{\mathscr{C}}_{\boldsymbol{X}^{[L]}} - \mathscr{C}_{\boldsymbol{X}^{[L]}}\|_{\mathcal{S}} = O_{\mathbb{P}}(L^{3/2}N^{-1/2}), \tag{4.5}$$

$$\|\hat{\mathscr{D}}_{\boldsymbol{X}^{[L]},\boldsymbol{X}} - \mathscr{D}_{\boldsymbol{X}^{[L]},\boldsymbol{X}}\|_{\mathcal{S}} = O_{\mathbb{P}}(LN^{-1/2}). \tag{4.6}$$

Remark 4.2. (a) This assumption is fulfilled for centered, $L^4_{\mathcal{H}}$-m-approximable processes, see Corollary 3.1 and Proposition 3.2 in [18], using the weak dependence concept of [11]. For $p \ge 1$, a process $(X_k)_{k\in\mathbb{Z}} \subset L^p_{\mathcal{H}}$ is called *$L^p_{\mathcal{H}}$-m-approximable* if for each k it holds $X_k = f(\epsilon_k, \epsilon_{k-1}, \ldots)$ for some i.i.d. process $(\epsilon_k)_{k\in\mathbb{Z}} \subset \mathcal{H}$ and a measurable function $f\colon \mathcal{H}^\infty \to \mathcal{H}$ such that $\sum_{m=1}^\infty \nu_p(X_m - X_m^{(m)}) < \infty$, where $\nu_p(\cdot) := (\mathbb{E}\|\cdot\|^p)^{1/p}$, and where $X_k^{(m)} := f(\epsilon_k, \epsilon_{k-1}, \ldots, \epsilon_{k-m+1}, \epsilon^{(m)}_{k-m}, \epsilon^{(m)}_{k-m-1}, \ldots)$ for all k, m, with $(\epsilon_j^{(m)})_j$ being independent copies of (ϵ_j) for all m.
(b) It is worthwhile to note that Assumption 3 can even be satisfied for not necessarily causal processes, for instance *linear processes* (LP) in $\mathcal{H}$

$$X_k = \sum_{\ell=-\infty}^{\infty} \phi_\ell(\varepsilon_{k-\ell}), \tag{4.7}$$

with $\sum_{\ell=-\infty}^\infty \|\phi_\ell\|_{\mathcal{L}} < \infty$. This is the case if the LP operators fulfill the stronger condition $\sum_{k=1}^\infty \sum_{\ell=k}^\infty [\|\phi_\ell\|_{\mathcal{L}} + \|\phi_{-\ell}\|_{\mathcal{L}}] < \infty$.

The covariance operator of an LP can be described in terms of the linear operators in (4.15) and error covariance operator $\mathscr{C}_\varepsilon$ through

$$\mathscr{C}_{\boldsymbol{X}} = \sum_{i=0}^{\infty} \phi_i \mathscr{C}_\varepsilon \phi_i^*. \tag{4.8}$$

Let $(\lambda_j, c_j)_{j\in\mathbb{N}} = (\lambda_j(L), c_j(L))_{j\in\mathbb{N}}$ and $(\hat\lambda_j, \hat c_j)_{j\in\mathbb{N}} = (\hat\lambda_j(L), \hat c_j(L))_{j\in\mathbb{N}}$ be the sequences of eigenvalue/eigenfunction pairs of the covariance operators $\mathscr{C}_{\boldsymbol{X}^{[L]}}$ and $\hat{\mathscr{C}}_{\boldsymbol{X}^{[L]}}$, respectively. According to [4], Lemma 4.2, it holds that

$$\sup_{j\in\mathbb{N}} |\hat{\lambda}_j - \lambda_j| \leq \|\hat{\mathscr{C}}_{\boldsymbol{X}^{[L]}} - \mathscr{C}_{\boldsymbol{X}^{[L]}}\|_{\mathcal{L}}. \tag{4.9}$$

Suitable estimators for the eigenfunctions c_j, which are unambiguous except for their sign, are $\hat{c}'_j := \operatorname{sgn}\langle \hat{c}_j, c_j\rangle \hat{c}_j$, with $\operatorname{sgn}(x) := 1_{[0,\infty)}(x) - 1_{(-\infty,0)}(x), x \in \mathbb{R}$. As covariance operators are positive semi-definite, injectivity of $\mathscr{C}_{\boldsymbol{X}^{[L]}}$ after Proposition 4.1 is equivalent to strict positivity. Hence, the eigenvalues satisfy $\lambda_1 \geq \lambda_2 \geq \cdots > 0$. If the eigenspaces associated to each eigenvalues λ_j are one-dimensional, according to [4], Lemma 4.3, it holds that

$$\|\hat{c}'_j - c_j\| \leq \frac{2\sqrt{2}}{\alpha_j} \|\hat{\mathscr{C}}_{\boldsymbol{X}^{[L]}} - \mathscr{C}_{\boldsymbol{X}^{[L]}}\|_{\mathcal{L}},$$

where $\alpha_1 := \lambda_1 - \lambda_2$, and $\alpha_j := \min(\lambda_{j-1} - \lambda_j, \lambda_j - \lambda_{j+1})$ for $j > 1$. Moreover, for all $k \in \mathbb{N}$ it holds with $\Lambda_k := \sup_{1\leq j\leq k}(\lambda_j - \lambda_{j+1})^{-1}$ that

$$\sup_{1\leq j\leq k} \|\hat{c}'_j - c_j\| \leq 2\sqrt{2}\,\Lambda_k \|\hat{\mathscr{C}}_{\boldsymbol{X}^{[L]}} - \mathscr{C}_{\boldsymbol{X}^{[L]}}\|_{\mathcal{L}}. \tag{4.10}$$

The estimates for the operators in the inverted representation are derived from the approximate Yule–Walker equation

$$\mathscr{D}_{\boldsymbol{X}^{[L]},\boldsymbol{X}} = \Psi_L \mathscr{C}_{\boldsymbol{X}^{[L]}} + \sum_{\ell>L} \psi_\ell \mathscr{C}^{1-\ell}_{\boldsymbol{X}^{[L]},\boldsymbol{X}}, \tag{4.11}$$

where the process $\boldsymbol{X}^{[L]} = (X^{[L]}_k)_k \subset \mathcal{H}^L$ is defined in (4.2),

$$\Psi_L := (\psi_1 \cdots \psi_L), \tag{4.12}$$

and where $\sum_{\ell>L} \psi_\ell \mathscr{C}^{1-\ell}_{\boldsymbol{X}^{[L]}\boldsymbol{X}}$ exists for all fixed $L \in \mathbb{N}$ as it converges in the operator norm topology. Based on a sample $X_1, \ldots, X_N$ from $\boldsymbol{X}$, we focus on estimating Ψ_L. This can be achieved by using Tychonoff regularization, with an estimator $\hat{\Psi}_L = \hat{\Psi}_L(K, N)$ of the form

$$\hat{\Psi}_L := \hat{\mathscr{D}}_{\boldsymbol{X}^{[L]},\boldsymbol{X}} \hat{\mathscr{C}}^{\dagger}_{\boldsymbol{X}^{[L]}} \coprod_{\hat{c}_1}^{\hat{c}_K}, \tag{4.13}$$

where $\hat{\mathscr{C}}^{\dagger}_{\boldsymbol{X}^{[L]}} := (\hat{\mathscr{C}}_{\boldsymbol{X}^{[L]}} + \theta_N \mathbb{I})^{-1}$ stands for the pseudo-inverse of $\hat{\mathscr{C}}_{\boldsymbol{X}^{[L]}}$, with $\mathbb{I}\colon \mathcal{H}^L \to \mathcal{H}^L$ denoting the identity map, and where $(\theta_N)_{N\in\mathbb{N}} \subset (0,\infty)$ is a tuning parameter sequence with $\theta_N \to 0$ as $N \to \infty$. Further, $\hat{\mathscr{C}}_{\boldsymbol{X}^{[L]}}$ and $\hat{\mathscr{D}}_{\boldsymbol{X}^{[L]},\boldsymbol{X}}$ are the estimates in (4.3)–(4.4), $K = K_N \in \mathbb{N}$ is a sequence satisfying $K_N \to \infty$ as $N \to \infty$, and $\hat{c}_1, \ldots, \hat{c}_K$ are the eigenfunctions associated to the first K largest eigenvalues $\hat{\lambda}_1 \geq \cdots \geq \hat{\lambda}_K$ of $\hat{\mathscr{C}}_{\boldsymbol{X}^{[L]}}$.

Assumption 4 *The eigenvalues $\lambda_j = \lambda_j(L)$ of $\mathscr{C}_{\boldsymbol{X}^{[L]}}$ satisfy $\lambda_j \neq \lambda_{j+1}$ for each $L = L_N$ and j.*

Assumption 5 *Each of the operators in* (4.1) *satisfy that* $\psi_i \in \mathcal{S}_{\mathcal{H}}$, *and* $\sum_{i=1}^{\infty} \|\psi_i\|_{\mathcal{S}}^2 < \infty$.

The method we use to quantify the error between $\hat{\Psi}_L$ and Ψ_L entails first decomposing Ψ_L into a suitable finite dimensional representation and a remainder term, and then evaluating the error of $\hat{\Psi}_L$ in estimating each component. In order to quantify the estimation of the remainder term, we employ a Sobolev condition akin to that used in [9] to derive precise asymptotic results for the estimation errors of densities in the deconvolution setting. To state the Sobolev type condition, let $(\mu_j, d_j)_{j\in\mathbb{N}}$ denote the eigenvalue/eigenfunction pairs corresponding to the covariance operator $\mathscr{C}_X$. Then, $(c_i \otimes d_j)_{i,j\in\mathbb{N}} = (c_i(L) \otimes d_j)_{i,j\in\mathbb{N}}$ is a CONS of $\mathcal{S}_{\mathcal{H}^L,\mathcal{H}}$, so, if $\Psi_L \in \mathcal{S}_{\mathcal{H}^L,\mathcal{H}}$, which holds if the ψ_i are Hilbert–Schmidt operators, $\Psi_L = \sum_i \sum_j \langle \Psi_L, c_i \otimes d_j\rangle_{\mathcal{S}} (c_i \otimes d_j)$.

Assumption 6 *For some* $\beta > 0$, *the limit* $S_\Psi(\beta) := \lim_{N\to\infty} S_{\Psi_{L_N}}(\beta) < \infty$ *exists, where for each* $L = L_N \in \mathbb{N}$ *holds,*

$$S_{\Psi_L}(\beta) = \sum_{i=1}^{\infty}\sum_{j=1}^{\infty} \langle \Psi_L, c_i \otimes d_j\rangle_{\mathcal{S}}^2 \,(1 + i^{2\beta} + j^{2\beta}) < \infty.$$

Theorem 4.1. *Under the given assumptions, and with* $K = K_N \to \infty$ *and* $\theta_N \to 0$ *be sequences with* $\theta_N = o(\lambda_K)$, $K^{1+\beta}\lambda_K^{-1}\Lambda_K L^2 = O(N^{1/2})$, $\sum_{\ell>L} \|\psi_\ell\|_{\mathcal{L}} = o(K\Lambda_K L^{3/2} N^{-1/2})$ *and* $\theta_N = o(K\Lambda_K L^2 N^{-1/2})$, *where* $\beta > 0$ *is defined in the last assumption we have*

$$\|\hat{\Psi}_L - \Psi_L\|_{\mathcal{S}} = O_{\mathbb{P}}(K^{-\beta}). \tag{4.14}$$

Remark 4.3. (a) Theorem 4.1 improves on comparable results in the literature in several ways. In contrast to [17], our result is stated for general, invertible linear processes without assuming that they originate from a specific model, and where the processes are allowed to attain values in arbitrary, separable Hilbert space rather than only in $L^2[0, 1]$. Most notably, we analyze the complete operators and state explicit asymptotic upper bounds, rather than either focusing on finite-dimensional projections which is commonly done in operator estimation, see [10] and [2], or omitting stating the rate when completely observing the operators, see [6] for fGARCH models, and [3] and [15] for invertible linear processes. Further, we measured the estimation errors even with the Hilbert–Schmidt instead of the weaker operator norm as in [3] and [15]. The Sobolev type condition we utilized in our estimation, is, as far as we are aware, in the context of operator estimation only used for fGARCH models in [17]. Moreover, we refrained from a simplifying convexity assumption for the eigenvalues.

4.2.1 Estimating the operators in the causal linear process representation

We now suppose $\boldsymbol{X} = (X_k)_{k\in\mathbb{Z}} \subset \mathcal{H}$ follows a causal LP, see (4.7) for general LP's, so

$$X_k = \varepsilon_k + \sum_{j=1}^{\infty} \phi_j(\varepsilon_{k-j}), \quad k \in \mathbb{Z}, \tag{4.15}$$

where $(\phi_j)_{j\in\mathbb{N}} \subset \mathcal{L}_{\mathcal{H}}$ are linear operators with $\sum_{j=1}^{\infty} \|\phi_j\|_{\mathcal{L}} < \infty$. Such an LP is invertible if it may be expressed as in (4.1), see [4]. Relating the estimates from an inverted representation of $\boldsymbol{X}$ to its causal linear representation can be done using the following proposition.

Proposition 4.1. *Under the given assumptions, we have that*
(a) *The covariance operator* $\mathscr{C}_{\boldsymbol{X}} : \mathcal{H} \to \mathcal{H}$ *is injective.*
(b) *The covariance operators* $\mathscr{C}_{\boldsymbol{X}[L]} : \mathcal{H}^L \to \mathcal{H}^L$ *are injective for all* L.
(c) *If* $\sum_{\ell=0}^{\infty} A_\ell(\varepsilon_{i-\ell}) = \sum_{\ell=0}^{\infty} B_\ell(\varepsilon_{i-\ell})$ *for some* $i \in \mathbb{N}$, *then,* $A_\ell = B_\ell$ *for all* $\ell \in \mathbb{N}$.
(d) *If* $\boldsymbol{X}$ *is a causal linear process, and if* $\sum_{\ell=0}^{\infty} A_\ell(X_{i-\ell}) = \sum_{\ell=0}^{\infty} B_\ell(X_{i-\ell})$ *for some* $i \in \mathbb{N}$, *then* $A_\ell = B_\ell$ *for all* $\ell \in \{0, 1, 2, \dots\}$.

For an invertible LP $\boldsymbol{X}$ in Assumption 1 such that

$$X_k = \varepsilon_k + \sum_{i=1}^{\infty} \psi_i(X_{k-i}), \text{ and } X_{k-i} = \varepsilon_{k-i} + \sum_{j=1}^{\infty} \phi_j(\varepsilon_{k-i-j}),$$

for all $k, i \in \mathbb{Z}$, we have with $\phi_0 := \mathbb{I}$ that

$$X_k = \varepsilon_k + \sum_{i=1}^{\infty} \Big[\sum_{j=1}^{i} \psi_j \phi_{i-j}\Big](\varepsilon_{k-i}), \quad k \in \mathbb{Z}.$$

This representation, the linear representation (4.15), and Proposition 4.1(c) yield

$$\phi_i = \sum_{j=1}^{i} \psi_j \phi_{i-j}, \quad i \in \mathbb{N}.$$

Hence, with $\hat{\psi}_j = \hat{\Psi}_L^{(j)}$ denoting the j'th component of the estimator $\hat{\Psi}_L$ in (4.13) for $\Psi_L = (\psi_1, \dots, \psi_L)$, a reasonable estimate for ϕ_i is $\hat{\phi}_i = \hat{\phi}_i(L, K)$, defined iteratively with $\hat{\phi}_0 := \mathbb{I}$, and

$$\hat{\phi}_i := \sum_{j=1}^{i} \hat{\psi}_j \hat{\phi}_{i-j}, \quad i \in \mathbb{N}. \tag{4.16}$$

By assuming that the operators satisfy a Sobolev condition, we obtain the following consistency result for the complete operators in the linear representation.

Proposition 4.2. *Under the assumptions of Theorem 4.1,*

$$\|\hat{\phi}_i - \phi_i\|_{\mathcal{S}} = O_{\mathbb{P}}(K^{-\beta}), \quad i \in \mathbb{N}.$$

Remark 4.4. The novelties of Proposition 4.2 are similar to those in Theorem 4.1, see Remark 4.3 (a). Here, too, in the linear representation, the processes can attain their values in a general, separable Hilbert space, they are not constrained to stem from a certain type of time series, and we estimated the complete operators with stating an explicit asymptotic upper bound for the estimation errors in the Hilbert–Schmidt norm. Such a result is to the best of our knowledge new. However, consistent estimates for the operators in the linear representation, measured in the operator norm, and without convergence rate, can be found in [3] and [15].

References

1. Aue, A., van Delft, A.: Testing for stationarity of functional time series in the frequency domain. Ann. Statist. **48**, 2505–2547 (2020)
2. Aue, A., Horváth, L., Pellatt, D.: Functional generalized autoregressive conditional heteroskedasticity. J. Time Series Anal. **38**, 3–21 (2017)
3. Aue, A., Klepsch, J.: Estimating functional time series by moving average model fitting. arXiv:1701.00770 (2017)
4. Bosq, D.: Linear Processes in Function Spaces. Lecture Notes in Statistics. New York: Springer (2000)
5. Caponera, A. and Panaretos, V.: On the rate of convergence for the autocorrelation operator in functional autoregression. Statist. Probab. Lett. **189**, 109575, (2022)
6. Cerovecki, C., Francq, C., Hörmann, S, Zakoïan, J.-M.: Functional GARCH models: The quasi-likelihood approach and its applications. J. Econometrics **209**, 353–375 (2019)
7. van Delft, A., Characiejus, V., Dette, H.: A nonparametric test for stationarity in functional time series. Statist. Sinica **31**, 1375–1395 (2021)
8. Düker, M.-C.: Limit theorems for Hilbert space-valued linear processes under long-range dependence. Stochastic Process. Appl. **128**, 1439–1465 (2018)
9. Hall, P. , Meister, A.: A ridge-parameter approach to deconvolution. Ann. Statist. **35**, 1535–1558 (2007)
10. Hörmann, S., Horváth, L., Reeder, R.: A functional version of the ARCH model. Econometric Theory **29**, 267–288 (2013)
11. Hörmann, S., Kokoszka, P.: Weakly dependent functional data. Ann. Statist. **38**, 1845–1884 (2010)
12. Horváth, L., Kokoszka, P.: Inference for Functional Data with Applications. New York: Springer (2012)
13. Horváth, L., Kokoszka, P., Rice, G. Testing stationarity of functional time series. J. Econometrics **179**, 66–82 (2014)
14. Hsing, T., Eubank, R.: Theoretical Foundations of Functional Data Analysis, with an Introduction to Linear Operators. West Sussex: Wiley (2015)
15. Klepsch, J., Klüppelberg, C, Wei, T.: Prediction of functional ARMA processes with an application to traffic data. Econom. Stat. **1**, 128–149 (2017)
16. Kuenzer, T.: Estimation of functional ARMA models. Bernoulli **30**, 117–142 (2024)
17. Kühnert, S.: Functional ARCH and GARCH models: A Yule-Walker approach. Electron. J. Stat. **14**, 4321–4360 (2020)
18. Kühnert, S.: Estimating lagged (cross-)covariance operators of L^p-m-approximable processes in Cartesian product Hilbert spaces. J. Time Series Anal.

19. Merlevède, F.: Central limit theorem for linear processes with values in a Hilbert space. Stochastic Process. Appl. **65**, 103–114 (1996)
20. Merlevède, F., Peligrad, M., Utev, S.: Sharp conditions for the CLT of linear processes in a Hilbert space. J. Theoret. Probab. **10**, 681–693 (1997)
21. Turbillon, C., Bosq, D., Marion, J.-M., Pumo, B.: Estimation du paramètre des moyennes mobiles Hilbertiennes. C. R. Math. Acad. Sci. Paris **346**, 347–350 (2008)

Chapter 5
Uniform Confidence Bands for Joint Angles Across Different Fatigue Phases

Patrick Bastian, Rupsa Basu and Holger Dette

Abstract We develop uniform confidence bands for the mean function of stationary time series as a post-hoc analysis of multiple change point detection in functional time series. In particular, the methodology in this work provides bands for those segments where the jump size exceeds a certain threshold Δ. In [1] such exceedences of Δ were related to fatigue states of a running athlete. The extension to confidence bands stems from an interest in understanding the range of motion (ROM) of lower-extremity joints of running athletes under fatiguing conditions. From a biomechanical perspective, ROM serves as a proxy for joint flexibility under varying fatigue states, offering individualized insights into potentially problematic movement patterns. The new methodology provides a valuable tool for understanding the dynamic behavior of joint motion and its relationship to fatigue.

5.1 Introduction

Fatigue detection in biomechanical data of the lower extremity joint angles may be recast as change point detection in a functional time series, as seen previously in [2] and [1]. Methods in these references provide statistical methodologies for segmenting data based on the tiredness of athletes, ultimately providing phases where the athlete may be considered to be rested, pre-fatigued or fully exhausted. In this paper will explicate some of the consequences of theses results by studying

Patrick Bastian
Ruhr Universität Bochum, Universitätsstraße 150, 44801 Bochum,
e-mail: patrick.bastian@rub.de

Rupsa Basu
Universität zu Köln, Universitätstraße 24, 50931 Köln, e-mail: rbasu@uni-koeln.de

Holger Dette (✉)
Ruhr Universität Bochum, Universitätsstraße 150, 44801 Bochum
e-mail: holger.dette@rub.de

© The Author(s), under exclusive license to Springer Nature Switzerland AG 2025
G. Aneiros et al. (eds.), *New Trends in Functional Statistics and Related Fields*,
Contributions to Statistics, https://doi.org/10.1007/978-3-031-92383-8_5

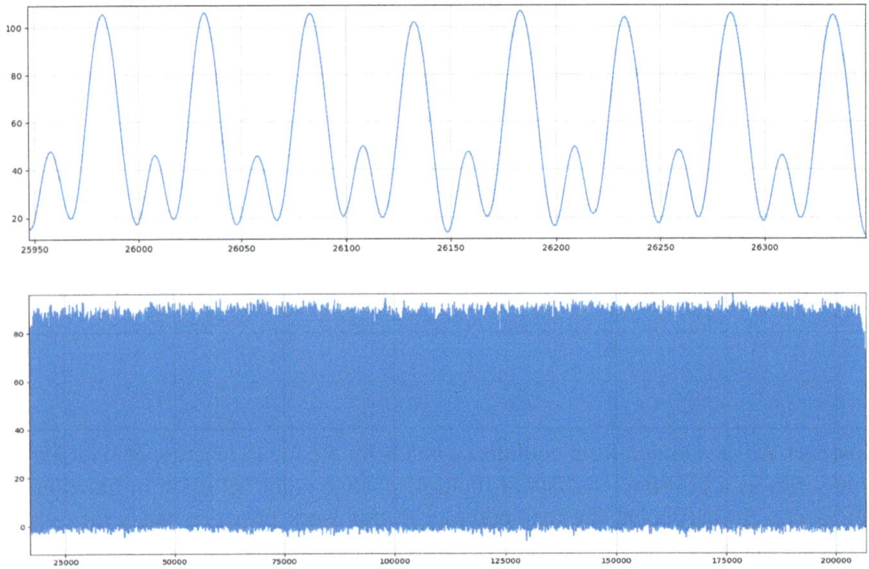

Fig. 5.1: Functional time series of knee angle data from runners. The upper panel shows 8 cycles of the data starting and ending with the contact of the same foot with the ground. The lower panel shows the full functional time series of knee angle data from the above runner with about 1200 cycles. Individual cycles not visible due to being condensed in the plot.

the fatigue attributed impact on the range of motion (ROM) within these phases. To achieve this, we construct confidence bands for the mean curves corresponding to each fatigue state.

The dataset motivating our work comprises of measurements of lower-extremity joint angles collected from running athletes subjected to a stress-inducing protocol specifically designed to simulate fatiguing conditions. Under fatigue, the body is expected to make adjustments to maintain endurance, resulting in changes in joint angles. In Fig. 5.1, we provide a snapshot of knee angle data from a single runner. Running, being a repetitive activity, naturally exhibits cycles, also called strides, which are clearly visible. In the upper part of the figure we see 8 strides, but the full data set consists of more than 1200 strides as shown in the lower part of Fig. 5.1.

Detecting fatigue related adaptations of the body can be considered as change detection problem for functional time series data. In this work, supposing that the locations of fatigue haven been identified, we address the problem of understanding the movement of the runner within the stable periods of no change by developing confidence bands for the mean function of the time series between two changes. Such bands are of interest to annotate the individual ROM of the joint angles of a particular running athlete, thereby obtaining a better understanding of the ROM during rested, pre-fatigue and fatigued stages of the run. In particular they serve as a powerful visual tool for the biomechanical scientist to establish whether the athlete under study is somehow undergoing movement that is abnormal as compared to her standard course of motion. Ultimately, this knowledge is beneficial for ensuring that injuries resulting from harmful movements during fatigued phases is curtailed timely, thereby ensuring enhanced longevity of the runners performance.

Summary of approach Consider that for a runner the lower extremity joint angle data from the knee[1] obtained over a course of the run are simply (discrete) realizations of an underlying function of time. Due to the cyclic nature of this data we have a long functional time series with each such function beginning and ending with the subsequent contact of the same foot with the ground. Each of these cycles, or strides in biomechanical terms, may be represented mathematically as an observation taking values in the space of continuous functions, say $C[0, 1]$, leading to a time series, say $X_{n,1}, ..., X_{n,n} \in C[0, 1]$, where n is the total number of strides. We model these observations in the form,

$$X_{n,j}(t) = \mu_{n,j}(t) + \epsilon_{n,j}(t) \quad j = 1, ..., n; \tag{5.1}$$

where $\{\epsilon_{n,j}\}_{j=1,...n}$ is triangular array whose rows are stationary and $\mu_{n,j} = \mathbb{E}[X_j] \in C[0, 1]$ denotes the underlying mean function of $X_{n,j}$. Notice here that the index j denotes the sequential nature of the cycles i.e., $j = 1$ is the joint angle data from the first stride followed by $j = 2$ and so on. Under this setup, detecting changes with statistical guarantees in the mean function $\mu_{n,j}$ then corresponds in biomechanical terms to detecting the onset of different fatigue states over the locations j in the average movement of the runner. In this work, we assume that $\mu_{n,j}$ is piecewise constant (in j) and that there are m (rescaled) locations of change denoted by $1 < \lfloor s_1 n \rfloor < \ldots < \lfloor s_m n \rfloor$, where $0 < s_1 < \ldots < s_m < 1$, where we define $s_0 = 0$ and $s_{m+1} = 1$.. More precisely, we assume that the mean function on the segment between $j = \lfloor ns_i \rfloor$ and $j = \lfloor s_{i+1} n \rfloor$ is given by

$$\mu_i(t) := \mu_{n,\lfloor s_i n \rfloor}(t) \ , \ \ i = \lfloor ns_i \rfloor, \ldots, \lfloor s_{i+1} n \rfloor - 1. \tag{5.2}$$

We are now interested in characterizing the range of motion in those intervals $[s_i, s_{i+1}]$ where the change at time s_i corresponds to a transition from one fatigue state to another.

This poses three problems. First we need to estimate the change points in such a functional time series, secondly we need to determine which of these change locations are relevant, i.e. correspond to transitions of the fatigue state. Thirdly, we then need to construct confidence bands for the associated segments. The first and second problems have been investigated in detail by [1]. In this note we will solve the third problem, i.e. we construct upper and lower confidence bands, respectively given by functions $t \to \hat{\mu}_i^+(t)$, $t \to \hat{\mu}_i^-(t)$ such that,

[1] Similar analysis from the hip and ankle- angle data may be pursued. For sake of brevity, we omit the study of these examples in this paper.

$$\lim_{n\to\infty} \mathbb{P}\Big(\cap_{i\in I} \big\{\text{for all } t \in [0,1] : \hat{\mu}_i^-(t) \le \mu_i(t) \le \hat{\mu}_i^+(t)\big\}\Big) = 1-\alpha, \quad (5.3)$$

where

$$I := \{0\} \cup \big\{1 \le i \le m \,\big|\, \|\mu_i - \mu_{i-1}\|_\infty > \Delta\big\} \quad (5.4)$$

is the set of changes of size larger than $\Delta > 0$. Here we include 0 for notational convenience of later statements. Δ denotes a threshold that can be chosen either automatically or by the practitioner and it serves to eliminate subtle changes in the data which may be due to external conditions like pedestrians, obstacles, or curvature on the path of the athlete. The main contribution of this note is the construction of a bootstrap procedure that allows us to compute the lower and upper confidence functions $\hat{\mu}_i^{\pm}, i \in I$ that fulfill (5.3), provided that estimators of the change points s_i and the set I are available.

With the additional contribution of this note we transform the biomechanical problem of studying the impact of fatigue on the range of motion into a three-step statistical problem of *(i)* detecting locations of change s_i in a functional time series, *(ii)* demarcating those changes $s_i, i \in I$ corresponding to relevant changes in the mean function, followed by *(iii)* constructing simultaneous confidence bands for the mean functions within the segments $[s_i, s_{i+1}], i \in I$.

5.2 Confidence bands for multiple means

Suppose there are a total of m change locations $s_1, ..., s_m$ whereof we have $k \le m$ locations of relevant change $s_{i_l}, l = 1, ..., k$ so that the set I in (5.4) is given by $I = \{0, i_1, ..., i_k\}$. We are interested in the $k+1$ segments of the functional time series given by $[0, s_{i_1}], [s_{i_1}, s_{i_1+1}], ..., [s_{i_k}, s_{i_k+1}]$ and their corresponding mean curves $\mu_0, \mu_{i_1}, ..., \mu_{i_k}$. Therefore, provided that the locations of the change points $s_i, i = 1, \dots, m$ and the indices $\{i_1, ..., i_k\}$ are known or can be estimated, the problem in (5.3) may be rephrased as obtaining confidence bands for the multivariate vector given by,

$$\mu = (\mu_0, \mu_{i_1}, ..., \mu_{i_k})^\top \in (C[0,1])^{k+1}\,. \quad (5.5)$$

The case $k = 0$ corresponding to no change point has already been treated in [4]. For the extension to the situation, where there are (relevant) change points, we require estimators for the the locations $s_1, ..., s_m$ of the changes, for their number m and for the set I. For this purpose we denote by $\hat{s}_i, i = 1, ..., \hat{m}$ the estimators from Algorithm 1 of [1], where $\hat{m}$ is the estimated number of change points and additionally let $\hat{I}$ denote the estimator of the set I of the relevant chnage points that can be obtained by Algorithm 2 in the same paper. Precise consistency statements for these estimates may be obtained in Theorem 4.1 and 4.2 of the given reference.

To obtain confidence bands we need to derive the asymptotic distribution of a suitable statistics such as

$$T_n = \max_{i_l \in \hat{I}} \sqrt{\hat{n}_{i_l}} \left\| \frac{\hat{\mu}_{i_l} - \mu_{i_l}}{\hat{\sigma}} \right\|_\infty , \tag{5.6}$$

where $\hat{n}_i = \lfloor n(\hat{s}_{i+1} - \hat{s}_i) \rfloor$, $\hat{\mu}_i = n_i^{-1} \sum_{k=\lfloor n\hat{s}_i \rfloor}^{\lfloor n\hat{s}_{i+1} \rfloor - 1} X_{n,k}$ and $\hat{\sigma}^2$ is an estimator of the long run variance function

$$\sigma^2 = \sum_{j\in\mathbb{Z}} \mathbb{E}[\epsilon_0 \epsilon_j],$$

which we define in equation (5.15) further below. One can show that it holds that $T_n \xrightarrow{d} T$ for some random variable T and we can then obtain the required confidence bands by defining,

$$\mu_i^{\pm}(t) = \hat{\mu}_i(t) \pm \hat{\sigma}(t) q_{1-\alpha/2} / \sqrt{n}, \quad i \in \hat{I}, \tag{5.7}$$

where $q_{1-\alpha}$ is the $(1-\alpha)$-quantile of T. The distribution of T however depends not only on the set I but also on the covariance structure of the functional time series and as such needs to also be estimated from the data. For this purpose, we will employ a multiplier bootstrap procedure.

We first construct a bootstrap version of T_n. To that end let $L = L(n)$ denote a block length and define the random variables

$$Y_{n,j} = X_{n,j} - \hat{\mu}_i, \qquad \hat{s}_i \le j/n < \hat{s}_{i+1} \tag{5.8}$$

$$\hat{\mu}_i^* = \hat{n}_i^{-1} \sum_{j=\lfloor n\hat{s}_i \rfloor}^{\lfloor n\hat{s}_{i+1} \rfloor - 1} \nu_j \Big(L^{-1/2} \sum_{l=0}^{L-1} Y_{n,j+l} \Big) \tag{5.9}$$

where $\{\nu_j\}_{j\in\mathbb{N}}$ is an iid sequence of standard normal random variables. The bootstrap test statistic is then given by

$$T_n^* = \max_{i\in\hat{I}} \sqrt{\hat{n}_i} \left\| \hat{\mu}_i^* / \hat{\sigma} \right\|_\infty . \tag{5.10}$$

From the bootstrap procedure, we obtain statistics $T_n^{*,(1)}, \dots, T_n^{*,(R)}$ and denote the $(1-\alpha)$-quantile by $q_{1-\alpha}^*$. For convenience we summarize the above procedure in Algorithm 1. Finally we obtain the desired confidence bands by defining,

$$\hat{\mu}_i^{\pm} = \hat{\mu}_i \pm \hat{\sigma}(t) \hat{q}_{1-\alpha/2}^* / \sqrt{\hat{n}_i} , \quad i \in \hat{I}. \tag{5.11}$$

Algorithm 1 Bootstrapping Quantiles

Compute the number $\hat{m}$ and all (ordered) change points $\hat{S} = \{\hat{s}_1, \ldots, \hat{s}_{\hat{k}}\}$ using BINSEG$(1, n, \xi_n)$ from [1]
Compute at significance level β the set of relevant change points $\hat{I}$ by Algorithm 2 from [1]
Compute the means $\hat{\mu}_i = \hat{n}_i^{-1} \sum_{j=\lfloor n\hat{s}_i \rfloor}^{\lfloor n\hat{s}_{i+1} \rfloor - 1} X_{n,j}$ for $i \in \hat{I}$
Compute the long run variance $\hat{\sigma}^2$ as given in (5.15)
Compute $Y_{n,j} = X_{n,j} - \hat{\mu}_i$ for $\hat{s}_i \leq j/n < \hat{s}_{i+1}$
Fix block length L, number of bootstrap replications R
for $r = 1, \ldots, R$ **do**
    Compute $\hat{\mu}_i^* = \hat{n}_i^{-1} \sum_{j=\lfloor n\hat{s}_i \rfloor}^{\lfloor n\hat{s}_{i+1} \rfloor} \nu_j \left(L^{-1/2} \sum_{l=0}^{L-1} Y_{n,j+l} \right)$ for $i \in \hat{I}$
    Compute $\hat{T}_n^* = \max_{i \in \hat{I}} \sqrt{\hat{n}_i} \left\| \hat{\mu}_i^* / \hat{\sigma} \right\|_\infty$
end for
Compute $\hat{q}_{1-\alpha}^* \leftarrow$ as the empirical $(1-\alpha)$-quantile of bootstrap sample $\hat{T}_n^{*,(1)}, \ldots, \hat{T}_n^{*,(R)}$.

Theorem 5.1. *Assume that Condition (A1)-(A4) in [3] hold. Let $\hat{q}_{1-\alpha}^*$ denote the $(1-\alpha)$ quantile obtained by Algorithm 1 where $\hat{I}$ is calculated at level β. Further assume that $\sigma(t)^2 > 0$ for all $t \in [0, 1]$ and that $\hat{\sigma}^2$ is a consistent estimator for σ^2. Then it holds, with $\hat{\mu}_i^{\pm}$ given as in* (5.11)*, that*

$$\liminf_{n\to\infty} \mathbb{P}\Big(\cap_{i\in I} \left\{ \text{for all } t \in [0,1] : \ \hat{\mu}_i^-(t) \leq \mu_i(t) \leq \hat{\mu}_i^+(t) \right\} \Big) \geq 1 - \alpha - \beta. \quad (5.12)$$

In the case that $\|\mu_{i+1} - \mu_i\|_\infty \neq \Delta$ for $1 \leq i \leq m$ we can strengthen this result to to

$$\lim_{n\to\infty} \mathbb{P}\Big(\cap_{i\in I} \left\{ \text{for all } t \in [0,1] : \ \hat{\mu}_i^-(t) \leq \mu_i(t) \leq \hat{\mu}_i^+(t) \right\} \Big) = 1 - \alpha. \quad (5.13)$$

Proof. We provide only a sketch of the proof for the sake of brevity, focusing on the second statement. The first follows by similar arguments. The proof proceeds in three steps, we first replace the change point estimators by their true values, then identify the relevant changes and finally employ weak invariance principles to establish the necessary asymptotic statements that yield validity of the bootstrap. In a bit more detail we

1. Use Theorem 4.1 from [1] to replace the estimated change points by the true change points in the definitions of $\hat{n}_i$ and $\hat{\mu}_i$.
2. Use Theorem 4.2 from [1] to show that $\hat{I} = I$ with probability $1 - o(1)$.
3. Observe that multiplication by σ, taking component wise maxima and the supremum norm are continuous functions and that by Lemma 1 (below) we may replace $\hat{\sigma}$ by σ. Then use Theorem 2.2 and 4.3 from [4] in combination with the continuous mapping theorem to show that T_n and finitely many copies $T_{n,i}^*$ of T_n^* converge jointly in distribution, more precisely

$$(T_n, T_n^{*,(1)}, \ldots, T_n^{*,(R)}) \xrightarrow{d} (T, T^{(1)}, \ldots, T^{(R)}), \quad (5.14)$$

   where T_i are independent copies of T.

Combining these steps immediately yields validity of the bootstrap scheme.

Now we provide the precise definition of the long run variance estimator $\hat{\sigma}^2$ and derive its consistency. Let

$$\hat{\sigma}^2 = \sum_{l=-c}^{c} \hat{\sigma}_l^2 K(l/c) \tag{5.15}$$

for some bandwidth parameter $c \to \infty$, kernel K and where

$$\hat{\sigma}_l^2 = \begin{cases} \frac{1}{n}\sum_{j=1}^{n-l} \big(X_j - \hat{\mu}^{(j)}\big)\big(X_{j+\ell} - \hat{\mu}^{(j)}\big), & l \geq 0, \\ \frac{1}{n}\sum_{j=1-l}^{n} \big(X_j - \hat{\mu}^{(j)}\big)\big(X_{j+\ell} - \hat{\mu}^{(j)}\big), & l < 0, \end{cases} \tag{5.16}$$

are estimators for the lag variances and $\hat{\mu}^{(j)} = \hat{\mu}_i$ for $\lfloor \hat{s}_i \rfloor \leq j < \lfloor \hat{s}_{i+1} \rfloor$. A symmetric function K is called kernel whenever $K(0) = 1, K(1) = 0$ and $K(x) = 0$ for $|x| > 1$.

Lemma 1 *Assume that Condition (A1)-(A4) in [3] hold, then the long run variance estimator defined in* (5.15) *is consistent, i.e.* $\|\hat{\sigma}^2 - \sigma^2\| = o_{\mathbb{P}}(1)$ *whenever the bandwidth* c *satisfies* $c \to \infty$ *and* $c^3/n \to 0$.

Proof. We will show the result for mean zero processes $\epsilon_1, \ldots, \epsilon_n$ satisfying Conditions (A1) to (A4) in [3] and the estimator $\hat{\sigma}^2 = \sum_{l=-c}^{c} \hat{\sigma}_l^2 K(l/c)$ where

$$\hat{\sigma}_l^2 = \begin{cases} \frac{1}{n}\sum_{j=1}^{n-l} \epsilon_j \epsilon_{j+l}, & l \geq 0, \\ \frac{1}{n}\sum_{j=1-l}^{n} \epsilon_j \epsilon_{j+l}, & l < 0, \end{cases} \tag{5.17}$$

and we define $\sigma_l^2 = \mathbb{E}[\epsilon_j \epsilon_{j+l}]$. The general result then follows by straightforward approximation arguments. Observe the following decomposition:

$$\begin{aligned} \hat{\sigma}^2 - \sigma^2 = &- \sum_{|j|>c} \sigma_j^2 + \sum_{j=-c}^{c} (\hat{\sigma}_j^2 - \sigma_j^2) \\ &- \sum_{j=-c}^{c} (\hat{\sigma}_j^2 - \sigma_j^2)(1 - K(j/c)) - \sum_{j=-c}^{c} \sigma_j^2 (1 - K(j/c)) \end{aligned}$$

By Lemma A from [5] we have that $\sup_{t\in[0,1]} \sigma_j^2(t) \lesssim \sqrt{\phi(j)}$ which is a summable sequence by (A4). Therefore the first summand is of order $o(1)$. The fourth summand converges to 0 by Lebesgue's Theorem. Finally, Arguments similar to those given on page 18 of the supplement of [4] and a union bound argument yield that

$$\max_{-c\leq j\leq c} \left\|\hat{\sigma}_j^2 - \sigma_j^2\right\|_\infty = O_{\mathbb{P}}(\sqrt{c/n}).$$

This implies that the second and third summand are of order $O_{\mathbb{P}}(\sqrt{c^3/n})$.

5.3 Application to sports data

In this section, we explore the application of the methodology presented above to biomechanical knee angle data collected under a fatigue protocol obtained from the collaboration project *Sports, Data, and Interaction*[2]. The data examples presented in this work are collected via marker-based optical motion capture systems placed in a laboratory while the athlete ran on a treadmill. Following relevant change detection to demarcate changes corresponding to fatigue we now study the range of motion of the athlete by means of the confidence bands we introduced. For a more detailed discussion of this dataset and the literature on fatigue-related change detection, we refer the interested reader to [1] and [2]. In all of the following data examples, Band 0 corresponds to the confidence bands around the mean curve μ_0 i.e, the mean in the segments $[0, s_1]$, Band 1 corresponds to mean μ_1, i.e. to the segment $[s_{i_1}, s_{i_1+1}]$ and so on.

5.3.1 I: Confidence bands for knee angles

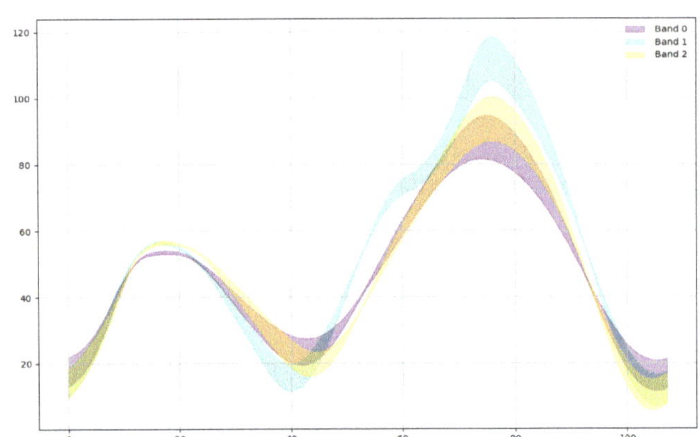

Fig. 5.2: Confidence bands pre- and post- each change for right knee angles from runner A. Total sample size $n = 1830$ and three phases are detected for a relevant size $\Delta = 6.6$.

In Fig. 5.2, we present the confidence bands for a novice runner under fatiguing conditions. They have about 2 years of running experience and their dominant leg is the right leg. To determine the value of Δ in a data-driven manner, we compute the average of the initial 5% of the data, denoted as $\hat{\mu}_{\text{initial}}$, and the average of the final

[2] http://www.sports-data-interaction.com/

5% of the data, denoted as $\hat{\mu}_{\text{final}}$. Then, we set $\Delta = \frac{\|\hat{\mu}_{\text{final}} - \hat{\mu}_{\text{initial}}\|_\infty}{3}$, similarly to [1]. For the case of this runner, we obtain $\Delta = 6.6$. Band 0 corresponds to the rest phase, Band 1 corresponds to the pre-fatigue phase and Band 2 finally corresponds to the fatigue phase. Notice that the intermediate shift upwards in the second peak of knee angles is an initial adjustment made by the runner to endure fatigue. Band 2 indicates that the final adjustment corresponds to lesser bending of the knee while the foot is in the air, which is clearly reflected by the large jump in the second peak of the bands. The lower and upper bands at each part of the cycle provide useful information on the range of motion of the knee movement through different segments of the run. We stress that the exceedance threshold Δ eliminates all segments of non-fatigue related deviations in the knee angles.

5.3.2 II: Comparison novice and experienced runner

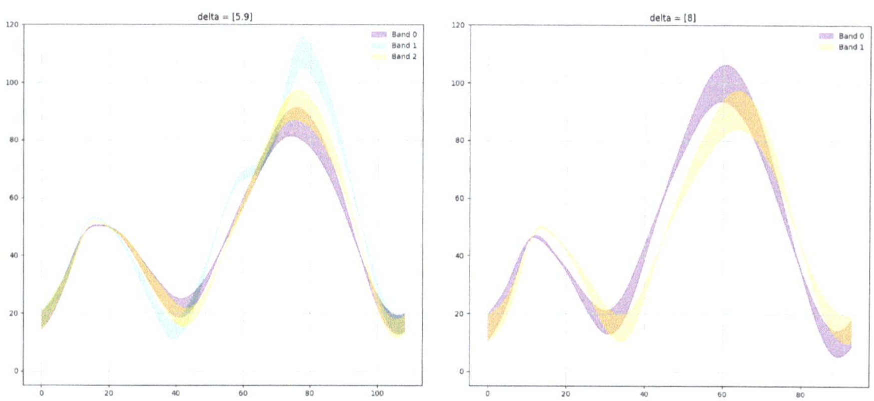

Fig. 5.3: Confidence bands from two runners. **Left**: (Novice-) runner A left knee angles for relevant size $\Delta = 5.9$ and **Right**: (experienced-) runner B left knee angles for relevant size $\Delta = 8$.

In Fig. 5.3 we compare two runners with different experience levels. On the left we have the same runner A as in Fig. 5.2 while on the right we have an experienced runner B with about 11 years of running experience and a dominant left leg. This time we consider the left leg for both runners, notice that runner B has only one relevant change of size $\Delta = 8$. This may be due to B being well trained, i.e. their body is used to fatiguing conditions and therefore needs to make fewer adjustments to endure accumulated fatigue. However, just as for the less experienced runner A, we can see that in the fatigue phase there is reduced bending of the knee at the second peak. This behaviour is consistent across runners even beyond the two cases we present here and is in accordance with prior studies in the biomechanical gait literature, see for example [6], where this is attributed to a protection mechanism of the runner.

Acknowledgements We thank R. van Middelaar and A. Balasubramaniam of the University of Twente., the Netherlands for the immense amount of work they put in providing high-quality data from running athletes. The authors gratefully acknowledge the support by the Deutsche Forschungsgemeinschaft (DFG), project number 511905296 titled: *Modeling functional time series with dynamic factor structures and points of impact* and TRR 391 *Spatio-temporal Statistics for the Transition of Energy and Transport*, project number 520388526.

References

1. Bastian, P., Basu, R., and Dette, H.: Multiple change point detection in functional data with applications to biomechanical fatigue data. The Annals of Applied Statistics, **18**(4), 3109 – 3129, (2024)
2. Basu, R. and Proksch, K.: Fatigue detection via sequential testing of biomechanical data using martingale statistic. (2023)
3. Dette, H. and Kokot, K.: Detecting relevant differences in the covariance operators of functional time series: a sup-norm approach. Annals of the Institute of Statistical Mathematics, **74**, 195–231, (2022)
4. Dette, H., Kokot, K., and Aue, A.: Functional data analysis in the Banach space of continuous functions. The Annals of Statistics, **48**(2), 1168 – 1192, (2020)
5. Yoshihara, K.-i.: Moment inequalities for mixing sequences. Kodai Mathematical Journal, **1**(2), 316 – 328, (1978)
6. Zandbergen, M. A., Marotta, L., Bulthuis, R., Buurke, J. H., Veltink, P. H., and Reenalda, J.: Effects of level running-induced fatigue on running kinematics: A systematic review and meta-analysis. Gait & posture, **99**, 60–75, (2023)

Chapter 6
Scalar on Shape Regression Using Function Data

Sayan Bhadra and Anuj Srivastava

Abstract Functional data contains two components: shape and phase. This paper is a part of functional data analysis (FDA), called *Shape-Based FDA*, that isolates and focuses on shapes of functions. Specifically, this paper focuses on Scalar-on-Shape (ScoSh) regression models that utilize shapes of predictor functions and treat their phases as free variables. ScoSh model optimizes over phases using the (nonparametric) Fisher-Rao inner product and nonlinear index functions to capture complex predictor-response relationships. The paper presents experiments involving COVID curves as predictors and related health outcomes as responses.

6.1 Introduction

Functional regression is a branch of functional data analysis (FDA) that deals with using functional variables in regression models as predictors, responses, or both. Specifically, in *Scalar-on-Function* (ScoF) models, some functions $\{f_i\}$ play the role of predictors, and some scalars $\{y_i\}$ are treated as responses. ScoF models have widespread applications across scientific domains and are natural extensions of the standard multivariate regression models to functional data. Our focus in this paper is on the *shapes* (also termed *amplitudes*) of functions rather than the full functions themselves. This focus is motivated, for example, by problems in neuroimaging where morphologies of anatomical parts are used to predict clinical measurements, such as disease progression or treatment effects. Accordingly, we develop a regression model, called *Scalar-on-Shape* (ScoSh), where the *shapes of functions* are treated as predictors for scalar clinical responses. For a detailed version of this article, please

Sayan Bhadra
Dept. of Statistics, Florida State University, e-mail: sbhadra@fsu.edu

Anuj Srivastava (✉)
Dept. of Statistics, Florida State University, e-mail: anuj@stat.fsu.edu

© The Author(s), under exclusive license to Springer Nature Switzerland AG 2025

G. Aneiros et al. (eds.), *New Trends in Functional Statistics and Related Fields*,
Contributions to Statistics, https://doi.org/10.1007/978-3-031-92383-8_6

refer to [2].

Shapes of Functions: Mathematically, *shape* is a property of data objects that is invariant to certain transformations (see [3] etc.). These transformations are called shape-preserving and are treated as nuisance variables in shape analysis. In case of 3D objects, these transformations are rigid rotations, translations, and global scaling. For parameterized data objects, such as curves or surfaces, their shapes are additionally invariant to the re-parameterizations. For scalar functions, $\mathcal{F} = \{f : [0,1] \to \mathbb{R}\}$, the notion of shape relates to the number and height of extremas (peaks and valleys), but is invariant to horizontal shifts in their locations. Changes in the locations of these points are effected by time warping functions (positive diffeomorphisms γ_i of the time domain $[0,1]$) and implemented using the transformation $f_i \mapsto f_i \circ \gamma_i$. This transformation are called *phase changes* and are considered nuisance or free variables in shape analysis of f_i's. In other words, a function f_i and its transformation $f_i \circ \gamma_i$, for any time warping γ, are deemed to have identical shape. Several methods (*e.g.*, [6, 7, 10]) have been developed to separate shape from phase components in functional data analysis. These methods are primarily based on aligning the geometric features (peaks and valleys) of $\{f_i\}$ using optimal time warping.

In ScoSh, we want to utilize the *shapes* of predictor functions $\{f_i\}$, rather than full functions, as predictors in regression models. One possibility is to use a preprocessing step where we separate phase and shape components using a past method, followed by a regression model that intakes only the shape components. However, a more comprehensive approach is to perform phase-shape separation inside the regression model. This way, the response variables $\{y_i\}$ can also contribute to the definition of phase. How can we develop such a model? We can do that by setting the phases $\{\gamma_i\}$ as free variables in a functional regression model. While previously $\{\gamma_i\}$ were selected to align peaks and valleys of $\{f_i\}$, for isolating shapes, now we can select $\{\gamma_i\}$ to help minimize the prediction errors of $\{y_i\}$. Consequently, the phase-shape separation takes place inside the regression model rather than as a preprocessing step. Treating phases as free variables and optimizing over them can enhance prediction performance and interpretation in a regression model.

The literature on such shape regression is scarce, but extensive research exists in some related areas. We discuss the related ScoF models briefly. [9] introduced the parametric functional linear model (FLM) that was extended ([1]) into an index FLM as:

$$y_i = \alpha + h(\langle \beta, f_i \rangle) + \epsilon_i, \; i = 1, 2, \ldots, n, \tag{6.1}$$

where the $y_i \in \mathbb{R}$ is the response, $f_h \in \mathcal{F}$ is the predictor function, $\alpha \in \mathbb{R}$ is an offset, $\langle \cdot, \cdot \rangle$ is the $\mathbb{L}^2$ inner-product, $\beta \in \mathcal{F}$ is regression coefficient function β, $h : \mathbb{R} \to \mathbb{R}$ is an index function, and ϵ_i is the observational noise. Several papers ([8, 4, 5]) introduced penalty terms to regularize the estimation of β.

We develop a Scalar-on-Shape (ScoSh) model that is fully invariant to phase changes in the predictors. It uses predictor phases as free variables and optimizes over them during model estimation.

6.2 Proposed Model: Scalar on Shape Regression

To focus on shapes of $\{f_i\}$, we need invariance to their phases, *i.e.*, replacing any f_i with $f_i \circ \gamma_i$ should not change the response y_i. In other words, γ_i's become free variables. To achieve this invariance, we use a nonparametric version of the Fisher-Rao metric (FRM) (see [10]). The calculation of a FRM utilizes the square-root velocity function (SRVF) of a function f_i, defined by $q(t) \triangleq \text{sign}(\dot{f}(t))\sqrt{|\dot{f}(t)|}$. The SRVF of a time-warped function $f_i \circ \gamma_i$ is given by $(q_i \circ \gamma_i)\sqrt{\dot{\gamma}_i}$, which we will denote by $(q_i \star \gamma_i)$. The Fisher-Rao inner product between two functions f_i, f_j is computed as the $\mathbb{L}^2$ inner product between their corresponding SRVFs, *i.e.*, $\langle q_i, q_j \rangle$. A critical property of FRM is its invariance to time warping, *i.e.*, $\langle q_i, q_j \rangle = \langle q_i \star \gamma, q_j \star \gamma \rangle$ for any time warping γ. With this notation, we utilize the term $\sup_{\gamma \in \Gamma} \langle \beta, q_i \star \gamma \rangle$ as a surrogate for $\langle \beta, f_i \rangle$ in Eqn. 6.1. Here Γ is the group of all-time warping functions. The invariance of FRM and the group structure of Γ results in the property: $\sup_{\gamma \in \Gamma} \langle \beta, q_i \star \gamma \rangle = \sup_{\gamma \in \Gamma} \langle \beta, (q_i \star \gamma_0) \star \gamma \rangle$, for any $\gamma_0 \in \Gamma$. Thus, this expression is truly invariant to the phase of f_i and depends only on its shape. To add flexibility to the model, we introduce two functions: (1) an index function $h : \mathbb{R} \to \mathbb{R}$, and (2) an offset function $g : \mathbb{R} \to \mathbb{R}$. We will assume that $h, g \in C(\mathbb{R}, \mathbb{R})$ (or just C). The overall model can now be stated as:

$$y_i = g(f_i(0)) + h\left(\sup_{\gamma_i \in \Gamma} \langle \beta, q_i \star \gamma_i \rangle \right) + \epsilon_i \ , \ i = 1, \ldots, n. \tag{6.2}$$

Here $q_i, \beta \in \mathbb{L}^2$, $\gamma_i \in \Gamma$, $g, h \in C$, and $\epsilon_i \in \mathbb{R}$ are *i.i.d.* from $N(0, \sigma^2)$. We will call this the Single-Index Scalar-on-Shape (SI-ScoSh) model. The parameters of this model are $\{\beta, h, g, \sigma^2\} \in \mathbb{L}^2 \times C \times C \times \mathbb{R}_+$. As a special case, we will also study when $h(x) = x$ and will term it the ScoSh model.

Next, we consider the estimation of model parameters from the observed data $\{(f_i, y_i) \in \mathcal{F} \times \mathbb{R},\ i = 1, 2, \ldots, n\}$. We pre-compute the SRVFs $\{q_i\} \in \mathbb{L}^2$ of the predictor functions $\{f_i\}$. Then, given the observations $\{(y_i, q_i, f_i(0)) \in \mathbb{R} \times \mathbb{L}^2, i = 1, \ldots, n\}$, the inference problem is to estimate the quantities h, g, β, and σ^2 from the data. To simplify estimation, we will express $\beta \in \mathbb{L}^2$ using a truncated orthogonal basis $\mathcal{B} = \{b_j, j = 1, \ldots, J\}$ according to: $\beta(t) = \sum_{j=1}^{J} c_j b_j(t)$. $\mathcal{B}$ can be either a predefined basis, *e.g.*, the Fourier basis, or can be extracted from the training dataset through functional PCA. Then, the maximum-likelihood estimates of h, g and $\mathbf{c} = \{c_j\}$ are given by:

$$(\widehat{\mathbf{c}}, \widehat{h}, \widehat{g}) = \underset{\mathbf{c} \in \mathbb{R}^J, h \in C(\mathbb{R},\mathbb{R}), g \in C(\mathbb{R},\mathbb{R})}{\arg\min} H(\mathbf{c}, g, h), \quad \text{where} \tag{6.3}$$

$$H(\mathbf{c}, g, h) \triangleq \left[\sum_{i=1}^{n} \left\{ y_i - g(f_i(0)) - h\left(\sup_{\gamma_i \in \Gamma} \left\langle \sum_{j=1}^{J} c_j b_j, (q_i \star \gamma_i) \right\rangle \right) \right\}^2 \right].$$

One can additionally impose a roughness penalty on β to control its smoothness if needed.

To minimize H with respect to g, h, and β, we use a coordinate-descent approach, optimizing one parameter at a time while fixing the others. Estimating σ^2 from the residual variance is straightforward and not discussed. Keeping h and g fixed, $\widehat{\beta}$ is estimated by minimizing $H(c, g, h)$ over $\mathbf{c}$ while optimizing over free variables γ_i. Keeping β and g fixed, we estimate h using a fitting procedure that uses either polynomials or splines. An estimate of g is obtained similarly keeping h and β fixed.

6.3 Experimental Results

Simulated Data Experiments

In this section, we simulate several datasets and use them to evaluate the proposed as well as some current models. We generate $f_i^0(t) = c_{i,1}\sqrt{2}\sin(2\pi t)+c_{i,2}\sqrt{2}\cos(2\pi t)$, where $c_{i,1}, c_{i,2} \sim \mathcal{N}(0,1)$ and we create predictors by perturbing f^0's with arbitrary γ_i's: $f_i(t) = f_i^0 \circ \gamma_i$, where $\gamma_i(t) = t + \alpha \cdot t(1-t)$, $t \in [0,1]$, $\alpha \sim U(-1,1)$. We calculate the SRVFs (q_i's) of these f_i's using $q_i = \text{sign}(\dot{f}_i(t))\sqrt{|\dot{f}_i(t)|}$. To define the coefficient function β, we use the first J elements of the Fourier basis $\{b_j\} = \{\sqrt{2}\cos(2\pi jx), \sqrt{2}\sin(2\pi jx), j = 1, 2 \ldots, J/2\}$ and fixed coefficients $\tilde{c}_0 = \{1, \cdots, 1\}$. We also use low-order polynomials h (listed below with the results) and a fixed $g(x) = x^2 - 1$. Responses y_i's are calculated by adding $\epsilon_i \sim \mathcal{N}(0, 0.01^2)$ as per Eqn. 6.2. For a sample size of $n = 100$, we randomly set 80% of the dataset for training and the rest for testing in a five-fold cross-validation.

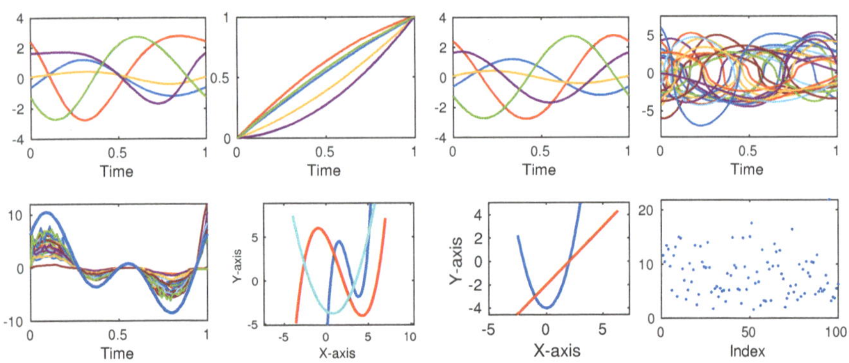

Fig. 6.1: Simulated Data. Top: From left to right, initial functions $\{f_i^0\}$ formed using a Fourier basis, random time-warpings $\{\gamma_i\}$, time-warped functions $\{f_i = f_i^0 \circ \gamma_i\}$, and their SRVFs $\{q_i\}$. Bottom: SRVF's after registering with a β (blue line), some index functions h, some polynomial offset functions g, and the responses $\{y_i\}$ generated after adding random noise.

Experiment 1: Prediction performance comparison with different parameter complexities

Table 6.1: Listing of various models studied in this paper. Here SI = Single Index. SI-ScoSh is the proposed model, SI-ScoF are past ideas. Here q_i is the SRVF of predictor f_i.

SI-Scalar-on-Shape (SI-ScoSh)	$y_i = g(f_i(0)) + h\left(\sup_{\gamma_i \in \Gamma} \langle \beta, q_i \star \gamma_i \rangle\right) + \epsilon_i$
SI-Scalar-on-Function (SI-ScoF)($\mathbb{L}^2$)	$y_i = c_i + h(\langle \beta, f_i \rangle) + \epsilon_i$
SI-Scalar-on-Function (SI-ScoF)(FR)	$y_i = g(f_i(0)) + h(\langle \beta, q_i \rangle) + \epsilon_i$

This section evaluates prediction performance R^2s of the models SI-ScoSh and SI-ScoF(FR) (Table 6.1) for different model parameters using the simulated data. Note that we consider two ScoF models: one based on $\mathbb{L}^2$ inner product and one based on FRM.

1. **Varying degrees of Index Function** h: Here we study how varying the degree of the index function h affects the prediction performance of the SI-ScoSh model and compare it with the SI-ScoF model. We generate data from a quadratic or cubic h_{true} (keeping other parameters fixed) and allow different polynomial degrees $(1-4)$ during estimating of h. From the prediction summaries presented in the top part of Table 6.2, one can see that higher-order polynomials improve prediction only up to the order of the true polynomial index function. In comparison, the SI-ScoF is not able to exhibit any significant R^2 for any h.

Table 6.2: Prediction performance comparison SI-ScoSh vs SI-ScoF for different complexities of one of $\widehat{h}, \widehat{\beta}$ and $\widehat{g}$ with the other two parameters fixed. The numbers are mean R^2s (with standard deviations in parenthesis) over five cross-validated sets. (quad=quadratic, cub=cubic)

		SI-ScoSh for varying degrees of h_{est}			SI-ScoF(FR)
h_{true}	quad	linear	quadratic	cubic	
		0.89(0.03)	0.98(0.01)	0.99(0.01)	0.6(0.3)
	cub	quadratic	cubic	quartic	
		0.73(0.16)	0.95(0.04)	0.96(0.03)	0.5(0.2)
		SI-ScoSh for varying # basis functions for β_{est}			SI-ScoF(FR)
β_{true}	J=4	J=3	J=4	J=6	
		0.93(.03)	0.99(.01)	0.98(.01)	0.62(.14)
	J=6	J=4	J=6	J=10	
		0.93(.03)	0.99(.01)	0.98(.01)	0.62(.14)
		SI-ScoSh			SI-ScoF(FR)
g_{true}	quad	0.98(0.01)			0.54 (0.2)
	cub	0.99(0.01)			0.12 (0.4)

2. **Varying complexities of** β: Next we compare prediction performances for a varying the number of basis elements to estimate β, while keeping h and g fixed. We construct β_{true} using 4 and 6 Fourier basis elements, and study the results. In the middle portion of Table 6.2, the resulting R^2's show that increasing $J(\beta_{est})$ beyond the order of $J(\beta_{true})$ does not enhance prediction performance in SI-ScoSh. For $J < J_{true}$, our model performs worse than $J \geq J_{true}$ as they fail to capture the shape

of the β_{true}. In comparison, SI-ScoF gives an inferior performance for all values of J.

3. **Estimation Error for** g: To compare the prediction performances of SI-ScoSH vs. SI-ScoF for different complexities of the offset function, we generate data with fixed h and β, setting g_{true} as quadratic or cubic. The prediction performances (see bottom part of Table 6.2) show the SI-ScoSh model's superiority over the SI-ScoF(*FR*) model.

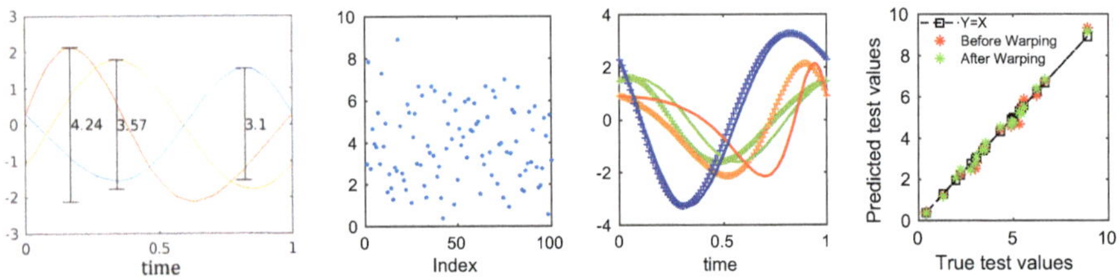

Fig. 6.2: Examples of f_is (left), noisy measurements of y_is (second), the original test set in solid lines and their perturbed version in hashed lines (third panel), and SI-ScoSh predicted $\widehat{y}_i$ plotted versus true y_is (last panel). Responses y_i are the max amplitude of predictors f_i.

Experiment 2: Evaluating Model Invariance to Random Phases

This experiment assesses the invariance of SI-ScoSh to phase variability in predictors. Here we design a response variable that is mathematically invariant to phase shifts using the generative model: $y_i = (\max(f_i(t)) - \min(f_i(t)) + \epsilon_i$, where $\epsilon_i \sim \mathcal{N}(0, 0.5)$, and the predictors $\{f_i\}$ are generated as in Sect. 6.3. Fig. 6.2 shows some predictors (first panel) and responses (second panel). A separate test set (third panel) is created by perturbing the training set predictors with random time warpings. Then, we apply SI-ScoSh and SI-ScoF(*FR*) to this noisy and time-warped data.

On this data, SI-ScoSh achieves $R^2 = 0.98$, while SI-ScoF(*FR*) has $R^2 < 0.1$. SI-ScoF(*FR*)'s lack of optimization over $\{\gamma_i\}$ results in inferior performance. The high performance of the ScoSh model (last column in Fig. 6.2) underscores its invariance to random phases of predictor functions.

Real Data Experiments:

In this section, we investigate the proposed SI-ScoSh model on several real datasets and compare the results with SI-ScoF models ($\mathbb{L}^2$ & *FR*). Two versions of SI-ScoSh have been implemented: one in which the index function is estimated by a polynomial, and the other in SVM regression.

1. **Spanish Weather Data**: This data consists of functional predictors for 73 Spanish weather stations. Each curve has 365 points, one for each day of the year. The value for each data is the average temperature on that day over several years from 1980 to 1993 (see left plot in Fig. 6.3). The scalar responses are the mean of temperatures over a future period (1994-2009) at each station. On this data,

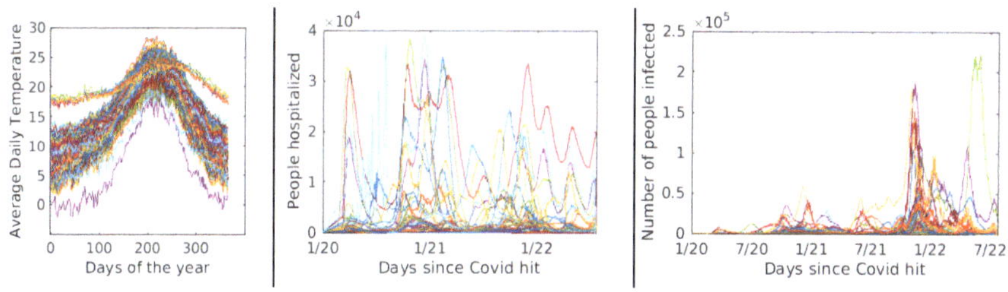

Fig. 6.3: From left to right are the predictor functions for *Spanish Weather*, *Covid Hospitalization* & *Covid Infection* datasets

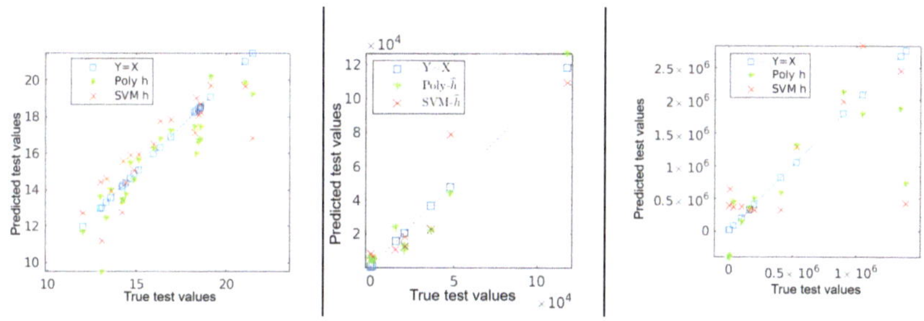

Fig. 6.4: From left to right are the SI-ScoSh models' predictions against true test values for the *Spanish Weather*, *Covid Hospitalization* & *Covid Infection* datasets

SI-ScoSh models have prediction R^2s around 0.9 (left plot Fig. 6.4) whereas SI-ScoF($\mathbb{L}^2$& FR) models have R^2s around 0.5.

2. **Covid Hospitalization Data**: This data contains daily COVID-19 hospitalizations across 31 European nations, which are used as predictors (middle plot in Fig. 6.3). The response variables y_i are the total deaths per country during the observation period. The SI-ScoSh model with a quadratic g, cubic h, and six Fourier basis elements β performed best (test set prediction $R^2 > 0.92$, see middle plot Fig. 6.4). The SI-ScoF ($\mathbb{L}^2$ & FR) models showed weaker predictions (R^2 less than 0.2).
3. **Covid Infection Data**: This dataset includes daily COVID-19 infection rates for 41 countries as predictors (right plot in Fig. 6.3), with the total number of hospitalizations as the response. The SI-ScoSh models predict test responses with $R^2 = 0.89$ (right plot Fig. 6.4), but the SI-ScoF(FR & $\mathbb{L}^2$) models show a weaker predictor-response relationship ($R^2 = 0.4$).

6.4 Conclusion

Functional data consists of phase and shape components, contributing differently to regression models. This paper introduces the SI-ScoSh model that focuses on the predictor shapes, treats phases as free variables, and optimizes phases to minimize prediction errors. In several experiments, it outperforms the classical ScoF models.

A ScoSh model is effective when phase components provide little information about the response. Such is often the case on image analysis and neuroimaging datasets. In contrast, when phase holds significant information, one can separate phases and shapes and use them both as predictors in a regression model.

References

1. Ait-Saïdi A., Ferraty F., Kassa R., and Vieu P.: Cross-validated estimations in the single-functional index model. Statistics, **42**(6), 475–494, (2008)
2. Bhadra S. and Srivastava A.: Scalar-on-shape regression models for functional data analysis. arXiv:2411.15326, (2024)
3. Dryden I. L. and Mardia K. V.: Statistical Shape Analysis, with Applications in R. Second Edition. John Wiley and Sons, Chichester, (2016)
4. James G.M., Wang J., and Zhu J.: Functional linear regression that's interpretable. The Annals of Statistics, **37**(5A), 2083 – 2108, (2009)
5. Ryung E.L. and Park B.U.: Sparse estimation in functional linear regression. Journal of Multivariate Analysis, **105**(1), 1–17, (2012)
6. Marron J.S., Ramsay J.O., Sangalli L.M., and Srivastava A.: Statistics of time warpings and phase variations. Electronic Journal of Statistics, **8**(2), 1697–1702, (2014)
7. Marron J.S., Ramsay J.O., Sangalli L.M., and Srivastava A.: Functional Data Analysis of Amplitude and Phase Variation. Statistical Science, **30**(4), 468 – 484, (2015)
8. Marx B.D. and Eilers P.H.C.: Generalized linear regression on sampled signals and curves: a p-spline approach. Technometrics, **41**(1), 1–13, (1999)
9. Ramsay J.O. and Silverman B.W.: Fitting differential equations to functional data: Principal differential analysis. Functional data analysis, 327–348, (2005)
10. Srivastava A. and Klassen E.P. Functional and shape data analysis, volume 1. Springer, (2016)

Chapter 7
A Comparison of Band-based Approaches to Functional Depth

Filip Bočinec, Erik Mendroš and Stanislav Nagy

Abstract Statistical depth functions have become essential tools for analyzing multivariate data, as they provide center-outward ordering of the data and thus measure centrality. Classical depth notions, such as Tukey depth and simplicial depth, are widely applied in various statistical tasks, including outlier detection and data visualization. With the growing demand for analyzing complex data structures, depth functions have been extended to functional data, where observations are treated as functions rather than finite-dimensional vectors. This article focuses on functional depths defined using bands formed by random functions. We study four such depth functions: (i) band depth, (ii) modified band depth, (iii) spherical depth, and (iv) modified spherical depth, the last of which is introduced here for the first time. We compare these notions with an emphasis on theoretical properties and evaluate their practical performance through an application to real-world data.

7.1 Introduction: Statistical Depth in Multivariate and Functional Data Analysis

Statistical depth functions are essential tools in multivariate data analysis [4], offering robust methods for ordering data. Classical notions such as the Tukey depth and the simplicial depth have been widely applied in tasks like outlier detection,

Filip Bočinec (✉)
Charles University, Faculty of Mathematics and Physics, Sokolovská 83, 186 75 Prague, Czech Republic, e-mail: bocinec@karlin.mff.cuni.cz

Erik Mendroš
Charles University, Faculty of Mathematics and Physics, Sokolovská 83, 186 75 Prague, Czech Republic, e-mail: erikmendros@gmail.com

Stanislav Nagy
Charles University, Faculty of Mathematics and Physics, Sokolovská 83, 186 75 Prague, Czech Republic, e-mail: nagy@karlin.mff.cuni.cz

© The Author(s), under exclusive license to Springer Nature Switzerland AG 2025

G. Aneiros et al. (eds.), *New Trends in Functional Statistics and Related Fields*, Contributions to Statistics, https://doi.org/10.1007/978-3-031-92383-8_7

classification, clustering, and other rank-based methods. These functions provide a center-outward ranking of data points and measure centrality with respect to multivariate datasets. Lately, the increasing complexity of modern datasets has driven the extension of depth functions beyond the multivariate framework.

Functional data analysis (FDA) deals with datasets where individual observations are functions, often continuous or sampled over a dense grid. Such data appear in applications like growth curve analysis, temperature records, stock market trends, and biomechanical measurements. Standard depth functions for multivariate data often face conceptual and computational challenges in the functional setting [8], leading to the development of depth notions specifically tailored to functional data. Functional depths [1] assess the centrality of entire trajectories of the functional data rather than of individual functional values, making them indispensable for tasks such as outlier detection and visualization [10] in FDA.

Among functional depth approaches, those based on bands formed by random functions have gained particular attention. The band depth and its extension, the modified band depth [6], evaluate how well a function is "covered" by bands generated from subsets of the data. Conversely, the spherical depth, originally developed for multivariate data [2], has been adapted to functional settings [7], where it assesses centrality based on distances in a specific function space. Despite sharing similarities, a systematic comparison of these functional depths and their properties is lacking.

This article seeks to address this gap by comparing four functional depths: (i) band depth (**BD**), (ii) modified band depth (**MBD**), (iii) spherical depth (**SD**), and (iv) modified spherical depth (**MSD**), the last depth being a newly proposed concept. In Sect. 7.2, we introduce these depth notions, including the novel **MSD**. Sect. 7.3 examines their theoretical properties, and Sect. 7.4 demonstrates their application to real-world data.

Notations. Let $\mathcal{F} = C([0,1])$ denote the space of continuous real functions defined on $[0,1]$ equipped with the supremum norm $\|\cdot\| = \|\cdot\|_\infty$. Throughout, we assume that all random elements are defined on a common probability space $(\Omega, \mathcal{A}, \mathbb{P})$. The set of all Borel probability measures on a metric space $\mathcal{S}$ is denoted by $\mathcal{P}(\mathcal{S})$. For a random element $X \sim P \in \mathcal{P}(\mathcal{F})$ and a measurable map $\varphi \colon \mathcal{F} \to \mathcal{S}$, we write $P_{\varphi(X)} \in \mathcal{P}(\mathcal{S})$ to denote the distribution of $\varphi(X)$. The symbol $\stackrel{d}{=}$ represents equality in distribution.

7.2 Band-based Functional Depths

In this section, we define the band depth and spherical depth, along with their modified versions. These statistical depths provide similar approaches to assess the centrality of a function within a distribution of functional data. They are all based on the geometric notion of a band of functions [6]; for $x_1, x_2 \in \mathcal{F}$, the *band* of functions delimited by x_1 and x_2 is

$$B(x_1, x_2) = \{y \in \mathcal{F} \mid \forall t \in [0,1], \min\{x_1(t), x_2(t)\} \le y(t) \le \max\{x_1(t), x_2(t)\}\}.$$

The band depth **BD** of $x \in \mathcal{F}$ is the probability that x falls into a band of two random functions from P. In contrast, the spherical depth **SD** relies on geometric properties of distances between functions, while also evaluating probabilities of x falling into certain bands. The modified versions of these depths address some limitations of **BD** and **SD**, enhancing their applicability to diverse FDA settings.

Band depth (BD). Using the notion of a band, López-Pintado and Romo [6] defined the (second-order[1]) *band depth* of $x \in \mathcal{F}$ with respect to (w.r.t.) $P \in \mathcal{P}(\mathcal{F})$ as

$$\mathcal{BD}(x; P) = \mathbb{P}(x \in B(X_1, X_2)),$$

where X_1, X_2 are two independent random functions sampled from P. From the definition, it follows that higher **BD** of x indicates that x is more likely to be included in the band generated by two random functions sampled from P and is, therefore, more representative or central w.r.t. P. The properties of **BD** have been well studied [1, 6, 9]. This depth has also found practical applications, such as in classifying functional observations [5].

Spherical depth (SD). The concept of spherical depth was first defined in a finite-dimensional setting. Elmore et al. [2] define the *spherical depth* of $\boldsymbol{x} \in \mathbb{R}^d$ w.r.t. $P \in \mathcal{P}(\mathbb{R}^d)$ as

$$\mathcal{SD}(\boldsymbol{x}; P) = \mathbb{P}(\|\boldsymbol{x} - (\boldsymbol{X}_1 + \boldsymbol{X}_2)/2\|_2 \le \|(\boldsymbol{X}_1 - \boldsymbol{X}_2)/2\|_2)$$

for $\boldsymbol{X}_1, \boldsymbol{X}_2 \sim P$ independent, and $\|\cdot\|_2$ the Euclidean norm on $\mathbb{R}^d$. As a natural extension of **SD** to functional data, for $x_1, x_2 \in \mathcal{F}$, we define the sphere[2] formed by x_1 and x_2 as

$$S(x_1, x_2) = \{y \in \mathcal{F} \mid \|y - (x_1 + x_2)/2\| \le \|(x_1 - x_2)/2\|\}.$$

In contrast with $B(x_1, x_2)$, the set $S(x_1, x_2)$ consists of functions that are at most $\|(x_1 - x_2)/2\|$-away from the averaged function $(x_1 + x_2)/2$. Therefore, the band $S(x_1, x_2)$ is of constant width in $t \in [0, 1]$. The *spherical depth* of $x \in \mathcal{F}$ w.r.t. $P \in \mathcal{P}(\mathcal{F})$ is defined as

$$\mathcal{SD}(x; P) = \mathbb{P}(x \in S(X_1, X_2)),$$

where X_1 and X_2 are two independent random functions sampled from P.

An important property of both **BD** and **SD** is that they can be computed very quickly, even in function (infinite-dimensional) spaces, because their sample versions are naturally U-statistics of order 2. As such, their computation requires $O(n^2)$ operations for datasets of size n in $\mathcal{F}$. This should be compared with the time

[1] López-Pintado and Romo [6] define the band depth of order J, $J \ge 2$, using bands formed by up to J functions. We consider only $J = 2$ due to its similarity with other depth concepts.

[2] Unlike **BD**, **SD** depends on the choice of the norm. Here, we use the supremum norm $\|\cdot\|$, but the theory can be developed for any other norm.

complexity $O(n^{d-1}\log(n))$ of the fastest known algorithm for computing the Tukey depth in $\mathbb{R}^d$, which grows exponentially in the dimension d [8].

Modified band depth (MBD). A problem with **BD** is that if two functions x_1 and x_2 cross at $t \in [0,1]$, the band $B(x_1, x_2)$ degenerates to a point at t. Consequently, for distributions with continuous marginals, no other function will be contained in $B(x_1, x_2)$ almost surely. To obtain a more flexible functional depth, López-Pintado and Romo [6] introduced the modified band depth. Instead of examining whether an entire function x lies within the band $B(x_1, x_2)$, the modified band depth measures the set of arguments $t \in [0,1]$ for which $x(t)$ lies between $x_1(t)$ and $x_2(t)$. Specifically, it takes the Lebesgue measure (denoted by λ) of the set

$$T_b(x; x_1, x_2) = \{t \in [0,1] \mid \min\{x_1(t), x_2(t)\} \leq x(t) \leq \max\{x_1(t), x_2(t)\}\}.$$

The *modified band depth* of $x \in \mathcal{F}$ w.r.t. $P \in \mathcal{P}(\mathcal{F})$ is

$$\mathcal{MBD}(x; P) = \mathbb{E}\lambda\left(T_b(x; X_1, X_2)\right),$$

where X_1 and X_2 are two independent random functions sampled from P.

Modified spherical depth (MSD). Motivated by how **MBD** improves on some properties of **BD**, we apply a similar adaptation to **SD**. Instead of considering whether $y \in \mathcal{F}$ is fully contained in the sphere $S(x_1, x_2)$, we measure the set

$$T_s(x; x_1, x_2) = \left\{t \in [0,1] \mid \left|x(t) - \frac{(x_1 + x_2)(t)}{2}\right| \leq \left\|\frac{x_1 - x_2}{2}\right\|\right\} \tag{7.1}$$

of those t where $x(t)$ is at most $\|(x_1 - x_2)/2\|$-away from the averaged function value $(x_1 + x_2)(t)/2$. The *modified spherical depth* of $x \in \mathcal{F}$ w.r.t. $P \in \mathcal{P}(\mathcal{F})$ is then

$$\mathcal{MSD}(x; P) = \mathbb{E}\lambda\left(T_s(x; X_1, X_2)\right),$$

where again $X_1, X_2 \sim P$ are independent. Comparing **MBD** and **MSD**, we naturally observe similarities. A major difference between these depths is that while **MBD** can be represented as a so-called depth of integrated type [9], this is not possible for **MSD**, because of the norm of the entire function $x_1 - x_2$ in the definition of the set T_s in (7.1). For this reason, while **MBD** can be computed in $O(n)$ operations for datasets of size n, the naive computation of the sample **MSD** costs $O(n^2)$, and it is not clear whether this can be improved substantially.

7.3 Theoretical Properties

We now analyze and compare the theoretical properties of the four functional depths from Sect. 7.2. The comparison is carried out w.r.t. a set of desirable properties that have been proposed in the literature [9], as they serve as a benchmark for depth notions. By $\mathcal{D}(\cdot;\cdot)\colon \mathcal{F} \times \mathcal{P}(\mathcal{F}) \to [0,\infty)$ we denote a general depth functional.

(P_1) **Nondegeneracy:** For any $P \in \mathcal{P}(\mathcal{F})$ it holds that

$$P\left(\left\{x \in \mathcal{F} \mid \mathcal{D}(x;P) > \inf_{y \in \mathcal{F}} \mathcal{D}(y;P)\right\}\right) > 0.$$

(P_2) **Scalar-affine invariance:** $\mathcal{D}(f(x); P_{f(X)}) = \mathcal{D}(x; P_X)$ holds for all $x \in \mathcal{F}$, $X \sim P_X \in \mathcal{P}(\mathcal{F})$ and f of the form $f(y) = ay + b$, where $y, b \in \mathcal{F}$ and $a \in \mathbb{R} \setminus \{0\}$.

(P_3) **Function-affine invariance:** $\mathcal{D}(f(x); P_{f(X)}) = \mathcal{D}(x; P_X)$ holds for all $x \in \mathcal{F}$, $X \sim P_X \in \mathcal{P}(\mathcal{F})$ and f of the form $f(y) = ay + b$, where $a, y, b \in \mathcal{F}$, $a(t) \neq 0$ for all $t \in [0, 1]$, and $ay \in \mathcal{F}$ is defined by $(ay)(t) = a(t)y(t)$ for $t \in [0, 1]$.

(P_4) **Maximality at a Gaussian process mean:** For P a zero-mean, stationary, continuous Gaussian process on $[0, 1]$, $\mathcal{D}(x; P) = \sup_{y \in \mathcal{F}} \mathcal{D}(y; P) > \inf_{y \in \mathcal{F}} \mathcal{D}(y; P)$, where $x \equiv 0$ is the constant zero mean function of P.

(P_5) **Maximality at center (central symmetry):** For any $P \in \mathcal{P}(\mathcal{F})$ centrally symmetric about the zero function $0 \in \mathcal{F}$ (i.e. $X \overset{d}{=} -X$ for $X \sim P$) we have that $\mathcal{D}(x; P) = \sup_{y \in \mathcal{F}} \mathcal{D}(y; P)$ if and only if $x \equiv 0$.

(P_6) **Maximality at center (halfspace symmetry):** For any $P \in \mathcal{P}(\mathcal{F})$ halfspace symmetric around the zero function $0 \in \mathcal{F}$,[3] we have that $\mathcal{D}(x; P) = \sup_{y \in \mathcal{F}} \mathcal{D}(y; P)$ if and only if $x \equiv 0$.

(P_7) **Monotonicity w.r.t. the deepest point:** For any $P \in \mathcal{P}(\mathcal{F})$ and $z \in \mathcal{F}$ such that $\mathcal{D}(z; P) = \sup_{y \in \mathcal{F}} \mathcal{D}(y; P)$ we have that $\mathcal{D}(z; P) > \inf_{y \in \mathcal{F}} \mathcal{D}(y; P)$ and $\mathcal{D}(x; P) \leq \mathcal{D}(z + \alpha(x - z); P)$ for all $x \in \mathcal{F}$ and $\alpha \in [0, 1]$.

(P_8) **Vanishing at infinity:** For all $P \in \mathcal{P}(\mathcal{F})$ we have that $\lim_{\|x\| \to \infty} \mathcal{D}(x; P) = 0$.

(P_9) **Semicontinuity in x:** For all $P \in \mathcal{P}(\mathcal{F})$, $\mathcal{D}(\cdot; P)$ is upper semi-continuous, meaning that whenever $x_n \to x$ as $n \to \infty$, we have $\limsup_{n \to \infty} \mathcal{D}(x_n; P) \leq \mathcal{D}(x; P)$.

(P_{10}) **Continuity in P:** For any $x \in \mathcal{F}$ and any $P_n \to P$ converging in the sense of weak convergence of measures in $\mathcal{P}(\mathcal{F})$, we have $|\mathcal{D}(x; P_n) - \mathcal{D}(x; P)| \to 0$ as $n \to \infty$.[4]

Table 7.1 compares the properties (P_1) – (P_{10}) for the four considered depths. It aims to provide a clear and concise overview of the theoretical properties for easy comparison across the studied depth notions. Note the blank spaces in Table 7.1 — the question of whether properties (P_1) and (P_4) hold true for **SD** and **MSD** appears to be an interesting open problem. Table 7.1 extends [9, Table 1], with the properties of **BD** and **MBD** taken from that paper. All results presented for **SD** and **MSD** are original, with some properties of **SD** also announced in [7]. Due to space limitations, the complete proofs of our theoretical results will be provided elsewhere.

From Table 7.1, several conclusions can be drawn. **BD** and **MBD** satisfy both types of affine invariance, while **SD** and **MSD** satisfy only scalar-affine invariance.

---

[3] See [3, Sect. 2.2] for the definition of halfspace symmetry in function spaces.

[4] Also the stronger requirement of uniform continuity of $\mathcal{D}$ in P is considered in [3], i.e., $\sup_{x \in \mathcal{F}} |\mathcal{D}(x; P) - \mathcal{D}(x; P_n)| \to 0$ as $n \to \infty$. In function spaces, however, this condition is rather difficult to establish, see [9].

Table 7.1: A table of results summarizing which of the theoretical properties from Sect. 7.3 are satisfied (indicated by ✓) or not (indicated by ×) for the considered functional depths

Properties		Depth functionals			
		Band	Spherical	Modified band	Modified spherical
		$\mathcal{BD}$	$\mathcal{SD}$	$\mathcal{MBD}$	$\mathcal{MSD}$
Nondegeneracy	(P_1)	×		✓	
Invariance	(P_2)	✓	✓	✓	✓
	(P_3)	✓	×	✓	×
Maximality at centre	(P_4)	×		✓	
	(P_5)	×	×[1]	✓[2]	×[3]
	(P_6)	×	×[1]	✓[2]	×[3]
Monotonicity w.r.t. median	(P_7)	×	×[1]	✓[2]	×[3]
Vanishing at infinity	(P_8)	✓	✓	×	×
Semicontinuity in x	(P_9)	✓	✓	✓	✓
Consistency/robustness	(P_{10})	×	✓[4]	✓	✓[5]

[1] Not satisfied in general. An open problem is whether this can hold under additional continuity conditions imposed on P.
[2] If every univariate marginal distribution of $X(t)$ for $X \sim P$ has a unique median.
[3] Does not hold even when the marginals $X(t)$ have absolutely continuous distributions with unique medians.
[4] For $\{P_n\}_{n=1}^{\infty}$ the sequence of empirical measures of random samples of size n from P, or for P satisfying an additional technical continuity condition that will be detailed elsewhere.
[5] For $X \sim P$ such that $X(t)$ has no atoms for each $t \in [0, 1]$, or for $\{P_n\}_{n=1}^{\infty}$ the sequence of empirical measures of random samples of size n from P.

Regarding maximality at the center for symmetric distributions, only **MBD** exhibits favorable behavior due to its nature as an integrated functional depth. The modified depths (**MBD** and **MSD**) do not vanish at infinity, as they are less affected by large local deviations. All depth functions are semi-continuous in x. We also demonstrated the consistency of **SD** and **MSD** and their continuity in P under additional technical conditions.

7.4 A Real-world Data Demonstration

A practical comparison of the four depths from Sect. 7.2 is now demonstrated in two real-world examples. Our first example is the classical tecator data from the R package fds, where each of the 100 observations represents a spectrometric curve corresponding to the absorbance of a meat sample measured at different wavelengths (ranging from 852 to 1050 nm). Most of these observations differ mainly by a vertical shift while maintaining a similar shape. Therefore, the bands used to define the given depths are nearly identical, which results in similar orderings of the curves, as shown in Fig. 7.1, where the 10% deepest and 10% least deep observations are highlighted.

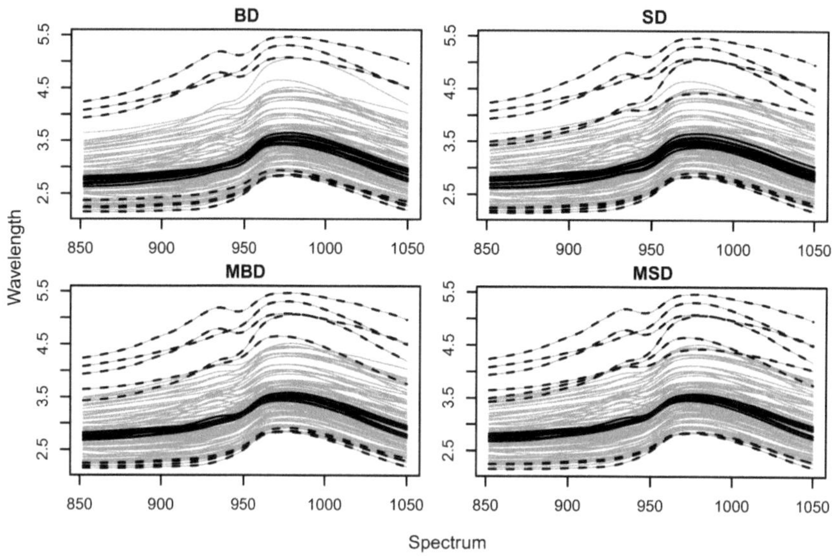

Fig. 7.1: Tecator data: 100 observations represented by a spectrometric curve corresponding to the absorbance of a meat sample measured at different wavelengths (from 852 to 1050 nm). For **BD**, **SD**, **MBD**, and **MSD**, the 10% deepest observations are highlighted with a solid line, while the 10% least deep observations are shown with a dashed line

The second example (Fig. 7.2) is the ECBYieldcurve data from R package fds, consisting of 100 daily yield curve spot rates from 29/12/2006 to 24/07/2009 for government bonds, nominal, and all triple-A-rated companies, with maturity terms ranging from 3 months, 6 months, and 1 to 30 years. These sample curves exhibit decreasing variance over time. Therefore the bands used to define the depths vary greatly in their width. From Fig. 7.2, we observe a difference in the orderings induced by the four depths. **SD** is strongly influenced by the initial high variance, causing it to detect as peripheral observations only those curves that deviate at the beginning of the domain. In contrast, **SD** is insensitive to observations that slip outside the band defined by the other curves only later in the domain. **MSD** eliminates this issue and behaves similarly to **MBD**.

Acknowledgements This research was supported by the Czech Science Foundation (project n. 24-10822S) and the ERC CZ grant LL2407 of the Ministry of Education, Youth and Sport of the Czech Republic.

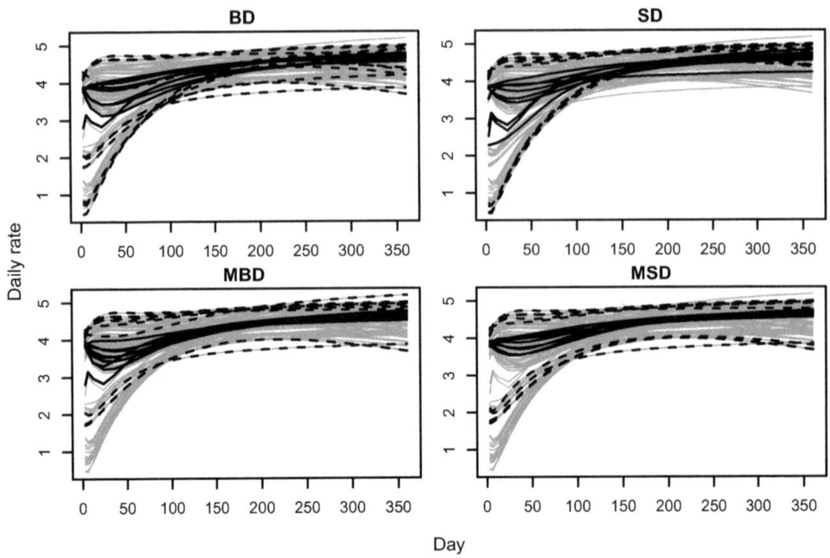

Fig. 7.2: ECBYieldcurve data: Provided by the European Central Bank, this dataset contains 100 daily yield curve spot rates from 29/12/2006 to 24/07/2009 for government bonds, nominal, and all triple-A-rated companies, with maturity terms of 3 months, 6 months, and 1 to 30 years. For **BD**, **SD**, **MBD**, and **MSD**, the 10% deepest observations are highlighted with a solid line, while the 10% least deep observations are shown with a dashed line

References

1. Chakraborty, A., Chaudhuri, P.: On data depth in infinite dimensional spaces. Ann. Inst. Statist. Math. **66**(2), 303–324 (2014)
2. Elmore, R. T., Hettmansperger, T. P., Xuan, F.: Spherical data depth and a multivariate median. In: Liu, R. Y., Serfling, R., Souvaine, D. L. (eds.) Data depth: robust multivariate analysis, computational geometry and applications, pp. 87–101. Amer. Math. Soc., Providence, RI (2006)
3. Gijbels, I., Nagy, S.: On a general definition of depth for functional data. Statist. Sci. **32**(4), 630–639 (2017)
4. Liu, R. Y., Parelius, J. M., Singh, K.: Multivariate analysis by data depth: descriptive statistics, graphics and inference. Ann. Statist. **27**(3), 783–858 (1999)
5. López-Pintado, S., Romo, J.: Depth-based classification for functional data. In: Liu, R.Y., Serfling, R., Souvaine, D.L. (eds.) Data depth: robust multivariate analysis, computational geometry and applications, pp. 103–119. Amer. Math. Soc., Providence, RI (2006)
6. López-Pintado, S., Romo, J.: On the concept of depth for functional data. J. Amer. Statist. Assoc. **104**(486), 718–734 (2009)
7. Mendroš, E., Nagy, S.: The spherical depth for functional data. In "New Trends in Functional Statistics and Related Fields", (G.Aneiros et al. Ed.s), Springer, (2025)
8. Mosler, K., Mozharovskyi, P.: Choosing among notions of multivariate depth statistics. Statist. Sci. **37**(3), 348–368 (2022)
9. Nagy, S., Gijbels, I., Omelka, M., Hlubinka, D.: Integrated depth for functional data: Statistical properties and consistency. ESAIM Probab. Stat. **20**, 95–130 (2016)
10. Nagy, S., Mrkvička, T., and Elías, A.: Which depth to use to construct functional boxplots? arXiv:2409.18603 (2024)

Chapter 8
Analysing the Complexity Mixture Structure of Daily Probability Densities of Bitcoin Returns

Enea G. Bongiorno, Lax Chan and Aldo Goia

Abstract This work deals with a sample of probability density functions calculated from a time series of high frequency Bitcoins returns. Taking advantage on the functional nature of these data, a study of the complexity, a concept tied to the minimum number of random sources used to define the underlying process, is performed. In particular a mixture of processes with different complexities is identified and studied through the use of the small-ball probability of the process. Based on a theoretical discussion about the complexity mixture process structure, a heuristic method is proposed to isolate the different components of the mixture.

8.1 Introduction

Studying and modeling returns of Bitcoin, one of the most well-known cryptocurrencies, is a critical task in finance, as it serves as the foundation for evaluating the prices of other financial contracts that use this returns as an underlying factor (see e.g. [8]) and, among other features, the probability density function (pdf) of returns plays an important role.

Consider the time series of Bitcoin returns observed from 7th of December 2020 to 11th of October 2022 with 1 minute frequency and, for each day, a kernel density approach is used to estimate the pdf of returns. As a result, a dataset of functional

Enea G. Bongiorno
Dipartimento di Studi per l'Economia e l'Impresa, Università del Piemonte Orientale, Via E. Perrone, 18, Novara, e-mail: enea.bongiorno@uniupo.it

Lax Chan (✉)
School of Mathematics, University of Birmingham, Edgbaston, Birmingham, B15 2TT, e-mail: l.chan.1@bham.ac.uk

Aldo Goia
Dipartimento di Studi per l'Economia e l'Impresa, Università del Piemonte Orientale, Via E. Perrone, 18, Novara, e-mail: aldo.goia@uniupo.it

© The Author(s), under exclusive license to Springer Nature Switzerland AG 2025

G. Aneiros et al. (eds.), *New Trends in Functional Statistics and Related Fields*, Contributions to Statistics, https://doi.org/10.1007/978-3-031-92383-8_8

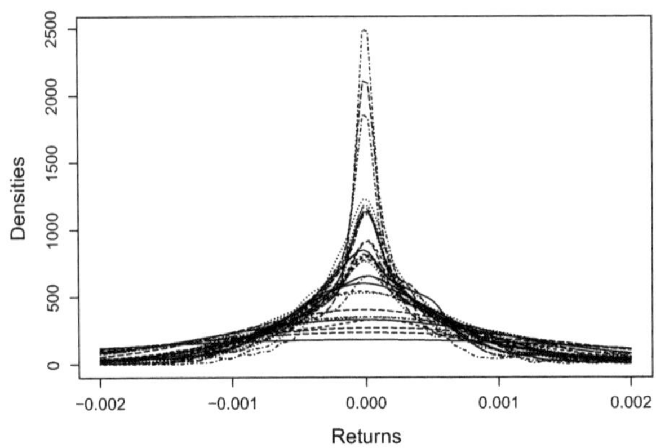

Fig. 8.1: A random selection of 30 estimated pdfs of Bitcoin returns

data is obtained: to appreciate the nature and the shape of them, a random selection of 30 pdfs is reproduced in Fig. 8.1.

Starting from this dataset, an estimate of the complexity can be obtained. Here, complexity refers to the degree of freedom, dimensionality, or the minimum number of random sources required to accurately describe the probability distribution of the underlying process. By understanding this measure, one could then aim to better model the dynamics of Bitcoin return and its implications for related financial instruments.

By an inspection of the plot, it is possible to perceive substantial differences between the represented curves in terms of symmetry, peakness, tails thickness: this suggests the possibility that there exists an underlying mixture of processes having different complexity that drives these aspects.

This problem is approached by exploiting some of the recent developments on complexity (see [2]) in the more general framework of functional data analysis (see [7, 10, 11, 12]).

The remaining of the paper is organised as follows: Sect. 8.2 introduces the notions of Small–Ball Probability, complexity and complexity mixture process; Sect. 8.3 provides the methods to determine the structure of a complexity mixture process; these concepts will finally be applied to analyse a functional time series of bitcoins density returns in Sect. 8.4.

8.2 Complexity Mixture Process

The problem described above is approached through the statistical notion of complexity which was introduced for the first time in [4]. This notion is strictly related to a factorization of the Small-Ball Probability (SmBP) of the process which is usually assumed to by-pass the lack of a dominant measure in the functional statistical or more general setting. In particular, given X a random element in a metric space $(\mathcal{F}, d)$, the SmBP at a point $\chi \in \mathcal{F}$ is the limiting behaviour of

$$\varphi(\chi, h) = \mathbb{P}(d(X, \chi) \leq h), \qquad h \to 0.$$

Roughly speaking, $\varphi(\chi, h)$ looks at how the probability of X concentrates around χ. As previously mentioned, to by-pass the requirement of the existence of a dominant measure, it is convenient to assume that

$$\varphi(\chi, h) \sim \psi(\chi)\phi(h), \qquad h \to 0,$$

with the identifiability condition $\mathbb{E}[\psi(X)] = 1$. In other words, such factorization is isolating the way the SmBP depends on the spatial variable χ and on the "volume" variable h. Moreover, it turns out that while the term $\psi(\chi)$ can be interpreted as a surrogate density and can be used to define modes or to implement (un–)supervised classification problems (see [3]), the term $\phi(h)$ carries the information about the complexity and/or dimensionality of the process (see [1, 4, 5, 6]).

In particular, in [2], the case in which X is a mixture whose components have different complexities and how this reflects on the volume term $\phi(h)$ that appears in the factorization above is studied. To be more precise, assume that $\mathcal{F}$ is endowed with a sum operation and consider

$$X(\omega, t) = \sum_{g=1}^{G} X_g(\omega, t)\, \mathbb{I}_{\Omega_g}(\omega), \qquad t \in [0, 1],$$

where $\{\Omega_g, g = 1, \ldots, G\}$ is a partition of the sample space Ω for which $\pi_g = \mathbb{P}(\Omega_g) > 0$, $g = 1, \ldots, G$, with $\sum_g \pi_g = 1$. Assume that:

- For any $g = 1, \ldots, G$, the minimum number of real random variables defining X_g is d_g and these are denoted $\tau_{g,j}$ (with $j = 1, \ldots, d_g$).
- The components X_g have distinct complexities and, without loss of generality, $1 \leq d_1 < \cdots < d_G < \infty$.
- For $g \in \{1, \ldots, G\}$, the random vector $\boldsymbol{\tau} = (\tau_{g,1}, \ldots, \tau_{g,d_g})$ satisfies some mild regularity assumptions.

Therefore, the SmBP of X exhibits a mixture form as well:

$$\varphi_X(x, h) = \sum_{g=1}^{G} \pi_g \varphi_g(x, h),$$

where $\varphi_g(x, h) = \mathbb{P}(\delta(X, x) \leq h \mid \Omega_g)$ is the SmBP conditioned on the g-th component for which it is assumed that $\varphi_g(x, h) \sim \psi_g(x)\, \phi_g(h)$, $h \to 0$, with $\mathbb{E}[\psi_g(X)\, |\Omega_g] = 1$, for any $g = 1, \ldots, G$ and where, it is assumed that, $\phi_g(h) \sim c_g h^{d_g}$, as h goes to zero, with c_g denoting a suitable constant.

In this setting, it can be shown (see [2]) that

$$\widetilde{\phi}(h) = \mathbb{E}[\varphi_X(X', h)] = \sum_{g=1}^{G} \left(\pi_g c_g h^{d_g} + o\left(h^{d_g}\right)\right) = \pi_1 c_1 h^{d_1} + o\left(h^{d_1}\right). \tag{8.1}$$

In other words, it is $\widetilde{\phi}$ that carries the information about the complexity of the mixture and, in particular, its leading term coincides with the complexity function of the component with the lowest complexity in the mixture.

The idea is, starting from Equation (8.1), to identify, extract and isolate the lowest component from the rest of the mixture. In doing so and iterating such a procedure, all the components of the mixture will be isolated and the structure will be completely determined. This is described in greater details in the next section.

8.3 Identifying a Complexity Mixture Process

Let $\chi = (\chi_1, \ldots, \chi_n)$ be a sample, and the restriction to two groups, i.e. $G = 2$ is exposed here. Consider the empirical estimator for $\varphi(x, h)$ and $\widetilde{\phi}$

$$\varphi_\chi(x, h) = \frac{1}{n} \sum_{1 \le i \le n} \mathbb{I}_{\{d(\chi_i, x) \le h\}}, \qquad \widetilde{\phi}_\chi(h) = \binom{n}{2}^{-1} \sum_{1 \le i < j \le n} \mathbb{I}_{\{d(\chi_i, \chi_j) \le h\}}, \tag{8.2}$$

where $x \in \mathcal{F}$ and $h > 0$. Then, an estimator of d_1 is

$$\widehat{d_1} = \arg\min_{d \in D} \Delta(\widetilde{\phi}_\chi, \phi_0), \tag{8.3}$$

where D is a suitable compact subset of $\mathbb{R}$ that one assumes to contain the true value of d_1 and $\Delta(\widetilde{\phi}_\chi, \phi_0) = 1 - \langle \widetilde{\phi}_\chi, \phi_0 \rangle^2 \|\widetilde{\phi}_\chi\|^{-2} \|\phi_0\|^{-2}$ with $\phi_0(h) = c_d h^d$, $d \in \mathbb{N}$ and where the integrals defining the inner product and norms are approximated numerically over a discretized mesh of points in a suitable interval $\mathcal{H}$ (see [4]). Theoretical properties of the nonparametric estimators (8.2) and (8.3) are derived in [6] in the i.i.d. case and in [2] under a β-mixing dependency structure.

Note that $\chi = \chi_1 \cup \chi_2$, with $\chi_1 \cap \chi_2 = \emptyset$ and where χ_g is the set of observed curves belonging to the mixture component X_g, $g = 1, 2$. Therefore, if one removes χ_1 from χ and computes the estimators (8.2) and (8.3) on $\chi \setminus \chi_1$, then obtains $\widehat{d_2}$, an estimate of the complexity index of the second component X_2. Moreover, an estimate of the mixture weight is given by $\widehat{\pi}_1 = n_1/n$, where $n_1 = \#\chi_1$.

The elements of χ_1 are identified through a Bayesian argument: let Y be a latent group variable given by $Y(\omega) = 1$ if $\omega \in \Omega_1$ and $Y(\omega) = 2$ otherwise, and x be an observation. If x belongs to the first component X_1 then accordingly to the naive Bayes classifier $\mathbb{P}(Y = 1 \mid X = x) \ge \mathbb{P}(Y = 2 \mid X = x)$. With a slight modification on the conditional event, one has that there exists $h_0 > 0$ such that for any $0 < h < h_0$ it holds

$$\mathbb{P}\left(Y=1 \mid \|X-x\| \leq h\right) \geq \mathbb{P}\left(Y=2 \mid \|X-x\| \leq h\right).$$

Thanks to the Bayes formula and because $\varphi_g\left(x, h\right) > 0$, the latter inequality becomes

$$\pi_1 \varphi_1\left(x, h\right) \geq \left(1-\pi_1\right) \varphi_2\left(x, h\right), \qquad 0 < h \leq h_0. \tag{8.4}$$

Intuitively, the latter is true since $(1-\pi_1)\varphi_2(x, h) = o(\varphi_1(x, h))$ when h tends to zero.

Therefore, the following heuristic method is introduced: fix h small enough to guarantee that Equation (8.4) hold true but large enough to avoid that the indicator function in (8.2) vanish systematically, if $\varphi_{\mathcal{X}}\left(\chi_i, h\right)$ is bounded away from zero ($\varphi_{\mathcal{X}}\left(\chi_i, h\right) > \rho$ with $\rho > 0$ being a suitable threshold; see later for practical details), then χ_i is assigned to $\mathcal{X}_1$, otherwise to $\mathcal{X}_2$. It is then possible to estimate d_2 by minimizing the dissimilarity measure $\Delta(\widetilde{\phi}_{\mathcal{X}_2}, \phi)$ where $\widetilde{\phi}_{\mathcal{X}_2}$ is the complexity mixture function estimated over $\mathcal{X}_2$. Some heuristic rules to select h and ρ following a data driven approach are illustrated in Sect. 8.4 below.

In principle, the procedure can be generalized to a mixture with more than two components by iteratively applying the method described above to the sub-samples $\mathcal{X}_{g:G} = \mathcal{X} \setminus \cup_{j=1}^{g-1} \mathcal{X}_j$ for $g = 2, \ldots, G$. The estimates $\widehat{d}_g$ are obtained by minimizing the dissimilarities $\Delta(\widetilde{\phi}_{\mathcal{X}_{g:G}}, \phi_0)$, where $\widetilde{\phi}_{\mathcal{X}_{g:G}}$ is computed based on the observations belonging to $\mathcal{X}_{g:G}$. Clearly, due to the nonparametric nature of the procedure, a sufficiently large sample size is required to ensure reliable estimates of the SmBP and the complexity indexes. As the procedure is iterated, the sub-samples become progressively smaller, which may lead to a potential decline in the quality of the resulting estimates.

8.4 Bitcoin Density Returns Data Analysis

Consider the pdfs of Bitcoin returns described in the Introduction. Each pdf is estimated by using a kernel density approach with a Gaussian kernel and the bandwidth selected according to the unbiased cross-validation approach (see [13]). To guarantee a common support for all these functional data, the pdfs are truncated over a common interval and discretized over a mesh of 100 equispaced points: the resulting dataset consists of 735 discretized functional data defined over $[-0.002, 0.002]$.

Since the nature of the data allows to assume the existence of an underlying complexity mixture, the algorithm proposed in Sect. 8.3 is performed. The first step is to select h and ρ. About the first parameter, after defining a suitable set of possible values and computing the corresponding empirical SmBPs for each element in the sample, the most interesting radius is the one that reveals the highest heterogeneity, producing a bimodality in the distribution of values of SmBPs for a fixed h and thus allowing to better identifying a partition. Here, the Sarle's coefficient is adopted to asses the bimodality (see [9]). The behaviour of the estimated SmBP varying h is displayed in the top-left panel of Fig. 8.2 where the red dashed vertical line indicates

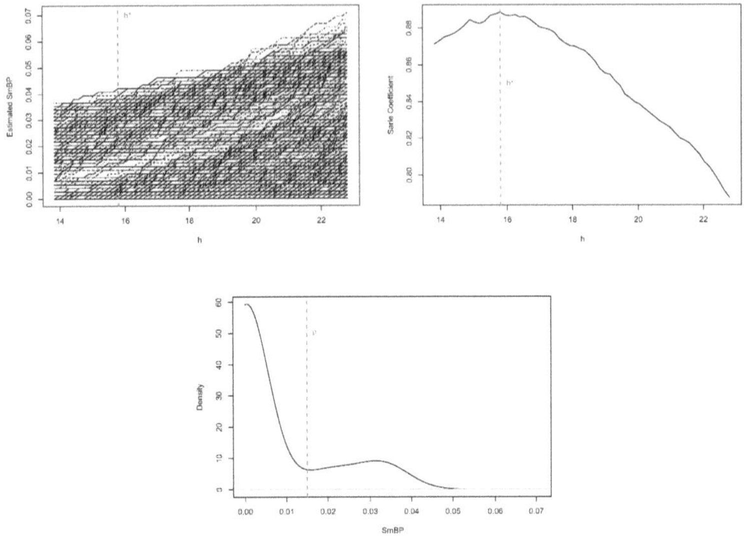

Fig. 8.2: Top-Left: Estimated SmBP varying the bandwidth h (the red dashed vertical line indicates the optimal bandwidth $h^\star$). Top-Right: Sarle coefficients varying the bandwidth. Bottom-Center: Estimated density of the SmBP evaluated at the optimal bandwidth (the red dashed horizontal line indicates the threshold chosen ρ)

the optimal value $h^\star = 15.7$ that maximizes the Sarle's coefficient (see the top-right panel of Fig. 8.2). At this point, the threshold ρ can be chosen: according to the shape of the distribution of SmBP values at $h^\star$, one can select $\rho = 0.015$ (see the panel at the bottom of Fig. 8.2).

At this stage it is possible to separate the two different groups according to their complexities. The first one contains the pdfs having a complexity index $\widehat{d_1} = 3$ and the second one the pdfs with complexity index $\widehat{d_2} = 6$. The averages of the pdfs of the two groups, calculated by using the geometric mean, are plotted in Fig. 8.3: at a first glance the mean distributions of the groups seem to center at zero, the first group has pdfs with a rather flat mean whereas the second one presents the highest level of peakness.

In order to explore more in detail the meaning of these differences, the main features of distributions in each groups are computed: in particular, one deals with the volatility (measured in term of standard deviation), the skewness and the kurtosis. The distributions of these indexes are plotted in Fig. 8.4. It emerges that the distributions in the first group present averagely rather high volatility, are rather symmetric and, on average, show lower kurtosis. The second group is connoted by pdfs showing asymmetry and the highest kurtosis and this can be explained the greater complexity detected.

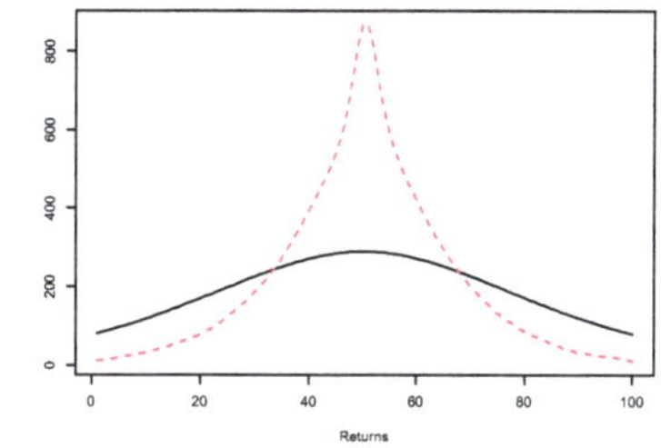

Fig. 8.3: Averages density in each group (Group 1: black, Group 2: dashed red)

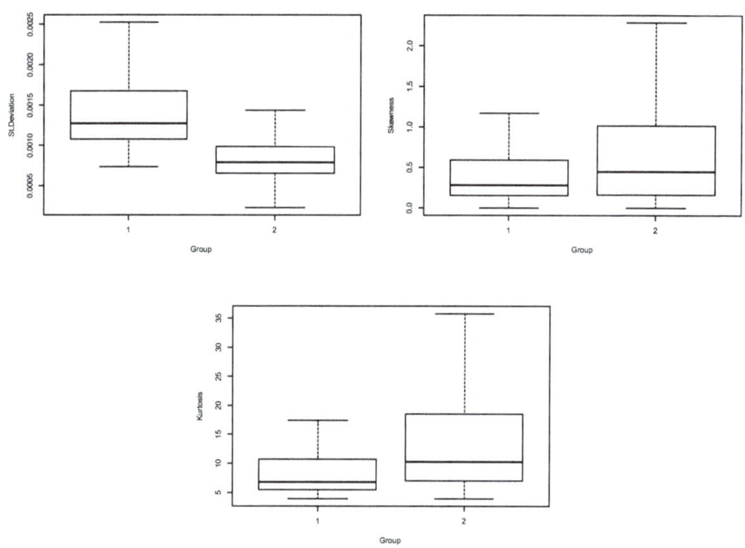

Fig. 8.4: Top-Left: Distributions of daily Standard deviation in each group. Top-Right: Distributions of daily Skewness in each group. Bottom-Center: Distributions of daily Kurtosis in each group

Acknowledgements E.G. Bongiorno, L. Chan and A. Goia are members of Gruppo Nazionale per l'Analisi Matematica, la Probabilità e le loro Applicazioni (GNAMPA) of the Istituto Nazionale di Alta Matematica (INdAM). This article is based upon work from COST Action HiTEc, CA21163, supported by COST (European Cooperation in Science and Technology). The authors thank Gianluca Fusai and Anna Maria Gambaro for some discussions they had on the financial use of bitcoin.

References

1. Bongiorno, E.G., Chan, L., Goia, A.: Detecting the complexity of a functional time series. Journal of Nonparametric Statistics, **36**(3), 600–622 (2024)
2. Bongiorno, E.G., Chan, L., Goia, A.: Complexity Mixture Processes on Metric Spaces. Submitted. (2025)
3. Bongiorno, E.G., Goia, A.: Classification methods for Hilbert data based on surrogate density. Computational Statistics & Data Analysis, **99**, 204–222 (2016)
4. Bongiorno, E.G., Goia, A., Vieu, P.: Evaluating the complexity of some families of functional data. SORT **42**(1), 27–44 (2018)
5. Bongiorno, E.G., Goia, A., Vieu, P.: Modeling Functional Data: a test procedure. Computational Statistics **34**, 451–468 (2019)
6. Bongiorno, E.G., Goia, A., Vieu, P.: Estimating the complexity index of functional data: some asymptotics. Statistics and Probability Letters, 161, article 108731 (2020)
7. Ferraty, F., Vieu, P.: Nonparametric functional data analysis. Theory and practice. Springer, New York, (2006)
8. Kaur, G.: Crypto derivatives 101: A beginner's guide on crypto futures, crypto options and perpetual contracts. (2024) Curated by Cointelegraph's editorial staff.
9. Knapp, T. R.: Bimodality revisited. Journal of Modern Applied Statistical Methods, 6(1):3. (2007)
10. Horvath, L., Kokoszka, P.: Inference for functional data with applications. Springer, New York, (2012)
11. Kokoszka, P., Reimherr, M.: Introduction to functional data analysis. CRC Press, Boca Raton, FL, (2017)
12. Ramsay, J.O., Silverman, B.W.: Functional data analysis. Second edition. Springer, New York, (2005)
13. Scott, W. D.: Multivariate density estimation: theory, practice and visualisation. Second edition. John Wiley & Son, Inc. (2015)

Chapter 9
Regularized Nonparametric Estimation of Covariance Kernels for High-Dimensional Interferometric Data

Teresa Bortolotti, Roberta Troilo, Alessandra Menafoglio and Simone Vantini

Abstract Sentinel-1 satellites offer extensive synthetic aperture radar data globally, revisiting locations every six days. Leveraging these data, differential interferometric processing techniques yield high-resolution ground displacement images that are accurate to millimeter precision, enabling comprehensive monitoring of large areas prone to environmental hazards. Nonetheless, challenges arise when single pixels or entire areas of the images have missing information consistently over time. As typical reconstruction techniques from functional and spatial statistics ground on the second-order structure of the target field, we address the challenge of estimating the spatial covariance operator in a highly non-stationary, non-parametric setting. By grounding in the theory of functional data analysis, we discuss a flexible, non-parametric methodology which accounts for the non-stationarity of the field and ensures continuity of the reconstructed operator through a second-order regularization. The methodology is showcased on ground displacement images collected in the Phlegraean Fields, Italy, a region vulnerable to seismic and bradyseismic activity.

Teresa Bortolotti (✉),
MOX - Department of Mathematics, Politecnico di Milano, Piazza Leonardo da Vinci 32, 20133, Milan, Italy, e-mail: `teresa.bortolotti@polimi.it`

Roberta Troilo,
MOX - Department of Mathematics, Politecnico di Milano, Piazza Leonardo da Vinci 32, 20133, Milan, Italy, e-mail: `roberta.troilo@polimi.it`

Alessandra Menafoglio,
MOX - Department of Mathematics, Politecnico di Milano, Piazza Leonardo da Vinci 32, 20133, Milan, Italy, e-mail: `alessandra.menafoglio@polimi.it`

Simone Vantini,
MOX - Department of Mathematics, Politecnico di Milano, Piazza Leonardo da Vinci 32, 20133, Milan, Italy, e-mail: `simone.vantini@polimi.it`

© The Author(s), under exclusive license to Springer Nature Switzerland AG 2025

G. Aneiros et al. (eds.), *New Trends in Functional Statistics and Related Fields*,
Contributions to Statistics, https://doi.org/10.1007/978-3-031-92383-8_9

9.1 Introduction

Sentinel-1 satellites continuously provide a huge volume of Synthetic Aperture Radar (SAR) data, guaranteeing global coverage with, in the best case, a revisit time down to six days. The Small Baseline Subsets (SBAS) methodology [1] takes advantage of SAR data collected sequentially over time to estimate the deformation time series of an observed area. Specifically, the SBAS algorithm builds a sequence of high-resolution images, each pixel of which containing the time series of ground displacement occurred in the corresponding portion of the ground (e.g., [3]). The scattering, absorption or reflection away from the sensor of radar signals – happening when specific spatial entities on the ground are radiated, such as water, vegetation, or rocks – cause the presence of missing values in some pixels of the SBAS-processed images. Additionally, the ground displacement measurements in the pixels are persistently missing in all images. For natural hazard monitoring purposes, reconstructing ground displacement at missing locations is of crucial importance.

As typical reconstruction techniques from functional (e.g., [6], [5]) and spatial statistics (e.g., Kriging) ground on the second-order structure of the target field, we address the challenge of estimating the covariance of the process recorded by the SBAS-processed images. The high non-stationarity of the process, the time dependence of the images and the presence of persistently missing values over replicates are the main factors responsible for the failure of existing statistical techniques to this aim.

We propose a novel nonparametric covariance reconstruction technique, rooted in the principles of functional data analysis (FDA, [7]). In this framework, the sequence of images is considered as a sequence of partially observed functional data with bidimensional domain. Modeling the images as functions is convenient to capture the natural smoothness of the recorded spatial phenomenon, and helps reduce the dimensionality of the problem due to the high resolution of the images. Estimating the covariance operator from partially observed functional fragments may notably pose challenges [4]. Specifically, relying on the classical empirical covariance kernel over the entire domain may lead to inconsistency issues, due to the unavailability of a sufficient number of curves being observed for every pair of domain locations, or even be infeasible, if no curve is observed for some couple of domain points. The work of [4] focuses on retrieving the covariance operator in the setting known as *banded regime* – where each functional sample is observed on a generic subinterval of the domain of fixed length δ, with δ smaller than the length of the full domain. They propose to reconstruct the empirical covariance kernel by solving a matrix completion problem, which aims to recover a low-rank matrix from a partially observed matrix. When the pattern of missing values is persistently missing over replicates, the low-rank constraint imposed by the method of [4] fails to produce smooth estimates of the covariance operator. We define the setting characterized by persistent patterns of missing values as *fragmented regime*, and propose a covariance kernel reconstruction technique for this setting, which aims to recover a regularized, fixed-rank matrix from a partially observed matrix.

9.2 Relevant Technical Background

We consider for simplicity functional data with one-dimensional domain. Note that all the methods discussed here can be easily extended to higher dimensional settings. Let $X_1, \dots, X_n$ be independent and identically distributed realizations of a random variable with values in the separable Hilbert space of square integrable functions on a bounded domain. Without loss of generality, we consider the space $L^2([0,1])$. Let $\mu : [0,1] \to \mathbb{R}$ and $\mathcal{R} : L^2([0,1]) \to L^2([0,1])$ denote the mean function and the covariance operator of the random function X_1, respectively. Recall that the covariance operator in L^2 is a kernel operator, namely

$$\mathscr{R}f = \mathbb{E}[\langle f, X_1 - \mu\rangle (X_1 - \mu)] = \int_0^1 r(\cdot, t) f(t) dt,$$

for any $f \in L^2([0,1])$, with $r(s,t)$ being the covariance kernel of the random function X_1, such that $r(s,t) = Cov(X_1(s), X_1(t))$ a.e. for $s, t \in [0,1]$.

In our framework, each X_i is observed on a subdomain $O \subset [0,1]$, where O can in principle be a union of subintervals. Let $M := [0,1] \backslash O$, and denote as X_{iO} and X_{iM} the observed and the missing parts of X_i, respectively. Then, note that $X_i(t)$ can be written as $X_i(t) = X_{iO}(t) 1_O(t) + X_{iM}(t) 1_M(t)$, where $1_O(t) := 1(t \in O)$ and $1_M(t) := 1(t \in M)$. When functional data are only partially observed, straightforward estimators of the mean and the covariance operator are no longer defined. The estimation of the mean can easily be obtained in the points of $[0,1]$ where the functional observations are observed. For locations where no observation is available, however, alternative techniques need to be employed.

The estimate of the covariance operator is directly derived from the estimate of its kernel. Inspired by [4], we define the covariance kernel estimator $r_n(s,t)$ as

$$r_n(s,t) := \frac{U(s,t)}{n} \sum_{i=1}^{n} \left((X_i(s) - \mu_n(s)) \, (X_i(t) - \mu_n(t)) \right),$$

where $(s,t) \in [0,1]^2$, $U(s,t) = 1_O(s) 1_O(t)$ and $\mu_n(t)$ is the mean estimator evaluated at location t. As a result, $r_n(s,t) = 0$ if $s, t \in M$. These correspond to locations where the covariance kernel needs to be reconstructed.

We consider the case in which functional data are only measured on an equally-spaced grid of points. Consequently, we represent each functional datum by its K evaluations at specific locations $(t_1, ..., t_K) \in T \subset O$, each corresponding to one successive point of the domain. As a consequence, the discretized covariance kernel can be synthesized by the $K \times K$ covariance matrix $R_n^K = \{r_n(t_j, t_l)\}_{j,l=1}^K$. Our purpose is to estimate R_n^K at its missing elements.

9.3 Regularized Covariance Kernel Estimation

Inspired by the work of [4], we propose a fixed-rank matrix completion problem with a Laplacian regularization. The Laplacian regularization encourages smoothness in the solutions by penalizing abrupt changes or oscillations in the reconstructed covariance kernel. This is particularly beneficial when the process underlying the functional data evolves with smooth variations. Since the Laplacian regularization term is calculated relying on the whole reconstructed kernel, we introduce a matrix of weights which modulates the impact that the reconstruction in the observed and in the missing cells has on the resulting penalization term. Note that this approach is consistent with that of [2].

Building on these considerations, the Laplacian regularization term Δ is designed to incorporate the structure of the domain using an adjacency-based Laplacian matrix, which captures the connectivity between domain points. A weighting matrix adjusts the contribution of the Laplacian term, assigning more importance to the observed cells while allowing a lower impact from the missing cells. This is achieved by defining the weighting matrix $P \in [0, 1]^{K\times K}$ as function of a parameter $m \in [0, 1]$, which effectively balances the impact from both parts of the matrix on the resulting regularization term. The weight given to the missing cells increases with m. At the extremes, $P(0)$ assigns weight 1 to observed cells and 0 to missing cells, while $P(1)$ does the opposite.

In the space of positive semi-definite matrices with fixed rank r^*, the optimization problem minimizes a trade-off between accurately reconstructing the observed entries of the covariance kernel and enforcing smoothness in the reconstruction. The trade-off between the two components is regulated by a parameter α, so that the solution to the problem is the matrix $\theta \in \mathbb{R}^{K\times K}$ that minimizes

$$\left\|P(0) \circ (R_n^K - \theta)\right\|_F^2 + \alpha \left\|P(m) \circ \Delta(\theta)\right\|_F^2 .$$

The rank r^* and the hyperparameters α and m are tuned according to the criteria established by conducting an extensive simulation study. In order to achieve a more rapid resolution, the optimization problem is solved by exploiting the positive-semidefiniteness of covariance matrices. In particular, θ is reparametrized as $\theta = \gamma\gamma^T$, with $\gamma \in \mathbb{R}^{r^*\times K}$. Then, the Broyden–Fletcher–Goldfarb–Shanno (BFGS) algorithm is used to solve the optimization problem for γ.

9.4 Simulation Study

The simulation study serves two main purposes: to test our method's performance and to identify criteria for selecting the rank of the solution and the hyperparameters α and m. By using simulated functional data generated from a Gaussian process with a Matérn covariance kernel, we tested different parameter combinations and evaluated reconstruction performance through three indicators: the reconstruction root mean squared error (RMSE), the RMSE over the observed cells ($\mathrm{RMSE_O}$), and the RMSE over the missing cells ($\mathrm{RMSE_M}$). Though RMSE is the quantity that we aim to minimize, $\mathrm{RMSE_O}$ is the only indicator we have access to in real applications.

The rank r^* is set as high as the available computational resources allow. In fact, additional results from the simulation study show that the choice of the optimal α and m is not affected by the rank, and that the RMSE of reconstruction decreases with r^*, stabilizing for r^* large. With fixed rank, the simulation study allow us to identify two distinct selection criteria for α and m.

For α, the $\mathrm{RMSE_O}$ curve shows a clear elbow point in correspondence of the value of α minimizing both the RMSE and the $\mathrm{RMSE_M}$. The elbow-behavior of $\mathrm{RMSE_O}$ is due to the trade-off in the optimization problem. In fact, small values of α prioritize information from observed cells, while larger values reduce its significance, resulting in an increase of the $\mathrm{RMSE_O}$. These results indicate that α should be chosen at the elbow before the abrupt increase of the $\mathrm{RMSE_O}$ curve, which is detectable even in real data scenarios.

The same argument does not apply to the selection of m. Although RMSE and $\mathrm{RMSE_M}$ show a global minimum in correspondence of the same m, the behaviour of the $\mathrm{RMSE_O}$ curve does not indicate where that point could be. Indeed, $\mathrm{RMSE_O}$ consistently decreases as m increases, indicating better estimation of observed cells when the optimization problem focuses on the first term. As a result, this does not provide a clear criterion for selecting m based solely on $\mathrm{RMSE_O}$. This leads us to identify the best m by relying on an alternative strategy based on cross-validation. Specifically, the strategy sequentially considers the rows and columns near to the already missing rows and columns of the covariance kernel, and fictitiously treat them as missing. Then, we select m minimizing the reconstruction RMSE made in the left-out cells.

Fixed all hyperparameters, the solution to the optimization problem discussed in Sect. 9.3 is a covariance matrix which effectively exploits the information carried by the observed cells, while guaranteeing continuity over the missing rows and columns of the matrix. Fig. 9.1 reports a simulated covariance reconstruction for specific choice of parameters.

9.5 Case Study

The proposed methodology is applied to SBAS-processed images relative to the area of Phlegraean Fields, Italy. The area of interest has the size of 101×101 pixels, each of which covering an area of 80 meters $\times$ 80 meters, and comprises $n = 391$ temporal observations. With the goal of reconstructing the incomplete temporal series of ground displacement, our initial step involves examining the independence among time instants. However, this condition is not met, as the temporal images clearly exhibit autocorrelation. Consequently, we preprocess our data to eliminate autocorrelation and retrieve independence within our problem framework.

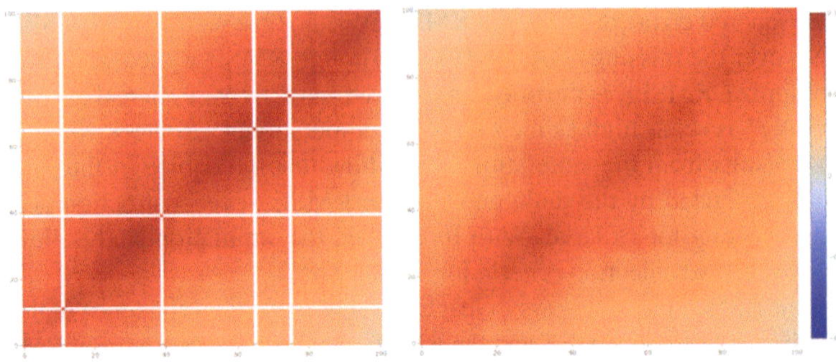

Fig. 9.1: Reconstruction of the Matérn covariance. The rank of the solution is $r^* = 48$, and the selected hyperparameters are $\alpha = 0.01$ and $m = 0.13$.

In order reconstruct the covariance of the image, we adapt the methodology presented in Sect. 9.3 to handle functions with a two-dimensional domain. Specifically, we extend the Laplacian term to take into consideration the four-dimensional proximity of elements in the four-dimensional covariance kernel. Drawing on a criterion established through the simulation study, we choose parameter α by looking at the curve of RMSE_O for some fixed values of m. Then, m is selected following the cross-validation scheme just described. Although the method is applied in the 4-dimensional setting, Fig. 9.2 displays a temptative visualization of the reconstructed empirical covariance kernel obtained via vectorization of three consecutive columns of the considered area.

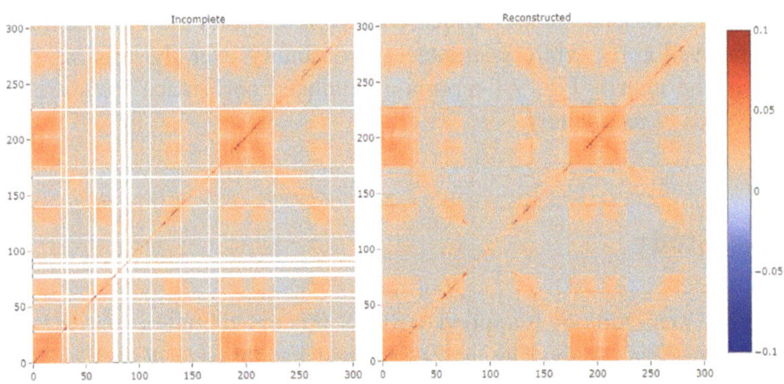

Fig. 9.2: Empirical covariance kernel (left) and reconstructed covariance kernel (right) for 79th, 80th and 81st column of the considered area. The rank of the solution is $r^* = 130$, and the selected hyperparameters are $\alpha = 0.01$ and $m = 0.1$.

9.6 Conclusions

We propose a nonparametric approach which effectively reconstructs the covariance kernel for partially observed functional data, where the pattern of missing values belongs to the *fragmented regime*. Inspired by the low-rank matrix completion technique proposed by [4], which minimizes the rank of the covariance matrix and preserves the values of the observed cells, we propose a new method which relies on a Laplacian regularization of a fixed-rank matrix. The Laplacian regularization promotes smoothness of the reconstructed covariance kernel across neighbouring cells. In its current form, the proposed method offers a practical solution for cases where missing values appear in isolated patterns within the functional data, resulting in narrow bands of missing values in the empirical covariance kernel. Extending the method to more general cases, where the fragmented regime leads to wider bands, requires a deeper theoretical investigation into the identifiability of the solution. Addressing this challenge will be the focus of future work. The performance of our method is showcased in a simulated scenario. Moreover, the simulation study provides criteria for determining the most suitable regularization parameter α and weight parameter m, demonstrating their efficacy. Finally, by applying our method to reconstruct the temporal series of ground displacement, we successfully achieve covariance reconstruction. The reconstructed covariance kernel could be coupled with any reconstruction method for partially observed functional data (e.g., [6], [5]) to reconstruct the ground displacement time series in the pixels where they are missing.

Acknowledgements This work is partially supported by ACCORDO Attuativo ASI-POLIMI "Attività di Ricerca e Innovazione" n. 2018-5-HH.0, collaboration agreement between the Italian Space Agency and Politecnico di Milano, and by the initiative "Dipartimento di Eccellenza 2023–2027", MUR, Italy, Dipartimento di Matematica, Politecnico di Milano.
The authors gratefully acknowledge the financial support of IREA-CNR (Istituto per il Rilevamento Elettromagnetico dell'Ambiente del Consiglio Nazionale delle Ricerche) for funding a PhD grant in cooperation with Politecnico di Milano.

References

1. Berardino, P., Fornaro, G., Lanari, R., Sansosti, E.: A new algorithm for surface deformation monitoring based on small baseline differential SAR interferograms. IEEE Transactions on Geoscience and Remote Sensing **40**(11), 2375–2383 (2002)
2. Bortolotti, T., Peli, R., Lanzano, G., Sgobba, S., Menafoglio, A.: Weighted Functional Data Analysis for the Calibration of a Ground Motion Model in Italy. Journal of the American Statistical Association, **119**(547), 1697–1708 (2024)
3. Casu, F., Manzo, M., Lanari, R.: A quantitative assessment of the SBAS algorithm performance for surface deformation retrieval from DInSAR data. Remote Sensing of Environment, **102**(3-4), 195–210 (2006).
4. Descary, M.-E., Panaretos, V. M.: Recovering Covariance from Functional Fragments. Biometrika **105**(4), 883–896 (2018)
5. Kneip, A., Liebl, D.: On the optimal reconstruction of partially observed functional data. The Annals of Statistics **48**(3), 1692–1717 (2020)

6. Kraus, D.: Components and completion of partially observed functional data. Journal of the Royal Statistical Society **77**(4), 777–801 (2015)
7. Ramsay, J. O., Silverman, B. W.: Functional Data Analysis. Springer, New York (2005)

Chapter 10
Statistical Properties of a Random Series Transmitted by Filtering

Alain Boudou and Sylvie Viguier-Pla

Abstract Considering a multidimensional stationary series, this work aims to present how the properties of periodicity in distribution and of ergodicity are stable by filtering.

10.1 Introduction

The property of weak stationarity of a multidimensional random series is transmitted by linear filtering (see, for example, [3], [1]). We know that the properties of periodicity in distribution and of ergodicity are stable by linear filtering for univariate random series ([6], [5]). This paper aims at developing a very general theoretical framework for which these properties stand. These properties become relevant for Principal Component Analysis in the frequency domain, as it is obtained by linear filtering. We can then envisage applications to functional data, in a variety of fields such as health, meteorology [2], fluid mechanics [4], etc. The random series is not necessarily assumed to have a spectral density and its spectrum may be any. It takes values in a separable Hilbert space, of any dimension, non necessarily finite.

The Hilbert spaces encountered in this text, H, H', H'', of type $L^2_H(\Omega, \mathcal{A}, P)$ (which we denote $L^2_H(\mathcal{A})$ when there is no ambiguity, and $L^2(\mathcal{A})$ when $H = \mathbb{C}$), are assumed to be separable. So they are equiped with an orthonormal basis and we may consider Hilbert-Schmidt operators. Of course, if $\mathcal{A}'$ is a sub-σ−field of $\mathcal{A}$, $L^2_H(\mathcal{A}')$ is separable as a closed sub-space of $L^2_H(\mathcal{A})$. If a Hilbert space is separable, it is also the case for any isometric Hilbert space. When X is an element of $\mathcal{L}^2_H(E, \tau, \eta)$, that is a map from E into H, measurable and of η−integrable squared norm, we note X

Alain Boudou
IMT, Université Paul Sabatier, Toulouse e-mail: palainboudou@gmail.com

Sylvie Viguier-Pla (✉)
LAMPS, Université de Perpignan via Domitia and IMT, Université Paul Sabatier, Toulouse
e-mail: viguier@univ-perp.fr

© The Author(s), under exclusive license to Springer Nature Switzerland AG 2025

G. Aneiros et al. (eds.), *New Trends in Functional Statistics and Related Fields*,
Contributions to Statistics, https://doi.org/10.1007/978-3-031-92383-8_10

or $\overline{X}$, its equivalence class, which is consequently an element of $L^2_H(E,\tau,\eta)$. When the context is obvious, we will not differentiate between the two notations.

Theorem 10.1. *If X is an element of $L^2_H(E,\tau,\eta)$, then:*
i) for any y of $\mathcal{L}^2(E,\tau,\eta)$, the map $\mathcal{X}(.)y(.)$ is measurable and of η−integrable norm;
ii) the map $\widetilde{X} : y \in L^2(E,\tau,\eta) \mapsto \int \mathcal{X}(.)y(.)d\eta(.) \in H$ is a Hilbert-Schmidt operator;
iii) for any h from H, $\widetilde{X}^* h$, element of $L^2(E,\tau,\eta)$, has as representative the map $\langle h, \mathcal{X}(.)\rangle$.

Definition 10.1. We say that a family $(X_n)_{n\in\mathbb{Z}}$ of elements of $L^2_H(\mathcal{A})$ is a H−stationary series when, for any pair (n,m) of elements of $\mathbb{Z}$, we have $\widetilde{X}_n \circ \widetilde{X}^*_m = \int \mathcal{X}_m(.) \otimes \mathcal{X}_n(.)\mathrm{d}P = \widetilde{X}_{n-m} \circ \widetilde{X}^*_0$.

Let us note that when $H = \mathbb{C}$ and when $E(X_n) = 0$, we find the classical definition of the weak stationarity: $\mathrm{cov}(X_n, X_m) = \mathrm{cov}(X_{n-m}, X_0)$.

In this text, we study the transmission by filtering of properties which are statistically of interest, as that of periodicity in distribution, strict stationarity, and ergodicity. As we situate our work in a multidimensional context, these results apply, for example, to Principal Components Analysis in the frequency domain.

We will often use the following result.

Theorem 10.2. *If T is a measurable map from (E,τ,η) into (E',τ'), then the map $\overline{f} \in L^2_H(E',\tau',T\eta) \mapsto \overline{f \circ T} \in L^2_H(E,T^{-1}\tau,\eta)$ is an isometry.*

Remark 10.1. $T\eta : A' \in \tau' \mapsto \eta T^{-1}A' \in [0;1]$. Of course, when E' is an Hilbert space, we speak about distribution.

10.2 The Spaces $H^{\mathbb{Z}}$ and $\mathcal{H}^{\mathbb{Z}}$

In this section, we define the σ−field $\mathcal{B}^H$ of subsets of $H^{\mathbb{Z}}$, and we study the maps $T^H_n : (h_p)_{p\in\mathbb{Z}} \in H^{\mathbb{Z}} \mapsto h_n \in H$ and $\theta^H_n : (h_p)_{p\in\mathbb{Z}} \in H^{\mathbb{Z}} \mapsto (h_{p+n})_{p\in\mathbb{Z}} \in H^{\mathbb{Z}}$, which allow us define later the trajectory of a process.

Lemma 10.1. *For any pair (n,m) of elements of $\mathbb{Z}$, we have $T^H_n \circ \theta^H_m = T^H_{n+m}$ and $\theta^H_n \circ \theta^H_m = \theta^H_{n+m}$.*

Let us now examine the σ−field $\mathcal{B}^H$ of subsets of $H^{\mathbb{Z}}$ generated by the family $\{(T^H_n)^{-1}B; (n,B) \in \mathbb{Z}\times\mathcal{B}_H\}$ ($\mathcal{B}_H$ is the Borel σ−field of H). This σ−field is sometimes named cylindric measurable σ−field.

Theorem 10.3. *i) The σ−field $\mathcal{B}^H$ is the smallest σ−field of subsets of $H^{\mathbb{Z}}$ which makes the maps T^H_n measurable;*
ii) a map f from (E,τ) in $H^{\mathbb{Z}}$ is measurable if and only if, for any n of $\mathbb{Z}$, $T^H_n \circ f$ is measurable;
iii) for any n of $\mathbb{Z}$, $(\theta^H_n)^{-1}\mathcal{B}^H = \mathcal{B}^H$.

The last point is easy to check: from $T_n^H \theta_m^H = T_{n+m}^H$, for any n of $\mathbb{Z}$, we can deduce, from i) and ii), that θ_m^H is measurable. So we can write that $(\theta_{-m}^H)^{-1}\mathcal{B}^H \subset \mathcal{B}^H$, so that $(\theta_m^H)^{-1}(\theta_{-m}^H)^{-1}\mathcal{B}^H \subset (\theta_m^H)^{-1}\mathcal{B}^H \subset \mathcal{B}^H$, that is $(\theta_m^H)^{-1}\mathcal{B}^H = \mathcal{B}^H$.

Let us now study a bijection from $H^{\mathbb{Z}}$ on $\mathcal{H}^{\mathbb{Z}}$, where $\mathcal{H}$ is the Hilbert space H^p, p integer greater or equal to 2. For this, let us consider the maps
$\overline{p} : n \in \mathbb{Z} \mapsto n - p[\frac{n}{p}] + 1 \in \{1, \ldots, p\}$, where $[x]$ is the integer part of x,
$K_j^H : h \in H \mapsto (\delta_{jl} h)_{l=1,\ldots,p} \in \mathcal{H}$,
$\mathcal{F} : h \in H^{\mathbb{Z}} \mapsto ((\sum_{j=1}^p K_j^H \circ T_{pn+j-1}^H) h)_{n\in\mathbb{Z}} \in \mathcal{H}^{\mathbb{Z}}$
and $\mathcal{L} : h \in (\mathcal{H})^{\mathbb{Z}} \mapsto ((K_{\overline{pn}}^H)^* \circ T_{[\frac{n}{p}]}^{\mathcal{H}} h)_{n\in\mathbb{Z}} \in H^{\mathbb{Z}}$.

We first remark that $(K_j^H)^* \circ K_{j'}^H = \delta_{jj'} I_H$, for any pair (j, j') of elements of $\{1, \ldots, p\}$ and that $\sum_{j=1}^p K_j^H \circ (K_j^H)^* = I_{\mathcal{H}}$.

If we note that $T_n^{\mathcal{H}} \circ \mathcal{F} = \sum_{j=1}^p K_j^H \circ T_{pn+j-1}^H$ and that $T_n^H \circ \mathcal{L} = (K_{\overline{pn}}^H)^* \circ T_{[\frac{n}{p}]}^{\mathcal{H}}$, for any n of $\mathbb{Z}$, we deduce that $\mathcal{F}$ and $\mathcal{L}$ are measurable. Moreover, for any n of $\mathbb{Z}$, we have $T_n^{\mathcal{H}} \circ \mathcal{F} \circ \mathcal{L} = \sum_{j=1}^p K_j^H \circ T_{pn+j-1}^H \circ \mathcal{L} = \sum_{j=1}^p K_j^H \circ (K_j^H)^* \circ T_n^{\mathcal{H}} = T_n^{\mathcal{H}}$, and $T_n^H \circ \mathcal{L} \circ \mathcal{F} = (K_{\overline{pn}}^H)^* \circ T_{[\frac{n}{p}]}^{\mathcal{H}} \circ \mathcal{F} = \sum_{j=1}^p (K_{\overline{pn}}^H)^* \circ K_j^H \circ T_{p[\frac{n}{p}]+j-1}^H = T_n^H$.

Theorem 10.4. *The maps $\mathcal{F}$ and $\mathcal{L}$ are inverse one each other: $\mathcal{F} \circ \mathcal{L} = I_{\mathcal{H}^{\mathbb{Z}}}$ and $\mathcal{L} \circ \mathcal{F} = I_{H^{\mathbb{Z}}}$. Moreover, $\mathcal{F} \circ \theta_p^H = \theta_1^{\mathcal{H}} \circ \mathcal{F}$ and $\theta_p^H \circ \mathcal{L} = \mathcal{L} \circ \theta_1^{\mathcal{H}}$.*

10.3 H−stationarity

We study now the various operators which we can associate with a H−stationary series $(X_n)_{n\in\mathbb{Z}}$ and define the derived filtered H'−stationary series.

Definition 10.2. We name shift operator of $(X_n)_{n\in\mathbb{Z}}$ any unitary operator U of $L^2(\Omega, \mathcal{A}, P)$ such that $U \circ \widetilde{X}_n^* = \widetilde{X}_{n+1}^*$, for any n of $\mathbb{Z}$.

Let us note that any H−stationary series $(X_n)_{n\in\mathbb{Z}}$ is associated with a shift operator. Indeed, for any pair $((n_1, h_1), (n_2, h_2))$ of elements of $\mathbb{Z} \times H$, we have
$\langle \widetilde{X}_{n_1}^* h_1, \widetilde{X}_{n_2}^* h_2 \rangle = \langle \widetilde{X}_{n_2} \circ \widetilde{X}_{n_1}^* h_1, h_2 \rangle = \langle \widetilde{X}_{n_2+1} \circ \widetilde{X}_{n_1+1}^* h_1, h_2 \rangle = \langle \widetilde{X}_{n_1+1}^* h_1, \widetilde{X}_{n_2+1}^* h_2 \rangle$.

So there exists an isometry (and only one) V from $\overline{\text{vect}}\{\widetilde{X}_n^* h; (n, h) \in \mathbb{Z} \times H\}$ on $\overline{\text{vect}}\{\widetilde{X}_{n+1}^* h; (n, h) \in \mathbb{Z} \times H\}$ such that $V(\widetilde{X}_n^* h) = \widetilde{X}_{n+1}^* h$, for any (n, h) of $\mathbb{Z} \times H$.

The following property comes from the relation $\overline{\text{vect}}\{\widetilde{X}_n^* h; (n, h) \in \mathbb{Z} \times H\} = \overline{\text{vect}}\{\widetilde{X}_{n+1}^* h; (n, h) \in \mathbb{Z} \times H\}$.

Theorem 10.5. *There exists a unitary operator V of $H' = \overline{vect}\{\widetilde{X}_n^* h; (n, h) \in \mathbb{Z} \times H\}$ such that $V(\widetilde{X}_n^* h) = \widetilde{X}_{n+1}^* h$, for any (n, h) of $\mathbb{Z} \times H$.*

Let P be the projector from $L^2(\mathcal{A})$ on H' and by L the canonical injection $y \in H' \mapsto y \in L^2(\mathcal{A})$. Then $U = P_{\perp} + L \circ V \circ L^*$ is a unitary operator of $L^2(\mathcal{A})$ such that $U(\widetilde{X}_n^* h) = \widetilde{X}_{n+1}^* h$, for any (n, h) of $\mathbb{Z} \times H$, so $U \circ \widetilde{X}_n^* = \widetilde{X}_{n+1}^*$, for any n of $\mathbb{Z}$.

The shift operator is not unique. Moreover, with a double induction, we can show that $U^n \circ \widetilde{X}_0^* = \widetilde{X}_n^*$, for any n of $\mathbb{Z}$. We name it shift operator because when $(X_n)_{n\in\mathbb{Z}}$ is unidimensional ($H = \mathbb{C}$) and taking real values, then $UX_n = X_{n+1}$.

From the shift operator U and $(X_n)_{n\in\mathbb{Z}}$, we define a whole family of shift operators. For this, we use the notion of ampliation of a bounded operator of $L^2(\mathcal{A})$.

Definition 10.3. We name ampliation of an element A of $L(L^2(\mathcal{A}))$ the map A^H defined by $K \in \sigma_2(L^2(\mathcal{A}), H) \mapsto K \circ A^* \in \sigma_2(L^2(\mathcal{A}), H)$.

Let us give some properties of the ampliation.

Theorem 10.6. *i) For any A of $L(L^2(\mathcal{A}))$, A^H is linear and bounded.*
ii) For any A of $L(L^2(\mathcal{A}))$, $(A^H)^* = (A^*)^H$.
iii) For any pair (A, B) of elements of $L(L^2(\mathcal{A}))$, $(A \circ B)^H = A^H \circ B^H$.
iv) $(I_{L^2(\mathcal{A})})^H = I_{\sigma_2}$.

By $\mathcal{I}_{H'}$ we denote the isometry $X' \in L^2_{H'}(\mathcal{A}) \mapsto \widetilde{X'} \in \sigma_2(L^2(\mathcal{A}), H')$. Let us consider $U_{H'} = \mathcal{I}^*_{H'} \circ U^{H'} \circ \mathcal{I}_{H'}$, which is clearly a unitary operator.

Theorem 10.7. *i) For any n of $\mathbb{Z}$, $U^n_H X_0 = X_n$.*
ii) For any (K, X') of $L(H', H'') \times L^2_{H'}(\mathcal{A})$, $U_{H''}(K \circ X') = K \circ (U_{H'} X')$.

Let us briefly examine the proves of these properties. They come from the following.

$$U^n_H X_0 = \mathcal{I}^*_H \circ (U^H)^n \circ \mathcal{I}_H X_0 = \mathcal{I}^*_H(\widetilde{X}_0 \circ U^{-n}) = \mathcal{I}^*_H(\widetilde{X}_n) = X_n;$$
$$U_{H''}(K \circ X') = K \circ \widetilde{X'} \circ U^{-1} = K \circ \widetilde{(U_{H'} X')} = K \circ \widetilde{(U_{H'} X')}.$$

These tools let us define the filter of $(X_n)_{n\in\mathbb{Z}}$.

Definition 10.4. A filter of $(X_n)_{n\in\mathbb{Z}}$ is a series of the type $(U^n_{H'} X')_{n\in\mathbb{Z}}$, where X' belongs to $L^2_{H'}(\mathcal{A})$.

Theorem 10.8. *A filter $(U^n_{H'} X')_{n\in\mathbb{Z}}$ of $(X_n)_{n\in\mathbb{Z}}$ is a H'−stationary series, and U is its shift operator.*

Indeed, for any n of $\mathbb{Z}$, we have $\widetilde{U^n_{H'} X'} = (U^{H'})^n \widetilde{X'} = \widetilde{X'} \circ U^{-n}$. We deduce that $(U^n_{H'} X')_{n\in\mathbb{Z}}$ is a H'−stationary series:

$$\widetilde{U^n_{H'} X'} \circ \widetilde{U^m_{H'} X'}^* = \widetilde{X'} \circ U^{-n} \circ U^m \circ \widetilde{X'}^* = \widetilde{X'} \circ U^{m-n} \circ \widetilde{X'}^* = \widetilde{U^{n-m}_{H'} X'} \circ \widetilde{U^0_{H'} X'}^*,$$

and that U is a shift operator of $(U^n_{H'} X')_{n\in\mathbb{Z}}$: $U^n \circ \widetilde{X'}^* = U^n \circ \widetilde{U^0_{H'} X'}^* = \widetilde{U^n_{H'} X'}^*$.

Such a filter can have the apearance of a moving average. Indeed, let $\{A_p; p \in \mathbb{Z}\}$ be a family of elements of $L(H, H')$ such that $\{A_p \circ X_{-p}; p \in \mathbb{Z}\}$ is a summable family of elements of $L^2_{H'}(\mathcal{A})$, of sum X'_0, we can affirm that $\{U^n_{H'}(A_p \circ X_{-p}); p \in \mathbb{Z}\}$ is a summable family of sum $U^n_{H'} X'_0$, because $U^n_{H'}$ is an isometry. For any p of $\mathbb{Z}$, we have $U^n_{H'}(A_p \circ X_{-p}) = A_p \circ X_{n-p}$, and so $U^n_{H'} X'_0 = \sum_{p\in\mathbb{Z}} A_p \circ X_{n-p}$.

10.4 Trajectory

Let $(\mathcal{X}_n)_{n\in\mathbb{Z}}$ be a series of elements of $\mathcal{L}^2_H(\mathcal{A})$ such that $\overline{\mathcal{X}_n} = X_n$, for any n of $\mathbb{Z}$.

Definition 10.5. We name trajectory of the H−stationary series $(X_n)_{n\in\mathbb{Z}}$ the map $\widehat{X} : \omega \in \Omega \mapsto (\mathcal{X}_n(\omega))_{n\in\mathbb{Z}} \in H^{\mathbb{Z}}$.

This trajectory may be considered as a random value, as it is measurable: $T_n^H \circ \widehat{X} = \mathcal{X}_n$. It is relevant to notice that if $(\mathcal{X}'_n)_{n\in\mathbb{Z}}$ is another series of elements of $\mathcal{L}^2_H(\mathcal{A})$ such that $\overline{\mathcal{X}'_n} = X_n$, for any n of $\mathbb{Z}$, that is such that $\mathcal{X}_n = \mathcal{X}'_n$ P−almost anywhere, then for P−almost any ω of Ω, $\widehat{X}\omega = \widehat{X}'\omega$, and then $\widehat{X}P = \widehat{X}'P$. Thus the image probability $\widehat{X}P$ does not depend on the choice of the representative $\mathcal{X}_n$.

Definition 10.6. We say that the H−stationary series $(X_n)_{n\in\mathbb{Z}}$ has got an order p periodicity in distribution when $\theta_p^H \widehat{X}P = \widehat{X}P$.

Especially, for any n of $\mathbb{Z}$, $\mathcal{X}_{n+p}(P) = T_{n+p}^H \widehat{X}P = T_n^H \theta_p^H \widehat{X}P = T_n^H \widehat{X}P = \mathcal{X}_n(P)$. When $p = 1$, we talk about strict stationarity. Classically, strict stationarity is defined differently, but in an equivalent way.

For any n of $\mathbb{Z}$, we have $(T_n^H)^{-1}\mathcal{B}_H \subset \mathcal{B}^H$, so $\mathcal{X}_n^{-1}\mathcal{B}_H = \widehat{X}^{-1}(T_n^H)^{-1}\mathcal{B}_H \subset \widehat{X}^{-1}\mathcal{B}^H$. Then the family $\{\mathcal{X}_n; n \in \mathbb{Z}\}$ is made of elements of $\mathcal{L}^2_H(\Omega, \widehat{X}^{-1}\mathcal{B}^H, P)$ and the family $\{X_n; n \in \mathbb{Z}\}$ of elements of $L^2_H(\Omega, \widehat{X}^{-1}\mathcal{B}, P)$. From Theorem 10.2, we can affirm the following.

Theorem 10.9. *The map* $T_{H'} : \overline{f} \in L^2_{H'}(H^{\mathbb{Z}}, \mathcal{B}^H, \widehat{X}P) \mapsto \overline{f \circ \widehat{X}} \in L^2_{H'}(\widehat{X}^{-1}\mathcal{B}^H)$ *is an isometry.*

10.5 Transmission of Strict Stationarity and Ergodicity by Filtering

We show here that the periodicity in distribution of order 1 and ergodicity are transmitted by filtering. Let $(X_n)_{n\in\mathbb{Z}}$ be a H−stationary series such that $\theta_1^H \widehat{X}P = \widehat{X}P$.

As $\theta_1^H \widehat{X}P = \widehat{X}P$ and $(\theta_1^H)^{-1}\mathcal{B}^H = \mathcal{B}^H$, Theorem 10.2 lets us establish the following.

Corollary 10.1. *The map* $V : \overline{t} \in L^2(H^{\mathbb{Z}}, \mathcal{B}^H, \widehat{X}P) \mapsto \overline{t \circ \theta_1^H} \in L^2(H^{\mathbb{Z}}, \mathcal{B}^H, \widehat{X}P)$ *is a unitary operator.*

We know that if T' belongs to $\mathcal{L}^2_{H'}(H^{\mathbb{Z}}, \mathcal{B}^H, \widehat{X}P)$, then $T' \circ \widehat{X}$ belongs to $\mathcal{L}^2_{H'}(\Omega, \widehat{X}^{-1}\mathcal{B}^H, P)$. So we have, between the two elements, the following relation.

Lemma 10.2. *For any* T' *of* $\mathcal{L}^2_{H'}(H^{\mathbb{Z}}, \mathcal{B}^H, \widehat{X}P)$, $\widetilde{\overline{T'}} = \widetilde{\overline{T' \circ \widehat{X}}} \circ T_{\mathbb{C}}$.

Indeed, $\langle h', T'(.)\rangle$ is a representative of $\widetilde{\overline{T'}}^* h'$, and $\langle h', T'(.)\rangle \circ \widehat{X}$ of $T_{\mathbb{C}}(\widetilde{\overline{T'}}^* h')$. But as $\langle h', (T' \circ \widehat{X})(.)\rangle$, that is $\langle h', T'(.)\rangle \circ \widehat{X}$ is a representative of $\widetilde{\overline{T' \circ \widehat{X}}}^* h'$, we can write $\widetilde{\overline{T' \circ \widehat{X}}}^* h' = T_{\mathbb{C}}(\widetilde{T'}^* h')$, for any h' of H', so $\widetilde{\overline{T' \circ \widehat{X}}}^* = T_{\mathbb{C}} \circ \widetilde{\overline{T'}}^*$ and then $\widetilde{\overline{T' \circ \widehat{X}}} \circ T_{\mathbb{C}} = \widetilde{\overline{T'}}$. In the particular case where $T' = T_n^H$, we have the following.

Corollary 10.2. *For any n of $\mathbb{Z}$, $\widetilde{\widetilde{T_n^H}} = \widetilde{X_n} \circ T_{\mathbb{C}}$.*

If T' belongs to $\mathcal{L}^2_{H'}(H^{\mathbb{Z}}, \mathcal{B}, \widehat{X}P)$, so it is for $T' \circ \theta_1^H$. The following property establishes a relation between the two elements.

Lemma 10.3. *For any T' of $\mathcal{L}^2_{H'}(H^{\mathbb{Z}}, \mathcal{B}^H, \widehat{X}P)$, $V \circ \widetilde{\overline{T'}}^* = \widetilde{\overline{T' \circ \theta_1^H}}^*$.*

The element $\widetilde{\overline{T'}}^* h'$, of $L^2(H^{\mathbb{Z}}, \mathcal{B}, \widehat{X}P)$, has as representative $\langle h', T'(.)\rangle$, so $V \circ \widetilde{\overline{T'}}^* h'$ has as representative $\langle h', T'(.)\rangle \circ \theta_1^H$.

As for $\widetilde{\overline{T' \circ \theta_1^H}}^* h'$, one of its representatives is $\langle h', T' \circ \theta_1^H(.)\rangle$, but as $\langle h', T'(.)\rangle \circ \theta_1^H = \langle h', T' \circ \theta_1^H(.)\rangle$, we deduce that $V \circ \widetilde{\overline{T'}}^* h' = \widetilde{\overline{T' \circ \theta_1^H}}^* h'$. As it is exact for any h' of H', we have $V \circ \widetilde{\overline{T'}}^* = \widetilde{\overline{T' \circ \theta_1^H}}^*$.

Lemma 10.3 and Corollary 10.2 let us find a shift operator of the $H-$stationary series $(X_n)_{n\in\mathbb{Z}}$.

Lemma 10.4. *$T_{\mathbb{C}} \circ V \circ T_{\mathbb{C}}^*$ is a shift operator of $(X_n)_{n\in\mathbb{Z}}$.*

Indeed, $T_{\mathbb{C}} \circ V \circ T_{\mathbb{C}}^* \circ \widetilde{X_n}^* = T_{\mathbb{C}} \circ V \circ \widetilde{\overline{T_n^H}}^* = T_{\mathbb{C}} \circ \widetilde{\overline{T_n^H \circ \theta_1^H}}^* = T_{\mathbb{C}} \circ \widetilde{\overline{T_{n+1}^H}}^* = \widetilde{X_{n+1}}^*$.

From these properties, we can prove the following.

Theorem 10.10. *For any T' of $\mathcal{L}^2_{H'}(H^{\mathbb{Z}}, \mathcal{B}^H, \widehat{X}P)$, $U_{H'}\overline{(T' \circ \widehat{X})} = \overline{T' \circ \theta_1^H \circ \widehat{X}}$.*

Indeed, $U_{H'}\overline{(T' \circ \widehat{X})} = \mathcal{I}_{H'}^* \circ (T_{\mathbb{C}} \circ V \circ T_{\mathbb{C}}^*)^{H'} \circ \mathcal{I}_{H'}\overline{T' \circ \widehat{X}} = \mathcal{I}_{H'}^*(\widetilde{\overline{T' \circ \widehat{X}}} \circ T_{\mathbb{C}} \circ V^{-1} \circ T_{\mathbb{C}}^*) = \mathcal{I}_{H'}^*(\widetilde{\overline{T'}} \circ V^{-1} \circ T_{\mathbb{C}}^*) = \mathcal{I}_{H'}^*(\widetilde{\overline{T' \circ \theta_1^H}} \circ T_{\mathbb{C}}^*) = \mathcal{I}_{H'}^*(\widetilde{\overline{T' \circ \theta_1^H \circ \widehat{X}}}) = \overline{T' \circ \theta_1^H \circ \widehat{X}}$.

With a double induction, we can generalize the previous result.

Theorem 10.11. *For any T' of $\mathcal{L}^2_{H'}(H^{\mathbb{Z}}, \mathcal{B}^H, \widehat{X}P)$ and for any n of $\mathbb{Z}$, we have $U_{H'}^n\overline{T' \circ \widehat{X}} = \overline{T' \circ \theta_n^H \circ \widehat{X}}$.*

Now we have got the necessary tools to examine the trajectory $\widehat{X'}$ of a filter $(U_{H'}^n X')_{n\in\mathbb{Z}}$ as a function of the trajectory $\widehat{X}$ of the $H-$stationary series $(X_n)_{n\in\mathbb{Z}}$.

Theorem 10.12. *There exists a measurable map F from $H^{\mathbb{Z}}$ in $H'^{\mathbb{Z}}$ such that*
i) $\widehat{X'} = F \circ \widehat{X}$;
ii) $\theta_1^{H'} \circ F = F \circ \theta_1^H$.

Let then $(U_{H'}^n X')_{n\in\mathbb{Z}}$ be a $H'-$stationary series, filter of $(X_n)_{n\in\mathbb{Z}}$. Let T' be an element of $\mathcal{L}^2_{H'}(H^{\mathbb{Z}}, \mathcal{B}^H, \widehat{X}P)$ such that $T_{H'}\overline{T'} = \overline{T' \circ \widehat{X}} = X'_0$.

Let us consider the map $F : h \in H^{\mathbb{Z}} \mapsto (T'\theta_n^H h)_{n\in\mathbb{Z}} \in H'^{\mathbb{Z}}$. This map is measurable, because for any n of $\mathbb{Z}$, we have $T_n^{H'} \circ F = T' \circ \theta_n^H$, and it is such that $\theta_1^{H'} \circ F = F \circ \theta_1^H$ ($F \circ \theta_1^H h = (T'\theta_n^H \theta_1^H h)_{n\in\mathbb{Z}} = (T'\theta_{n+1}^H h)_{n\in\mathbb{Z}} = \theta_1^{H'}(T'\theta_n^H h)_{n\in\mathbb{Z}} = \theta_1^{H'} F h$).

For any n of $\mathbb{Z}$, we have $U_{H'}^n X'_0 = U_{H'}^n\overline{T' \circ \widehat{X}} = \overline{T' \circ \theta_n^H \circ \widehat{X}}$.

So $T' \circ \theta_n^H \circ \widehat{X}$, element of $\mathcal{L}^2_{H'}(\Omega, \widehat{X}^{-1}\mathcal{B}^H, P)$, is a representative of $U^n_{H'}X'_0$, element of $L^2_{H'}(\Omega, \widehat{X}^{-1}\mathcal{B}^H, P)$. The trajectory $\widehat{X'}$ of $(U^n_{H'}X'_0)_{n\in\mathbb{Z}}$ is then such that $\widehat{X'}(\omega) = (T'\theta_n^H \widehat{X}\omega)_{n\in\mathbb{Z}} = F\widehat{X}\omega$, for any ω of Ω, so we have $\widehat{X'} = F \circ \widehat{X}$.

This factorization lets us prove the strict stationarity of the filtered series.

Corollary 10.3. $\theta_1^{H'}\widehat{X'}P = \widehat{X'}P$.

Indeed, $\theta_1^{H'}\widehat{X'}P = \theta_1^{H'}F\widehat{X}P = F\theta_1^H\widehat{X}P = F\widehat{X}P = \widehat{X'}P$.

We recall that a H−stationary series $(X_n)_{n\in\mathbb{Z}}$ is said to be ergodic when, for any B of $\mathcal{B}^H$, $P(\widehat{X}^{-1}B\Delta(\theta_1^H)^{-1}\widehat{X}^{-1}B) = 0$ implies either $\widehat{X}PB = P\widehat{X}^{-1}B = 0$, either $\widehat{X}PB = 1$, where Δ is the symmetric difference.

Theorem 10.13. *If $(X_n)_{n\in\mathbb{Z}}$ is ergodic, so it is for $(U^n_{H'}X'_0)_{n\in\mathbb{Z}}$.*

For any B' of $\mathcal{B}^{H'}$, we have $\widehat{X'}P(B'\Delta(\theta_1^{H'})^{-1}B') = F\widehat{X}P(B'\Delta(\theta_1^{H'})^{-1}B') = \widehat{X}P(F^{-1}(B'\Delta(\theta_1^{H'})^{-1}B')) = \widehat{X}P(F^{-1}B'\Delta F^{-1}(\theta_1^{H'})^{-1}B') = \widehat{X}P(F^{-1}B'\Delta(\theta_1^{H})^{-1}F^{-1}B')$.

If $\widehat{X'}P(B'\Delta(\theta_1^{H'})^{-1}B') = 0$, then $\widehat{X}P(F^{-1}B'\Delta(\theta_1^{H})^{-1}F^{-1}B') = 0$.

Either $\widehat{X}PF^{-1}B' = 0$, and then $0 = F\widehat{X}PB' = \widehat{X'}PB'$, either $\widehat{X}PF^{-1}B' = 1$, and then $1 = \widehat{X'}P(F^{-1}B') = F\widehat{X}PB' = \widehat{X'}PB'$.

10.6 Deployment

Definition 10.7. We name deployment of order p of the H−stationary series $(X_n)_{n\in\mathbb{Z}}$ the series $(Y_n)_{n\in\mathbb{Z}} = (\sum_{j=1}^p K_j^H \circ X_{pn+j-1})_{n\in\mathbb{Z}}$.

It is easy to establish the following.

Theorem 10.14. *The deployment of order p of a H−stationary series is a $\mathcal{H}$−stationary series.*

From $Y_n(\omega) = \sum_{j=1}^p K_j^H \circ X_{pn+j-1}(\omega) = \sum_{j=1}^p K_j^H \circ T_{pn+j-1}X(\omega)$, we deduce, with obvious notation, that

$$\widehat{Y}(\omega) = (Y_n(\omega))_{n\in\mathbb{Z}} = ((\textstyle\sum_{j=1}^p K_j^H \circ T_{pn+j-1})X(\omega))_{n\in\mathbb{Z}} = \mathcal{F} \circ \widehat{X}(\omega).$$

Hence the relations between the trajectories of $(X_n)_{n\in\mathbb{Z}}$ with that of its deployment of order p $(Y_n)_{n\in\mathbb{Z}}$:

$$\widehat{Y} = \mathcal{F} \circ \widehat{X} \text{ and } \mathcal{L} \circ \widehat{Y} = \widehat{X}.$$

Theorem 10.15. *If U is a shift operator of $(X_n)_{n\in\mathbb{Z}}$, then U^p is a shift operator of $(Y_n)_{n\in\mathbb{Z}}$ issued from a deployment of order p.*

Indeed, $U^p \circ \widetilde{Y_n}^* = U^p \circ \sum_{j=1}^p \widetilde{X}^*_{pn+j-1} \circ (K_j^H)^* = \sum_{j=1}^p \widetilde{X}^*_{pn+p+j-1} \circ (K_j^H)^* = \sum_{j=1}^p \widetilde{X}^*_{p(n+1)+j-1} \circ (K_j^H)^* = \widetilde{Y}^*_{n+1}$.

Let us consider a H'−stationary series $(X'_n)_{n\in\mathbb{Z}}$, filtered from $(X_n)_{n\in\mathbb{Z}}$, $X'_n = U^n_{H'}X'_0$. With obvious notation, $(K_j^{H'}h' = (\delta_{jl}h')_{l=1,\dots,p})$, its deployment of order p is the series $(Y'_n)_{n\in\mathbb{Z}} = (\sum_{j=1}^p K_j^{H'} \circ X'_{pn+j-1})_{n\in\mathbb{Z}}$.

Theorem 10.16. *The $\mathcal{H}'$–stationary series ($\mathcal{H}' = H'^p$) $(U^n_{\mathcal{H}'}Y'_0)_{n\in\mathbb{Z}}$ is a filtered series of the $\mathcal{H}$–stationary series $(Y_n)_{n\in\mathbb{Z}}$.*

Indeed, for any n of $\mathbb{Z}$, we have $U^n_{\mathcal{H}'}Y'_0 = (\mathcal{I}^*_{\mathcal{H}'}\circ(U^p)^{\mathcal{H}'}\circ\mathcal{I}_{\mathcal{H}'})^n Y'_0 = \mathcal{I}^*_{\mathcal{H}'}\circ((U^p)^{\mathcal{H}'})^n\circ \widetilde{Y'_0} = \mathcal{I}^*_{\mathcal{H}'}((U^p)^n)^{\mathcal{H}'}\widetilde{Y'_0} = \mathcal{I}^*_{\mathcal{H}'}\widetilde{Y'_0}\circ U^{-pn} = \mathcal{I}^*_{\mathcal{H}'}\widetilde{Y'_n} = Y'_n$.

So the deployment $(Y'_n)_{n\in\mathbb{Z}}$ is equal to $(U^n_{\mathcal{H}'}Y'_0)_{n\in\mathbb{Z}}$, that is a filtered of the deployment $(Y_n)_{n\in\mathbb{Z}}$. We summarize these results in Fig. 10.1.

$$
\begin{array}{ccc}
(X_n)_{n\in\mathbb{Z}} & \xrightarrow{\text{deployment}} & (Y_n)_{n\in\mathbb{Z}} \\
\Big\downarrow \text{filter} & & \Big\downarrow \text{filter} \\
(X'_n)_{n\in\mathbb{Z}} & \xrightarrow{\text{deployment}} & (Y'_n)_{n\in\mathbb{Z}}
\end{array}
$$

Fig. 10.1: Summary of relations deployment-filtering

10.7 Trasmission of the Periodicity in Distribution

We assume now that $(X_n)_{n\in\mathbb{Z}}$ is periodic of order p in distribution, so $\theta^H_p\widehat{X}P = \widehat{X}P$.

The trajectory $\widehat{Y}$ of the deployment of order p is such that $\widehat{Y} = \mathcal{F}\circ\widehat{X}$. So we have $\theta^{\mathcal{H}}_1\widehat{Y}P = \theta^{\mathcal{H}}_1\mathcal{F}\widehat{X}P = \mathcal{F}\theta^H_p\widehat{X}P = \mathcal{F}\widehat{X}P = \widehat{Y}P$. This means that the series $(Y_n)_{n\in\mathbb{Z}}$, deployment of order p of $(X_n)_{n\in\mathbb{Z}}$, is periodic in distribution of order 1, and then so it is for the filtered $(Y'_n)_{n\in\mathbb{Z}}$ of $(Y_n)_{n\in\mathbb{Z}}$ (the periodicity in distribution of order 1 is transmitted by filtering). So we have $\theta^{\mathcal{H}'}_1\widehat{Y'}P = \widehat{Y'}P$ (with obvious notation, $\mathcal{F}'$ is a map from $H'^{\mathbb{Z}}$ into $\mathcal{H}'^{\mathbb{Z}}$ and $\mathcal{L}'$ its reverse). So the trajectory $\widehat{X'}$ of $(X'_n)_{n\in\mathbb{Z}}$ is such that $\widehat{X'} = \mathcal{L}'\circ\widehat{Y'}$. Then we have $\theta^{H'}_p\widehat{X'}P = \theta^{H'}_p\mathcal{L}'\widehat{Y'}P = \mathcal{L}'\theta^{\mathcal{H}'}_1\widehat{Y'}P = \mathcal{L}'\widehat{Y'}P = \widehat{X'}P$.

As a conclusion, the filtered series $(X'_n)_{n\in\mathbb{Z}}$ is periodic of order p in distribution. The periodicity in distribution of order p is transmitted by filtering. More generally, we can prove that the strict stationarity is also transmitted by filtering for any series $(X_g)_{g\in G}$, where G is an abelian locally compact group (e.g. $\mathbb{R}^k, \mathbb{Z}^k$).

References

1. Boudou, A., Dauxois, J.: Principal component analysis for a stationary random function defined on a locally compact abelian group, J. Multivariate Anal. **51** pp1-16 (1994)
2. Boudou, A., Viguier-Pla, S.: On proximity between PCA in the frequency domain and usual PCA, Statistics **40**:5 pp447-464 (2006)
3. Brillinger, D. R.: Times Series: Theory and Methods. Springer, New-York (1981)
4. Nekkanti, A., and Schmidt, O.T.: Frequency–Time Analysis, Low-Rank Reconstruction and Denoising of Turbulent Flows Using SPOD. Journal of Fluid Mechanics, **256**, A26 (2021)
5. Papoulis, A., Pillai, S.U.: Probability, Random Variables, and Stochastic Processes. Fourth Edition, McGraw-Hill (2002)
6. Priestley, M. B.: Spectral Analysis and Time Series. Academic Press (1981)

Chapter 11
Multi-Object Regression: A Linear Framework via Partial Least Squares

Robert Cantwell and John Aston for the Alzheimer's Disease Neuroimaging Initiative*

Abstract Modern statistical problems regularly involve data collected with more structure than simple scalars, for example functional or image data. Accompanying these more exotic data types has been the generalisation of statistical methods to analyse data which take values in almost arbitrary Hilbert spaces. Yet to date, nearly all regression approaches treat a single, often specific, data type within a single Hilbert space. This work considers a linear framework for statistical analysis of data objects naturally represented in a variety of Hilbert spaces, using latent space projections. As part of this framework, we generalise the method of Partial Least Squares and demonstrate its use on data related to estimating early Alzheimer's disease.

11.1 Introduction

Modern data collection now encompasses information collected in a variety of formats, from classic scalar observables, to pointwise observations of functional data, or data collected as images or graphs. Statistical methods have been developed or generalised to treat 'exotic' (non-scalar) data, especially in the infinite-dimensional case of functional data, and extend to almost arbitrary Hilbert spaces [6]. However, there is little that explicitly deals with regression problems for data objects across

Robert Cantwell (✉)
Statistical Laboratory, University of Cambridge, UK e-mail: rc936@cam.ac.uk

John Aston
Statistical Laboratory, University of Cambridge, UK e-mail: jada2@cam.ac.uk

* Data used in preparation of this article were obtained from the Alzheimer's Disease Neuroimaging Initiative (ADNI) database (adni.loni.usc.edu). As such, the investigators within the ADNI contributed to the design and implementation of ADNI and/or provided data but did not participate in the analysis or writing of this report. A complete listing of ADNI investigators can be found at: http://adni.loni.usc.edu/wp-content/uploads/how_to_apply/ADNI_Acknowledgement_List.pdf

© The Author(s), under exclusive license to Springer Nature Switzerland AG 2025

G. Aneiros et al. (eds.), *New Trends in Functional Statistics and Related Fields*, Contributions to Statistics, https://doi.org/10.1007/978-3-031-92383-8_11

differing Hilbert spaces. While multi-object data may be considered jointly in a direct sum Hilbert space, an extended framework is required to consider the decomposition of joint dimension reduction procedures, for example if one wishes to be able to test the inclusion of individual data objects in the modelling procedure.

In functional data analysis, a common approach is Principal Component Analysis (PCA) which guarantees analytically tractable results from the natural symmetry of the covariance. However, allowing for very high- or possibly infinite-dimensional spaces in a regression setting does not guarantee that the first dominant modes of predictor variation are also the dominant modes of covariation. We use this opportunity to consider the extension of Partial Least Squares (PLS) to a general Hilbert space setting building on ideas in [8], but which could be interchanged for any linear dimension reduction scheme. This framework establishes preliminary modelling considerations which can be extended to multi-object settings. For example, how might one perform object selection for a regression model projected onto a latent subspace in a way that can compare with data collection costs or communicate the extent of joint information, notions similar to multivariate work by [7] and [9].

In setting up a framework for generalised PLS, we apply our methodology to heterogeneous data objects collected to inform analysis and diagnosis of Alzheimer's disease (AD). We use data obtained from the Alzheimer's Disease Neuroimaging Initiative (ADNI) database (`adni.loni.usc.edu`). The original goal of ADNI was to test whether serial magnetic resonance imaging (MRI) combined with information from other biological markers and clinical and neuropsychological assessment can be combined to measure the progression of mild cognitive impairment and early AD. Information is collected in the form of MRI (tensor images), physiological variables (e.g. age) and also cognitive scores measured over time (treated as sparse observations of functional data). We estimate the joint relationship between these measured data objects and the level of tau protein measured in the brains, treated as a proxy measurement for the possible onset of AD, being correlated with the development of neurofibrillary tangles [5].

11.2 Partial Least Squares Methodology

Consider a linear regression problem where the explanatory variables come as observations of r objects that lie in various different separable Hilbert spaces $\mathcal{Z}_1, \ldots, \mathcal{Z}_r$, each equipped with its own inner product. The model considered is

$$Y = \sum_{k=1}^{r} \mathcal{B}_k Z_k + \varepsilon, \tag{11.1}$$

where $\mathcal{B}_k$ is a linear operator from $\mathcal{Z}_k$ to the response space $\mathcal{Y}$. A conventional linear dimension reduction scheme would consider that in each of their respective spaces, we can write

$$Z_k = \sum_j \xi_{kj}\phi_{kj}, \quad Y = \sum_j \gamma_j \psi_j, \tag{11.2}$$

for deterministic bases $\{\phi_{kj}\}_j, \{\psi_j\}_j, k = 1, \ldots, r$ and the ξ_{kj}, γ_j random coefficients containing the distributional information for the Z_k's and Y. Considering a sample setting with sample size n, and implicitly assuming finite dimension of all spaces ($\dim(\mathcal{Z}_k) = p_k$, $\dim(\mathcal{Y}) = q$) only in this sentence, one could conceive that by stacking the ξ_{ikj} into the matrix $\mathbf{\Xi} \in \mathbb{R}^{n\times\sum_{k=1}^r p_k}$ and γ_{ij} into $\mathbf{\Gamma} \in \mathbb{R}^{n\times q}$, we have the equivalent multivariate linear regression problem

$$\mathbf{\Gamma} = \mathbf{\Xi}\mathbf{B} + \mathbf{E}. \tag{11.3}$$

However, where the dimension of the data spaces exceeds the sample size, n, including infinite dimensional spaces, the least squares problem becomes ill-specified. Under the assumption that the regression relation is, at least mostly, contained within a relatively low-dimensional subspace, then truncating the basis expansion after the first L basis elements could capture the desired relationship. Linear dimension reduction methods including PCA and PLS select the projection subspace basis as directions which iteratively maximise an objective function designed to capture the most relevant information at each step. PCA selects a subspace basis as the directions along which variation *within* a data object is maximised. Here, we focus on PLS, which instead selects directions that maximise the joint covariation *between* the predictor data objects and the response object. Indeed, with possibly infinite-dimensional settings, there is no guarantee that the first few directions of maximal variation within the predictors are those driving the variation within the response. PLS has evolved in a multivariate setting from its origins in econometric and chemometric literature attributed to Hermann Wold [11], [4], [10], and was generalised to the functional predictor setting [8]. For the results contained here, we assume the data objects all have trace class covariance operators independent of the sample size.

11.2.1 PLS: computing maximal covariance directions

To select maximal covariance directions, PLS seeks directions $\tilde{\phi} \in \mathcal{X}$ and $\tilde{\psi} \in \mathcal{Y}$ which are the maximisers

$$(\tilde{\phi}, \tilde{\psi}) = \operatorname{argmax}_{\substack{\phi\in\mathcal{X},\psi\in\mathcal{Y}\\ \|\phi\|=\|\psi\|=1}} \left[\operatorname{Cov}\left(\langle\phi, X\rangle, \langle\psi, Y\rangle\right)\right]^2. \tag{11.4}$$

These are the directions in $\mathcal{X}$ and $\mathcal{Y}$ along which the predictors and response covary the most. To relate this back to PCA, one would take $Y = X$ and $\psi = \phi$, and seek maximal directions in $\mathcal{X}$ which maximise the variation within the predictors $\mathcal{X}$. As we are measuring linear dependence in relation to the assumed linear model (11.1), it suffices to consider $\mathcal{X} = \bigoplus_{k=1}^r \mathcal{Z}_k$ the direct sum space of the predictor data spaces with the natural inner product the sum of the respective inner products. This allows for joint covariance maximisation over all objects as opposed to conducting separate dimension reduction schemes for each object.

The directions maximising (11.4) (sometimes called the Tucker Criterion) can be given as the top singular vectors of a linear operator acting from $\mathcal{X} \to \mathcal{Y}$. We define the cross-covariance operators

$$C_{XY} := \mathbb{E}\left((X - \mu_X) \otimes (Y - \mu_Y)\right), \tag{11.5}$$
$$C_{YX} := \mathbb{E}\left((Y - \mu_Y) \otimes (X - \mu_X)\right), \tag{11.6}$$

where $\mu_X = \mathbb{E}(X)$, $\mu_Y = \mathbb{E}(Y)$ and the outer product acts $y \otimes x : \mathcal{X} \to \mathcal{Y}$ with

$$(y \otimes x)\, v = \langle v, x\rangle y. \tag{11.7}$$

Proposition 11.1. *The maximal covariance directions $\tilde{\phi}, \tilde{\psi}$ satisfying* (11.4) *are the top left ($\phi_1 \in \mathcal{X}$) and right ($\psi_1 \in \mathcal{Y}$) singular vectors of C_{XY} respectively. Equivalently, $\tilde{\phi}$ is the top eigenvector of the self-adjoint operator $U_X = C_{XY} \circ C_{YX}$ and $\tilde{\psi}$ the top eigenvector of $U_Y = C_{YX} \circ C_{XY}$.*

This follows as a generalisation of a similar result given in the functional setting in [8]. An important consideration in these linear dimension reduction schemes is then how one wishes to compute subsequent directions. There is an 'orthogonalisation' step to compute subsequent maximal covariance directions, however this orthogonalisation step is not directly in the spaces $\mathcal{X}$, $\mathcal{Y}$. The purpose of finding these objective-maximising directions is to then consider the projection of the data along these directions, that is $t = \langle\tilde{\phi}, X\rangle$ and $u = \langle\tilde{\psi}, Y\rangle$, and use the information contained in these random coordinates ('scores') as a reduced representation of the full objects X, Y. As such, the desired orthogonality should actually be orthogonality of the scores, which as random objects linearly orthogonal is interpreted as uncorrelated. That is, given $t_1 = \langle\tilde{\phi}_1, X\rangle$ we aim seek $\tilde{\phi}_2$ as a maximiser of (11.4) constrained that that $\mathbb{C}\mathrm{ov}(t_2, t_1) = 0$, where $t_2 = \langle\tilde{\phi}_2, X\rangle$. In this interpretation, having accounted for the linear dependence between X and Y captured in directions $\tilde{\phi}_1$ and $\tilde{\psi}_1$, we seek a direction with new linear *information* than that already captured, which may well lie along directions in the data space not orthogonal to those already found. This can be represented as an orthogonalisation constraint of subsequent directions $\tilde{\phi}$ with respect to the modified inner product $\langle\cdot,\cdot\rangle_{C_{XX}}$, where

$$\langle u, v\rangle_{C_{XX}} = \langle u, C_{XX} v\rangle, \tag{11.8}$$

for C_{XX} the covariance operator associated with X, e.g. letting $Y = X$ in C_{XY}.

In PCA, which seeks eigenvectors of C_{XX}, this is clear why the subsequent directions are orthogonal. However, for other dependence maximisation criteria, such as PLS, the directions are not orthogonal, and thus PLS directions are not able to all be recovered from the eigendecomposition of a single operator, instead requiring an iterative approach to computation. One can instead focus on directly estimating the scores in a sample setting $t = \langle\tilde{\phi}, X\rangle$, which it can be shown are given by the top eigenvector of what we call the Gram operator product

$$\mathrm{K}_X\mathrm{K}_Y : \mathcal{T}_X \to \mathcal{T}_X, \tag{11.9}$$
$$\mathrm{K}_X\, s = \langle \mathbb{E}(sX), X\rangle, \tag{11.10}$$
$$\mathrm{K}_Y\, t = \langle \mathbb{E}(tY), Y\rangle. \tag{11.11}$$

where the individual operators act between the random Hilbert spaces

$$\mathcal{T}_X = \{t \in \mathbb{R} \mid t = \langle \phi, X\rangle \text{ for some non-random } \phi \in \mathcal{X}\}, \tag{11.12}$$
$$\mathcal{T}_Y = \{s \in \mathbb{R} \mid s = \langle \psi, Y\rangle \text{ for some non-random } \psi \in \mathcal{Y}\}, \tag{11.13}$$

with inner product $\langle t, u\rangle = \mathbb{E}(tu)$. While somewhat abstract at population level, in a sample setting these operators are approximated by the $n \times n$ Gram matrices $(\mathbf{K}_X)_{ij} = \langle X_i, X_j\rangle / n$, and similarly for $\mathbf{K}_Y$. In a sample setting, we can directly deal with matrix algebra in $\mathbb{R}^{n\times n}$, rather than covariance operators in the different data spaces (which with high-dimensional objects can become significantly cheaper).

Proposition 11.2. *The scores corresponding to the maximal covariance directions under a PLS scheme at step h can be computed as the top eigenvector t_h of the $n \times n$ matrix product $\mathbf{K}_{X^{(h)}}\mathbf{K}_Y$, where*

$$\mathbf{K}_{X^{(h)}} = (\mathbf{I} - \mathbf{T}_{h-1}\mathbf{T}_{h-1})\mathbf{K}_X(\mathbf{I} - \mathbf{T}_{h-1}\mathbf{T}_{h-1}) \tag{11.14}$$

is the joint left and right orthogonal projection of $\mathbf{K}_X$ relative to the subspace spanned by all previous scores t_l, $l = 1, \ldots, h-1$ as the columns of $\mathbf{T}_{h-1}$.

These Gram operators, and estimating matrices relate to and extend the notions of the 'Escoufier operators' discussed in the functional setting in [8]. It is worth noting that while we seek orthogonal vectors t_l being the top eigenvectors of the matrix product, one cannot simply take the full eigendecomposition of $\mathbf{K}_X\mathbf{K}_Y$, which as an asymmetric matrix will not guarantee orthogonal eigenvectors, hence why the iterative orthogonalisation is explicitly necessary. Indeed, this results in the generalised Non-Linear Iterative PArtial Least Squares (NIPALS) algorithm (Alg. 2), which in the context of high-dimensional data is computationally cheaper than the equivalent power method application using the cross-covariance operators from Proposition 11.1.

Algorithm 2 NIPALS for arbitrary data objects

$t = (1, 0, \ldots, 0)^\top \in \mathbb{R}^n$
while $\|\Delta t\|_{\mathbb{R}^n} \geq \varepsilon$ **do**
  $t \leftarrow \mathbf{K}_X\mathbf{K}_Y t \,/\, \|\mathbf{K}_X\mathbf{K}_Y t\|$
end while
$\mathbf{K}_X \leftarrow (\mathbf{I} - tt^\top)\mathbf{K}_X(\mathbf{I} - tt^\top)$
return t
repeat L times as number of desired components.

In a multi-object setting, one can compute the joint kernel matrix as the sum of the respective object kernel matrices, i.e. $\mathbf{K}_X = \sum_k \mathbf{K}_{Z_k}$. Of course, in dealing with the covariance, scaling is of important consideration, as variables measured on a large scale relative to others. Here we do not consider this in great depth, opting for scaling each constituent matrix by its estimated variance $\mathrm{tr}(\mathbf{K}_{Z_k})$.

Under a dimension reduction scheme, the issue of model selection becomes more challenging, yet with heterogeneous data objects, often more interesting. This framework, treating a linear dimension reduction case across multi-object data provides the groundwork upon which issues of model selection may be considered, for example in settings where there are data collection or computation costs, raising questions about how much *unique* information a particular object may provide. While not explored in this paper, the use of a joint dimension reduction criterion (such as in (11.4)) allows one to develop a set of nested models, upon which the individual contribution of a particular data object may be measured or tested given the best relevant information extracted already extracted from other objects. Of course, the influence of the original data space's dimension is also an important consideration, similar to how the number of predictors for a model appears in the Akaike information criterion [1].

11.2.2 Convergence and estimation

Under a sample setting, covariance operators, their eigenvectors and the scores are estimated, provided $\mathbb{E}(\|X\|^2\|Y\|^2) < \infty$, by the natural sample covariance operator

$$\hat{C}_{XY} = \frac{1}{n-1}\sum_{i=1}^{n}(X_i - \bar{X}) \otimes (Y_i - \bar{Y}) \tag{11.15}$$

which converges to C_{XY} with rate $O_P(n^{-1/2})$, and similarly for $\hat{C}_{YX}$. By a similar application of the delta method [3] or Theorem 9.1.3 of [6], we also obtain the same convergence rate for the eigenvectors $\hat{\phi}_h$, and consequently a similar rate of convergence for the empirical distribution of the sample score vectors $\hat{t}_h$ to the distribution of the random score $t_h = \langle\tilde{\phi}_h, X\rangle$.

As for the linear regression operator $\mathscr{B}$, we approximate it with the truncated form $\mathscr{B}_{L,M}$, being the projection onto the first L (M) PLS directions in $\mathcal{X}$ ($\mathcal{Y}$). Unable to express PLS as a single eigendecomposition of an operator in $\mathcal{X}$ under the natural inner product, we cannot use Mercer's theorem to guarantee the proportion of covariation explained decaying with the top eigenvalue at each iteration. Nonetheless, similar analytical methods guarantee under fairly regular conditions the rate of convergence of the truncated estimator decreases in similar fashion to sequence of eigenvalues estimated as the iteratively maximal values of (11.4).

Finally, the estimation itself of the coefficients of $\mathscr{B}$ by the coefficients $\mathbf{B}$ in the reduced regression problem (11.3) hold to the rates corresponding to the multivariate method used, such as least squares or penalised least squares (e.g. ridge regression). However, some care must be taken to understand the bias of estimation due to the error being no longer independent of the predictors (cf. [2]). In the simple case, one could consider the estimator $\hat{\mathbf{B}} = \mathbf{T}^\top\mathbf{U}$, being the least squares estimator from (11.3).

11.3 Simulation and Results

We test the generalised procedure for PLS regression on a synthetic dataset consisting of a functional response $Y \in L^2([0,1])$ built from the first 3 Fourier basis functions predicted by a vector object $Z_1 \in \mathbb{R}^4$, a matrix object $Z_2 \in \mathbb{R}^{3\times 3}$ and a functional object $Z_3 \in L^2([0,2])$. We use a sample size of $n = 1000$ and consider natural inner products in these spaces. For the reduced regression estimation, we just consider the least squares estimator of the coefficient operator's coordinates on the truncated subspace. Using a scree-plot selection criterion, we select to use $L = 4$ predictor scores and $M = 3$ response scores, as the response directions $\{\psi_m\}_m$ only span a 3-dimensional subspace up to random noise, as expected from the construction. Considered across 100 resampled experiments, PLS (compared to PCA) achieves an average mean squared error (MSE) of 0.38 (PCA: 0.53) and average mean squared prediction error on 1000 new data points of 0.45 (PCA: 0.54).

Considering the ADNI data, we use a randomly selected subsample of 85 subjects and estimate the level of tau protein based on subject age, cognitive score trajectories constructed as functions from sparse cognitive score measurements via sparse fPCA [12], and Magnetic Resonance Imaging (MRI) brain scans treated as 3-dimensional tensor images with a score at each point indicating the grayscale intensity in the image. For each space, we use here the standard inner product, however it should be noted that the consideration of the appropriate inner product for the data space is important. For example, were the MRI images to be diffusion MRI images measuring positive definite matrices at each voxel, then a transformed inner product to account for the positive definiteness would be more appropriate. The standard inner products suffice for this modelling exercise.

The selected PLS model uses 9 components and obtained a MSE of 0.023 (R-squared: 0.977). Fig. 11.1 plots the proportion each data object contributes to the implicit projection directions in the product space for the first 9 PLS components. As expected, age, being a scalar is only prominent in the first two directions, with the majority of information quickly dominated in later directions by the MRI data.

As may also be expected with a relatively small sample size, a degree of overfitting occurs. We observe an out-of-sample MPSE of 0.785, on 21 new data points, likely due to maximal covariation directions in a low sample distorted by spurious covariation in the training MRI data. Sample size aside, this also suggests that an object selection criterion may need to penalise the dimension of original object data spaces, as larger degrees of freedom increase the likelihood of noise disrupting a dimension reduction scheme, for example, a form of generalised AIC.

Acknowledgements The authors would like to acknowledge Nina Dekoninck Bruhin for her assistance with the ADNI data, in particular the use of sparse fPCA for cognitive scores. Data collection and sharing for the ADNI is funded by the National Institute on Aging (National Institutes of Health Grant U19AG024904).

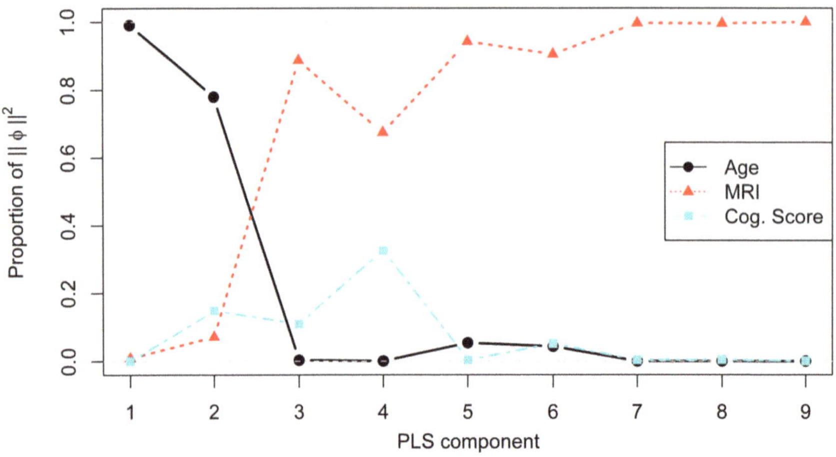

Fig. 11.1: Proportional contribution to each PLS component by each data object.

References

1. Akaike, H.: A new look at the statistical model identification. IEEE Trans. on Auto. Contr. **19**, 716–723 (1974)
2. Aue, A., Hörmann, S., Horváth, L., Hušková, M.: Dependent functional linear models with applications to monitoring structural change. Stat. Sin. **24**, 1043–1073, (2014)
3. Cupidon, J., Gilliam, D., Eubank, R., Ruymgaart, F.: The delta method for analytic functions of random operators with application to functional data. Bernoulli **13**, 1179–1194, (2007)
4. Garthwaite, P.H.: An interpretation of Partial Least Squares. J. Am. Stat. Assoc. **89**, 122–127, (1994)
5. Jack, C.R., Knopman, D.S., Jagust, W.J., Shaw, L.M., Aisen, P.S., Weiner, M.W., Petersen, R.C., Trojanowski, J.Q.: Hypothetical model of dynamic biomarkers of the Alzheimer's pathological cascade. Lancet Neurol. **9**, 119–128, (2010)
6. Hsing, T., Eubank, R.: Theoretical Foundations of Functional Data Analysis, with an Introduction to Linear Operators. John Wiley & Sons Ltd. (2015)
7. Lock, E., Hoadley, K., Marron, J.S., Nobel, A.: Joint and Individual Variation Explained (JIVE) for Integrated Analysis of Multiple Data Types. Ann. App. Stat. **7**, 523–542, (2013)
8. Preda, C., Saporta, G.: PLS regression on a stochastic process. Comput. Stat. Data Anal. **48**, 149–158, (2005)
9. Prothero, J., Jiang, M., Hannig, J., Tran-Dinh, Q., Ackerman, A., Marron, J.S.: Data integration via analysis of subspaces (DIVAS). TEST, **33**, 633–674, (2024)
10. Rosipal, R., Krämer, N.: Overview and Recent Advances in Partial Least Squares. In: Saunders, C., Grobelnik, M., Gunn, S., Shawe-Taylor, J. (eds.) Subspace, Latent Structure and Feature Selection. 2005. Lecture Notes in Computer Science **3940** pp. 34–51. Springer, Berlin, Heidelberg (2006)
11. Wold, H.: Nonlinear Iterative Partial Least Squares (NIPALS) Modelling: Some Current Developments. In: Krishnaiah, P. (ed.) Multivariate Analysis-III pp. 383–407. Academic Press Inc. (1973)
12. Yao, F., Müller, H.-G., Wang, J.-L.: Functional Data Analysis for Sparse Longitudinal Data. J. Am. Stat. Assoc. **100**, 577–590, (2005)

Chapter 12
Monitoring the Covariance of Multichannel Profiles

Christian Capezza, Davide Forcina, Antonio Lepore, Biagio Palumbo

Abstract This paper proposes a control chart for detecting covariance shifts in multichannel profiles, which are a specific type of multivariate functional data characterized by multiple interdependent channels with shared structural patterns. Such profiles are common in modern industrial systems, where monitoring covariance is critical for ensuring product quality and process stability. Existing methods focus primarily on monitoring mean profiles but often struggle with detecting complex covariance shifts due to the high dimensionality and unknown shift patterns. The proposed control chart leverages penalized likelihood ratio tests with varying penalty parameters to effectively identify diverse covariance shift patterns. Covariance monitoring is recast as detecting changes in a precision matrix by representing the conditional dependence structure among multichannel profiles through functional graphical models. The superior performance of the proposed control chart compared to state-of-the-art approaches and its practical applicability are illustrated through a simulation study and a case study.

12.1 Introduction

Advancements in sensing technologies have significantly improved statistical process monitoring (SPM) in modern manufacturing, enabling the analysis of increasingly complex datasets. A key tool in SPM is the control chart, which monitors process stability by detecting shifts in behavior over time. Control charts are typically applied in two phases: Phase I identifies a dataset representative of the in-control (IC) process state, while Phase II prospectively monitors new data for out-of-control (OC) conditions [14]. The performance of these methods is commonly measured

Christian Capezza (✉), Davide Forcina, Antonio Lepore, Biagio Palumbo
University of Naples Federico II, Piazzale Tecchio 80, 80125, Naples, Italy,
e-mail: christian.capezza@unina.it, davide.forcina@unina.it
e-mail: antonio.lepore@unina.it, biagio.palumbo@unina.it

© The Author(s), under exclusive license to Springer Nature Switzerland AG 2025
G. Aneiros et al. (eds.), *New Trends in Functional Statistics and Related Fields*,
Contributions to Statistics, https://doi.org/10.1007/978-3-031-92383-8_12

by the average run length (ARL), the average number of observations needed to detect an OC process state or signal false alarms under IC conditions. The ARL corresponding to no process shift is usually referred to as IC ARL or ARL_0.

In practical applications, product quality is best characterized by continuous and smooth functions over some domain, e.g., time or space, commonly referred to as functional data or profiles. Functional data analysis (FDA) [15] provides powerful tools for modeling such data and calls for novel and sophisticated methods in profile monitoring [9]. A particularly relevant case arises when functional data are collected from multiple sensors, also referred to as channels, that simultaneously measure the same or related process variables, exhibiting shared patterns across channels. This setting results in multichannel profiles, which are frequently encountered in real-world applications [10, 2, 7].

While existing research has made significant progress in monitoring the mean of multichannel profiles [11, 18, 16], monitoring changes in process covariance, which reflects variability between profiles, has received less attention. Covariance monitoring is crucial, as shifts in the relationships between profiles often signal process deterioration even when the mean remains unchanged.

For instance, in a roasting process, shifts in temperature profile relationships across chambers may indicate faults, impacting the final quality of the product [17]. Similar challenges arise in financial markets, where shifts in dependencies among multichannel profiles, such as intraday prices, influence asset management [1], or in seismic monitoring, where changes in profile relationships may reflect changes in volcanic dynamics [23].

Functional graphical models provide an interpretable framework for monitoring the covariance by representing the individual channels as nodes and their conditional dependencies as edges. Several methods have been proposed for estimating functional graphical models, including functional graphical lasso [13], functional neighborhood selection [22], Bayesian inference frameworks [24], and partial separability approaches [21]. Despite these advances, covariance monitoring for multichannel profiles remains largely unexplored, with only a few studies addressing this issue [19, 16]. To the best of our knowledge, apart from the method proposed by [19], only a brief subsection in [16] touches on monitoring covariance in multichannel profiles, as the article primarily focuses on detecting mean shifts. Moreover, [19] proposed a Shewhart-type control chart, i.e., a control chart that assesses the process state based only on current data, only suitable for the detection of large, transient process shifts [14]. Furthermore, the proposed chart has a limited ability to detect sparse covariance shifts, i.e., changes occur in only a subset of the covariance parameters [20]. In the multivariate scalar case, significant research has focused on monitoring sparse covariance shifts [6, 8, 17], but these methods are not designed for multichannel profiles. Nevertheless, they face challenges in selecting penalty parameters and managing parameter uncertainty, which can affect their performance [5].

To address these challenges, this paper proposes an adaptive control chart for the online monitoring of covariance in multichannel profiles, referred to as ACOVCC. The method combines multichannel functional principal component analysis (MFPCA) to extract features with penalized likelihood ratio tests under varying penalty parameters, enabling flexible and robust detection of diverse shift patterns. Simulation studies compare the proposed method with state-of-the-art approaches, and a case study on multichannel temperature profiles in a roasting machine illustrates its practical utility.

12.2 Methodology

The proposed ACOVCC is designed to monitor covariance shifts in multichannel profiles. It is based on functional graphical models and MFPCA to extract meaningful features from high-dimensional data and detect changes in the relationships between profiles. This is briefly illustrated in Sect. 12.2.1, while Sect. 12.2.2 shows the monitoring procedure.

12.2.1 Modeling Multichannel Profile Relationships

The first step is to model the relationships between the multichannel profiles $\boldsymbol{X}(t) = (X_1(t), \ldots, X_p(t))^T$, with $t \in [0, 1]$. These profiles are treated as functional data, where each channel represents a variable that evolves over time or space. The mean function $\boldsymbol{\mu}(t) = (\mu_1(t), \ldots, \mu_p(t))^T$ captures the central tendency, while deviations from the mean reflect random variation. The profiles are modeled as $\boldsymbol{X}(t) = \boldsymbol{\mu}(t) + \boldsymbol{Y}(t)$, where $\boldsymbol{Y}(t)$ represents the stochastic error.

The multivariate FPCA [3] is based on the full multivariate functional covariance. In contrast, we use MFPCA [11], which employs a pooled covariance function that accounts for the specific structure of multichannel profiles, enabling the capture of shared variability across channels. Specifically, the error term $\boldsymbol{Y}(t)$ is decomposed as $\boldsymbol{Y}(t) = \sum_{k=1}^{\infty} \boldsymbol{\xi}_k v_k(t)$, where $\boldsymbol{\xi}_k = (\xi_1, \ldots, \xi_p)^T$ are the principal component scores, and $v_k(t)$ are orthonormal eigenfunctions derived from the pooled covariance function $c(s,t) = \sum_{j=1}^{p} \mathbb{E}[Y_j(s)Y_j(t)]$. In practice, this decomposition is truncated to a finite number of components, K, such that $\boldsymbol{Y}(t) \approx \sum_{k=1}^{K} \boldsymbol{\xi}_k v_k(t)$ captures the essential variation.

Under the assumption that the multichannel profiles follow a multivariate Gaussian process, when the process is IC, the scores $\boldsymbol{\xi}_k$ have a covariance structure described by the precision matrix $\boldsymbol{\Theta}_0$, the inverse of the covariance matrix $\boldsymbol{\Sigma}_0$. Functional graphical models estimate the conditional dependencies among the profiles, represented by the sparsity pattern of $\boldsymbol{\Theta}_0$. These dependencies can be represented as a graph, where the edges correspond to the significant conditional relationships between the profiles.

12.2.2 Monitoring Procedure

In the monitoring phase, newly collected multichannel profiles $\boldsymbol{X}_n(t)$, $n = 1, 2, \ldots$, are projected onto the functional components identified in Phase I to obtain new scores, $\boldsymbol{\xi}_{n,k} = \int_0^1 (\boldsymbol{X}_n(t) - \boldsymbol{\mu}(t)) v_k(t) dt$. The goal is to detect deviations in the covariance structure, corresponding to shifts in the precision matrix $\boldsymbol{\Theta}_0$.

A moving average statistic accumulates historical information for enhanced sensitivity. Specifically, $\boldsymbol{S}_{n,k} = (1-\rho)\boldsymbol{S}_{n-1,k} + \rho \boldsymbol{\xi}_{n,k} \boldsymbol{\xi}_{n,k}^T$ tracks the covariance structure over time, where $\rho \in (0, 1)$ is a weighting parameter. A penalized likelihood ratio test is used to compare IC and OC states. Given the precision matrix $\boldsymbol{\Theta}_1$ corresponding to the newly collected multichannel profiles, the test evaluates the null hypothesis $H_0 : \boldsymbol{\Theta}_1 = \boldsymbol{\Theta}_0$ against the alternative $H_1 : \boldsymbol{\Theta}_1 \neq \boldsymbol{\Theta}_0$. The test statistic is defined as $\Lambda_\gamma = \ell_{1\gamma} - \ell_0$, where ℓ_0 and $\ell_{1\gamma}$ represent the log-likelihoods under H_0 and H_1, respectively. The inclusion of the penalty parameter γ ensures the test adapts to diverse covariance shift scenarios by controlling the complexity of the model.

Since shift patterns are often unknown, we combine multiple penalized likelihood tests using varying penalty parameters $\gamma \in \Gamma$. To achieve this, we employ the nonparametric combination methodology [12], which aggregates p-values from individual tests. This approach enhances adaptability, allowing the method to detect a wide range of potential covariance shifts. The overall monitoring statistic is defined as $\Lambda = -2 \log(\min_{\gamma \in \Gamma} p_\gamma)$, where p_γ is the p-value associated with the test for each penalty parameter. The control chart triggers a signal when the monitoring statistic Λ exceeds a control limit h, determined based on the desired ARL_0.

12.3 Simulation Study

To evaluate the performance of the proposed ACOVCC to detect an OC scenario, inspired by [22], IC multichannel profile observations are generated according to the following procedure. Let

$$X_{ij}(t) = \boldsymbol{c}_{ij}^T \boldsymbol{\phi}(t), \quad i = 1, \ldots, N, \quad j = 1, \ldots p, \quad t \in [0, 1], \tag{12.1}$$

be the multichannel profile observations, where $\boldsymbol{c}_{ij} \in \mathbb{R}^M$ and $\boldsymbol{\phi}(t)$ is a vector containing the first $M = 5$ Fourier basis functions. The vector of coefficients $\left(\boldsymbol{c}_{i1}^T, \ldots, \boldsymbol{c}_{ip}^T\right)^T \in \mathbb{R}^{pM}$ follows a multivariate Gaussian distribution with zero mean and covariance matrix $\boldsymbol{\Sigma}_0 = \boldsymbol{\Theta}_0^{-1}$. Let $\boldsymbol{A} \in \mathbb{R}^{M \times M}$ be a tridiagonal matrix with $\boldsymbol{A}_{k,k} = 1$, $\boldsymbol{A}_{k,k+1} = \boldsymbol{A}_{k+1,k} = 0.6$ and $\boldsymbol{A}_{k,k+2} = \boldsymbol{A}_{k+2,k} = 0.3$. The precision matrix $\boldsymbol{\Theta}_0$ is generated as a block-banded precision matrix with blocks given as $\boldsymbol{\Theta}_{0j,j} = \boldsymbol{A}$, $\boldsymbol{\Theta}_{0j,j+1} = \boldsymbol{\Theta}_{0j+1,j} = 0.6\boldsymbol{A}$ and $\boldsymbol{\Theta}_{0j,j+2} = \boldsymbol{\Theta}_{0j+2,j} = 0.3\boldsymbol{A}$. All remaining blocks are set to zero. We fix $N = 500$, $p = 20$, while each random function is observed on a grid with 100 equally spaced time points in $[0, 1]$. Then, for each time point, the observed values are generated using Equation (12.1) and adding noise

$\varepsilon_{ij}(t) \sim N\left(0, 0.5^2\right)$. The proposed ACOVCC is compared with two state-of-the-art methods: the hierarchical graphical model (HGM) [19] and the multichannel profile monitoring approach of [16], referred to as Ren. These methods aim to identify a shift in the relationships between profiles. Four increasing severity levels (SL) are explored, each corresponding to a progressively larger covariance shift. For each SL, 100 simulation runs are performed. The number of functional principal components is chosen so that the fraction of variance explained is at least 95%. In Phase II, a testing set of 1000 OC observations is randomly generated, and the performance of the control chart is evaluated by means of ARL as a function of SL. When the process is IC, ARL should be as close as possible to ARL_0, which is set equal to 100, while it should be as small as possible when the process is OC.

As explained above, the simulation results are shown in Fig. 12.1, which shows the mean ARL as a function of SL for each monitoring method. The proposed ACOVCC outperforms all competing methods in terms of ARL, especially for moderate shift sizes. HGM performs the worst, relying on single observations without leveraging historical data. Ren performs better than HGM because of its use of historical data, but does not match the adaptability of ACOVCC in detecting diverse shift patterns.

It is worth noting that when the process is IC, all methods exhibit ARL values that are larger than or equal to the pre-specified ARL_0 100, demonstrating their reliability in maintaining false alarm rates under IC conditions. However, ACOVCC demonstrates a more consistent decrease in ARL as the severity of the shift increases, indicating its adaptability to various scenarios and its ability to maintain a high detection power across a wide range of shift magnitudes. The observed performance differences emphasize the importance of using penalized likelihood ratio tests, which allow the model to detect subtle changes in the underlying structure of multichannel profiles while adapting to varying levels of sparsity in covariance shifts. The nonparametric combination framework further enhances ACOVCC's flexibility in

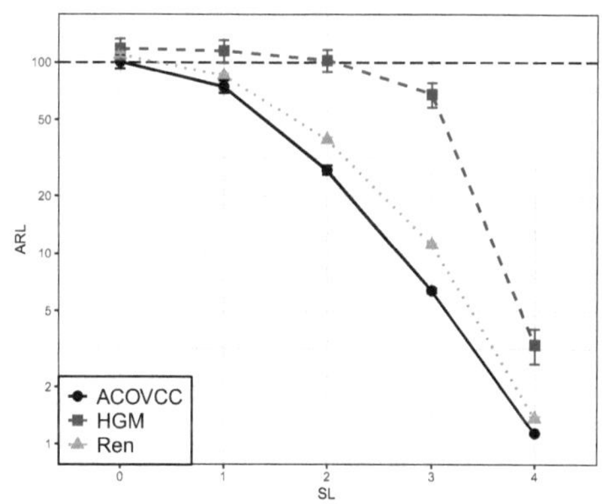

Fig. 12.1: Mean ARL achieved in Phase II by ACOVCC, HGM, and Ren in the Simulation study.

handling various shift patterns, a critical advantage over more rigid approaches like HGM and Ren.

12.4 Case Study

The practical applicability of the ACOVCC is demonstrated through a case study on the monitoring of a roasting process. Data are publicly available at `https://www.kaggle.com/datasets/podsyp/production-quality?select=data_X.csv`. The roasting machine consists of five chambers, each containing a set of three sensors placed in specific locations away from each other to obtain multichannel temperature profiles. As a result, 15 temperature profiles ($p = 15$) can be collected in one roasting cycle, which lasts one hour. The data is collected at a frequency of one temperature data point per minute, providing 60 data points in one profile. The temperature profile dynamics are crucial for the quality of the final product, which is measured by overall scores that evaluate the industry standards of size and hardness of products and are available for this case study. Products with quality scores smaller than the 10th percentile are considered defective units. Accordingly, the remaining products are considered non-defective. Then, the corresponding multichannel profiles were split into training and tuning sets and used in Phase I. The former is used to estimate the model parameters, while the latter is used to calculate the control limit for an $ARL_0 = 100$. Moreover, to compare the performance of the ACOVCC with that of the competing methods presented in Sect. 12.3, 100 OC sequences were obtained by sampling the OC multichannel profile observations and monitored by each method. Then, the ARL is calculated as the mean of the run lengths for each sequence. Fig. 12.2 shows the results achieved in the case study, where the proposed ACOVCC method outperforms all competing methods.

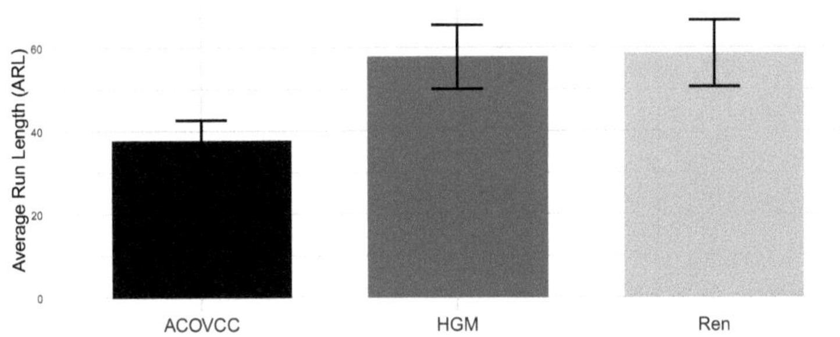

Fig. 12.2: Mean ARL achieved in Phase II by ACOVCC, HGM, and Ren in the case study.

12.5 Conclusions

This paper introduces a novel control chart for monitoring covariance shifts in multichannel profiles, leveraging multichannel functional principal component analysis and functional graphical models. The proposed method effectively adapts to various shift patterns in high-dimensional data by integrating penalized likelihood ratio tests with a nonparametric combination methodology. The methodology addresses the challenges of detecting sparse and structured covariance changes and provides a scalable framework suitable for modern industrial systems.

The proposed control chart demonstrates several advantages over the existing methods. It handles the high-dimensionality of multichannel profiles, capturing between-profile relationships while maintaining computational feasibility. Integrating a moving covariance framework enhances sensitivity to small or persistent shifts, while the use of penalized likelihood estimation ensures robust detection of changes in the precision matrix.

Future research could explore applications of the proposed method beyond manufacturing. Fields such as finance, environmental monitoring, and healthcare often involve multichannel profiles, where shifts in covariance structures play a critical role. Demonstrating the versatility of the approach in these domains could further highlight its practical utility and adaptability to diverse real-world scenarios.

Acknowledgements This work was supported by Piano Nazionale di Ripresa e Resilienza (PNRR) - Missione 5 Componente 2, Investimento 1.3-D.D. 1551.11-10-2022, PE00000004 within the Extended Partnership MICS (Made in Italy - Circular and Sustainable) and by the MOST - Sustainable Mobility National Research Center and received funding from the European Union Next-GenerationEU (PIANO NAZIONALE DI RIPRESA E RESILIENZA (PNRR) - MISSIONE 4 COMPONENTE 2, INVESTIMENTO 1.4 - D.D. 1033 17/06/2022, CN00000023). The computing resources and related technical support used for this work have been provided by the CRESCO/ENEAGRID High Performance Computing infrastructure and its staff [4]. CRESCO/ENEAGRID High Performance Computing infrastructure is funded by ENEA, the Italian National Agency for New Technologies, Energy and Sustainable Economic Development and by Italian and European research programmes, see `http://www.cresco.enea.it/english` for information.

References

1. Avanesov V., Buzun N.: Change-point detection in high-dimensional covariance structure. Electronic Journal of Statistics **12**(2), 3254–3294, (2018)
2. Capezza C., Centofanti F., Lepore A., Palumbo B.: Robust multivariate functional control chart. Technometrics **66**(4), 531–547, (2024)
3. Chiou J.M.,Chen Y.T., Yang Y.F.: Multivariate functional principal component analysis: A normalization approach. Statistica Sinica pp 1571–1596, (2014)
4. Iannone F., Ambrosino F., Bracco G., De Rosa M., Funel A., Guarnieri G., Migliori S., Palombi F., Ponti G., Santomauro G., Procacci P.: CRESCO ENEA HPC clusters: a working example of a multifabric GPFS Spectrum Scale layout. In: 2019 International Conference on High Performance Computing Simulation (HPCS), pp 1051–1052, (2019)

5. Jensen W.A., Jones-Farmer L.A., Champ C.W., Woodall W.H.: Effects of parameter estimation on control chart properties: a literature review. Journal of Quality technology **38**(4), 349–364, (2006)
6. Li B., Wang K., Yeh A.B.: Monitoring the covariance matrix via penalized likelihood estimation. IIE Transactions **45**(2), 132–146, (2013)
7. Li Y., Zhou Q., Huang X., Zeng L.: Pairwise estimation of multivariate gaussian process models with replicated observations: Application to multivariate profile monitoring. Technometrics **60**(1), 70–78, (2018)
8. Maboudou-Tchao E.M., Diawara N.: A lasso chart for monitoring the covariance matrix. Quality Technology & Quantitative Management **10**(1), 95–114, (2013)
9. Noorossana R., Saghaei A., Amiri A.: Statistical analysis of profile monitoring. John Wiley & Sons, (2011)
10. Paynabar K., Jin J., Pacella M.: Monitoring and diagnosis of multichannel nonlinear profile variations using uncorrelated multilinear principal component analysis. IIE Transactions **45**(11), 1235–1247, (2013)
11. Paynabar K., Zou C., Qiu P.: A change-point approach for phase-I analysis in multivariate profile monitoring and diagnosis. Technometrics **58**(2), 191–204, (2016)
12. Pesarin F., Salmaso L.: Finite-sample consistency of combination-based permutation tests with application to repeated measures designs. Journal of Nonparametric Statistics **22**(5), 669–684, (2010)
13. Qiao X., Guo S., James G.M.: Functional graphical models. Journal of the American Statistical Association **114**(525), 211–222, (2019)
14. Qiu P.: Introduction to statistical process control. CRC press, (2014)
15. Ramsay J., Silverman B.: Functional data analysis. Springer, (2005)
16. Ren H., Chen N., Wang Z.: Phase-II monitoring in multichannel profile observations. Journal of Quality Technology **51**(4), 338–352, (2019)
17. Shen X., Tsung F., Zou C.: A new multivariate ewma scheme for monitoring covariance matrices. International Journal of Production Research **52**(10), 2834–2850, (2014)
18. Wang Y., Mei Y., Paynabar K.: Thresholded multivariate principal component analysis for phase i multichannel profile monitoring. Technometrics **60**(3), 360–372, (2018)
19. Wu H., Zhang C., Li Y.F.: Monitoring heterogeneous multivariate profiles based on heterogeneous graphical model. Technometrics **64**(2), 210–223, (2022)
20. Yeh A.B., Li B., Wang K.: Monitoring multivariate process variability with individual observations via penalised likelihood estimation. International Journal of Production Research **50**(22), 6624–6638, (2012)
21. Zapata J., Oh S.Y., Petersen A.: Partial separability and functional graphical models for multivariate gaussian processes. Biometrika **109**(3), 665–681, (2022)
22. Zhao B., Zhai P.S., Wang Y.S., Kolar M.: High-dimensional functional graphical model structure learning via neighborhood selection approach. Electronic Journal of Statistics **18**(1), 143–188, (2024)
23. Zhao Y., Li Y., Wu N.: Distributed acoustic sensing vertical seismic profile data denoiser based on convolutional neural network. IEEE Transactions on Geoscience and Remote Sensing **60**, 1–11, (2020)
24. Zhu H., Strawn N., Dunson D.B.: Bayesian graphical models for multivariate functional data. Journal of Machine Learning Research **17**(204), 1–27, (2016)

Chapter 13
Statistical Modeling of Categorical Trajectories with Multivariate Functional Data Approaches

Hervé Cardot and Caroline Peltier

Abstract There are many examples in which the statistical units of interest are samples of a continuous time categorical random process, taking values in a finite state space. Without loosing any information, we associate to each state a binary random function, taking values in $\{0, 1\}$, and turn the problem of statistical modeling of a categorical process into a multivariate functional data analysis issue. The (multivariate) covariance operator has nice interpretations in terms of departure from independence of the joint probabilities and the multivariate functional principal components are simple to interpret. Under a weak hypothesis assuming only continuity in probability of the $0 - 1$ trajectories, it is simple to build consistent estimators of the covariance kernel and perform multivariate functional PCA. The trajectories being piecewise constant, with a finite number of jumps, this a rare case in functional data analysis in which the trajectories can be observed exhaustively. The approach is illustrated on a data set of sensory perceptions, considering different gustometer-controlled stimuli experiments. It can be easily extended to analyze experiments, such as temporal check-all-that-apply, in which two states or more can be present at the same time or to analyze multivariate categorical trajectories.

13.1 Introduction

A lot of attention has been paid in the statistical literature over the last decades to develop tools dedicated to the analysis and modeling of functional data, considering

Hervé Cardot (✉)
Université Bourgogne Europe, Institut de Mathématiques de Bourgogne, UMR UBE-CNRS 5584, Dijon, France, e-mail: `herve.cardot@ube.fr`

Caroline Peltier
Inrae, Centre des Sciences du Goût et de l'Alimentation, UMR CNRS-INRAE-Institut Agro, Dijon, France and Probe Research Infrastructure, Chemosens facility, CNRS-INRAE, Dijon, France e-mail: `caroline.peltier@inrae.fr`

© The Author(s), under exclusive license to Springer Nature Switzerland AG 2025
G. Aneiros et al. (eds.), *New Trends in Functional Statistics and Related Fields*,
Contributions to Statistics, https://doi.org/10.1007/978-3-031-92383-8_13

random trajectories defined on an interval of time $[0, T]$ and taking values in $\mathbb{R}$, at each $t \in [0, T]$. Much less attention has been given to the case in which the trajectories take values in a finite set of elements that are not necessarily numbers, that is to say when $\mathbb{R}$ is replaced by the finite set $\mathcal{S} = \{S_1, \ldots, S_q\}$, with cardinality$q$.

There are many examples in which the statistical units of interest are samples of a continuous time random categorical process : in a pioneer work on demographic studies, Deville [7] extended the notion of multiple correspondence analysis to continuous time correspondence analysis in order to analyze the time evolution of the "marital status" of a sample of women over the period $[1962, 1975]$. The trajectories related to the marital status take values in a state space $\mathcal{S}$ with $q = 4$ states (divorced, married, single, widowed). Recent examples of statistical analysis of individual categorical trajectories are found in food science, a domain in which it is of great interest to get information about the temporal perception of aliments to understand the perception mechanisms. Of particular interest is the Temporal Dominance of Sensations (TDS) approach which consists in choosing sequentially attributes, among a list composed of q predefined items, describing a food product over tasting. The chosen states correspond to the most striking perception at a given time and the results of TDS experiments, after time normalization, can be represented via barplots as in Fig. 13.1. The technique developed by [7], also called qualitative harmonic analysis (see [8]) or categorical functional data analysis (CFDA) is now available in **R**, package `cfda` (see [16]) and can be used to analyze such kind of data. Even if the CFDA approach can be very powerful, encoding categorical trajectories into a sequence of real components, the interpretation of the results in terms of individual trajectories often may be delicate (see [15] for an illustration in sensory analysis).

Another natural approach is based on Markov and semi-Markov processes (see [14] for an introduction and [13] for an illustration in sensory analysis), which provide a simple representation of results via the graph of transitions between states. Considering parametric distributions for the sojourn time in the different states also allows to deal with maximum likelihood estimation techniques, permitting to build two-sample tests to compare two food products or different categories of consumers (see [4]) as well as model-based clustering, considering mixtures of semi-Markov processes (see [3]). A major drawback of this approach is the Markovian assumption which is often too simplistic with real data.

We can associate to each state S_j, for $j = 1, \ldots, q$, a random trajectory $X_j(t), t \in [0, T]$, taking value 1 when state S_j is observed at time t and zero else. Another possible way of analyzing such data is to consider that at each instant t of time we have to deal with compositional data (see [1] for a seminal paper and [9] for a recent review). A major difficulty in that case is that, at each instant t, $q - 1$ individual trajectories have value 0, so that usual logarithmic transforms cannot apply directly.

We consider in this paper a way of dealing with categorical functional data based on multivariate functional principal component analysis of $\mathbf{X}(t) = (X_1(t), \ldots, X_q(t))$ we the aim of finding a small dimension function space that can give the best approximation to the individual 0-1 trajectories. We show that it leads both to results that can be simply interpreted and powerful tools able to reduce effectively the dimension of categorical functional data in a finite dimension vector space. More details can be found in [5].

13.2 Notations and Mathematical Framework

To describe continuous time categorical trajectories, we consider the random element $Y = \{Y(t),\ t \in [0,T]\}$, with $Y(t) \in \mathcal{S} = \{S_1, \dots, S_q\}$ for all $t \in [0,T]$, and define, for $(s,t) \in [0,T] \times [0,T]$ and $(j,\ell) \in \{1,\dots,q\} \times \{1,\dots,q\}$, the joint probability $p_{j\ell}(s,t) = \mathbb{P}\left[Y(s) = S_j \text{ and } Y(t) = S_\ell\right]$. We denote by $p_j(t) = \mathbb{P}\left[Y(t) = S_j\right]$, the probability of being in state S_j at time t.

We associate to the random element Y, q trajectories X_j, $j = 1, \dots, q$, related to the occurrence of state S_j, and defined as follows, $X_j(t) = 1$ if $Y(t) = S_j$ and 0 otherwise. We have $\mathbb{E}[X_j(t)] = \mathbb{E}[\mathbf{1}_{Y(t)=S_j}] = p_j(t)$ and $\mathbb{E}[X_j(s)X_\ell(t)] = \mathbb{E}[\mathbf{1}_{Y(s)=S_j}\mathbf{1}_{Y(t)=S_\ell}] = p_{j\ell}(s,t)$. We suppose in the following that hypothesis $\mathbf{H}_1$ is satisfied:

$$\mathbf{H}_1: \quad \lim_{h\to 0} \mathbb{P}[Y(t) \neq Y(t+h)] = 0, \quad \forall t \in [0,T],$$

ensuring that the trajectories X_j, $j = 1, \dots, q$ are continuous in probability. Hypothesis $\mathbf{H}_1$ prevents them to have too many jumps. Remarking that $|X_j(t)| \le 1$ almost

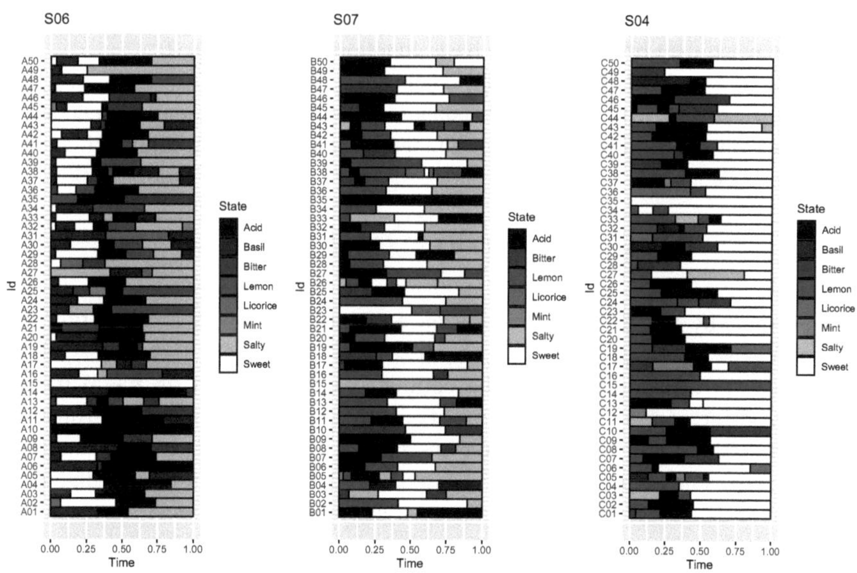

Fig. 13.1: TDS bandplot for $n = 150$ tasting experiments, considering three different gustometer-controlled stimuli, S06, S07 and S04.

surely, we deduce, by Theorem 1.3.6 in [17], that the trajectories are also continuous in the L^2 sense when $\mathbf{H}_1$ is true, that is to say,

$$\lim_{h \to 0} \mathbb{E}[(X_j(t) - X_j(t+h))^2] = 0, \quad \forall t \in [0,T],\ \forall j \in \{1,\ldots,q\}.$$

Define the covariance functions $\gamma_j(s,t) = \mathbb{C}ov(X_j(s), X_j(t))$ and, for $\ell \neq j$, $\gamma_{j\ell}(s,t) = \mathbb{C}ov(X_j(s), X_\ell(t))$. We remark that $\gamma_j(t,t) = p_j(t)(1 - p_j(t))$ and $\gamma_{j\ell}(s,t) = p_{j\ell}(s,t) - p_j(s)p_\ell(t)$, so that the terms out of the diagonal, when $s \neq t$, can be related to a departure from independence.

Proposition 13.1. *Under hypothesis $\mathbf{H}_1$, for all $j \in \{1,\ldots,q\}$,*

- *p_j (resp. γ_j) is uniformly continuous on $[0,T]$ (resp. on $[0,T] \times [0,T]$)*
- *$\gamma_{j\ell}$ is uniformly continuous on $[0,T] \times [0,T]$, for all $\ell \neq j$.*

The proof is a consequence of Theorem 1.3.6 in [17] and Theorem 7.3.2 in [12].

Note that the binary trajectories X_j take values in the Hilbert space $L^2[0,T]$ equipped with the usual inner product, denoted by $\langle .,. \rangle$ and norm $\|.\|$. As in [6], we consider $\mathbb{H} = L^2[0,T] \times \cdots \times L^2[0,T]$, the Hilbert space of q dimensional vectors of functions in $L^2[0,T]$, equipped with the inner product $\langle \mathbf{f}, \mathbf{g} \rangle_{\mathbb{H}} = \sum_{j=1}^q w_j \langle f_j, g_j \rangle$, where the weights w_j are strictly positive. The choice of the value of w_j can be important in practice and is discussed in [5].

Define $\mathbf{X} = (X_1, \ldots, X_q) \in \mathbb{H}$ and denote by $\boldsymbol{\gamma}(s,t)$ the corresponding multivariate (weighted) covariance function, with elements $\left(w_\ell \gamma_{j\ell}(s,t)\right)_{j,\ell}$. We deduce from Proposition 13.1 and the multivariate version of Mercer's theorem (see [6] or [11]) that there exist a set of orthonormal basis functions $\boldsymbol{\phi}_r(s) = (\phi_{r1}(s), \ldots, \phi_{rq}(s))$ in $\mathbb{H}$, $r = 1, 2, \ldots$ and corresponding eigenvalues $\lambda_1 \geq \lambda_2 \geq \cdots \geq 0$ such that $\boldsymbol{\gamma}(s,t) = \sum_{r \geq 1} \lambda_r \boldsymbol{\phi}_r(s) \boldsymbol{\phi}_r(t)^\top$. At the state level, this expansion can be seen as an additive expansion of the departure from independence of the joint probabilities (see [5] for details),

$$p_{j\ell}(s,t) - p_j(s)p_\ell(t) = \sum_{r \geq 1} \lambda_r \phi_{rj}(s) \phi_{r\ell}(t).$$

We denote by $\boldsymbol{\Gamma} : \mathbb{H} \to \mathbb{H}$ the covariance operator of $\mathbf{X}$ whose kernel function is $\boldsymbol{\gamma}(s,t)$. The eigenvalues satisfy $\sum_{r \geq 1} \lambda_r = \mathbb{E}\left[\|\mathbf{X} - \mathbf{p}\|_{\mathbb{H}}^2\right] \leq T \max_j w_j$, where $\mathbf{p}(t) = (p_1(t), \ldots, p_q(t)) \in \mathbb{H}$. The optimal linear expansion of $\mathbf{X} - \mathbf{p}$ in a k dimensional vector space of $\mathbb{H}$, in terms of quadratic mean, is given by the truncated Karhunen-Loève expansion, so that $\mathbf{X} \approx \mathbf{p} + \sum_{r=1}^k \langle \mathbf{X} - \mathbf{p}, \boldsymbol{\phi}_r \rangle_{\mathbb{H}} \boldsymbol{\phi}_r$ if k is large enough. Our aim is to estimate $\mathbf{p}$ and $\boldsymbol{\phi}_r$, for $r = 1, \ldots,$ in order to be able to capture the main variations of a categorical random function Y in a small dimension vector space.

13.3 Sampling and Estimators

Suppose now we have a sample of n categorical trajectories $Y_1, \ldots, Y_n$ observed on $[0, T]$. We define $\mathbf{X}_i = (X_{i1}, \ldots, X_{iq})$ the vector of the q corresponding binary trajectories, with $X_{ij}(t) = \mathbf{1}_{\{Y_i(t)=S_j\}}$. These trajectories are piecewise constant, with a finite number of jumps. It is a rare case in functional data analysis in which the trajectories can be observed exhaustively, that is to say for all time point $t \in [0, T]$.

We consider, for all s and t in $[0, T]$, the empirical probabilities $\widehat{p}_j(t) = n^{-1}\sum_{i=1}^n X_{ij}(t)$ and $\widehat{p}_{j\ell}(s,t) = n^{-1}\sum_{i=1}^n X_{ij}(s)X_{i\ell}(t)$, and define the estimators of the covariance functions $\widehat{\gamma}_j(t,t) = \widehat{p}_j(t)(1 - \widehat{p}_j(t))$ and $\widehat{\gamma}_{j\ell}(s,t) = \widehat{p}_{j\ell}(s,t) - \widehat{p}_j(s)\widehat{p}_\ell(t)$. We then deduce estimators $\widehat{\boldsymbol{\gamma}}$ of the kernel covariance function, and $\widehat{\boldsymbol{\Gamma}}$ of the covariance operator of $\boldsymbol{X}$. Noting that $\|\mathbf{X}\|_{\mathbb{H}}^2 \leq T \max_j w_j$, all the moments of $\|\mathbf{X}\|_{\mathbb{H}}$ are finite and we can directly apply Theorems 8.1.1 and 8.1.2 in [12] to get:

Proposition 13.2. *Suppose that hypothesis* $\mathbf{H}_1$ *is fulfilled, as* n *tends to infinity,* $\widehat{\mathbf{p}} \to \mathbf{p}$ *and* $\widehat{\boldsymbol{\gamma}} \to \boldsymbol{\gamma}$ *almost surely, and* $\sqrt{n}\left(\widehat{\mathbf{p}} - \mathbf{p}\right)$ *and* $\sqrt{n}\left(\widehat{\boldsymbol{\Gamma}} - \boldsymbol{\Gamma}\right)$ *are asymptotically Gaussian.*

We can deduce from Proposition 13.2 (see *e.g* Theorem 9.1.3 in [12]) the consistency and asymptotic normality of the estimators of the eigenfunctions $\widehat{\boldsymbol{\phi}}_r(t)$, $r = 1, \ldots, q$.

13.4 An Illustration in Sensory Analysis

The dataset used for this illustration was derived from a publicly available source [2] and the interested readers can consult [5] for more details. Fifty panelists participated in the study, during which they were instructed to click on the sensation they perceived as dominant in real-time from a list of descriptors while tasting controlled temporal solutions delivered by a gustometer. Three sensory signals were tested by the panelists: S04 (Lemon, followed by Acid, and finally Sweet), S06 (Sweet, Acid, and finally Salty, with a continuous hint of Basil), and S07 (Acid, followed by Sweet, and finally Salty, with a continuous hint of Lemon). The descriptor list included Acid, Sweet, Lemon, Basil, and Salty, alongside distractors such as Mint, Licorice and Bitter, so that the categorical process Y has $q = 8$ states. The overall number of experiments is $n = 150$ and time has been normalized to be $[0, 1]$ for each experiment, resulting in the observed categorical trajectories drawn in Fig. 13.1.

The estimated mean probability of perception $\widehat{p}_j(t)$ are drawn in Fig. 13.2, for the states j for which $\int_0^1 \widehat{p}_j(t)dt \geq 0.05$. A multivariate FPCA of the trajectories $\mathbf{X}_i, i = 1, \ldots, n$, has been performed with library `MFPCA` in **R** (see [10]) considering equal weights $w_1 = \ldots = w_q = 1$. The estimated principal component scores $\langle X_i - \widehat{\mathbf{p}}, \widehat{\boldsymbol{\phi}}_r \rangle_{\mathbb{H}}$, for $r = 1, 2$ are drawn in Fig. 13.3, whereas the main variations around the mean functions, for the first dimension, is drawn in Fig. 13.4. We only consider

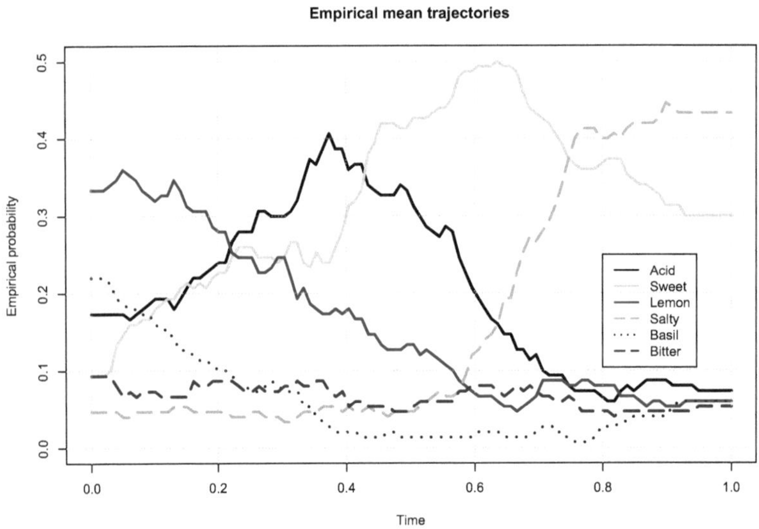

Fig. 13.2: Empirical probabilities $\widehat{p}_j(t)$, $t \in [0, 1]$. Only the states j whose average probability of occurrence is larger than 0.05 are presented

the components j with the largest variations, that is to say with the largest values of $\|\phi_{rj}\|$, $j = 1, \ldots, q$. The results are rather simple to interpret. For example, the black dots in Fig. 13.3, corresponding to the S04 experiment, are characterized by a first principal component taking positive values, related to a high probability of occurrence of Lemon and a small probability of Acid at the beginning of the period, and a high probability of occurrence of Sweet and a small probability of Salty at the end of the period.

13.5 Concluding Remarks

The new way, presented in this work, of reducing the dimension in a vector space of a sample of categorical trajectories provides a simple tool to describe and reduce the dimension of such kind of data. It can be easily extended to multivariate categorical trajectories, considering simultaneously, in our example, the three experiments made on the same panelists, that is to say $\mathbf{Y} = (Y_{S04}, Y_{S06}, Y_{S07})$. It is also direct to apply the proposed methodology to other protocols such as Temporal Check-All-That-Apply, in which $X_j(t)$ and $X_\ell(t)$, $\ell \neq j$, can be both equal to one at the same time t, whereas this would require to increase considerably to 2^q the number states with the CFDA or Markov processes approaches. Finally, this methodology can be useful to build predictive models, such as scalar-on-categorical functional data

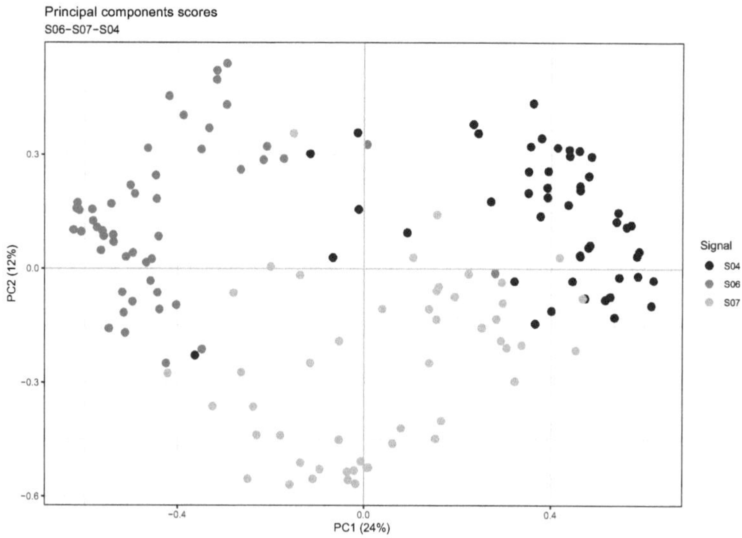

Fig. 13.3: Estimated principal component scores. Different labels are used to distinguish the observations according the set of gustomer-controlled stimuli

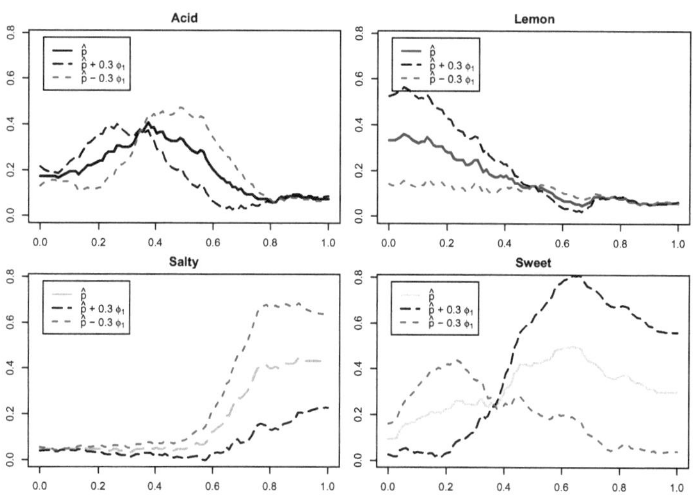

Fig. 13.4: Variations around the probability of occurrence related to the first component of the Karhunen Loève expansion, for the most important categories

regression models, permitting to use categorical trajectories as explanatory variables in statistical models.

Acknowledgements The Institut de Mathématiques de Bourgogne (UMR UB-CNRS 5584) receives support from the EIPHI Graduate School (contract ANR-17-EURE-0002).

References

1. Aitchison, J.: Principal component analysis of compositional data. Biometrika, **70**, 57-65 (1983)
2. Béno, N., Nicolle, L., Visalli, M.: A dataset of consumer perceptions of gustometer-controlled stimuli measured with three temporal sensory evaluation methods. Data in Brief, **48**, 109271, (2023)
3. Cardot, H., Frascolla, C., Schlich, P., Visalli, M. : Estimating finite mixtures of semi-Markov chains: an application to the segmentation of temporal sensory data. J. R. Stat. Soc., Ser. C, Appl. Stat. **68**, 1281–1303, (2019)
4. Cardot, H., Frascolla, C.: Hypothesis testing for Panels of Semi-Markov Processes with parametric sojourn time distributions. J. Stat. Plann. Inference, **228**, 59–79, (2024)
5. Cardot, H., Peltier, C.: Statistical modeling of categorical trajectories with multivariate functional principal components. Arxiv 2502.09986, (2025)
6. Chiou, J.M., Chen, Y.T., Yang, Y.F. : Multivariate functional principal component analysis: a normalization approach. Statistica Sinica, **24**, 1571–1596, (2014)
7. Deville, J.C.: Analyse des données chronologiques qualitatives, Annales de l'INSEE, **45**, 45-104, (1982)
8. Deville, J.C., Saporta, G.: Analyse harmonique qualitative, Data analysis and informatics, Proc. int. Symp., Versailles, 375-389, (1980)
9. Greenacre, M.: Compositional Data Analysis. Annu. Rev. Stat. Appl., **8**, 271–299, (2021)
10. Happ, C.: MFPCA Multivariate Functional Principal Component Analysis. R package version 1.3-10, (2022)
11. Happ, C., Greven, S.: Multivariate functional principal component analysis for data observed on different (dimensional) domains. J. Am. Stat. Assoc., **113**, 649–659, (2018)
12. Hsing, T., Eubank, R. : Theoretical foundations of functional data analysis, with an introduction to linear operators. Wiley Ser. Probab. Stat. Hoboken, NJ: John Wiley & Sons (2015)
13. Lecuelle, G., Visalli, M., Cardot, H., Schlich, P.: Modeling temporal dominance of sensations with semi-Markov chains. Food Quality and Preferences, **67**, 59-66, (2018).
14. Limnios, N., Oprişan, G.: Semi-Markov processes and reliability. Stat. Ind. Technol., Basel: Birkhäuser (2001)
15. Peltier, C., Visalli, M., Schlich, P. and Cardot, H.: Analyzing Temporal Dominance of Sensations data with Categorical Functional Data Analysis, Food Quality and Preference, 109, (2023)
16. Preda C., Grimonprez Q., Vandewalle V.: Categorical Functional Data Analysis. The cfda R Package. Mathematics **9**(23),3074, (2021)
17. Serfling, R.J.: Approximation theorems of mathematical statistics, Wiley Ser. Probab. Math. Stat., John Wiley & Sons, Hoboken, NJ (1980)

Chapter 14
Markov Switching Tensor Regressions

Roberto Casarin, Radu Craiu and Qing Wang

Abstract A new flexible tensor-on-tenor regression model that accounts for latent regime changes is proposed. The coefficients are driven by a common hidden Markov process that addresses structural breaks to enhance the model flexibility and preserve parsimony. A new soft PARAFAC hierarchical prior is introduced to achieve dimensionality reduction while preserving the structural information of the covariate tensor. The proposed prior includes a new multi-way shrinking effect to address over-parametrization issues while preserving interpretability and model tractability. An efficient MCMC algorithm is introduced based on random scan Gibbs and back-fitting strategy. The model framework's effectiveness is illustrated using financial and commodity market volatility data. The proposed model exhibits superior performance compared to the current benchmark, Lasso regression.

14.1 Introduction

As data grow in volume and complexity, it is increasingly common to record them as high-dimensional arrays or tensors. Such structures appear in many applications and fields, such as neuroimaging [8, 11], biostatistics, financial networks [2], or even more generally, in time series [3]. People are often interested in characterizing the relationship between a tensor predictor and a scalar outcome [8] or tensor outcome [15]. Tensor regression has been studied extensively in a linear model framework. Nevertheless, a common challenge within the framework of regression is model

Roberto Casarin
Ca' Foscari University of Venice, Italy, e-mail: r.casarin@unive.it

Radu Craiu
University of Toronto, Canada, e-mail: radu.craiu@utoronto.ca

Qing Wang (✉)
Ca' Foscari University of Venice, Italy, e-mail: qing.wang@unive.it

© The Author(s), under exclusive license to Springer Nature Switzerland AG 2025

G. Aneiros et al. (eds.), *New Trends in Functional Statistics and Related Fields*,
Contributions to Statistics, https://doi.org/10.1007/978-3-031-92383-8_14

misspecification. One of the sources of model misspecification is the presence of dynamic regimes, which naturally call for models with time-varying parameters.

In this paper, we assume a Hidden Markov chain dynamics for the regression coefficients because it allows for model parsimony while preserving a high level of flexibility compared to other time-varying parameter models. The model extends the soft tensor linear regression of [8, 14, 15] to an HMM or MS framework to accommodate structural breaks. In our multi-equation setting, the latent process is common to several tensor regression equations involving different response variables and, possibly, different sets of covariates. Using a common latent process facilitates the integration of information about the latent process from multiple outcomes and robustifies the estimation accounting for structural breaks. Regarding inference, we consider a Bayesian inference procedure combined with an efficient Gibbs sampler, which reduces computational costs and improves scalability [5].

14.2 A Markov-switching Tensor Regression Model

In our Markov Switching Tensor Regression Model (MSTR), we assume a system of N equations with time-varying parameters

$$\mathbf{y}_t = \boldsymbol{\mu}(s_t) + X_t \times_{1:M} \boldsymbol{B}(s_t) + \Sigma^{1/2}(s_t)\boldsymbol{\varepsilon}_t, \tag{14.1}$$

$t = 1, \dots, T$, where $\times_{m:n}$ denotes the tensor contract product along the modes from m to n [10], $\mathbf{y}_t = (y_{1t}, \dots, y_{Nt})'$ is the collection of response variables across equations, X_t is a $p_1 \times \dots \times p_{M-1}$ covariate tensor, $\boldsymbol{B}(s_t) = (B_1(s_t), \dots, B_N(s_t))$ is a M-modes coefficient tensor of size $p_1 \times \dots \times p_{M-1} \times N$ and $B_\ell(s_t)$ is a $p_1 \times \dots \times p_{M-1}$ coefficient tensor, $\boldsymbol{\mu}(s_t) = (\mu_1(s_t), \dots, \mu_N(s_t))'$ an intercept vector, $\Sigma^{1/2}(s_t)$ denotes the Cholesky's decomposition of the positive definite $N \times N$ covariance matrix $\Sigma(s_t)$, $\boldsymbol{\varepsilon}_t \sim \mathcal{N}_N(0, I_N)$, i.i.d. for $t = 1, \dots, T$.

The latent process $\{s_t, t = 1, \dots, T\}$ is a K-state homogeneous Markov chain with transition probability $\mathbb{P}(s_t = j | s_{t-1} = i) = p_{ij}$, $i, j = 1, \dots, K$ and the tensor regression parametrization used is

$$\mu_\ell(s_t) = \sum_{k=1}^{K} \mu_{\ell k}\mathbb{I}(s_t = k), B_\ell(s_t) = \sum_{k=1}^{K} B_{\ell k}\mathbb{I}(s_t = k), \Sigma(s_t) = \sum_{k=1}^{K} \Sigma_k\mathbb{I}(s_t = k).$$

See [7] and [4] for alternative coefficient parameterisations.

Since, in many applications, the number of covariates in Eq. 14.1 is large, a dimensionality reduction strategy is needed. In this paper, we follow [14] and [2, 3] and consider a low-rank representation combined with a hierarchical prior distribution. The hierarchical prior allows for shrinking effects in the coefficient matrices $B_{\ell k}$, $k = 1, \dots, K$, and the low-rank representation induces further shrinking effects along different modes. To simplify exposition, we assume $M = 3$, and the number of elements in each mode is p_1, p_2 and N. In our PARAFAC representation, the state-specific coefficient matrix is written as follows:

$$B_{\ell k} = \sum_{d=1}^{D} B_{\ell,1,k}^{(d)} \circ B_{\ell,2,k}^{(d)}, \tag{14.2}$$

where $\circ$ is the element-by-element Hadamard product and $B_{\ell,m,k}^{(d)}$ $m = 1, 2$ are multiplicative factors as in [14]. The soft PARAFAC prior distribution includes three stages. At the first stage, we assume an inverse gamma prior distribution $\mathcal{IG}(a_\sigma, b_\sigma)$ for $\sigma_{\ell,k}^2$ and a matrix-variate normal distribution for the coefficient tensor:

$$B_{\ell,m,k}^{(d)} \sim \mathcal{MN}_{p_1,p_2}\left(G_{\ell,m,k}^{(d)}, \tau_{\ell,k}\kappa_{\ell,m,k}^2\zeta_{\ell,k}^{(d)} I_{p_1}, I_{p_2}\right), \tag{14.3}$$

where $\mathcal{MN}_{p_1,p_2}(\mathbf{M}, \mathbf{U}, \mathbf{V})$ denotes the matrix-variate normal distribution with $p_1 \times p_2$ mean matrix $\mathbf{M}$, $p_1 \times p_1$ row covariance matrix $\mathbf{U}$ and $p_2 \times p_2$ column covariance matrix $\mathbf{V}$. See [9]. The location matrix $G_{\ell,m,k}^{(d)}$ is parametrized as follows:

$$G_{\ell,m,k}^{(d)} = \begin{cases} \boldsymbol{\gamma}_{\ell,1,k}^{(d)} \otimes \boldsymbol{\iota}_{p_2}, & \text{if } m = 1, \\ \boldsymbol{\iota}_{p_1} \otimes \boldsymbol{\gamma}_{\ell,2,k}^{(d)}, & \text{if } m = 2, \end{cases}$$

where $\otimes$ denotes the *outer product*, $\boldsymbol{\iota}_n = (1, \ldots, 1)'$ is the n-dimensional unit vector, $\boldsymbol{\gamma}_{\ell,1,k}^{(d)}$ and $\boldsymbol{\gamma}_{\ell,2,k}^{(d)}$ are the PARAFAC margins, which are vectors of sizes p_1 and p_2, respectively. In the conditional mean of the factors $B_{\ell,k}^{(d)}$, we have

$$\mathbb{E}\left(B_{\ell,k} \mid \boldsymbol{\gamma}_{\ell,1,k}^{(d)}, \boldsymbol{\gamma}_{\ell,2,k}^{(d)}\right) = \sum_{d=1}^{D} \mathbb{E}\left(B_{\ell,1,k}^{(d)}\right) \circ \mathbb{E}\left(B_{\ell,2,k}^{(d)}\right) = \sum_{d=1}^{D} \left(G_{\ell,1,k}^{(d)} \circ G_{\ell,2,k}^{(d)}\right)$$
$$= \sum_{d=1}^{D} (\boldsymbol{\gamma}_{\ell,1,k}^{(d)} \circ \boldsymbol{\iota}_{p_1}) \otimes (\boldsymbol{\gamma}_{\ell,2,k}^{(d)} \circ \boldsymbol{\iota}_{p_2}) = \sum_{d=1}^{D} \boldsymbol{\gamma}_{\ell,1,k}^{(d)} \otimes \boldsymbol{\gamma}_{\ell,2,k}^{(d)}.$$

In the second stage, we assume that the margins from the PARAFAC decomposition follow a multivariate normal distribution:

$$\boldsymbol{\gamma}_{\ell,m,k}^{(d)} \sim \mathcal{N}_{p_m}(\mathbf{0}, \tau_{\ell,k}\zeta_{\ell,k}^{(d)} W_{\ell,m,k}^{(d)}), \tag{14.4}$$

and assume the distributions are centred around the null vector and have random scales to allow for shrinkage at different levels. At the third stage, we borrow from [13] and specify the priors for scale parameters to induce shrinkage across PARAFAC components and fibers:

$$\tau_{\ell,k} \sim \mathcal{G}a(a_\tau, b_\tau), \quad \kappa_{\ell,m,k}^2 \sim \mathcal{G}a(a_\kappa, b_\kappa), \quad w_{\ell,m,j_m,k}^{(d)} \sim \mathcal{E}xp((\lambda_{\ell,m,k}^{(d)})^2/2),$$
$$\lambda_{\ell,m,k}^{(d)} \sim \mathcal{G}a(a_\lambda, b_\lambda), \quad \left(\zeta_{\ell,k}^{(1)}, \ldots, \zeta_{\ell,k}^{(D)}\right) \sim \mathcal{D}ir(\alpha/D, \ldots, \alpha/D),$$

where $\mathcal{Ga}(a,b)$, $\mathcal{Exp}(\lambda)$ and $\mathcal{Dir}(\nu_1,\dots,\nu_D)$ denote the Gamma, Exponential and Dirichlet distributions, respectively. Compared to [14] our prior assumes the global scale $\tau_{\ell,k}$ contributes not only to the variance of $\boldsymbol{\gamma}^{(d)}_{\ell,m,k}$ but also to that of one of the tensor coefficients $B^{(d)}_{\ell,m,k}$. This allows for stronger shrinkage effects and a full factorization of the prior variance, as detailed below. The matrices $W^{(d)}_{\ell,m,k} = \mathrm{diag}(w^{(d)}_{\ell,1,1,k},\dots,w^{(d)}_{\ell,m,j_m,k},\dots,w^{(d)}_{\ell,M,p_M,k})$, where $j_m = 1,\dots,p_m$ denotes the j_mth element along mode m, are the row-specific parameters that shrink the individual elements of the margins. Together with the prior on $\lambda^{(d)}_{\ell,m,k}$, they lead to an adaptive LASSO-type penalty on $\boldsymbol{\gamma}^{(d)}_{\ell,m,k}$ ([1]). The parameter $\zeta^{(d)}_{\ell,k}$ is component-specific and allows a subset of the D components to contribute substantially to the PARAFAC approximation while leaving the values of other components close to zero. The transition probabilities $(p_{1k},\dots,p_{Kk})$ are assumed to follow a Dirichlet distribution:

$$(p_{1k},\dots,p_{Kk}) \sim \mathcal{Dir}(\nu_1,\dots,\nu_K). \tag{14.5}$$

Dimensionality reduction is achieved by noticing that the number of equation- and state-specific coefficients needed to be estimated reduces from $p_1 \times p_2$ to $D(p_1+p_2)$ by applying a low-rank approximation. The choice of rank D for the soft PARAFAC decomposition of the tensor coefficient can lead to significant changes in computational costs, with a higher value of D triggering drastic increases in computational time. However, the increase in D doesn't necessarily guarantee a vast boost in inferential performance. Intuitively, the soft PARAFAC can expand away from the low-rank hard PARAFAC structure and achieve a higher-rank representation of the tensor coefficient. See [5] for further discussion.

14.3 Posterior Approximation

In this section, we assume tensor-valued covariates and denote with $B^{(d)}_{\ell,m,\tilde{j}_m,k}$ the j_mth slice of tensor $B^{(d)}_{\ell,m,k}$ along the mode m, where $\tilde{j}_m = (:,\cdots,:,j_m,:,\cdots,:)$ and with $B^{(d)}_{\ell,m,\tilde{j}_m,k}$ the $p_1 \times \cdots \times p_{m-1} \times p_{m+1} \times \cdots \times p_M$ tensor with $M-1$ modes. In the case of $M = 2$, introduced in the previous section, $B^{(d)}_{\ell,m,k}$ is a matrix. The slice $B^{(d)}_{\ell,m,\tilde{j}_m,k}$ is the j_1th row of $B^{(d)}_{\ell,m,k}$ when $m = 1$ with $\tilde{j}_1 = (j_1, j)$, $j = 1,\dots,p_2$ or the j_2th column of $B^{(d)}_{\ell,m,k}$ when $m = 2$ with $\tilde{j}_2 = (i, j_2)$, $i = 1,\dots,p_1$, respectively. Thus, choosing $j_1 = i$ and $j_2 = j$ and the jth and ith elements of $\tilde{j}_1$ and $\tilde{j}_2$, respectively one get the coefficient $B_{\ell,k,ij} = \sum_{d=1}^{D} B^{(d)}_{\ell,1,(i,j),k} B^{(d)}_{\ell,2,(i,j),k}$. For $\ell = 1,\dots,N$, $k = 1,\dots,K$, $m = 1,\dots,M$, $d = 1,\dots,D$ and $j_m = 1,\dots,p_m$ define the $q_m \times 1$ vector $\boldsymbol{\beta}^{(d)}_{\ell,m,j_m,k} = \mathrm{vec}(B^{(d)}_{\ell,m,\tilde{j}_m,k})$, with $q_m = \prod_{l\neq m} p_l$, obtained by stacking vertically all 1-mode fibers of the tensor following a lexicographic order of the indexes. We further define the collections $\boldsymbol{\beta}_k = (\boldsymbol{\beta}_{1k},\dots,\boldsymbol{\beta}_{Nk})$ and $\boldsymbol{\gamma}_k = (\boldsymbol{\gamma}_{1k},\dots,$

$\gamma_{Nk})$, with $\boldsymbol{\beta}_{\ell,k} = (\boldsymbol{\beta}^{(1)}_{\ell,1,1,k}, \ldots, \boldsymbol{\beta}^{(d)}_{\ell,m,j_m,k}, \ldots, \boldsymbol{\beta}^{(D)}_{\ell,M,p_M,k})'$ and $\boldsymbol{\gamma}_{\ell,k} = (\gamma^{(1)}_{\ell,1,1,k}, \ldots, \gamma^{(d)}_{\ell,m,j_m,k}, \ldots, \gamma^{(D)}_{\ell,M,p_M,k})'$, where $\gamma^{(d)}_{\ell,m,j_m,k}$ is the j_mth entry, $j_m = 1, \ldots, p_m$, of the PARAFAC marginal vector $\boldsymbol{\gamma}^{(d)}_{\ell,m,k}$ defined in Sect. 14.2.

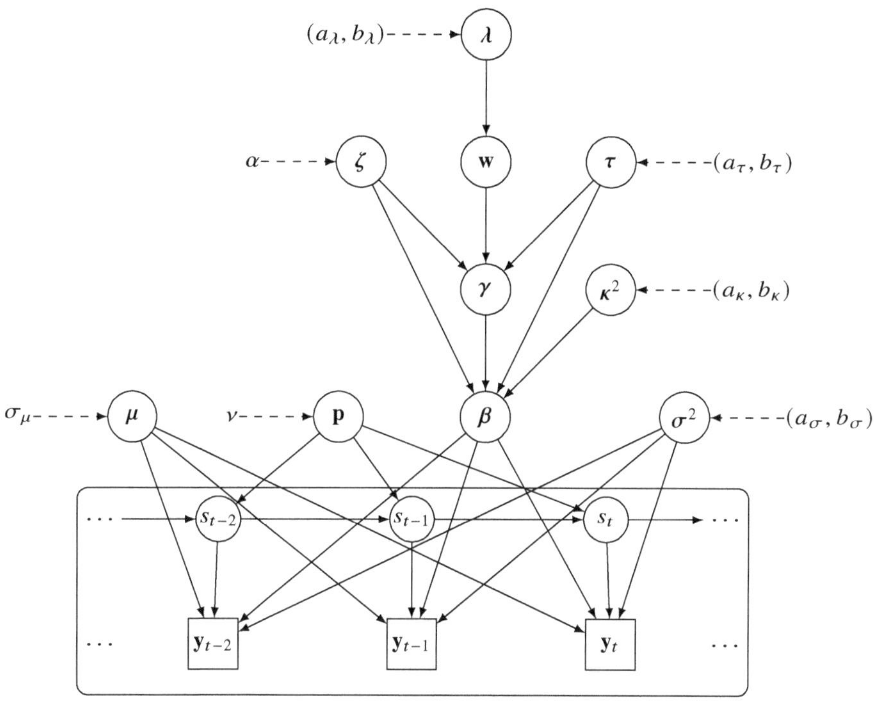

Fig. 14.1: Directed Acyclic Graph of the Bayesian Markov-switching Matrix Regression model. It exhibits the hierarchical structure of the observations $\mathbf{y}_t$ (boxes), the latent state variables s_t (grey circles), the parameters $\beta^{(d)}_{\ell,m,j_m,k}$, $\mu_{\ell,k}$, p_{ik} and $\sigma^2_{\ell,k}$, the hyper-parameters of the first stage $\gamma^{(d)}_{\ell,m,j_m,k}$ and $\kappa^2_{\ell,m,k}$, the second stage $\tau_{\ell,k}$, $\zeta^{(d)}_{\ell,m,k}$ and $w^{(d)}_{\ell,m,j_m,k}$ and the third stage $\lambda^{(d)}_{\ell,m,k}$ (white circles). The directed arrows show the conditional independence structure of the model

We summarize our Bayesian model in the Directed Acyclic Graph (DAG) representation of Fig. 14.1 where $\mathbf{p} = (\mathbf{p}_1, \ldots, \mathbf{p}_K)$ is the collection of transition probabilities, $\boldsymbol{\beta} = (\boldsymbol{\beta}_1, \ldots, \boldsymbol{\beta}_K)$, $\boldsymbol{\gamma} = (\boldsymbol{\gamma}_1, \ldots, \boldsymbol{\gamma}_K)$, $\boldsymbol{\sigma}^2 = (\boldsymbol{\sigma}^2_1, \ldots, \boldsymbol{\sigma}^2_K)$, and $\boldsymbol{\mu} = (\boldsymbol{\mu}_1, \ldots, \boldsymbol{\mu}_K)$ denote the collections across equations and states of the regression coefficients, the PARFAC factors, the error scale parameters and intercepts, respectively, where $\boldsymbol{\mu}_k = (\mu_{1,k}, \ldots, \mu_{N,k})'$, $\boldsymbol{\sigma}^2_k = (\sigma^2_{1,k}, \ldots, \sigma^2_{N,k})'$ and $\mathbf{p}_k = (p_{1k}, \ldots, p_{Kk})'$. See [5] for the derivation and further details on the sampling methods.

14.4 Empirical Application

To illustrate our MSTR model with $K = 3$ regimes and rank $D = 2$ (MSTR(3,2) in the following), we study the relationship between the daily volatility index of the US market, also known as VIX and the crude oil ETF oil volatility index (OVX) with several other financial indicators. This is motivated by the fact that VIX has been recognized as the key measure of the market's expectations and sentiments, and predicting VIX is crucial for developing investment strategies.

A vector of the log-VIX averages at different window sizes of $\{1, 5, 10, 22, 66\}$ days is usually employed to study the long-range dependence in VIX data [6]. This choice mirrors daily, weekly, bi-weekly, monthly and quarterly components and returns the family of heterogeneous autoregressive (HAR) processes. The model specification is shown in (14.6) and (14.7).

$$\text{VIX}_t = \mu_1(s_t) + \left\langle B_1(s_t), \begin{pmatrix} \text{SP}_{t-1} & \dots & \text{SP}_{t-h} & \dots & \text{SP}_{t-44} \\ \text{ER}_{t-1} & \dots & \text{ER}_{t-h} & \dots & \text{ER}_{t-44} \\ \text{Oil}_{t-1} & \dots & \text{Oil}_{t-h} & \dots & \text{Oil}_{t-44} \\ \text{OVX}_{t-1} & \dots & \text{OVX}_{t-h} & \dots & \text{OVX}_{t-44} \end{pmatrix} \right\rangle + \sigma_1(s_t)\epsilon_{1t}, \quad (14.6)$$

$$\text{OVX}_t = \mu_2(s_t) + \left\langle B_2(s_t), \begin{pmatrix} \text{SP}_{t-1} & \dots & \text{SP}_{t-h} & \dots & \text{SP}_{t-44} \\ \text{ER}_{t-1} & \dots & \text{ER}_{t-h} & \dots & \text{ER}_{t-44} \\ \text{Oil}_{t-1} & \dots & \text{Oil}_{t-h} & \dots & \text{Oil}_{t-44} \\ \text{VIX}_{t-1} & \dots & \text{VIX}_{t-h} & \dots & \text{VIX}_{t-44} \end{pmatrix} \right\rangle + \sigma_2(s_t)\epsilon_{2t}. \quad (14.7)$$

Left plots in **Fig.** 14.2 show that the MSTR(3,2) in-sample fitting is better than the one of the ordinary least squares and linear LASSO. The stepwise lines in the right plots show that by taking advantage of the Hidden Markov process, the MSTR(3, 2) model detects three oil volatility regimes: low volatility (regime 1), moderately high volatility (regime 2) and high volatility (regime 3). The regime identification follows from the prior identifying restriction $\mu_{11} < \mu_{12} < \mu_{13}$.

Fig. 14.3 provides the coefficient estimates of the Markov-switching Tensor Regression model MSTR(3, 2). In each plot the pairs of coefficients $(B_\ell(1), B_\ell(2))$ and $(B_\ell(1), B_\ell(3))$ are depicted (dots) for the VIX equation ($\ell = 1$, left plots) and OVX equation ($\ell = 2$, right plots).

The regime separation can be further described by inspecting the estimated effects of h-day log-return of oil prices (Oil_{t-h}, $h = 1, \dots, 44$) and S&P 500 (SP_{t-h} $h = 1, \dots, 44$) on VIX (blue dots, left) and OVX (red dots, right). The dots in the plots correspond to the values of parameters in different pairs of volatility regimes. The 70% HPD regions (grey ellipses) provide evidence of coefficient heterogeneity across regimes (asymmetric effects), equations (market asymmetry) and lags (long-term effects). The asymmetric effects in the coefficients are more substantial across regime 1 and 3 than regime 1 and 2. Moreover, there is evidence of a more balanced impact of the h-day S&P 500 log-returns on both markets (VIX and OVX) compared to the impacts of oil prices. There is also evidence of non-negligible long-term effects of oil prices on the stock market (dark blue dots in the left-bottom plot of Panel a).

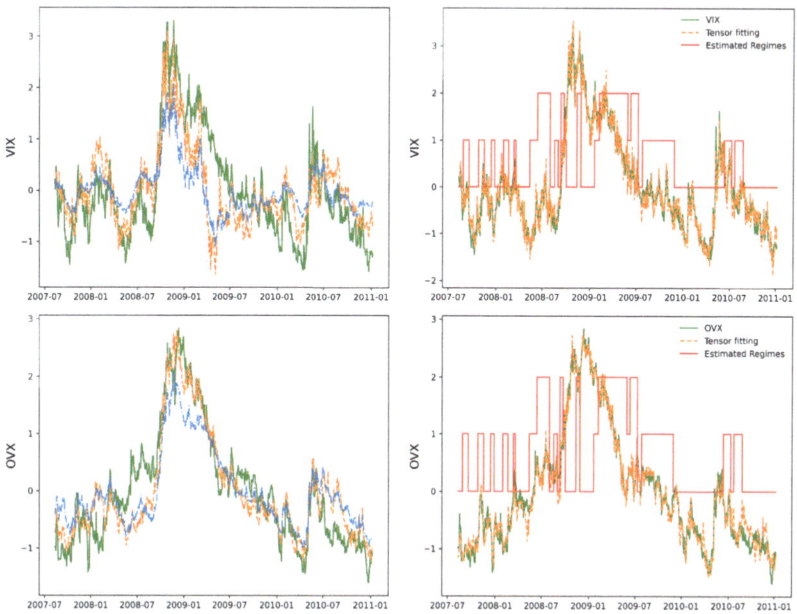

Fig. 14.2: In-sample fitting versus the actual data. Left: Least Squares (orange dashed) and LASSO (blue dashed) fitting. Right: Markov-Switching Tensor Regression model MSTR$(3,2)$ fitting (orange dashed) and estimated hidden states (red solid). For all plots, the green solid line represents the actual data for the VIX and VOX.

14.5 Conclusion

This paper proposes a Markov Switching Tensor-on-Tensor Regression Model (MSTR) for high dimensional data where a common hidden Markov chain process drives dependencies between equations and allows for regime changes and time-varying coefficients. A low-rank representation of the tensor coefficient is used to achieve dimensionality reduction. A hierarchical prior distribution introduces further shrinkage effects in the regression coefficients. An efficient MCMC sampler based on Random Partial Scan Gibbs and a back-fitting strategy is introduced. We illustrate our MSTR with a real-world application to oil and stock market volatility data. Our Bayesian MSTR model outperforms competing models and can capture structural changes in the parameters by identifying distinct volatility regimes. The MSTR can also capture the heterogeneity and asymmetric effects in the coefficient across the equations.

(a) Effects of Oil_{t-h}

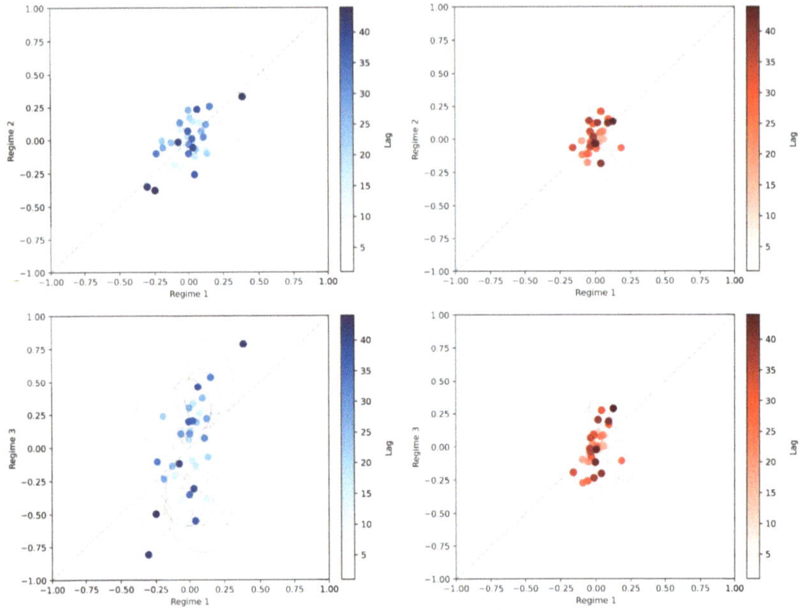

(b) Effects of SP_{t-h}

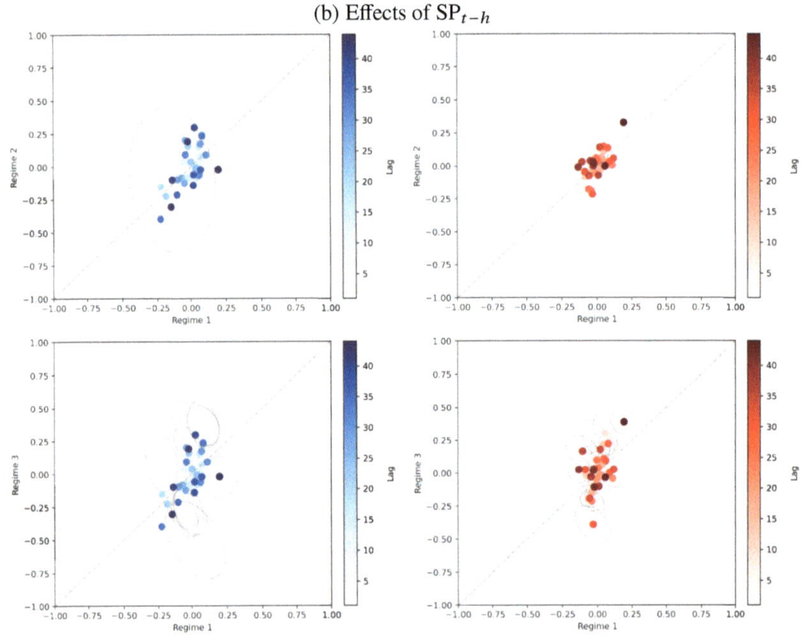

Fig. 14.3: Markov-switching Tensor Regression model MSTR(3, 2) coefficient estimates. Effects of h-day Oil (panel a) and S&P 500 (panel b) log-returns on VIX (left column) and OVX (right column) for $h \in \{1, \ldots, 44\}$. In each plot, lighter and darker colours represent smaller and larger h, respectively, and grey ellipses exhibit significant asymmetric effects across regimes (70% Highest Posterior Density regions).

References

1. Armagan, A., Dunson, D.B., Lee, J.: Generalized double Pareto shrinkage. Statistica Sinica **23**(1), 119 (2013)
2. Billio, M., Casarin, R., Iacopini, M.: Bayesian Markov-switching tensor regression for time-varying networks. J. Am. Stat. Assoc. **119**(545), 109–121 (2024)
3. Billio, M., Casarin, R., Iacopini, M., Kaufmann, S.: Bayesian dynamic tensor regression. J. Bus. Econ. Stat. **41**(2), 429–439 (2023)
4. Casarin, R., Sartore, D., Tronzano, M.: A Bayesian Markov-switching correlation model for contagion analysis on exchange rate markets. J. Bus. Econ. Stat. **36**(1), 101–114 (2018)
5. Casarin, R., Craiu, R.V., Wang, Q.: Markov Swithching Multiple-equation Tensor Regression. J. Multivar. Anal., (2025)
6. Fernandes, M., Medeiros, M.C., Scharth, M.: Modeling and predicting the CBOE market volatility index. J. Bank. Finance **40**, 1–10 (2014)
7. Frühwirth-Schnatter, S.: Finite Mixture and Markov Switching Models. Springer (2006)
8. Guha, S., Rodriguez, A.: Bayesian regression with undirected network predictors with an application to brain connectome data. J. Am. Stat. Assoc. **116**(534), 581–593 (2021)
9. Gupta, A.K., Nagar, D.K.: Matrix Variate Distributions. Chapman and Hall/CRC, Boca Raton (2018)
10. Kolda, T.G., Bader, B.W.: Tensor decompositions and applications. SIAM Rev. **51**, 455–500 (2009)
11. Spencer, D., Guhaniyogi, R., Shinohara, R., Prado, R.: Bayesian tensor regression using the Tucker decomposition for sparse spatial modeling. arXiv:2203.04733 (2022).
12. Wang, K., Xu, Y.: Bayesian tensor-on-tensor regression with efficient computation. Stat. Interface **17**, 199 (2024)
13. Guhaniyogi, R., Qamar, S., Dunson, D.B.: Bayesian tensor regression. J. Mach. Learn. Res. **18**(79), 1–31 (2017)
14. Papadogeorgou, G., Zhang, Z., Dunson, D.B.: Soft tensor regression. J. Mach. Learn. Res. **22**, 1–53 (2021)
15. Wang, K., Xu, Y.: Bayesian tensor-on-tensor regression with efficient computation. Stat. Interface **17**, 199 (2024)

Chapter 15
Functional Data Depth for the Analysis of Earth Surface Temperatures

Michele Cavazzutti, Eleonora Arnone, Ying Sun, Marc G. Genton and Laura M. Sangalli

Abstract In this work, we introduce an integrated depth measure for functional data defined over complex multidimensional domains. We consider functional data whose discrete realizations are irregularly spaced, and may be available only over portions of the domain. To address this issue, we propose an integrated depth based on a Voronoi tessellation of the multidimensional domain. This approach ensures favorable statistical properties for the proposed depth, as well as computational efficiency, enabling the analysis of large-scale functional datasets. We validate our proposal with the study of air temperatures across the Earth surface, as provided by the CESM2 Large Ensemble Community Project. The proposed depth is able to capture the increase in global temperatures since the 1980s, coherently with global warming.

15.1 Introduction

The data depth is a measure of *centrality* of a datum with respect to a distribution. At first, it was introduced to establish an inward-outward ordering, and therefore

Michele Cavazzutti (✉)
Politecnico di Milano, Milan, Italy, e-mail: `michele.cavazzutti@polimi.it`

Eleonora Arnone
Università di Torino, Turin, Italy, e-mail: `eleonora.arnone@unito.it`

Ying Sun
King Abdullah University of Science and Technology, Thuwal, KSA,
e-mail: `ying.sun@kaust.edu.sa`

Marc G. Genton
King Abdullah University of Science and Technology, Thuwal, KSA,
e-mail: `marc.genton@kaust.edu.sa`

Laura M. Sangalli
Politecnico di Milano, Milan, Italy, e-mail: `laura.sangalli@polimi.it`

© The Author(s), under exclusive license to Springer Nature Switzerland AG 2025

G. Aneiros et al. (eds.), *New Trends in Functional Statistics and Related Fields*,
Contributions to Statistics, https://doi.org/10.1007/978-3-031-92383-8_15

a ranking, in the context of multivariate data [18, 9, 19]. Starting from the early 2000s [5], it was also generalized to analyze functional data [10, 11]. In addition, the depth-induced rankings have allowed the solution of various statistical problems, ranging from visualization and outlier detection [7, 16, 17] to hypothesis testing, as well as clustering and classification [9].

The majority of the existing literature on functional data depth focuses on functional data defined over one-dimensional domains. Only recently some authors have started to focus on the problem of providing a data depth for functional data defined over multidimensional domains, mainly considering simple two-dimensional planar domains and hypothesizing the functional data observed on regular uniform grids [6, 12]. However, many applications of interest concern data observed over complex multidimensional domains and with measurement locations that are irregular in space; furthermore, observations may be affected by some severe missing data patterns. These problems may be handled naturally within the framework of functional integrated depth, that has been recently extended also to partially observed functional data over one-dimensional domains [3, 4]. However, the discretization of the integrated depth needs to be handled cautiously, to guarantee favorable consistency properties for real data sample. In this work, we propose to discretize the integrated depth based on a Voronoi tessellation of the domain, which implicitly takes into account the complex geometry of the domain, and also permits to correctly treat irregular spatial measurements. The use of the Voronoi tessellation is also advantageous in terms of computational efficiency. Indeed, while the existing functional depth discretizations require the ranking (sorting) of the functional evaluations in each measurement location [1, 13, 3], we only need to rank the evaluations in the Voronoi seeds, significantly reducing the number of computations compared to the full set of locations.

We illustrate our method with the analysis of Earth's surface average yearly temperatures over the period 1850-2100, as provided by the CESM2 Large Ensemble Project [15]. The CESM2 Large Ensemble (LENS2) consists of a set of temperature simulations at a 1-degree spatial resolution, covering the period 1850-2100 under CMIP6 historical and SSP370 future radiative forcing scenarios. In this context, we employ the proposed depth to study the evolution of average annual temperatures between 1850 and 2014. Fig. 15.1 shows the average annual temperatures for the years 1850 and 2014. By applying the proposed depth, we observe that temperature functions have exhibited increasingly extreme behaviors since the 1980s. This finding aligns with established knowledge about global warming, and demonstrates the effectiveness of our method.

15.2 Integrated Functional Depth for Multidimensional Domains

Let $D : \mathbb{R} \times \Gamma \to [0, 1]$ be a univariate depth that satisfies the conditions expressed in [13], where Γ is the space of distributions on $\mathbb{R}$. These conditions are required to guarantee the good properties of the overall integrated functional depth, and are

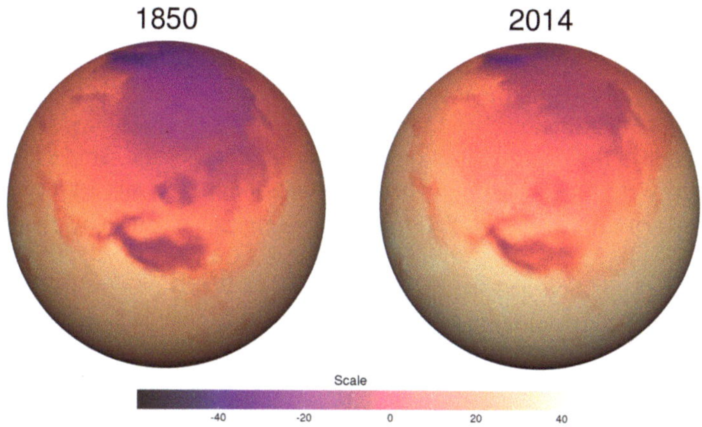

Fig. 15.1: Visualization of the annual average temperature (°C) field in 1850 and 2014 from the CESM2 Large Ensemble Project

encountered by the main univariate depths of interest, such as the halfspace depth [18] and the simplicial depth [9]. Let $\Omega \subset \mathbb{R}^d$ be the support of the functional data, with $d = 2, 3$. In the application to temperature data introduced in Sec. 15.1, Ω is the Earth surface, a two-dimensional manifold embedded in $\mathbb{R}^3$. Consider a stochastic process $X : \Omega \to \mathbb{R}$ with law P. Let $P_{\boldsymbol{p}}$ be the marginal distribution of $X(\boldsymbol{p})$, $\forall \boldsymbol{p} \in \Omega$.

We consider a natural extension of the definition of the weighted integrated functional depth, originally developed for a one-dimensional domain in [1]. Let ϕ be a positive continuous function on Ω, $\phi(\boldsymbol{p}) > 0 \; \forall \boldsymbol{p} \in \Omega$. We define the following weighting function:

$$w_\phi(\boldsymbol{p}) = \frac{\phi(\boldsymbol{p})}{\int_\Omega \phi(\boldsymbol{s}) \mathrm{d}\boldsymbol{s}}. \tag{15.1}$$

We can thus define the Weighted Integrated Functional Depth (WIFD) for multi-dimensional domains as follows. The Weighted Integrated Functional Depth of the continuous function $x : \Omega \to \mathbb{R}$, with respect to P, is given by

$$\mathrm{WIFD}(x, P) = \int_\Omega D(x(\boldsymbol{p}), P_{\boldsymbol{p}}) w_\phi(\boldsymbol{p}) \mathrm{d}\boldsymbol{p}. \tag{15.2}$$

The weight function is used to weight differently different portions of the domain. For instance, in the analysis of temperature across the Earth surface, we know that the variability is reduced in water areas with respect to the inland, due to the water's thermal inertia, and we may wish to weight more either water or land areas, depending on the target of the analysis. Notice that the weight function can also be used to address partial observability in the functional data, as done by [3, 4] in the context of univariate functional data over one-dimensional domains.

15.2.1 Depth for Discrete Measurements

In practice the distribution P is not known in advance, and needs to be estimated with a sample distribution. Let $\{X_i : \Omega \to \mathbb{R}\}_{i=1,\dots,n}$ be a set of i.i.d. stochastic processes drawn from the distribution of interest P. Let $\hat{P}$ be the sample distribution that assigns weight $\frac{1}{n}$ to each X_i, where n is the sample size, and let also $\hat{P}_{\boldsymbol{p}}$ be its marginal distribution at $\boldsymbol{p}$. The WIFD sample estimate is obtained by plugging $\hat{P}$ in (15.2).

Both the population and the sample versions of WIFD require the availability of continuous functional observations. However, functional data are available only through discrete measurements, and the integral in (15.2) needs to be approximated with a discrete sum. In the existing literature for one-dimensional domains, this problem has so far been addressed by assuming a common, regular measurement grid [14, 12], therefore weighting each point equally. From a practical point of view, this approach is equivalent to the vectorization of the functional data, therefore discarding the information related to the spatial displacement of the measurements. When dealing with multidimensional domains, the so obtained depth would be problematic for various reasons. First, the curse of dimensionality imposes a significantly larger number of measurement locations to observe each function with minimal accuracy. Even if the functions were observed on a common set of locations — which is not always the case — the computational burden of ranking functions for every location quickly becomes prohibitive. Additionally, the measurement locations might follow a highly irregular pattern, depending on the specific problem under consideration. In such cases, assigning equal weights to all locations can disproportionately amplify the influence of areas with a higher density of observations, skewing the overall results. These factors collectively pose significant challenges in accurately and efficiently determining functional depth measures for multidimensional domains. Such difficulties can be solved by means of an appropriate discretization of the integral.

In this work, we consider a Voronoi tessellation of the multidimesional domain Ω; see, e.g., [2]. Voronoi tessellations can represent accurately complex multidimensional geometries, including non-convex domains; moreover, they can be easily obtained as the dual representation of triangular meshes, as shown in Fig. 15.2 for a simple two-dimensional domain.

In order to approximate the depth's integral, we represent the discrete realization of the functional data with a step function, that takes as constant value, in each Voronoi cell, the average of the datum's measurements in the cell itself. Thus, the point-wise depth in (15.2) is substituted in each cell by the univariate depth of n averages, one for each functional datum. As a result, the discrete version of (15.2) is the sum of the univariate depths computed in each cell multiplied by the Lebesgue measure of the cell itself.

The approach can naturally handle different measurement grids across different statical units, as well as irregular density of the measurement locations, without overemphasis on the regions of the domain with more observations. For instance, in the application to temperature data, the functions are measured on a regular longitude-latitude grid (of size 55296), leading to much denser locations near the

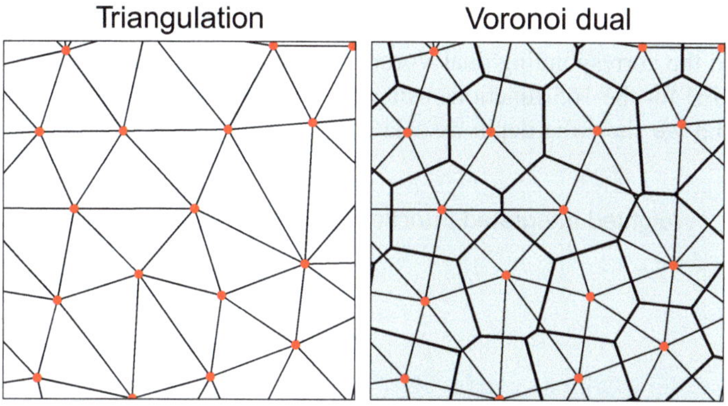

Fig. 15.2: Example of domain discretization in two dimensions. In the left panel, we show a detail of the triangulation of a two-dimensional planar domain. In the right panel, we show instead the corresponding dual Voronoi tessellation, where the edges of the Voronoi cells are depicted with thicker linewidth with respect to the edges of the triangulation

poles with respect to the equator zone. Using a regular Voronoi tessellation of the globe, instead of the point-wise observational grid, we avoid over-weighting the contribution of the poles with respect to areas close to the equator. In addition, the computational cost related to this formula is now depending on the number of Voronoi cells, and no longer on the number of locations, allowing the analysis of large data sets.

15.3 Application to LENS-CESME Temperature Data

The CESM2 Large Ensemble Community Project (LENS2) [15] is a publicly accessible collection of climate model simulations designed to enhance knowledge of internal climate variability and climate change. Among the various simulations available, we are interested in analyzing the set of surface temperatures in the period 1850-2014. The data sample consists of 165 annual temperature simulations observed at 1-degree spatial resolution, obtained under CMIP6 historical radiative forcing scenarios. An example of such functions is depicted in Fig. 15.1. Given that global warming has become to be evident from the late 1980s, we expect the temperature functions to become more and more extreme after that time as whole functional objects across the Earth surface. Therefore, we seek for validation from our methods by applying the proposed depth to the aforementioned functional temperatures. In this context, the domain is a two-dimensional sphere embedded in a three-dimensional space. As anticipated above, the temperature surfaces are measured on a regular longitude-latitude grid with 55296 locations.

We discretize the domain using a uniform triangulation of the domain of 13969 nodes, and the corresponding dual Voronoi tessellation. We then compute the proposed WIFD for the 165 functions using the univariate simplicial depth for D [9]. In Fig. 15.3 we show the depth associated with each annual temperature function

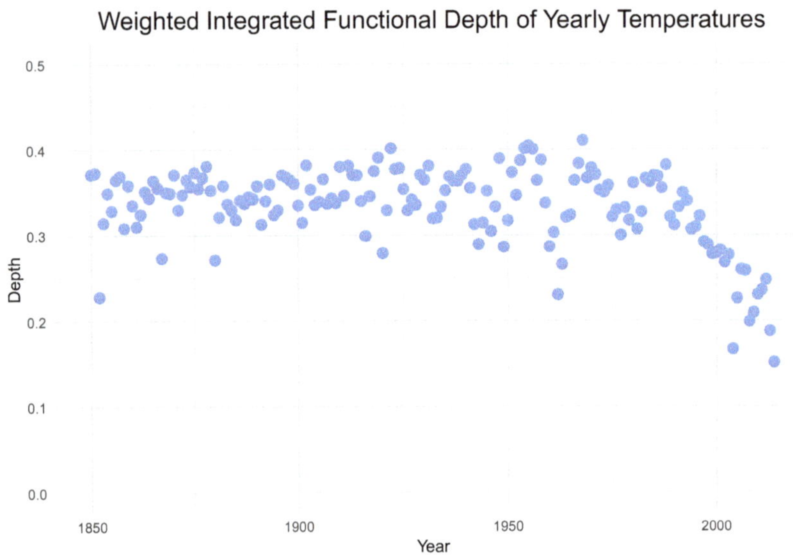

Fig. 15.3: WIFD computed with point-wise simplicial depth D for the Yearly Temperatures in the period 1850-2014. Notably, the depth seems to decrease with time starting from 1980s on, and the great majority of the most extreme years is clearly clustered in the last decade

against the year. The data shows a noticeable decrease in depth starting from the 1980s, signaling a trend of increasingly extreme air temperatures over time. This result corroborates knowledge about global warming.

15.4 Conclusion

In this work we illustrated a functional depth for data defined on multidimensional domains. Our proposal handles the difficulties related to the complexity of the domain by means of a Voronoi-based integration. We apply the proposed depth to study the global surface temperatures estimates for the years 1850-2014, showing that the climate conditions have started becoming extreme in the last three decades.

There are several directions of research that could be explored, starting from this work. One possible extension would be to exploit the weight function in (15.1) to handle functional data that are only partially observable, similarly to what is done in [3, 4] in the one-dimensional domain setting. Another natural extension of the proposed depth goes toward functional vectors.

Acknowledgements L.M. Sangalli acknowledges the PRIN 2022 project CoEnv - Complex Environmental Data and Modeling (CUP 2022E3RY23), funded by the European Union – NextGenerationEU programme, and by the Italian Ministry for University and Research. L.M. Sangalli also acknowledges the project GRINS - Growing Resilient, INclusive and Sustainable (GRINS PE00000018 – CUP D43C22003110001), funded by the European Union - NextGenerationEU programme. The views and opinions expressed are solely those of the authors and do not necessarily reflect those of the European Union, nor can the European Union be held responsible for them. M. Cavazzutti and L.M. Sangalli acknowledge MUR, Dipartimento d'Eccellenza 2022-2027, Dipartimento di Matematica, Politecnico di Milano. M. Cavazzutti also acknowledges financial support from KAUST as a visiting student.

References

1. Claeskens G., Hubert M., and Slaets L. and Vakili K.: Multivariate Functional Halfspace Depth. Journal of the American Statistical Association, **109**(505), 411-423 (2014)
2. De Berg M.: Computational geometry: algorithms and applications. Springer Science & Business Media (2000)
3. Elías A., Jiménez R., Paganoni A. M., Sangalli L. M.: Integrated Depths for Partially Observed Functional Data. Journal of Computational and Graphical Statistics, **32**(2), 341–352 (2022)
4. Elías A., Nagy S.: Statistical properties of partially observed integrated functional depths. Test, to appear (2024)
5. Fraiman R., Muniz G.: Trimmed means for functional data. Test, **10**(2), 419–440 (2001)
6. Genton M. G., Johnson C., Potter K., Stenchikov G., Sun, Y.: Surface boxplots. Stat, **3**(1), 1-11 (2014)
7. Hyndman R. J., Shang H. L.: Rainbow plots, bagplots, and boxplots for functional data. Journal of Computational and Graphical Statistics, **19**(1), 29–45 (2010)
8. Kokoszka P., Reimherr M.: Introduction to Functional Data Analysis. Chapman and Hall/CRC, New York (2017)
9. Liu R. Y.: On a notion of data depth based on random simplices. The Annals of Statistics, **18**(1), 405–414 (1990)
10. López-Pintado S., Romo J.: On the Concept of Depth for Functional Data. Journal of the American Statistical Association, **104**(486), 718–734 (2009)
11. López-Pintado S., Romo J.: A half-region depth for functional data. Computational Statistics & Data Analysis **55**(4), 1679–1695 (2011)
12. López-Pintado S., Wrobel J.: Robust non-parametric tests for imaging data based on data depth. Stat, **6**, 405–419 (2017)
13. Nagy S., Gijbels I., Omelka M. and Hlubinka D.: Integrated depth for functional data: statistical properties and consistency. ESAIM: Probability and Statistics, **20**, 95-130 (2016)
14. Nagy S. and Gijbels I. and Hlubinka D.: Weak convergence of discretely observed functional data with applications. Journal of Multivariate Analysis, **146**, 46-62 (2016)
15. Rodgers K. B., Lee S.-S., Rosenbloom N., Timmermann A., Danabasoglu G., Deser C., Edwards J., Kim J.-E., Simpson I. R., Stein K., Stuecker M. F., Yamaguchi R., Bódai T., Chung E.-S., Huang L., Kim W. M., Lamarque J.-F., Lombardozzi D. L., Wieder W. R., Yeager S. G.: Ubiquity of human-induced changes in climate variability. Earth Sysem Dynamics, 12, 1393–1411 (2021)
16. Sun Y., Genton M. G.: Functional boxplots. Journal of Computational and Graphical Statistics, **20**(2), 316–334 (2011)

17. Sun Y., Genton M. G.: Adjusted functional boxplots for spatio-temporal data visualization and outlier detection. Environmetrics, **23**(1), 54–64 (2012)
18. Tukey, J. W.: Proceedings of the 1974 International Congress of Mathematicians.Vol. 2, Canadian Mathematical Congress, Vancouver, pp. 523–531 (1975)
19. Zuo, Y., Serfling, R.: General notions of statistical depth function. The Annals of Statistics, **28**(2), 461–482 (2000)

Chapter 16
Improving Finite Samples Performances in Nonparametric Functional Regression by Using Weighted Pseudo-Metrics

Lax Chan, Laurent Delsol and Aldo Goia

Abstract The choice of the pseudo-metric plays a crucial role in functional nonparametric regression as it has a direct impact on the type of models one considers. In this work, the idea of weighted pseudo-metric is introduced and discussed. The effects and the performances of the choices of pseudo-metric used are evalulated by means of a Monte Carlo study.

16.1 Introduction

Starting from the seminal work [5], the interest towards nonparametric regression in functional statistics has grown both in theoretical and applied frameworks. In particular, a lot of studies have been devoted to the so-called scalar on functional model where a real random variable is modeled by using a functional covariate (see e.g. the collection [1], the monograph [6] or the review [7]).

One of the main problems in functional data analysis is the selection of a good (pseudo-)metric as a distance measure between curves (see e.g. [3, 8]). In the regression framework, the pseudo-metric, entering in the estimation procedure, captures the information contained in the explanatory curve by extracting the informative features of the explanatory curve. Hence a good selection criterion is useful to build more accurate estimates, also when the sample size is rather small.

Lax Chan
School of Mathematics, University of Birmingham, Edgbaston, Birmingham, B15 2TT, e-mail: l.chan.1@bham.ac.uk

Laurent Delsol (✉)
Institut Denis Poisson, Université d'Orléans, Rue de Chartres, 45067, Orleans, France, e-mail: Laurent.Delsol@univ-orleans.fr

Aldo Goia
Dipartimento di Studi per l'Economia e l'Impresa, Università del Piemonte Orientale, Via E. Perrone, 18, Novara, e-mail: aldo.goia@uniupo.it

© The Author(s), under exclusive license to Springer Nature Switzerland AG 2025
G. Aneiros et al. (eds.), *New Trends in Functional Statistics and Related Fields*, Contributions to Statistics, https://doi.org/10.1007/978-3-031-92383-8_16

The selection of the pseudo-metric in the regression context is therefore a crucial step; until now it is based on some heuristic rules based on cross-validation approaches and a systematic theory on this topic is not yet provided.

This paper aims to make progress on this open problem by illustrating how the use of weighted pseudo-metric can improve the quality of the estimate. After presenting the model and its estimate in Sect. 16.2, the effects due to the introduction of a weighted pseudo-metric are investigated from a theoretical point of view in Sect. 16.3 in the same spirit as in [2, 4]. The practical implementation of the technique is discussed in Sect. 16.4 and a simulation study concludes the paper (see Sect. 16.5) showing good performances in comparison with other approaches.

16.2 The Model and the Estimate

Consider the following regression model:

$$Y = r(X) + \epsilon, \tag{16.1}$$

where Y is a real-valued random variable, X is a random object lying in the space $\mathcal{F}$ of the real-valued functions defined on a compact set $\mathcal{T} \subset \mathbb{R}$ which can be equipped with a family of pseudo-metrics d, r is a real-valued regression operator, and the residual ϵ satisfies $\mathbb{E}[\epsilon|X] = 0$.

Given a sample $(X_i, Y_i)_{1\leq i\leq n}$ drawn from (X,Y), in the nonparametric case, it is usual to make the kernel estimator of the regression operator r depending on a pseudo-metric d as follows (see e.g. [6]):

$$\widehat{r}_d(x) = \begin{cases} \sum_{i=1}^n Y_i \dfrac{K\left(\frac{d(X_i,x)}{h}\right)}{\sum_{i=1}^n K\left(\frac{d(X_i,x)}{h}\right)} & \text{if } \sum_{i=1}^n K\left(\frac{d(X_i,x)}{h}\right) > 0, \\ 0 & \text{else,} \end{cases}$$

where h is a positive bandwidth, K a (positive) kernel function, and x a given element of $\mathcal{F}$.

Suppose that only some parts of the random curve X are relevant in defining the link with the response Y. In order to try to improve the precision of the estimation (in a finite dimensional sample point of view), an *informative support selection* approach can be performed by using a suitable weighted pseudo-metric $d_{w,n}$ that may depend on n.

The resulting estimator writes:

$$\widehat{r}_{d_{w,n},h}(x) = \begin{cases} \sum_{i=1}^n Y_i \dfrac{K\left(\frac{d_{w,n}(X_i,x)}{h}\right)}{\sum_{i=1}^n K\left(\frac{d_{w,n}(X_i,x)}{h}\right)} & \text{if } \sum_{i=1}^n K\left(\frac{d_{w,n}(X_i,x)}{h}\right) > 0, \\ 0 & \text{else.} \end{cases} \tag{16.2}$$

As an example, if X is m times differentiable, the following weighted pseudo-metric can be considered:

$$d_{m,w}(y,x)^2 = \int_{\mathcal{T}} \left(y^{(m)}(t) - x^{(m)}(t)\right)^2 w(t)\,dt, \tag{16.3}$$

where $x^{(m)}$ stands for the m-th derivative of x (with $x^{(0)} = x$), and the weight w is a density function, i.e. $w(t) \geq 0$, $\int_{\mathcal{T}} w(t)\,dt = 1$.

16.3 The Weighted Pseudo-metric Effect

The pseudo-metric is a usual and relevant parameter in nonparametric kernel regression with functional covariate (see e.g. [6]) to extract the informative part of the explanatory curve. However, up to our knowledge, only a few comments have been done on the theoretical effect on using a pseudo-metric instead of a metric. This point is discussed in the current section. To make the notations simpler, note $p = d_{w,n}$ hereafter.

16.3.1 The General Case

Firstly define the relation $\mathcal{R}_p$ on $\mathcal{F} \times \mathcal{F}$ induced by the pseudo-metric:

$$x\mathcal{R}_p y \Leftrightarrow \{p(x,y) = 0\}.$$

It is easy to prove it is an equivalence relation on $\mathcal{F}$. Hence we can define for any element $x \in \mathcal{F}$ its equivalence class

$$C_p(x) = \{y \in \mathcal{F},\ p(x,y) = 0\}.$$

The first important fact is that the kernel estimator given by (16.2) is constant on every equivalence class and does not actually estimate $r(x)$ but

$$r_p(x) = \left(\mathbb{E}[Y|p(X,x)]\right)\big|_{p(X,x)=0}. \tag{16.4}$$

Indeed, the bias term (with respect to r) is usually obtained from the Taylor expansion of $\phi_p(u) = \mathbb{E}[r(X) - r(x)|p(X,x)]\big|_{p(X,x)=u}$ around 0 (see for instance [4]), where the assumption $\phi_p(0) = 0$ is commonly used. However, it comes quite easily

$$\phi_p(0) = \left(\mathbb{E}[Y - r(x)|p(X,x)]\right)\big|_{p(X,x)=0} = r_p(x) - r(x).$$

Hence if $r(x) \neq r_p(x)$ we have an additional term in the bias. Moreover, the variance dominant term depends on

$$r_{2,p}(x) = \left(\mathbb{E}[Y^2|p(X,x)]\right)\Big|_{p(X,x)=0},$$

that is usually different of $r_2(x) = r^2(x) + \sigma^2(x)$ (used in [4]), where σ^2 is given by $\sigma^2(X) = \mathbb{E}[\epsilon^2|X]$. Moreover, following the same ideas than [4, Theorem 1], one may expect to get, under appropriate assumptions,

$$\mathbb{E}[\widehat{r}_p(x)]-r(x) = \mathbb{E}[\widehat{r}_p(x)]-r_p(x)+\Delta r_p(x) = C_{1,p}h+\Delta r_p(x)+o(h)+O\left(\frac{1}{nF_p(h)}\right),$$

and

$$Var(\widehat{r}_p(x)) = \frac{C_{2,p}}{nF_p(h)} + o\left(\frac{1}{nF_p(h)}\right),$$

where $F_p(h) = \mathbb{P}\left(p(X,x) \leq h\right)$ and $C_{1,p}$ and $C_{2,p}$ are explicit constants depending on x.

16.3.2 Extracted Features

Let's focus now on a specific case where everything may be written in a easier way. Assume the pseudo-metric writes in the following way:

$$\forall x, y \in \mathcal{F},\ p(x,y) = p_0(g_p(x), g_p(y)), \tag{16.5}$$

where p_0 is a metric and $g_p : \mathcal{F} \mapsto \mathcal{F}_p = g_p(\mathcal{F})$ is an operator. Assumption (16.5) holds for instance, when the pseudo-metric is given by (16.3), with p_0 is the $\mathcal{L}^2$ metric and $x_p = x^{(m)}\sqrt{w}$. Then for any $x \in \mathcal{F}$, the element $x_p := g_p(x)$ corresponds to the features of x extracted by p and $g_p^{-1}(x_p) = C_p(x)$. In that case $\widehat{r}_{p,h}(x)$ is in fact designed for the regression model induced on $\mathcal{F}_p$:

$$Y = r_p(X_p) + \epsilon_p, \tag{16.6}$$

where $r_p(X_p) = \mathbb{E}[Y|X_p]$ and $\epsilon_p = Y - r_p(X)$. In other words, $\widehat{r}_{p,h}$ does not actually estimate r but r_p. The following properties explicit some links between the regression models (16.1) and (16.6), when g_p is measurable:

- $\mathbb{E}[r(X)|X_p] = \mathbb{E}[\mathbb{E}[Y|X]|X_p] = \mathbb{E}[\mathbb{E}[Y|X,X_p]|X_p] = r_p(X_p)$
- $\mathbb{E}[\epsilon|X_p] = \mathbb{E}[Y - r(X)|X_p] = 0$
- $\forall x \in \mathcal{F},\ \mathbb{E}[Y|p(X,x)] = \mathbb{E}[\mathbb{E}[Y|X_p]|p_0(X_p,x_p)] = \mathbb{E}[r_p(X_p)|p_0(X_p,x_p)]$

and the last equality links (16.4) and (16.6).

16.3.3 Comments

Let's take time to discuss the impact of the pseudo-metric p from the content of the previous sections. On the one hand, we have explained the use of p instead of a metric may be useful to extract interesting features of the explanatory curve and induce a kind of semi-parametric model (16.6) which we can expect to be more relevant (see Sect. 16.3.2). On the other hand, this may lead to a biased estimator whenever $\Delta r_p(x) = r_p(x) - r(x) \neq 0$ (see Sect. 16.3.1). Hence the choice of p is crucial and has to to be done in a relevant way.

Assume first p does not depend on n. If $\Delta r_p(x) = 0$, the pseudo-metric extracts the relevant features of X related to Y and the estimator is asymptotically unbiased and has usual properties. If $\Delta r_p(x) \neq 0$, the pseudo-metric forgets some relevant features of X related to Y and the estimator is asymptotically biased. However, for finite sample size, the semi-parametric model may lead to a better estimator than the non-parametric one, for instance when $\Delta r_p(x)$ is quite small and $\mathcal{F}_p$ is more concentrated than what it would be for a metric.

Assume now p may depend on n. Then $\Delta r_p(x)$, $\mathcal{F}_p$, $C_{1,p}$ and $C_{2,p}$ depend on n, and the asymptotic properties of the kernel estimator will be impacted by their asymptotic behaviour. Again, for finite sample size, the semi-parametric model may lead to a better estimator than the non-parametric one.

For all these reasons, the selection of a relevant pseudo-metric p from the data is crucial and interesting. Indeed, the selected p should provide a better estimation and help to identify the features of X relevant to estimate $Y\cdot$

16.4 Weighted Pseudo-metric Selection

The aim is to find the optimal weight $w^\star \in \left\{w > 0, \int w = 1\right\}$ and the optimal bandwidth $h^\star$ so that the mean square error,

$$MISE(\widehat{r}_{w,h}) = \mathbb{E}\left[\left(\widehat{r}_{w,h}(x) - r(x)\right)^2 \gamma(x)\right]$$

is minimized, where γ is a known positive operator with compact support.

In practice, since the MISE is unknown, and to avoid overfitting, we consider cross-validation criterion as proxy. The problem is then to find:

$$\left(\widehat{w}, \widehat{h}\right) = \arg \min_{w \in D_n, h \in H_n} \left\{ \frac{1}{n} \sum_{i=1}^{n} \left(Y_i - \widehat{r}_{w,h}^{-i}(X_i)\right)^2 \gamma(X_i) + pen_n(w) \right\},$$

where $\widehat{r}_{w,h}^{-i}$ is computed without (X_i, Y_i), D_n is a suitable family of pdfs, H_n a compact subset of $\mathbb{R}$, and $pen_n(w)$ a penalty term that may be added to constrain w.

For practical purposes, one can represent $\log w$ by using B-splines of order d and k internal knots, then Equation (16.3) can be rewritten as:

$$d_{m,w}(y,x)^2 = \int_{\mathcal{T}} \left(y^{(m)}(t) - x^{(m)}(t)\right)^2 \frac{\exp\left(\sum_{j=1}^{k+d+1} \alpha_j B_j(t)\right)}{\int_{\mathcal{T}} \exp\left(\sum_{j=1}^{k+d+1} \alpha_j B_j(t)\right) dt} dt. \quad (16.7)$$

The algorithm to estimate the optimal weight $\widehat{w}$ is based on the selection of the optimum vector $(\widehat{\alpha}_1, \ldots, \widehat{\alpha}_{k+d+1})$ that may be done using a parametric optimization procedure since no constrains have to be imposed. Note that the expression of w is related to the well-known centered *log-ratio tranformation* (*clr*) applied to pdfs when one deals with these kind of functional objects.

16.5 A Numerical Example

This section is devoted to show how the proposed technique can improve the performances in the nonparametric setting through a simulation study.

Consider a sample of $n \in \mathbb{N}$ trajectories X_i of a standard Brownian motion on $[0, 1]$, discretized over a grid of 100 equispaced points $0 \leq t_1 < \cdots < t_{100} \leq 1$, and the regression model

$$Y_i = \int_{0.2}^{0.3} \sin(\pi X_i(t))dt + \int_{0.8}^{0.9} \cos(\pi X_i(t))\, dt + \epsilon_i \qquad i = 1, \ldots, 200, \quad (16.8)$$

where the errors ϵ_i are i.i.d. $\mathcal{N}(0, \sigma)$ and σ is γ times the standard deviation of $r(X)$ (with $\gamma > 0$), in order to control the signal-to-noise ratio. Note that the model is fully nonlinear and only the segments of the curves defined over the intervals $[0.2, 0.3]$ and $[0.8, 0.9]$ are useful to explain the response.

In this numerical study, the following experimental settings are used: $n = 50, 100, 200$ and $\delta = 0.2, 0.5$ that correspond to a squared correlation ratio equals to 0.96 and 0.8 respectively. The performances are evaluated in terms of mean square error (MSE) calculated starting from 500 Monte Carlo with training-sets generated according to (16.8) and evaluated over a test-set of size $n_{test} = 50$. The results are compared with them obtained by using the Nadaraya-Watson nonparametric approach with the standard unweighted $\mathcal{L}^2$ pseudo-metric and a semi-parametric approch, namely the single functional index model (SFIM). To better appreciate the quality of the performances, the estimates based on the nonparametric approach restricted to the informative supports for the predictor are derived and the corresponding MSEs are computed.

For what concerns the implementation of the estimation procedure illustrated in Sect. 16.4, cubic splines with $k = 8$ internal knots are used and, for the sake of simplicity, the penalization term in not included in the minimization criterion.

The obtained results are collected in Table 16.1: they show that the proposed methodology produces a substantial increment in terms of quality of prediction over the classical unweighted approach and the semi-parametric one in all the considered experimental settings. Note that the performances are not so far from the case where the informative support is assumed known: the introduction of a weighted pseudo-metric allows to better describe the link between the covariate and the response through an informative support selection.

δ	n	Weighted $\mathcal{L}^2$ pseudo-metric	Standard $\mathcal{L}^2$ pseudo-metric	SFIM	$\mathcal{L}^2$ pseudo-metric over segments
	50	0.4168	0.4795	0.5014	0.3666
0.2	100	0.3840	0.5178	0.5867	0.3637
	200	0.3779	0.5118	0.5924	0.3596
	50	0.5761	0.6135	0.6909	0.5423
0.5	100	0.6292	0.7147	0.8241	0.6143
	200	0.4634	0.5483	0.6213	0.4371

Table 16.1: MSEs of four different approaches varying δ and n. The results displayed are multiplied by a factor of 100 for ease of comparison

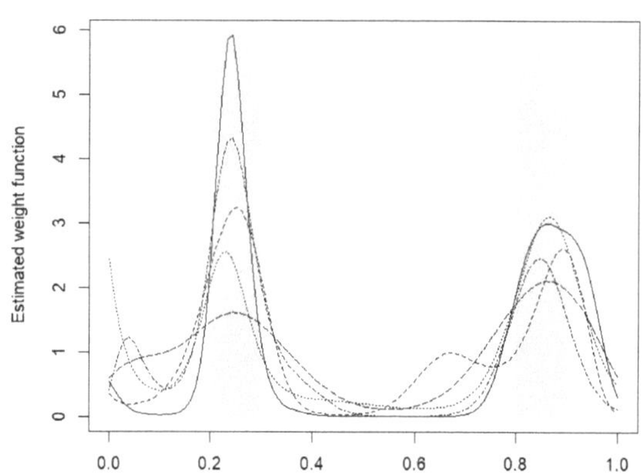

Fig. 16.1: Estimates of the weight functions over five samples and the relevant parts of the curves (in light purple).

To appreciate how the proposed methodology is able to identify the informative support, the estimates of the weight functions obtained starting from five samples of size $n = 200$ are plotted in Fig. 16.1. For the purpose of clarity, the parts of the curves that are relevant in explaining the response in the regression model are visualized in light purple.

Acknowledgements L. Chan and A. Goia are members of Gruppo Nazionale per l'Analisi Matematica, la Probabilità e le loro Applicazioni (GNAMPA) of the Istituto Nazionale di Alta Matematica (INdAM).

References

1. Aneiros, G., Horová, I., Hušková, M., Vieu, P. (Eds.): Functional and High-Dimensional Statistics and Related Fields. Springer, (2020)
2. Chagny, G., Roche, A.: Adaptive estimation in the functional nonparametric regression model. Journal of Multivariate Analysis, **146**, 105–118, (2016)
3. Chen, H., Philip T., Reiss, P.T., Tarpey T.: Optimally weighted L^2 distance for functional data, Biometrics, **70**(3), 516–25, (2014)
4. Ferraty, F., Mas, A., Vieu, P.: Advances on nonparametric regression for functional variables. Australian and New Zealand Journal of Statistics, **49**(3), 267–286, (2007)
5. Ferraty, F., Vieu, P.: The Functional Nonparametric Model and Application to Spectrometric Data, Computational Statistics, **17**(4), 545–564, (2002)
6. Ferraty, F., Vieu, P.: Nonparametric functional data analysis. Theory and practice. Springer, New York, (2006)
7. Ling, N., Vieu, P.: Nonparametric modelling for functional data: selected survey and tracks for future. Statistics, **52**(4), 934—949, (2018)
8. Romano, E., Diana, A., Miller, C., O'Donnell, R.: Optimally weighted L^2 distances for spatially dependent functional data, Spatial Statistics, **39**, 100468, (2020)

Chapter 17
Smoothing with Nonlinear Partial Differential Equation Regularization

Aldo Clemente, Alessandro Palummo, Eleonora Arnone and Laura M. Sangalli

Abstract This work proposes a novel smoothing technique for functional data observed over space and time, that penalizes the misfit with respect to a time-dependent nonlinear partial differential equation. This method extends existing physics-informed statistical models, so far explored in the case of penalty terms involving linear partial differential equations. Their use in the regularizing term of a smoothing model broaden extensively the scope of this method, but introduces new challenges in both theory and numerical implementation. Through a simulation study, we show that the proposed method outperforms simpler models, which are unable to incorporate complex physics information in the statistical model. These results demonstrate the potential of combining advanced physical knowledge with statistical modelling, paving the way for future applications to real-world data, as briefly discussed in the context of neuroimaging data.

Aldo Clemente (✉)
Politecnico di Milano, Piazza Leonardo da Vinci, 32, 20133, Italy
e-mail: aldo.clemente@polimi.it

Alessandro Palummo
Politecnico di Milano, Piazza Leonardo da Vinci, 32, 20133, Italy
e-mail: alessandro.palummo@polimi.it

Eleonora Arnone
Università degli Studi di Torino, Corso Unione Sovietica, 218 b, Italy
e-mail: eleonora.arnone@unito.it

Laura M. Sangalli
Politecnico di Milano, Piazza Leonardo da Vinci, 32, 20133, Italy
e-mail: laura.sangalli@polimi.it

© The Author(s), under exclusive license to Springer Nature Switzerland AG 2025
G. Aneiros et al. (eds.), *New Trends in Functional Statistics and Related Fields*,
Contributions to Statistics, https://doi.org/10.1007/978-3-031-92383-8_17

17.1 Introduction

There is a vast literature concerning smoothing with roughness penalties involving differential operators. In this framework, classical methods usually include isotropic linear operators in the penalty term (see, e.g., [7, 22, 23, 18, 16, 27]). Specifically, these methods typically penalize the roughness of the estimate by penalizing the second derivative of the unknown function they aim to estimate. In the same framework, we find the family of physics-informed statistical models, discussed, e.g., in [20, 5, 19], where the regularization term can involve an anisotropic and non-stationary linear Partial Differential Equation (PDE), encoding problem-specific information about the phenomenon under study. This class of models can handle functional data observed over multidimensional spatial domains, possibly displaying complex geometries, such as two-dimensional manifold domains [9], and non-convex three-dimensional domains [3]. Moreover, the class of models has been recently enriched by [4] to handle spatiotemporal data.

While linear PDE-based regularization has proven effective, many real-world phenomena exhibit nonlinear dynamics that cannot be adequately captured by linear models. This work extends Physics-Informed functional data models by considering a nonlinear PDE in the penalty term to account for more realistic and complex phenomena. Nonlinear PDEs model a variety of complex phenomena across different scientific and engineering domains. As an example, the Fisher-KPP equation, named after Fisher, Kolmogorov, Petrovsky, and Piskunov, is commonly used to model population growth and wave propagation (see, e.g.,[10]). There is a vast literature concerning smoothing with roughness penalties involving differential operators. In this framework, classical methods usually include isotropic linear operators in the penalty term (see, e.g., [7, 22, 23, 18, 16, 27]). Specifically, these methods typically penalize the roughness of the estimate by penalizing the second derivative of the unknown function they aim to estimate. In the same framework, we find the family of physics-informed statistical models, discussed, e.g., in [20, 5, 19], where the regularization term can involve an anisotropic and non-stationary linear Partial Differential Equation (PDE), encoding problem-specific information about the phenomenon under study. This class of models can handle functional data observed over multidimensional spatial domains, possibly displaying complex geometries, such as two-dimensional manifold domains [9], and non-convex three-dimensional domains [3]. Moreover, the class of models has been recently enriched by [4] to handle spatiotemporal data.

While linear PDE-based regularization has proven effective, many real-world phenomena exhibit nonlinear dynamics that cannot be adequately captured by linear models. This work extends Physics-Informed functional data models by considering a nonlinear PDE in the penalty term to account for more realistic and complex phenomena. Nonlinear PDEs model a variety of complex phenomena across different scientific and engineering domains. As an example, the Fisher-KPP equation, named after Fisher, Kolmogorov, Petrovsky, and Piskunov, is commonly used to model population growth and wave propagation (see, e.g., [10]).

The use of nonlinear PDEs in the regularization term of physics-informed functional data analysis models extensively broadens their applicability. However, introducing nonlinear PDEs also raises significant theoretical and computational challenges, such as ensuring the identifiability of the solution and handling increased model complexity. In cases where physical information is only partially available, one could adopt a parameter-cascading approach, as explored by [21], to estimate unknown PDE coefficients. The use of nonlinear PDEs in the regularization term of physics-informed functional data analysis models extensively broadens their applicability. However, introducing nonlinear PDEs also raises significant theoretical and computational challenges, such as ensuring the identifiability of the solution and handling increased model complexity. In cases where physical information is only partially available, one could adopt a parameter-cascading approach, as explored by [21], to estimate unknown PDE coefficients.

17.2 Smoothing with Nonlinear PDEs

Let $\mathcal{D} \subset \mathbb{R}^d$, $d \leq 3$, be a bounded domain, and let $[0, T] \subset \mathbb{R}$ be a time interval. We consider functional data defined over the support $\mathcal{D} \times [0, T]$. For each functional realization f, we only observe a noisy and discrete realization. Specifically, let $\{y_{ij}\}_{ij}$ be the value of the functional datum observed at the spatial location $\mathbf{p}_i \in \mathcal{D}$, for $i = 1, \cdots, n$, and time instant $t_j \in [0, T]$, for $j = 1, \cdots, m$. We assume the following model:

$$y_{ij} = f(\mathbf{p}_i, t_j) + \varepsilon_{ij},$$

where $f : \mathcal{D} \times [0, T] \to \mathbb{R}$ is the unknown spatiotemporal function to be estimated, and ε_{ij} are uncorrelated errors with zero mean and constant finite variance. We propose to estimate the unknown field f by minimizing the following functional:

$$\sum_{i=1}^{n} \sum_{j=1}^{k} \left(y_{ij} - f(\mathbf{p}_i, t_j)\right)^2 + \lambda \int_0^T \int_{\mathcal{D}} (\partial_t f + \mathcal{L}[f] - u)^2. \qquad (17.1)$$

The first term in (17.1) is the sum-of-squared errors and measures the model's adherence to the data, and the second term is the regularization term, which encodes physic-specific information about the problem at hand through the time-dependent PDE $\partial_t f + L[f] = u$, where $\mathcal{L}[f]$ is a nonlinear differential operator and u is the forcing term.

The estimation problem (17.1) does not admit a solution in a closed form. Different numerical discretization strategies can be considered to find a numerical approximation to the problem. In this work, we first linearize the infinite-dimensional problem. Then, we rely on the Finite Element Method to discretize the problem in space, and we employ a semi-implicit Euler scheme to advance in time.

The two following sections briefly discuss a simulation study and an application to neuroimaging data, using the Fisher-KPP equation.

17.3 Simulation Study

This section compares the proposed method with Space-Time regression with non-linear PDE regularization (STR-PDE-NL) to the method with Space-Time regression with linear PDE regularization (STR-PDE) proposed by [4].

We consider functional data generated by the solution to the Fisher-KPP equation, an equation extensively used to model population growth and wave propagation. Specifically, let $\mathcal{D} = [-2.5, 2.5]^2$ and $[0, 1]$ be the spatial domain and the temporal domain of interest, respectively. Consider the following Fisher-KPP equation:

$$\begin{cases} \frac{\partial f}{\partial t} - \mu \Delta f = r f(1-f) & \text{in } \mathcal{D} \times T \\ f(\mathbf{p}, 0) = \mathbb{I}_{\{x^2+y^2 \leq 1\}} & \text{in } \mathcal{D} \\ f(\mathbf{p}, t) = 0 & \text{on } \partial\mathcal{D} \times T, \end{cases} \tag{17.2}$$

where $\mu = 0.1$ is the diffusion coefficient and $r = 3.0$ is the reaction coefficient. This differential problem (17.2) exhibits travelling wave solutions, which are not usually recovered analytically.

We discretize the spatial domain $\mathcal{D}$ using a mesh with 289 nodes, and we divide the interval T into 10 subintervals of equal dimension $\Delta t = 0.1$. We consider as test function a reference solution to the problem (17.2), provided by the numerical software `FreeFem++`[12]. The test function is shown in the second column of Fig. 17.2 at different time instants. To generate the data for this simulation study, we evaluate the test function at different $n = 100, 250, 500, 1000$ locations uniformly sampled over the spatial domain $\mathcal{D}$, for each $t_j = j\Delta t, j = 1, \cdots, m$, and by adding to the test function independent Gaussian noises with standard deviation equal to 0.05, i.e. about the 5% of the range of the test field. We repeat the data generation 30 times, for each sample size n.

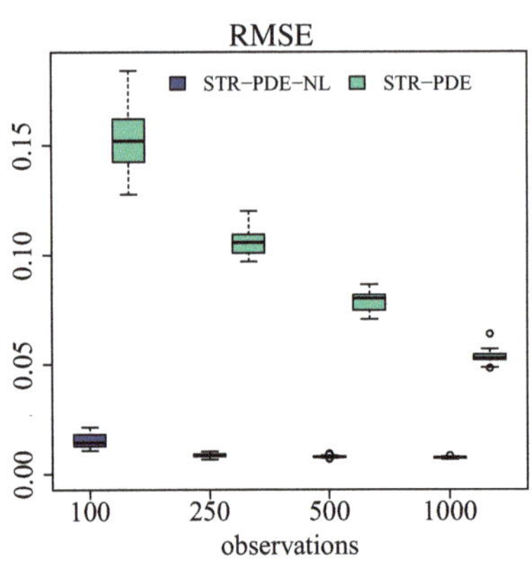

Fig. 17.1: Boxplots of the L^2 errors across 30 simulation repetitions for the STR-PDE-NL method (blue) and STR-PDE method (green) as the number of observations increases.

Fig. 17.2 shows the data in the first column, the test field in the second column, and the estimate provided by the two competing methods in the third and fourth columns, respectively. From a qualitative point of view, the proposed STR-PDE-NL method recovers the overall shape of the test field better than the competing method. The previous considerations are confirmed by Fig. 17.1, which shows the boxplots of the L^2 errors as the sample size increases. These results highlight the advantages of incorporating the complex physics of the phenomenon under study, motivating the application of the proposed model to real-world data.

17.4 Application to Neuroimaging Data

The proposed method offers fascinating new opportunities for the analysis of neuroimaging data.

Advances in neuroscience have led to the development of neuroimaging techniques, such as structural and functional Magnetic Resonance Imaging (MRI) and Positron Emission Tomography (PET), which are used to monitor the progression of neurological diseases. In particular, structural and functional MRI assess brain connectivity and atrophy [1]. On the other hand, PET provides pathological information about the brain. PET has been used to identify aggregates of toxic proteins, such as Amyloid-β (Aβ) and Tau, which are associated with the development of neurological diseases, such as Alzheimer's disease. This is achieved using, for instance, Aβ-imaging tracers and Tau-imaging tracers [2, 13]. Fig. 17.3 shows an example of an image scan recorded by PET.

From a differential equation modelling perspective, the Fisher-KPP equation is one of the simplest nonlinear PDEs that has been considered to model the diffusion and spreading of such toxic proteins in the human brain (see, e.g., [17, 14, 24, 8, 26, 11, 15]). In recent years, various numerical methods have been developed to find a discrete solution to the aforementioned differential problem (see, e.g., [25, 6, and references therein]). These studies offer mathematical models for the progression and development of the disease.

The statistical model proposed in this work would permit to validate the mathematical models cited above, through the analysis of real neuroimaging data, obtained by the modalities mentioned above. Integrating real data with these advanced physical models mentioned above can indeed lead to significant advances of knowledge in this field.

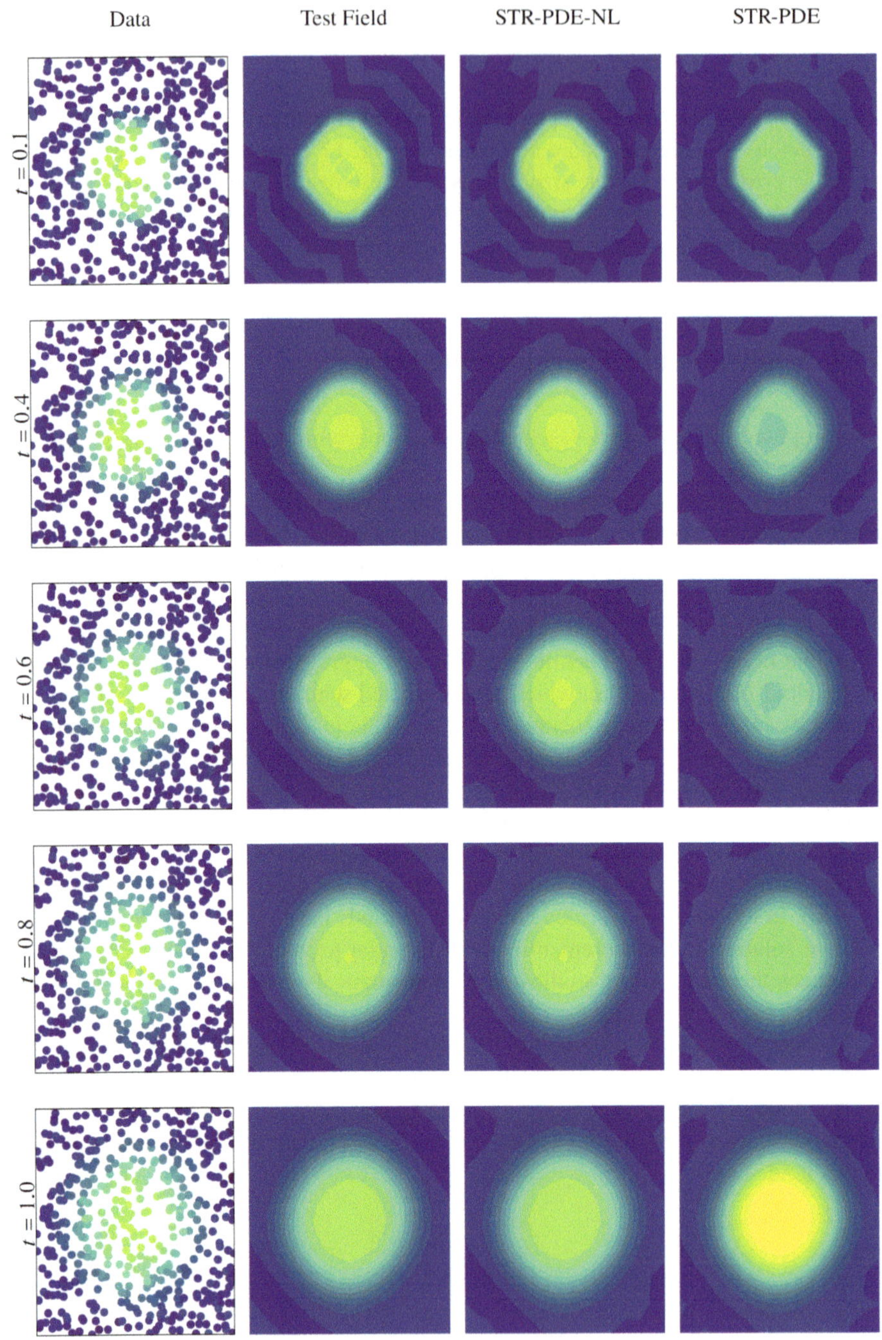

Fig. 17.2: From left to right: synthetic data, test function, estimates provided by the proposed Space-Time regression with nonlinear PDE regularization (STR-PDE-NL) and by Space-Time regression with linear PDE regularization (STR-PDE) proposed by [4].

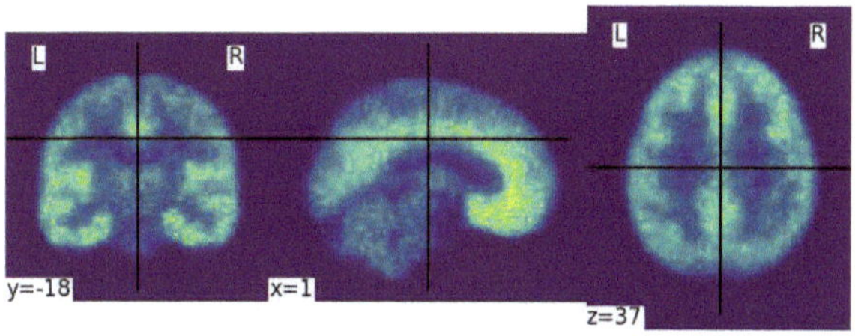

Fig. 17.3: Example of an image provided by PET modality. PET provides pathological information about the brain by detecting radiation from a radioactive tracer.

References

1. Agosta, F., Galantucci, S., and Filippi, M.: Advanced magnetic resonance imaging of neurodegenerative diseases. Neurological sciences, **38**(1), 41–51, (2017)
2. Anand, K. and Sabbagh, M.: Amyloid imaging: poised for integration into medical practice. Neurotherapeutics, **14**(1), 54–61, (2017)
3. Arnone, E., Negri, L., Panzica, F., and Sangalli, L. M.: Analyzing data in complicated 3d domains: Smoothing, semiparametric regression, and functional principal component analysis. Biometrics, **79**(4), 3510–3521, (2023)
4. Arnone, E., Sangalli, L. M., and Vicini, A.: Smoothing spatio-temporal data with complex missing data patterns. Statistical Modelling, **23**(4), 327–356, (2023)
5. Azzimonti, L., Sangalli, L. M., Secchi, P., Domanin, M., and Nobile, F.: Blood flow velocity field estimation via spatial regression with pde penalization. Journal of the American Statistical Association, **110**(511), 1057–1071, (2015)
6. Corti, M., Bonizzoni, F., Quarteroni, A. M., Antonietti, P. F., et al.: Discontinuous galerkin methods for fisher–kolmogorov equation with application to α-synuclein spreading in parkinson's disease. Computer Methods in Applied Mechanics and Engineering, **417**, 116450, (2023)
7. Cox, D. D: Multivariate smoothing spline functions. SIAM Journal on Numerical Analysis, **21**(4), 789–813, (1984)
8. Cruz-Haces, M., Tang, J., Acosta, G., Fernandez, J., and Shi, R.: Pathological correlations between traumatic brain injury and chronic neurodegenerative diseases. Translational neurodegeneration, **6**, 1–10, (2017)
9. Ettinger, B., Perotto, S., and Sangalli, L. M.: Spatial regression models over two-dimensional manifolds. Biometrika, **103**(1), 71–88, (2016)
10. Fisher, R. A.: The wave of advance of advantageous genes. Annals of eugenics, **7**(4), 355–369, (1937)
11. Fornari, S., Schäfer, A., Jucker, M., Goriely, A., and Kuhl, E.: Prion-like spreading of alzheimer's disease within the brain's connectome. Journal of the Royal Society Interface, **16**(159), 20190356, (2019)
12. Hecht, F.: New development in freefem++. J. Numer. Math., **20**(3-4), 251–265, (2012)
13. Jie, C. V., Treyer, V., Schibli, R., and Mu, L.: Tauvid™: the first fda-approved pet tracer for imaging tau pathology in alzheimer's disease. Pharmaceuticals, **14**(2), 110, (2021)
14. Jucker, M. and Walker, L. C.: Self-propagation of pathogenic protein aggregates in neurodegenerative diseases. Nature, **501**(7465), 45–51, (2013)
15. Kevrekidis, P., Thompson, T. B., and Goriely, A.: Anisotropic diffusion and traveling waves of toxic proteins in neurodegenerative diseases. Physics Letters A, **384**(36), 126935, (2020)

16. Lai, M.-J. and Schumaker, L. L.: Spline functions on triangulations. Number 110. Cambridge University Press. (2007)
17. Prusiner, S. B.: Prions. Proceedings of the National Academy of Sciences, **95**(23), 13363–13383, (1998)
18. Ramsay, J. O. and Silverman, B. W.: Applied functional data analysis: methods and case studies, volume 77. Springer, (2002)
19. Sangalli, L. M.: Spatial regression with partial differential equation regularisation, International Statistical Review, **89**, 505–531, (2021)
20. Sangalli, L. M., Ramsay, J. O., and Ramsay, T. O.: Spatial spline regression models. Journal of the Royal Statistical Society: Series B (Statistical Methodology), **75**(4), 681–703, (2013)
21. Tomasetto, M., Arnone, E., and Sangalli, L. M.: Modeling anisotropy and non-stationarity through physics-informed spatial regression, Environmetrics, **35**(8):e2889, (2024)
22. Wahba, G.: Spline interpolation and smoothing on the sphere. SIAM Journal on Scientific and Statistical Computing, **2**(1), 5–16, (1981)
23. Wahba, G.: Spline models for observational data, volume 59 of CBMS-NSF Regional Conference Series in Applied Mathematics. Society for Industrial and Applied Mathematics (SIAM), Philadelphia, PA, (1990)
24. Watts, J. C., Condello, C., Stöhr, J., Oehler, A., Lee, J., DeArmond, S. J., Lannfelt, L., Ingelsson, M., Giles, K., and Prusiner, S. B.: Serial propagation of distinct strains of aβ prions from alzheimer's disease patients. Proceedings of the National Academy of Sciences, **111**(28), 10323–10328, (2014)
25. Weickenmeier, J., Jucker, M., Goriely, A., and Kuhl, E.: A physics-based model explains the prion-like features of neurodegeneration in alzheimer's disease, parkinson's disease, and amyotrophic lateral sclerosis. Journal of the Mechanics and Physics of Solids, **124**, 264–281, (2019)
26. Weickenmeier, J., Kuhl, E., and Goriely, A.: Multiphysics of prionlike diseases: Progression and atrophy. Physical review letters, **121**(15), 158101, (2018)
27. Wood, S. N., Bravington, M. V., and Hedley, S. L.: Soap film smoothing. Journal of the Royal Statistical Society Series B: Statistical Methodology, **70**(5), 931–955, (2008)

Chapter 18
Functional Principal Component Analysis for Bivariate Densities and their Orthogonal Decomposition

Adéla Czolková, Karel Hron and Sonja Greven

Abstract Embedding bivariate probability density functions in Bayes spaces enables their orthogonal decomposition into independent and interactive parts, the former can be further decomposed into orthogonal geometric marginals. After performing the clr (centered logratio) transformation, densities can be analysed as functional data in L_0^2 space. In this paper, we focus on functional principal component analysis (FPCA) and its use for bivariate densities as well as for the vector of orthogonal densities from their decomposition. We show that performing FPCA on the original bivariate densities is equivalent to performing multivariate FPCA on the decomposed densities (the vector of the interactive part and geometric marginals). Moreover, eigenfunctions and scores decompose accordingly, and this allows to identify which parts of the decomposition contribute the most to the variation of the densities. The theoretical results are complemented by an illustration on an empirical data set.

18.1 Bayes Spaces

Bayes spaces provide a geometric representation for probability density functions (PDFs) which possess the property of scale invariance, i.e. given the domain Ω, two positive functions (densities) $f(x)$ and $g(x)$, $x \in \Omega$, which are proportional if $g(x) = cf(x)$ for a positive real number c, carry the same relative information [2, 3]. This is a direct consequence of scale invariance of the associated measures μ and ν

Adéla Czolková (✉)
Palacký University, 17. listopadu 12, Olomouc, Czech Republic,
e-mail: adela.czolkova01@upol.cz

Karel Hron
Palacký University, 17. listopadu 12, Olomouc, Czech Republic, e-mail: karel.hron@upol.cz

Sonja Greven
Humboldt-Universität zu Berlin, Unter den Linden 6, Berlin, Germany,
e-mail: sonja.greven@hu-berlin.de

© The Author(s), under exclusive license to Springer Nature Switzerland AG 2025

G. Aneiros et al. (eds.), *New Trends in Functional Statistics and Related Fields*,
Contributions to Statistics, https://doi.org/10.1007/978-3-031-92383-8_18

such that their densities are $f = \frac{\mathrm{d}\mu}{\mathrm{d}\lambda}$ and $g = \frac{\mathrm{d}\nu}{\mathrm{d}\lambda}$ for a σ-finite positive real-valued reference measure λ (here the Lebesgue measure).

Given a reference measure λ, the Bayes space $\mathcal{B}(\lambda)$ is defined as a space of equivalence classes of densities (densities are equivalent, $f =_{\mathcal{B}} g$, if they are proportional). In this space, we can define two operations – perturbation $\oplus$ and powering $\odot$,

$$(f \oplus g) =_{\mathcal{B}} f \cdot g, \quad (\alpha \odot f) =_{\mathcal{B}} f^{\alpha}, \quad \text{for } f, g \in \mathcal{B}(\lambda),\ \alpha \in \mathbb{R}, \tag{18.1}$$

where the index in $=_{\mathcal{B}}$ means equivalence of the resulting density up to scale, i.e. without changing the relative information which it contains [2, 3].

Suppose finite λ and define subspace $\mathcal{B}^2(\lambda)$ of space $\mathcal{B}(\lambda)$ as those elements that are square-integrable log-densities with respect to λ, i.e.

$$\mathcal{B}^2(\lambda) = \left\{ f \in \mathcal{B}(\lambda) \,\middle|\, \int_{\Omega} |\ln(f)|^2 \,\mathrm{d}\lambda < +\infty \right\}. \tag{18.2}$$

By defining the inner product

$$\langle f, g \rangle_{\mathcal{B}^2(\lambda)} = \frac{1}{2\lambda(\Omega)} \int_{\Omega} \int_{\Omega} \ln \frac{f(s)}{f(t)} \ln \frac{g(s)}{g(t)} \,\mathrm{d}\lambda(s)\,\mathrm{d}\lambda(t), \tag{18.3}$$

we obtain a Hilbert space structure of $\mathcal{B}^2(\lambda)$. The inner product also implies the norm and the distance

$$\|f\|_{\mathcal{B}^2(\lambda)} = \sqrt{\langle f, f \rangle_{\mathcal{B}^2(\lambda)}}, \quad \mathrm{d}_{\mathcal{B}^2(\lambda)}(f, g) = \|f \ominus g\|_{\mathcal{B}^2(\lambda)}, \tag{18.4}$$

where $f \ominus g = f \oplus [(-1) \odot g]$ is a perturbation-subtraction.

The densities are not usually processed directly in the Bayes space, but they are mapped into a subset of the $L^2(\lambda)$ space (space of real-valued square-integrable functions), where the standard methods of functional data analysis can be used [2, 3]. For this purpose, we define the centered logratio transformation, clr : $\mathcal{B}^2(\lambda) \to L^2(\lambda)$, as

$$\mathrm{clr}(f) = \ln f - \frac{1}{\lambda(\Omega)} \int_{\Omega} \ln f \,\mathrm{d}\lambda, \quad f \in \mathcal{B}^2(\lambda). \tag{18.5}$$

This transformation is an isomorphism (in particular, it is a linear mapping) [2], and for $f \in \mathcal{B}^2(\lambda)$, we get $\mathrm{clr}(f)\colon \Omega \to \mathbb{R}$. Moreover, the zero-integral constraint,

$$\int_{\Omega} \mathrm{clr}(f) \,\mathrm{d}\lambda = 0, \tag{18.6}$$

holds so that $\mathrm{clr}(f)$ is in the subspace of $L^2(\lambda)$, denoted $L^2_0(\lambda)$ and referred to as clr space in the context of Bayes spaces.

The inverse clr transformation, clr^{-1}, is defined as $\exp\{\mathrm{clr}(f)\}$.

In the case of bivariate densities, the domain Ω is a Cartesian product of two intervals, Ω_X and Ω_Y, i.e. $\Omega = \Omega_X \times \Omega_Y$, and a reference measure is a product measure $\lambda = \lambda_X \times \lambda_Y$. The Bayes spaces $\mathcal{B}^2(\lambda)$, $\mathcal{B}^2(\lambda_X)$, $\mathcal{B}^2(\lambda_Y)$ can be defined in a similar manner as stated above in (18.2) as can the clr transformation (18.5) and the clr spaces $L_0^2(\lambda)$, $L_0^2(\lambda_X)$, $L_0^2(\lambda_Y)$ [2].

The space $\mathcal{B}^2(\lambda_X)$ (and also $\mathcal{B}^2(\lambda_Y)$) can be easily embedded in $\mathcal{B}^2(\lambda)$ by defining subspaces $\mathcal{B}_X^2(\lambda)$ and $\mathcal{B}_Y^2(\lambda)$ of $\mathcal{B}^2(\lambda)$ as

$$\mathcal{B}_X^2(\lambda) = \left\{ f \in \mathcal{B}^2(\lambda) \,\middle|\, f =_{\mathcal{B}} f_X \cdot 1, f_X \in \mathcal{B}^2(\lambda_X), 1 \in \mathcal{B}^2(\lambda_Y) \right\}, \quad (18.7)$$

$$\mathcal{B}_Y^2(\lambda) = \left\{ f \in \mathcal{B}^2(\lambda) \,\middle|\, f =_{\mathcal{B}} 1 \cdot f_Y, 1 \in \mathcal{B}^2(\lambda_X), f_Y \in \mathcal{B}^2(\lambda_Y) \right\}, \quad (18.8)$$

where 1 is a neutral element in the respective space. Moreover, it can be shown that $\mathcal{B}_X^2(\lambda)$ and $\mathcal{B}_Y^2(\lambda)$ are orthogonal [2], which is important for the following orthogonal decomposition. A similar procedure can be applied to $L_0^2(\lambda_X)$ and $L_0^2(\lambda_Y)$ to gain the orthogonal subspaces $L_{0,X}^2(\lambda)$ and $L_{0,Y}^2(\lambda)$ of $L_0^2(\lambda)$ [2].

To keep in line with the geometry of Bayes spaces, we reformulate the standard marginals (called arithmetic marginals in this context) to the geometric marginal densities [2, 3] as

$$f^{\mathrm{X}} = \exp\left\{\mathrm{clr}(f^{\mathrm{X}})\right\} = \exp\left\{\frac{1}{\lambda_Y(\Omega_Y)} \int_{\Omega_Y} \mathrm{clr}(f)\, \mathrm{d}\lambda_Y\right\}, \quad (18.9)$$

$$f^{\mathrm{Y}} = \exp\left\{\mathrm{clr}(f^{\mathrm{Y}})\right\} = \exp\left\{\frac{1}{\lambda_X(\Omega_X)} \int_{\Omega_X} \mathrm{clr}(f)\, \mathrm{d}\lambda_X\right\}; \quad (18.10)$$

they can be considered as univariate densities, i.e. elements of $\mathcal{B}^2(\lambda_X)$ or $\mathcal{B}^2(\lambda_Y)$, as well as bivariate densities, i.e. elements of $\mathcal{B}_X^2(\lambda)$ or $\mathcal{B}_Y^2(\lambda)$. Then the geometric marginals can also be viewed as orthogonal projections of the density f onto the subspaces $\mathcal{B}_X^2(\lambda)$ and $\mathcal{B}_Y^2(\lambda)$. Similarly for clr marginals $\mathrm{clr}(f^{\mathrm{X}})$ and $\mathrm{clr}(f^{\mathrm{Y}})$ which fulfill a zero-integral constraint so they are elements of $L_0^2(\lambda_X)$ and $L_0^2(\lambda_X)$; but can also be considered elements of $L_{0,X}^2(\lambda)$ and $L_{0,Y}^2(\lambda)$, respectively.

The independent part of the density $f \in \mathcal{B}^2(\lambda)$ is defined as a perturbation of geometric marginals,

$$f^{\mathrm{ind}} = f^{\mathrm{X}} \oplus f^{\mathrm{Y}}, \quad (18.11)$$

and is an element of a subspace $\mathcal{B}_{\mathrm{ind}}^2(\lambda) = \mathcal{B}_X^2(\lambda) \oplus \mathcal{B}_Y^2(\lambda)$ of space $\mathcal{B}^2(\lambda)$.

The complementary part of the density f is called the interactive part and is defined as

$$f^{\mathrm{int}} = f \ominus f^{\mathrm{ind}} = f \ominus f^{\mathrm{X}} \ominus f^{\mathrm{Y}}. \quad (18.12)$$

It can be considered as an element of a subspace $\mathcal{B}_{\mathrm{int}}^2(\lambda)$ of $\mathcal{B}^2(\lambda)$. It holds that $\mathcal{B}_{\mathrm{ind}}^2(\lambda)$ and $\mathcal{B}_{\mathrm{int}}^2(\lambda)$ are orthogonal so that f^{ind} and f^{int} are orthogonal [2, 3]. Note that in the truly independence case, i.e. when the interactive part equals the uniform density, both arithmetic and geometric marginals coincide as expected.

This orthogonal decomposition, $f = f^{\mathrm{int}} \oplus f^{\mathrm{X}} \oplus f^{\mathrm{Y}}$, can be formulated in clr space as well [2, 3], then

$$\mathrm{clr}(f) = \mathrm{clr}(f^{\mathrm{int}}) + \mathrm{clr}(f^{\mathrm{X}}) + \mathrm{clr}(f^{\mathrm{Y}}), \quad (18.13)$$

where the parts of the density f are elements of corresponding subspaces of $L^2_0(\lambda)$, namely $L^2_{0,\text{int}}(\lambda), L^2_{0,\text{X}}(\lambda), L^2_{0,\text{Y}}(\lambda)$.

For simplicity, we omit λ from the notation in the following sections.

18.2 FPCA for Bivariate Densities

Let $f_1, \ldots, f_N$ be a sample of bivariate probability density functions (PDFs) defined on a compact domain $\Omega = \Omega_X \times \Omega_Y \subset \mathbb{R}^2$. We consider these PDFs as elements of the Bayes space $\mathcal{B}^2$ on Ω, and suppose these densities to be centered; if they are not, we can always center them, i.e. $\tilde{f}_i = f_i \ominus \overline{f},\ i = 1, \ldots, N,\ \overline{f} = \frac{1}{N} \odot \bigoplus_{i=1}^{N} f_i$.

We aim to find the main modes of variability of the PDFs (principal components) [4] which are orthogonal elements $\{\phi_j\}_{j\geq 1},\ \phi_j \in \mathcal{B}^2$ that maximize the following objective functional

$$\frac{1}{N}\sum_{i=1}^{N}\langle f_i, \phi_j\rangle^2_{\mathcal{B}^2} \quad \text{subject to } \|\phi_j\|_{\mathcal{B}^2} = 1 \text{ and } \langle \phi_j, \phi_k\rangle_{\mathcal{B}^2} = 0,\ k < j,\ j \geq 2. \tag{18.14}$$

The solution of this maximization can be found as the eigenfunctions of the sample covariance operator $V\colon \mathcal{B}^2 \longrightarrow \mathcal{B}^2$ that, to each $f \in \mathcal{B}^2$, assigns

$$Vf = \frac{1}{N} \odot \bigoplus_{i=1}^{N} \langle f_i, f\rangle_{\mathcal{B}^2} \odot f_i. \tag{18.15}$$

The corresponding eigenvalues are $\{\rho_j\}_{j\geq 1},\ \rho_1 \geq \rho_2 \geq \cdots \geq 0$.

For computation of the eigenfunctions (functional principal components), we use the clr transformation to map this problem into L^2_0 space [4]. Then we maximize the functional

$$\frac{1}{N}\sum_{i=1}^{N}\langle \text{clr}(f_i), \psi_j\rangle^2_{L^2_0} \quad \text{subject to } \|\psi_j\|_{L^2_0} = 1 \text{ and } \langle \psi_j, \psi_k\rangle_{L^2_0} = 0,\ k < j,\ j \geq 2, \tag{18.16}$$

where $\psi_j = \text{clr}(\phi_j) \in L^2_0$ and also $\phi_j = \text{clr}^{-1}(\psi_j) \in \mathcal{B}^2$.

The solution is a collection $\{\psi_j\}_{j\geq 1}, \psi_j \in L^2_0$, of the eigenfunctions of the sample covariance operator $V_{\text{clr}}\colon L^2_0 \to L^2_0$ that, to each $f \in L^2_0$, assigns

$$V_{\text{clr}} f = \frac{1}{N}\sum_{i=1}^{N}\langle \text{clr}(f_i), f\rangle_{L^2_0} \cdot \text{clr}(f_i). \tag{18.17}$$

For FPCA scores it then holds

$$\xi_{ij} = \langle \text{clr}(f_i), \psi_j\rangle_{L^2_0} = \langle f_i, \phi_j\rangle_{\mathcal{B}^2}, \quad j \geq 1,\ i = 1, \ldots, N \tag{18.18}$$

and each PDF can be expressed, for $(s,t) \in \Omega,\ i = 1, \dots, N$, as

$$f_i(s,t) = \bigoplus_{j=1}^{\infty} \xi_{ij} \odot \phi_j(s,t) \quad \text{or} \quad \mathrm{clr}(f_i)(s,t) = \sum_{j=1}^{\infty} \xi_{ij}\psi_j(s,t). \tag{18.19}$$

18.3 Multivariate FPCA for Orthogonal Parts of Bivariate Densities

All PDFs can be decomposed into orthogonal parts, cf. equations (18.11)–(18.13):

$$f_i = f_i^{\mathrm{int}} \oplus f_i^{\mathrm{X}} \oplus f_i^{\mathrm{Y}} \quad \text{and} \quad \mathrm{clr}(f_i) = \mathrm{clr}(f_i^{\mathrm{int}}) + \mathrm{clr}(f_i^{\mathrm{X}}) + \mathrm{clr}(f_i^{\mathrm{Y}}). \tag{18.20}$$

We consider vectors of densities $\boldsymbol{f}_i = (f_i^{\mathrm{int}}, f_i^{\mathrm{X}}, f_i^{\mathrm{Y}}) \in \mathcal{B}^2_{\mathrm{int}} \times \mathcal{B}^2_{\mathrm{X}} \times \mathcal{B}^2_{\mathrm{Y}} \subset \left[\mathcal{B}^2\right]^3$, $i = 1, \dots, N$, and we would like to find the main modes of variability of these vectors [1] in a similar manner as for the sample of PDFs in Sect. 18.2.

To obtain the principal components (PCs), we maximize the following functional

$$\begin{gathered}\frac{1}{N}\sum_{i=1}^{N}\langle \boldsymbol{f}_i, \boldsymbol{\phi}_j\rangle^2_{[\mathcal{B}^2]^3} = \frac{1}{N}\sum_{i=1}^{N}\left(\langle f_i^{\mathrm{int}}, \phi_j^{\mathrm{int}}\rangle^2_{\mathcal{B}^2} + \langle f_i^{\mathrm{X}}, \phi_j^{\mathrm{X}}\rangle^2_{\mathcal{B}^2} + \langle f_i^{\mathrm{Y}}, \phi_j^{\mathrm{Y}}\rangle^2_{\mathcal{B}^2}\right) \\ \text{subject to } \|\boldsymbol{\phi}_j\|_{[\mathcal{B}^2]^3} = 1 \text{ and } \langle \boldsymbol{\phi}_j, \boldsymbol{\phi}_k\rangle_{[\mathcal{B}^2]^3} = 0,\ k < j,\ j \geq 2,\end{gathered} \tag{18.21}$$

where $\boldsymbol{\phi}_j = (\phi_j^{\mathrm{int}}, \phi_j^{\mathrm{X}}, \phi_j^{\mathrm{Y}}) \in \mathcal{B}^2_{\mathrm{int}} \times \mathcal{B}^2_{\mathrm{X}} \times \mathcal{B}^2_{\mathrm{Y}} \subset \left[\mathcal{B}^2\right]^3$.

The PCs are eigenfunctions of the sample covariance operator $\boldsymbol{V}\colon [\mathcal{B}^2]^3 \to [\mathcal{B}^2]^3$ that, to each $\boldsymbol{f} \in [\mathcal{B}^2]^3$, assigns

$$\boldsymbol{V}\boldsymbol{f} = \frac{1}{N} \odot \bigoplus_{i=1}^{N} \langle \boldsymbol{f}_i, \boldsymbol{f}\rangle_{[\mathcal{B}^2]^3} \odot \boldsymbol{f}_i, \tag{18.22}$$

where the operations $\oplus$ and $\odot$ are defined component-wise [1].

After clr transformation of the PCA problem (18.21), we maximize

$$\begin{gathered}\frac{1}{N}\sum_{i=1}^{N}\langle \mathbf{clr}(\boldsymbol{f}_i), \boldsymbol{\psi}_j\rangle^2_{[L^2_0]^3} \\ \text{subject to } \|\boldsymbol{\psi}_j\|_{[L^2_0]^3} = 1 \text{ and } \langle \boldsymbol{\psi}_j, \boldsymbol{\psi}_k\rangle_{[L^2_0]^3} = 0,\ k < j,\ j \geq 2\end{gathered} \tag{18.23}$$

where $\boldsymbol{\psi}_j = \mathbf{clr}(\boldsymbol{\phi}_j) \in L^2_{0,\mathrm{int}} \times L^2_{0,\mathrm{X}} \times L^2_{0,\mathrm{Y}} \subset [L^2_0]^3$ and $\boldsymbol{\phi}_j = \mathbf{clr}^{-1}(\boldsymbol{\psi}_j) \in \mathcal{B}^2_{\mathrm{int}} \times \mathcal{B}^2_{\mathrm{X}} \times \mathcal{B}^2_{\mathrm{Y}} \subset \left[\mathcal{B}^2\right]^3$. Clr transformation $\mathbf{clr}$ and its inverse $\mathbf{clr}^{-1}$ are defined component-wise, so we use (18.5) for each component [1].

Then, to each $\boldsymbol{f} \in [L^2_0]^3$, the transformed sample covariance operator $\boldsymbol{V}_{\mathbf{clr}} : [L^2_0]^3 \to [L^2_0]^3$ assigns

$$\boldsymbol{V}_{\mathbf{clr}}\boldsymbol{f} = \frac{1}{N}\sum_{i=1}^{N}\langle \mathbf{clr}(\boldsymbol{f}_i), \boldsymbol{f}\rangle_{[L_0^2]^3} \cdot \mathbf{clr}(\boldsymbol{f}_i). \tag{18.24}$$

The PCs $\boldsymbol{\phi}_j,\ j \geq 1$ are orthonormal, and for each $j \geq 1$, $\phi_j^{\text{int}}, \phi_j^{\text{X}}, \phi_j^{\text{Y}}$ are mutually orthogonal. However, $\{\phi_j^{\text{int}}\}_{j\geq 1}$ alone are not orthonormal and neither orthogonal, the same applies for $\{\phi_j^{\text{X}}\}_{j\geq 1}$ and $\{\phi_j^{\text{Y}}\}_{j\geq 1}$. Analogously, $\{\boldsymbol{\psi}_j\}_{j\geq 1}$ has the same properties.

For vectors of PDFs, it holds that

$$\boldsymbol{f}_i(s,t) = \bigoplus_{j=1}^{\infty} \zeta_{ij} \odot \boldsymbol{\phi}_j(s,t) \quad \text{and} \quad \mathbf{clr}(\boldsymbol{f}_i)(s,t) = \sum_{j=1}^{\infty} \zeta_{ij}\boldsymbol{\psi}_j(s,t) \tag{18.25}$$

for $(s,t) \in \Omega,\ i = 1,\dots,N$, where $\zeta_{ij} = \langle \mathbf{clr}(\boldsymbol{f}_i), \boldsymbol{\psi}_j\rangle_{[L_0^2]^3} = \langle \boldsymbol{f}_i, \boldsymbol{\phi}_j\rangle_{[\mathcal{B}^2]^3}$ are the respective scores.

Now we show that using this approach, we are able to obtain the same results as in Sect. 18.2 – this means the same scores and loadings. Moreover, hereby we get an orthogonal decomposition of the loadings, and the scores can also be decomposed as shown later. We formulate the following theorem. It is stated in the clr space; however, below we discuss that a direct formulation in the Bayes space is also possible.

Theorem 18.1. *Let $\{\psi_j^0\}_{j\geq 1}$ be the principal components maximizing the functional in (18.16), and let $\boldsymbol{\psi}_j = (\psi_j^{\text{int}}, \psi_j^{\text{X}}, \psi_j^{\text{Y}}),\ j \geq 1$ be the vectors of principal components maximizing (18.23). Then $\psi_j^0 = \psi_j^{\text{int}} + \psi_j^{\text{X}} + \psi_j^{\text{Y}}$ for all $j \geq 1$.*

Proof. For all $j \geq 1$, let $\psi_j^* \in L_0^2$ maximize (18.16) and $\tilde{\boldsymbol{\psi}}_j = (\tilde{\psi}_j^{\text{int}}, \tilde{\psi}_j^{\text{X}}, \tilde{\psi}_j^{\text{Y}}) \in L_{0,\text{int}}^2 \times L_{0,\text{X}}^2 \times L_{0,\text{Y}}^2 \subset [L_0^2]^3$ maximize (18.23). We show the equivalence of the maximizers and the corresponding constraints in (18.16) and (18.23), and therefore the equality $\psi_j^* = \tilde{\psi}_j^{\text{int}} + \tilde{\psi}_j^{\text{X}} + \tilde{\psi}_j^{\text{Y}},\ \forall j \geq 1$.

First, we show the equality of the inner products in (18.16) and (18.23). For any $u, v \in L_0^2$, $u = u^{\text{int}} + u^{\text{X}} + u^{\text{Y}}$, $v = v^{\text{int}} + v^{\text{X}} + v^{\text{Y}}$, and $\boldsymbol{u}, \boldsymbol{v} \in L_{0,\text{int}}^2 \times L_{0,\text{X}}^2 \times L_{0,\text{Y}}^2$ with $\boldsymbol{u} = (u^{\text{int}}, u^{\text{X}}, u^{\text{Y}}),\ \boldsymbol{v} = (v^{\text{int}}, v^{\text{X}}, v^{\text{Y}})$, we can write

$$\langle u, v\rangle_{L_0^2}^2 = \langle u^{\text{int}} + u^{\text{X}} + u^{\text{Y}}, v^{\text{int}} + v^{\text{X}} + v^{\text{Y}}\rangle_{L_0^2}^2 = \tag{18.26}$$

$$= \langle u^{\text{int}}, v^{\text{int}}\rangle_{L_0^2}^2 + \langle u^{\text{X}}, v^{\text{X}}\rangle_{L_0^2}^2 + \langle u^{\text{Y}}, v^{\text{Y}}\rangle_{L_0^2}^2 = \langle \boldsymbol{u}, \boldsymbol{v}\rangle_{[L_0^2]^3}^2, \tag{18.27}$$

where the equality between (18.26) and (18.27) holds because of the orthogonality of the spaces $L_{0,\text{int}}^2, L_{0,\text{X}}^2, L_{0,\text{Y}}^2$.

Thus, with $\psi_j = \psi_j^{\text{int}} + \psi_j^{\text{X}} + \psi_j^{\text{Y}},\ \boldsymbol{\psi}_j = (\psi_j^{\text{int}}, \psi_j^{\text{X}}, \psi_j^{\text{Y}}),\ j \geq 1$, it holds

$$\|\psi_j\|_{L_0^2} = 1 \quad \Leftrightarrow \quad \|\boldsymbol{\psi}_j\|_{[L_0^2]^3} = 1, \tag{18.28}$$

$$\langle \psi_j, \psi_k\rangle_{L_0^2} = 0 \quad \Leftrightarrow \quad \langle \boldsymbol{\psi}_j, \boldsymbol{\psi}_k\rangle_{[L_0^2]^3} = 0 \tag{18.29}$$

and

$$\psi_j^* = \operatorname{argmax} \frac{1}{N}\sum_{i=1}^{N}\langle \operatorname{clr}(f_i), \psi_j\rangle_{L_0^2}^2 \quad \Leftrightarrow \quad \boldsymbol{\psi}_j^* = \operatorname{argmax} \frac{1}{N}\sum_{i=1}^{N}\langle \mathbf{clr}(\boldsymbol{f}_i), \boldsymbol{\psi}_j\rangle_{[L_0^2]^3}^2, \tag{18.30}$$

as

$$\frac{1}{N}\sum_{i=1}^{N}\langle \operatorname{clr}(f_i), \psi_j\rangle_{L_0^2}^2 = \frac{1}{N}\sum_{i=1}^{N}\langle \mathbf{clr}(\boldsymbol{f}_i), \boldsymbol{\psi}_j\rangle_{[L_0^2]^3}^2, \tag{18.31}$$

where (18.28), (18.29), (18.31) follow from the equality of the inner products.

As $\tilde{\boldsymbol{\psi}}_j$ was the maximizer of the right-hand side of (18.30) under the same constraints, altogether, this shows that $\psi_j^* = \tilde{\psi}_j^{\text{int}} + \tilde{\psi}_j^{\text{X}} + \tilde{\psi}_j^{\text{Y}},\ j \geq 1$.

The equivalence is shown only in L_0^2 space, but it can also be shown to hold in the Bayes space $\mathcal{B}^2$, where $\phi_j = \phi_j^{\text{int}} \oplus \phi_j^{\text{X}} \oplus \phi_j^{\text{Y}},\ j \geq 1$.

From the above equalities, it is obvious that for the scores it holds

$$\begin{aligned}\xi_{ij} &= \langle \operatorname{clr}(f_i), \psi_j\rangle_{L_0^2} = \langle \operatorname{clr}(f_i^{\text{int}}), \psi_j^{\text{int}}\rangle_{L_0^2} + \langle \operatorname{clr}(f_i^{\text{X}}), \psi_j^{\text{X}}\rangle_{L_0^2} + \langle \operatorname{clr}(f_i^{\text{Y}}), \psi_j^{\text{Y}}\rangle_{L_0^2} = \\ &= \langle \mathbf{clr}(\boldsymbol{f}_i), \boldsymbol{\psi}_j\rangle_{[L_0^2]^3} = \zeta_{ij}\end{aligned} \tag{18.32}$$

and also

$$\begin{aligned}\xi_{ij} &= \langle f_i, \phi_j\rangle_{\mathcal{B}^2} = \langle f_i^{\text{int}}, \phi_j^{\text{int}}\rangle_{\mathcal{B}^2} + \langle f_i^{\text{X}}, \phi_j^{\text{X}}\rangle_{\mathcal{B}^2} + \langle f_i^{\text{Y}}, \phi_j^{\text{Y}}\rangle_{\mathcal{B}^2} = \\ &= \langle \boldsymbol{f}_i, \boldsymbol{\phi}_j\rangle_{[\mathcal{B}^2]^3} = \zeta_{ij}\end{aligned} \tag{18.33}$$

for $j \geq 1,\ i = 1, \ldots, N$. There we can also see the decomposition of the scores into the sum of three elements, each corresponding to one of the parts from the orthogonal decomposition of PDFs.

When putting it all together, we get

$$\begin{aligned}f_i(s,t) &= f_i^{\text{int}}(s,t) \oplus f_i^{\text{X}}(s,t) \oplus f_i^{\text{Y}}(s,t) = \\ &= \bigoplus_{j=1}^{\infty} \xi_{ij} \odot \left(\phi_j^{\text{int}}(s,t) \oplus \phi_j^{\text{X}}(s,t) \oplus \phi_j^{\text{Y}}(s,t)\right).\end{aligned} \tag{18.34}$$

18.4 Application

In this section, we will focus on analysing an empirical data set concerning flight delays from the three New York airports (Newark, J. F. Kennedy, La Guardia) in 2013. These flights were divided into several groups according to months (January to December 2013) and the three airports – 36 groups. For each group, the bivariate density (PDF) of departure and arrival delay was obtained using kernel density estimation, then all the densities were transformed using clr transformation and decomposed into their orthogonal parts – geometric marginals and the interactive part. After this preprocessing procedure, both approaches to FPCA (described in Sect. 18.2 and Sect. 18.3) were performed on the densities of flight delays. The resulting scores and their decomposition are displayed in Fig. 18.1 and Fig. 18.2.

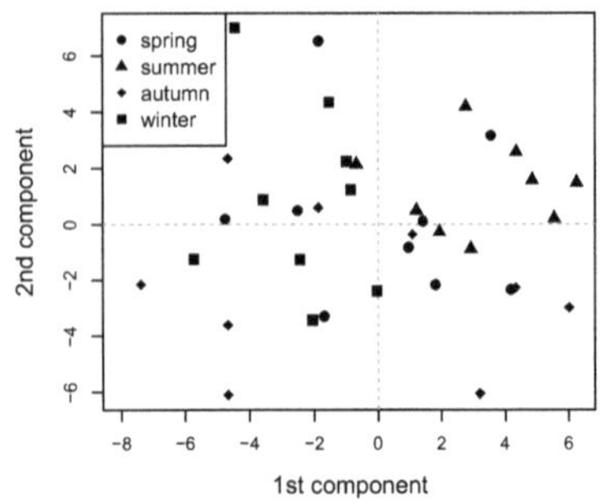

Fig. 18.1 Scores corresponding to the first two principal components $\psi_1, \psi_2 \in L_0^2$ (or $\phi_1, \phi_2 \in \mathcal{B}^2$). We can see that it is possible to distinguish the scores of summer and winter flights (or more precisely flights in warm and cold months).

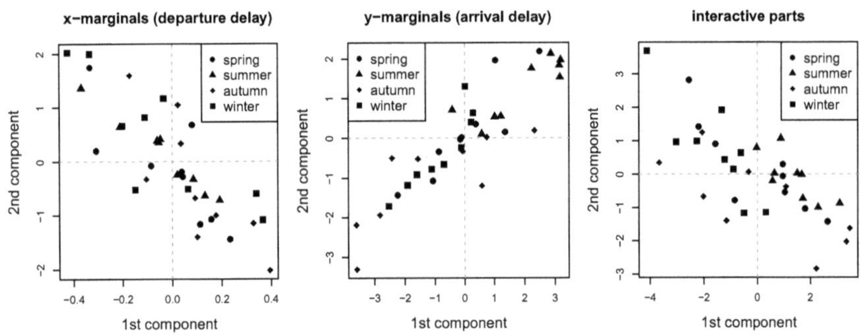

Fig. 18.2: The decomposed scores. It is visible that the distinction between summer and winter flights is mostly related to the interactive parts of densities and y-marginals (arrival delay). For example, while looking back to the original data (especially, PDFs of flight arrival delays), we would see that the more delayed flights are at the arrival, the higher the scores corresponding to the first component ψ_1^Y of the y-marginals are.

References

1. Filzmoser, P., Hron, K., Menafoglio, A.: Logratio Approach to Distributional Modeling. In Daouia, A., Ruiz-Gazen, A. (eds.): Advances in Contemporary Statistics and Econometrics: Festschrift in Honor of Christine Thomas–Agnan, Cham: Springer International Publishing 451–470, (2021)
2. Genest, C., Hron, K., Nešlehová, J. G.: Orthogonal decomposition of multivariate densities in Bayes spaces and relation with their copula–based representation. Journal of Multivariate Analysis **198**, 105228, (2023)
3. Hron, K., Machalová, J., Menafoglio, A.: Bivariate densities in Bayes spaces: orthogonal decomposition and spline representation. Statistical Papers **64**, 1629–1667, (2023)
4. Hron, K., Menafoglio, A., Templ, M., Hrůzová, K., Filzmoser, P.: Simplical principal component analysis for density functions in Bayes spaces. Computational Statistics and Data Analysis **94**, 330–350, (2016)

Chapter 19
Estimating Multiple Quantile Surfaces: A Penalized Functional Approach

Marco F. De Sanctis, Ilenia Di Battista, Eleonora Arnone, Cristian Castiglione, Mauro Bernardi, Francesca Ieva and Laura M. Sangalli

Abstract The simultaneous estimation of multiple quantiles is essential for analyzing phenomena with complex distributional characteristics, which standard parametric assumptions can not describe, and where the interest lies in tail behavior, rather than in the mean. In this work in particular, we are interested in estimating multiple quantile surfaces, which is crucial when investigating environmental phenomena. For instance, in the analysis of particulate matter (PM_{10}) concentrations, researchers are particularly interested in tail events, which have severe consequences for human health. Estimating simultaneously multiple quantile levels is crucial for preserving monotonicity, which ensures consistent and reliable estimates. Moreover, the resulting monotone quantile surfaces allow for a fully nonparametric reconstruction of the probability density function of the variable of interest, at any location of the domain. This work introduces a novel methodology, based on a penalized functional approach, that permits to estimate simultaneously multiple quantile surfaces. The method is developed within the framework of physics-informed models, where the minimization functional includes a regularization term based on a Partial Differential Equation, that embeds physical knowledge on the phenomenon under study. In the

Marco F. De Sanctis (✉), Ilenia Di Battista, Francesca Ieva, Laura M. Sangalli
MOX, Dipartimento di Matematica, Politecnico di Milano, Piazza Leonardo Da Vinci 32, Milano, 20133, Italy e-mail: marcofrancesco.desanctis@polimi.it
e-mail: ilenia.dibattista@polimi.it; francesca.ieva@polimi.it
e-mail: laura.sangalli@polimi.it

Eleonora Arnone
Dipartimento di Management, Università degli Studi di Torino, Corso Unione Sovietica, 218 bis, Torino, 10134, Italy e-mail: eleonora.arnone@unito.it

Cristian Castiglione
Bocconi Institute for Data Science and Analytics, Bocconi University, Via Röntgen 1, Milano, 20136, Italy e-mail: cristian.castiglione@unibocconi.it

Mauro Bernardi
Dipartimento di Scienze Statistiche, Università degli Studi di Padova, Via C. Battisti 241, 35121, Padova, Italy e-mail: mauro.bernardi@unipd.it

© The Author(s), under exclusive license to Springer Nature Switzerland AG 2025
G. Aneiros et al. (eds.), *New Trends in Functional Statistics and Related Fields*, Contributions to Statistics, https://doi.org/10.1007/978-3-031-92383-8_19

case of PM_{10} analysis, this enables for instance the integration in the statistical model of information concerning wind stream over the region, a critical factor in the dispersion of air pollutants. A second regularization term is used to enforce monotonicity of the quantiles. The resulting monotone quantile surfaces facilitates the development of data analysis tools, such as spatial exceedance probability maps, which may provide policymakers with robust insights to inform strategies for mitigating the harmful effects of air pollution.

19.1 Introduction

Given a set of realizations of a real-valued random variable, scattered across a bounded spatial domain, we consider the problem of estimating conditional quantiles of the unknown distribution of data. Estimating quantile surfaces provides a flexible approach to model and analyze complex spatial phenomena, marked by skewness, heteroscedasticity and kurtosis. For these phenomena, central tendency measures are insufficient to capture key aspects of the underlying distribution, making quantile-based approaches better suited for their analysis.

Specifically, statistical models for estimating multiple quantiles have been recently gaining increasing success in the spatial regression context (see, e.g., [11, 15, 5]). These models allow to analyze the whole conditional distribution of a variable of interest, by estimating an entire sequence of quantiles, while ensuring their monotonicity. This approach offers critical insights into the variability of the phenomenon. Moreover, by targeting a sequence of quantiles in the tail of the distribution, we can reconstruct tail features of the phenomenon, which are the main interest for several applications.

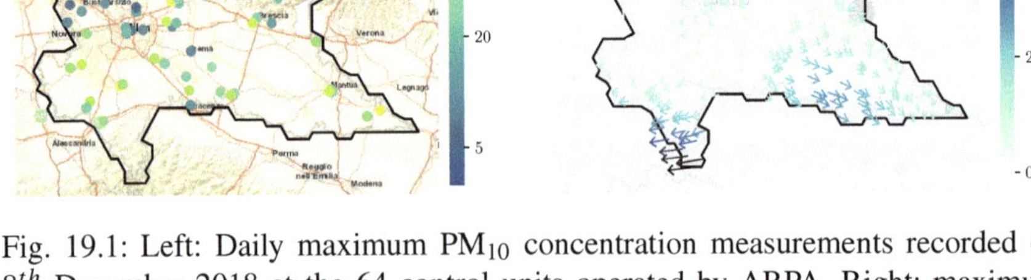

Fig. 19.1: Left: Daily maximum PM_{10} concentration measurements recorded on 8th December 2018 at the 64 control units operated by ARPA. Right: maximum wind speed measurements in Lombardy, collected over 121 ARPA meteorological stations, on 8th December 2018.

For instance, scientists monitoring the air quality of a territory are primarily concerned with extremely high levels of pollutants, aiming to assess their impact on public health. Fig. 19.1 displays the daily maximum concentrations of particulate matter (PM_{10}), an highly toxic pollutant for the human health. Data have been collected by Agenzia Regionale per la Protezione dell'Ambiente (ARPA) at 64 monitoring stations, placed throughout the Lombardy region. As we can see from the figure, certain areas exhibit unusually high PM_{10} concentrations, which can be effectively identified by estimating quantiles of the right tail of the distribution.

However, modeling the spatial pattern of this kind of data is a challenging task. In fact, PM_{10} values are characterized by strong non-stationarities due to the influence of external physical sources, such as wind streams, which may induce non-stationary and localized effects. Incorporating such dynamics into statistical modeling is essential to accurately capture the spatial structure of the phenomenon.

In this work, we propose a novel method for estimating multiple quantile surfaces, based on a penalized functional approach typical of functional data analysis (see, e.g., [14, 7, 9]). The estimation procedure is formulated as an optimization problem involving a composite loss functional defined over the multidimensional domain of interest. This functional combines a data-fitting term with two additional regularization penalties, one aimed at incorporating prior information about the underlying physics of the phenomenon under study, and the other needed to enforce monotonicity of the quantiles. In particular, following the blueprint of [2], we use a regularizing term based on a Partial Differential Equation (PDE), defined over the spatial domain, and encoding the available physical information. This allows us to include physical information, such as the wind currents, in the statistical model. To enforce the monotonicity of the quantiles, instead, we use a L^1-type penalty for quantile crossing, following the discussion in [12]. By leveraging these functional penalties, the model incorporates both statistical and physical information, yielding a hybrid framework that is well-suited to address challenges posed by spatial quantile estimation.

19.2 Semiparametric Physics-Informed Multiple Quantile Model

For a bounded domain $\mathcal{D} \subset \mathbb{R}^2$, we consider realizations $\{y_i\}_{i=1}^n$ of a real-valued random variable Y, along with a set of q covariates $\{\mathbf{x}_i\}_{i=1}^n$, observed at the spatial locations $\{\mathbf{p}_i\}_{i=1}^n \subset \mathcal{D}$. The objective is to estimate multiple quantile surfaces of Y, defined as $Q_{Y|\mathbf{p}_i,\mathbf{x}_i}(\alpha_j) = \inf\{y \in \mathbb{R} : \mathbb{P}(Y \leq y \mid \mathbf{p}_i, \mathbf{x}_i) \geq \alpha_j\}$, for a sequence of r probability levels $\{\alpha_j\}_{j=1}^r \subset (0, 1)$. We propose the following semiparametric model for the conditional quantiles of Y:

$$Q_{Y|\mathbf{p}_i,\mathbf{x}_i}(\alpha_j) = \mathbf{x}_i^\top \boldsymbol{\beta}_j + f_j(\mathbf{p}_i), \quad i = 1, \dots, n, \quad j = 1, \dots, r, \tag{19.1}$$

where $\boldsymbol{\beta}_j \in \mathbb{R}^q$ quantifies the linear effect of the covariates on the α_j-th quantile, and $f_j : \mathcal{D} \to \mathbb{R}$ is a nonparametric function that captures the spatial pattern of the α_j-th quantile.

Setting $\boldsymbol{\beta} = (\boldsymbol{\beta}_1, \dots, \boldsymbol{\beta}_r)^\top$ and denoting by $\boldsymbol{f}$ the collection of functions $f_1, \dots, f_r$ defined over the domain $\mathcal{D}$, our goal is to estimate the unknowns $(\boldsymbol{\beta}, \boldsymbol{f})$, ensuring the monotonicity of the quantiles $\{Q_{Y|\mathbf{p}_i,\mathbf{x}_i}(\alpha_j)\}_{j=1}^r$. To this end, generalizing the approach of [4] for single quantile estimation, we propose to minimize the penalized loss functional

$$J(\boldsymbol{\beta}, \boldsymbol{f}) = \mathcal{L}(\boldsymbol{\beta}, \boldsymbol{f}) + \sum_{j=1}^{r} \lambda_j P(f_j) + \gamma\, C(\boldsymbol{\beta}, \boldsymbol{f}), \tag{19.2}$$

which consists of three main components: a data loss $\mathcal{L}$, which quantifies the discrepancy between the estimated quantiles and the observed data; a smoothing penalty term P, that promotes smoothness in the nonparametric fields $\{f_j\}_{j=1}^r$; finally, a crossing penalty C, which penalizes violations of the monotonicity property among the quantile estimates.

The three terms of the estimation functional (19.2) are defined as follows. The data loss term is given by

$$\mathcal{L}(\boldsymbol{\beta}, \boldsymbol{f}) = \frac{1}{n} \sum_{i,j=1}^{n,r} \rho_{\alpha_j}(y_i - \mathbf{x}_i^\top \boldsymbol{\beta}_j - f_j(\mathbf{p}_i)), \tag{19.3}$$

where $\rho_\alpha(t) := 0.5|t| + (\alpha - 0.5)t$ represents the pinball loss function [8]. The roughness penalties $P(f_j)$ are defined as

$$P(f_j) = \int_{\mathcal{D}} (Lf_j(\mathbf{p}))^2 \, d\mathbf{p},$$

where Lf_j is the misfit of a PDE over the spatial domain $\mathcal{D}$, that incorporates some problem-specific knowledge about the physics of the phenomenon under study. Here, L represents a linear second-order elliptic differential operator defined as

$$Lf = -\nabla \cdot (K \nabla f) + \mathbf{b} \cdot \nabla f + cf, \tag{19.4}$$

where $K \in \mathbb{R}^{2\times 2}$ is a symmetric positive definite matrix, which models the anisotropy of the underlying process, $\mathbf{b} \in \mathbb{R}^2$ controls potential unidirectional effects, and $c \in \mathbb{R}$ regulates the shrinkage of the solution towards the null function. The coefficients K, $\mathbf{b}$, c can vary spatially, allowing for non-stationary modeling. The PDE is coupled with homogeneous Neumann boundary conditions, $K \nabla f \cdot \nu = 0$ on $\partial\mathcal{D}$, even though more general boundary conditions could be considered (see, e.g., [1]). The weight of the roughness penalties $P(f_j)$ in the objective functional J is regulated by the hyperparameters $\{\lambda_j\}_{j=1}^r$, tuned on the basis of the Generalized Cross-Validation criterion. Finally, $C(\boldsymbol{\beta}, \boldsymbol{f})$ in (19.2) is the crossing penalty, that quantifies the crossing between two consecutive quantiles at each spatial location, and is defined as

$$C = \sum_{i,j=1}^{n,r-1} \max\{0, \varepsilon - \mathbf{x}_i^\top (\boldsymbol{\beta}_{j+1} - \boldsymbol{\beta}_j) - (f_{j+1}(\mathbf{p}_i) - f_j(\mathbf{p}_i))\}, \text{ where } \varepsilon > 0.$$

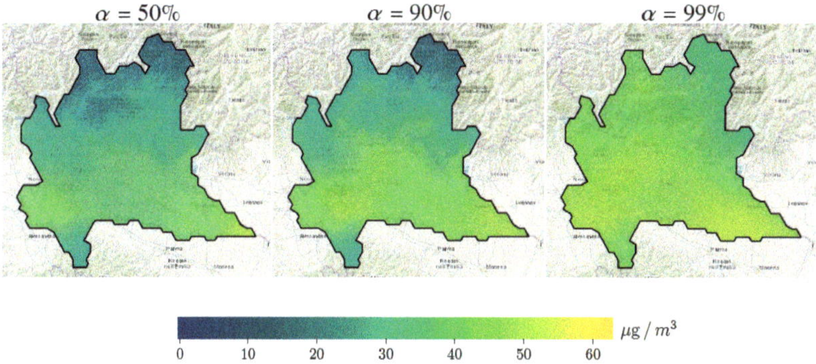

Fig. 19.2: Fitted 50%, 90% and 99% semiparametric quantile maps of the PM_{10} concentration in Lombardy region.

It should be pointed out that the functional (19.2) cannot be minimized analytically, due to the non-quadratic terms ρ_α and C. To address this issue, we resort to an iterative algorithm grounded in the Maximization-Minimization framework (see, e.g., [10]). The infinite-dimensional solution is then discretized using the Finite Element Method, which provides an efficient and flexible numerical scheme for solving the problem (see [16]).

19.3 Application to Air Quality Data

We briefly present an application of the proposed model to analyze PM_{10} concentrations in the Lombardy region. Specifically, we focus on the maximum PM_{10} measurements recorded on 8th December 2018 by the 64 ARPA sensors distributed throughout the region. The data, shown in Fig. 19.1, are publicly available through Regione Lombardia[1].

The purpose of this study is to estimate a fine sequence of quantiles along the entire distribution of the pollutant, with particular emphasis on the high probability levels associated with the most severe health risks. We include the altitude as covariate in the model. Moreover, we fully exploit the flexibility of the model by specifying a transport term within the regularization component in (19.4). This term encodes physical knowledge about wind patterns in the region and is optimized via parameter cascading (see [13, 18, 3, 17]), through a two-step procedure as proposed by [4], starting from the wind data collected by ARPA, and shown in Fig. 19.1.

Fig. 19.2 shows the quantile surfaces of levels 50%, 90% and 99% of the PM_{10} concentration in Lombardy. The highest levels of pollutant concentration are reached in the southern parts of the Po Valley, for each of the quantile levels. Interestingly, the results indicate that the most concerning PM_{10} levels are associated with industrial

[1] Open Data: Regione Lombardia (2024). Available at: `https://www.dati.lombardia.it/stories/s/Meteo-inquinamento-aria-e-acqua-e-altri-dati-da-AR/auv9-c2sj`

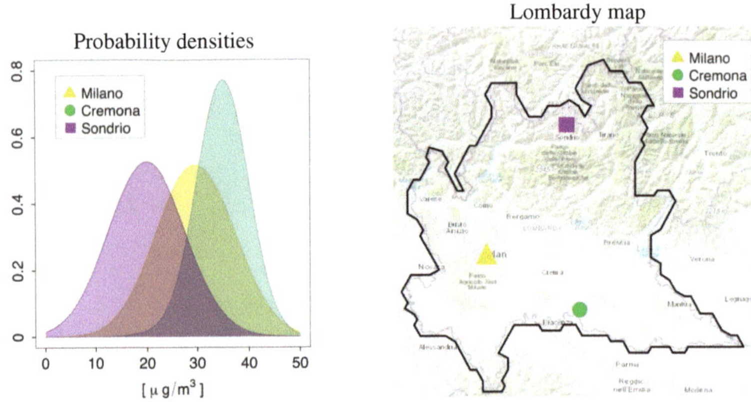

Fig. 19.3: Left: PM_{10} probability density functions in Milano (yellow), Cremona (green), and Sondrio (purple), obtained from the monotone sequence of estimated quantiles. Right: Lombardy map illustrating the three analyzed locations.

sites, around the areas of Lodi, Cremona and Mantova, rather than in large city centers, like the metropolitan city of Milano. This observation underscores the dominant role of industrial activities in PM_{10} emissions, which aligns with findings that PM_{10} comprises a variety of particulate sources, not only including car exhausts, but also generated by agricultural operations and industrial processes [2].

Moreover, as expected, the altitude exhibits a mitigating effect on PM_{10} pollution across all quantile levels. These effects are statistically significant on the mean, as demonstrated through appropriate Wald tests detailed in [6].

Owing to the monotonicity of the spatial quantile estimates, we can carry out additional analyses that are directly relevant for policymakers. The accuracy of these quantile estimates has been extensively evaluated in appropriate simulation studies, demonstrating the advantage of our method compared to existing approaches in the spatial quantile regression literature. Firstly, we reconstruct the PM_{10} probability density function, in a fully nonparametric manner, focusing on three diverse locations across Lombardy: Milano, the region's largest metropolitan hub; Cremona, a key industrial center; and Sondrio, a small mountain town in the northern part of the region. These probability density functions are shown in Fig. 19.3. We can see that the industrial pole of Cremona exhibits the highest values of PM_{10} concentration.

Secondly, we generate spatial maps illustrating the probability of exceeding alarming PM_{10} levels, providing practical tools for policymakers to assess the risk faced by citizens in specific areas. Fig. 19.4 shows the exceedance probability for a threshold of $30\,\mu g/m^3$. The analysis confirms the significant risks in the industrial areas of Cremona, Lodi, Pavia and Mantova, in contrast to the mountain provinces of Sondrio, Como and Lecco, where the exceedance probability is nearly zero.

[2] California Environmental Protection Agency – Inhalable Particulate Matter and Health (PM2.5 and PM10).

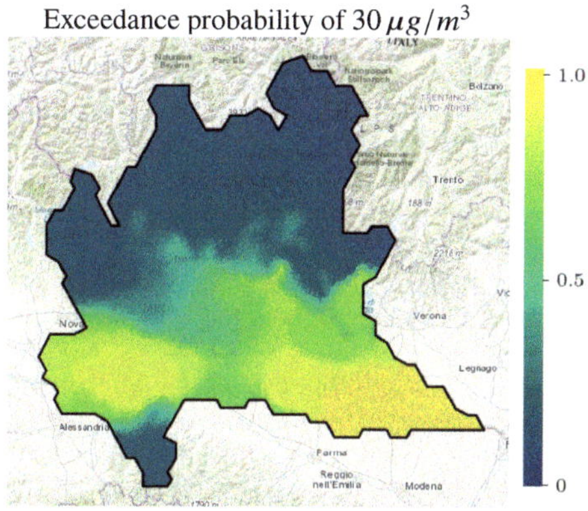

Fig. 19.4: Risk of exceeding the threshold of $30\mu g/m^3$ PM_{10} across Lombardy.

In conclusion, the analysis of the spatial distribution of the PM_{10} field reveals a clear pattern, with the highest concentrations observed in the southern parts of the Po Valley and significantly lower values in the mountainous regions. These findings highlight the need for targeted mitigation strategies focusing on industrial emissions to effectively address the high PM_{10} levels in these areas. Building on this spatial analysis, a promising direction for future research involves extending the proposed statistical model to include a temporal dimension. This would enable the estimation of multiple spatio-temporal quantile surfaces, advancing the approach discussed in [1] for linear regression settings.

19.4 Future Directions

A promising direction for future research involves extending our model to better accommodate nonlinear covariate effects. Indeed, while our current approach already includes a nonlinear component through the nonparametric term, additional flexibility could further enhance the model's adaptability to complex spatial dependencies. For instance, in the study of the air pollution problem presented in this work, the considered covariates exhibit strong spatial correlation, and incorporating more sophisticated nonlinear structures into the model could improve estimation accuracy.

Acknowledgements Research funded by the European Union - NextGenerationEU, in the framework of the projects GRINS - Growing Resilient, INclusive and Sustainable (GRINS PE00000018 – CUP D43C22003110001), and PRIN2022 CoEnv - Complex Environmental Data and Modeling (CUP 2022E3RY23). The views and opinions expressed are solely those of the authors and do not necessarily reflect those of the European Union, nor can the European Union be held responsible for them. We also acknowledge the MUR research project Dipartimento di Eccellenza 2023-2027, Dipartimento di Matematica, Politenico di Milano.
Contribution Marco F. De Sanctis and Ilenia Di Battista equally contributed to the work.

References

1. Arnone, E., Sangalli, L. M., Vicini, A.: Smoothing spatio-temporal data with complex missing data patterns. Statistical Modelling, **23**(4), 327–356 (2023)
2. Azzimonti, L., Sangalli, L. M., Secchi, P., Domanin, M., Nobile, F.: Blood flow velocity field estimation via spatial regression with PDE penalization. Journal of the American Statistical Association, **110(511)**, 1057–1071 (2015)
3. Bernardi, M.S., Carey, M., Ramsay, J., Sangalli, L.M.: Modeling spatial anisotropy via regression with partial differential regularization. Journal of Multivariate Analysis, **167**, 15–30 (2018)
4. Castiglione, C., Arnone, E., Bernardi, M., Farcomeni, A., Sangalli, L.M.: PDE-regularised spatial quantile regression. Journal of Multivariate Analysis, **205**, (2024)
5. Das, P., Ghosal, S.: Analyzing ozone concentration by Bayesian spatio-temporal quantile regression. Environmetrics, **28**(4), e2443 (2017)
6. Ferraccioli, F., Sangalli, L.M., Finos, L.: Nonparametric tests for semiparametric regression models. TEST, **32**(3), 1106—1130 (2023)
7. Vieu, P., Ferraty, E.: Nonparametric Functional Data Analysis: Theory and Practice. Springer New York, (2006)
8. Koenker, R., Bassett J.G.: Regression quantiles. Econometrica: journal of the Econometric Society, **46**, 33–50 (1978)
9. Kokoszka, P., Reimherr M.: Introduction to functional data analysis. CRC Press, (2017)
10. Lange, K.: MM optimization algorithms. Society for Industrial and Applied Mathematics, (2016)
11. Liu, Y., Wu, Y.: Simultaneous multiple non-crossing quantile regression estimation using kernel constraints. Journal of Nonparametric Statistics, **23**(2), 415–437 (2011)
12. Nocedal, J., Wright, S. J.: Numerical optimization. Springer, New York (2006)
13. Ramsay, J.O., Hooker, G., Campbell, D., Cao, J.: Parameter Estimation for Differential Equations: a Generalized Smoothing Approach. Journal of the Royal Statistical Society Series B: Statistical Methodology, (2007)
14. Ramsay, J.O., Silverman, B.W.: Functional data analysis. Springer, (2006)
15. Reich, B. J., Fuentes, M., Dunson, D. B.: Bayesian spatial quantile regression. Journal of the American Statistical Association **106**(493), 6–20 (2011)
16. Sangalli, L. M., Ramsay, J.O., Ramsay, T. O.: Spatial spline regression models. Journal of the Royal Statistical Society Series B: Statistical Methodology, **75**(4), 681–703 (2013)
17. Tomasetto, M., Arnone, E., Sangalli, L. M.: Modeling Anisotropy and Non-Stationarity Through Physics-Informed Spatial Regression. Environmetrics, (2024).
18. Zhang, X., Cao, J., Carroll, R. J.: Estimating Varying Coefficients for Partial Differential Equation Models. Biometrics, (2017).

Chapter 20
Neural Drift Estimation for Ergodic Diffusions: Nonparametric Analysis and Numerical Exploration

Simone Di Gregorio and Francesco Iafrate

Abstract We take into consideration generalization bounds in [3] for the problem of the estimation of the drift component for ergodic stochastic differential equations, when the estimator is a ReLU neural network and the estimation is nonparametric with respect to the statistical model. We show a practical way to enforce the theoretical estimation procedure, enabling inference on noisy and rough functional data. Results are shown for a simulated Itô-Taylor approximation of the sample paths.

20.1 Neural Nonparametric Drift Estimation for Ergodic Diffusions

The problem of statistical modeling for multi-dimensional ergodic diffusions from discrete observations has a long-standing literature that largely focused on parametric approaches to the estimation of the coefficient set [4, 5]. In this work, we pick a different approach stemming from the analysis in [3], focusing on neural network estimators of the drift coefficient and without parametric assumptions on the statistical model. This nonparametric approach is relevant because, under mild regularity conditions, it enables practical and flexible estimation for an underlying functional variable with a non-trivial noise dynamic, from which we sample points. Moreover, the generality, rough path features and analytical tractability of diffusions make them a standard tool for modeling structured and irregular time sequences as random functions. While finance is the standard application, as stochastic calculus provides its

Simone Di Gregorio (✉)
Department of Computer, Control and Management Engineering Antonio Ruberti, Sapienza University of Rome, Italy, e-mail: simone.digregorio@uniroma1.it

Francesco Iafrate
Department of Mathematics, University of Hamburg, Bundestr. 55 20146 Hamburg, Germany, e-mail: francesco.iafrate@uni-hamburg.de

© The Author(s), under exclusive license to Springer Nature Switzerland AG 2025

G. Aneiros et al. (eds.), *New Trends in Functional Statistics and Related Fields*, Contributions to Statistics, https://doi.org/10.1007/978-3-031-92383-8_20

mathematical framework, these methods are also valuable for investigating data in broader contexts, such as biological or geological data analysis [1, 2].

The formal setting is the following. Let $\{W_t\}_{t\in[0,T]}$ be a d-dimensional Brownian motion adapted to an underlying filtered probability space, $\sigma : \mathbb{R}^d \to \mathbb{R}^{d\times m}$ and $b : \mathbb{R}^d \to \mathbb{R}^d$, which for our purposes we assume as globally Lipschitz (**B1**):

$$dX_t = b(X_t)dt + \sigma(X_t)dW_t,\ X_0 = \eta \tag{20.1}$$

According to standard results, under **B1**, a strong solution $\{X_t\}_{t\in[0,T]}$ to the above stochastic differential equation (SDE) always exists. We also require an exponential mixing condition (**B2**): $\exists\, C_\beta, C'_\beta : \beta_X(t) \le C'_\beta \exp\left(-C_\beta t\right)$. β_X is the β-strong mixing function of the solution. A useful sufficient condition [6] for **B2** is that we require that η has a moment generating function, that $\sigma\sigma^\top$ is bounded and bounded away from zero and that $x^\top b(x) \le -r\|x\|_2^\alpha\ \forall x$ s.t. $\|x\|_2 \ge M_0$, with $M_0 \ge 0,\ \alpha \ge 1,\ r > 0$.

The estimation target is b, with σ not entering the analysis. The estimation is restricted to a parallelotope, which can be taken without loss of generality as $[0,1]^d$: one may pick a generic parallelotope and scale and shift accordingly. The target is thus a specific restricted component of the drift, $f_0 := b_i \mathbb{1}_{[0,1]^d}$, for $i \in \{1,\dots,d\}$.

Following [3], we formalize a neural network as a tuple $(L, \mathbf{p}, \{W_i\}_{i=0}^L, \{\mathbf{v}_j\}_{j=1}^L)$, and the non-linearity is taken to be the *Rectified Linear Unit* (ReLU), i.e. $\rho_{v_i}(x_i) := \max(0, x - v_i)$, where v_i is the i-th entry of a generic shift vector $\mathbf{v}$. Every tuple like the above induces a map f in the following standard way:

$$\begin{aligned} f(x) &:= W_L \rho_{\mathbf{v}_L} W_{L-1} \rho_{\mathbf{v}_{L-1}} \dots W_1 \rho_{\mathbf{v}_1} W_0 x, \\ & \quad x \in \mathbb{R}^d,\ W_l \in \mathbb{R}^{p_{l+1}\times p_l},\ p_0 = d,\ p_{L+1} = 1 \\ \rho_{\mathbf{v}}(y) &:= \begin{pmatrix} \rho_{v_1}(y_1) \\ \vdots \\ \rho_{v_r}(y_r) \end{pmatrix},\ y \in \mathbb{R}^r \end{aligned} \tag{20.2}$$

The hypothesis class of the estimator we consider is:

$$\begin{aligned} \mathfrak{F}(L, \mathbf{p}, s, F) := \Bigg\{ & f\mathbb{1}_{[0,1]^d} : f \text{ follows (20.2)},\ \max_{j=0,\dots,L} \|W_j\|_\infty \vee \|\mathbf{v}_j\|_\infty \le 1, \\ & \|W_0\|_0 + \sum_{j=1}^{L} (\|W_j\|_0 + \|\mathbf{v}_j\|_0) \le s,\ \|f\mathbb{1}_{[0,1]^d}\|_\infty \le F \Bigg\} \end{aligned} \tag{20.3}$$

Given a constant sampling interval δ and a number of points N, we observe a sample of the solution $\{X_{k\delta}\}_{k=0}^N$. The idea is defining a response variable for learning f_0 as $Y_{k\delta} := \frac{1}{\delta}\left(X^i_{(k+1)\delta} - X^i_{k\delta}\right)$. The almost sure rough path properties of the SDE solution and the exploding variance of the difference quotients $Y_{k\delta}$ present a challenge in this context, see the analysis in Subsect. 20.3.2. One may connect this reasoning to the standard statistical learning scenario, showing that the drift function in $X_{k\delta}$ can be seen as an increasingly precise (as $\delta \downarrow 0$) approximation of the regression function for the prediction of $Y_{k\delta}$. This in turn heuristically justifies a mean square empirical risk minimization, which targets f_0 through the proxy $Y_{k\delta}$. This leads to a training loss and a generalization risk for the estimator $\hat{f}_N$ given by:

$$Q_N(f) := \frac{1}{N} \sum_{k=0}^{N-1} (Y_{k\delta} - f(X_{k\delta}))^2 \tag{20.4}$$

$$\mathcal{R}_N(\hat{f}_N, f_0) := \mathbb{E}\left[\frac{1}{N} \sum_{k=0}^{N-1} \left(\hat{f}_N(X'_{k\delta}) - f_0(X'_{k\delta})\right)^2\right] \tag{20.5}$$

where $\{X'_t\}_{t\in[0,N\delta]}$ is an independent copy of the solution to (20.1).

In [3] the authors prove that under **B1**, **B2**, with $F \geq \|f_0\|_\infty$, any estimator taking values in $\mathfrak{F}(L, \mathbf{p}, s, F)$ satisfies the following risk inequality, for all $\tau > 0$:

$$\begin{aligned}\mathcal{R}_N(\hat{f}_N, f_0) \leq & \tau \left(\Psi_N^{\mathfrak{F}(L,\mathbf{p},s,F)}(\hat{f}_N) + \inf_{f\in\mathfrak{F}(L,\mathbf{p},s,F)} \|f - f_0\|\right) + \\ & + C_\tau F^2 \left(\frac{s(L\log s + \log(N\delta)) \cdot \log(N\delta)}{N\delta} + \delta\right)\end{aligned} \tag{20.6}$$

where in the above $\Psi_N^{\mathfrak{F}(L,\mathbf{p},s,F)}$ is the expected optimization error of the fitting procedure with respect to the considered class $\mathfrak{F}$, while C_τ is a constant depending on τ, the specification of **B1** and **B2** and the $\|\cdot\|_\infty$ norm of the coefficients over the estimation set. The last term - which represents the estimation variance - vanishes under the standard asymptotic regime [4], i.e. $T = N\delta \uparrow \infty$ and $\delta \downarrow 0$.

20.2 Building Regular Networks

The problem of enforcing the regularity specified in (20.3) is relevant. Clearly, we could just impose an appropriate hard norm constraint on blocks of the weight set and take s as the total cardinality of the parameter set. For example, for a specific parameter vector, constraining the estimate inside a ℓ^2 ball of unitary radius would suffice to enforce each entry being bounded by 1. This clearly strongly limits the flexibility of the model leading to increased depth or width to reduce the approximation error. We aim for a much more compact representation of the networks while preserving the same regularity as the one needed for the risk inequality in (20.6) to hold. The following lemma is Proposition A.3 from [7], with an additional remark from [3]:

Lemma 20.1. *For any ReLU neural network function $g : \mathbb{R}^d \to \mathbb{R}$ with parameters bounded by B, there exists another ReLU network ϕ such that $\phi(x) = g(x)\,\forall x \in \mathbb{R}^d$ and such that the weights of ϕ are uniformly bounded by 1. Additionally, with L, s and p respectively the depth, the number of non-null parameters and the maximum* *width of g, the depth of the new network* ϕ *is* $L' = \lceil(\log B + 5)L\rceil$, *its width* p' *is bounded by* $\max\{3, p\}$ *and the number of active parameters is at most* $s' = 2s + 12L'$.

Therefore, we get a logarithmic growth in B for all the complexity metrics that appear in (20.6). What this suggests is that if we are able to enforce any uniform bound on the weight set we can still get that the produced estimator $\hat{f}_N \in \mathfrak{F}(L', \mathbf{p}', s', F)$, just by considering the regular version of the estimator stemming from Lemma 20.1. The next Lemma shows that a bounding constant F always exists as soon as we impose, while optimizing, a norm constraint on each element of the parameter set.

Lemma 20.2. *Let* K *be any set of indexes, let* $\{(L, \mathbf{p}, \{W_{ik}\}_{i=0}^{L}, \{\mathbf{v}_{jk}\}_{j=1}^{L})\}_{k\in K}$ *be a set of neural network tuples with input in* $\mathbb{R}^d$ *for which* $\exists D > 0 : \|W_{ik}\| \leq D, \|\mathbf{v}_{jk}\| \leq D \,\forall i \in \{0, \dots, L\}, j \in \{1, \dots, L\}, k \in K$. *Let* $\{g_k\}_{k\in K}$ *be the set of functions which are induced by each of the networks above. Then we have that* $\exists F \geq 0 : \|g_k \mathbb{1}_A\|_\infty \leq F \,\forall\, k \in K$, *where* $A \subset \mathbb{R}^d$ *and is compact.*

The result can be proved with induction over composition and sub-multiplicativity of matrix norms. By combining Lemma 1 and Lemma 2 we get the following Theorem:

Theorem 20.1. *Let the underlying diffusion process satisfy* ***B1*** *and* ***B2***, *and let* $\hat{f}_N$ *be a feedforward ReLU neural network estimator with weights bounded by an arbitrary constant B. Then the estimator is bounded on* $[0, 1]^d$ *by a constant F whose existence is assured through Lemma 20.2 and thus satisfies - if* $F \geq \|f_0\|_\infty$ *- the risk inequality in (20.6), with the configuration specified by Lemma 1.*

Notice that the optimization error and the approximation error in (20.6) need to be taken w.r.t. $\mathfrak{F}(L', \mathbf{p}', s', F)$. These results enable much more flexibility when tuning the optimization procedure: for example, in our case, setting $L = 2$ and 32 as hidden dimensionality, we enforce the required regularity conditions just by imposing $\|\mathbf{v}_l\|_2 \leq \sqrt{p_l}$ for $l \in \{1, 2\}$ and $\|W_{l_{j,\cdot}}\|_2 \leq \sqrt{p_l}$ for $j \in \{1, \dots, p_{l+1}\}$, $l \in \{0, 1, 2\}$, which is the ℓ^2 bound we would immediately get if all the entries of each row in each weight component were bounded by 1. This still clearly implies that the $\|\cdot\|_\infty$ norms in (20.3) are bounded, but by $\max_{l\in\{0,\dots,L\}} \sqrt{p_l}$ and not by 1. The approach of constraining the ℓ^2 norm of the weight vector of each linear combination is a common heuristic idea when training neural networks [8], being used to drive regularization on the estimator. The ℓ^2 constraint is easily enforced when training with stochastic gradient descent on the loss from (20.4), simply by rescaling after each update if needed.

Further developments. Other approaches to neural network regularization with statistical guarantees have been studied e.g. in [11]. Future developments of our work can be an extension of such results to our setting.

20.3 Numerical Monte Carlo Testing

In order to check the validity of the theoretical results we introduce a testing framework based on the SDE model:

$$dY_t = \begin{bmatrix} -\alpha_1 Y_{t,1} + c_1\alpha_2 \left(\sin\left(\frac{Y_{t,2}}{c_2}\right) + 2\right) \\ c_2\alpha_3 \left(\cos\left(\frac{Y_{t,1}}{c_1}\right) + 2\right) - \alpha_4 Y_{t,2} \end{bmatrix} dt + \begin{bmatrix} \beta_1 c_1 s\left(\frac{Y_{t,1}}{c_1}\right) + c_1\beta_3 & 0 \\ 0 & \beta_2 c_2 s\left(\frac{Y_{t,2}}{c_2}\right) + c_2\beta_3 \end{bmatrix} dW_t \tag{20.7}$$

This model is an extension of the one proposed in [9] to the case of non-constant diffusion. In the above, $\alpha_1 = 1, \alpha_2 = 2, \alpha_3 = 2, \alpha_4 = 1, \beta_1 = 0.5, \beta_2 = 0.5, \beta_3 = 0.1$ are hyperparameters that specify the probabilistic behavior of the solution; specifically, it is crucial that α_1 and α_4 are negative and that the diagonal of the diffusion coefficient remains bounded away from zero, something that is achieved with the β_3 offset. $c_1 = \frac{1}{6}$ and $c_2 = \frac{1}{5}$ are instead much less relevant and, as anticipated before, are just constants that results from scaling the SDE solution for it to reside with high probability mostly in $[0, 1]^2$.

20.3.1 Numerical Setup

With a deterministic initial condition (in our case $Y_0 = 0$) and our hyperparameter choice, it can be shown that (20.7) satisfies the needed ergodic conditions from [6]; additionally, this SDE model satisfies the linear growth and Lipschitz conditions on the first Itô-Taylor coefficients [10], which in turn allow an approximation of the sample paths that is L^1 convergent to the solution, the so-called Milstein scheme.

We fix a time approximation mesh $\Delta = 10^{-3}$. The sampling of the solution is induced by skipping points evenly in the time discretization. This allows us to mitigate the spurious error induced by the numerical computation of the sample paths of the solution to (20.7). Having fixed Δ, the sampling δ appearing in (20.6) is determined by the number of skipped points, which we call *skip* in the following.

The actual implementation of the neural network, according to the configuration specified in Sect. 20.2, is coded in Python with the Tensorflow framework. The number of epochs for the (mini-batch) Adam stochastic gradient descent routine is set fixed to 200, while the number of Monte Carlo iterations is set fixed to 50. After each training routine and evaluation on the test trajectory, we save the error metric contained in the expectation in (20.5) for Monte Carlo approximation.

Notice also that we build a single network to estimate both components together, something that differs from the treatment in [3]: the idea is that we can split the network in two by taking slices of the last dense layer. This induces two networks that both satisfy the regularity we have been describing, sharing the same parameter set except for the last linear map.

20.3.2 Analysis of Results

In Fig. 20.1, we show a single Monte Carlo iteration and get some visual insights. We train and test with two different time horizons $T = 10, 100$, $\delta = 0.02$, corresponding to *skip* = 20. The impact of T on improving the fit arises. This can also be seen from the final results in Table 20.2. This is expected, since the increase in T leads to the exploration of the high-probability sets under the ergodic measure of the process. For low T, ergodic behavior cannot help much, although Fig. 20.1.a shows that a crude modeling of the relevant section of the drift surface can still be achieved. This is also not surprising due to the exponentially fast total variation convergence of the Markov kernels to the ergodic measure, which is again assured here [6].

In the interpretation of Fig. 20.1 it is important to note that the target of our estimate is the drift evaluated on the random trajectory, not the drift itself uniformly over the $[0, 1]^2$ domain. This means that we cannot expect to be able to estimate state space sets that are visited with low probability. This immediate intuition is actually shown in practice with another experiment, which is highlighted in Fig. 20.2. The idea is that upward scaling the diffusion coefficient should increase the noise component of the process, leading it to a more uniform coverage of the considered box domain.

In Fig. 20.2 we show the variance and the bias of (a slice of) the estimated surface $\hat{b}_j(y_1, y_2)$, $j = 1, 2$. The Monte Carlo simulation supports this idea of the drift being learned non-homogeneously, with a strong estimation bias in some regions. This effect is indeed mitigated when increasing the magnitude of the noise stemming from the diffusion component.

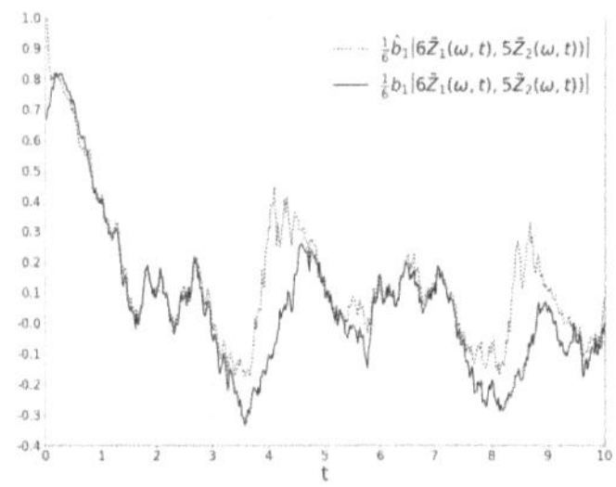

First component of the drift, $T = 10$

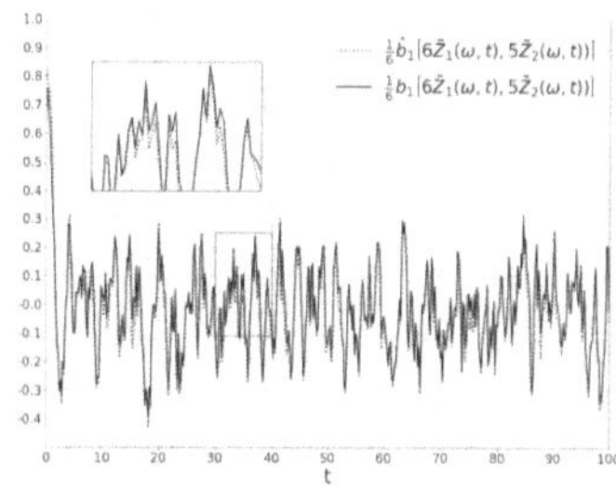

First component of the drift, $T = 100$

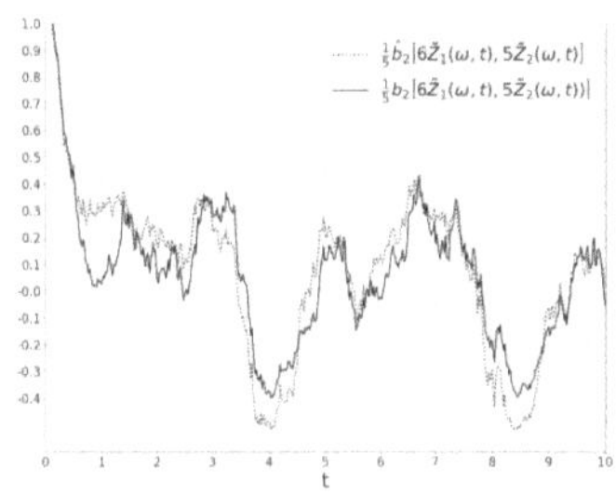

Second component of the drift, $T = 10$

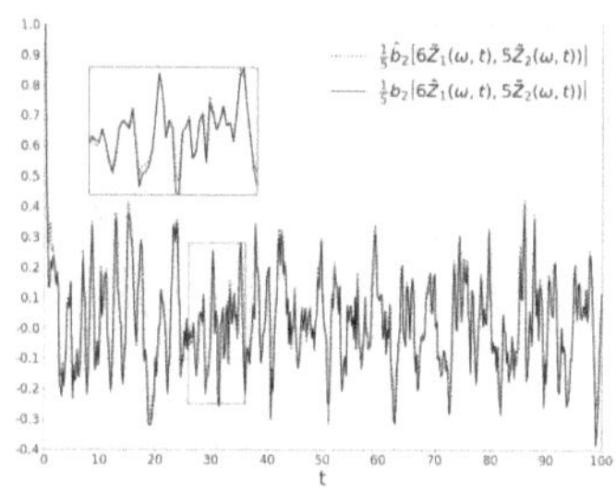

Second component of the drift, $T = 100$

(a) Drift trajectories for $T = 10$

(b) Drift trajectories for $T = 100$

Fig. 20.1: Actual and modeled trajectories for the drift components evaluated on a sample path of the approximation of the test process, $\{\tilde{Z}_t\}_t$; 20 skipped observations and $T = 10$ (left) and $T = 100$ (right). b_1 and b_2 in the plot legends are the components from the non-scaled SDE (i.e. $c_1 = c_2 = 1$) from (20.6)

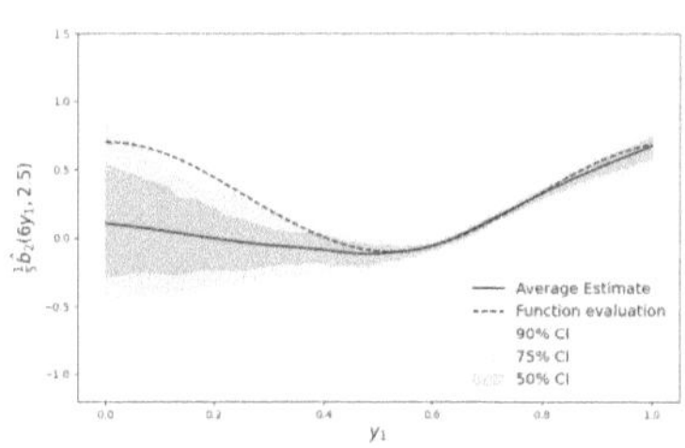

(a) Average estimate, confidence bands and actual evaluation for the second component of the drift, when fixing the second of the two input variables

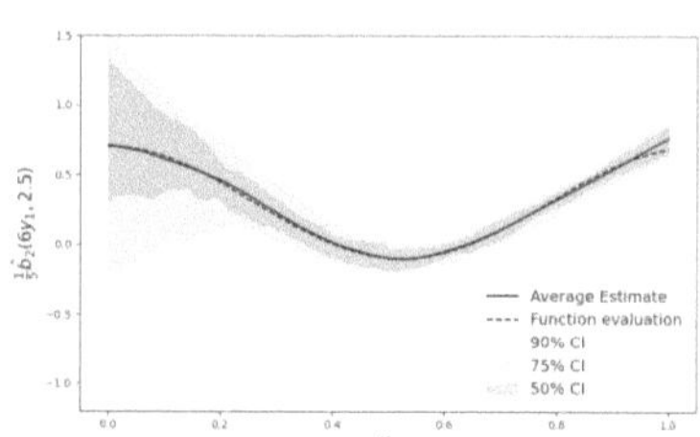

(b) Average estimate, confidence bands and actual evaluation for the second component of the drift, when fixing the second input variable and with diffusion scaled by 2.5

Fig. 20.2: A 50 iterations Monte Carlo experiment: we compare the function surface with the estimated one, across two SDE configurations where for one we are scaling the diffusion component. For the fit, the configuration is the one of Fig. 20.1.b

Table 20.2 gives an overall view of the performance in terms of the risks involved. Apart from the obvious relevance of T, the analysis highlights a counter-intuitive behavior with the decreasing δ (i.e. decreasing *skip* variable). Specifically, *Test MSE* seems not to benefit from low values of δ, after decreasing the *skip* variable past 100.

This would seem in contrast with the finite-sample valid bound in (20.6). Nevertheless, we stress that the bound **may not be tight** in our setting due to unspecified universal constants. This non-tightness may be caused by the model being moderately parameterized. This in turn may cause overfitting problems arising when fitting the difference quotients, since a vanishing δ increases the variance of the response (recall that the underlying process is not differentiable). The error on the quotients (last column in Table 20.2) is compatible with the irreducible error from Table 20.1, with the discrepancy between the two indeed increasing for *skip* < 100, with the fit starting to capture noise. This explains the non-monotonic in δ behaviour of *Train MSE*.

It is apparent that the difference between *Train MSE* and *Test MSE* also increases, meaning that the estimated surface is greedily representing the drift evaluated in the points of the sample. Our hypothesis is that this happens since increasing δ leads to datasets with observations that are close to each other in the state space, with the result similar to increasing the training iterations, by continuity of the drift trajectory.

Table 20.1: Average MSE (scaled by 10^3 and averaged across the two components together, as in Table 20.2) between the difference quotients and the actual drift evaluation, computed with 1000 Monte Carlo iterations for trajectories with $T = 100$

Estimate of the irreducible error ($\times 10^3$)				
skip = 200	skip = 100	skip = 50	skip = 20	skip = 10
48.297	104.348	218.657	562.421	1136.375

Table 20.2: The MSEs are computed with 50 Monte Carlo iterations. *Test MSE* is the estimate of the risk (20.5), *Train MSE* is the same but with evaluation on the training process and *Quotients MSE* is the error on the training difference quotients from (20.4)

skip	**T**	**Test MSE** ($\times 10^3$)	**Train MSE** ($\times 10^3$)	**Quotients MSE** ($\times 10^3$)
200	10	**16.353** (1.204)	14.417 (0.629)	49.800 (1.281)
200	25	**8.627** (0.439)	7.818 (0.266)	48.627 (0.700)
200	50	**4.557** (0.282)	4.177 (0.237)	46.461 (0.528)
200	100	**2.269** (0.107)	2.193 (0.093)	47.032 (0.279)
100	10	**10.995** (0.575)	10.148 (0.441)	101.165 (1.463)
100	25	**4.648** (0.353)	4.305 (0.195)	101.971 (0.870)
100	50	**2.625** (0.110)	2.538 (0.104)	101.546 (0.605)
100	100	**2.013** (0.140)	1.786 (0.107)	104.299 (0.444)
50	10	**11.773** (0.737)	10.295 (0.447)	215.721 (2.014)
50	25	**5.190** (0.498)	3.900 (0.182)	215.832 (1.340)
50	50	**3.430** (0.324)	2.849 (0.180)	216.440 (0.851)
50	100	**2.328** (0.162)	2.083 (0.135)	218.162 (0.745)
20	10	**12.450** (1.155)	8.667 (0.416)	552.487 (3.776)
20	25	**6.529** (0.494)	5.128 (0.327)	552.572 (2.368)
20	50	**3.706** (0.222)	2.838 (0.139)	559.533 (1.628)
20	100	**2.570** (0.232)	2.072 (0.155)	561.386 (0.914)
10	10	**13.894** (1.597)	8.527 (0.384)	1101.550 (5.237)
10	25	**7.061** (0.807)	4.996 (0.299)	1121.530 (2.998)
10	50	**4.309** (0.430)	3.268 (0.256)	1129.400 (1.940)
10	100	**3.706** (0.426)	2.375 (0.157)	1131.730 (1.327)

References

1. Zhu, B., Song, P.X.K. and Taylor, J.M.: Stochastic functional data analysis: A diffusion model-based approach. Biometrics, **67**(4), 1295-1304, (2011)
2. Hirano, S.: Source time functions of earthquakes based on a stochastic differential equation. Scientific reports, **12**(1), 3936, (2022)
3. Oga, A. and Koike, Y.: Drift estimation for a multi-dimensional diffusion process using deep neural networks. Stochastic Processes and their Applications **170**, 104240, (2024)
4. Kessler, M.: Estimation of an ergodic diffusion from discrete observations. Scandinavian Journal of Statistics **24**(2), 211-229, (1997)
5. De Gregorio, A. and Iafrate, F.: Regularized bridge-type estimation with multiple penalties. Annals of the Institute of Statistical Mathematics **73**(5), 921-951, (2021)
6. A. Yu. Veretennikov: Bounds for the mixing rate in the theory of stochastic equations. Theory of Probability & Its Applications **32**(2), 273–281, (1988)
7. Elbrächter, D., Perekrestenko, D., Grohs, P. and Bölcskei, H.: Deep neural network approximation theory. IEEE Transactions on Information Theory **67**(5), 2581-2623 (2021)
8. Srivastava, N., Hinton, G., Krizhevsky, A., Sutskever, I. and Salakhutdinov, R.: Dropout: a simple way to prevent neural networks from overfitting. The journal of machine learning research **15**(1), 1929-1958, (2014)
9. Uchida, M., and Nakahiro Y.: Adaptive estimation of an ergodic diffusion process based on sampled data. Stochastic Processes and their Applications **122**(8), 2885-2924, (2012)

10. Kloeden, P.E. and Platen, E.: Stratonovich and Itô stochastic taylor expansions. Mathematische Nachrichten **151**(1), 33–50, (1991)
11. Taheri, M., Xie, F. and Lederer, J.: Statistical guarantees for regularized neural networks. Neural Networks, **142**, 148-161, (2021)

Chapter 21
Amplitude-Invariant Functional Motif Discovery

Jacopo Di Iorio, Marzia A. Cremona and Francesca Chiaromonte

Abstract A recent problem of interest in Functional Data Analysis (FDA), which generalizes clustering, concerns the discovery of functional motifs in a single curve or in a set comprising multiple curves. Functional motifs are defined as distinctive patterns that are repeated – with a certain amount of noise – within the curve(s) under study. Two methods; namely, *probKMA* [3] and *funBIalign* [6], have recently been proposed to tackle this problem. Although they are effective in discovering motifs whose occurrences have similar shape and allow occurrences to be shifted vertically, these methods cannot deal with motifs composed of portions sharing the same shape but having different amplitudes. In this work, we introduce a new definition of motifs based on a multiplicative model that includes this more challenging scenario, and we extend *funBIalign* to discover amplitude-invariant functional motifs. Performance is assessed through extensive simulations.

Jacopo Di Iorio
Dept. of Quantitative Theory & Methods, Emory University, 36 Eagle Row, Atlanta, 30322, GA, USA, e-mail: jacopo.di.iorio@emory.edu

Marzia A. Cremona (✉)
Dept. of Operations and Decision System, Université Laval, 2325 Rue de la Terrasse, Québec, G1V0A6, Québec, Canada
CHU de Québec – Université Laval Research Center, 1050 Chemin Sainte-Foy, Québec, G1S4L8, Québec, Canada e-mail: marzia.cremona@fsa.ulaval.ca

Francesca Chiaromonte
Dept. of Statistics, Pennsylvania State University, 326 Thomas Building, University Park, PA, 16802, US
Institute of Economics and L'EMbeDS, Sant'Anna School of Advanced Studies, Piazza Martiri della Libertà 33, Pisa, 56127, PI, Italy e-mail: fxc11@psu.edu

© The Author(s), under exclusive license to Springer Nature Switzerland AG 2025
G. Aneiros et al. (eds.), *New Trends in Functional Statistics and Related Fields*, Contributions to Statistics, https://doi.org/10.1007/978-3-031-92383-8_21

21.1 Introduction

While the problem of motif discovery in sequences of symbols has been widely studied in the "omics" sciences – where motifs are repeated sequences of DNA nucleotides or protein amino-acids (see, e.g., [5]) – its continuous counterpart has only recently been tackled in the Functional Data Analysis (FDA) literature. Intuitively, functional motifs are patterns that are repeated (with noise) in different portions of a single curve, or of a set of curves. Functional motif discovery generalizes the problems of clustering misaligned curves (see, e.g., [10, 12]) and of local clustering of functional data (see, e.g., [11, 1]); indeed, motifs are local patterns which can be misaligned in the different curves. In addition, certain curves could present multiple functional motifs in different positions of their domain (hence belonging to multiple local clusters), while other curves may contain occurrences of only one motif (hence belonging to a single cluster), and others could show no motif at all.

To the best of our knowledge, only two FDA methods have been developed to discover functional motifs: probabilistic K-means with local alignment (*probKMA*; [3]) and *funBIalign* [6]. The former is based on a local K-means algorithm, while the latter exploits hierarchical clustering of curve portions. Importantly, these methods consider functional motifs whose occurrences have both a similar shape and level, or a similar shape but different levels (i.e., they allow for vertical shift among occurrences of the same motif), but they are not suitable for discovering motifs whose occurrences have different amplitudes. Amplitude-invariant motifs are of interest in many real-data applications in which the amplitude of the repeated shapes is proportional to the level of the curve. For example, in a price time series where the average price increases with time, it is intuitive that a pattern could have a smaller amplitude when the price is lower [4]. *ProbKMA* has recently been generalized to discover amplitude-invariant functional motifs using a normalization of curve portions to $[0, 1]$. In this work, we formally define amplitude-invariant functional motifs using a multiplicative model, we introduce the functional Virtual Error (fVE) to evaluate motifs, and we extend *funBIalign* to discover this type of motifs.

21.2 Methodology

21.2.1 Model-based definition of amplitude-invariant functional motifs

We consider a set of $N \geq 1$ curves $f_i(t) : [0, T_i] \longrightarrow \mathbb{R}$. Let l be the length of the functional motif and L a sub-interval of $[0, T_i]$ of length l. We define $f_{i,L}(t)$ as the portion corresponding to the restriction of the i-th curve $f_i(t)$ to L, after shifting it to the interval $[0, l]$, in symbols $f_{i,L}(t) := f_i|_L \circ h(t)$ with $t \in [0, l]$, where h is the shift transformation from $[0, l]$ to L. Finally, we define I as the set of all portions of length l obtained from all N curves $I := \{p(t) = f_{i,L}(t) \mid i = 1, \dots, N \text{ and } L \subseteq [0, T_i] \text{ with } |L| = l\}$.

In this framework, [6] provides a definition of functional motifs based on an additive model which considers a mean level, a portion-specific adjustment and a t-varying adjustment. We generalize this definition to the following multiplicative model, which allows the occurrences of a functional motif to have different amplitudes.

Definition 21.1. An *amplitude-invariant functional motif* Q is a collection $I_Q \subset I$ of $n_Q > 1$ curve portions (called *motif occurrences*) $p_k^Q(t)$, $k = 1, \dots, n_Q$ such that

$$p_k^Q(t) = \mu^Q + \alpha_k^Q + \gamma_k^Q \beta^Q(t) + \varepsilon_k(t) \quad \forall t \in [0, l] \tag{21.1}$$

where μ^Q is the motif mean level, α_k^Q and γ_k^Q are the shift and amplitude portion-specific adjustments, $\beta^Q(t)$ is the t-varying adjustment, and $\varepsilon_k(t)$ is the error term.

Similarly to [6], we require $\sum_{k=1}^{n_Q} \alpha_k^Q = 0$, $\int_0^l \beta^Q(t)dt = 0$, and $\mathbb{E}[\varepsilon_k(t)] = 0$ for $\forall k = 1, \dots, n_Q$ and $\forall t \in [0, l]$. In addition, in order to obtain unique and identifiable parameters, we impose $\prod_{k=1}^{n_Q} \gamma_k^Q = 1$.

Note that our definition of amplitude-invariant functional motif generalizes the definition of functional motif in [6], which corresponds to the case of $\gamma_k^Q = 1$, $\forall k = 1, \dots, n_Q$. Since our definition is inspired by the biclustering literature (see, e.g., [8]), in line with this literature we say that a functional motif is "ideal" when there is no noise, i.e. $\varepsilon_k(t) = 0$. In this case, a functional motif is composed of portions that share exactly the same shape up to a multiplicative factor (an example is provided in Fig. 21.1).

Fig. 21.1: Example of ideal ($\varepsilon_k(t) = 0$) amplitude-invariant functional motif: all the portions share exactly the same shape up to a multiplicative factor.

21.2.2 *Evaluation of Amplitude-Invariant Functional Motifs and Dissimilarity Between Curve Portions*

FunBIalign [6] employs a functional Mean Squared Residue score (fMSR), or functional H-score, to evaluate the quality of functional motifs, i.e. their coherence to the (unknown) additive models which define them. The non-functional version of this evaluation score is used by the most popular biclustering algorithm [2] to assess the coherence of the rows (genes) and columns (experiments) of a gene expression bicluster, considering only shifted patterns. Several scores have been proposed in the biclustering literature to identify biclusters presenting both shifted and scaling patterns; that is, biclusters defined by multiplicative models. Among these scores, the Virtual Error (VE) [7] was introduced to measure how the rows (genes) of a biclusters follow a general trend among the columns (experiments). Here, in order to evaluate the quality of an amplitude-invariant functional motif Q, we employ the following functional version of the Virtual Error (fVE):

$$fVE(Q) = \frac{1}{l}\frac{1}{n_Q}\sum_{k=1}^{n_Q}\int_0^l \left|\tilde{p}_k^Q(t) - \bar{\tilde{p}}^Q(t)\right| dt\,, \tag{21.2}$$

where $\tilde{p}_k^Q(t)$ indicates the k-th standardized portion (i.e. the portion minus its mean value, divided by its standard deviation) and $\bar{\tilde{p}}^Q(t)$ is the the functional mean of all standardized portions, i.e. $\bar{\tilde{p}}^Q(t) = \frac{1}{n_Q}\sum_{k=1}^{n_Q}\tilde{p}_k^Q(t)$ (the virtual pattern). To identify amplitude-invariant functional motifs, we look for motifs with low fVE.

Finally, we use the fVE to define the dissimilarity between two curve portions $p_1(t)$ and $p_2(t)$ in $[0, l]$ as

$$d_{\mathbf{fVE}}(p_1, p_2) = \frac{1}{l}\int_0^l \left|\tilde{p}_1^Q(t) - \tilde{p}_2^Q(t)\right| dt\,, \tag{21.3}$$

which is proportional to the fVE of the amplitude-invariant functional motif composed by the two portions $fVE\left(\{p_1(t), p_2(t)\}\right)$.

21.2.3 *The* **funBIalign** *Algorithm*

To discover amplitude-invariant functional motifs, we employ the *funBIalign* algorithm developed in [6], which consists of an agglomerative hierarchical clustering run with complete linkage, followed by a post-processing procedure designed to sort the motifs and retain only the most interesting. While in the original algorithm the hierarchical clustering is based on the fMSR dissimilarity, here we employ the fVE dissimilarity of equation 21.3. The algorithm requires as input the curves $f_i(t)$, $i = 1, \ldots, N$ ($N \geq 1$) evaluated over an equally spaced grid, the discretized length ℓ of the motifs to be discovered (i.e., their length expressed in number of grid points) and their minimum cardinality n_{min}. In the following, we summarize the four steps of the *funBIalign* algorithm (for details, see [6]).

1. *Portion creation and alignment.* We consider all portions of length ℓ obtained from all curves, and align them so that their domain starts at $t = 0$.
2. *Hierarchical clustering based on the fVE dissimilarity.* For each pair of curve portions, we compute the fVE dissimilarity using equation 21.3. If two portions are "acolytes", i.e. if they share 50% or more of their grid points, we add to their dissimilarity the maximum observed dissimilarity value, say M. This makes acolytes "artificially dissimilar" and prevents them from being clustered together. We then employ the dissimilarities to perform an agglomerative hierarchical clustering with complete linkage, and we cut the resulting dendrogram where it displays its longest branches (these correspond to the merger of acolytes).
3. *Collection of candidate functional motifs.* For every sub-tree obtained after cutting the dendrogram, we select one or multiple nodes comprising at least n_{min} portions as candidate functional motifs (see [6] for details on this selection).
4. *Post-processing.* We sort all candidate motifs according to their fVE, or to a combination of their fVE and their cardinality. Finally, we screen motifs starting from the top of the ranking; a motif is filtered out if it is composed of acolytes of a higher ranking motif, and retained otherwise. This produces the final list of discovered motifs.

We observe that the use of complete linkage in step 2 of our algorithm is needed in order to guarantee that the longest dendrogram branches correspond to the merger of acolytes. Other linkages (e.g., single or average) would not prevent acolytes to be clustered together, making the dendrogram cut less straightforward and requiring a complex post-processing to remove acolytes from candidate motifs.

21.3 Simulations

We evaluate the performance of our modified *funBIalign* by applying it to smooth curves embedding multiple occurrences of amplitude-invariant functional motifs. We simulate data using the model proposed by [3], which defines a smooth curve as $f(t) = \sum_{\gamma=1}^{\Gamma} c_\gamma \phi_\gamma(t)$ where $\{\phi_\gamma\}_{\gamma=1}^{\Gamma}$ is a B-spline basis of order ν with $\Gamma - \nu + 2$ equally spaced knots, and $\{c_\gamma\}_{\gamma=1}^{\Gamma}$ are B-splines coefficients. Motifs are defined by fixing the values of the B-spline coefficients that correspond to them. We consider B-splines of order $\nu = 3$ and knots spaced at intervals of length 10, and for each simulation we generate a single smooth curve of length 10,000, evaluated over a grid of 10,001 equally spaced points. Within each curve, we embed the same number (either 6, 8, or 10) of occurrences of four distinct motifs, each of length 50 (i.e., $\ell = 51$ points). The B-splines coefficients are randomly sampled from a Beta$(0.45, 0.45)$ distribution, and rescaled to the range $[-15, 15]$. Each motif occurrence is then modified to add a vertical shift, as well as a multiplicative factor. In particular, we add a vertical shift randomly sampled from a $U(-10, 10)$ to all

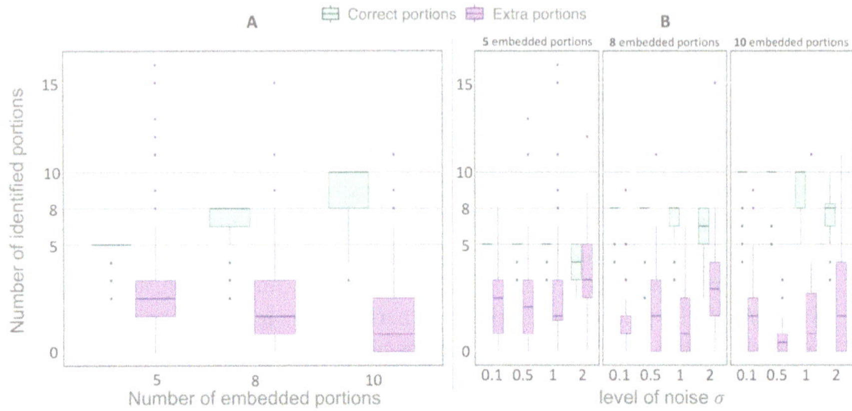

Fig. 21.2: **A.** Performance of *funBIalign* for all simulations where motifs have n_Q = 6, 8, or 10 occurrences. The algorithm is run with minimum cardinalities n_{min} = 5, 6, 7, 8, 9, 10 and results are pooled. For each number of occurrences, the green and pink boxplots (left and right) represent correctly identified portions and extra portions, respectively. **B.** Performance of *funBIalign* in the identification of motifs with n_Q = 6, 8, or 10 occurrences (left, central, and right panel, respectively) and σ = 0.1, 0.5, 1, or 2. Results obtained by running the algorithm with minimum cardinalities n_{min} = 5, 6, 7, 8, 9, 10 are pooled. For each number of occurrences, the green and pink boxplots (left and right) represent correctly identified portions and extra portions, respectively.

the B-spline coefficients corresponding to a motif occurrence, and then we multiply them by a multiplicative factor randomly sampled from a $U[0.05, 4]$. Finally, we add noise sampled from a $N(0, \sigma^2)$ to each motif coefficient. Note that a motif generated in this way deviates from the additive model in equation 21.1, since the noise is added to the B-spline coefficients instead than to the motif itself; this poses a greater challenge to our methodology. We consider three alternative numbers of occurrences for the motifs (n_Q = 6, 8 or 10). For each n_Q we construct 10 simulation sets of 10 curves with four distinct amplitude-invariant motifs each, with n_Q occurrences and with noise level $\sigma = 0.1$, $\sigma = 0.5$, $\sigma = 1$, and $\sigma = 2$, respectively. We run *funBIalign* with input $\ell = 51$ (the true length of the motifs) and $n_{min} = 5, 6, \ldots, 10$ (the minimum cardinalities). We rank the candidate motifs using their fVE (see Step 4 of the *funBIalign* algorithm), and report the discovered motifs which are most similar to the embedded motifs.

Fig. 21.2A. summarizes the performance of *funBIalign* across the 10 simulations. For each of the 3 alternative numbers of occurrences (n_Q = 6, 8, or 10), two boxplots are displayed, each comprising $4 \times 6 \times 10 = 240$ values: for each of the four motifs in each of the 10 simulations and across the 6 different n_{min} runs, we count the number of correctly identified portions (i.e. portions embedded in the curve and correctly

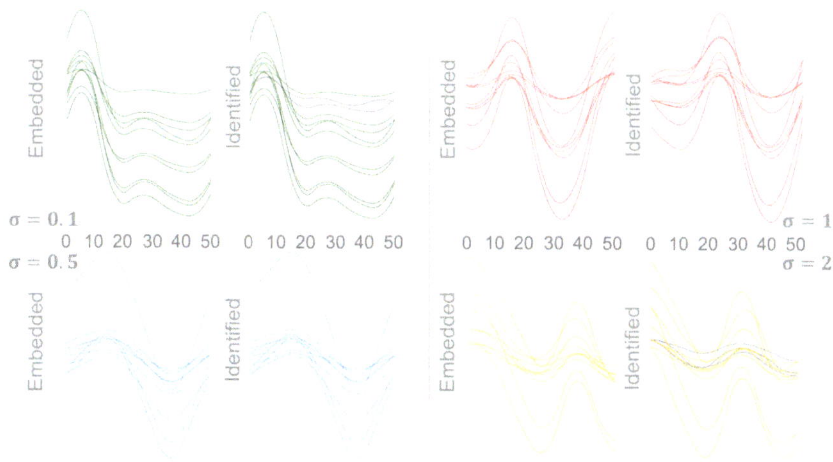

Fig. 21.3: Details on the motifs identified in the simulation set n.2 with $n_Q = 10$ obtained by running *funBIalign* with $n_{min} = 8$. All the 10 embedded portions were correctly identified, except for the motif with $\sigma = 2$, where one portion is missing. The identified extra portions (not intentionally embedded in the curve; in black), visually resembles occurrences of the motif.

identified by the algorithm; left boxplot) and the number of extra portions (i.e. portions that were not embedded in the curve, but were identified by the algorithm; right boxplot). We can observe that *funBIalign* is quite effective in identifying all embedded motif occurrences. However, some extra portions are usually identified – though these are usually very similar to the embedded portions (see Fig. 21.3).

Fig. 21.2B. displays the performance of *funBIalign* in the identification of motifs with $n_Q = 6, 8$, or 10 occurrences (left, central, and right panel, respectively) and $\sigma = 0.1, 0.5, 1$, or 2. We can note, as expected, that both the number of extra portions and the number of portions that are not identified increase with the level of noise and with the number of occurrences embedded in the motif. However, in terms of correctly identified portions, performance remains excellent even at high noise levels.

Finally, Fig. 21.3 provides, as an example, details on the motifs identified in the simulation set n.2 with $n_Q = 10$ obtained by running *funBIalign* with $n_{min} = 8$. We can observe that the extra portions, i.e. the portions that were not intentionally embedded in the curve but were identified by *funBIalign* as motif occurrences, albeit by chance, do resemble the other occurrences of the motif. Hence, they cannot really be considered as "false positives".

21.4 Conclusions

This work extends the recently developed functional motif discovery method *funBIalign* [6] to identify amplitude-invariant functional motifs in a single curve or in a set of curves. We define this type of motifs using a multiplicative model and we propose a novel score inspired by the biclustering literature, namely the functional Virtual Error (fVE), to discover and evaluate motifs. As in the original *funBIalign* algorithm, our extended version is particularly easy to use; it only requires as input, in addition to the curve(s) themselves, the length of the motifs and their minimum number of occurrences. Our simulations demonstrate the good performance of the method across different levels of noise and motif cardinalities.

Acknowledgements M.A.C. acknowledges the support of the Natural Sciences and Engineering Research Council of Canada (NSERC) and of Faculty of Business Administration, Université Laval. The work of F.C. was partially supported by the Huck Institutes of the Life Sciences at Penn State, the L'EMbeDS Department of Excellence of the Sant'Anna School of Advanced Studies, and the SMaRT COnSTRUCT project (CUP J53C24001460006, as part of FAIR, PE0000013, CUP B53C22003630006, Italian National Recovery and Resilience Plan funded by NextGenerationEU).

References

1. Centofanti, F., Lepore, A., Palumbo, B.: Sparse and smooth functional data clustering. Stat. Pap. **65**(2), 795-825 (2024).
2. Cheng, Y., Church, GM.: Biclustering of expression data. In: Proceedings of the 8th International Conference on Intelligent Systems for Molecular Biology, La Jolla, CA, pp. 93–103 (2000).
3. Cremona, M.A., Chiaromonte, F.: Probabilistic K-means with local alignment for clustering and motif discovery in functional data. JCGS **32**(3), 1119–1130 (2023)
4. Cremona, M.A., Doroshenko, L., Severino, F.: Functional motif discovery in stock market prices. SSRN 4642040 (2023)
5. Das, M. K., Dai, H. K.: A survey of DNA motif finding algorithms. BMC Bioinform. **8**, 1-13 (2007)
6. Di Iorio, J., Cremona, M.A., Chiaromonte, F.: funBIalign: a hierachical algorithm for functional motif discovery based on mean squared residue scores. Stat. Comput. **35**, 11 (2025)
7. Divina, F., Pontes, B., Giráldez, R., Aguilar-Ruiz, J. S.: An effective measure for assessing the quality of biclusters. Comput. Biol. Med. **42**(2), 245-256 (2012).
8. Madeira, S. C., Oliveira, A. L.: Biclustering algorithms for biological data analysis: a survey. TCBB **1**(1), 24-45 (2004)
9. Pontes, B., Girldez, R., Aguilar-Ruiz, J.S.: Quality measures for Gene expression biclusters. PLoS ONE **10**(3): e0115497 (2015)
10. Sangalli, L. M., Secchi, P., Vantini, S., Vitelli, V.: K-mean alignment for curve clustering. CSDA **54**(5), 1219-1233 (2010)
11. Vitelli, V.: A novel framework for joint sparse clustering and alignment of functional data. JNPS **36**(1), 182-211 (2024)
12. Welbaum, A., Qiao, W.: Mean shift-based clustering for misaligned functional data. CSDA, 108107 (2025)

Chapter 22
Testing the Significance of Covariates in Nonparametric Regression without the Curse of Dimensionality

Daniel Diz-Castro, Manuel Febrero-Bande and Wenceslao González-Manteiga

Abstract Significance tests are invaluable tools for detecting redundant covariates in regression, enabling the proposal of simpler, more parsimonious models. This is particularly relevant in nonparametric regression with high dimensional or functional data due to the effects of the curse of dimensionality on estimation and inference procedures. A new kernel-based test, designed to be robust against the curse of dimensionality, is presented in this contribution. This proposal is related to two of the most prominent approaches in the literature on significance tests: smoothing and empirical process-based tests. We provide a result establishing the asymptotic behavior of the proposed test statistic under independence and weak dependence assumptions.

22.1 Introduction

It is well known that nonparametric estimation and inference procedures tend to suffer as the dimensionality of the data increases (see e.g. [18]). This issue has motivated the development of multiple variable selection procedures and, in the particular context of regression modelling, the derivation of covariate significance tests which provide, by construction, model simplification mechanisms with a controlled type I

Daniel Diz-Castro (✉)
Department of Statistics, Mathematical Analysis and Optimization, University of Santiago de Compostela, Rúa Lope Gómez de Marzoa s/n, 15782 Santiago de Compostela, Spain
e-mail: danieldiz.castro@usc.es

Manuel Febrero-Bande
Department of Statistics, Mathematical Analysis and Optimization, University of Santiago de Compostela, Rúa Lope Gómez de Marzoa s/n, 15782 Santiago de Compostela, Spain
e-mail: manuel.febrero@usc.es

Wenceslao González-Manteiga
Department of Statistics, Mathematical Analysis and Optimization, University of Santiago de Compostela, Rúa Lope Gómez de Marzoa s/n, 15782 Santiago de Compostela, Spain
e-mail: wenceslao.gonzalez@usc.es

© The Author(s), under exclusive license to Springer Nature Switzerland AG 2025

G. Aneiros et al. (eds.), *New Trends in Functional Statistics and Related Fields*,
Contributions to Statistics, https://doi.org/10.1007/978-3-031-92383-8_22

error.
We are concerned in this work with the problem of testing the significance of covariates in nonparametric regression models that may involve some functional variables. In particular, given a response $Y \in \mathcal{Y}$ and a pair of covariates $X \in \mathbb{R}^d$ and $W \in \mathcal{W}$, where $\mathcal{Y}$ and $\mathcal{W}$ are two separable Hilbert spaces, we are interested in the hypothesis test

$$\mathcal{H}_0 : \; g(X,W) = m(X) \text{ a.s.} \quad \text{vs.} \quad \mathcal{H}_1 : \; \mathbb{P}(g(X,W) \neq m(X)) > 0, \tag{22.1}$$

where $m(X) = \mathbb{E}[Y|X]$ and $g(X,W) = \mathbb{E}[Y|(X,W)]$ are both unspecified; that is, we want to test the significance of W in the nonparametric regression model of Y on (X,W).

The literature on this type of covariate significance tests is extensive for regression models with finite-dimensional covariates and responses. See, for example, [7, 2, 10, 5, 11, 21, 20, 3] and the references therein. For models involving functional data, there are also some proposals for no-effect tests which, in our notation, focus on testing whether $\mathbb{E}[Y|W] = \mathbb{E}[Y]$ holds almost surely (see e.g. [14, 15, 12, 9]). However, for the particular test in (22.1), we are only aware of the papers by [13], where the asymptotic theory for the completely nonparametric case is not developed, and by [6], which formally requires to know the distribution function of (X,W). Moreover, in both proposals, the curse of dimensionality hinders the theoretical power porperties of their tests.

In this contribution we present a covariate significance test free from the curse of dimensionality which is able to handle both independent and weakly dependent data and it is based on recent developments in the theory of kernel-based characterizations of null conditional expectations.

22.2 Connecting Smoothing and Empirical Process-Based Tests

For simplicity of exposition, let us assume that X is an absolutely continuous random vector with density function f. Let Z and $\mathcal{Z}$ denote the pair (X,W) and the product space $\mathbb{R}^d \times \mathcal{W}$, respectively. Moreover, for any measurable function $u : \mathcal{Z} \to \mathbb{R}$ which is positive (almost everywhere) in the support of Z, let $I(u)$ be defined as

$$I(u) = \mathbb{E}[u(Z)\langle Y - m(X), \mathbb{E}[Y - m(X)|Z]\rangle_{\mathcal{Y}}] = \mathbb{E}[u(Z)||\mathbb{E}[Y - m(X)|Z]||^2_{\mathcal{Y}}] \geq 0.$$

It is easy to verify that $\mathcal{H}_0$ holds if and only if $I(u) = 0$ and, hence, an estimator of $I(u)$ can be used to test $\mathcal{H}_0$. Since m is unknown, it also needs to be estimated nonparametrically, usually by means of the Nadaraya-Watson estimator $\hat{m}_h$ defined as

$$\hat{m}_h(x) = \sum_{k=1}^{n} \frac{K_h(x - X_k)}{n\hat{f}_h(x)} Y_k,$$

where $\hat{f}_h(x) = \sum_{k=1}^{n} K_h(x - X_k)/n$, for some rescaled kernel, $K_h(\cdot) = K(\cdot/h)/h^d$, depending on a smoothing bandwidth parameter $h > 0$. Similarly, $\mathbb{E}[Y - m(X)|Z]$

is estimated by another kernel smoother. This is the fundamental idea behind the family of smoothing-based tests which encompasses most proposals in the literature (see e.g. [7, 10, 6, 21]). The main advantage of this family of tests is the standard Gaussian limiting distribution of the (conveniently normalized) test statistics. However, smoothing-based tests are specially affected by the curse of dimensionality due to the double smoothing required to estimate $\mathbb{E}[Y - m(X)|Z]$ or similar, alternative, quantities.

A different approach, adopted in [2, 5], comes from the empirical process literature for goodness-of-fit. Given some set $\mathcal{T}$ and a measurable function $v : \mathcal{Z} \times \mathcal{T} \to \mathbb{R}$ let $\tilde{I}(v,t)$ be defined as

$$\tilde{I}(v,t) = \mathbb{E}\left[v(Z,t)f(X)(Y - m(X))\right],$$

for all $t \in \mathcal{T}$. If $\mathcal{H}_0$ holds, then $\tilde{I}(v,t) = 0$ for almost all t. The reciprocal also holds under some conditions over $\mathcal{T}$ and v (see e.g. [17]). Some classical examples are the family of complex exponentials [1] and indicator functions [19] where $v(Z,t) = e^{-i\langle Z,t\rangle}$ and $v(Z,t) = \mathbb{1}_{\{Z\leq t\}}$, respectively, considering $\mathcal{Z} = \mathcal{T} = \mathbb{R}^p$ for some $p > d$. Plugging-in the estimators $\hat{m}_h$ and $\hat{f}_h$ in the empirical analogous of $\tilde{I}(v,t)$, we obtain an empirical process driven by the nonparametric residuals $\{\hat{f}_h(X_i)(Y_i - \hat{m}_h(X_i))\}_{i=1}^n$:

$$\hat{\tilde{I}}(v,t) = \frac{1}{n}\sum_{i=1}^{n} v(Z_i,t)\hat{f}_h(X_i)(Y_i - \hat{m}_h(X_i)).$$

Thus, a consistent test can be constructed based on a continuous functional of $\hat{\tilde{I}}(v,t)$ such as a $L_q(\mathcal{T})$ norm for some $0 < q \leq \infty$. The limit distribution of this type of test statistics is generally nonpivotal and resampling schemes are needed for calibration. By avoiding the double smoothing issue mentioned before, this approach yields tests with better theoretical power properties than the smoothing-based tests although, for small sample sizes, the former generally does not outperform the latter.

To introduce our proposal, let c be a positive definite kernel defined on $\mathcal{Z} \times \mathcal{Z}$. Then, c generates a reproducing kernel Hilbert space C, as the closure of the linear span of $\{c(z,\cdot) : z \in \mathcal{Z}\}$, and it holds that $c(z,z') = \langle c(z,\cdot), c(z',\cdot)\rangle_C$ for all $z, z' \in \mathcal{Z}$. Furthermore, let $\{\Phi_j\}_{j=1}^{\infty}$ be an orthonormal basis of $\mathcal{Y}$. It is easy to verify that

$$\tilde{I}(c,z) = \sum_{j=1}^{\infty} \left\langle \mathbb{E}\left[c(Z,\cdot)f(X)\langle Y - m(X), \Phi_j\rangle_{\mathcal{Y}}\right], c(z,\cdot)\right\rangle_C \Phi_j$$

and, moreover, $\tilde{I}(c,z) = 0$ for all $z \in \mathcal{Z}$ if and only if the squared norm of $\mathbb{E}\left[c(Z,\cdot)f(X)\langle Y - m(X), \Phi_j\rangle\right]$ is zero for all j. Assuming $\tilde{I}(c,z)$ characterizes $\mathcal{H}_0$, it follows that

$$\mathcal{H}_0 \iff 0 = \sum_{j=1}^{\infty} \left\| \mathbb{E}\left[c(Z,\cdot) f(X) \langle Y - m(X), \Phi_j \rangle_{\mathcal{Y}}\right] \right\|_C^2$$
$$= \mathbb{E}\left[c(Z,Z')\langle f(X)(Y-m(X)), f(X')(Y'-m(X'))\rangle_{\mathcal{Y}}\right], \quad (22.2)$$

given two i.i.d. (independent and identically distributed) copies (Z,Y) and (Z',Y'). Taking into account the results in [12, 9], the equivalence in (22.2) holds for Hilbert space-valued data when $c(z,z') = ||z - z'||$ and also for any bounded characteristic kernel c (see e.g. [22]). As a natural estimator of the term on the right-hand side of (22.2), when m and f are unknown, we suggest:

$$V_n = \frac{1}{n^2} \sum_{i,j=1}^{n} c(Z_i, Z_j) \langle \hat{f}_h(X_i)(Y_i - \hat{m}_h(X_i)), \hat{f}_h(X_j)(Y_j - \hat{m}_h(X_j)) \rangle_{\mathcal{Y}},$$

which is a fourth order V-statistic with a (fairly complicated) sample size dependent kernel. Furthermore the structure of V_n is very similar to that of the smoothing-based estimators of $I(u)$ in [7, 10, 6, 21], with the main difference of the usage of a positive definite kernel c instead of a second smoothing kernel depending on a bandwidth parameter converging to zero.

22.3 Assumptions

To derive the limiting behavior of V_n we impose the following model-free assumptions:

(A1) K is a product kernel verifying $K(x) = \prod_{j=1}^{d} K^u(x^j)$ for all $x = (x^1, \ldots, x^d)$, where K^u is a bounded function with support in $[-1, 1]$ and symmetric with respect to 0 such that, for some natural number $\nu \geq 2$,

$$\int K^u(t) t^i dt = \mathbb{1}_{\{i=0\}}, \quad 0 \leq i \leq \nu - 1.$$

(A2) $\{h(n)\}_{n=1}^{\infty}$, where $h(n)$ is simply denoted as h, is a deterministic sequence of positive bandwidths, such that $h \to 0$ and $nh^d \to \infty$ as $n \to \infty$.

(A3) c is characteristic kernel. Furthermore, it is positive definite, bounded, continuous and symmetric in its two arguments.

Assumptions (A1) and (A2) are common in the context of specification tests where kernel smoothing is involved. Assumption (A3) establishes some restrictions on the learning kernel c used to characterise the null hypothesis through the equivalence in (22.2). Some examples are the Gaussian and Laplace kernels, defined as $c(z,z') = \exp\left(-||z - z'||_{\mathcal{Z}}^2/\sigma_c^2\right)$ and $c(z,z') = \exp\left(-||z - z'||_{\mathcal{Z}}/\sigma_c^2\right)$, respectively, for some $\sigma_c^2 > 0$.

Denote the regression model error by $\mathcal{E} = Y - g(Z)$, verifying $\mathbb{E}[\mathcal{E}|Z] = 0$. Denote also $b(x,z') = \mathbb{E}[c(Z,z')|X = x]$. Let us assume that there exists some sample $\{(Z_i, \mathcal{E}_i)\}_{i=1}^{n}$ and we observe $\{(Z_i, Y_i)\}_{i=1}^{n}$. The sample is either i.i.d. or a partial sample path of a strictly stationary, absolutely regular process $\{(Z_i, \mathcal{E}_i)\}_{i=-\infty}^{\infty}$ identically distributed as (Z, Y); that is,

$$\beta_k = \mathbb{E}\left[\sup_{A\in\sigma_k^{\infty}} \left|\mathbb{P}\left(A|\sigma_{-\infty}^{0}\right) - \mathbb{P}(A)\right|\right] \xrightarrow{k\to\infty} 0,$$

where the σ-algebras generated by $\{(Z_i, \mathcal{E}_i)\}_{i=-\infty}^{0}$ and $\{(Z_i, \mathcal{E}_i)\}_{i=k}^{\infty}$ are denoted by $\sigma_{-\infty}^{0}$ and σ_k^{∞}, respectively.

Besides assumptions (A1)-(A3), we also need to impose some technical conditions where, for the sake of notational simplicity, we omit references to the spaces where inner products are computed. In the following, δ is a nonnegative real number.

(T1) $\mathbb{E}[||\mathcal{E}||^{4+2\delta} + ||m(X)||^{4+2\delta} + ||g(Z)||^{2+\delta}] < \infty$.

(T2) As a function of $x \in \mathrm{supp}(\mathbb{P}^X)$, both the density f and the regression function m are $\mathbb{P}^X$ almost surely $\nu + 1$ times continuously differentiable in a neighborhood of x. Furthermore, for $a \in \{m, f\}$ it holds that

$$\sum_{k=1}^{\nu} \sum_{j_1,\ldots,j_k=1}^{d} \mathbb{E}\left[\left\|\frac{\partial^k}{\partial x^{j_1}\cdots\partial x^{j_k}} a(X)\right\|^{4+2\delta}\right] < \infty.$$

In addition, there will be some $h_1 > 0$ such that the following quantity is finite:

$$\mathbb{E}\left[\sup_{||R||\le h_1} \sum_{j_1,\ldots,j_{\nu+1}=1}^{d} \left\|\frac{\partial^{\nu+1}}{\partial x^{j_1}\cdots\partial x^{j_{\nu+1}}} a(X+R)\right\|^{4+2\delta} \mathbb{1}_{\{X+R\in\mathrm{supp}(\mathbb{P}^X)\}}\right].$$

(T3) For almost all $z' \in \mathrm{supp}(\mathbb{P}^Z)$, $b(x, z')$ is $\mathbb{P}^X$ almost surely continuously differentiable with respect to $x \in \mathrm{supp}(\mathbb{P}^X)$, in a neighborhood of x. Given two i.i.d. copies Z and Z', there will be some $h_2 > 0$ such that,

$$\mathbb{E}\left[\sup_{||R||\le h_2} \sum_{j=1}^{d} \left|\frac{\partial}{\partial x^j} b(X+R, Z)\right|^{2+\delta} \mathbb{1}_{\{X+R\in\mathrm{supp}(\mathbb{P}^X)\}}\right] < \infty.$$

(T4) The density function f is bounded.

(T5) $\sum_{k=1}^{\infty} k\beta_k^{\frac{\delta}{2+\delta}} < \infty$.

(T6) For all $i > 1$, the conditional distribution function of X_i given X_1 is absolutely continuous with density function $f_{i|1}$. There is some $h_3 > 0$ such that,

$$\sup_{1<i<\infty} \mathbb{E}\left[\sup_{||R||\le h_3} \left|f_{i|1}(X_1 + R|X_1)\right|^2 \mathbb{1}_{\{X_1+R\in\mathrm{supp}(\mathbb{P}^{X_i|X_1})\}}\right] < \infty.$$

(T7) For all $i > 1$, denote $s_i(X_i, X_1) = \mathbb{E}[(||\mathcal{E}_i||^{2+\delta} + ||\mathcal{E}_1||^{2+\delta})|X_i, X_1]$. There is some $h_4 > 0$ such that

$$\sup_{1<i<\infty} \mathbb{E}\left[\sup_{||R||\leq h_4} |s_i(X_1+R, X_1)|^2 \, \mathbb{1}_{\{X_1+R\in \mathrm{supp}(\mathbb{P}^{X_i|X_1})\}} \right] < \infty.$$

Condition (T1) imposes some finite moments restrictions on the regression errors and the conditional expectation of Y. Conditions (T2)-(T3) bound, $\mathbb{P}^X$ almost surely, the moments of the Taylor expansions of m, f and b. Condition (T4) is analytically convenient to simplify the derivation of theoretical results but it can be replaced by the finiteness of $\mathbb{E}[|f(X)|^{4+2\delta}]$ and $\mathbb{E}[||g(Z)||^{4+2\delta}]$. Conditions (T5)-(T7) are required to bound the variance of certain terms in a Hoeffding-type decomposition of V_n in the absolutely regular case.

22.4 Asymptotics

Denote by $\mathcal{HS} = C \otimes \mathcal{Y}$ the Hilbert space of tensor products $\zeta \otimes y : \mathcal{Y} \to C$, where $\zeta \in C$ and $y \in \mathcal{Y}$, such that $\zeta \otimes y(y') = \zeta\langle y, y'\rangle$, endowed with the following inner product structure: $\langle \zeta \otimes y, \zeta' \otimes y'\rangle = \langle \zeta \otimes \zeta'\rangle\langle y \otimes y'\rangle$ (see e.g. [8]). Moreover, denote by $b(x,\cdot) = \mathbb{E}[c(Z,\cdot)|X = x]$, $L_h(x) = \mathbb{E}[K_h(X - X')(m(X) - m(X'))|X = x]$ for any X, X' i.i.d and $B(x) = \lim_{h\to 0} L_h(x)/h^\nu$, the latter of which is well defined for all $x \in \mathrm{supp}\left(\mathbb{P}^X\right)$ under (A1)-(A2) and (T1)-(T2).

Theorem 22.1. *Assume (A1)-(A3) hold. In addition, suppose one of the following three sets of conditions is verified:*

- $\{(Z_i, \mathcal{E}_i)\}_{i=1}^n$ *is an i.i.d. sample and (T1)-(T4) hold for* $\delta = 0$.
- $\{(Z_i, \mathcal{E}_i)\}_{i=1}^n$ *is a partial sample path of an absolutely regular process, (T1)-(T5) hold for some* $\delta > 0$ *and* $nh^{2d} \to \infty$.
- $\{(Z_i, \mathcal{E}_i)\}_{i=1}^n$ *is a partial sample path of an absolutely regular process, (T1)-(T7) hold for some* $\delta > 0$ *and* $nh^{d\left(1+\frac{\delta}{2+\delta}\right)} \to \infty$.

Then:

1. V_n *converges to* $\mathbb{E}\left[c(Z, Z')\langle f(X)(Y - m(X)), f(X')(Y' - m(X'))\rangle\right] \geq 0$ *in probability.*
2. *Under* $\mathcal{H}_0$*, assuming* $nh^{2\nu} \to \infty$,

$$\frac{\sqrt{n}}{h^\nu}(V_n - \mu_h) \xrightarrow{\mathcal{D}} N(0, 4S_0),$$

where $\xrightarrow{\mathcal{D}}$ *means convergence in distribution and* μ_h *and* S_0 *are defined as*

$$\mu_h = ||\mathbb{E}[c(Z,\cdot) \otimes L_h(X)]||^2 = O(h^{2\nu}),$$

$$S_0 = \sum_{k=-\infty}^{\infty} \mathbb{E}\left[\langle U_0, \mathbb{E}[c(Z,\cdot) \otimes B(X)]\rangle \langle U_k, \mathbb{E}[c(Z,\cdot) \otimes B(X)]\rangle\right],$$

where $U_k = (c(Z_k,\cdot) - b(X_k,\cdot)) \otimes f(X_k)\mathcal{E}_k$ *for all* k.

3. Under $\mathcal{H}_0$, assuming $nh^{2\nu} \to M \geq 0$, $M < \infty$,

$$nV_n \xrightarrow{\mathcal{D}} ||\sqrt{M}\mathbb{E}[c(Z,\cdot) \otimes B(X)] + N_{\mathcal{HS}}||^2$$

where $N_{\mathcal{HS}}$ is a Gaussian random variable taking values in $\mathcal{HS}$ with zero mean and covariance operator S defined by

$$\langle S(\zeta \otimes y), \zeta' \otimes y' \rangle = \sum_{k=-\infty}^{\infty} \mathbb{E}\left[\langle U_0, \zeta \otimes y \rangle \langle U_k, \zeta' \otimes y' \rangle\right].$$

Theorem 22.1 establishes the consistency of a test based on rejecting $\mathcal{H}_0$ for large enough $\sqrt{n}(V_n - \mu_h)/h^\nu$ or nV_n (depending on the rate of convergence of h to zero) although obtaining a $\sqrt{n/h^\nu}$-consistent estimator for μ_h is quite difficult. Furthermore, consider sequences of local alternatives in Pitman's sense of the form

$$\mathcal{H}_{1n}^\gamma : \ g(Z) = m(X) + \frac{1}{n^\gamma}\tilde{g}(Z) \ \text{a.s.},$$

where $\mathbb{E}[\tilde{g}(Z)|X] = 0$ a.s. and $\mathbb{E}[||\tilde{g}(Z)||^2] < \infty$ for some $\gamma > 0$. We have found that, in general, only the test under the bandwidth regime $nh^{2\nu} \to 0$ can detect $\mathcal{H}_{1n}^\gamma$ for any value of $\gamma \leq 1/2$, making that version of the test theoretically comparable to the empirical process-based proposals in [2, 5] and superior to the smoothing-based tests.

The limit distribution of the test statistic when $nh^{2\nu} \to 0$ is nonpivotal and has an intrincated form of dependence on the data distribution. However, the test can be consistently calibrated via wild bootstrap (see e.g. [16]).

22.5 Discussion

The asymptotic results presented in Sect. 22.4 indicate that the proposed methodology is an useful tool to test $\mathcal{H}_0$ against $\mathcal{H}_1$. In addition, with an adequate choice of the smoothing bandwidth h (such that $nh^{2\nu} \to 0$), the test is free from the curse of dimensionality, since it is able to detect local alternatives converging to the null hypothesis at the rate of $n^{-1/2}$, independently of the data dimension. This results hold for independent data and sample paths of absolutely regular processes, although some preliminar results suggest that the test may also be applied under more general dependence conditions (see e.g. [4]).

The test methodology can be easily extended to regression models where X contains a discrete component. It is also possible to handle Hilbert-valued X, under some restrictions on the dependence between X and W.

Acknowledgements The authors acknowledge support from project PID2020-116587GB-I00 funded by MCIN/AEI/10.13039/501100011033 and the Competitive Reference Groups 2021-2024 (ED431C 2021/24) from the Xunta de Galicia.

References

1. Bierens, H.J.: Consistent model specification tests. Journal of Econometrics **20**(1), 105–134 (1982)
2. Chen, X., Fan, Y.: Consistent hypothesis testing in semiparametric and nonparametric models for econometric time series. Journal of Econometrics **91**(2), 373–401 (1999)
3. Dai, B., Shen, X., Pan, W.: Significance tests of feature relevance for a black-box learner. IEEE Transactions on Neural Networks and Learning Systems **35**(2), 1898–1911 (2024)
4. Dehling, H., Sharipov, O.S., Wendler, M.: Bootstrap for dependent Hilbert space-valued random variables with application to von Mises statistics. Journal of Multivariate analysis **133**, 200–215 (2015)
5. Delgado, M.A., González-Manteiga, W.: Significance testing in nonparametric regression based on the bootstrap. The Annals of Statistics **29**(5), 1469–1507 (2001)
6. Delsol, L., Ferraty, F., Vieu, P.: Structural test in regression on functional variables. Journal of Multivariate Analysis **102**(3), 422–447 (2011)
7. Fan, Y., Li, Q.: Consistent model specification tests: Omitted variables and semiparametric functional forms. Econometrica **64**(4), 865–890 (1996)
8. Gretton, A., Bousquet, O., Smola, A., Schölkopf, B.: Measuring statistical dependence with Hilbert-Schmidt norms. In: Jain, S., Simon, H.U., Tomita, E. (eds) Algorithmic Learning Theory, pp. 63-77. Springer, Berlin (2005)
9. Lai, T., Zhang, Z., Wang, Y.: A kernel-based measure for conditional mean dependence. Computational Statistics & Data Analysis **160**, 107246 (2021)
10. Lavergne, P., Vuong, Q.: Nonparametric significance testing. Econometric Theory **16**(4), 576–601 (2000)
11. Lavergne, P., Maistre, S., and Patilea, V.: A significance test for covariates in nonparametric regression. Electronic Journal of Statistics **9**(1), 643–678 (2015)
12. Lee, C.E., Zhang, X., Shao, X.: Testing conditional mean independence for functional data. Biometrika **107**, 331–346 (2020)
13. Maistre, S., Patilea, V.: Testing for the significance of functional covariates. Journal of Multivariate Analysis **179**, 104648 (2020)
14. Patilea, V., Sánchez-Sellero, C., Saumard, M.: Testing the predictor effect on a functional response. Journal of the American Statistical Association **111**(516), 1684–1695 (2016)
15. Patilea, V., Sánchez-Sellero, C.: Testing for lack-of-fit in functional regression models against general alternatives. Journal of Statistical Planning and Inference **209**, 229–251 (2020)
16. Shao, X.: The dependent wild bootstrap. Journal of the American Statistical Association **107**, 218–235 (2010)
17. Stinchcombe, M., White, H.: Consistent specification testing with nuisance parameters present only under the alternative. Econometric Theory **14**(3), 295–325 (1998)
18. Stone, C.J.: Optimal rates of convergence for nonparametric estimators. The Annals of Statistics **8**(6), 1348–1360 (1980)
19. Stute,W.: Nonparametric model checks for regression. The Annals of Statistics **25**, 613–641. (1997)
20. Williamson, B.D., Gilbert, P.B., Simon, N.R., Carone, M.: A general framework for inference on algorithm-agnostic variable importance. Journal of the American Statistical Association **118**(543), 1645–1658 (2022)
21. Zhu, X. and Zhu, L.: Dimension reduction-based significance testing in nonparametric regression. Electronic Journal of Statistics **12**(1), 1468–1506 (2018)
22. Ziegel, J., Ginsbourger, D., Dümbgen, L.: Characteristic kernels on Hilbert spaces, Banach spaces, and on sets of measures. Bernoulli **30**(2), 1441–1457 (2024)

Chapter 23
Gaussian Process Methods for Covariate-Based Intensity Estimation

Patric Dolmeta and Matteo Giordano

Abstract We study nonparametric Bayesian inference for the intensity function of a covariate-driven point process. We extend recent results from the literature, showing that a wide class of Gaussian priors, combined with flexible link functions, achieve minimax optimal posterior contraction rates. Our result includes widespread prior choices such as the popular Matérn processes, with the standard exponential (and sigmoid) link, and implies that the resulting methodologically attractive procedures optimally solve the statistical problem at hand, in the increasing domain asymptotics and under the common assumption in spatial statistics that the covariates are stationary and ergodic.

23.1 Introduction

In the statistical analysis of many point patterns, a central issue is to determine the influence of covariates on the point distribution. We refer to [2], and references therein, for an overview of the problem and applications. The established framework to tackle this scenario is to model the point pattern via a Cox process [4], namely a 'doubly-stochastic' point process with covariate-dependent intensity function, cf. Sect. 23.2.1 below. The resulting statistical problem is then to estimate the intensity from observations of the points and covariates.

There is a vast literature on the parametric approach to this issue. For example the log-Gaussian Cox model [17] is widely used; further see the monograph [6]. Existing nonparametric frequentist approaches largely rely on kernel-type estimators, e.g. [2]. These methods were first shown to be asymptotically consistent in a seminal paper

Patric Dolmeta
Bocconi University, Via Sarfatti, 25 Milan, Italy e-mail: patric.dolmeta@unibocconi.it

Matteo Giordano (✉)
University of Turin, Corso Unione Sovietica 218/bis, Turin, Italy
e-mail: matteo.giordano@unito.it

© The Author(s), under exclusive license to Springer Nature Switzerland AG 2025
G. Aneiros et al. (eds.), *New Trends in Functional Statistics and Related Fields*,
Contributions to Statistics, https://doi.org/10.1007/978-3-031-92383-8_23

by Guan [13], under the typical 'large domain' framework of spatial statistics and the assumption that the covariates are stationary and ergodic.

On the other hand, nonparametric Bayesian intensity estimation has so far been considered almost exclusively in models without covariates; see [17, 1, 3, 14, 7] for methodology and theoretical results, as well as for further references. The first frequentist asymptotic analysis of posterior distributions for covariate-driven point processes was developed in the recent paper by Giordano et al. [10], in an analogous increasing domain framework as the one considered in [13]. Among their results, they showed that truncated Gaussian wavelet priors, combined with suitable positive link functions, achieve minimax-optimal rates of posterior contraction towards the ground truth, in L^1-distance.

In this paper, we build on the investigation of [10], obtaining in the same setting optimal posterior contraction rates for a much wider class of (rescaled) Gaussian priors, including widespread choices such as the popular Matérn processes (e.g. [8, Sect. 11.4.4]). See Theorem 23.3. The proof of our result follows the general program set forth in [10], based on the derivation of preliminary convergence properties in a covariate-dependent metric, cf. (23.7), to be combined with concentration inequalities for integral functional of stationary processes. To obtain the required uniformity over the prior support beyond the specific wavelet structure considered in [10], we then employ techniques from the statistical theory of inverse problems, e.g. [11, 12, 18].

Another novelty in our result is that, compared to [10], we also significantly weaken the assumptions on the link function, which we only require to be locally Lipschitz (and bijective). Among the others, this allows to incorporate into the procedure the standard exponential link, as well as the sigmoid one, which had previously been employed for efficient posterior sampling in non-covariate-based intensity estimation [1]; see also [14]. Thus, while theoretical, our result has also potential interesting methodological implications. We discuss these and other related research question in the concluding discussion in Sect. 23.3.

23.2 Posterior Contraction Rates for Covariate-Based Intensity Estimation

23.2.1 Observation Model

Throughout, we take $\mathbb{R}^D$ ($D \in \mathbb{N}$) to be the ambient space. Let the covariates be given by a d-dimensional ($d \in \mathbb{N}$) random field $Z := (Z(x),\ x \in \mathbb{R}^D)$, and let $(\mathcal{W}_n)_{n\in\mathbb{N}}$ be a sequence of compact sets in $\mathbb{R}^D$ satisfying $\mathcal{W}_n \subseteq \mathcal{W}_{n+1}$ for all n.

For some $n \in \mathbb{N}$, the data consist in an observed realisation of the covariates over the window $\mathcal{W}_n$, denoted by $Z^{(n)} := (Z(x),\ x \in \mathcal{W}_n)$, and of a random point process $N^{(n)}$ on $\mathcal{W}_n$ which, conditionally given $Z^{(n)}$, is of inhomogeneous Poisson type with intensity

$$\lambda_\rho^{(n)}(x) := \rho(Z(x)), \qquad x \in \mathcal{W}_n, \tag{23.1}$$

for some unknown bounded function $\rho : \mathbb{R}^d \to [0, \infty)$. Formally, we may write

$$N^{(n)} \stackrel{d}{=} \{X_1, \dots, X_{N_n}\}, \quad N_n | Z^{(n)} \sim \text{Po}(\Lambda_\rho^{(n)}), \quad X_i | Z^{(n)} \stackrel{\text{iid}}{\sim} \frac{\lambda_\rho^{(n)}(x)dx}{\Lambda_\rho^{(n)}}, \tag{23.2}$$

with $\Lambda_\rho^{(n)} := \int_{\mathcal{W}_n} \lambda_\rho^{(n)}(x)dx$. In other words, $N^{(n)}$ is a Cox process on $\mathcal{W}_n$ directed by the random measure $\lambda_\rho^{(n)}(x)dx$ [4]. The statistical problem is to estimate non-parametrically ρ from data $(Z^{(n)}, N^{(n)})$.

We will denote by $P_\rho^{(n)}$ the joint law of $(N^{(n)}, Z^{(n)})$. By standard theory for Poisson processes, e.g. [15], $P_\rho^{(n)}$ is absolutely continuous with respect to the law $P_1^{(n)}$ corresponding to the homogeneous case, with log-likelihood equal to

$$l_n(\rho) := \log \frac{dP_\rho^{(n)}}{dP_1^{(n)}}(D^{(n)}) = \int_{\mathcal{W}_n} \log(\rho(Z(x)))dN^{(n)}(x) - \int_{\mathcal{W}_n} \rho(Z(x))dx, \tag{23.3}$$

cf. [15, Theorem 1.3]. We will write $E_\rho^{(n)}$ for the expectation with respect to $P_\rho^{(n)}$.

23.2.2 Assumptions on the Covariates

We consider the important case where the covariates are given by a Gaussian process, according to the following assumption. Under the latter, the point process (23.2) is then a non-parametric generalisation of the celebrated log-Gaussian Cox process [17], wherein ρ in (23.1) takes the parametric form $\rho(z) = \exp(\beta^T z)$ for some $\beta \in \mathbb{R}^d$. Gaussian processes are among the most ubiquitously employed models for spatial random fields in statistical applications, e.g. [5, Sect. 2.3].

Condition 23.1 *Let* $\tilde{Z}^{(h)} := (\tilde{Z}^{(h)}(x),\ x \in \mathbb{R}^D)$, *for* $h = 1, \dots, d$, *be independent, almost surely locally bounded, centred and stationary Gaussian processes with integrable covariance functions. Further assume without loss of generality that* $\text{Var}[Z^{(h)}(x)] = 1$ *for all* h *and* x. *Let* Z *be given by*

$$Z(x) := [\phi(\tilde{Z}^{(1)}(x)), \dots, \phi(\tilde{Z}^{(d)}(x))], \qquad x \in \mathbb{R}^D, \tag{23.4}$$

where ϕ *is the standard normal cumulative distribution function.*

Under Condition 23.1, $\tilde{Z}(x) := (\tilde{Z}^{(1)}(x), \dots, \tilde{Z}^{(d)}(x)) \sim N_d(0, 1)$ for all $x \in \mathbb{R}^D$, which amounts to the common practice of standardising the covariates, cf. [13, Sect. 3.2.1]. The component-wise application of ϕ in (23.4) is without loss of generality in view of its invertibility. It can be thought of as a convenient 'pre-processing' step that implies, in particular, that Z takes values in the compact set $[0, 1]^d$. The integrability of the covariances is a mild requirement that is satisfied under the (often realistic, e.g. [5, Sec. 2.3]) assumption that the correlation between covariates at distant locations decays sufficiently fast. This condition implies that Z is ergodic, which, similarly to [13], crucially enables in our analysis a sufficient 'accumulation of information' from distinct points in the observation window with similar covariate values, allowing for consistent inference in the increasing domain asymptotics.

23.2.3 Prior Specification

Gaussian processes are among the most universal models for prior distributions in function spaces, and have been successfully employed in related nonparametric Bayesian intensity estimation problems; see [17, 1, 14] and references therein.

In the present setting, under Condition 23.1, we place on the unknown function $\rho : [0, 1]^d \to [0, \infty)$ in (23.1) an (n-dependent) log-Gaussian prior Π_n, constructed starting from a base distribution that we require to satisfy the following mild regularity condition.

Condition 23.2 *Let $\tilde{\Pi}$ be a centred Gaussian probability measure on the Banach space $C([0, 1]^d)$ that is supported on a linear subspace of $C^1([0, 1]^d)$. Further assume that, for some $\alpha > 0$, the reproducing kernel Hilbert space (RKHS) $\tilde{\mathcal{H}}$ of $\tilde{\Pi}$ is equal to the Sobolev space $H^\alpha([0, 1]^d)$, with RKHS norm satisfying $\|\cdot\|_{\mathcal{H}} \simeq \|\cdot\|_{H^\alpha}$.*

Condition 23.2 is satisfied by a wide variety of Gaussian processes, including by stationary ones with polynomially-tailed spectral measures (e.g. the popular Matérn processes, cf. [8, Sect. 11.4.4]), as well as by series priors defined on bases spanning the scale of Sobolev spaces (e.g. [8, Sect. 11.4.3]).

Given $\tilde{\Pi}$ satisfying Condition 23.2, we then construct the prior Π_n as the law of the random function

$$R_n(z) := \exp\left(n^{-d/(4\alpha+2d)} \tilde{W}(z)\right), \qquad z \in [0, 1]^d, \qquad \tilde{W} \sim \tilde{\Pi}. \tag{23.5}$$

The introduction of the exponential link in (23.5) is motivated by the positivity constraint on ρ. This particular choice was made for concreteness, and we note that any other locally Lipschitz and bijective link function may be used in the result to follow, including for example the sigmoid one used in [1, 14]. The n-dependent rescaling of the base Gaussian prior is a common technique in the statistical theory for inverse problems, e.g. [11, 12, 18]. In the proof, it implies a bound for the norm of the draws from Π, which is key to obtaining certain uniform concentration inequalities, going beyond the specific wavelet structure considered in [10]. Non-asymptotically, the rescaling amounts to a simple adjustment of the covariance function of $\tilde{\Pi}$, and does not require further tuning in practice.

23.2.4 Main Result

In our main result, we characterise the speed of asymptotic concentration of the posterior distributions $\Pi_n(\cdot|N^{(n)}, Z^{(n)})$ arising from the log-Gaussian priors (23.5), under the frequentist assumption that the data $(N^{(n)}, Z^{(n)})$ have been generated by some fixed ground truth ρ_0. We work in the (widely adopted in spatial statistics) increasing domain asymptotics, namely $\mathrm{vol}(\mathcal{W}_n) \to \infty$ as $n \to \infty$, which reflects the common situation where a single realisation of the points and covariates is observed over a large window. Similarly to [13], our proof requires $\mathcal{W}_n$ to grow uniformly in all directions, and for concreteness we restrict to square domains

$$\mathcal{W}_n = \left[-\frac{1}{2}n^{1/D}, \frac{1}{2}n^{1/D}\right]^D, \tag{23.6}$$

whence vol $= n$. Extensions to other domains containing the above sets is straightforward.

Theorem 23.3. *Let $\rho_0 = \exp \circ w_0$, for some $w_0 \in C^\beta([0,1]^d) \cap H^\beta([0,1]^d)$ and some $\beta > \min(1, d/2)$. Consider data $(N^{(n)}, Z^{(n)}) \sim P^{(n)}_{\rho_0}$ arising from the observation model (23.2) with $\rho = \rho_0$, $\mathcal{W}_n$ as in (23.6) and with Z a random field satisfying Condtion 23.1. Let the prior Π_n be as in (23.5), with $\tilde{\Pi}$ satisfying Condition 23.2 with $\alpha = \beta$. Then, for $M > 0$ large enough, as $n \to \infty$,*

$$E^{(n)}_{\rho_0}\left[\Pi_n\left(\rho : \|\rho - \rho_0\|_{L^1} > Mn^{-\beta/(2\beta+d)} \Big| N^{(n)}, Z^{(n)}\right)\right] \to 0.$$

Theorem 23.3 asserts that the posterior distribution $\Pi_n(\cdot|N^{(n)}, Z^{(n)})$ asymptotically concentrates its mass over small neighbourhoods of the ground truth, whose L^1-radius shrinks at rate $n^{-\beta/(2\beta+d)}$. The latter is known to be minimax for β-smooth intensities, e.g. [15], implying the optimality of Theorem 23.3.

The proof of the result, presented in Sect. 23.4, is based on recent work by Giordano et al. [10], who developed a general program to derive posterior contraction rates in covariate-based intensity estimation. In particular, the argument also implies that $\Pi(\cdot|N^{(n)}, Z^{(n)})$ concentrates at the same speed also in the 'empirical' (i.e. covariate-dependent) L^1-metric

$$\frac{1}{n}\|\lambda^{(n)}_\rho - \lambda^{(n)}_{\rho_0}\|_{L^1(\mathcal{W}_n)} = \frac{1}{|\mathcal{W}_n|}\int_{\mathcal{W}_n} |\rho(Z(x)) - \rho_0(Z(x))| dx, \tag{23.7}$$

cf. (23.8) below. This implies that the proposed nonparametric Bayesian methodology also optimally solves the 'prediction' problem of inferring the overall spatial intensity function $\lambda^{(n)}_\rho$ in (23.1).

23.3 Concluding Discussion

In this article, we have considered Gaussian process methods for estimating the intensity function of a covariate-driven point process. In a realistic increasing domain framework with ergodic covariates, our main result, Theorem 23.3, shows that the posterior distributions arising from a large class of (rescaled) Gaussian priors concentrates at optimal rate around the ground truth. This is based on recent work by [10], and extends a result obtained in the latter reference for the specific case of truncated Gaussian wavelet priors.

We conclude mentioning some related interesting research questions. Firstly, Theorem 23.3 is non-adaptive in that the prior specification under which optimal rates are obtained requires knowledge of the regularity β of the ground truth. Extensions to adaptive version of the rescaled priors (23.5) in the present setting is of primary interest, but is known to be an involved issue in the broader literature, cf. [11, 12].

Implementation of posterior inference with the considered Gaussian priors (as well as the other priors studied in [10]) is also of great importance. In ongoing work, we are exploring extensions to the problem at hand of the existing efficient posterior sampling methodologies for non-covariate-based Bayesian intensity estimation [1].

Lastly, beyond the large domain framework considered in the present article, 'infill asymptotics' are also of interest in spatial statistics. For covariate-based intensity estimation, this setting can be formulated as having access to repeated (and possibly independent) realisations of the covariates and the points over a fixed observation window. We refer to [10] for possible extensions of the proof techniques employed here to such situations. This is beyond the scope of the present article, but represents an interesting avenue for future research.

23.4 Proof

We employ techniques from [10], to which we refer for the necessary background. We begin proving posterior contraction in the empirical distance (23.7). This is then combined with an exponential concentration inequality for $n^{-1}\|\lambda_\rho^{(n)} - \lambda_{\rho_0}^{(n)}\|_{L^1(\mathcal{W}_n)}$ around its 'ergodic average' $\|\rho - \rho_0\|_{L^1}$, which allows to extend the obtained rates to the latter metric, concluding the proof.

Set $\varepsilon_n := n^{-\beta/(2\beta+d)}$. For $\tilde{\Pi}$ satisfying Condition 23.2, let $\tilde{W} \sim \tilde{\Pi}$ and denote by $\tilde{\Pi}_n$ the law of the random function $(\sqrt{n}\varepsilon_n)^{-1}\tilde{W} = n^{-d/(4\alpha+2d)}\tilde{W}$. Using standard techniques from the posterior contraction rate theory for (rescaled) Gaussian priors, e.g. [11, 12, 18], for all $K_1 > 0$ there exist a sufficiently large constant $M_1 > 0$ and a set

$$\mathcal{F}_n \subset \left\{ f \in C^1([0,1]^d) : \|f\|_{C^1} \le M_1 \right\}$$

such that, for all n large enough,

$$\tilde{\Pi}_n(\mathcal{F}_n^c) \le e^{-K_1 n \varepsilon_n^2}; \qquad \log N(\varepsilon_n; \mathcal{F}_n, \|\cdot\|_{L^\infty}) \lesssim n\varepsilon_n^2.$$

See e.g. Lemma 5 in [12] and its proof. Set $\mathcal{R}_n := \{\exp\circ f,\ f \in \mathcal{F}_n\}$. Then, by construction of Π_n, cf. (23.5), the first inequality in the last display implies that $\Pi_n(\mathcal{R}_n^c) \le e^{-K_1 n\varepsilon_n^2}$. Further, since $\mathcal{F}_n$ is contained in a sup-norm ball, $\|\exp\circ f_1 - \exp\circ f_2\|_{L^\infty} \lesssim \|f_1 - f_2\|_{L^\infty}$ for all $f_1, f_2 \in \mathcal{F}_n$. It follows that $\log N(\varepsilon_n; \mathcal{R}_n, \|\cdot\|_{L^\infty}) \lesssim \log N(\varepsilon_n; \mathcal{F}_n, \|\cdot\|_{L^\infty}) \lesssim n\varepsilon_n^2$. Lastly, using Corollary 2.6.18 of [9], since $w_0 \in H^\beta([0,1])^d = \mathcal{H}$ by assumption, we have

$$\tilde{\Pi}_n(w : \|w - w_0\|_{L^\infty} \le \varepsilon_n) \ge e^{-\frac{1}{2}\|w_0\|_{\mathcal{H}}^2 n\varepsilon_n^2}\tilde{\Pi}(w : \|w\|_\infty \le \sqrt{n}\varepsilon_n^2),$$

which, by the metric entropy estimate in Theorem 4.3.36 of [9] and the centred small ball inequality in Theorem 1.2 of [16], is lower bounded by $e^{-\frac{1}{2}\|w_0\|_{\mathcal{H}}^2 n\varepsilon_n^2} e^{-c_1 n\varepsilon_n^2} = e^{-c_2 n\varepsilon_n^2}$ for some $c_1, c_2 > 0$. Arguing as in the proof of Theorem 3.2 in [10], this implies that for all sufficiently large $M_2 > 0$, as $n \to \infty$,

$$E_{\rho_0}^{(n)}\left[\Pi(\mathcal{S}_n | N^{(n)}, Z^{(n)})\right] \to 1, \tag{23.8}$$

where $\mathcal{S}_n := \left\{\rho \in \mathcal{R}_n : n^{-1}\|\lambda_\rho^{(n)} - \lambda_{\rho_0}^{(n)}\|_{L^1(\mathcal{W}_n)} \le M_2\varepsilon_n\right\}$. Now for $M > 0$ to be chosen below, set $\mathcal{T}_n := \left\{\rho \in C^1([0,1]^d) : \|\rho - \rho_0\|_{L^1} \le M\varepsilon_n\right\}$. The proof is then concluded by showing that $E_{\rho_0}^{(n)}[\Pi(\mathcal{T}_n^c | N^{(n)}, Z^{(n)})] \to 0$ as $n \to \infty$. To do so, note that by (23.8),

$$\begin{aligned}\Pi(\mathcal{T}_n^c | N^{(n)}, Z^{(n)}) &= \Pi(\mathcal{T}_n^c \cap \mathcal{S}_n | N^{(n)}, Z^{(n)}) + o_{P_{\rho_0}^{(n)}}(1)\\ &= \frac{\int_{\mathcal{T}_n^c \cap \mathcal{S}_n} e^{l_n(\rho) - l_n(\rho_0)} d\Pi(\rho)}{\int_{C([0,1]^d)} e^{l_n(\rho) - l_n(\rho_0)} d\Pi(\rho)} + o_{P_{\rho_0}^{(n)}}(1).\end{aligned}$$

Denote by D_n the denominator in the previous display. The proof of Theorem 3.2 in [10] shows that $P_{\rho_0}^{(n)}(D_n \le e^{-K_1 n\varepsilon_n^2}) = o(1)$ as $n \to \infty$ for some constant $K_1 > 0$, so that by Fubini's theorem,

$$\begin{aligned}&E_{\rho_0}^{(n)}\left[\Pi(\mathcal{T}_n^c | N^{(n)}, Z^{(n)})\right]\\ &\le e^{K_1 n\varepsilon_n^2} \int_{\mathcal{T}_n^c \cap \{\rho \in \mathcal{R}_n : \|\rho\|_{C^1} \le M_1\}} \Pr(\|\lambda_\rho^{(n)} - \lambda_{\rho_0}^{(n)}\|_{L^1(\mathcal{W}_n)} \le M_2 n\varepsilon_n) d\Pi(\rho) + o(1).\end{aligned}$$

Fix any $\rho \in \mathcal{T}_n^c \cap \{\rho \in \mathcal{R}_n : \|\rho\|_{C^1} \le M_1\}$. Then, if $\|\lambda_\rho^{(n)} - \lambda_{\rho_0}^{(n)}\|_{L^1(\mathcal{W}_n)} \le M_2 n\varepsilon_n$, necessarily

$$\Delta^{(n)}(\rho) := \|\rho - \rho_0\|_{L^1} - \frac{1}{n}\|\lambda_\rho^{(n)} - \lambda_{\rho_0}^{(n)}\|_{L^1(\mathcal{W}_n)} > (M - M_2)\varepsilon_n \ge \frac{M_2}{2}\varepsilon_n$$

upon taking $M > 2M_2$. For all such M, it follows that the expectation of interest is upper bounded by

$$e^{K_1 n\varepsilon_n^2}\int_{\mathcal{T}_n^c\cap\{\rho\in\mathcal{R}_n:\|\rho\|_{C^1}\le M_1\}} \Pr(\Delta^{(n)}(\rho) > M_2\varepsilon_n/2)d\Pi(\rho) + o(1).$$

The concentration inequality in Proposition D.1 of [10], applied with $f := |\rho - \rho_0| - n^{-1}\|\lambda_\rho^{(n)} - \lambda_{\rho_0}^{(n)}\|_{L^1(\mathcal{W}_n)}$, for $\rho \in C^1([0,1]^d)$, whose (weak) gradient satisfies $\|\nabla f\|_{L^\infty} \le \|\rho\|_{C^1} + \|\rho_0\|_{C^1} \le 2M_1$ now gives that

$$\sup_{\rho\in\mathcal{T}_n^c\cap\{\rho\in\mathcal{R}_n:\|\rho\|_{C^1}\le M_1\}} \Pr(\Delta^{(n)}(\rho) > M_2\varepsilon_n/2) \le e^{-K_2(M_2)^2 n\varepsilon_n^2}$$

for some $K_2 > 0$. The claim follows taking $M_2 > 0$ large enough and combining the last two displays. □

Acknowledgements M.G. has been partially supported by MUR, PRIN project 2022CLTYP4.

References

1. Adams, R. P., Murray, I., and MacKay, D. J. C.: Tractable nonparametric bayesian inference in poisson processes with gaussian process intensities. In Proceedings of the 26th Annual International Conference on Machine Learning (New York, NY, USA, 2009), ICML '09, Association for Computing Machinery, pp. 9–16.
2. Baddeley, A., Chang, Y.-M., Song, Y., and Turner, R.: Nonparametric estimation of the dependence of a spatial point process on spatial covariates. Stat. Interface 5, **2**, 221–236 (2012)
3. Belitser, E., Serra, P., and van Zanten, H.: Rate-optimal Bayesian intensity smoothing for inhomogeneous Poisson processes. J. Statist. Plann. Inference 166, 24–35 (2015)
4. Cox, D. R.: Some statistical methods connected with series of events. J. Roy. Statist. Soc. Ser. B **17** 129–157; discussion, 157–164 (1955)
5. Cressie, N.: Statistics for spatial data. John Wiley & Sons (2015)
6. Diggle, P. J.: Statistical analysis of spatial and spatio-temporal point patterns, third ed., vol. 128 of Monographs on Statistics and Applied Probability. CRC Press, Boca Raton, FL (2014)
7. Donnet, S., Rivoirard, V., Rousseau, J., and Scricciolo, C.: Posterior concentration rates for counting processes with Aalen multiplicative intensities. Bayesian Anal. **12**(1) 53–87, (2017)
8. Ghosal, S., and van der Vaart, A. W.: Fundamentals of Nonparametric Bayesian Inference. Cambridge University Press, New York, (2017)
9. Giné, E., and Nickl, R.: Mathematical foundations of infinite-dimensional statistical models. Cambridge University Press, New York (2016)
10. Giordano, M., Kirichenko, A., and Rousseau, J.: Nonparametric bayesian intensity estimation for covariate-driven inhomogeneous point processes. arXiv preprint 2312.14073 (2023)
11. Giordano, M., and Nickl, R.: Consistency of Bayesian inference with Gaussian process priors in an elliptic inverse problem. Inverse Problems **36**(8), 085001, 35 (2020)
12. Giordano, M., and Ray, K.: Nonparametric Bayesian inference for reversible multidimensional diffusions. Ann. Statist. **50**(5), 2872–2898 (2022)
13. Guan, Y.: On consistent nonparametric intensity estimation for inhomogeneous spatial point processes. J. Amer. Statist. Assoc. **103**(483), 1238–1247 (2008)
14. Kirichenko, A., and van Zanten, H.: Optimality of Poisson processes intensity learning with Gaussian processes. J. Mach. Learn. Res. **16**, 2909–2919 (2015)
15. Kutoyants, Y. A.: Statistical inference for spatial Poisson processes, Lecture Notes in Statistics. Springer-Verlag, New York, 1998.
16. Li, W. V., and Linde, W.: Approximation, metric entropy and small ball estimates for Gaussian measures. Ann. Probab. **27**(3), 1556–1578 (1999)
17. Møller, J., Syversveen, A. R., and Waagepetersen, R. P.: Log Gaussian Cox processes. Scand. J. Statist. **25**(3), 451–482 (1998)
18. Nickl, R.: Bayesian non-linear statistical inverse problems. EMS press, (2023)

Chapter 24
Spectral Analysis of Multidimensional Thermal Fields

Mélanie Dreina, Sylvie Viguier-Pla and Stéphane Abide

Abstract The study of multidimensional data along time often needs data reduction methods to reduce dimension, and to study some particular phenomena. In the context of fluid mechanics, we propose to compare a version of Functional Principal Components Analysis (FPCA), named Proper Orthogonal Decomposition (POD), with two methods based on the spectral decomposition of data Fourier transform, the Spectral Proper Orthogonal Decomposition (SPOD) and Principal Components Analysis in the Frequency Domain (PCAFD). In this context, both POD and SPOD have been proposed, while PCAFD has been newly applied to this domain. Thus, we provide a discussion on the contribution of PCAFD to deal with multiscale physics.

24.1 Introduction

Simulation of multidimensional data, such as in fluid mechanics, or observation of such data, such as ocean temperature, leads to the production of a large amount of information. Therefore, dimension reduction is of major importance to be able to carry out fine analyses of the underlying physical phenomena. Naturally, Principal Components Analysis (PCA) is the basic method, which has been declined by several approaches since the last decade. The pioneer work concerning fluid mechanics has

Mélanie Dreina (✉)
LAMPS, Université de Perpignan, 52 avenue Paul Alduy 66000 Perpignan Cedex 9, France
e-mail: melanie.dreina@univ-perp.fr

Sylvie Viguier-Pla
LAMPS, Université de Perpignan, 52 avenue Paul Alduy 66000 Perpignan Cedex 9, France & IMT, Université Paul Sabatier, 118 route de Narbonne 66860 Toulouse Cedex 4, France
e-mail: viguier@univ-perp.fr

Stéphane Abide
LJAD, Université Côte d'Azur, 28 avenue Valrose 06108 Nice Cedex 02, France
e-mail: Stephane.ABIDE@univ-cotedazur.fr

© The Author(s), under exclusive license to Springer Nature Switzerland AG 2025

G. Aneiros et al. (eds.), *New Trends in Functional Statistics and Related Fields*,
Contributions to Statistics, https://doi.org/10.1007/978-3-031-92383-8_24

been introduced by Lumley (1970), with Proper Orthogonal Decomposition (POD), which can be apparented with Functional Principal Components Analysis (FPCA). The Spectral POD (SPOD), also proposed by Lumley (2007), which performs the Fourier transform on data before POD has been brought to the fore by Towne et al. (2018), and put into practice by various authors, such as Schmidt and Colonius (2020), Xiao et al. (2021) or Nekkanti and Schmidt (2021). More recently the Dynamic Mode Decomposition (DMD) proposed by Schmid (2010), is performed by a time-domain PCA, with time-dependent assumptions. We will focus our work on PCA in the frequency domain, introduced by Brillinger (1981) generalised to a wide context by Boudou (1995), performed for a periodic flow by Boudou et al. (2004), and generalised for a cyclostationary flow by Boudou and Viguier-Pla (2020). This technique of using reduction methods in the frequency domain is knowing a large interest, as it is better appropriate for time-dependent series than working in the time domain (see also Hörmann et al, 2015).

In this study, we propose to compare results from POD, SPOD and an improved method of PCA in the frequency domain, that we name PCAFD, which is not restrictive on the structure of the multidimensional signal spectrum. In this work, we first present the data of interest. Secondly, we present the three compared methods, POD, SPOD and PCAFD. In the third part, we apply the three methods to our simulated data. Finally, for each method, we show the phenomenon it reveals at each step and we compare the qualities of reconstruction.

24.2 Description of the Data

The comparison of POD, SPOD and PCAFD methods is carried out on the spatio-temporal series based on a natural convection flow temperature field (Sergent et al., 2013, Trias et al., 2007). In particular, we simulate the thermal coupling between a fluid and a solid wall, imposing continuity of the temperature field at the fluid/solid interface. The data considered here is therefore a sampling of simulations $\{u(t,x_1,x_2);(t,x_1,x_2) \in \mathbb{R}_+ \times [a;b] \times [c;d]\}$, where t is time index, x_1 and x_2 are coordinates of the point where the random process u is determined by Direct Numerical Simulation (DNS). For our application, we study 100 snapshots of 24×22 points each. A snapshot, i.e. an image at a time index t, and time series for three points over time are given in Fig. 24.1. Details of the DNS solver used in this work are presented in Abide et al. (2017, 2018).

24.3 The Methods of Dimension Reduction

Proper Orthogonal Decomposition. Let $\{u(t,x)\}$ be a stochastic process defined on $\mathbb{R}\times\mathbb{R}^p$, as for example a random p−dimensional field observed along the time. The POD is the search of a deterministic function $\phi(x)$ that best approximates the stochastic function in average. In practice, it consists of considering a grid of p points $(x_1,\ldots,x_p)$, and measures of u on each point at times $t_1,\ldots,t_n$. The method

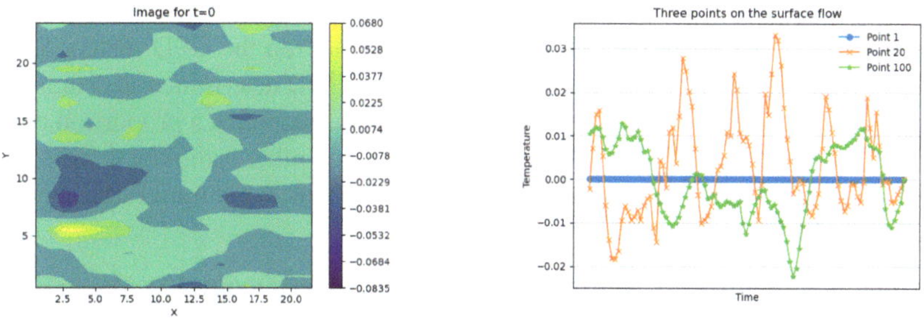

Fig. 24.1: First snapshot and time series of the temperature fluctuations

is implemented by PCA of the matrix $U = (u(t_i, x_j))_{i=1,\dots,n; j=1,\dots,p}$. Briefly, PCA is processed by a decomposition of $R = U_r^T U_r / n$, where U_r is the column-wise centered and reduced matrix U. So we have $R = B \Lambda B^T$, where B is composed of the eigenvectors and Λ of the eigenvalues of R. Each principal component, i.e. each column of $C = U_r B$, of such PCA is named a mode.

Spectral Proper Orthogonal Decomposition. SPOD is designed for statistically stationary flows, from which it is aimed to extract coherent structures. The eigenvalues of the Cross-Spectral Density (CSD) matrix at each frequency represent the energy associated with each mode at a given frequency. We perform SPOD using the open access Python script proposed and successfully applied to several examples by He et al. (2021) [https://github.com/HexFluid/spod_python (accessed December 2024)]. We describe here the main steps of this method. First it builds a matrix with the spatio-temporal data. Let the vector $u_k \in \mathbb{R}^p$ be the kth time snapshot after subtracting the time-averaged data. The chronologically sorted spatio-temporal data matrix is:

$$Q = U^T = [u_1, u_2, \dots, u_n] \in \mathbb{R}^{p \times n},$$

where n is the number of snapshots. Secondly, the data matrix is splitted into N_b blocks using the Welch periodogram method and the discrete Fourier transform is applied to each block to pass into the frequency domain with N_f frequencies. At this stage, to prevent loss of precision due to spectral leakage, each data block is processed with a Hamming window and then overlapped with neighbouring blocks. The jth block matrix is given by

$$\hat{Q}^{(j)} = \left[\hat{u}_1^{(j)}, \hat{u}_2^{(j)}, \dots, \hat{u}_{N_f}^{(j)}\right] \in \mathbb{C}^{p \times N_f}.$$

Then, according to the frequency, the matrices are reshaped so that the matrix for the kth frequency is

$$\hat{Q}_k = \left[\hat{q}_k^{(1)}, \hat{q}_k^{(2)}, \dots, \hat{q}_k^{(N_b)}\right] \in \mathbb{C}^{p \times N_b}.$$

The weighted cross-spectral density (CSD) matrix for the kth frequency, denoted as S_k is obtained as follows:

$$S_k = \frac{1}{N_b} W^{1/2} \hat{Q}_k^* \hat{Q}_k W^{1/2} \in \mathbb{C}^{N_b \times N_b},$$

where W is the weight matrix for scaling the various flow variables (we chose $W = I$). Finally, the eigen-decomposition is performed on the CSD matrix S_k for each frequency. Similarly to other versions of POD, SPOD determines an orthogonal basis for the data, meaning that a subset of these modes captures a proportion of the total energy (variance) within the data compared to any other orthogonal basis. The function used for the reconstruction is based on Nekkanti & Schmidt (2021), and is available in the Python script above mentioned.

Principal Components Analysis in the Frequency Domain. Let $(X_n)_{n\in\mathbb{Z}}$ be a stationary p−dimensional random time series. The PCAFD of $(X_n)_{n\in\mathbb{Z}}$ is the search of a q−dimensional series ($q < p$) $(X'_n)_{n\in\mathbb{Z}}$, stationarily correlated with $(X_n)_{n\in\mathbb{Z}}$, and a filtering $X''_n = \sum_m A_m X'_{n-m}$, where the A_m's are linear applications from $\mathbb{C}^q$ in $\mathbb{C}^p$, such that $\|X_n - X''_n\|$ (which equals $\|X_0 - X''_0\|$, as $(X_n)_{n\in\mathbb{Z}}$ and $(X''_n)_{n\in\mathbb{Z}}$ are stationary and stationarily correlated) is as small as possible.

The X_n's of the stationary series $(X_n)_{n\in\mathbb{Z}}$, are p−dimensional random vectors: $X_n = \left(x_n^1, \ldots, x_n^p\right)^T$. The stationarity is assumed in a broad sense, that is $\mathbb{E}(X_n^T \overline{X_m}) = \mathbb{E}({X_{n-m}}^T \overline{X_0})$ for any pair (n, m) of elements from $\mathbb{Z}$. It is equivalent with the usual second order stationarity of each of its components $(x_n^i)_{n\in\mathbb{Z}}$ and with the pairwise correlated stationarity: $\mathbb{E}(x_n^i \overline{x}_m^j) = \mathbb{E}(x_{n-m}^i \overline{x}_0^j)$ for any (n, m, i, j) from $\mathbb{Z} \times \mathbb{Z} \times \{1, ..., p\} \times \{1, ..., p\}$.

We assume that the conditions are satisfied for the existence of the spectral density,

$$(2\pi)^{-1} \sum_{n\in\mathbb{Z}} e^{-i.n} \mathbb{E}\, X_n^T \overline{X_0}.$$

Theoretically, the PCAFD needs to process the PCA of $(2\pi)^{-1} \sum_{n\in\mathbb{Z}} e^{-i\lambda n} \mathbb{E}\, X_n^T \overline{X_0}$, for each λ from $[-\pi, \pi[$, this means an infinity of PCA's. We overcome this difficulty by a discretization of the spectrum $[-\pi, \pi[$. More precisely, if k is an integer, we consider the measurable application from $[-\pi, \pi[$ into itself:

$$f_k = \sum_{l=-k}^{k-1} \frac{\pi l}{k} 1_{B_{lk}},$$

where $B_{-k,k} = \{-\pi\}$, $B_{lk} =]\frac{\pi l}{k} - \frac{\pi}{k}, \frac{\pi l}{k}]$ for $l = -k+1, \ldots, -1$, $B_{0k} =]-\frac{\pi}{k}, \frac{\pi}{k}[$, and $B_{lk} = [\frac{\pi l}{k}, \frac{\pi l}{k} + \frac{\pi}{k}[$ for $l = 1, \ldots, k-1$. The PCAFD can be approximated by a spectral decomposition of each spectral density M_{lk} defined on B_{lk}; $l = -k + 1, \ldots, k-1$. The matrices M_{lk} can be estimated by

$$(2\pi m)^{-1} \sum_{u=1}^{m} \sum_{v=1}^{m} (\int_{B_{lk}} e^{i\lambda(u-v)}\, d\lambda) X_v^T \overline{X_u}.$$

Let $(X'_n)_{n\in\mathbb{Z}}$ be the q−dimensional solution of the q−order PCAFD of $(X_n)_{n\in\mathbb{Z}}$. This series is of the form $X'_n = \sum_{m\in\mathbb{Z}} C'_m X_{n-m}$. It can be approximated via the discretization of the spectrum, by the series

$$X'^k_n = \sum_{m\in\mathbb{Z}} C'_{m,k} X_{n-m},$$

where

$$C'_{m,k} = (2\pi)^{-1} \sum_{l=-k+1}^{k-1} (\int_{B_{lk}} e^{i\lambda m}\, d\lambda) \sum_{j=1}^{q} F_j{}^T \overline{A_{jlk}},$$

F_j being the jth vector of the canonical basis of $\mathbb{C}^q$, and A_{jlk} being the jth unitary eigenvector of M_{lk}.

The reconstructed series is then $(X''^k_n)_{n\in\mathbb{Z}}$, which can be writen

$$X''^k_n = \sum_{m\in\mathbb{Z}} C''_{m,k} X'^k_{n-m} = \sum_{m\in\mathbb{Z}} D_{m,k} X_{n-m},$$

where

$$C''_{m,k} = (2\pi)^{-1} \sum_{l=-k+1}^{k-1} \left(\int_{B_{lk}} e^{i\lambda m}\, d\lambda\right) \sum_{j=1}^{q} A_{jlk}{}^{T} \overline{F_j},$$

and

$$D_{m,k} = (2\pi)^{-1} \sum_{l=-k+1}^{k-1} \left(\int_{B_{lk}} e^{i\lambda m}\, d\lambda\right) \sum_{j=1}^{q} A_{jlk}{}^{T} \overline{A_{jlk}}.$$

Of course, the greater k is, the closest the approximated PCAFD is to the theoretical PCAFD defined above.

We can examine the norms of the $C'_{m,k}$, which are high when the autocorrelation of order m is high, what happens, for example, when the series is periodic of period m. We can also compare the series before and after PCAFD, for various dimensions q of the reconstruction.

24.4 Results and Discussion

For our data, $n = 100$ and $p = 22 \times 24$, so the initial dimension is $min(n, p) = 100$.

Analysis with POD. We examine the modes of this analysis, which match with the principal components in usual PCA. In Fig. 24.2, the reconstruction is very slightly improved using 2 dimensions. At least, the essential of the variations is returned, that is those for median temperature.

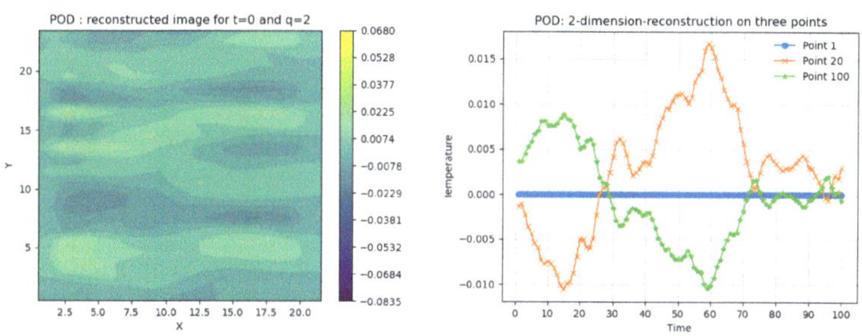

Fig. 24.2: POD. Reconstruction of image at $t = 0$ and of three points variations with two modes

Analysis with SPOD. In Fig. 24.3, we can see that the most reconstructed variations are those for extreme temperatures at $t = 0$. The variations of points 20 and 100 are slightly more complex than those from POD, but the same variations are retrieved first.

Analysis with PCAFD. Fig. 24.4 shows the first snapshot and trajectories of three points for a two-dimensional reconstruction. By comparison with the initial first snapshot, we can recognise the main variations of the flow. As for the trajectories, PCAFD retrieves more complexity than the previous methods. It takes into account more frequencies in the first modes.

Now we can focus on the possible periodicities in the signal, by looking at the norms of the coefficients of the reconstruction of (X_n). In Fig. 24.5, the norms of the

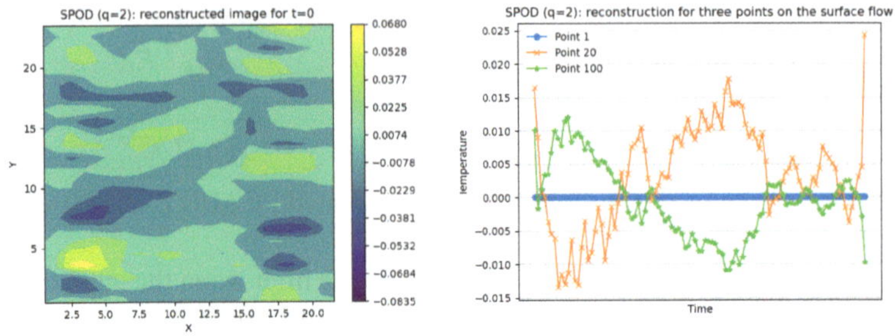

Fig. 24.3: SPOD. Reconstruction of image at $t = 0$ and of three points variations with two modes

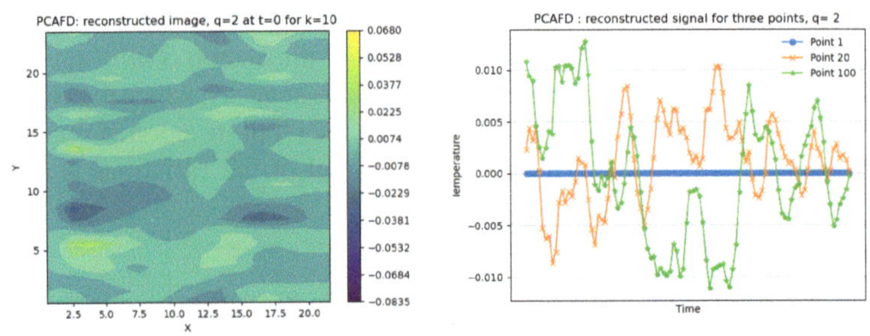

Fig. 24.4: PCAFD, $k = 10$. Reconstruction on dimension 2

C'_{lk} show some phenomena with a period around 26. So we guess that any frequency of the form $p\frac{2\pi}{26}$ in $[-\pi; \pi[$ is of high energy. The frequency $\frac{4\pi}{26}$ falls in the interval B_{lk}, where $l = 1$ for $k = 10$. In the second diagram of Fig. 24.5, we show the three selected trajectories for frequencies in $B_{1,10}$ and $q = 2$. Note that this period depends on the time discretisation. This technique offers a way to analyze what is specific to a given frequency, when the bandwidth is small enough.

24.5 Comparisons

One way to assess the efficiency of decomposition method relies on its ability to reconstruct the initial signal with few modes. To this end, we evaluate the error in reconstruction with respect to the mode numbers. Fig. 24.6 presents the relative error computed for the three methods POD, SPOD and PCAFD. One can note that PCAFD is able to reconstruct the data with fewer modes than the other methods. In this way, PCAFD overcomes the POD and SPOD in its ability to retrieve data. Moreover, the control parameter k greatly improves the decomposition efficiency. When k is small, the number of subdivisions of the frequency spectrum is small, so

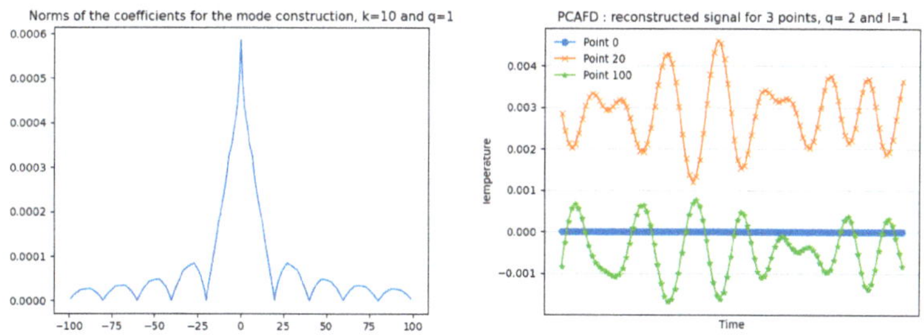

Fig. 24.5: PCAFD, $k = 10$. Norms of the C'_{lk} and reconstruction for 3 points on $B_{1,10}$

few frequencies are taken into account. The higher k is, the higher is the number of considered potential frequencies. The method PCAFD has been performed with $k = 10$ and $k = 20$. This comparison of two values of k illustrates the fact that the higher k is, the smaller the error is, for a fixed value of q. As POD and SPOD present similar errors in the first dimensions, SPOD tends to be better with dimension getting higher. PCAFD has more little errors, and the quality of reconstruction is almost perfect as soon as the dimension reaches $q = 10$ when $k = 10$, and $q = 6$ when $k = 20$.

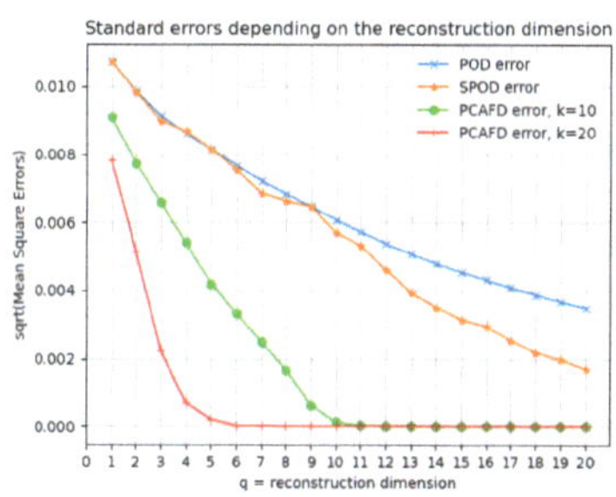

Fig. 24.6: Standard deviation of errors of reconstruction for dimensions 1 to 20

24.6 Conclusion

The PCAFD sounds interesting for several purposes in fluid mechanics. The summary needs few modes to give good quality of reconstruction compared to POD and SPOD. We can analyse the coefficients of the reconstruction for information above the periodic parts of the signal, and we can select part of the spectrum for the extraction of some particular phenomena. Moreover, Boudou and Viguier-Pla (2006) have investigated the conditions where PCA and PCAFD give the same results. This condition is the independence of data from time, and a consequence of this independence is that PCAFD and POD become equivalent. The differences between POD, SPOD and PCAFD results gives indications of how time dependent the data are. PCAFD is newly compared to SPOD, which is supposed to proceed with the same way of dealing with the frequency domain, on data simulated from fluid mechanics models. However, we must also analyse the computational efficiency of each method, and the ability of these methods to apply to large volumes of data. As PCAFD has got longer execution time, one of the challenges is to adapt its algorithms to this context.

References

1. Abide, S. and Binous, M.S. and Zeghmati, B.: An efficient parallel high-order compact scheme for the 3D incompressible Navier-Stokes equations. International Journal of Computational Fluid Dynamics, **31** 4-5, 214-229 (2017)
2. Abide, S., Viazzo, S. and Raspo, I.: Higher-order compact scheme for high-performance computing of stratified rotating flows. Computers & Fluids, **174**, 300-310 (2018)
3. Boudou, A.: Mise en œuvre de l'analyse en composantes principales d'une série stationnaire multidimensionnelle. Publications de l'Institut de Statistique de l'Université de Paris, XXXIX, fasc.1, 89-104 (1995)
4. Boudou, A., Caumont, O., Viguier-Pla, S.: Principal components analysis in the frequency domain. COMPSTAT 2004-Proceedings in Computational Statistics, Physica, Heidelberg, 729-736 (2004)
5. Boudou, A., Viguier-Pla, S.: Principal components analysis of a Cyclostationary Random Function. IWFOS 2020-Special issue, 43-50. Springer, Heideberg (2020)
6. Brillinger, D.R.: Time Series: Theory and Methods., Springer, New-York (1981)
7. Hörmann, S., Kidziński, L., Hallin, M.: Dynamic Functional Principal Components. Journal of the Royal Statistical Society. Series B: Statistical Methodology **77**(2), 319-348. (2002)
8. He, X., Fang, Z., Rigas, G., Vahdati, M.: Spectral proper orthogonal decomposition of compressor tip leakage flow. Physics of Fluids vol. 33, 105105 (2021)
9. Lumley, J.L.: Stochastic Tools in Turbulence. Courrier Corporation. Initially published in 1970 in Academic Press, New-York (1970, 2007)
10. Nekkanti, A., and Schmidt, O.T.: Frequency–Time Analysis, Low-Rank Reconstruction and Denoising of Turbulent Flows Using SPOD. Journal of Fluid Mechanics, **926**, A26 (2021)
11. Schmidt, O.T., Colonius, T.: Guide to Spectral Proper Orthogonal Decomposition. AIAA Journal, **58**, 3, 1023-1033 (2020)
12. Schmid, P.J.: Dynamic mode decomposition of numerical and experimental data. Journal of Fluid Mechanics, **656**, 5-28 (2010)
13. Sergent, A., Xin, Shihe, Joubert, P., Le Quéré, P., Salat, J., Penot, F.: Resolving the stratification discrepancy of turbulent natural convection in differentially heated air-filled cavities - Part I: Reference solutions using Chebychev spectral methods. International Journal of Heat and Fluid Flow, **39**, 1-14 (2013)
14. Towne, A., Schmidt, O.T., Colonius, T.: Spectral proper orthogonal decomposition and its relationship to dynamic mode decomposition and resolvent analysis. Journal of Fluid Mechanics, **847**, 821-867 (2018)
15. Trias, F.X., Soria, M., Oliva, A., Pérez-Segarra, C.D.: Direct numerical simulations of two- and three-dimensional turbulent natural convection flows in differentially heated cavity of aspect ratio 4. Journal of Fluid Mechanics, **586**, 259-293 (2007)

Chapter 25
Forecasting Dynamic Factor Scores by UNALSE Spectral Density Matrix Estimator

Matteo Farnè and Xuanye Dai

Abstract In this short paper, we propose a new generalized dynamic factor model estimator using the UNALSE spectral density matrix estimator, which is obtained by nuclear norm plus ℓ_1 norm penalization. In this way, the matrix of factor loadings is optimally estimated in the minmax sense, as are the factor scores by both Bartlett's and Thomson's method. This approach makes it possible to counteract the impossibility of obtaining the same estimates with classical dynamic principal component analysis, due to the fact that the estimated idiosyncratic covariance matrix is not invertible. The theoretical background is presented and mathematical results are announced to demonstrate the usefulness and novelty of the new approach. In particular, the optimality properties of the new estimates of factor loadings and scores are highlighted.

25.1 Theoretical Background

FHLR2000 [5] believes that a factor model must have two features. First, it must be dynamic, since business cycle problems are typically dynamic problems. Second, it must take into account the cross-correlation of the idiosyncratic component, since orthogonality is unrealistic for most purposes in high dimensions. Consequently, FHLR2000 proposes a *generalized dynamic factor model* (GDFM), allowing the observed variables to be affected by infinite lagged values of latent factors. In particular, FHLR2000 assumes the observation X_{it} to be decomposed into a sum of two mutually orthogonal (all leads, all lags) unobservable components: the common component χ_{it} and the idiosyncratic component ξ_{it} (which represent the share of

Matteo Farnè (✉)
University of Bologna, Via delle Belle Arti 41, Bologna, Italy
e-mail: matteo.farne@unibo.it

Xuanye Dai
University of Florence, Viale Morgagni 59, Florence, Italy, e-mail: xuanye.dai@unifi.it

© The Author(s), under exclusive license to Springer Nature Switzerland AG 2025

G. Aneiros et al. (eds.), *New Trends in Functional Statistics and Related Fields*,
Contributions to Statistics, https://doi.org/10.1007/978-3-031-92383-8_25

X_{it} explained and not explained by the common factors, respectively). The common component is driven by a finite number q of common dynamic factors, which summarize the information from a large number of variables into a small number of factors. The idiosyncratic component is allowed to show mild cross-correlation, like the approximate factor model proposed by [3], capturing the information not explained by the factors. Thus, the GDFM can be written as

$$X_{it} = \chi_{it} + \xi_{it},$$

where, denoting the $p \times p$ spectral density matrices of the p-dimensional processes $X_t = (X_{1t}, \ldots, X_{pt})^\top$, $\chi_t = (\chi_{1t}, \ldots, \chi_{pt})^\top$, and $\xi_t = (\xi_{1t}, \ldots, \xi_{pt})^\top$ as $\Sigma^X(\theta)$, $\Sigma^\chi(\theta)$, and $\Sigma^\xi(\theta)$ respectively, the assumptions below hold (see FHLR2005 [6]):

- Assumption (A1). The process χ_t is second order stationary, so that for any $i, j \in \mathbb{N}, t \in \mathbb{Z}$ and $k \in \mathbb{Z}$, the covariance $\mathrm{Cov}(\chi_{it}, \chi_{j(t-k)})$ exists, is finite, and does not depend on t.
- Assumption (A2). For any $i \in \mathbb{N}$ and $t \in \mathbb{Z}, \mathbb{E}[\chi_{it}] = 0$ and $\mathrm{Var}(\chi_{it}) > 0$.
- Assumption (A3). For all $p \in \mathbb{N}$, the spectral measure of X_t is absolutely continuous with respect to the Lebesgue measure on $[-\pi, \pi]$, that is, X_t has a spectral density matrix $\Sigma^X(\theta)$ with entries $\sigma_{ij}(\theta)$, for $i, j = 1, \ldots, p$.

The estimation approach of FHLR2000 starts from a consistent estimator of the spectral density matrix based on principal components in the frequency domain, that is, from the q leading dynamic principal components obtained by the smoothed periodogram (see [2], Chap.9) computed at each Fourier frequency, to estimate the low-rank component χ_{it} of the GDFM.

FHLR2005 combines the advantages of [5] and [7] to propose a new estimator of dynamic factor scores. FHLR2005 divides the estimation process in two steps: first, they use the approach of FHLR2000 to obtain the estimates of the auto-covariance matrices of the common and idiosyncratic components. Second, they apply them to obtain a new estimate of the current value and of the projection of future values of the common component.

The smoothed periodogram $\widetilde{\Sigma}^X(\theta_h)$ is defined as

$$\widetilde{\Sigma}^X(\theta_h) = \frac{1}{2\pi} \sum_{k=-\infty}^{\infty} K\left(\frac{k}{M_T}\right) \widetilde{\Gamma}^X(k) e^{-ik\theta_h},$$
$$\theta_h = \frac{h\pi}{M_T}, \quad |h| \leq M_T,$$

where $\widetilde{\Gamma}^X(k) = T^{-1} \sum_{t=1}^{T-|k|} X_t X_{t+k}^\top$ and $K(\cdot)$ is a suitable kernel function with M_T being the associated smoothing parameter. For each Fourier frequency θ_h, the dynamic eigenvalues and eigenvectors are obtained by $\widetilde{\Sigma}^X(\theta_h)$. To transform back into the time domain, the eigenvectors are derived using the inverse Fourier transform. Thus, we have

$$\widehat{p}_j(L) = \frac{1}{2\pi} \sum_{k=-\infty}^{\infty} \left[\int_{-\pi}^{\pi} \widehat{p}_j(\theta) e^{ik\theta} d\theta\right] L^k,$$

where $\widehat{p}_j(\theta) = (\widehat{p}_{1j}(\theta), \ldots, \widehat{p}_{pj}(\theta))$ is the jth row eigenvector. The top q dynamic principal component vector $\widehat{\chi}_t$ is given in FHLR2000 by $\widehat{P}^\top(L)\widehat{P}(L)X_t$. However, this estimate requires a two-sided filter to be computed, which may be problematic at the sample extremes and cannot be used for prediction.

Therefore, in FHLR2005 the estimator of the spectral density matrix of the common component is given by

$$\widetilde{\Sigma}^{\chi}(\theta) = \widehat{\mathbf{P}}_q(\theta)\widehat{\Lambda}_q(\theta)\widehat{\mathbf{P}}_q^\top(\theta) = \widehat{\lambda}_1\widehat{p}_1^\top(\theta)\widehat{p}_1(\theta) + \ldots + \widehat{\lambda}_q\widehat{p}_q^\top(\theta)\widehat{p}_q(\theta),$$

where $\widehat{\Lambda}_q = \text{diag}(\widehat{\lambda}_1 \geq \widehat{\lambda}_2 \geq \cdots \geq \widehat{\lambda}_{\text{q}} > 0)$ is a $q \times q$ diagonal matrix containing the largest q dynamic eigenvalues on the diagonal and $\widehat{\mathbf{P}}_q(\theta) = (\widehat{p}_1(\theta), \ldots, \widehat{p}_q(\theta))$ is a $p \times q$ matrix containing corresponding eigenvectors. Subsequently, the spectral density matrix of the idiosyncratic component can be estimated as $\widetilde{\Sigma}^{\xi}(\theta) = \widetilde{\Sigma}^{X}(\theta) - \widetilde{\Sigma}^{\chi}(\theta)$, and we have

$$\widetilde{\Sigma}^{\xi}(\theta) = \widehat{\mathbf{P}}_{p-q}(\theta)\widehat{\Lambda}_{p-q}(\theta)\widehat{\mathbf{P}}_{p-q}^\top(\theta) = \widehat{\lambda}_{q+1}\widehat{p}_{q+1}^\top(\theta)\widehat{p}_{q+1}(\theta) + \ldots + \widehat{\lambda}_p\widehat{p}_p^\top(\theta)\widehat{p}_p(\theta),$$

where $\widehat{\Lambda}_{p-q} = \text{diag}(\widehat{\lambda}_{\text{q+1}} \geq \widehat{\lambda}_{\text{q+2}} \geq \cdots \geq \widehat{\lambda}_{\text{p}} > 0)$ is a $(p-q) \times (p-q)$ diagonal matrix containing the remaining $p-q$ dynamic eigenvalues on the diagonal and $\widehat{\mathbf{P}}_{p-q}(\theta) = (\widehat{p}_{q+1}(\theta), \ldots, \widehat{p}_p(\theta))$ is a $p \times (p-q)$ matrix containing corresponding eigenvectors.

At this stage, it is possible to estimate the covariance matrices of χ_t and ξ_t by inverse Fourier transform as follows:

$$\widetilde{\Gamma}_k^{\chi} = \int_{-\pi}^{\pi} \widetilde{\Sigma}^{\chi}(\theta)e^{ik\theta}d\theta, \tag{25.1}$$

and

$$\widetilde{\Gamma}_k^{\xi} = \int_{-\pi}^{\pi} \widetilde{\Sigma}^{\xi}(\theta)e^{ik\theta}d\theta. \tag{25.2}$$

Following FHLR2005 , the common component can then be estimated by solving the following generalized PCA problem:

$$\widehat{Z}_j = \arg\max_{a\in\mathbb{R}^p} a^\top\widetilde{\Gamma}_0^{\chi}a \quad \text{s.t.} \quad a^\top\widetilde{\Gamma}_0^{\xi}a = 1 \quad \text{and} \quad a^\top\widetilde{\Gamma}_0^{\xi}\widehat{Z}_m = 0, \quad 1 \leq m \leq j-1,$$

for $j = 1, \ldots, q$, where the generalized eigenvectors associated with the generalized eigenvalues of $\widetilde{\Gamma}_0^{\chi}$ are denoted by $\widehat{Z}_j$, and the first q generalized principal components of X_t are obtained as

$$\widehat{Z}^\top X_t = \left(\widehat{Z}_1^\top X_t \quad \widehat{Z}_2^\top X_t \quad \cdots \quad \widehat{Z}_q^\top X_t\right)^\top.$$

FHLR2005 notes the fact that the covariance between χ_t and $\widehat{Z}^\top X_t$ equals the covariance between χ_t and $\widehat{Z}^\top \chi_t$. Thus, the estimators of the projection can be written in vector form, as

$$\widehat{\chi}_t = \left[\widetilde{\Gamma}_0^{\chi} \widehat{Z} \left(\widehat{Z}^\top \widetilde{\Gamma}_0^{X} \widehat{Z} \right)^{-1} \right] \left[\widehat{Z}^\top X_t \right]. \tag{25.3}$$

Since they use Generalized PCA to forecast the common component and assume $\widehat{Z}^\top \widehat{Z} = I_q$, which is typical of PCA, we can rewrite Equation (25.3) as

$$\begin{aligned}\widehat{\chi}_t &= \left[\widehat{Z} \widetilde{\Lambda}_0^{\chi} \widehat{Z}^\top \widehat{Z} \left(\widehat{Z}^\top \widetilde{\Gamma}_0^{X} \widehat{Z} \right)^{-1} \right] \left[\widehat{Z}^\top X_t \right] \\ &= \left[\widehat{Z} \widetilde{\Lambda}_0^{\chi} \left(\widehat{Z}^\top \widetilde{\Gamma}_0^{X} \widehat{Z} \right)^{-1} \right] \left[\widehat{Z}^\top X_t \right],\end{aligned}$$

where $\widetilde{\Lambda}_0^{\chi}$ is the diagonal matrix containing the eigenvalues of $\widetilde{\Gamma}_0^{\chi}$ explained by each principal component of the common component. The forecast of the common component h steps ahead, for $h \in \mathbb{Z}^+$, is given by

$$\widehat{\chi}_{T+h|T} = \left[\widetilde{\Gamma}_h^{\chi} \widehat{Z} \left(\widehat{Z}^\top \widetilde{\Gamma}_0^{X} \widehat{Z} \right)^{-1} \right] \left[\widehat{Z}^\top X_T \right].$$

25.2 Estimating Factor Loadings and Scores

Let us estimate the factor loadings matrix as $\widehat{B} = \widehat{Z}(\widehat{\Lambda}_0^{\chi})^{1/2}$, which is typical in the statistical literature. Importantly, the idiosyncratic covariance matrix $\widetilde{\Gamma}_0^{\xi}$ is not invertible, as its rank is $p - q$. This prevents the use of Bartlett's or Thomson's method to estimate dynamic factor scores, since for the observation $t, t \in \{1, \ldots, T\}$, they are defined as follows: $\widehat{f}_t^{\mathrm{B}} = (\widehat{B}^\top (\widetilde{\Gamma}_0^{\xi})^{-1} \widehat{B})^{-1} \widehat{B}^\top (\widetilde{\Gamma}_0^{\xi})^{-1} X_t$, $\widehat{f}_t^{\mathrm{T}} = (I_q + \widehat{B}^\top (\widetilde{\Gamma}_0^{\xi})^{-1} \widehat{B})^{-1} \widehat{B}^\top (\widetilde{\Gamma}_0^{\xi})^{-1} X_t$.

To solve these drawbacks, [1] proposes a new high-dimensional spectral density estimator, named ALgebraic Spectral Estimator (ALSE), which is consistent when the dimension p and the sample size T diverge to infinity. The low-rank estimate of ALSE is positive semidefinite, and its rank is the true rank q with high probability. Furthermore, the ALSE residual estimate is positive definite, and its sparsity pattern is the true one with high probability, while FHLR estimate would be positive semidefinite. The ALSE estimate $\widehat{\Sigma}^X(\theta)$ is also positive definite. [1] regularizes $\widetilde{\Sigma}^X(\theta)$ by minimizing a quadratic loss under a nuclear norm plus l_1 norm penalty, getting at each $\theta \in [-\pi, \pi]$ the ALSE as $\widehat{\Sigma}^X(\theta) = \widehat{\Sigma}^{\chi}(\theta) + \widehat{\Sigma}^{\xi}(\theta)$, with

$$\begin{aligned}(\widehat{\Sigma}^{\chi}(\theta), \widehat{\Sigma}^{\xi}(\theta)) =& \arg\min_{\underline{\Sigma}^{\chi}(\theta) \succeq 0,\quad \underline{\Sigma}^{\xi}(\theta) \succ 0,\quad \underline{\Sigma}^{\chi}(\theta) + \underline{\Sigma}^{\xi}(\theta) \succ 0} \\ & \frac{1}{2} \| \widetilde{\Sigma}^X(\theta) - (\underline{\Sigma}^{\chi}(\theta) + \underline{\Sigma}^{\xi}(\theta)) \|_{\mathrm{F}}^2 + \psi \| \underline{\Sigma}^{\chi}(\theta) \|_* + \rho \| \underline{\Sigma}^{\xi}(\theta) \|_1,\end{aligned}$$

where $\widetilde{\Sigma}^X(\theta)$ is the smoothed periodogram estimator, $\underline{\Sigma}^\chi(\theta)$ and $\underline{\Sigma}^\xi(\theta)$ indicate generic $p \times p$ matrices, ψ and ρ are positive threshold parameters, $\|\underline{\Sigma}^\chi(\theta)\|_* = \mathrm{tr}(\underline{\Sigma}^\chi(\theta))$, and $\|\underline{\Sigma}^\xi(\theta)\|_1 = \sum_{i,j}^p |\underline{\Sigma}_{ij}^\xi(\theta)|$.

The following theorem summarizes a number of properties of the ALSE estimator, proved in [1], regarding the common, idiosyncratic, and overall spectral density matrix estimated by the ALSE procedure, as well as the inverses of the latter two, and their minimal eigenvalues, providing a set of simplified conditions which ensure that dynamic factor loadings and scores estimated by the ALSE approach are consistent for all $p \in \mathbb{N}$ as $T \to \infty$.

Theorem 25.1. *For all $p \in \mathbb{N}$, assume that: (i) the q nonzero eigenvalues of $\Sigma^\chi(\theta)$ are such that $p^{-1}\lambda_j(\Sigma^\chi(\theta))$ is finite and has a lower bound strictly greater than zero for all $j = 1, \ldots, q$; (ii) $\|\Sigma^\xi(\theta)\|_{0,v} = \sum_{i,j=1}^p \mathbb{1}(\sigma_{ij}^\xi(\theta) \neq 0)$ is bounded; (iii) $\{\chi_{it}, t \in \mathbb{Z}, i = 1, \ldots, p\}$ and $\{\epsilon_{it}, t \in \mathbb{Z}, i = 1, \ldots, p\}$, are Gaussian processes. Let us define $m_p = \max_{|h| \leq M_T} \|\Sigma^\xi(\theta_h)\|_{0,v}$. Then, there exist positive reals κ_1, κ_2, κ_3, κ_4, and κ_5, which are independent of p and T, such that, as $T \to \infty$, the following hold for all $p \in \mathbb{N}$:*

1. $\mathcal{P}\left(\max_{|h|\leq M_T} \frac{1}{p}\|\widehat{\Sigma}^\chi(\theta_h) - \Sigma^\chi(\theta_h)\|_2 \leq \kappa_1\sqrt{\frac{M_T \log(M_T p)}{T}}\right) \to 1;$
2. $\mathcal{P}\left(\max_{|h|\leq M_T} \|\widehat{\Sigma}^\xi(\theta_h) - \Sigma^\xi(\theta_h)\|_2 \leq \kappa_2 m_p\sqrt{\frac{M_T \log(M_T p)}{T}}\right) \to 1;$
3. $\mathcal{P}\left(\max_{|h|\leq M_T} \frac{1}{p}\|\widehat{\Sigma}^X(\theta_h) - \Sigma^X(\theta_h)\|_2 \leq \kappa_1\sqrt{\frac{M_T \log(M_T p)}{T}}\right) \to 1;$
4. $\mathcal{P}(\min_{|h|\leq M_T} \lambda_p(\widehat{\Sigma}^\xi(\theta_h)) \geq \kappa_3) \to 1;$
5. $\mathcal{P}\left(\max_{|h|\leq M_T} \|(\widehat{\Sigma}^\xi(\theta_h))^{-1} - (\Sigma^\xi(\theta_h))^{-1}\|_2 \leq \kappa_4 m_p\sqrt{\frac{M_T \log(M_T p)}{T}}\right) \to 1;$
6. $\mathcal{P}(\min_{|h|\leq M_T} \lambda_p(\widehat{\Sigma}^X(\theta_h)) \geq \kappa_3) \to 1;$
7. $\mathcal{P}\left(\max_{|h|\leq M_T} \frac{1}{p}\|(\widehat{\Sigma}^X(\theta_h))^{-1} - (\Sigma^X(\theta_h))^{-1}\|_2 \leq \kappa_5\sqrt{\frac{M_T \log(M_T p)}{T}}\right) \to 1.$

At this stage, we note that in high-dimensional scenarios, when p is large and the latent eigenvalues are spiked, the singular value thresholding procedure may lead to the over-shrinkage of the eigenvalues of $\widetilde{\Sigma}^\chi(\theta)$, which can introduce some bias, so that the optimality of ALSE cannot be guaranteed. To address this, [1] proposes the UNshrunk ALgebraic Spectral Estimator (UNALSE), to un-shrink the eigenvalues of the estimated low-rank component. UNALSE is the closest (according to the Frobenius norm) estimator to the smoothed periodogram pre-estimator.

Let us define the spectral decomposition of $\widehat{\Sigma}^\chi(\theta_h)$ as $\widehat{Z}(\theta_h)\widehat{D}(\theta_h)\widehat{Z}^\top(\theta_h)$. For any $\theta_h = \frac{h\pi}{M_T}$, $|h| \leq M_T$, the resulting UNshrunk ALgebraic Spectral Estimator (UNALSE) is defined as:

$$\widehat{\Sigma}^\chi_{\mathrm{UNALSE}}(\theta_h) = \widehat{Z}(\theta_h)\left(\widehat{D}(\theta_h) + \psi I_r\right)\widehat{Z}^\top(\theta_h),$$
$$\mathrm{diag}(\widehat{\Sigma}^\xi_{\mathrm{UNALSE}}(\theta_\mathrm{h})) = \mathrm{diag}(\widehat{\Sigma}(\theta_\mathrm{h})) - \mathrm{diag}(\widehat{\Sigma}^\chi_{\mathrm{UNALSE}}(\theta_\mathrm{h})),$$

$$\text{off-diag}(\widehat{\Sigma}^{\xi}_{\text{UNALSE}}(\theta_h)) = \text{off-diag}(\widehat{\Sigma}^{\xi}(\theta_h)),$$
$$\widehat{\Sigma}^{X}_{\text{UNALSE}}(\theta_h) = \widehat{\Sigma}^{\chi}_{\text{UNALSE}}(\theta_h) + \widehat{\Sigma}^{\xi}_{\text{UNALSE}}(\theta_h).$$

We now focus on the estimation of dynamic factor scores. The estimation procedure involves two steps. First, periodogram smoothing of the spectral density matrix derives a pre-estimator for the dynamic eigenvalues and eigenvectors in the frequency domain using UNALSE. Second, the inverse Fourier transform transitions these eigenvectors back to the time domain. This estimation guarantees that the resulting residual covariance matrix is positive definite, facilitating the use of Bartlett's method to estimate factor scores.

Therefore, in practice, we compute the UNALSE estimates of Γ_0^{χ} and Γ_0^{ξ} following (25.1) and (25.2), respectively. We thus obtain $\widehat{\Gamma}^{\chi}_{0\text{UNALSE}}$ and $\widehat{\Gamma}^{\xi}_{0\text{UNALSE}}$. Once we summarize the top q eigenvalues and eigenvectors of $\widehat{\Gamma}^{\chi}_{0\text{UNALSE}}$ in the matrix $\widehat{Z}_{\text{UNALSE}}(\widehat{\Lambda}^{\chi}_{0\text{UNALSE}})\widehat{Z}^{\top}_{\text{UNALSE}}$, we can follow [4] to estimate $\widehat{B}_{\text{UNALSE}} = \widehat{Z}_{\text{UNALSE}}(\widehat{\Lambda}^{\chi}_{0\text{UNALSE}})^{1/2}$, as [1] also adopts this solution for the approximate dynamic factor model. Then, we define Bartlett factor score estimates for the observation $t, t \in \{1, \dots, T\}$, as follows:

$$\widehat{f}_t^{\text{B}} = (\widehat{B}^{\top}_{\text{UNALSE}}(\widehat{\Gamma}^{\xi}_{0\text{UNALSE}})^{-1}\widehat{B}_{\text{UNALSE}})^{-1}\widehat{B}^{\top}_{\text{UNALSE}}(\widehat{\Gamma}^{\xi}_{0\text{UNALSE}})^{-1}X_t,$$

and the true Bartlett factors are defined as

$$f_t^{\text{B}} = (B^{\top}_{\text{UNALSE}}(\Gamma^{\xi}_{0\text{UNALSE}})^{-1}B_{\text{UNALSE}})^{-1}B^{\top}_{\text{UNALSE}}(\Gamma^{\xi}_{0\text{UNALSE}})^{-1}X_t.$$

Let us define the matrix varieties $\widehat{\mathcal{L}}(q)$ and $\widehat{\mathcal{S}}_{\text{diag}}$ as

$$\widehat{\mathcal{L}}(q) = \{\widehat{\Gamma}_0^{\chi} \in \mathbb{R}^{p\times p} \mid \widehat{\Gamma}_0^{\chi} = \widehat{Z}_{\text{UNALSE}}\widehat{D}\widehat{Z}^{\top}_{\text{UNALSE}},\ \widehat{D} \in \mathbb{R}^{q\times q}\text{diagonal}\},$$
$$\widehat{\mathcal{S}}_{\text{diag}} = \{\widehat{\Gamma}_0^{\xi} \in \mathbb{R}^{p\times p} \mid \text{diag}(\widehat{\Gamma}_0^{\chi}) + \text{diag}(\widehat{\Gamma}_0^{\xi}) = \text{diag}(\widehat{\Gamma}_0^{\text{X}}),$$
$$\text{off-diag}(\widehat{\Gamma}_0^{\xi}) = \text{off-diag}(\widehat{\Gamma}_0^{\xi}), \widehat{\Gamma}_0^{\chi} \in \widehat{\mathcal{L}}(q)\}.$$

The following result for Bartlett's factor scores and commonalities based on UNALSE spectral density matrix estimates holds.

Corollary 25.1. *Let $\boldsymbol{H}_q$ be any orthogonal matrix of $q \times q$ and $\widehat{\mathcal{L}}(q)$ be the low-rank matrix variety spanned by $\widehat{Z}_{\text{UNALSE}}$. Under the assumptions of Theorem 25.1, then*

$$\arg\min \max_{\widehat{B}, \widehat{\Gamma}_0^{\chi} = \widehat{B}\widehat{B}^{\top} \in \widehat{\mathcal{L}}(q), \widehat{\Gamma}_0^{\xi} \in \widehat{\mathcal{S}}_{\text{diag}}} \|\widehat{B} - B\boldsymbol{H}_q\| = \widehat{B}_{\text{UNALSE}}.$$

Corollary 25.1 ensures that the UNALSE loadings matrix is the closest possible to the true one in Euclidean norm into the low rank matrix variety with rank q spanned by $\widehat{Z}_{\text{UNALSE}}$.

In this paper, we show that, under the assumptions of Theorem 25.1, for $t \in \{1, \dots, T\}$ the minmax

$$\min_{\widehat{B},\widehat{\Gamma}_0^{\chi}=\widehat{B}\widehat{B}^{\top}\in\widehat{\mathcal{L}}(q),\widehat{\Gamma}_0^{\xi}\in\widehat{\mathcal{S}}_{\text{diag}}} \max \|\widehat{f}_t^{\mathrm{B}} - \mathbf{H}_q f_t^{\mathrm{B}}\|$$

is achieved for $\widehat{B} = \widehat{B}_{\text{UNALSE}}$ and $\widehat{\Gamma}_0^{\xi} = \widehat{\Gamma}_{0\text{UNALSE}}^{\xi}$, and similarly the minmax

$$\min_{\widehat{B},\widehat{\Gamma}_0^{\chi}=\widehat{B}\widehat{B}^{\top}\in\widehat{\mathcal{L}}(q),\widehat{\Gamma}_0^{\xi}\in\widehat{\mathcal{S}}_{\text{diag}}} \max \|\widehat{f}_t^{\mathrm{T}} - \mathbf{H}_q f_t^{\mathrm{T}}\|$$

(we leave the formal proofs to future publications).

These results mean that Bartlett's and Thomson's factor scores obtained by the UNALSE estimates of the covariance matrices provide the tightest error bounds in Euclidean norm within the low rank variety spanned by $\widehat{Z}_{\text{UNALSE}}$ and the sparse matrix variety with the same sparsity pattern as the ALSE/UNALSE residual covariance matrix estimate, keeping fixed the diagonal elements of the overall covariance matrix estimate.

25.3 Numerical Example

In this section, we briefly describe a simulation study to show the effectiveness of ALSE and UNALSE estimates of dynamic factor scores by Bartlett's and Thomson's method. We generate the data according to a VMA(1) and a VAR(1) model which respects all the assumptions of Theorem 25.1, following Sect. D.1 and Sect. D.2 in the Appendix of [1], respectively. For each of the two model types, we consider three cases: case 1 ($p = 100, T = 1000$), case 2 ($p = 150, T = 150$), and case 3 ($p = 150, T = 100$ for VAR(1), $p = 200, T = 100$ for VMA(1)). For each of the $N =$ 100 trials, we estimate $\widehat{f}_t^{\mathrm{T}}$, $\widehat{f}_{t+1}^{\mathrm{T}}$ by FHLR2005, ALSE and UNALSE, and measure $\mathrm{RMSE}^{\mathrm{T}} = \sqrt{\frac{1}{\mathrm{qTN}}\|\widehat{\mathrm{f}}_\tau^{\mathrm{T}} - \mathrm{f}_\tau\|_2^2}$, $\mathrm{MAE}^{\mathrm{T}} = \frac{1}{\mathrm{qTN}}\|\widehat{\mathrm{f}}_\tau^{\mathrm{T}} - \mathrm{f}_\tau\|_1$ for in-sample estimation ($\tau =$ t), and $\mathrm{RMSE}^{\mathrm{T}} = \sqrt{\frac{1}{\mathrm{q(T-1)N}}\|\widehat{\mathrm{f}}_\tau^{\mathrm{T}} - \mathrm{f}_\tau\|_2^2}$ and $\mathrm{MAE}^{\mathrm{T}} = \frac{1}{\mathrm{q(T-1)N}}\|\widehat{\mathrm{f}}_\tau^{\mathrm{T}} - \mathrm{f}_\tau\|_1$ for one-step-ahead prediction ($\tau = t + 1$).

Tables 25.1 and 25.2 contain the results for the two types of models, the three settings, and the three methods. We can observe that the three methods perform in a very similar way. Importantly, ALSE/UNALSE also systematically provides the correct estimate of the dynamic rank. In addition, unlike FHLR2005, ALSE/UNALSE also allows to estimate dynamic factor scores by Bartlett's methods ($\widehat{f}_t^{\mathrm{B}}$, $\widehat{f}_{t+1}^{\mathrm{B}}$), performing equally well. We leave to future research to adapt POET estimator [4] to the spectral density matrix, and to monitor its performance for the relative Bartlett's estimators $\widehat{f}_t^{\mathrm{B}}$, $\widehat{f}_{t+1}^{\mathrm{B}}$.

Table 25.1: In-sample estimation results ($\tau = t$, Thomson's method): RMSE and MAE averaged across 100 trials for FHLR2005, ALSE and UNALSE.

Scenario	FHLR2005		UNALSE		ALSE	
	RMSE	MAE	RMSE	MAE	RMSE	MAE
VAR(1) case 1 ($p = 100, T = 1000$)	0.1022	0.1292	0.1028	0.1299	0.1033	0.1305
VAR(1) case 2 ($p = 150, T = 150$)	0.0840	0.1058	0.0849	0.1069	0.0857	0.1079
VAR(1) case 3 ($p = 150, T = 100$)	0.0867	0.1093	0.0880	0.1109	0.0893	0.1125
VMA(1) case 1 ($p = 100, T = 1000$)	0.0842	0.1065	0.0845	0.1069	0.0848	0.1072
VMA(1) case 2 ($p = 150, T = 150$)	0.0722	0.0917	0.0728	0.0924	0.0733	0.0930
VMA(1) case 3 ($p = 200, T = 100$)	0.0582	0.0737	0.0590	0.0747	0.0597	0.0756

Table 25.2: One-step ahead prediction results ($\tau = t + 1$, Thomson's method): RMSE and MAE averaged across 100 trials for FHLR2005, ALSE and UNALSE.

Scenario	FHLR2005		UNALSE		ALSE	
	RMSE	MAE	RMSE	MAE	RMSE	MAE
VAR(1) case 1 ($p = 100, T = 1000$)	0.0200	0.0253	0.0202	0.0254	0.0203	0.0253
VAR(1) case 2 ($p = 150, T = 150$)	0.0155	0.0194	0.0157	0.0196	0.0159	0.0199
VAR(1) case 3 ($p = 150, T = 100$)	0.0166	0.0210	0.0170	0.0215	0.0174	0.0220
VMA(1) case 1 ($p = 100, T = 1000$)	0.0347	0.0438	0.0350	0.0441	0.0352	0.0444
VMA(1) case 2 ($p = 150, T = 150$)	0.0280	0.0355	0.0288	0.0360	0.0286	0.0360
VMA(1) case 3 ($p = 200, T = 100$)	0.0210	0.0264	0.0213	0.0269	0.0216	0.0272

References

1. Barigozzi, M., Farnè, M.: An algebraic estimator for large spectral density matrices. Journal of the American Statistical Association, **119**(545), 498–510, (2024)
2. Brillinger, D. R.: Time series: Data analysis and theory. Holden-Day, San Francisco (1981)
3. Chamberlain, G., Rothschild, M.: Arbitrage, factor structure, and mean-variance analysis on large asset markets. Econometrica, **51**(5), 1281–1304, (1983)
4. Fan, J., Liao, Y., Mincheva, M.: Large covariance estimation by thresholding principal orthogonal complements. Journal of the Royal Statistical Society, **75**, 603–680, (2013)
5. Forni, M., Hallin, M., Lippi, M., Reichlin, L.: The generalized dynamic-factor model: Identification and estimation. The Review of Economics and Statistics, **82**(4), 540–554, (2000)
6. Forni, M., Hallin, M., Lippi, M., Reichlin, L.: The generalized dynamic factor model. Journal of the American Statistical Association, **100**(471), 830–840, (2005)
7. Stock, J. H., Watson, M. W.: Macroeconomic forecasting using diffusion indexes. Journal of Business and Economic Statistics, **20**(2), 147–162, (2002)

Chapter 26
Testing for Linearity and Independence in Scalar-on-Function Regression with Responses Missing at Random by Generalized Distance Covariance

Manuel Febrero-Bande, Pedro Galeano and Wenceslao González-Manteiga

Abstract Scalar-on-function regression models are used to predict a scalar response based on a functional covariate. A procedure for jointly testing the linearity of the relationship between the scalar response and the functional covariate and the independence between the functional covariate and the error term when some responses are Missing at Random (MAR) is proposed. The method uses generalized distance covariance (GDC) to detect dependency between the functional covariate and residuals from a functional linear model fit. The null distribution of the test statistic is calibrated using residual bootstrap. Accurate slope estimation in the functional linear model is crucial for handling MAR responses. Three estimation methods based on Functional Principal Components (FPCs) are evaluated: (*i*) the simplified method, which omits pairs with missing responses; (*ii*) the imputed method, which fills in missing responses to increase data usage; and (*iii*) the inverse probability weighted method, which adjusts for the probability of missing responses. Cross-validation is used to select the optimal number of FPCs for each method.

26.1 Introduction

In scalar-on-function regression, a scalar response y depends on a functional covariate χ through the model $y = m(\chi) + e$, where $m(\chi) := \mathbb{E}[y|\chi]$ is the regression

---

Manuel Febrero-Bande (✉)
Departamento de Estadística, Análisis Matemático y Optimización, Universidade de Santiago de Compostela, e-mail: manuel.febrero@usc.es

Pedro Galeano
Departamento de Estadística, Universidad Carlos III de Madrid,
e-mail: pedro.galeano@uc3m.es

Wenceslao González-Manteiga
Departamento de Estadística, Análisis Matemático y Optimización, Universidade de Santiago de Compostela e-mail: wenceslao.gonzalez@usc.es

© The Author(s), under exclusive license to Springer Nature Switzerland AG 2025

G. Aneiros et al. (eds.), *New Trends in Functional Statistics and Related Fields*,
Contributions to Statistics, https://doi.org/10.1007/978-3-031-92383-8_26

operator of y on χ, and e is an error term independent of χ with zero mean and finite variance σ_e^2. Given an iid random sample $\{(\chi_i, y_i)\}_{i=1}^n$ from (χ, y), both centered, testing mainly focuses on the form of the regression operator m, with particular interest in whether it is linear. Thus, the omnibus hypothesis H_0 assumes that m is linear, meaning $m \in \mathcal{L} := \{\langle \cdot, \beta \rangle : \beta \in \mathbb{H}\}$, where $\mathbb{H}$ is a real separable Hilbert space equipped with an inner product $\langle \cdot, \cdot \rangle$ and its associated norm $\| \cdot \|$, and $\beta \in \mathbb{H}$ is the unspecified functional slope. Then, if H_0 is true, the relationship between y and χ is captured by the Functional Linear Model with Scalar Response (FLMSR):

$$y = \langle \chi, \beta \rangle + e. \tag{26.1}$$

Several recent papers have introduced different goodness-of-fit tests for H_0. For instance, [1] proposed a test based on a Cramér–von Mises-like norm, which integrates the marked empirical processes indexed by the projected functional covariate along a set of functional directions, while [2] used randomly selected directions for computational efficiency. In another approach, [3] jointly tested linearity and independence between the functional covariate and the error term in (26.1) using generalized distance covariance, as defined by [4]. For alternatives and related advances in goodness-of-fit tests for functional regression models, see [5] and [6].

Random samples may suffer from missing data, particularly Missing at Random (MAR) responses. For instance, [7] encountered missing sunshine duration data due to unavailable cloud coverage recordings, and [8] faced missing brain activity data in their study of major depressive disorder. While multiple regression with MAR responses has been extensively studied, e.g., [9], [10], [11], [12], [13] and [14], the functional covariate setting is less explored. Assuming a FLMSR, [15] proposed a simplified estimator for the functional slope using only complete pairs while [7] introduced an imputed method that re-estimates the slope using both complete and imputed pairs. More recently, [16] proposed an inverse probability weighted method to account for the missing response mechanism. These authors have also proposed a goodness-of-fit test for linearity in FLMSRs with MAR responses. The test statistic is based on a marked empirical process and its null distribution is calibrated using wild bootstrap.

This paper aims to simultaneously test two hypotheses: (i) linearity of the scalar-on-function regression operator m; and (ii) independence between the covariate χ and the error term e, in the presence of MAR responses. Formally, the hypotheses are expressed as:

$$H_0 : m \in \mathcal{L} := \{\langle \cdot, \beta \rangle : \beta \in \mathbb{H}\} \qquad \text{and} \qquad \chi \perp e \tag{26.2}$$

versus

$$H_1 : \text{ Otherwise.} \tag{26.3}$$

The approach builds upon the work of [17], [18] and [3] for fully observed data, which highlighted the challenge of separately testing goodness-of-fit and independence. [17] focused on real-valued covariates, using the Hilbert-Schmidt independence criterion, while [3] extended this to functional covariates using semimetrics on the covariate and residual spaces. The proposed test involves fitting the FLMSR, computing the generalized distance covariance between covariates and residuals, and using residual bootstrap to determine the p-value. In the context of MAR responses, accurate functional slope estimation is crucial for test performance, so the simplified, imputed and inverse probability weighted methods are briefly reviewed.

The paper is structured as follows: Sect. 26.2 discusses MAR assumptions and estimators for the functional slope. Sect. 26.3 details the testing procedure and bootstrap resampling. Finally, Sect. 26.4 provides with some conclusions and extensions.

26.2 The FLMSR with MAR Responses

This section describes the main assumptions underlying the FLMSR when responses are MAR and introduces suitable estimators for the functional slope β. While $\mathbb{H}$ can be any real separable Hilbert space, a common choice is $\mathbb{H} = L^2[a,b]$, where, for a function $\eta : [a,b] \to \mathbb{R}$, the norm is defined by $\|\eta\| := \langle \eta, \eta \rangle^{1/2} < \infty$, and for any two functions $\eta, \varphi \in L^2[a,b]$, their inner product is $\langle \eta, \varphi \rangle := \int_a^b \eta(t)\,\varphi(t)\,dt$. The functional covariate χ is assumed to be a zero-mean functional random variable in $\mathbb{H}$ with a covariance operator Γ defined as

$$\Gamma[\eta] := \mathbb{E}\left[(\chi \otimes \chi)(\eta)\right] := \mathbb{E}\left[\langle \chi, \eta \rangle \chi\right], \quad \text{for any } \eta \in \mathbb{H}, \tag{26.4}$$

where Γ has positive eigenvalues $a_1 > a_2 > \cdots > 0$ associated with orthonormal eigenfunctions $\psi_1, \psi_2, \ldots$, such that $\Gamma[\psi_k] = a_k \psi_k$, for $k \in \mathbb{N}$.

The FLMSR establishes that the functional covariate χ is related to the zero-mean scalar response y via a scalar-on-function regression model as in (26.1) provided that the following conditions are met (see, [19]):

A1. χ and y satisfy $\sum_{k=1}^{\infty} \frac{1}{a_k^2} \langle \mathbb{E}[\chi y], \psi_k \rangle^2 < \infty$;
A2. $\mathrm{Ker}(\Gamma) = \{0\}$.

Under these assumptions, χ and β can be expanded using the eigenfunctions of Γ as follows:

$$\chi = \sum_{k=1}^{\infty} s_k \psi_k \quad \text{and} \quad \beta = \sum_{k=1}^{\infty} b_k \psi_k,$$

where $s_k := \langle \chi, \psi_k \rangle$ are uncorrelated functional principal component (FPC) scores with $\mathbb{E}[s_k] = 0$ and $\mathbb{V}\mathrm{ar}[s_k] = a_k$, and $b_k := \langle \beta, \psi_k \rangle$ are deterministic coefficients. The FLMSR (26.1) can then be written as

$$y = \sum_{k=1}^{\infty} b_k s_k + e, \tag{26.5}$$

which serves at the basis for estimating β.

Additionally, it is assumed that some responses y may be missing under a MAR mechanism, meaning the probability of a missing response depends on χ rather than y. i.e.:

$$\mathbb{P}(r = 1|y, \chi) = \mathbb{P}(r = 1|\chi) =: p(\chi), \qquad (26.6)$$

where r is a Bernoulli variable indicating whether y is observed ($r = 1$) or not ($r = 0$), and $p : \mathbb{H} \to [0, 1]$ is an unspecified function of χ. The MAR assumption allows the missing responses to be predicted based on χ.

The MAR-adjusted testing procedure introduced in Sect. 26.3 for testing (26.2) vs (26.3) relies on estimating β from the FLMSR (26.1). Assuming an i.i.d. sample $\{(\chi_i, y_i, r_i)\}_{i=1}^n$ is drawn from the joint distribution of (χ, y, r), the simplified, imputed, and inverse probability weighted estimators, as described in [7] and [16], are reviewed.

The simplified estimator uses only complete pairs of response and FPC scores to estimate the coefficients $\{b_k\}_{k=1}^{K_S}$ in (26.5). Let $i_S := \{i : r_i = 1\}_{i=1}^n$ denote the indices of observed responses, and $n_S := \#\{i_S\}$ be their count. The covariance operator Γ from (26.4) is estimated using the sample covariance operator of $\{\chi_i\}_{i \in i_S}$, given by

$$\widehat{\Gamma}_S[\eta] := \frac{1}{n_S} \sum_{i \in i_S} \langle \chi_i, \eta \rangle \chi_i, \quad \text{for any } \eta \in \mathbb{H},$$

that has nonnegative eigenvalues $\widehat{a}_{1,S} \geq \widehat{a}_{2,S} \geq \cdots \geq 0$, where $\widehat{a}_{k,S} = 0$ for $k > n_S$, and corresponding orthonormal eigenfunctions $\widehat{\psi}_{1,S}, \widehat{\psi}_{2,S}, \ldots$ If $\widehat{s}_{i,k,S} := \langle \chi_i, \widehat{\psi}_{k,S} \rangle$, for $i \in i_S$ and $k \in \mathbb{N}$ are the sample FPC scores then the simplified estimator of β is:

$$\widehat{\beta}_S := \sum_{k=1}^{K_S} \widehat{b}_{k,S} \widehat{\psi}_{k,S}, \qquad (26.7)$$

where $\widehat{b}_{1,S}, \ldots, \widehat{b}_{K_S,S}$ are OLS estimators of $b_1, \ldots, b_{K_S}$:

$$\widehat{b}_{k,S} := \frac{1}{n_S \widehat{a}_{k,S}} \sum_{i \in i_S} y_i \widehat{s}_{i,k,S}, \quad \text{for } k = 1, \ldots, K_S.$$

The cutoff K_S is selected via leave-one-out cross validation from $\{1, \ldots, K_S^{\max}\}$, where $K_S^{\max}$ retains only FPCs capturing significant variability in the covariate.

The imputed estimator leverages the MAR assumption to predict missing responses based on covariates making use of the equality $\mathbb{E}[ry + (1 - r)\langle \chi, \beta \rangle] = \mathbb{E}[y]$ that justifies the completed sample $\left\{(\chi_i, y_{i,I}, r_i)\right\}_{i=1}^n$ to estimate β, where $y_{i,I} := r_i y_i + (1 - r_i)\left\langle \chi_i, \widehat{\beta}_S \right\rangle$, for $i = 1, \ldots, n$. The covariance operator Γ is estimated using the sample covariance operator of $\{\chi_i\}_{i=1}^n$:

$$\widehat{\Gamma}_C[\eta] := \frac{1}{n} \sum_{i=1}^n \langle \chi_i, \eta \rangle \chi_i, \quad \text{for any } \eta \in \mathbb{H},$$

that has eigenvalues $\widehat{a}_{1,C} \geq \widehat{a}_{2,C} \geq \cdots \geq 0$, where $\widehat{a}_{k,C} = 0$, for $k > n$, and orthonormal eigenfunctions $\widehat{\psi}_{1,C}, \widehat{\psi}_{2,C}, \ldots$ If $\widehat{s}_{i,k,C} := \langle \chi_i, \widehat{\psi}_{k,C} \rangle$ are the sample FPC scores, the imputed estimator of β is:

$$\widehat{\beta}_I := \sum_{k=1}^{K_I} \widehat{b}_{k,I} \widehat{\psi}_{k,C}, \tag{26.8}$$

where $\widehat{b}_{1,I}, \ldots, \widehat{b}_{K_I,I}$ are OLS estimators of $b_1, \ldots, b_{K_I}$:

$$\widehat{b}_{k,I} := \frac{1}{n\widehat{a}_{k,C}} \sum_{i=1}^{n} y_{i,I} \widehat{s}_{i,k,C}, \quad \text{for } k = 1, \ldots, K_I.$$

The cutoffs K_S and K_I are selected by leave-one-out cross validation from $\{1, \ldots, K_C^{\max}\} \times \{1, \ldots, K_C^{\max}\}$. As shown in [7], the imputed estimator $\widehat{\beta}_I$ may have a lower MSE than $\widehat{\beta}_S$ if cutoffs are selected jointly.

The inverse probability weighted estimator accounts for missingness probability when imputing missing responses. Based on the equality $\mathbb{E}[(r/p(\chi))\, y + (1 - r/p(\chi)) \langle \chi, \beta \rangle] = \mathbb{E}[y]$ the completed sample $\left\{(\chi_i, y_{i,W}, r_i)\right\}_{i=1}^n$ is used to estimate β, where $y_{i,W} := (r_i/\widehat{p}(\chi_i))\, y_i + (1 - r_i/\widehat{p}(\chi_i)) \langle \chi_i, \widehat{\beta}_S \rangle$, and $\widehat{p}(\cdot)$ is the Nadaraya–Watson estimator of the MAR mechanism (26.6):

$$\widehat{p}(\chi) := \sum_{i=1}^{n} \frac{K_h(\|\chi - \chi_i\|)}{\sum_{l=1}^{n} K_h(\|\chi - \chi_l\|)} r_i,$$

where $K_h(\cdot) := K(\cdot/h)/h$ is a kernel function with bandwidth h, selected by leave-one-out cross validation. The inverse probability weighted estimator of β is:

$$\widehat{\beta}_W := \sum_{k=1}^{K_W} \widehat{b}_{k,W} \widehat{\psi}_{k,C}, \tag{26.9}$$

where $\widehat{b}_{1,W}, \ldots, \widehat{b}_{K_W,W}$ are the OLS estimators of $b_1, \ldots, b_{K_W}$. The cutoffs K_S and K_W are selected as for the imputed estimator $\widehat{\beta}_I$.

26.3 The Testing Procedure

This section outlines the procedure for testing the null hypothesis (26.2) against the alternative hypothesis (26.3) using the random sample $\{(\chi_i, y_i, r_i)\}_{i=1}^n$ and the generalized distance covariance. The concept of generalized distance covariance is briefly reviewed next, adapted for the specific case in question.

Generalized distance covariance is a powerful tool for testing the independence between two random objects, as explored by [20] and [4], and rooted in the work of [21] and [22] on distance covariance and correlation. Specifically, the generalized distance covariance between a functional predictor χ and an error term e is defined as

$$\begin{aligned}\mathcal{V}^2_{\rho_{\mathbb{H}},\rho_{\mathbb{R}}}(\chi,e) = \mathbb{E}_{\chi,e}\left[\mathbb{E}_{\chi',e'}\left[\rho_{\mathbb{H}}(\chi,\chi')\rho_{\mathbb{R}}(e,e')\right]\right] + \\ + \mathbb{E}_{\chi}\left[\mathbb{E}_{\chi'}\left[\rho_{\mathbb{H}}(\chi,\chi')\right]\right]\mathbb{E}_{e}\left[\mathbb{E}_{e'}\left[\rho_{\mathbb{R}}(e,e')\right]\right] - \\ - 2\mathbb{E}_{\chi,e}\left[\mathbb{E}_{\chi'}\left[\rho_{\mathbb{H}}(\chi,\chi')\right]\mathbb{E}_{e'}\left[\rho_{\mathbb{R}}(e,e')\right]\right],\end{aligned} \tag{26.10}$$

where (χ', e') is an independent copy of (χ, e), and $\rho_{\mathbb{H}}$ and $\rho_{\mathbb{R}}$ are semimetrics on $\mathbb{H}$ and $\mathbb{R}$, respectively. These are symmetric, i.e., $\rho_{\mathbb{H}}(\chi,\chi') = \rho_{\mathbb{H}}(\chi',\chi)$ and $\rho_{\mathbb{R}}(e,e') = \rho_{\mathbb{R}}(e',e)$, and equal zero only when $\chi = \chi'$ and $e = e'$. The conditions $\mathrm{E}\left[\rho^2_{\mathbb{H}}(\chi,\chi)\right] < \infty$ and $\mathrm{E}\left[\rho^2_{\mathbb{R}}(e,e)\right] < \infty$ ensure the definition is meaningful. Under a strong negative type condition (see [20]), $\mathcal{V}^2_{\rho_{\mathbb{H}},\rho_{\mathbb{R}}}(\chi,e) = 0$ if and only if χ and e are independent.

Given that under the null hypothesis (26.2) the predictor χ and error e are independent, the objective is to measure the deviation of the sample with MAR responses, $\{(\chi_i, y_i, r_i)\}_{i=1}^n$, from (26.2) using the generalized distance covariance (26.10). This involves the three considered estimators for the functional slope β, leading to several sets of residuals that provide the basis for constructing test statistics. Specifically, the estimators $\widehat{\beta}_{\mathcal{S}}$ in (26.7), $\widehat{\beta}_{\mathcal{I}}$ in (26.8), and $\widehat{\beta}_{\mathcal{W}}$ in (26.9), yield residuals with respect the observed responses defined as

$$\widehat{e}_{i,\mathcal{S}} = y_i - \left\langle \chi_i, \widehat{\beta}_{\mathcal{S}} \right\rangle, \quad \widehat{e}_{i,\mathcal{I}} = y_i - \left\langle \chi_i, \widehat{\beta}_{\mathcal{I}} \right\rangle, \quad \widehat{e}_{i,\mathcal{W}} = y_i - \left\langle \chi_i, \widehat{\beta}_{\mathcal{W}} \right\rangle, \quad \text{for } i \in i_{\mathcal{S}}, \tag{26.11}$$

respectively. Additionally, the imputed response estimators, $\widehat{\beta}_{\mathcal{I}}$ in (26.8) and $\widehat{\beta}_{\mathcal{W}}$ in (26.9) yield residuals with respect to the set of imputed responses $\left\{y_{i,\mathcal{I}}\right\}_{i=1}^n$ and $\left\{y_{i,\mathcal{W}}\right\}_{i=1}^n$ given by

$$\widetilde{e}_{i,\mathcal{I}} = y_{i,\mathcal{I}} - \left\langle \chi_i, \widehat{\beta}_{\mathcal{I}} \right\rangle, \quad \widetilde{e}_{i,\mathcal{W}} = y_{i,\mathcal{W}} - \left\langle \chi_i, \widehat{\beta}_{\mathcal{W}} \right\rangle, \quad \text{for } i = 1, \ldots, n, \tag{26.12}$$

respectively.

Next, if $k_{ij} = \rho_{\mathbb{H}}(\chi_i, \chi_j)$ and $l_{ij} = \rho_{\mathbb{R}}(\widehat{e}_i, \widehat{e}_j)$, for $i, j \in i_{\mathcal{S}}$, where $\widehat{e}_i$ represents any type of residual in (26.11), the natural estimator of the generalized distance covariance $\mathcal{V}^2_{\rho_{\mathbb{H}},\rho_{\mathbb{R}}}(\chi,e)$ in (26.10) based on the set $\{(\chi_i, \widehat{e}_i)\}_{i \in i_{\mathcal{S}}}$ and that serves for evaluating (26.2) versus (26.3) is the test statistic

$$\mathcal{T}_{\rho_{\mathbb{H}},\rho_{\mathbb{R}}}(\chi_i, \widehat{e}_i) = \frac{1}{n^2_{\mathcal{S}}} \sum_{i,j \in i_{\mathcal{S}}} k_{ij} l_{ij} + \frac{1}{n^4_{\mathcal{S}}} \sum_{i,j,q,r \in i_{\mathcal{S}}} k_{ij} l_{qr} - \frac{2}{n^3_{\mathcal{S}}} \sum_{i,j,q \in i_{\mathcal{S}}} k_{ij} l_{iq}. \tag{26.13}$$

In addition, if $k_{ij} = \rho_{\mathbb{H}}(\chi_i, \chi_j)$ and $l_{ij} = \rho_{\mathbb{R}}(\widetilde{e}_i, \widetilde{e}_j)$, for $i, j = 1, \ldots, n$, where $\widetilde{e}_i$ represents any type of residual in (26.12), the estimator of $\mathcal{V}^2_{\rho_{\mathbb{H}},\rho_{\mathbb{R}}}(\chi,e)$ based on the set $\{(\chi_i, \widetilde{e}_i)\}_{i=1}^n$ is the test statistic

$$\mathcal{T}_{\rho_{\mathbb{H}},\rho_{\mathbb{R}}}(\chi_i, \widetilde{e}_i) = \frac{1}{n^2} \sum_{i,j=1}^{n} k_{ij} l_{ij} + \frac{1}{n^4} \sum_{i,j,q,r=1}^{n} k_{ij} l_{qr} - \frac{2}{n^3} \sum_{i,j,q=1}^{n} k_{ij} l_{iq}. \tag{26.14}$$

For any semimetrics $\rho_{\mathbb{H}}$ and $\rho_{\mathbb{R}}$, a residual bootstrap procedure calibrates the distribution of the statistics (26.13) and (26.14) under the null hypothesis (26.2). Below is an overview of the testing procedure, with $\widehat{\beta}$ representing any of $\widehat{\beta}_S$, $\widehat{\beta}_I$, or $\widehat{\beta}_{\mathcal{W}}$.

Generalized distance covariance testing procedure

Let $\{(\chi_i, y_i, r_i)\}_{i=1}^n$ be an i.i.d. sample from (χ, y, r), where $y = m(\chi) + e$ and $r|\chi \sim \text{Ber}(p(\chi))$ is the MAR operator. The procedure to test $H_0 : m \in \mathcal{L} := \{\langle \cdot, \beta \rangle : \beta \in \mathbb{H}\}$ and $\chi \perp e$, against H_1 : Otherwise, with the sample $\{(\chi_i, y_i, r_i)\}_{i=1}^n$ is as follows:

1. Compute the estimator $\widehat{\beta}$ as described in Sect. 26.2.
2. Compute the set of residuals $\{\widehat{e}_i\}_{i \in i_S}$ in (26.11) associated to $\widehat{\beta}$. Furthermore, compute residuals $\{\widetilde{e}_i\}_{i \in i_S}$ in (26.12) if used for the test.
3. Compute the statistic $\mathcal{T}_{\rho_{\mathbb{H}},\rho_{\mathbb{R}}}(\chi_i, \widehat{e}_i)$ in (26.13) or the statistic $\mathcal{T}_{\rho_{\mathbb{H}},\rho_{\mathbb{R}}}(\chi_i, \widetilde{e}_i)$ in (26.14).
4. Bootstrap resampling: For $b = 1, \ldots, B$, do:

    a. Construct the set $\{e_i^{*b}\}_{i \in i_S}$ chosen randomly with replacement from $\{\widehat{e}_i\}_{i \in i_S}$.
    b. Define a bootstrap sample $\{(\chi_i, y_i^{*b}, r_i)\}_{i=1}^n$, where $y_i^{*b} = \langle \chi_i, \widehat{\beta} \rangle + e_i^{*b}$, for $i \in i_S$.
    c. Obtain the estimator $\widehat{\beta}^{*b}$ with the bootstrap sample $\{(\chi_i, y_i^{*b}, r_i)\}_{i=1}^n$.
    d. Obtain the set of residuals $\{\widehat{e}_i^{*b}\}_{i \in i_S}$ in (26.11) associated with $\widehat{\beta}^{*b}$. Furthermore, compute residuals $\{\widetilde{e}_i^{*b}\}_{i=1}^n$ in (26.12) if used for the test.
    e. Compute the statistic $\mathcal{T}_{\rho_{\mathbb{H}},\rho_{\mathbb{R}}}(\chi_i, \widehat{e}_i^{*b})$ in (26.13) or the statistic $\mathcal{T}_{\rho_{\mathbb{H}},\rho_{\mathbb{R}}}(\chi_i, \widetilde{e}_i^{*b})$ in (26.14).

5. Estimate the p-value of the test with $\#\{\mathcal{T} \leq \mathcal{T}_{\rho_{\mathbb{H}},\rho_{\mathbb{R}}}(\chi_i, \widehat{e}_i^{*b})\}/B$ or with $\#\{\mathcal{T} \leq \mathcal{T}_{\rho_{\mathbb{H}},\rho_{\mathbb{R}}}(\chi_i, \widetilde{e}_i^{*b})\}/B$.

26.4 Conclusions

Some aspects not covered in this article will be addressed in subsequent papers. For example, we have conducted simulations that shows that the proposed methodology is more powerful when used in conjunction with the residuals defined in (26.12) from the imputed estimator (26.8). Furthermore, we have analyzed the real data example discussed in [7] and [16], finding results that align closely with their findings. Importantly, the proposed methodology exhibits flexibility and can be extended to multivariate settings. Future research will delve into alternative approaches, such as projection-based techniques, as highlighted in the work of [1] and [2]. These efforts aim to further refine and expand the applicability of our methodology.

Acknowledgements The authors acknowledge financial support by MCIN/AEI/10.13039/501100011033: the first and third authors from grant PID2020-116587GB-I00, and the second author from grant PID2022-138114NB-I00.

References

1. García-Portugués, E., González-Manteiga, W., Febrero-Bande, M.: A goodness-of-fit test for the functional linear model with scalar response. J. Comput. Graph. Stat. **23**, 761–778 (1995)
2. Cuesta-Albertos, J. A., García-Portugués, E., González-Manteiga, W., Febrero-Bande, M.: Goodness-of-fit tests for the functional linear model based on randomly projected empirical processes. Ann. Stat., **47**, 439–467 (2019)
3. Lai, T., Zhang, Z., Wang, Y.: Testing for independence and goodness-of-fit jointly for functional linear models. J. Korean. Stat. Soc., **50**, 380–402 (2021)
4. Sejdinovic, D., Sriperrumbudur, B., Gretton, A., Fukumizu, K. Equivalence of distance-based and rkhs-based statistics in hypothesis testing. Ann. Stat., **41**, 2263–2291 (2013)
5. González-Manteiga, W.: A review on specification tests for models with functional data. Span. J. Stat., **4**, 9–40 (2023)
6. González-Manteiga, W., Crujeiras, R. M., García-Portugués, E.: A review of goodness-of-fit tests for models involving functional data. In: Balakrishnan, N., Gil, M. A., Martín, N., Morales, D., Pardo, M. C. (eds.) Trends in Mathematical, Information and Data Sciences, pp. 349–358. Springer, Cham (2023)
7. Febrero-Bande, M., Galeano, P., González-Manteiga, W.: Estimation and prediction for the functional linear model with scalar response with responses missing at random. Comput. Stat. Data Anal., **131**, 91–103 (2019)
8. Ciarleglio, A., Petkova, E., Harel, O.: Elucidating age and sex-dependent association between frontal EEG asymmetry and depression: An application of multiple imputation in functional regression. J. Am. Stat. Assoc., **117**, 12–26 (2022)
9. González-Manteiga, W., Pérez-González, A.: Goodness-of-fit tests for linear regression models with missing response data. Can. J. Stat., **34**, 149–170 (2006)
10. Sun, Z., Wang, Q.: Checking the adequacy of a general linear model with responses missing at random. J. Stat. Plan. Inference, **139**, 3588–3604 (2009)
11. Li, X.: Lack-of-fit testing of a regression model with response missing at random. J. Stat. Plan. Inference, **142**, 155–170 (2012)
12. Sun, Z., Chen, F., Zhou, X., Zhang, Q.: Improved model checking methods for parametric models with responses missing at random. J. Multivar. Anal., **154**, 147–161 (2009)
13. Zheng, S.-J., Gao, S.-Y., Sun, Z.-H.: Projection-based consistent test for linear regression model with missing response and covariates. Acta. Math. Appl. Sin., **36**, 917–935 (2020)
14. Pérez-González, A., Cotos-Yáñez, T. R., González-Manteiga, W., Crujeiras, R. M.: Goodness-of-fit tests for quantile regression with missing responses. Stat. Pap., **62**, 1231–1264 (2021)
15. Crambes, C., Henchiri, Y.: Regression imputation in the functional linear model with missing values in the response. J. Stat. Plan. Inference, **201**, 103–119 (2019)
16. Febrero-Bande, M., Galeano, P., García-Portugués, E., González-Manteiga, W.: Testing for linearity in scalar-on-function regression with responses missing at random. Comput. Stat., **39**, 3405-3429 (2024)
17. Sen, A., Sen, B.: Testing independence and goodness-of-fit in linear models. Biometrika, **101**, 927–942 (2014)
18. Xu, K., He, D.: Omnibus model checks of linear assumptions through distance covariance. Statistica Sinica, **31**, 1055–1079 (2021)
19. Cardot, H., Mas, A., Sarda, P.: CLT in functional linear regression models. Probab. Theory Related Fields, **138**, 325–361 (2007)
20. Lyons, R.: Distance covariance in metric spaces. Ann. Probab., **41**, 3284-3305 (2013)
21. Székely, G. J., Rizzo, M. L., Bakirov, N. K.: Measuring and testing dependence by correlation of distances. Ann. Statist., **35**, 2769–2794 (2007)
22. Székely, G. J., Rizzo, M. L.: Brownian distance covariance. Ann. Appl. Stat., **3**, 1236–1265 (2009)

Chapter 27
A Novel Spectral Density Operator Approach to Unveil Dynamic Time Dependencies in Multivariate Long-Term ECGs

Antonino Gagliano, Chiara Di Maria, Gianluca Sottile, Sarah Beutler-Traktovenko, Luigi Augugliaro and Valeria Vitelli

Abstract Electrocardiograms (ECGs) are essential tools for assessing cardiac health. The extraction of meaningful features from multivariate ECG data is critical for identifying potential cardiac anomalies based on inter-channel dynamics. This work introduces an innovative approach rooted in functional data analysis (FDA) aimed at analyzing the temporal and frequency-domain dynamics of multivariate ECG signals. Considering each patient's heartbeats as a sample observation of a multivariate functional time series, this method employs the spectral density operator to reveal dynamic time-dependencies among ECG channels across frequency bands. Through the analysis of ECG data of a subject experiencing detachment from the self, this approach demonstrates its potential to uncover clinically relevant patterns that correlate with the patient's symptomatology.

27.1 Introduction

Electrocardiograms (ECGs) are essential diagnostic tools that record the electrical activity of the heart over time, providing critical insights into cardiac health. The standard ECG waveform comprises distinct components: the P wave, which represents initial atrial depolarization, the QRS complex, corresponding to the subsequent

Antonino Gagliano (✉), Chiara Di Maria, Gianluca Sottile, Luigi Augugliaro
Department of Economics, Business and Statistics, University of Palermo, Italy,
e-mail: antonino.gagliano02@unipa.it, chiara.dimaria@unipa.it
e-mail: gianluca.sottile@unipa.it, luigi.augugliaro@unipa.it

Sarah Beutler-Traktovenko
Department of Psychology, Friedrich Schiller University of Jena, Jena, Germany,
e-mail: sarah.beutler@uni-jena.de

Valeria Vitelli
Oslo Centre for Biostatistics and Epidemiology, Departiment of Biostatistics, University of Oslo, Norway, e-mail: valeria.vitelli@medisin.uio.no

© The Author(s), under exclusive license to Springer Nature Switzerland AG 2025

G. Aneiros et al. (eds.), *New Trends in Functional Statistics and Related Fields*,
Contributions to Statistics, https://doi.org/10.1007/978-3-031-92383-8_27

ventricular depolarization, and the T wave, which indicates ventricular repolarization. Accurate interpretation of these features is vital for diagnosing various cardiac conditions [12]. ECG data are collected using single or multiple electrodes, and in the latter case they are referred to as *multi-channel* ECGs.

Over the years, extensive research has focused on developing advanced techniques for ECG analysis. This analysis is generally divided into three main stages: data cleaning, feature extraction and classification. Among these, the feature extraction phase is particularly critical, as it transforms raw ECG signals into meaningful inputs that are essential for detecting potential anomalies in the patient's ECG.

Various feature extraction methods have been applied to ECG signals. The domain of analysis can be used [14] to broadly categorize them in time-domain approaches [5], frequency-domain techniques [1], time–frequency domain methods [8, 3], decomposition-based techniques [7] and deep learning models [4]. Recent work explores the relationships between ECG channels to develop advanced feature extraction methods, particularly for tasks such as disease classification. For example, M. Sepahvand (2021) [13] employs wavelet-based analysis to decompose ECG signals into frequency bands, used to calculate within- and cross-correlation matrices for estimating multichannel dependencies. However, their approach primarily focuses on static dependencies within predefined frequency bands, neglecting the dynamic temporal evolution of inter-channel relationships.

To address this problem, we propose a novel approach to extract features from ECG data, which is rooted in functional data analysis (FDA) of multivariate time series. By treating each heartbeat as a single multivariate functional observation, and by modeling the time series of heartbeats, we model the temporal dynamics of multichannel ECG data through the spectral density operator [6, 11]. While the cross-covariance operator captures static dependencies between ECG channels at specific time lags, the spectral density operator extends this analysis into the frequency domain, providing a dynamic representation of how these dependencies evolve over time across different frequency bands. The efficacy of the proposed methodology is illustrated through a case study of a patient experiencing a sensation of "detachment" from the self, showcasing how the spectral density operator can reveal dynamic dependencies between ECG channels that are correlated with the symptomatology.

27.2 Methods

Let us consider a weakly stationary bivariate functional time series $\{(X_t, Y_t)\}_{t\in\mathbb{Z}}$, where X_t and Y_t represent two ECG channels, modeled as random elements in $L^2([0,1],\mathbb{R})$. For a zero-mean process, the cross-covariance operator C_h^{XY} at lag h is defined as:

$$C_h^{XY}(x)(u) = \int_0^1 c_h^{XY}(u,v)x(v)\,dv, \tag{27.1}$$

where $c_h^{XY}(u, v) = \text{cov}\{X_{t+h}(u), Y_t(v)\}$ denotes the cross-covariance kernel. Under stationarity, $c_h^{XY}(u, v)$ depends only on the lag h and not on the time index t, thus capturing the dependency structure between lagged observations of X_t and Y_t. However, equation (27.1) primarily reflects localized temporal interactions, while it does not fully account for long-range dependencies or oscillatory behaviors typical of multivariate functional time series.

To overcome these limitations, we can instead consider the *cross-spectral density operator*, which extends the analysis to the frequency domain. The cross-spectral density operator $\mathcal{F}_\theta^{XY}$ is defined as:

$$\mathcal{F}_\theta^{XY}(x)(u) = \frac{1}{2\pi} \sum_{h \in \mathbb{Z}} C_h^{XY}(x)(u) \exp(-\mathrm{i}h\theta), \tag{27.2}$$

where $\theta \in [-\pi, \pi]$ represents the frequency, and i denotes the imaginary unit. This operator encapsulates the entire family of lagged covariance operators $\{C_h^{XY} : h \in \mathbb{Z}\}$ by decomposing their contributions across different frequencies through a *Fourier transform*. Equation (27.2) therefore provides a comprehensive representation of the dynamic temporal dependencies between channels X_t and Y_t across frequency bands.

27.3 Dataset Description and Analysis

The dataset used in this study consists of 3-channel ECG data collected over a 12-hour period for eight subjects using a portable device. These recordings provide a comprehensive view of the subjects electrophysiological dynamics, capturing signals across three distinct channels with 250 Hz frequency. The dataset includes also information on the onset and duration of a specific symptomatology described as "detachment" from the self, recorded by a click provided by each subject during the recording. Detachment can be described as a subjective experience of an altered state of consciousness, characterized by "a sense of separation from certain aspects of everyday experience" [2]. Fig. 27.1 illustrates the heartbeats of a subject for channels 1 and 3: the grey curves represent individual heartbeats, while the red curve shows the mean heartbeats.

Preprocessing and smoothing

The initial step was to select a single subject for analysis, and segment their ECG into individual heartbeats. Based on the subject's click indicating the moment they experienced the symptom, 27 regular beats and 27 symptomatic beats were selected. Preprocessing was performed using the *Neurokit2* package in Python [9], and consisted of three main phases. The first phase divided the time series into windows. In the second phase, the signal was cleaned to remove artifacts and noise using a high-pass *Butterworth* filter (0.5 Hz) for low-frequency noise and a notch filter to suppress specific interferences, such as 50 Hz powerline noise. The third and final phase fo-

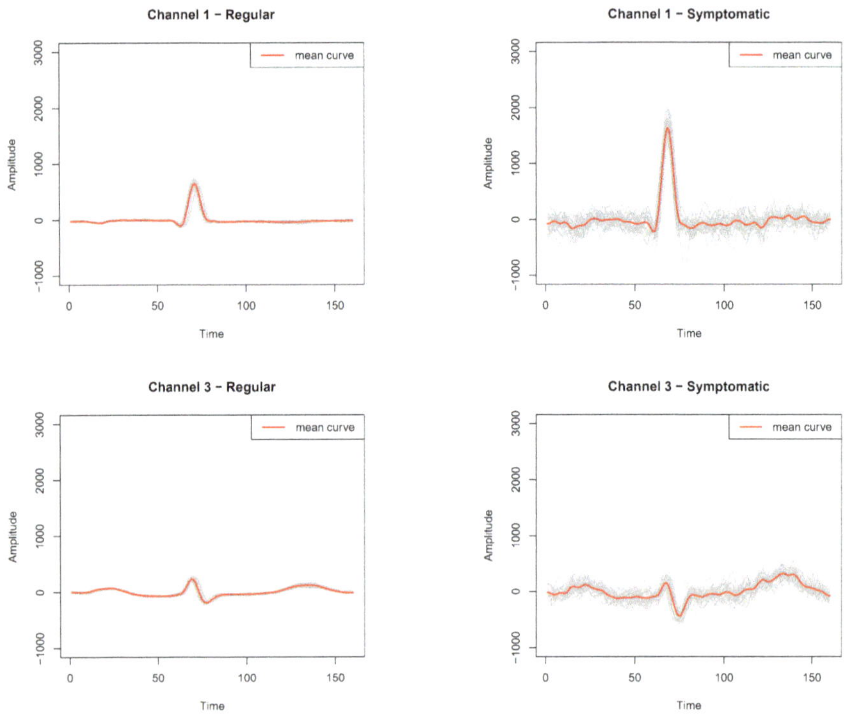

Fig. 27.1: Raw heartbeats for channel 1 (top) and channel 3 (bottom) in both regular (left) and symptomatic (right) conditions. The grey curves represent individual heartbeats, while the red curve represents the mean heartbeat

cused on identifying the $P-QRS-T$ segments to extract individual heartbeats. This step began with the detection of the R point, corresponding to the local maximum of the ECG signal within the temporal window. Subsequently, the remaining morphological points were identified using the discrete wavelet transform (DWT) [10]. Once all individual heartbeats were extracted, they were interpolated onto a uniform grid of 160 time points, representing the estimated average beat length. However, to fully utilize the functional characteristics of the data, it was necessary to convert the signal into a continuous representation. To this end, penalized B-spline smoothing was applied, with the number of bases and the penalization parameter selected using Generalized Cross-Validation (GCV).

Cross-Covariance Analysis

Considering individual heartbeats as functional time series observations, the dependencies between channels were initially analyzed using the functional cross-covariance operator defined in equation (27.1). Fig. 27.2 (a) and (b) illustrates the spectrogram of the cross-covariance operator between channels 1 and 3, representing the pair of channels exhibiting the most pronounced differences in the temporal dynamics between the physiological and pathological conditions, compared to other channel pairs. The spectrogram is obtained by fixing specific matrix coordinates and plotting the variation of the operator's values across all lags, thereby reducing the three-dimensional operator into a two-dimensional visual representation. In the stable condition (Fig. 27.2 (a)), covariance values are concentrated around lag $h = 0$, highlighting a stable relationship and high temporal synchrony between the signals. This symmetry is indicative of normal cardiac activity, with minimal dependence on lagged components beyond zero.

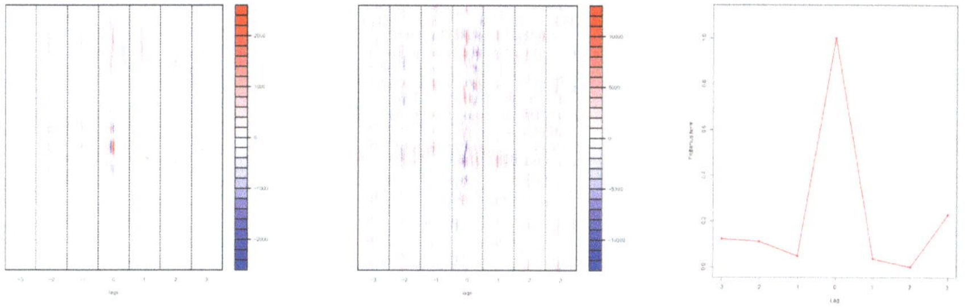

(a) Regular condition (b) Symptomatic condition (c) Frobenius distance

Fig. 27.2: Comparison of cross-covariance operators of Channels 1 and 3 in the two conditions: Regular condition (a) and Symptomatic condition (b). Panel (c) shows the Frobenius distance between the two cross-covariance operators for each lag

Conversely, in the symptomatic condition (Fig. 27.2 (b)), the covariance becomes less regular, with delayed components playing a more prominent role. This suggests that the interactions between channels are no longer confined to a local and stable relationship around lag zero but become more diffuse and less predictable. Such changes reflect an alteration in the temporal dynamic of the signals, likely influenced by pathological processes associated with the patient's symptomatic condition. However, this representation has certain limitations: the covariance operator, being calculated as a function of lag h, captures the linear dependency between signals at specific delays, i.e., an instantaneous covariance structure for lagged pairs of functional time series. It does not account for global temporal dynamics over time, therefore it limits the ability to describe complex temporal dependencies particularly in symptomatic ECG conditions, where signals often exhibit less stable behavior due to potential atrioventricular contractions or dilations that may occur over short or long periods.

Spectral Density Analysis

To address these limitations, spectral density (in equation (27.2)) leverages the Fourier transform of the covariance structure across temporal lags to capture the full temporal dependency characterizing the ECG data. Representing this temporal dependency within the frequency domain allows us to distinguish between slower variations in the signal (low frequencies), often associated with global changes or baseline conditions, and faster variations (high frequencies), which may correspond to specific cardiac phenomena.

Fig. 27.3 shows that, in the regular condition (a), the spectral density between channels 1 and 3 is concentrated at lower frequencies, indicating that their temporal dependency structure is dominated by slow and stable variations. This pattern is consistent with regular cardiac activity and suggests well-synchronized inter-channel interactions without abrupt changes or irregularities. In contrast, in symptomatic conditions (b), the spectral density exhibits a more complex and dynamic pattern, with values distributed across a broader frequency range. This indicates that the channels show temporal dependencies even at higher frequencies, reflecting faster and more unstable variations. Such a shift suggests an alteration in the rhythmic consistency of the cardiac cycle, a key feature of the patient's pathological condition.

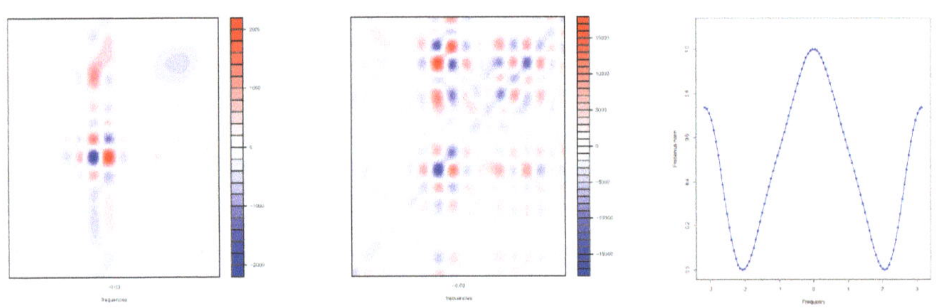

(a) Regular condition (b) Symptomatic condition (c) Frobenius distance

Fig. 27.3: Comparison of cross spectral density operators of Channels 1 and 3 in the two conditions: Regular condition (a) and Symptomatic condition (b). Panel (c) shows the Frobenius distance between the two cross spectral density operators for each frequency

Comparison by Frobenius Norm

To validate the insights provided by visual inspection of the spectral density operator, the Frobenius norm was employed as a quantitative measure of the differences between the two patient conditions. Specifically, for the cross-covariance operator, the norm was computed between matrices at each lag h, while for the spectral density operator, it was evaluated across frequencies (see Fig. 27.2(c) and Fig. 27.3(c)). The

Frobenius norm for matrices $A, B \in \mathbb{R}^{m \times n}$ is defined as:

$$\|A_k - B_k\|_F = \sqrt{\sum_{i=1}^{m} \sum_{j=1}^{n} \left(a_{ij}^{(k)} - b_{ij}^{(k)}\right)^2},$$

where A_k and B_k represent the symptomatic and regular matrices corresponding to a specific k, which denotes either a frequency (for spectral density) or a lag (for cross-covariance), depending on the operator being analyzed.

The results confirmed that the spectral density operator provides a more comprehensive characterization of inter-channel dependencies. While the cross-covariance operator primarily captures direct and immediate relationships between ECG channels, it shows limited sensitivity to interactions involving delayed signals at specific temporal lag. Expanding the analysis by integrating information across multiple lags, the spectral density operator reveals dependencies that evolve dynamically over time and across different frequency bands, with pronounced variations not only in the low frequency bands but also in the higher frequency corresponding to the rapid oscillations characteristic of the patient's symptomatic condition. This highlights its ability to uncover complex temporal dynamics that remain hidden in a purely lag-based framework.

27.4 Conclusion

This study introduces a novel approach to feature extraction from multivariate ECG data using the spectral density operator. By modeling each heartbeat as a sample realization of a functional time series, the proposed framework not only captures static inter-channel dependencies but also uncovers the dynamic temporal evolution of these relationships across varying frequency bands.

The case study of a patient experiencing "detachment" from the self demonstrates that the spectral density operator is a more insightful tool compared to traditional cross-covariance analysis. While cross-covariance provides a localized view of temporal dependencies at specific lags, it is limited in capturing the broader oscillatory dynamics of the signals. In contrast, the spectral density operator dynamically encapsulates these local temporal dependencies in the frequency domain. This dual-dimensional analysis enables a deeper characterization of both the intensity of inter-channel relationships and the specific frequency scales at which these interactions occur. Specifically, in our application, the patient's regular condition was characterized by a concentration of density in the low-frequency band, reflecting a stable and predictable cardiac signal. Conversely, the symptomatic condition exhibited increased dispersion in higher frequency bands, indicative of irregularities and heightened complexity in inter-channel interactions. These findings suggest that the patient's sensation of detachment correlates with increased instability or altered dynamics between channels.

Looking ahead, applying this framework to larger datasets and diverse clinical scenarios could further validate its robustness and establish it as a key tool for advancing cardiac diagnostics.

References

1. Benitez, D., Gaydecki, P. A., Zaidi, A., Fitzpatrick, A. P.: The use of the Hilbert transform in ECG signal analysis. Comput. Biol. Med. **31**(5), 399–406 (2001)
2. Butler, C., Dorahy, M. J., Middleton, W.: The Detachment and Compartmentalization Inventory (DCI): An assessment tool for two potentially distinct forms of dissociation. J. Trauma Dissociation. **20**(5), 526–547 (2019)
3. Dash, S. K., Rao, G. S.: ECG arrhythmia detection using Choi-Williams time-frequency distribution and artificial neural network. Int. J. Adv. Res. Comput. Commun. Eng. (2016)
4. Diker, A., Avci, E.: Feature extraction of ECG signal by using deep feature. In: 2019 7th International Symposium on Digital Forensics and Security (ISDFS), pp. 1–6. IEEE (2019)
5. Elgendi, M.: Fast QRS detection with an optimized knowledge-based method: evaluation on 11 standard ECG databases. PLoS ONE. **8**(9), e73557 (2013).
6. Hörmann, S., Kidziński, L., Hallin, M.: Dynamic functional principal components. J. R. Stat. Soc. Ser. B Stat. Methodol. **77**(2), 319–348 (2015)
7. Hu, Y., Zhao, Y., Liu, J., Pang, J., Zhang, C., Li, P.: An effective frequency-domain feature of atrial fibrillation based on time-frequency analysis. BMC Med. Inform. Decis. Mak. **20**(1), 308 (2020).
8. Huang, K., Zhang, L.: Cardiology knowledge-free ECG feature extraction using generalized tensor rank one discriminant analysis. EURASIP J. Adv. Signal Process. (2014)
9. Makowski, D., Pham, T., Lau, Z. J., Brammer, J. C., Lespinasse, F., Pham, H., Schölzel, C., Chen, S. A.: NeuroKit2: A Python toolbox for neurophysiological signal processing. Behav. Res. Methods. **53**(4), 1689–1696 (2021).
10. Mallat, S. G.: Multifrequency channel decompositions of images and wavelet models. IEEE Trans. Acoust. Speech Signal Process. **37**(12), 2091–2110 (1989)
11. Panaretos, V. M., Tavakoli, S.: Fourier analysis of stationary time series in function space. Ann. Stat. **41**, 568–603 (2013)
12. Sattar, Y., Chhabra, L.: Electrocardiogram. Treasure Island (FL): StatPearls Publishing (2022)
13. Sepahvand, M., Abdali-Mohammadi, F.: A novel multi-lead ECG personal recognition based on signals functional and structural dependencies using time-frequency representation and evolutionary morphological CNN. Biomed. Signal Process. Control. **68**, 102766 (2021)
14. Singh, A. K., Krishnan, S.: ECG signal feature extraction trends in methods and applications. BioMedical Engineering OnLine. **22**(1), 22 (2023)

Chapter 28
Comparison of Deep Learning Methods for Functional Data

Nouhaila Goujili, Matthieu Saumard and Maher Jridi

Abstract In this study, we explore the use of deep learning methods to perform functional regression models, highlighting their potential in analyzing and predicting functional data. Recent advancements in neural operators have paved the way for models like the Fourier Neural Operator (FNO) and Neural Operator Flows (OpFlow), which are capable of learning complex mappings between infinite-dimensional function spaces. Additionally, we discuss the Function Direct Neural Network model (FDNN), Deep Functional Multiple index Models (DFMM) and Adaptive Functional Neural Network (AdaFNN) to increase the flexibility of functional regression tasks. We present a comparative analysis of these deep functional networks, on simulations and on two case studies: Tecator and Building Occupancy datasets.

28.1 Introduction

Functional Data Analysis (FDA) provides a framework for analyzing continuous stochastic processes, as extensively explored in [1], and further developed in the

Nouhaila Goujili (✉)
L@bISEN, Vision-AD, ISEN Yncréa Ouest, 33 Chemin du Champ de Manoeuvres, 44470 Carquefou, France
Kardham Group, 10 rue Débarcadère, 75017 Paris, France
e-mail: nouhaila.goujili@isen-ouest.yncrea.fr, ngoujili@kardham.com

Matthieu Saumard
L@bISEN, Vision-AD, ISEN Yncréa Ouest, 20 rue Cuirassé Bretagne, 29200 Brest, France
e-mail: matthieu.saumard@isen-ouest.yncrea.fr

Maher Jridi
L@bISEN, Vision-AD, ISEN Yncréa Ouest, 33 Chemin du Champ de Manoeuvres, 44470 Carquefou, France
e-mail: maher.jridi@isen-ouest.yncrea.fr

© The Author(s), under exclusive license to Springer Nature Switzerland AG 2025

G. Aneiros et al. (eds.), *New Trends in Functional Statistics and Related Fields*, Contributions to Statistics, https://doi.org/10.1007/978-3-031-92383-8_28

context of non-parametric methods in [2]. Recent advancements in FDA have introduced novel methods, such as functional neural networks and neural operators, which extend FDA to non-linear and deep learning frameworks. These models have shown significant promise in applications requiring high flexibility and precision, such as predicting time series, modeling spatial-temporal dynamics, and solving complex regression tasks [3, 4].

In this paper, we explore and compare four deep learning models for functional data analysis. Specifically, we focus on Function Direct Neural Networks (**FDNN**)[5] presented in **Sect.** 28.2.1, Neural Operator Flows (**OpFlow**)[4] in **Sect.** 28.2.2, Deep Functional Multiple Index Models (**DFMM**)[6] in **Sect.** 28.2.3, and the Adaptive Functional Neural Network (**AdaFNN**)[7] in **Sect.** 28.2.4. We evaluate these methods on simulations in **Sect.** 28.3.1 and on two case studies – the Tecator and Building Occupancy datasets – in **Sect.** 28.3.2. Our findings highlight the transformative potential of these methods in addressing real-world challenges in functional data analysis.

28.2 Evolution Towards Non-Linear and Functional Deep Models

A notable advancement in the field of functional neural networks is the introduction of neural operators [4], which allow learning mappings between infinite-dimensional function spaces. Unlike standard neural networks that operate on finite-dimensional vectors, neural operators work directly on continuous functions. This makes them well-suited for complex tasks such as solving partial differential equations and modeling complex physical systems [8]. Among the classes of neural operators, Fourier neural operators (FNOs) stand out for their ability to leverage spectral representations to solve complex problems. By using Fourier transformation, FNOs can efficiently capture global information and handle high-dimensional problems. Their performance surpasses traditional numerical methods in prediction tasks in physical domains, such as fluid dynamics and weather forecasting [8]. In the remainder of this section, we present the four methods evaluated: FDNN, OpFlow, DFMM and AdaFNN.

28.2.1 Functional Direct Neural Network (FDNN)

Functional neural networks (FNNs) are an extension of classical neural networks, specifically designed to handle data where inputs or outputs are continuous functions rather than finite-dimensional vectors, as explained in [9]. This capability represents a significant improvement over traditional functional data analysis models, such as single index models, which often limit interactions between the temporal or spatial points of functional data. FNNs can manage the temporal or spatial structure of the data, which is essential for modeling phenomena evolving over time or space.

In [5], the authors introduce a functional neural network featuring continuous hidden layers, referred to as the Function Direct Neural Network (FDNN). This model uses continuous neurons and refer to a hidden layer with multiple continuous neurons as a continuous hidden layer. The kth continuous neuron in the lth continuous hidden layer is given by:

$$H_k^{(l)}(s) = \sigma\left(b_k^{(l)}(s) + \sum_{j=1}^{J} \int w_{(j,k)}^{(l)}(s,t) H_j^{(l-1)}(t)dt\right), \tag{28.1}$$

and the real output is given by:

$$\hat{Y} = H^{(L+1)} = \sigma\left(b^{(L+1)} + \sum_{j=1}^{J} \int w_j^{(L+1)}(t) H_j^{(L)}(t)dt\right), \tag{28.2}$$

where $l = 1, 2, 3, \ldots, L$, and L is the number of continuous hidden layers, $b_k^{(l)}(s) \in \mathcal{L}^2(T)$ is the unknown intercept function, $w_{(j,k)}^{(l)}(s,t)$ is the bivariate parameter function for the kth continuous neuron in the lth hidden layer coming from the jth continuous neuron of the $(l-1)^{\text{th}}$ hidden layer, $w_j^{(L+1)}(t)$ is a parameter function for the output layer coming from the jth continuous neuron of the $(L)^{\text{th}}$ hidden layer, and $\sigma(\cdot)$ is a non-linear activation function. The different parameters are adjusted by a back propagation algorithm described in [5].

28.2.2 Neural Operator Flows (OpFlow)

In [4], the authors introduced Neural Operator Flows (OpFlow), an infinite-dimensional extension of normalizing flows. OpFlow functions as an invertible operator that transforms a potentially unknown data function space into a Gaussian process, enabling precise likelihood estimation for functional point evaluations. The structure of the model is based on flow-based networks, combining spectral layers, integral operators, and affine couplings [10, 11]. These components are organized to perform a bijective transformation between simple distributions (e.g., Gaussian) and complex target distributions. Structural elements include 1D spectral layers and functions for noise injection or covariance computation. The architecture is built around an OpFlow model integrating multiple bijections (such as Affine Coupling), explicit data handling in the spectral domain, and a mechanism to optimize the log-probabilities of samples. The model is designed to model complex relationships in continuous functions using spectral integration techniques and deep networks.

28.2.3 Deep Functional Multiple Index Models (DFMM)

The article [6] is designed to classify emotions in speech based on MFCCs that can be regarded as a multivariate functional object. The transformation module is made of a stack of N transformer encoders with self-attention and a feed-forward network. The second module is the neural network of the paper [7] which is generalized to adapt to the multivariate context. In few words, the paper of [7] proposes to integrate the data by an adaptive function by a numerical scheme. So, the outputs of this second module are simply the L^2 product between each component of the transformed variables and an adaptive function. The third module of our network is a classical fully connected layer. The four modules of our new network architecture are:

1. The transformation module is made of N transformer encoders.
2. The Deep Functional Network that outputs the scores of multivariate functional variable.
3. Fully Connected Layers to reduce the dimension.
4. A regularisation module with a L^2 loss on the acceleration of parameter to produce the output.

There are three losses, L^1 and L^2 loss of all AdaFNN submodules as in [7] and the regularisation loss (ie. L^2 of the acceleration of the parameter to output the final Y_{n+1}).

28.2.4 Adaptive Functional Neural Network (AdaFNN)

In [7], the authors introduce the Adaptive Functional Neural Network (AdaFNN), a model designed to process functional data by exploiting an adaptive base layer. This architecture is based on a combination of adaptive integration methods and deep neural networks, enabling flexible representation of functional data. The model is broken down into several stages. Firstly, an adaptive basis layer is used to transform input functions into a discrete representation using a set of bases optimized during training. Then, a neural network is employed to capture the non-linear relationships between these representations. Regularization is a central element of the model, with penalization on successive derivatives of the basis coefficients to control complexity and promote smoothness of the representations.

28.3 Empirical Results

In this section, we evaluate and compare the performance of the four models — FDNN, OpFlow, DFMM and AdaFNN — across various simulation scenarios presented in [5], as well as benchmarking with two real world data sets.

Table 28.1: Comparison of the RMSE across methods under various settings involving functional predictors and a continuous response.

Mapping	Linear	Multiple Index	Quadratic
FDNN	**1.019**	**1.100**	**1.092**
OpFlow	1.064	1.206	1.118
DFMM	1.230	2.307	2.167
AdaFNN	1.214	2.902	2.603

28.3.1 Simulation Study

As in [5], we simulate $n = 1500$ i.i.d. random curves $\{X_1(t), \ldots, X_n(t)\}$ from a Gaussian process with mean 0 and covariance

$$C_X(t, s) = \frac{\sigma^2}{\Gamma(\nu)2^{\nu-1}} \left(\frac{\sqrt{2\nu}|t-s|}{\rho} \right)^{\nu} K_\nu \left(\frac{\sqrt{2\nu}|t-s|}{\rho} \right), \tag{28.3}$$

which is the Matérn covariance function, where K_ν represents the modified Bessel function of the second kind. The parameters are set as $\rho = 0.5$, $\nu = 5/2$, and $\sigma^2 = 1$. These curves are evaluated at $m = 200$ equally spaced time points over the interval $[0, 1]$. We replicated the three simulation scenarios from [5], considering a continuous response and a single predictor function: *Linear*, *Multiple index* and the *Quadratic* scenario.

We evaluate the root mean squared error (RMSE) of the prediction for the continuous response. In **Table 28.1**, we summarize the performance of each method across different simulation scenarios with a continuous response. FDNN offers the best performance for each configuration. In particular, for the *Linear* case, FDNN achieves an RMSE of 1.019, slightly lower than that of OpFlow (1.064), and significantly lower than that of DFMM (1.230). FDNN's advantage is reinforced in the *Multiple Index* and *Quadratic* situations: FDNN maintains RMSE values around 1.10, while OpFlow ranges between 1.118 and 1.206, and DFMM reaches significantly higher values (2.307 and 2.167).

28.3.2 Real Data

In this section, we present two real case studies to evaluate the performance of the different approaches introduced in **Sect. 28.2**. We begin by analyzing the Tecator dataset, an example of scalar regression using functional variables. We then study a dataset relating to the occupancy of an office building, where the response variable is functional.

Table 28.2: Models Performance on RMSE with Tecator data

Model	FDNN	**OpFlow**	DFMM	AdaFNN
RMSE	6.48	**1.16**	5.60	11.19

28.3.2.1 Tecator Data

In this section, we analyze and compare the performance of the four methods presented in **Sect.** 28.2 on the Tecator dataset. This dataset comprises 216 samples, each represented by a spectrometric curve capturing absorbance values at 100 wavelengths (ranging from 852 nm to 1050 nm, with 2 nm increments). Each sample is also associated with its fat content, determined through analytical chemical processing. The dataset is divided into 176 samples for training (including 40 for validation) and 40 samples reserved for testing. We adapted the **OpFlow** model to apply it to this case study "Tecator" data. In this case, the response variable to predict is scalar, we adjusted the Fourier Neural Operator (FNO) for scalar-on-function regression. The application of the **DFMM** model on "Tecator" data is tailored to predict a scalar variable from spectral data. Here, the inputs consist of discrete spectrum values, and the target is a scalar value (e.g., fat content). We adapted the model in this case to combine FeedForward networks with transformer layers. Additionally, hyperparameter tuning is performed using Optuna [12] to fine-tune model parameters and optimize prediction accuracy.

To evaluate forecasting and regression models, we use two main metrics: **MAPE** (Mean Absolute Percentage Error) and **RMSE** (Root Mean Squared Error).

In the case of the Tecator data, it is a regression problem because it involves modeling a functional relationship between functional explanatory variables (spectra at 100 wavelengths) and continuous quantitative response variables (such as fat content). We present the RMSE metric of the different models compared in **Table 28.2**.

The results in **Table 28.2** demonstrate that the OpFlow model significantly outperforms FDNN, DFMM and AdaFNN. The low RMSE (**1.16**) value indicates that OpFlow is better suited to model spectral data, probably due to its ability to learn complex relationships in functional spaces using spectral representations.

28.3.2.2 Building Occupancy Data

We applied our models to real data collected from a software publisher offering products for trading, risk management, and transaction processing for participants in financial markets. The data can be found in our GitHub repository[1]. This data was gathered at their Paris site, a 9-story office building with employees distributed

[1] github.com/goujilinouhaila-coder/Prediction-of-Employee-Presence-in-Smart-Workplaces

Table 28.3: Models Performance on MAPE with Occupancy data

Model	FDNN	OpFlow	**DFMM**	AdaFNN
MAPE	6.15%	9.10%	**5.86%**	6.54%

across 4 divisions (teams). The building is equipped with access control doors, where employees use their badges to enter the premises. We collected this badge data over a one-year period, from September 2022 to September 2023. In **Fig. 28.1**, we show an overview of the functional venue data in our real-life case study, segmented by day of the week, from 1 (Monday) to 5 (Friday).

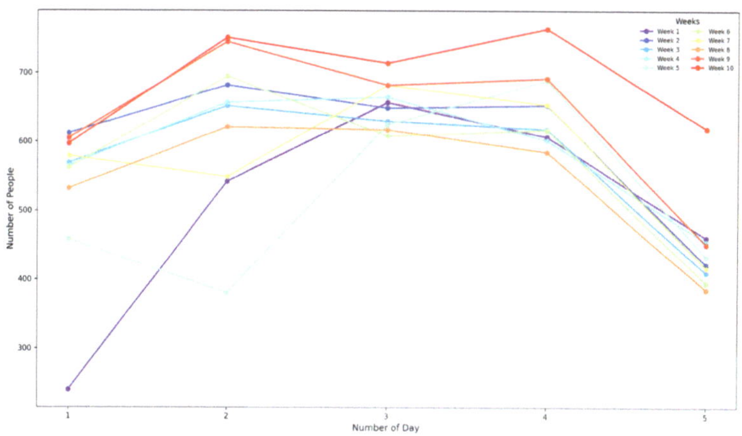

Fig. 28.1: Functional time series plot of number of people by day of week

To apply the **FDNN** model presented in **Sect.** 28.2.1, we adapted it to our case study "Occupancy" data set , in order to predict a functional response variable rather than a scalar one.

For the **OpFlow** model described in **Sect.** 28.2.2, the adaptations focus on the target variable, which is functional, and the structure of the model is specifically designed for time-series prediction on a single feature.

As shown in **Table 28.3**, DFMM achieves the best performance on the Building Occupancy dataset, with the lowest MAPE (**5.86%**). This suggests that its architecture is particularly effective in capturing temporal dependencies and modeling functional response variables. Although FDNN also performs well, its slightly higher MAPE (**6.15%**). In contrast, OpFlow's relatively higher MAPE (**9.10%**) highlights a limitation in handling occupancy data, potentially due to the simplified spectral layers not being well-suited for capturing time-series patterns.

28.4 Conclusion

Our comparative analysis, applying the FDNN, OpFlow, DFMM and AdaFNN models to two diverse datasets –Tecator and Building Occupancy data– highlights the strengths and trade-offs of each approach. Although OpFlow excels in tasks requiring precise modeling of numerical and scalar predictions, such as spectral data regression, DFMM proves effectiveness in predicting functional outcomes, as evidenced by its performance on the Building Occupancy dataset. Future exploration could include adaptations to incorporate hybrid features, for example combining the strengths of transformer networks with spectral representations.

References

1. Ramsay, J.O., Silverman, B.W.: Functional Data Analysis, 2nd edn. Springer, New York (2005).
2. Ferraty, F., Vieu, P.: Nonparametric Functional Data Analysis: Theory and Practice. Springer Series in Statistics. Springer, New York (2006).
3. Rossi, F., Delannay, N., Conan-Guez, B., Verleysen, M.: Representation of functional data in neural networks. Neurocomputing, **64**, 183–210 (2005).
4. Shi, Y., Gao, A.F., Ross, Z.E., Azizzadenesheli, K.: Universal Functional Regression with Neural Operator Flows. arXiv:2404.02986 (2024).
5. Rao, A. R., Reimherr, M.: Nonlinear Functional Modeling Using Neural Networks. Journal of Computational and Graphical Statistics, **32**(4), 1248–1257 (2023).
6. Saumard, M., El Haj, A., Napoleon, T.: Deep functional multiple index models with an application to SER. arXiv:2403.17562 (2024).
7. Yao, J., Mueller, J., Wang, J.-L.: Deep Learning for Functional Data Analysis with Adaptive Basis Layers. arXiv:2106.10414 (2021).
8. Li, Z., Kovachki, N., Azizzadenesheli, K., Liu, B., Bhattacharya, K., Stuart, A., Anandkumar, A.: Fourier Neural Operator for Parametric Partial Differential Equations. arXiv:2010.08895 (2021).
9. Rossi, F., Conan-Guez, B.: Functional multi-layer perceptron: a non-linear tool for functional data analysis. Neural Networks, **18**(1), 45–60 (2005).
10. Zhou, Q., Du, C., Li, D., Wang, H., Liu, J.K., He, H.: Neural Encoding and Decoding With a Flow-Based Invertible Generative Model. IEEE Transactions on Cognitive and Developmental Systems **15**(2), 724–736 (2023).
11. Ping, W., Peng, K., Zhao, K., Song, Z.: WaveFlow: A Compact Flow-based Model for Raw Audio. In: Proceedings of the 37th International Conference on Machine Learning (ICML), PMLR, pp. 7706–7716 (2020).
12. Akiba, T., Sano, S., Yanase, T., Ohta, T., Koyama, M.: Optuna: A Next-generation Hyperparameter Optimization Framework. Proceedings of the 25th ACM SIGKDD International Conference on Knowledge Discovery & Data Mining, 2623–2631 (2019).
13. Rasmussen, C.E., Williams, C.K.I.: Gaussian Processes for Machine Learning. The MIT Press, Cambridge (2005).
14. Wang, Q., Zheng, S., Farahat, A., Serita, S., Saeki, T., Gupta, C.: Multilayer Perceptron for Sparse Functional Data. In: 2019 International Joint Conference on Neural Networks (IJCNN), pp. 1–10. IEEE (2019).
15. Gertheiss, J., Rügamer, D., Liew, B.X.W., Greven, S.: Functional Data Analysis: An Introduction and Recent Developments. Biometrical Journal, **66**(7) (2024).

Chapter 29
The Common Support Function with Applications

Nicolás Hernández and Stanislav Nagy

Abstract Advances in information technology have amplified the need for sophisticated tools to analyze complex functional data. Functional Data Analysis meets this demand by emphasizing the selection of key features. This process reduces the data's dimensionality, making it easier to interpret and draw meaningful conclusions. This paper introduces the common support function, a novel method that leverages quantile and expectile estimation to identify subdomains where two stochastic processes differ the most. By capturing differences in mean and variance, the method highlights regions of minimal overlap between classes, enabling effective domain selection and shape analysis. This tool is applied to real ECG data to identify critical segments associated with myocardial infarction. We were able to pinpoint subintervals that align with medically relevant features such as the QRS complex and ST segments.

29.1 Introduction

Advances in information technology have led to an immense increase in the volume and complexity of data, creating an urgent need for sophisticated analytical tools to interpret data from random processes, such as functional data. One of the main objectives of Functional Data Analysis is feature selection, since it allows the reduction of the dimension of functional data, facilitating the use of inferential techniques for this kind of data. Recent studies explore feature selection in two primary forms: (i) *variable selection*, which implies selecting individual points within the domain of functions, as in [5, 4] and references therein; or (ii) *domain selection*, selecting

Nicolás Hernández (✉)
Queen Mary University of London, School of Mathematical Sciences, Mile End Rd, London E1 4NS, UK, e-mail: n.hernandez@qmul.ac.uk

Stanislav Nagy
Charles University, Faculty of Mathematics and Physics, Sokolovská 83, 186 75 Prague, Czech Republic, e-mail: nagy@karlin.mff.cuni.cz

© The Author(s), under exclusive license to Springer Nature Switzerland AG 2025

G. Aneiros et al. (eds.), *New Trends in Functional Statistics and Related Fields*, Contributions to Statistics, https://doi.org/10.1007/978-3-031-92383-8_29

intervals within the domain of the functions, which capture essential information for the desired inference; see [15, 16] and [6].

In domain selection, the learning problem is to determine which parts of the domain T show the greatest differences between two stochastic processes $X(t)$ and $Y(t)$. Identifying these regions provides valuable insights into the underlying differences between the groups, enabling us to focus the analysis where it matters most.

To address this need, we present the *common support function*, a new object that incorporates quantile and expectile estimation to capture the marginal distributions of functional data in each class. Through this function, we can uncover specific subdomains $[a, b] \subset T$ where the overlap in support between two classes is minimal, highlighting the areas where the processes differ the most.

This approach has practical implications not only for descriptive analysis and visualization, but also for prediction and inference. In supervised learning, for instance, reducing redundant information between classes can improve model performance, reducing testing error rates. Additionally, domain selection can increase the power of two-sample tests by minimizing the overlap in support between classes.

A key advantage of the common support function is that it incorporates information from the mean and variance of the functional data simultaneously, capturing differences in both moments across classes. In its simplest form, the marginal distributions are estimated independently at each point t.

The common support curves are introduced in Sect. 29.2. They are applied to real data analysis of ECG segments in Sect. 29.3, and some concluding remarks are given in Sect. 29.4.

29.2 The Common Support Curve

Write $\mathcal{F}$ for the space of continuous real functions defined on an interval $T = [0, 1]$. The set of all (Borel) probability measures on a measurable space $\mathcal{S}$ is denoted by $\mathcal{P}(\mathcal{S})$. All random variables are defined on a common probability space $(\Omega, \mathcal{A}, \mathsf{P})$.

Let $P \in \mathcal{P}(\mathcal{F})$, and let $\alpha \in [0, 1/2]$ be given. Let $X \sim P$ be a random function with distribution P. Denote by $F_{X(t)}$ the distribution function of the random variable $X(t)$, that is

$$F_{X(t)}(u) = \mathsf{P}(X(t) \leq u) \quad \text{for } u \in \mathbb{R},$$

and let $F^{-1}_{X(t)}(v) = \inf\left\{u \in \mathbb{R} \colon F_{X(t)}(u) \geq v\right\}$ be the associated quantile function.

29.2.1 *α-Support and the Common α-Support Function*

The *α-support* $S_\alpha(P)$ of the distribution P is defined as the set of functions $x \in \mathcal{F}$ whose graph $G(x) = \left\{(t, x(t))^\mathsf{T} \colon t \in [0, 1]\right\} \subset \mathbb{R}^2$ is contained in the coordinate-wise α-inter-quantile region given by

$$\left\{(t,y)^\top : t \in [0,1] \text{ and } F^{-1}_{X(t)}(\alpha) \le y \le F^{-1}_{X(t)}(1-\alpha)\right\}. \qquad (29.1)$$

For $\alpha = 1/2$, the elements of the α-support $S_{1/2}(P)$ are exactly the cross-sectional median functions of P, that is, functions that to each $t \in [0,1]$ assign the median of the random variable $X(t)$. The α-support is monotonous in α in the sense that for $\alpha_1 < \alpha_2$ we have $S_{\alpha_2}(P) \subseteq S_{\alpha_1}(P)$, and as $\alpha \to 0$, this set approximates the (cross-sectional) support of P.

The set $S_\alpha(P)$ has a close connection to the upper-level set of the so-called infimal depth (also called φ-depth [9], or extremal depth [12]) that can be found in the literature. It is also intimately related to the concept of envelope testing [11]. For a simple example of the α-support regions of two random samples of functions, see the top left panel of Fig. 29.1. As can be seen, the α-support regions describe the overall course and dispersion of the functional data quite well.

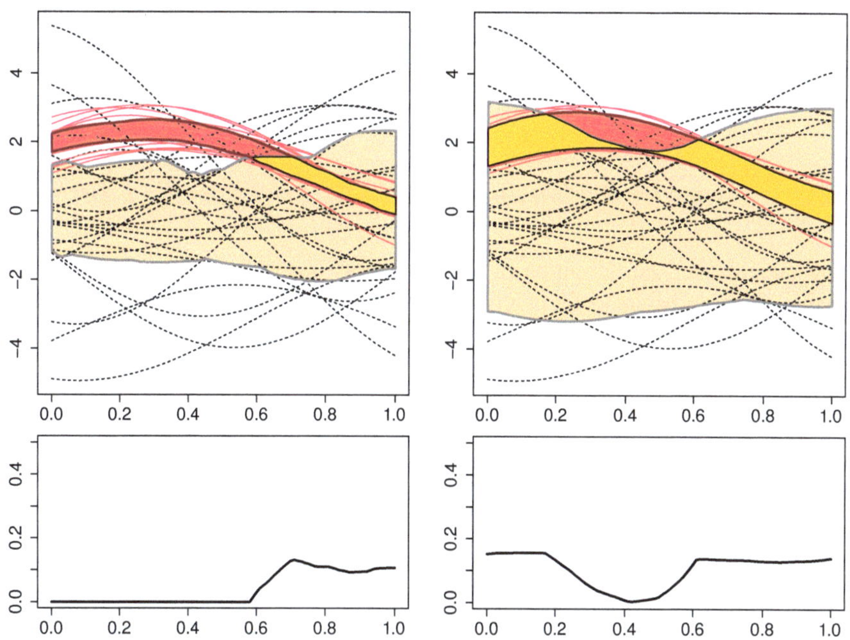

Fig. 29.1: The α-support/expectile regions and the common α-support/expectile functions of two random samples (dashed black and solid red curves, respectively) for $\alpha = 1/5$. The α-support regions are in the left-hand panels, the α-expectile regions are in the right-hand panels. The intersection of the two α-bands is in gold. The standardized width of the golden regions is measured by the common α-support/expectile functions displayed in the bottom panels

The *common α-support* function $s_\alpha(t; P, Q) = s_\alpha(t)$ of two distributions $P, Q \in \mathcal{P}(\mathcal{F})$ is defined as

$$s_\alpha(t) = \frac{\max\left\{\min\left\{F^{-1}_{X(t)}(1-\alpha), F^{-1}_{Y(t)}(1-\alpha)\right\} - \max\left\{F^{-1}_{X(t)}(\alpha), F^{-1}_{Y(t)}(\alpha)\right\}, 0\right\}}{\left(F^{-1}_{X(t)}(1-\alpha) - F^{-1}_{X(t)}(\alpha)\right) + \left(F^{-1}_{Y(t)}(1-\alpha) - F^{-1}_{Y(t)}(\alpha)\right)},$$

whenever the denominator is positive; if it is not, we define

$$s_\alpha(t; P, Q) = \mathbb{I}\left(F^{-1}_{X(t)}(\alpha) = F^{-1}_{Y(t)}(\alpha)\right).$$

In both these formulas, $X \sim P$ and $Y \sim Q$. The common α-support evaluates, for each time point $t \in [0, 1]$, the standardized degree of overlap between the marginal distributions of $X(t)$ and $Y(t)$. Certainly, $s_\alpha(t; P, Q) \in [0, 1]$. It equals 0 if the α-inter-quantile regions of $X(t)$ and $Y(t)$ do not overlap, and is 1 if and only if the two α-inter-quantile regions are the same. For an example of the α-support curve, see the bottom left panel of Fig. 29.1.

The integral of the common α-support $\int_0^1 s_\alpha(t; P, Q)\,\mathrm{d}t$ interprets as the overall level of similarity of the regions determined by the α- and $(1-\alpha)$-quantiles given by distributions P and Q.

29.2.2 α-Expectile Region and the Common α-Expectile Function

Using quantiles in the definition of the α-support is straightforward and easy to interpret. On the other hand, for practical purposes, quantiles are rather "non-smooth" functionals, and other approaches may be useful for summarizing the location/spread of the marginal distributions. To do so, we consider in this section the collection of expectiles [13, 8]. For simplicity, throughout this section, assume that the expected values of all considered random variables are well-defined and finite.

For $\alpha \in (0, 1)$, recall that the α-expectiles[1] are for a real-valued random variable $R \in \mathcal{P}(\mathbb{R})$ with distribution function G

$$E_R^-(\alpha) = \frac{1}{\alpha}\int_0^\alpha G^{-1}(u)\,\mathrm{d}u. \tag{29.2}$$

For $\alpha = 1$, we obtain

$$E_R^-(1) = \int_0^1 G^{-1}(u)\,\mathrm{d}u = \int_{-\infty}^{\infty} s\,\mathrm{d}G(s) = \mathsf{E}\,R,$$

the expected value of R, as follows by a simple change of variables. For $\alpha < 1$, the expectile can be interpreted as the conditional expectation of R given that $R \leq G^{-1}(\alpha)$. From this, we see that the expectiles are monotone in the sense that for

[1] In the literature, the quantity (29.2) is sometimes called integrated quantile, or expected shortfall, and the term expectile is used for a closely related (but different) concept, see [8]. Here, we use the term expectile for brevity.

$\alpha_1 < \alpha_2$ we have $E_R^-(\alpha_1) \leq E_R^-(\alpha_2)$, and as $\alpha \to 0$, $E_R^-(\alpha)$ converges to the lower endpoint of the support of R. As a major difference from the quantiles, expectiles are known to be sensitive to the shape of the (lower) tail of the distribution of R.

We see that the expectiles (29.2) of R always take values smaller than the expectation $\mathsf{E}\,R$. We thus call them also the lower expectiles, and consider the complementary notion of the upper α-expectiles, given by $E_R^+(\alpha) = -E_{-R}^-(\alpha)$, that is the (lower) α-expectile of random variable $-R$ with a negative sign. Then we have, similarly to the α-inter-quantile region $\left[G^{-1}(\alpha), G^{-1}(1-\alpha)\right]$ that

$$E_R^-(\alpha) \leq \mathsf{E}\,R \leq E_R^+(\alpha) \quad \text{for each } \alpha \in (0,1].$$

As for the quantiles, also the family of the upper and the lower expectiles characterizes the distribution of R uniquely [8] (given that $\mathsf{E}\,R$ exists).

For functional data, we introduce the *α-expectile region* of $P \in \mathcal{P}(\mathcal{F})$ as the set of all functions $x \in \mathcal{F}$ whose graph lies entirely in the coordinate-wise α-inter-expectile region given by

$$\left\{(t, y)^{\mathsf{T}} : t \in [0,1] \text{ and } E_{X(t)}^-(\alpha) \leq y \leq E_{X(t)}^+(\alpha)\right\}.$$

Further, the *common α-expectile* function of $X \sim P \in \mathcal{P}(\mathcal{F})$ and $Y \sim Q \in \mathcal{P}(\mathcal{F})$ is defined as

$$e_\alpha(t; P, Q) = \frac{\max\left\{\min\left\{E_{X(t)}^+(\alpha), E_{Y(t)}^+(\alpha)\right\} - \max\left\{E_{X(t)}^-(\alpha), E_{Y(t)}^-(\alpha)\right\}, 0\right\}}{\left(E_{X(t)}^+(\alpha) - E_{X(t)}^-(\alpha)\right) + \left(E_{Y(t)}^+(\alpha) - E_{Y(t)}^-(\alpha)\right)},$$

whenever the denominator is positive; if it is not, we again define

$$e_\alpha(t; P, Q) = \mathbb{I}\left(E_{X(t)}^-(\alpha) = E_{Y(t)}^-(\alpha)\right).$$

An example of the application of α-expectile regions and the common α-expectile function is in the right-hand panels of Fig. 29.1. As can be seen from that figure, compared to the quantile-based characteristics, the expectile-based bands are wider, and considerably "smoother". That is a consequence of the smoothing effect resulting from taking the conditional expectation in the tail of the distribution when considering an expectile. Similar features can also be observed theoretically, for instance, it can be proved that the expectile-based characteristics have better finite-sample consistency properties than the quantile-based ones. Due to space restrictions, a rigorous statement of this result will appear elsewhere.

29.3 Real Data Application: Identifying ECG Segments

From a medical perspective, when a patient arrives at the emergency room during a cardiac arrest, one of the primary diagnostic tools available is the ECG. This makes it an invaluable resource for immediate decision-making and for assessing both the causes and severity of cardiac conditions [10]. However, analyzing ECG data presents significant challenges, particularly due to its high dimensionality. Consequently, studying the local features of ECG signals plays a vital role in medicine: (i) improving early *diagnosis* of cardiac diseases by identifying abnormal patterns in the cardiac cycle and (ii) *understanding the causes and effects* of heart conditions through the study of anomalous time intervals, see [6].

In this context, the marginal common support curve can be used for domain selection to identify ECG segments by defining a threshold δ such that $e_\alpha(t; P, Q)$ or $s_\alpha(t; P, Q) \leq \delta$. This approach identifies sub-intervals within the domain where the two processes show less common α-support. The threshold δ can be defined considering some prior knowledge or via cross-validation. These selected domains can then be utilized to improve the accuracy of classification models and increase the power of two-sample tests.

In the left-hand panel of Fig. 29.2, we see the ECG dataset [14], available in the UEA & UCR[2] repository, contains 200 signals sampled at 96 equally spaced time points. Each observation represents the electrical activity of the heart recorded during a single heartbeat. The dataset is divided into two groups: 133 signals from normal heartbeats and 67 signals from myocardial infarctions.

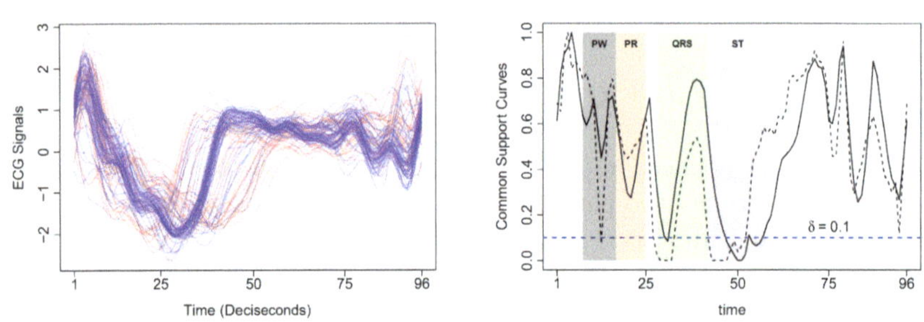

Fig. 29.2: **Left panel:** ECG signals corresponding to healthy (red lines —) and myocardial infarction (blue lines —) signals. **Right panel:** α-support/expectile curves in (- - -) and (—) respectively, for $\alpha = 0.2$, along with several relevant ECG segments in colors. The P-wave (PW) represents atrial depolarization, and its abnormalities.

Given the threshold $\delta = 0.1$, the selected domains (right-hand panel of Fig. 29.2) correspond to grid points 28–31 and 43–52 (in deciseconds) where normal heartbeats

[2] `http://timeseriesclassification.com/index.php`

and myocardial infarction signals show significant differences. These differences partially align with the QRS complex (28–42) and ST segments (42–58) deciseconds (approximately). The QRS complex, representing ventricular depolarization, is crucial for identifying ventricular arrhythmias [2]. Finally, the ST segment, an indicator of early ventricular repolarization, is a key marker in detecting acute myocardial infarction (ST elevation) or ischemia and hypokalemia (ST depression), playing a significant role in cardiovascular disease outcome studies [7]. There are also two other sections, that can be identified which play an important role. The P-wave (PW) represents atrial depolarization, and its abnormalities. From a medical standpoint, an abnormal P-wave can suggest atrial enlargement [3]. The PR segment represents the time between atrial depolarization and ventricular depolarization. Abnormalities in the timing of the PR segment can indicate pathology, such as heart block and ischemia [1].

29.4 Conclusion and Further Research

In this paper we introduce the common support function as a novel method for identifying subdomains where two stochastic processes, $X(t)$ and $Y(t)$, differ the most. By leveraging on quantile and expectile estimation, this approach captures differences in both the mean and variance of functional data, pinpointing regions with minimal overlap between classes. This tool addresses key challenges in functional data analysis, such as *Domain Selection*.

In our application to real ECG data, we use the marginal common support curve to effectively identify critical ECG segments associated with myocardial infarction. By setting a threshold $\delta = 0.1$, we detected subintervals where normal and pathological heartbeats significantly differ, particularly aligning with the medically important QRS complex and ST segments.

As a future research avenue, we will extend this approach to incorporate information from other points in the domain, resulting in a multivariate version of the common support function. This extension is particularly valuable for performing shape analysis in the functional data setting.

Acknowledgements The research of S. Nagy was supported by the Czech Science Foundation (project n. 24-10822S) and the ERC CZ grant LL2407 of the Ministry of Education, Youth and Sport of the Czech Republic.

References

1. Alnsasra, H., Ben-Avraham, B., Gottlieb, S., Ben-Avraham, M., Kronowski, R., Iakobishvili, Z., Goldenberg, I., Strasberg, B., and Haim, M.: High-grade atrioventricular block in patients with acute myocardial infarction. Insights from a contemporary multi-center survey. *Journal of Electrocardiology* **51**(3), 386–391 (2018)

2. Atwood, D. and Wadlund, D. L.: ECG interpretation using the CRISP method: a guide for nurses. *AORN journal* **102**(4), 396–408 (2015)
3. Baranchuk, A. and Bayés de Luna, A.: The P-wave morphology: what does it tell us? *Herzschrittmachertherapie & Elektrophysiologie* **26**(3), 192–199 (2015)
4. Berrendero, J. R., Cuevas, A., and Torrecilla, J. L.: Variable selection in functional data classification: a maxima-hunting proposal. *Statist. Sinica* **26**(2), 619–638 (2016)
5. Fraiman, R., Gimenez, Y., and Svarc, M.: Feature selection for functional data. *J. Multivariate Anal.* **146**, 191–208 (2016)
6. Hernández, N. and Martos, G.: Domain selection for Gaussian process data: An application to electrocardiogram signals. *Biometrical Journal* **66**(8), e70018 (2024)
7. Morris, N. P., and Body, R.: The De Winter ECG pattern: morphology and accuracy for diagnosing acute coronary occlusion: systematic review. *European Journal of Emergency Medicine* **24**(4), 236–242 (2017)
8. Mosler, K.: *Multivariate dispersion, central regions and depth: The lift zonoid approach*, volume 165 of *Lecture Notes in Statistics*. Springer-Verlag, Berlin (2002)
9. Mosler, K.: Depth statistics. In Becker, C., Fried, R., and Kuhnt, S., editors, *Robustness and complex data structures*, pages 17–34. Springer, Heidelberg (2013)
10. Mullainathan, S. and Obermeyer, Z.: Solving medicine's data bottleneck: Nightingale open science. *Nature Medicine* **28**(5), 897–899 (2022)
11. Myllymäki, M., Mrkvička, T., Grabarnik, P., Seijo, H., and Hahn, U.: Global envelope tests for spatial processes. *J. R. Stat. Soc. Ser. B. Stat. Methodol.* **79**(2), 381–404 (2017)
12. Narisetty, N. N. and Nair, V. N.: Extremal depth for functional data and applications. *J. Amer. Statist. Assoc.* **111**(516), 1705–1714 (2016)
13. Newey, W. K. and Powell, J. L.: Asymmetric least squares estimation and testing. *Econometrica* **55**(4), 819–847 (1987)
14. Olszewski, R. T.: Generalized feature extraction for structural pattern recognition in time-series data. Ph.D. thesis, Carnegie Mellon University, Pittsburgh PA, School of Computer Science (2001)
15. Pini, A. and Vantini, S.: Interval-wise testing for functional data. *J. Nonparametr. Stat.* **29**(2), 407–424 (2017)
16. Vitelli, V.: A novel framework for joint sparse clustering and alignment of functional data. *J. Nonparametr. Stat.* **36**(1), 182–211 (2024)

Chapter 30
Multivariate Densities in Bayes Spaces: The Novel Concept of Marginals and Its Implications

Karel Hron

Abstract With the collection of large amounts of data and the need to aggregate them into meaningful domains, there is an increasing interest in probability density functions (PDFs) as data objects and their statistical processing using functional data analysis (FDA) methods. When PDFs are embedded as scale-invariant positive functions in Bayes space with Hilbert space structure, it is possible to analyse them with popular methods of FDA due to isometric isomorphism with L^2 space. Moreover, if the PDFs are multivariate, it is possible to decompose them orthogonally into independent and interactive parts, the latter capturing interactions between all possible combinations of variables. The key aspect here is the appropriate reformulation of the marginals into so-called geometric marginals, which are orthogonal projections of the multivariate information contained in the distribution into the univariate case. In this paper we focus on discussing the role and interpretation of geometric marginals and highlight their main implications.

30.1 Bayes Spaces

When dealing with probability distributions, typically represented by probability density functions (PDFs), the natural goal is to filter out interactions from the independent part, commonly understood as the product of the marginals. A classical result is Sklar's theorem [5], which states that for a PDF $f(x_1, \ldots, x_d)$ of a random vector $(X_1, \ldots, X_d)$ in the domain $I = [a_1, b_1] \times \ldots, [a_d, b_d]$, marginal PDFs $f_1(x_1), \ldots, f_d(x_d)$ and marginal distribution functions $F_1(x_1), \ldots, F_d(x_d)$ we can write

Karel Hron (✉)
Department of Mathematical Analysis and Applications of Mathematics, Faculty of Science, Palacký University Olomouc, 17. listopadu 12, 771 46 Olomouc, Czech Republic
e-mail: karel.hron@upol.cz

© The Author(s), under exclusive license to Springer Nature Switzerland AG 2025

G. Aneiros et al. (eds.), *New Trends in Functional Statistics and Related Fields*, Contributions to Statistics, https://doi.org/10.1007/978-3-031-92383-8_30

$$f(x_1,\ldots,x_d)=f_1(x_1)\cdot\ldots\cdot f_d(x_d)\cdot c(F_1(x_1),\ldots,F_d(x_d)), \tag{30.1}$$

where c is the density of a copula

$$C(u_1,u_2,\ldots,u_d)=\mathsf{P}(U_1\le u_1,U_2\le u_2,\ldots,U_d\le u_d)$$

for $U_i=F_i(X_i)$ are the uniform random variables corresponding to the marginals $F_i(x_i)$ of X_i, $i=1,\ldots,d$. While the standard (arithmetic) marginal PDFs occur in (30.1), further decomposition of c in the original domain is not easily possible. Moreover, it is not specified whether there is a geometric relation between the marginal PDFs $f_1,\ldots,f_d$ and the "interaction component" c.

All this can be achieved by embedding f in the Bayes space setting. Let $(\Omega,\mathcal{A})$ be a d-dimensional product space and let λ be a finite product reference measure. Specifically, suppose that for $i\in D$, $(\Omega_i,\mathcal{A}_i)$ is a measurable space and λ_i is a finite, positive, real-valued measure on it. Set

$$\Omega=\Omega_1\times\cdots\times\Omega_d,\quad \mathcal{A}=\mathcal{A}_1\otimes\cdots\otimes\mathcal{A}_d,\quad \lambda=\lambda_1\otimes\cdots\otimes\lambda_d.$$

For arbitrary non-empty $I\subseteq D$, let

$$\Omega_I=\times_{i\in I}\Omega_i,\quad \mathcal{A}_I=\bigotimes_{i\in I}\mathcal{A}_i,\quad \lambda_I=\bigotimes_{i\in I}\lambda_i.$$

The need to have the product space is the only restriction we need to take into account. For the reference measure, the Lebesgue measure is the first choice for FDA, together with the usual case of the bounded domain, but more general settings can also be considered. For technical details in this section we refer to [2].

In Bayes space, PDFs are not considered as constrained positive functional data, but rather as *scale invariant* positive functions that can be arbitrarily rescaled without loss of information – of course, the unit integral representation with the probabilistic interpretation of PDFs is the most popular, see [6] for details and formal definition. Note that the concept of *proportionality* of PDFs is commonly used in Bayesian statistics, but as we will see, it will be crucial for further considerations.

The Bayes space of PDFs whose logarithm is square integrable is denoted as $\mathcal{B}^2(\lambda)$. Then for any two PDFs $f,g\in\mathcal{B}^2(\lambda)$ and $\alpha\in\mathbb{R}$ the operations of perturbation and powering,

$$f\oplus g=_{\mathcal{B}^2(\lambda)} fg,\quad \alpha\odot f=_{\mathcal{B}^2(\lambda)} f^{\alpha}, \tag{30.2}$$

(by convention $f\ominus g=_{\mathcal{B}^2(\lambda)} f\oplus[(-1)\odot g]$) and the Bayes inner product

$$\langle f,g\rangle_{\mathcal{B}^2(\lambda)}=\frac{1}{2\lambda(\Omega)}\int_\Omega\int_\Omega \ln\left\{\frac{f(\omega_1)}{f(\omega_2)}\right\}\ln\left\{\frac{g(\omega_1)}{g(\omega_2)}\right\}d\lambda(\omega_1)d\lambda(\omega_2),$$

can be defined; the lower index in $=_{\mathcal{B}^2(\lambda)}$ means that the right hand side of the equations in (30.2) can be arbitrarily rescaled without altering the relative information that the resulting density in $\mathcal{B}^2(\lambda)$ contains. The Bayes inner product induces a distance and a norm in the usual way. For any $f,g\in\mathcal{B}^2(\lambda)$, one sets

$$d_{\mathcal{B}^2(\lambda)}(f,g) = ||f \ominus g||_{\mathcal{B}^2(\lambda)}, \quad ||f||_{\mathcal{B}^2(\lambda)} = \sqrt{\langle f, f\rangle_{\mathcal{B}^2(\lambda)}}.$$

The Bayes space equipped with operations of perturbation and powering and the Bayes inner product has the structure of a separable Hilbert space. Perturbation and powering stand for addition of functions and multiplication by a scalar in the usual L^2 space. The one-to-one correspondence of the Bayes space $\mathcal{B}^2(\lambda)$ with

$$L_0^2(\lambda) = \left\{ f : \Omega \to \mathbb{R} \middle| \int_\Omega f^2 d\lambda < \infty, \int_\Omega f d\lambda = 0 \right\},$$

i.e., a subspace of the classical Hilbert space $L^2(\lambda)$ of square integrable functions with respect to λ, where standard methods of FDA can be applied, is achieved through the centred logratio (clr) transformation defined as

$$\mathrm{clr}(f) = \ln(f) - \frac{1}{\lambda(\Omega)} \int_\Omega \ln(f) d\lambda.$$

The inverse clr transformation for any function $f^c \in L_0^2(\lambda)$ is then defined as

$$\mathrm{clr}^{-1}(f^c) =_{\mathcal{B}^2(\lambda)} \exp(f^c).$$

Obviously, the scale invariance of the output can be used to arbitrarily rescale the resulting positive function to any integral constraint.

30.2 Key Concept: Geometric Marginals

As stated at the beginning of Sect. 30.1, the consideration of classical arithmetic marginals for the decomposition of the multivariate density f has its limitations. We will instead introduce the concept of geometric marginals, which is a *generalisation* of the former and allows a much deeper insight into the dependence structure of PDFs.

Let's consider $\mathcal{B}_I^2(\lambda)$ to be a subspace of $\mathcal{B}^2(\lambda)$ with PDFs of arguments with indices in I only. For $I \subsetneq D$, the *I-th geometric marginal* is

$$f_{I,g} = \exp\left\{ \frac{1}{\lambda_{D\setminus I}(\Omega_{D\setminus I})} \int_{\Omega_{D\setminus I}} \ln(f) d\lambda_{D\setminus I} \right\}. \tag{30.3}$$

and its clr transformation

$$\mathrm{clr}(f_{I,g}) = \frac{1}{\lambda_{D\setminus I}(\Omega_{D\setminus I})} \int_{\Omega_{D\setminus I}} \mathrm{clr}(f) d\lambda_{D\setminus I}.$$

The logarithm of f in $f_{I,g}$ can be replaced by its clr transformation due to properties of the exponential (and the proportionality of densities in the Bayes space framework). Nevertheless, the form in which the geometric marginal is defined in (30.3) clearly indicates its origin as a generalisation of the geometric mean of positive data. As a special case, for $I = \{i\}, i = 1, \ldots, d$ the *i-th geometric marginal* is

$$f_{i,g} = \exp\left\{\frac{1}{\lambda_{D\setminus\{i\}}(\Omega_{D\setminus\{i\}})}\int_{\Omega_{D\setminus\{i\}}}\ln(f)d\lambda_{D\setminus\{i\}}\right\}. \tag{30.4}$$

The I-th geometric marginal is the unique *orthogonal projection* of $f(\equiv f_\mu)$ onto $\mathcal{B}^2_I(\lambda)$, a property which has no counterpart in the concept of (classical) arithmetic marginals. And this is also a clear interpretive advance of the geometric marginals. In the case of true independence, when f is a product of (arithmetic) marginals, the i-th geometric and the corresponding arithmetic marginals coincide. Consequently, from this point of view, arithmetic marginals can be considered as their geometric counterparts *under the assumption of independence*. Geometric marginals also contain (univariate) information from the dependence structure. Consequently, the i-th geometric marginals of f and of f^*, where $D^* \subset D$, are generally different. There is also another seemingly confusing point. Looking at the decomposition (30.1), the i-th geometric marginal can be formulated as

$$f_{i,g} = \exp\left\{\frac{1}{\lambda_{D\setminus\{i\}}(\Omega_{D\setminus\{i\}})}\int_{\Omega_{D\setminus\{i\}}}\left[\ln\{c(F_1,\ldots,F_d)\} + \sum_{j=1}^{d}\ln(f_j)\right]d\lambda_{D\setminus\{i\}}\right\}.$$

Since for any distinct $i, j \in D$ the integral of $\ln(f_j)$ over $\Omega_{D\setminus\{i\}}$ is a constant, we find that $f_{i,g} \propto f_i g_i$, where

$$g_i = \exp\left\{\frac{1}{\lambda_{D\setminus\{i\}}(\Omega_{D\setminus\{i\}})}\int_{\Omega_{D\setminus\{i\}}}\ln\{c(F_1,\ldots,F_d)\}d\lambda_{D\setminus\{i\}}\right\} \tag{30.5}$$

depends on both the copula and the marginals $F_1, \ldots, F_d$. Consequently, if there are parameters specifically related to some (arithmetic) marginal, they will propagate through the dependence structure to other geometric marginals. However, this can be seen as an inherent property of the dependence structure rather than a deficiency of the geometric marginals: *i-th geometric marginals are those that capture all the univariate information about i-th variable from the multivariate structure of f*. We illustrate this with two examples for $d = 2$. The first is an extension of the illustrative example from [3]:

Example 30.1 (truncated bivariate normal). We consider a zero-mean bivariate Gaussian density $\mathcal{N}_2(\mu, \Sigma)$ with respect to the (product) Lebesgue measure $\lambda[I] = \lambda[I_1] \times \lambda[I_2]$, truncated on a rectangular domain $I = I_1 \times I_2 \subset \mathbb{R}^2$, with $I_1 = I_2 = [-T, T]$, $T = 5$. The PDF is defined, for $\mathbf{x} = (x, y) \in I$, as

$$f(x,y) =_{\mathcal{B}^2(\lambda)} \exp\left\{\mathbf{x}^\top\Sigma^{-1}\mathbf{x}\right\} = \exp\left\{-\frac{1}{2(1-\rho^2)}\left[\frac{x^2}{\sigma_1^2} - 2\rho\frac{xy}{\sigma_1\sigma_2} + \frac{y^2}{\sigma_2^2}\right]\right\}, \tag{30.6}$$

where $\sigma_i^2 = \Sigma_{ii}$ and $\rho \in [-1, 1]$ is the correlation coefficient. Clearly, for $\rho = 0$ the case of independence is achieved. The clr transformation of f yields

$$\mathrm{clr}(f)(x,y) = -\frac{1}{2(1-\rho^2)}\left[\frac{x^2}{\sigma_1^2} - 2\rho\frac{xy}{\sigma_1\sigma_2} + \frac{y^2}{\sigma_2^2}\right] + \frac{T^2}{6(1-\rho^2)}\left(\frac{1}{\sigma_1^2} + \frac{1}{\sigma_2^2}\right). \tag{30.7}$$

The X-th geometric marginal is derived from (30.7) as

$$\mathrm{clr}(f_X)(x) = -\frac{1}{2(1-\rho^2)} \cdot \frac{x^2}{\sigma_1^2} + \frac{T^2}{6(1-\rho^2)} \cdot \frac{1}{\sigma_1^2}, \qquad x \in I_1$$

and eventually back-transformed to the original space as

$$f_X(x) =_{\mathcal{B}^2(\lambda_X)} \exp\left[-\frac{1}{2(1-\rho^2)}\frac{x^2}{\sigma_1^2}\right]$$

(analogously for the Y-th marginal). Obviously, the parameter ρ for $\rho \neq 0$ propagates to the respective geometric marginals, which are again truncated normal PDFs (see Fig. 30.1), and reduces their variance. In turn, it is interesting to analyse what can be observed from the perturbation difference $f_{X,g} \ominus f_X$, which results in

$$f_{X,g} \ominus f_X(x) =_{\mathcal{B}^2(\lambda_X)} \exp\left[-\frac{\rho^2}{2(1-\rho^2)}\frac{x^2}{\sigma_1^2}\right]$$

and indicates how much univariate information about X interferes with the dependence structure (see Fig. 30.1). Starting from the uniform PDF for $\rho = 0$ as expected, which plays the role of the origin for the Lebesgue reference [6], the effect of increasing ρ can be clearly observed. On the other hand, geometric marginals are not affected by the parameter σ_2 of the marginal distribution of Y, unlike the situation in the next example from [2].

Example 30.2 (bivariate beta). Consider the following three-parameter bivariate beta distribution density. For arbitrary $\alpha_0, \alpha_1, \alpha_2 \in (0, \infty)$ and all $x_1, x_2 \in (0, 1)$, let

$$f_\mu(x_1, x_2) = \frac{1}{B(\alpha_0, \alpha_1, \alpha_2)} \frac{x_1^{\alpha_1 - 1}(1-x_1)^{\alpha_0+\alpha_2-1}x_2^{\alpha_2-1}(1-x_2)^{\alpha_0+\alpha_1-1}}{(1-x_1x_2)^{\alpha_0+\alpha_1+\alpha_2}},$$

where $B(\alpha_0, \alpha_1, \alpha_2) = \Gamma(\alpha_0)\Gamma(\alpha_1)\Gamma(\alpha_2)/\Gamma(\alpha_0 + \alpha_1 + \alpha_2)$ is the generalised beta function. The univariate marginals are beta distributions with parameters (α_1, α_0) and (α_2, α_0), respectively, which again correspond to geometric marginals under the assumption of independence, i.e. for $\alpha_0 + \alpha_1 + \alpha_2 \to 0$. This distribution was analyzed in detail from the Bayes spaces perspective in [2]. Here we just note that the corresponding geometric marginals are then given, for all $x_1, x_2 \in (0, 1)$ and $j \in \{1, 2\}$, by

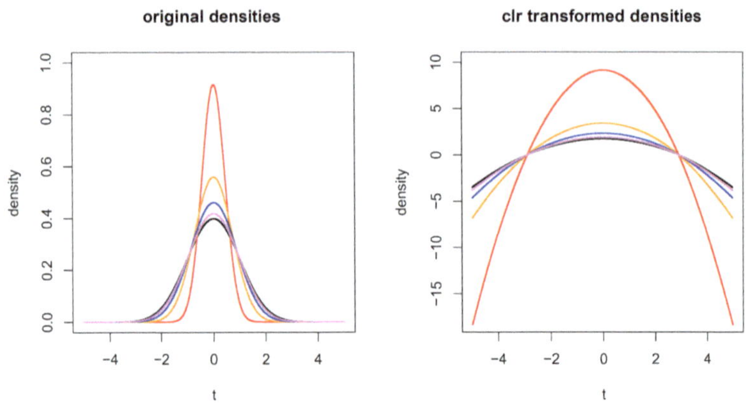

Fig. 30.1: Geometric marginals of X in the truncated bivariate normal PDF with $\sigma_1 = 1$. Colours stand for values of the parameter ρ: $\rho = 0$ (black), $\rho = 0.3$ (violet), $\rho = 0.5$ (blue), $\rho = 0.7$ (orange) and $\rho = 0.9$ (red).

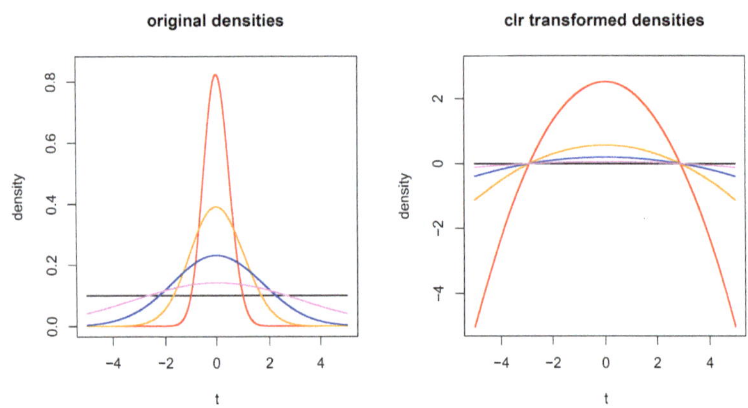

Fig. 30.2: Difference between geometric and arithmetic marginals of X in the truncated bivariate normal PDF with $\sigma_1 = 1$. Colours stand for values of the parameter ρ: $\rho = 0$ (black), $\rho = 0.3$ (violet), $\rho = 0.5$ (blue), $\rho = 0.7$ (orange) and $\rho = 0.9$ (red).

$$\operatorname{clr}(f_{\mu,j})(x_j) = (\alpha_j - 1)\ln(x_j) + (\alpha_0 + \alpha_j - 1)\ln(1 - x_j) \\ + (\alpha_0 + \alpha_1 + \alpha_2)\left\{\frac{\pi^2}{6} + \frac{(1 - x_1)\ln(1 - x_1)}{x_1}\right\} - 2,$$

i.e. they actually contain parameters of the latter marginal distribution.

Example 30.3. In practice, the perturbation difference between geometric and arithmetic marginals can result in PDFs that are far from the usual unimodal form, see Fig. 30.3. This can be nicely illustrated with RKP data from [4], where concentrations of the risk elements Cu, Pb and Zn were post-stratified according to 76 administrative districts in the Czech Republic. In the case of Pb (its arithmetic marginals are shown in Fig. 6 of [4]), while for the majority of districts the difference between the marginals is manifested in heavier tails (warm colours), this is rather not the case in the cluster of districts with prevalent industrial contamination, where changes can be observed in the more central parts of the domain.

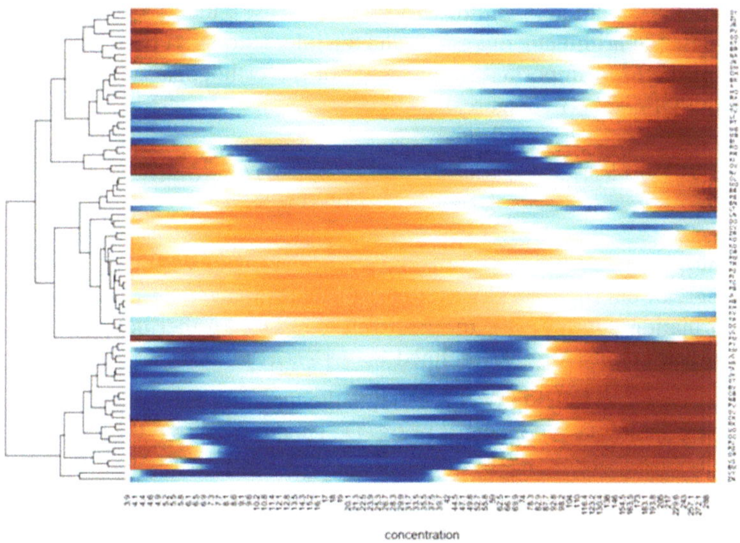

Fig. 30.3: PDFs of perturbation difference between geometric and arithmetic marginals of Pb concentrations, row-wise per district, showing the distribution of concentrations in a colour scale in individual districts and clustered accordingly.

30.3 Implications of Geometric Marginals

The concept of geometric marginals has many geometric and probabilistic implications which have no counterparts in the case of arithmetic marginals. In [2] the Hoeffding-Sobol decomposition was used to derive the following:

Theorem 30.1 (Decomposition of multivariate densities). *For any* $f \in \mathcal{B}^2(\lambda)$,

$$f = f_{\text{ind}} \oplus \bigoplus_{I \subseteq D, |I| \geq 2} f_{I,\text{int}}$$

where the so-called independence and interaction parts are respectively given by

$$f_{\text{ind}} = \bigoplus_{i=1}^{d} f_i, \quad f_{I,\text{int}} = \bigoplus_{J \subseteq I, J \neq \emptyset} \{(-1)^{|I \setminus J|}\} \odot f_J.$$

and all components of the decomposition are orthogonal to each other.

In other words, it is possible to extract the "independent" case of the product of univariate marginals (i-th geometric marginals, $i = 1, \ldots, d$), which are also all mutually orthogonal, and then use I-th geometric marginals to specify from bivariate interactions up to a PDF representing mutual interactions of all variables. Obviously, in the truly independent case (in the probabilistic sense), the independent part becomes a product of arithmetic marginals and the interaction part becomes a uniform PDF. One of the most important implications of the orthogonal decomposition of multivariate PDFs is the Pythagorean Theorem,

$$\|f\|^2_{\mathcal{B}^2(\lambda)} = \|f_{\text{ind}}\|^2_{\mathcal{B}^2(\lambda)} + \sum_{I \subseteq D, |I| \geq 2} \|f_{I,\text{int}}\|^2_{\mathcal{B}^2(\lambda)}, \tag{30.8}$$

where $\|f_{\text{ind}}\|^2_{\mathcal{B}^2(\lambda)} = \|f_1\|^2_{\mathcal{B}^2(\lambda)} + \cdots + \|f_d\|^2_{\mathcal{B}^2(\lambda)}$. Since the Bayes norm can serve as a scalar measure of information, [1], (30.8) can also be interpreted as a decomposition of the information conveyed by the multivariate PDF. Furthermore, let the total variance of a *random* PDF f is defined as $\text{var}_{\mathcal{B}^2(\lambda)}(f) = \|f \ominus \mu\|^2_{\mathcal{B}^2(\lambda)}$, where $\mu(x_1, \ldots, x_d) = \text{clr}^{-1}[\text{Eclr}(f(x_1, \ldots, x_d))]$ is the mean function of f (in the Bayes spaces sense); analogously for any $f_{I,g}$, $I \subsetneq D$. Then similar to (30.8),

$$\text{var}_{\mathcal{B}^2(\lambda)}(f) = \text{var}_{\mathcal{B}^2(\lambda)}(f_{\text{ind}}) + \sum_{I \subseteq D, |I| \geq 2} \text{var}_{\mathcal{B}^2(\lambda)}(f_{I,\text{int}}), \tag{30.9}$$

which sheds new light on dimension reduction in a sample of PDFs.

References

1. Egozcue J.J., Pawlowsky-Glahn V.: Evidence functions: A compositional approach to information. Statistics and Operations Research Transactions **42**, 101–124 (2018)
2. Genest, C., Hron, K., Nešlehová, J.: Orthogonal decomposition of multivariate densities in Bayes spaces and relation with their copula-based representation. Journal of Multivariate Analysis **198**, 105228 (2023)
3. Hron, K., Machalová, J., Menafoglio, A.: Bivariate densities in Bayes spaces: orthogonal decomposition and spline representation. Statistical Papers **64**, 1629–1667 (2023)
4. Matys Grygar, T., Radojičić, U., Pavlů, I., Greven, S., Nešlehová, J.G., Tůmová, Š., Hron, K.: Exploratory functional data analysis of multivariate densities for the identification of agricultural soil contamination by risk elements. Journal of Geochemical Exploration **259**, 107416 (2024)
5. Sklar, A.: Fonctions de répartition à dimensions et leurs marges. Annales de l'ISUP **8**, 229–231 (1959)
6. van den Boogaart, K.G., Egozcue, J.J., Pawlowsky-Glahn, V.: Bayes Hilbert spaces. Australian & New Zealand Journal of Statistics **56**, 171–194 (2014)

Chapter 31
Functional K Sample Problem via Multivariate Optimal Measure Transport-Based Permutation Test

Šárka Hudecová, Daniel Hlubinka and Zdeněk Hlávka

Abstract The null hypothesis of equality of distributions of functional data coming from K samples is considered. The proposed test statistic is multivariate and its components are based on pairwise Cramér von Mises comparisons of empirical characteristic functionals. The significance of the test statistic is evaluated via the novel multivariate permutation test, where the final single p-value is computed using the discrete optimal measure transport. The methodology is illustrated by real data on cumulative intraday returns of Bitcoin.

31.1 Introduction

Functional Data Analysis (FDA) is widely applied in various practical applications where the observed data can be represented as (possibly discretized) functions. This includes situations where the variable of interest is observed over a continuous time domain or over a high-frequency discrete time domain; setups often arising in economics, finance, biomedicine, engineering, and many other fields. FDA has become a very popular approach to analyzing high-dimensional datasets in past few

Šárka Hudecová (✉)
Charles University, Faculty of Mathematics and Physics, Department of Probability and Mathematical Statistics, Sokolovská 83, Prague, Czech Republic,
e-mail: `hudecova@karlin.mff.cuni.cz`

Daniel Hlubinka
Charles University, Faculty of Mathematics and Physics, Department of Probability and Mathematical Statistics, Sokolovská 83, Prague, Czech Republic,
e-mail: `hlubinka@karlin.mff.cuni.cz`

Zdeněk Hlávka
Charles University, Faculty of Mathematics and Physics, Department of Probability and Mathematical Statistics, Sokolovská 83, Prague, Czech Republic,
e-mail: `hlavka@karlin.mff.cuni.cz`

© The Author(s), under exclusive license to Springer Nature Switzerland AG 2025
G. Aneiros et al. (eds.), *New Trends in Functional Statistics and Related Fields*,
Contributions to Statistics, https://doi.org/10.1007/978-3-031-92383-8_31

decades, [14, 18]. One of the basic statistical problems in FDA is the functional K sample comparison, see [2, 3, 6, 22] and further references therein. Most studies focus on the null hypothesis of equality of means, resulting in a functional version of analysis of variance (ANOVA). However, the equality of covariance operators has also been investigated, [8]. In this contribution, we are concerned with a more strict null hypothesis that the whole underlying distributions in the K samples are equal.

In many FDA tests, the resulting test statistic is univariate and its (asymptotic) distribution is rather complex. It is, therefore, common to apply a permutation test, because it requires minimal distributional assumptions and leads to reasonable results even in small samples. The significance is then evaluated using so called permutation distribution obtained by random rearrangements of the data. The corresponding p-value is computed as a proportion of permutation test statistics that are more extreme than the value computed from the original data.

Unfortunately, the permutation principle cannot be easily generalized for multivariate test statistics, mainly due to the lack of ordering in dimension $d \geq 2$. A standard tool here is to compute p-values for each component of the vector test statistic (so called partial p-values) and then to combine them, [15, Chapter 6]. Recently, [11] proposed a multivariate permutation test based on discrete optimal measure transport, [16, 20], which is referred to in the following as the OMT permutation test.

Let $X_{j,1}, \ldots, X_{j,n_j}$ be independent and identically distributed functional observations coming from a distribution with a characteristic functional φ_j, $j = 1, \ldots, K$, and let the K samples be independent. Consider the null hypothesis of equality of distributions

$$\mathcal{H}_0 : \varphi_1 = \cdots = \varphi_K \tag{31.1}$$

against a general alternative. For $K = 2$, [12] propose a Cramér von Mises type test statistic for $\mathcal{H}_0$ based on the empirical characteristic functionals. For $K > 2$, the authors recommend to reformulate $\mathcal{H}_0$ using $d = \binom{K}{2}$ pairwise comparisons, i.e.

$$\mathcal{H}_0^{(j,l)} : \varphi_j = \varphi_l, \qquad \text{for } 1 \leq j < l \leq K. \tag{31.2}$$

The test statistic is then computed for each pair (j, l), $j < l$, leading to d-dimensional vector test statistic $\boldsymbol{T} = (T_1, \ldots, T_d)^\top$. Subsequently, the permutation test is applied to the univariate statistic $\max_{1 \leq j \leq d} T_j$. It is clear that some information can be lost by the transformation of the multivariate test statistic $\boldsymbol{T}$ to its maximum. Hence, in this contribution, we follow the latter approach for $K > 2$ samples, but the significance of $\boldsymbol{T}$ is evaluated using the OMT permutation test. This approach not only uses the information from all the elements of $\boldsymbol{T}$, but also allows for an interpretation of the contributions of the individual pairwise comparisons to the rejection of the null hypothesis. That is, if the null hypothesis $\mathcal{H}_0$ is rejected, the OMT permutation test detects the pairs (j, l) that contribute the most to the rejection. The benefits of this procedure are illustrated on a real dataset on cumulative intraday returns of Bitcoin prices.

The paper is organized as follows. The considered multivariate test statistic $\boldsymbol{T}$ is introduced in Sect. 31.2. Sect. 31.3 describes the OMT permutation test and discusses related issues. The real data application is provided in Sect. 31.4.

31.2 Multivariate Test Statistic for the *K* Sample Problem

A functional random variable X is a random element taking values in the Hilbert space $\mathcal{X} = \mathcal{L}_2[0,1]$ of square integrable functions on $[0,1]$ such that $\mathsf{E}\int_0^1 X^2(t)\mathrm{d}t < \infty$. The space $\mathcal{X}$ is equipped with an inner product $\langle u, v\rangle = \int_0^1 u(t)v(t)\mathrm{d}t$, $u, v \in \mathcal{X}$. A distribution of X is fully described by a characteristic functional $\varphi(w) = \mathsf{E}\exp\{\mathrm{i}\langle w, X\rangle\}$ for $w \in \mathcal{X}$, where $\mathrm{i} = \sqrt{-1}$.

Consider first the two sample problem, i.e. $K = 2$. Then (31.1) reduces to $\mathcal{H}_0^{(1,2)}$: $\varphi_1 = \varphi_2$. [12] proposed a test statistic for $\mathcal{H}_0^{(1,2)}$ based on a comparison of the empirical characteristic functionals of the two samples

$$S_{1,2} = \int_{\mathcal{X}} |\widehat{\varphi}_1(w) - \widehat{\varphi}_2(w)|^2 \mathrm{d}Q, \tag{31.3}$$

where $\widehat{\varphi}_j$ is the empirical characteristic functional in the j-th sample, defined as

$$\widehat{\varphi}_j(w) = \frac{1}{n_j}\sum_{i=1}^{n_j} \exp\left\{\mathrm{i}\langle w, X_{j,i}\rangle\right\}, \quad j = 1, 2, \tag{31.4}$$

and Q is a centered Gaussian measure on $\mathcal{X}$ with a covariance operator with a kernel function $v : [0,1]^2 \to \mathbb{R}$. In practice, the functions $X_{j,i}$, $i = 1, \ldots, n_j$, $j = 1, \ldots, K$, are recorded on a finite set of points $t_1 < \cdots < t_J$ from $[0,1]$, and therefore, the inner products in (31.4) need to be replaced by a suitable Riemann approximation of the integral. Furthermore, it suffices to specify v on the measurement points, via a matrix $\boldsymbol{V} = \left(v(t_i, t_j)\right)_{i,j=1}^{J}$. Further details on the numerical computation of $S_{1,2}$ and discussions on the choice of $\boldsymbol{V}$ can be found in [12].

Consider now the general problem with $K > 2$, the null hypothesis $\mathcal{H}_0$ in (31.1) and the pairwise partial hypothesis in (31.2). For each pair (j, l), $1 \le j < l \le K$, let $S_{j,l}$ be the two sample test statistic defined in (31.3) for samples j and l. It follows from (31.3) that large values of $S_{j,l}$ indicate that the distribution in samples j and l may differ. Hence, define the test statistic

$$\boldsymbol{T} = (S_{1,2}, S_{1,3}, \ldots, S_{K-1,K})^\top = (T_1, \ldots, T_d)^\top,$$

i.e. $\boldsymbol{T}$ is a vector with elements $S_{j,l}$ ordered according to the lexicographic ordering of (j, l). Then $\boldsymbol{T}$ takes values in $[0, \infty)^d$ and large values of its components indicate violation of the corresponding pairwise null hypothesis.

Remark 31.1. It is clear that the dimension d of $\boldsymbol{T}$ growths rapidly with the number of samples K. Remark that the dimension can be reduced, if required, by considering only some of the pairwise comparisons. For instance, $\mathcal{H}_0$ holds if and only if $\varphi_1 = \varphi_j$ for all $j = 2, \ldots, K$. Therefore, one can consider only $\mathcal{H}_0^{(1,j)}$ for $j = 2, \ldots, K$ and get $\boldsymbol{T}$ of dimension $K - 1$.

31.3 Permutation Test Based on the Optimal Measure Transport

Let

$$\mathcal{Z} = (X_{1,1}, \ldots, X_{1,n_1}, \ldots, X_{K,1}, \ldots, X_{K,n_K})$$

be the ordered list of the pooled data. Denote as $\boldsymbol{T}_0$ the d-dimensional test statistic $\boldsymbol{T}$ computed as described in Sect. 31.2 for the original data $\mathcal{Z}$. A permutation version of $\boldsymbol{T}$ is obtained as follows: The elements of $\mathcal{Z}$ are randomly permuted leading to a new ordered list $\mathcal{Z}^* = (Z^*_{1,1}, \ldots, Z^*_{1,n_1}, \ldots, Z^*_{K,1}, \ldots, Z^*_{K,n_K})$. The test statistic $\boldsymbol{T}$ is then computed for the corresponding K samples, with the j-th sample being $Z^*_{j,1}, \ldots, Z^*_{j,n_j}$. This procedure is repeated independently B times, leading to permutation test statistics $\boldsymbol{T}_1, \ldots, \boldsymbol{T}_B$.

The OMT permutation test from [11] is based on the discrete L_2 optimal measure transport of the set $\mathcal{T} = \{\boldsymbol{T}_0, \boldsymbol{T}_1, \ldots, \boldsymbol{T}_B\}$ of $B+1$ points in $\mathbb{R}^d$ to a specified grid set $\mathcal{G}$ of $B+1$ points in the unit ball $\{\boldsymbol{x} \in \mathbb{R}^d : \|\boldsymbol{x}\| \leq 1\}$, whose choice is discussed later. Namely, the optimal mapping F^* is defined as a bijection from $\mathcal{T}$ to $\mathcal{G}$ that minimizes the quadratic loss $\sum_{i=0}^{B} \|F(\boldsymbol{T}_i) - \boldsymbol{T}_i\|^2$. Note that computation of the discrete OMT F^* is a standard optimalization task, see [16, Chapter 3]. In R, [17], it can be computed using package `clue`, [13].

The permutation p-value is computed as the relative frequency of permutation statistics that are "more extreme" than $\boldsymbol{T}_0$ from the center-outward perspective (i.e. that are more distant from the center), that is

$$\widehat{p} = \frac{1}{B+1}\left(1 + \sum_{b=1}^{B} \mathbf{1}\Big\{\|F^*(\boldsymbol{T}_b)\| \geq \|F^*(\boldsymbol{T}_0)\|\Big\}\right),$$

where $\mathbf{1}\{\cdot\}$ is the indicator function. If the grid set $\mathcal{G}$ is specified as (31.5) below, one can use also a p-value $\widetilde{p} = 1 - \|F^*(\boldsymbol{T}_0)\|$.

Apart from the single final p-value, the OMT permutation test provides also an interpretation of the partial contributions of the individual pairwise comparisons. The quantity $(1 - \widetilde{p})^2$ can be interpreted as an overall non-conformity score that measures the deviation of the data from the null hypothesis, and $\mathcal{H}_0$ is rejected on level $\alpha \in (0, 1)$ if and only if the non-conformity score exceeds $(1 - \alpha)^2$. Let $F^*(\boldsymbol{T}_0) = \big(F_1(\boldsymbol{T}_0), \ldots, F_d(\boldsymbol{T}_0)\big)^\top$. Then

$$(1-\widehat{p})^2 = \|F^*(\boldsymbol{T}_0)\|^2 = \sum_{j=1}^{d} F_j^*(\boldsymbol{T}_0)^2,$$

so the value $F_j^*(\boldsymbol{T}_0)^2$ can be interpreted as an absolute contribution of the j-th component of $\boldsymbol{T}$ (corresponding to j-th pairwise comparison) to the rejection of the composite null hypothesis $\mathcal{H}_0$. Denote as $D_j = F_j^*(\boldsymbol{T}_0)/\|F^*(\boldsymbol{T}_0)\|$. Then $\sum_{j=1}^{d} D_j^2 = 1$, and the vector $(D_1^2, \dots, D_d^2)^\top$ can be interpreted as a vector of relative contributions of the individual pairwise comparisons. See [11] for examples and more details.

The computation of the OMT permutation p-value requires a choice of the grid $\mathcal{G}$. In the current context, each component of $\boldsymbol{T}$ takes values in $\mathbb{R}^+$ with large values indicating violation of the corresponding partial pairwise hypothesis. Therefore, in view of Sect. 3.2 in [11], a suitable grid set $\mathcal{G}$ is a subset of $\{\boldsymbol{x} : \boldsymbol{x} \in [0,1]^d, \|\boldsymbol{x}\| \le 1\}$. In view of [11], it is beneficial to specify B such that $B + 1 = n_R \cdot n_S$ for positive integers n_R and n_S and compute the grid points $\mathcal{G} = \{\boldsymbol{g}_{ij}\}_{i,j=1}^{n_R,n_S}$ as

$$\boldsymbol{g}_{ij} = \frac{i}{n_R+1}\boldsymbol{s}_j \quad i = 1, \dots, n_R,\ j = 1, \dots, n_S, \tag{31.5}$$

where $\boldsymbol{s}_1, \dots, \boldsymbol{s}_{n_S}$ are distinct directional vectors from the set $\mathcal{S}^+ = \{\boldsymbol{x} \in [0,1]^d : \|\boldsymbol{x}\| = 1\}$, distributed as uniformly as possible over $\mathcal{S}^+$. These directions can be obtained from a sequence $\{\boldsymbol{x}_i\}_{i=1}^{n_S}$ of low discrepancy points in $[0,1]^{d-1}$ as $\boldsymbol{s}_j = \tau(\boldsymbol{x}_j)$, where τ is a mapping from $[0,1]^{d-1}$ to $\mathcal{S}^+$ such that $\tau(\boldsymbol{X})$ has a uniform distribution on $\mathcal{S}^+$ whenever $\boldsymbol{X}$ has a uniform distribution in $[0,1]^{d-1}$. The mapping τ can be derived analogously as in [5, Sect. 1.5.3], see also Sect. 31.4 for the application with $d = 3$. Remark that the minimal p-value $\widehat{p}$ that can be obtained by the OMT permutation test with grid from (31.5) is $1/n_R$. For instance, if $n_R = 20$, then $1/n_R = 0.05$ and results with this p-value have to be interpreted as significant.

31.4 Real Data Application

As an illustration of the proposed methodology, we analyze the intraday Bitcoin prices from 2022. The data come from a larger database available at `kaggle.com` repository[1]. We consider intraday data with frequency 20 min (so there are $J = 72$ observations each day) and calculate the logarithmic cumulative intraday returns (CIR).

We are concerned with the question whether the behavior of CIR differs for working and weekend days. In particular, we consider the cumulative intraday returns for Mondays (sample 1), Wednesdays (sample 2) and Saturdays (sample 3), and test $\mathcal{H}_0$ for $K = 3$. The three samples, of sizes $n_1 = 52$, $n_2 = 52$, and $n_3 = 53$, respectively, are shown in Fig. 31.1.

[1] `https://www.kaggle.com/datasets/jkraak/bitcoin-price-dataset`

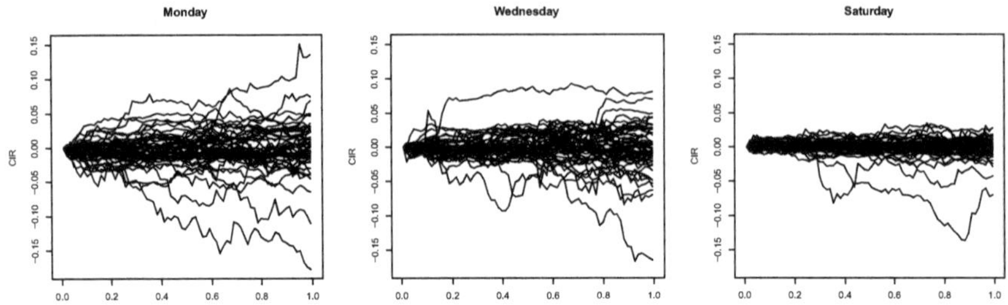

Fig. 31.1: Logarithmic cumulative intraday returns for Bitcoin in 2022: Mondays (52 curves), Wednesdays (52 curves), and Saturdays (53 curves).

Remark that the intraday Bitcoin prices for successive days are inherently dependent due to various market dynamics and external factors influencing price trends. Moreover, they are unlikely to be identically distributed because of long-term trends and evolving patterns. The latter issue is effectively addressed by transforming the data to CIR. The potential dependence between successive days is also partially mitigated by the CIR transformation. Furthermore, the selection of non-consecutive days (e.g., Monday, Wednesday, Saturday) reduces any lingering dependence between the observed data points. Consequently, any residual dependence between these chosen days can be reasonably assumed negligible.

The trivariate test statistic $\boldsymbol{T}$, whose components correspond to pairwise comparisons of samples (1,2), (1,3) and (2,3), respectively, was computed as described in Sect. 31.2. Based on empirical evidence, the authors from [12] recommend to choose the matrix $\boldsymbol{V}$ as data dependent, namely as an approximation to the inverse of the sample covariance matrix, because this choice leads to generally reasonable results under the null as well as against various alternatives. Following this recommendation, the following test statistics were considered:

– $\boldsymbol{T}^{\text{inv}}$ for $\boldsymbol{V}$ being the approximate inverse to the sample covariance matrix computed from all data (without distinguishing the samples) and
– $\boldsymbol{T}^{\text{inv.pool}}$ for $\boldsymbol{V}$ being the approximate inverse to the pooled sample covariance matrix.

In both cases, the approximated inverse is computed from 9 largest eigenvalues. The permutation test statistics were calculated for $B = 999$. The grid $\mathcal{G}$ in $\mathbb{R}^3$ is constructed as (31.5) with $n_R = 40$ and $n_S = 25$. The sequence $\{\boldsymbol{x}_i\}_{i=1}^{n_S}$ is taken as a Halton sequence in $[0,1]^2$, see [9], obtained in R using package `randtool`, [4]. For $d = 3$, the desired transformation from $\boldsymbol{x}_j$ to $\boldsymbol{s}_j$ is via mapping $\tau = (\tau_1, \tau_2, \tau_3)^\top$ defined as $\tau_1(x_1, x_2) = 1 - x_1$, $\tau_2(x_1, x_2) = \sqrt{2x_1 - x_1^2}\cos(\pi x_2/2)$, and $\tau_3(x_1, x_2) = \sqrt{2x_1 - x_1^2}\sin(\pi x_2/2)$.

Fig. 31.2 shows the test statistic $\boldsymbol{T}^{\text{inv}}$ computed from the original data together with its B permutation replicas, and the grid set $\mathcal{G}$ with the transformation $F^*(\boldsymbol{T}_0^{\text{inv}})$.

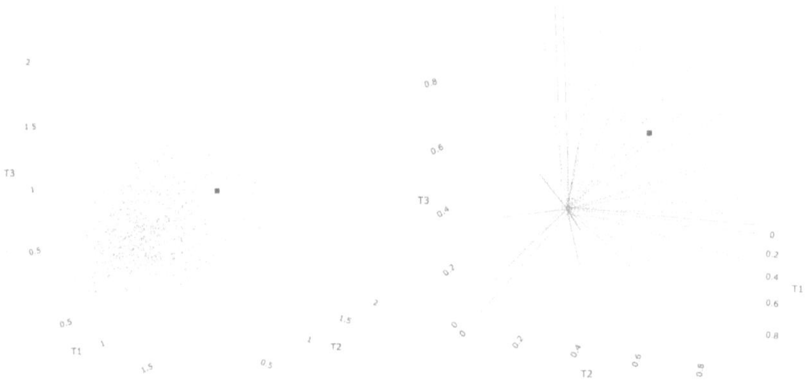

Fig. 31.2: Left panel: The test statistic $\boldsymbol{T}^{\text{inv}}$ (larger square) together with B permutation replicas (small circles). All values multiplied by 10^4. Right panel: The grid $\mathcal{G}$ with highlighted point $F^*(\boldsymbol{T}_0^{\text{inv}})$.

It is visible that $F^*(\boldsymbol{T}_0^{\text{inv}})$, and subsequently $\boldsymbol{T}_0^{\text{inv}}$, is one of the most extreme points, so the obtained p-value is the minimal possible and significant. Namely, we get $\widehat{p} = 0.025$ and $\widetilde{p} = 0.024$. The vector of individual contributions is $(D_1^2, D_2^2, D_3^2) = (0.353, 0.305, 0.343)^\top$, so it indicates that all three pairwise comparisons contribute equally to the rejection of the null hypothesis. The conclusions for the test statistic $\boldsymbol{T}^{\text{inv.pool}}$ are very similar, so these are not discussed in detail.

Table 31.1 presents p-values of various functional ANOVA tests for equality of mean functions, computed with the help of package `fdANOVA`, [7]. In contrast to the results of our tests, all the tests for equality of mean function lead to highly non-significant p-values. This suggests that the violation of $\mathcal{H}_0$ is not due to the difference in means, but rather due to some other distributional aspects. Fig. 31.1 suggests that the difference in the three groups can be related to different covariance structures. Hence, we consider also the null hypothesis $\widetilde{\mathcal{H}}_0 : C_1 = C_2 = C_3$, where C_j is the covariance operator in sample j. This hypothesis can be reformulated in terms of three pairwise comparisons. For each pair, the operators can be compared using the test statistic based on the square-root distance, proposed by [1]. This procedure results in a trivariate test statistic that can be evaluated via the OMT permutation test. Using again $B = 999$ and the same grid set, we get $\widehat{p} = 0.05$ and $\widetilde{p} = 0.049$. Note that due to the discreteness of possible p-values for the OMT permutation test, these values need to be interpreted as significant. The corresponding vector of individual contributions is $(0.062, 0.909, 0.028)^\top$, which reveals that $\widetilde{\mathcal{H}}_0$ is rejected mainly due to the difference in covariance operators of Mondays and Saturdays.

Finally note that our conclusions about differences in individual days are in agreement with some other empirical studies that examined differences in Bitcoin trading, see [10] and references therein.

Table 31.1: Results for various functional ANOVA tests for equality of mean functions for the considered three samples. FP is permutation test based on basis function representation from [6], CS is L_2-norm-based parametric bootstrap test for heteroscedastic samples from [3]. L2B is L_2-norm-based test with bias-reduced method of estimation, while L2b is L_2-norm-based bootstrap test, see [22]. FB is F-type test with bias-reduced method of estimation, see [19, 21], and Fb is F-type bootstrap test, see [22].

Test	FP	CS	L2B	L2b	FB	Fb
p-value	0.830	0.790	0.772	0.775	0.775	0.789

Acknowledgements The research was supported by the Czech Science Foundation project GAČR No. 25-15844S.

References

1. Cabassi A., Pigoli D., Secchi P. and Carter P.A.: Permutation tests for the equality of covariance operators of functional data with applications to evolutionary biology. Electron. J. Stat., **11**(2), 3815–3840, (2017)
2. Cuesta-Albertos J. A. and Febrero-Bande M.: A simple multiway ANOVA for functional data. TEST, **19**(3), 537–557, (2010)
3. Cuevas A., Febrero M. and Fraiman R.: An ANOVA test for functional data. Comput. Statist. Data Anal., **47**(1), 111–122, (2004)
4. Dutang C. and Savicky P.: randtoolbox: Generating and Testing Random Numbers, (2022) R package version 2.0.3.
5. Fang K.-T. and Wang Y.: Number-theoretic Methods in Statistics. Chapman & Hall, (1994)
6. Górecki T. and Smaga Ł.: A comparison of tests for the one-way ANOVA problem for functional data. Comput. Statist., **30**(4), 987–1010, (2015)
7. Górecki T. and Smaga Ł.: fdANOVA: Analysis of Variance for Univariate and Multivariate Functional Data, (2018) R package version 0.1.2.
8. Guo J., Zhou B., and Zhang J.-T.: New tests for equality of several covariance functions for functional data. J. Amer. Statist. Assoc., **114**(527), 1251–1263, (2019)
9. Halton J. H.: On the efficiency of certain quasi-random sequences of points in evaluating multi-dimensional integrals. Numer. Math., **2**, 84–90, (1960)
10. Hansen P. R., Kim C., and Kimbrough W.: Periodicity in cryptocurrency volatility and liquidity. J. Financ. Econom., **22**(1), 224–251, (2024)
11. Hlávka Z., Hlubinka D., and Hudecová Š.: Multivariate quantile-based permutation tests with application to functional data. J. Comput. Graph. Stat., doi:10.1080/10618600.2024.2444302 (2024)
12. Hlávka Z., Hlubinka D., and Koňasová K.: Functional ANOVA based on empirical characteristic functionals. J. Multivariate Anal., **189**, 104878, (2022)
13. Hornik K.: A CLUE for CLUster Ensembles. J. Stat. Softw., **14**(12), (2005)
14. Horváth L., and Kokoszka P.: Inference for Functional Data with Applications. Springer-Verlag, New York, (2012)
15. Pesarin F.: Multivariate permutation tests: with applications in biostatistics. Wiley, (2001)
16. Peyré G., Cuturi M., et al. Computational optimal transport: With applications to data science. Found. Trends Mach. Learn. g, **11**(5-6), 355–607, (2019)

17. R Core Team. R: A Language and Environment for Statistical Computing. R Foundation for Statistical Computing, Vienna, Austria, (2022)
18. Ramsay J. O., and Silverman B. W.: Functional Data Analysis. Springer Series in Statistics. Springer New York, (2013)
19. Shen Q., and Faraway J.: An F test for linear models with functional responses. Statistica Sinica, pages 1239–1257, (2004)
20. Villani C.: Topics in optimal transportation, volume 58. American Mathematical Soc., (2021)
21. Zhang J.-T.: Statistical inferences for linear models with functional responses. Statistica Sinica, pages 1431–1451, (2011)
22. Zhang J.-T.: Analysis of Variance for Functional Data. CRC Press, New York, (2013)

Chapter 32
Sequential Monitoring for Detection of Breaks in Panel Data

Marie Hušková and Charl Pretorius

Abstract Procedures for the detection of changes in panel data typically concern offline procedures, i.e., all observations are available at the beginning of the statistical analysis and testing is done retrospectively. The present paper deals with online procedures for detecting a change in the mean of panel data arriving sequentially, with the total number of observations eventually received being random. The procedures are developed using principles of offline detection procedures [5] in combination with those used in online procedures [8]. Some limit properties of the proposed procedures are presented, as well as simulation results exhibiting desirable finite-sample properties. The focus is on high-dimensional panel models, which are high dimensional time series with special properties often encountered in econometrics and financial applications.

32.1 Introduction

We deal with the problem of detection of a change in the mean of panel data when training data (of size m) without a change in the mean are available and each observation has N components. Specifically, we study the panel model

$$Y_{i,t} = \mu_{i,t} + \boldsymbol{\lambda}_i^\top \boldsymbol{f}_t + e_{i,t}, \quad i = 1, \ldots, N, \quad t = 1, 2, \ldots, \tag{32.1}$$

where $\mu_{i,t}$ denotes the mean of panel i at time t, $\boldsymbol{f}_t$ is an r-dimensional sequence of unobserved common factors on which the panel means depend through unknown

Marie Hušková (✉)
Department of Probability and Mathematical Statistics, Charles University, Sokolovská 83, Prague 18675, Czech Republic, e-mail: huskova@karlin.mff.cuni.cz

Charl Pretorius
Centre for Business Mathematics and Informatics, North-West University, Potchefstroom, South Africa, e-mail: charl.pretorius@nwu.ac.za

© The Author(s), under exclusive license to Springer Nature Switzerland AG 2025
G. Aneiros et al. (eds.), *New Trends in Functional Statistics and Related Fields*,
Contributions to Statistics, https://doi.org/10.1007/978-3-031-92383-8_32

factor loadings λ_i, and $e_{i,t}$ are the errors which are assumed to have zero mean. It is supposed that the first m observations are available and it is assumed that $\mu_{i,1} = \cdots = \mu_{i,m} = \mu_{i,0}$, $i = 1, \ldots, N$. These m observations are referred to as the training sample. Monitoring starts at time $m+1$ and continues until time $m + T_m$. Define $\boldsymbol{\mu}_t = (\mu_{1,t}, \ldots, \mu_{N,t})^\top$, $t = 0, 1, \ldots$.

We want to test the hypothesis

$$H_0 : \boldsymbol{\mu}_{m+t} = \boldsymbol{\mu}_0 \text{ for all } t = 1, \ldots, T_m,$$

against the alternative

$$\begin{aligned} H_1 &: \text{there exists a } k^0 > 0, \boldsymbol{\delta} \neq \mathbf{0} \text{ such that } \boldsymbol{\mu}_{m+t} = \boldsymbol{\mu}_0,\ 0 < t < k^0, \\ &\text{and } \boldsymbol{\mu}_{m+t} = \boldsymbol{\mu}_0 + \boldsymbol{\delta},\ k^0 \leq t \leq T_m, \end{aligned}$$

for a prior chosen T_m.

Panel models are applicable in econometrics and financial applications where high-dimensional time series are often encountered. Sparked by [2], several works on change-point detection have appeared in this context. We mention specifically the works of [1], [5], and [8]. These ideas have also been extended in several directions, such as the case of dynamic panel models [10].

Several sequential monitoring procedures for the finite-sample case have been proposed and studied. [6] study the case of monitoring for changes in a dynamic linear model. Other recently proposed sequential change-point detection procedures include that of [11] and [12]. On the other hand, sequential monitoring procedures for panel data are scarce.

Nowadays, many studies also consider self-normalizing procedures which do not rely on nuisance parameters (typically long-run variances) that are difficult to estimate [15, 16, 13]. [4] propose a self-normalizing statistic that can be used to sequentially monitor for changes in general time series models.

The remainder of this paper is structured as follows. We first state certain technical assumptions providing a framework for studying the theoretical properties of our test procedures proposed in Sect. 32.2. The results of a short simulation study are presented in Sect. 32.3, illustrating that a bootstrap-based implementation of the procedures has favorable finite-sample properties.

Technical Assumptions

(A1) The sequences of errors $\{e_{i,t}; -\infty < t < \infty\}$, $1 \leq i \leq N$, are strictly stationary and mutually independent. There exist finite constants $c_1, c_2, \Delta > 0$ and $\kappa > 4$ such that

$$\mathrm{E}\, e_{i,0} = 0, \quad \mathrm{E}\,|e_{i,0}|^{\kappa+\Delta} < \infty, \quad \text{and} \quad c_1 \leq \lim_{T\to\infty} \mathrm{E}\left(\frac{1}{\sqrt{T}}\sum_{t=1}^{T} e_{i,t}\right)^2 \leq c_2.$$

(A2) For each $i = 1, \ldots, N$, the error sequence $\{e_{i,t}, -\infty < t < \infty\}$ is assumed to be strong mixing with mixing coefficient $\alpha(\cdot)$, independent of i, satisfying $M_{\rho,\Delta}(\alpha) < \infty$ for some even integer $\rho \geq \kappa$, where

$$M_{\rho,\Delta}(\alpha) = \sum_{k=1}^{\infty}(k+1)^{\rho-2}\alpha(k)^{\Delta/(\rho+\Delta)}.$$

(A3) The common factor sequence $\{\boldsymbol{f}_t; -\infty < t < \infty\}$ is strictly stationary and independent of $\{e_{i,t}; 1 \leq i \leq N, -\infty < t < \infty\}$, $\mathrm{E}\,\boldsymbol{f}_0 = \mathbf{0}$, $\mathrm{E}\,\|\boldsymbol{f}_0\|^{\kappa+\Delta} < \infty$, and $\lim_{T\to\infty}\mathrm{E}(T^{-1}\sum_{t=1}^{T}\boldsymbol{f}_t\boldsymbol{f}_t^{\top}) = \mathbf{0}$.

(A4) $\limsup_{N\to\infty}\max_{1\leq i\leq N}\|\boldsymbol{\lambda}_{i,N}\| < \infty$ and $\lim_{N\to\infty}N^{-1/2}\sum_{i=1}^{N}\|\boldsymbol{\lambda}_{i,N}\|^2 = 0$, where $\boldsymbol{\lambda}_i = \boldsymbol{\lambda}_{i,N}$, $i = 1, \ldots, N$.

(A5) $\left(\dfrac{N}{m^2} + \dfrac{N}{T_m^2}\right) \to 0$ as $\min(N, m) \to \infty$.

Assumption (A4) states that the influence of the common factors is asymptotically negligible. The case of asymptotically influential common factors is in preparation.

32.2 Procedures

Using the ideas proposed in [5], [9], and [8], we consider the statistics

$$\begin{aligned} Z_{k,m}^{(1)} &= \frac{1}{\sqrt{N}}\sum_{i=1}^{N}\left\{\frac{1}{\sigma_i^2}W_{i,m,k}^2 - \frac{k(m+k)}{m}\right\}, \\ Z_{k,m}^{(2)} &= \frac{1}{\sqrt{N}}\sum_{i=1}^{N}\frac{1}{\sigma_i^2}\left\{W_{i,m,k}^2 - \frac{k(m+k)}{m^2}W_{i,\lfloor m/2\rfloor,m-\lfloor m/2\rfloor}^2\right\}, \end{aligned} \tag{32.2}$$

$k = 1, \ldots, T_m - m$, where $W_{i,m,k} = \sum_{j=m+1}^{m+k}(Y_{i,j} - \overline{Y}_{i,1:m})$, $\overline{Y}_{i,1:m} = m^{-1}\sum_{j=1}^{m}Y_{i,j}$, and σ_i^2 are appropriately chosen scaling constants. Note that the $Z_{k,m}^{(1)}$ are motivated by [5] while the $Z_{k,m}^{(2)}$ are motivated by [7] and [9].

Theorem 32.1. *Suppose that $T_m/m \to L$ as $m \to \infty$ for some fixed $L \in (0, \infty)$. Under Assumptions (A1)–(A5),*

$$\left\{\frac{Z_{k,m}^{(1)}}{mg(m,k)}; k = 1, \ldots, T_m\right\} \xrightarrow{\mathcal{D}[0,1]} \{\Gamma(u); u \in (0, L)\},$$

where $g(m,k) = 1 + k/m$ and $\Gamma(\cdot)$ is a zero-mean Gaussian process with

$$\mathrm{Cov}(\Gamma(u), \Gamma(v)) = 2u^2\frac{1+v}{1+u}, \qquad 0 < u \leq v < L.$$

It can be shown that Theorem 32.1 remains true if the constants σ_i^2 in (32.2) are replaced by estimators $\widehat{\sigma}_i^2$ satisfying $\sup_i |\widehat{\sigma}_i^2 - \sigma_i^2| = o_{\mathrm{P}}(1)$; cf. [5].

One can also define self-normalizing test procedures as follows:

$$Z_{k,m}^{(3)} = \frac{1}{\sqrt{N}} \sum_{i=1}^{N} \frac{1}{V_{i,m}} \left\{ W_{i,m,k}^2 - \frac{k(m+k)}{m^2} W_{i,\lfloor m/2 \rfloor, m-\lfloor m/2 \rfloor}^2 \right\},$$

$$Z_{k,m}^{(4)} = \frac{1}{\sqrt{N}} \sum_{i=1}^{N} \frac{1}{V_{i,m}} W_{i,m,k}^2,$$

where $V_{i,m} = m^{-2} \sum_{t=1}^{m} (\sum_{j=1}^{t} (Y_{i,j} - \overline{Y}_{i,1:m}))^2$.

The monitoring procedure continues until time $m+T_m$, or until the null hypothesis is rejected at time $m + S_m^{(\ell)}$ if $S_m^{(\ell)} \le T_m$, where

$$S_m^{(\ell)} = \min \left\{ k : \frac{|Z_{k,m}^{(\ell)}|}{mg(m,k)} > c_\alpha^{(\ell)} \right\}, \quad g(m,k) = \frac{m+k}{m}, \quad \ell = 1, \ldots, 4,$$

with $c_\alpha^{(\ell)}$ chosen such that the false rejection rate is α. Specifically, $c_\alpha^{(\ell)}$ is defined as the minimum value of c satisfying

$$\mathop{\mathrm{P}}_{H_0} \left(\sup_k \frac{|Z_{k,m}^{(\ell)}|}{mg(m,k)} > c \right) \le \alpha. \tag{32.3}$$

32.3 Simulation Study

The bootstrap scheme is motivated by the definition of c_α in (32.3). Let $Z_{k,m} = Z_{k,m}(\{Y_{i,t}\})$ stand for any $Z_{k,m}^{(\ell)}$, $\ell = 1, \ldots, 4$, calculated using the observations $Y_{i,t}$, $i = 1, \ldots, N$, $t = 1, \ldots, m + T_m$.

We employ a bootstrap algorithm adapted from [14] to obtain critical values of the tests. For $i = 1, \ldots, N$, and $t = 1, \ldots, m$, define the centered training observations $\widehat{\eta}_{i,t} = Y_{i,t} - \overline{Y}_{i,1:m}$. To implement the algorithm, we apply the information criteria proposed by [3] to the panels of centered observations $\widehat{\eta}_{i,t}$ to obtain an estimate $\widehat{r}$ of the number of common factors r and estimates $\widehat{\boldsymbol{\lambda}}_i$ and $\widehat{\boldsymbol{f}}_t$ of $\boldsymbol{\lambda}_i$ and $\boldsymbol{f}_t$, respectively. Define the residuals $\widehat{e}_{i,t} = \widehat{\eta}_{i,t} - \widehat{\boldsymbol{\lambda}}_i^\top \widehat{\boldsymbol{f}}_t$. To generate one bootstrap realization of $\max_k |Z_{k,m}|/(mg(m,k))$, we proceed as follows:

1. Generate T_m cross-sectional bootstrap observations $(e_{1,m+t}^*, \ldots, e_{N,m+t}^*)$, $t = 1, \ldots, T_m$, from the centered *training* observations $(\widehat{e}_{1,t}, \ldots, \widehat{e}_{N,t})$, $t = 1, \ldots, m$.
2. For each (i,t), generate $\xi_{i,t}$ independently of all other quantities with $\mathrm{E}(\xi_{i,t}) = 0$ and $\mathrm{Var}(\xi_{i,t}) = 1$.
3. For each $i = 1, \ldots, N$, construct the bootstrap errors $\tilde{e}_{i,1}^*, \ldots, \tilde{e}_{i,m+T_m}^*$, where

$$\tilde{e}^*_{i,t} = \begin{cases} \xi_{i,t}\widehat{e}_{i,t} & \text{for } t \leq m, \\ \xi_{i,t}e^*_{i,t} & \text{for } t > m. \end{cases}$$

4. Generate a $(m+T_m)\times\widehat{r}$ matrix $\boldsymbol{f}^*_t$ from a multivariate normal distribution having the same long-run covariance structure as estimated by $\widehat{\boldsymbol{f}}_t$.
5. Construct a bootstrap sample consisting of observations $Y^*_{i,t} = \widehat{\boldsymbol{\lambda}}_i^\top \boldsymbol{f}^*_t + \tilde{e}^*_{i,t}$.
6. For each $k = 1, \ldots, T_m$, calculate the bootstrap statistic $Z^*_{k,m} = Z_{k,m}(\{Y^*_{i,t}\})$.
7. Calculate $U^*_m = \max_k |Z^*_{k,m}|/(mg(m,k))$.

Steps 1 to 7 can be repeated many times, say B, to obtain bootstrap realizations U^*_m, of which the ordered values are denoted by $U^*_m(1) < \cdots < U^*_m(B)$. An estimate of the critical value c_α is given by $U^*_m(\lfloor(1-\alpha)B\rfloor)$.

In the simulations, we use the naive (i.i.d.) bootstrap in Step 1. In Step 2, we make use of the two-point distribution defined by $\mathrm{P}(\xi_{i,t} = -\frac{1}{2}(\sqrt{5}-1)) = q$ and $\mathrm{P}(\xi_{i,t} = \frac{1}{2}(\sqrt{5}+1)) = 1-q$, where $q = (\sqrt{5}+1)/2\sqrt{5}$.

Numerical Results

Training samples of size $m \in \{100, 200\}$ are considered and incoming observations are monitored sequentially until the critical value is exceeded or $T_m = Lm$ additional observations have been received, $L \in \{0.5, 1, 2, 4\}$. Observations are generated from the model in (32.1) with $N \in \{100, 200\}$ panels. A constant change of $\delta_i \in \{0, 0.1\}$ is introduced in all panels at time $m + \frac{1}{2}T_m$, i.e., halfway during monitoring.

Table 32.1 shows the empirical rejection rates in the case where the observations are generated from the model in (32.1) with errors $e_{i,t} = \phi e_{i,t-1} + \nu_{i,t}$, where $\nu_{i,t}$ are independent $\mathrm{N}(0,1)$ random variables, and no common factors (i.e. $\boldsymbol{\lambda}_i^\top \boldsymbol{f}_t = 0$). For the autoregressive parameter we consider the choices $\phi = 0, 0.3$.

Under the null hypothesis, as m increases, the nominal significance level of 5% is approached. There is a gain in power under alternatives as the magnitude of the change δ_i or m, the size of the training sample, is increased.

Although our theoretical results are for the case where the cross-dependence across panels is negligible in the sense of (A4), we briefly demonstrate the finite-sample behavior in the case of stronger cross-dependence. Specifically, for each t, we sample $\boldsymbol{f}_t = f_t$ in (32.1) independently from a univariate $\mathrm{N}(0,1)$ distribution and choose factor weights $\boldsymbol{\lambda}_i = \lambda_i = N^{-1/4}$ for all $i = 1, \ldots, N$. Table 32.2 shows the empirical rejection rates for this case. Similar observations can be made as in the case of no cross-dependence, but with a slight decrease in power under alternatives.

Acknowledgements The research of the first author was supported by grant GAČR-23-05737S.

Table 32.1: Rejection percentages of the procedures based on $Z_{k,m}^{(j)}$, $j = 1, \ldots, 4$, in the case of a change of size δ_i in all panels with no common factors. Significance level 0.05

				$Z_{k,m}^{(1)}$ m			$Z_{k,m}^{(2)}$ m			$Z_{k,m}^{(3)}$ m			$Z_{k,m}^{(4)}$ m		
δ_i	ϕ	N	L	100	200	400	100	200	400	100	200	400	100	200	400
0.0	0.0	100	0.5	4.6	5.4	5.2	4.9	4.8	5.4	5.1	4.5	5.3	4.7	5.3	5.5
			1.0	5.0	3.9	5.1	5.4	4.3	5.1	5.2	4.2	5.3	5.3	4.5	4.0
			2.0	5.5	4.8	4.3	4.6	4.7	4.2	4.9	4.7	4.5	4.5	4.5	4.6
			4.0	5.6	5.3	5.0	5.5	5.4	5.5	5.0	5.2	5.8	5.3	5.0	5.3
			8.0	5.0	4.9	4.8	5.1	4.6	5.1	5.4	5.2	5.0	5.5	4.8	4.9
		200	0.5	4.7	5.5	5.2	5.3	4.9	5.3	5.0	5.0	5.1	5.2	5.5	4.6
			1.0	6.2	5.4	5.0	5.5	5.5	5.6	5.7	5.6	5.2	5.9	5.4	5.1
			2.0	4.5	4.5	5.1	5.2	4.7	5.6	5.2	4.7	5.5	4.6	4.9	5.3
			4.0	4.7	5.1	4.8	5.1	5.0	5.2	5.3	4.8	4.7	5.1	4.9	5.1
			8.0	5.4	4.9	5.2	5.6	5.4	4.9	5.8	5.3	5.2	5.4	4.7	5.6
	0.3	100	0.5	4.0	4.7	5.5	7.5	5.7	5.7	8.5	6.1	6.1	4.1	4.1	5.8
			1.0	4.9	5.4	6.0	8.7	6.7	6.0	9.1	6.5	5.8	4.9	5.1	5.3
			2.0	5.9	5.0	6.0	10.4	5.7	6.1	11.7	6.2	6.1	5.6	4.9	6.0
			4.0	6.5	5.6	5.7	10.0	6.9	6.2	10.7	7.7	6.4	5.9	5.3	5.3
			8.0	5.8	5.4	5.5	10.3	7.3	5.7	10.8	7.2	5.8	6.2	5.0	5.4
		200	0.5	3.9	4.2	5.7	9.8	6.8	6.4	10.6	7.3	6.5	4.0	4.3	5.6
			1.0	4.6	5.3	5.3	10.3	7.1	6.0	10.8	7.9	5.9	5.0	5.3	4.6
			2.0	5.5	6.3	6.2	11.4	8.9	5.5	11.8	9.0	5.7	6.0	5.6	5.2
			4.0	5.4	5.8	6.9	11.9	7.6	5.6	12.7	8.1	5.8	6.1	6.3	5.1
			8.0	6.7	6.9	6.2	13.0	7.7	6.4	13.3	8.4	7.1	6.0	6.1	5.7
0.1	0.0	100	0.5	11.7	26.2	62.7	9.7	18.3	42.6	10.2	19.0	45.4	9.5	21.2	44.4
			1.0	17.7	42.4	86.7	15.8	28.3	63.6	16.3	30.3	65.8	15.9	33.3	69.7
			2.0	24.7	61.0	97.7	19.2	41.4	83.7	19.2	43.4	86.1	21.6	47.9	88.4
			4.0	32.7	73.3	99.5	26.0	54.7	92.9	27.5	54.9	94.4	26.9	59.8	95.2
			8.0	37.7	80.1	99.8	28.6	59.2	95.6	29.9	61.0	96.3	32.5	67.8	97.3
		200	0.5	16.9	40.1	86.0	13.9	29.2	67.0	14.1	31.3	71.3	13.9	30.7	72.8
			1.0	25.8	65.4	98.6	21.6	45.7	87.5	21.9	47.5	90.4	21.0	50.0	92.0
			2.0	38.2	86.1	100.0	29.9	65.1	99.4	31.1	67.3	99.6	32.6	71.2	99.3
			4.0	50.1	93.5	100.0	38.0	78.1	99.6	40.4	79.9	99.7	40.1	84.2	99.9
			8.0	58.1	96.9	100.0	43.4	85.5	99.9	43.9	86.6	100.0	47.5	89.8	100.0
	0.3	100	0.5	6.8	12.9	27.4	11.1	12.7	19.2	11.8	13.4	20.4	6.7	11.0	17.4
			1.0	8.8	20.4	43.6	14.8	18.4	31.1	14.4	20.3	33.9	9.2	15.4	29.6
			2.0	14.0	27.4	65.5	18.2	24.2	42.3	18.1	25.7	44.8	12.4	22.1	43.7
			4.0	16.2	36.3	76.0	20.2	28.1	55.0	20.9	30.2	56.3	13.9	27.4	56.3
			8.0	19.3	41.6	83.6	22.9	32.1	61.5	23.6	33.8	63.1	17.3	29.3	63.8
		200	0.5	9.4	18.6	45.6	16.1	18.7	31.0	17.0	18.7	33.9	9.2	14.4	26.3
			1.0	11.8	29.4	68.6	19.2	27.5	47.8	20.4	29.3	50.3	10.0	23.9	48.9
			2.0	17.0	45.5	90.0	24.6	35.1	67.8	26.5	38.7	71.9	15.1	32.0	67.3
			4.0	23.7	57.0	95.1	33.3	43.8	77.3	35.0	45.7	80.8	21.7	43.6	78.7
			8.0	27.5	63.2	98.8	32.0	49.0	86.3	34.8	51.3	87.5	24.1	46.9	89.1

Table 32.2: Rejection percentages of the procedures based on $Z^{(j)}_{k,m}$, $j = 1, \ldots, 4$, in the case of a change of size δ_i in all panels and strong cross-dependence across panels Significance level 0.05

				$Z^{(1)}_{k,m}$ m			$Z^{(2)}_{k,m}$ m			$Z^{(3)}_{k,m}$ m			$Z^{(4)}_{k,m}$ m		
δ_i	ϕ	N	L	100	200	400	100	200	400	100	200	400	100	200	400
0.0	0.0	100	0.5	2.4	4.8	3.9	3.7	4.6	4.5	2.6	4.1	4.3	3.3	3.7	4.5
			1.0	3.8	6.4	4.4	4.1	5.7	4.3	4.1	5.3	4.4	4.1	6.4	5.2
			2.0	3.5	4.4	2.3	4.2	3.9	3.5	2.9	3.7	3.1	4.0	3.6	3.7
			4.0	3.4	4.4	3.8	3.7	4.2	3.6	3.6	3.1	3.4	4.3	4.8	3.5
			8.0	4.0	4.2	4.6	3.8	4.8	4.4	3.8	3.4	4.2	3.7	5.1	4.3
		200	0.5	4.2	3.9	3.0	5.6	5.2	3.9	5.5	4.8	3.0	5.6	4.9	4.3
			1.0	2.8	4.0	3.9	2.6	4.3	4.5	2.7	3.5	4.4	2.4	4.1	4.3
			2.0	3.3	4.0	5.1	5.3	3.3	4.7	3.2	4.7	4.3	5.4	4.1	4.6
			4.0	3.8	5.7	3.7	3.9	5.9	3.3	3.9	3.9	3.8	4.0	6.0	3.4
			8.0	2.5	4.0	5.0	2.9	4.4	5.7	3.2	3.7	5.7	3.3	4.0	5.2
	0.3	100	0.5	0.9	2.8	1.3	5.5	5.2	3.0	1.0	2.2	1.6	5.7	4.8	2.2
			1.0	1.6	1.3	1.2	5.9	1.9	2.6	1.4	0.9	1.0	6.5	2.0	1.9
			2.0	2.1	2.0	2.5	6.3	4.3	5.3	1.4	1.7	1.3	6.0	4.7	4.2
			4.0	1.3	1.8	1.3	4.3	3.7	1.5	1.5	1.5	0.7	4.4	3.7	1.3
			8.0	1.8	2.7	1.9	5.0	4.1	3.9	0.9	2.2	1.6	5.2	4.9	3.3
		200	0.5	1.4	1.3	1.6	4.8	3.8	3.0	1.3	1.0	1.7	4.8	3.6	3.0
			1.0	2.0	1.6	2.3	6.8	4.5	4.3	2.1	0.7	1.3	6.1	3.7	4.6
			2.0	2.1	1.6	2.8	6.7	6.0	5.0	1.9	2.5	2.3	6.5	5.4	3.8
			4.0	1.4	1.6	2.8	6.1	4.6	3.1	2.0	2.6	1.6	4.8	4.7	3.1
			8.0	2.2	3.4	2.2	6.0	6.0	3.2	1.6	3.2	1.0	6.3	7.1	2.8
0.1	0.0	100	0.5	7.8	15.4	31.5	8.7	13.6	28.8	6.6	12.1	26.1	8.4	13.4	27.3
			1.0	12.0	23.3	40.6	12.7	22.0	39.7	9.1	20.0	36.4	11.8	22.4	38.3
			2.0	13.7	38.3	58.1	15.9	31.9	52.7	12.4	27.4	47.0	14.9	31.6	52.6
			4.0	20.2	37.3	65.8	17.6	35.1	61.5	16.3	33.0	54.9	17.8	35.7	58.5
			8.0	23.7	47.3	71.3	21.9	42.2	64.7	20.8	39.1	60.4	19.8	39.5	63.7
		200	0.5	13.4	24.5	46.1	12.3	21.9	42.2	11.3	19.2	37.8	12.0	20.7	40.7
			1.0	15.7	30.2	61.3	17.3	31.6	58.5	13.8	29.1	54.7	17.2	30.9	57.5
			2.0	25.3	43.5	76.6	24.6	39.6	71.7	20.6	37.0	66.5	24.0	38.4	68.4
			4.0	24.6	50.7	83.2	25.0	47.5	79.7	23.1	43.6	75.7	23.8	45.9	79.9
			8.0	27.3	58.3	89.9	25.0	53.3	84.9	23.9	51.2	84.4	23.8	52.5	85.6
	0.3	100	0.5	3.0	8.4	16.3	7.5	10.5	12.4	3.4	5.9	8.9	7.7	9.5	10.9
			1.0	4.8	12.3	26.7	7.9	13.8	23.8	3.6	7.7	17.5	6.3	13.0	22.8
			2.0	4.8	16.4	36.4	10.2	17.2	29.8	5.1	13.1	25.8	9.8	15.0	30.4
			4.0	9.9	19.1	43.3	12.5	23.9	37.7	8.1	15.1	28.5	11.6	22.4	36.5
			8.0	9.3	25.4	54.4	13.7	22.0	49.1	7.3	17.7	40.2	14.3	21.9	46.9
		200	0.5	4.3	10.1	23.3	9.8	15.1	20.6	4.0	6.8	16.2	8.8	13.0	17.8
			1.0	8.2	18.8	39.3	14.5	19.5	37.0	6.5	14.6	25.5	12.9	19.0	32.8
			2.0	9.0	23.9	49.5	16.6	26.0	46.8	9.0	18.5	40.6	15.8	24.9	44.4
			4.0	11.4	32.3	66.8	18.0	31.0	62.4	9.3	24.6	53.7	17.9	28.6	60.5
			8.0	16.8	36.2	73.4	23.1	31.5	66.3	12.3	28.8	60.4	20.5	30.5	66.3

References

1. Antoch, J., Hanousek, J., Horváth, L., Hušková, M., and Wang, S.:Structural breaks in panel data: Large number of panels and short length time series. Econometric Reviews **38**, 828–855 (2019)
2. Bai, J.: Common breaks in means and variances for panel data. Journal of Econometrics.**157**, 8–92 (2010)
3. Bai, J. and Ng, S.. Determining the number of factors in approximate factor models: Econometrica, 70(1):191–221 (2002)
4. Chan, N. H., Ng, W. L., and Yau, C. Y.: A self-normalized approach to sequential change-point detection for time series. Statistica Sinica **31**(1) 491–517 (2021)
5. Horváth, L. and Hušková, M.: Change-point detection in panel data. Journal of Time Series Analysis, **33** 631–648 (2012)
6. Horváth, L., Liu, Z., and Lu, S.: Sequential monitoring of changes in dynamic linear models, applied to the us housing market. Econometric Theory. **38**(2), 209–272 (2022)
7. Horváth, L., Liu, Z., Rice, G., and Zhao, Y.: Detecting common breaks in the means of high dimensional cross-dependent panels. The Econometrics Journal. **25**, 362–383 (2022)
8. Horváth, L. and Rice, G.: Change Point Analysis for Time Series. Springer Series in Statistics. Springer, Cham (2024)
9. Hušková, M. and Pretorius, C. : Detection of changes in panel data models with stationary regressors. Recent Advances in Econometrics and Statistics: Festschrift in Honour of Marc Hallin. Springer, Cham, 305– 324 (2024)
10. Jo, M. and Lee, S.: On cusum test for dynamic panel models. Statistical Methods & Applications. **30**,515–542 (2021)
11. Kirch, C. and Weber, S.: Modified sequential change point procedures based on estimating functions. Electronic Journal of Statistics. **12**(1), 1579 – 1613 (2018)
12. Lee, S., Meintanis, S. G., and Pretorius, C.: Monitoring procedures for strict stationarity based on the multivariate characteristic function. Journal of Multivariate Analysis. **189** 104892 (2022)
13. Pešta, M. and Wendler, M.: Nuisance-parameter-free changepoint detection in non-stationary series. TEST. **29**(2), 379–408(2019)
14. Prášková, Z.: Testing structural breaks in large dynamic panel models. Talk presented at PROBASTAT 2024, Smolenice, Slovakia (2024)
15. Shao, X. and Zhang, X.: Testing for change points in time series. Journal of the American Statistical Association **151**, 1228–1240 (2010)
16. Zhang, T. and Lavitas, L.: Unsupervised self-normalized change-point testing for time series. Journal of the American Statistical Association. **113**(522), 637–648 (2018)

Chapter 33
Robust Penalized Splines for Location Estimation from Discretely Sampled Functional Data

Ioannis Kalogridis and Stefan Van Aelst

Abstract We study a broad family of lower-rank spline estimators for the mean of a process X based on discretely sampled functional data observed with measurement error. We derive its rates of convergence under general assumptions and show that the parametric rate n^{-1} is attainable, even under discretely sampled data with heavy-tailed distributions of either the functional variable or the measurement error.

33.1 Introduction

Consider a process $\{X(t), t \in [0,1]\}$ with a well-defined mean function, that is, $\sup_{t\in[0,1]} \mathbb{E}\{|X(t)|\} < \infty$. Let $\{X_i\}_{i=1}^n$ denote independent and identically distributed copies of X. Our goal is to robustly estimate the mean function $\mu(t) = \mathbb{E}\{X(t)\}$ from the discretely sampled data

$$Y_{ij} = X_i(T_j) + \zeta_{ij}, \quad (j = 1, \ldots, m; i = 1, \ldots, n), \tag{33.1}$$

where $\{T_j\}_{j=1}^m$ are sampling points and $\{\zeta_{ij}\}_{i=1,j=1}^{n,m}$ are random noise variables. This setting has been considered among others by [1, 9, 4]. The latter authors proposed a broad class of robust smoothing spline estimators for the estimation of μ from the partially observed data Y_{ij} and demonstrated the minimax rate optimality of these estimators.

A drawback of the estimators of [4] is their dimensionality: these spline estimators have knots at every discretization point T_j, which makes them m-dimensional. For densely sampled functional data with large m, these estimators are therefore computationally intensive. To remedy this drawback, we introduce in this contri-

Ioannis Kalogridis (✉)
University of Glasgow, UK, e-mail: ioannis.kalogridis@glasgow.ac.uk

Stefan Van Aelst
KU Leuven, Belgium, e-mail: stefan.vanaelst@kuleuven.be

© The Author(s), under exclusive license to Springer Nature Switzerland AG 2025
G. Aneiros et al. (eds.), *New Trends in Functional Statistics and Related Fields*,
Contributions to Statistics, https://doi.org/10.1007/978-3-031-92383-8_33

bution a family of lower rank penalized spline estimators with flexible choices of knots and penalties. In particular, fix a value $K \in \mathbb{N}_+$ and define the interior knots $0 = x_0 < x_1, \ldots, x_K < x_{K+1} = 1$, which do not need to be discretization points T_j. For any fixed integer $p \in \mathbb{N}_+$, let $S_K^p([0,1])$ denote the set of spline functions on $[0,1]$ of order p with knots at the x_i. For $p = 1$, $S_K^1([0,1])$ is the set of step functions with jumps at the knots while for $p \geq 2$

$$S_K^p([0,1]) = \{s \in C^{p-2}([0,1]) : s(x) \text{ is a polynomial of degree } (p-1) \text{ on each } [x_i, x_{i+1}]\}.$$

It is well-known that the dimension of $S_K^p([0,1])$ is $K + p$ (see, e.g., [2]). For some choice of $r < p$, we propose to estimate $\mu(t) = \mathbb{E}\{X(t)\}$ by

$$\widehat{\mu}_n = \underset{g \in S_K^p([0,1])}{\operatorname{argmin}} \left[\frac{1}{nm} \sum_{i=1}^{n} \sum_{j=1}^{m} \rho\left(Y_{ij} - g(T_j)\right) + \lambda \int_{[0,1]} \left| g^{(r)}(t) \right|^2 \mathrm{d}t \right], \tag{33.2}$$

for a convex loss function $\rho : \mathbb{R} \to \mathbb{R}_+$. The parameter $\lambda \geq 0$ regulates the trade-off between smoothness, measured by the squared integrated rth derivative, and fidelity to the data. Common choices are $p = 4$ and $r = 2$, i.e. cubic splines with a penalty on their second derivative. For the standard choice $\rho(x) = x^2$, this estimator has been considered by [9], but for other ρ-functions, such as $\rho(x) = |x|$, (33.2) yields novel estimators whose theoretical properties have not yet been studied.

33.2 Rates of convergence

We establish rates of convergence of the estimator $\widehat{\mu}_n$ towards the mean function $\mu(t)$ in the $\mathcal{L}^2([0,1])$-norm $\|\cdot\|$, i.e., $\|g\|^2 = \int_0^1 |g(t)|^2 \mathrm{d}t$. In what follows we assume that the measurement errors ζ_{ij} have been generated by a process ζ such that $\zeta_{ij} = \zeta_i(T_j)$. We can then define the corresponding error processes

$$\epsilon_i(t) = X_i(t) - \mu(t) + \zeta_i(t), \quad (i = 1, \ldots, n).$$

and write $\epsilon_{ij} = \epsilon_i(T_j)$. We use the following assumptions.

(A1) $\rho : \mathbb{R} \to \mathbb{R}_+$ is absolutely continuous and convex with derivative ψ existing almost everywhere.

(A2) The error processes $\epsilon_i(\omega, t) : (\Omega, \mathcal{A}, \mathbb{P}) \times [0,1] \to \mathbb{R}$, $i = 1, \ldots, n$, are independent and identically distributed.

(A3) There exist finite constants κ and M_1 such that for all $|y| < \kappa$,

$$|\psi(x+y) - \psi(x)| \leq M_1, \quad \forall x \in \mathbb{R}.$$

(A4) There exists a finite constant M_2 such that

$$\sup_{1\le j\le m} \mathbb{E}\left\{\left|\psi(\epsilon_{ij}+u)-\psi(\epsilon_{ij})\right|^2\right\} \le M_2|u|, \text{ as } u \to 0.$$

(A5) $\sup_{j\le m}\mathbb{E}\{|\psi(\epsilon_{ij})|^2\} < \infty$, $\mathbb{E}\{\psi(\epsilon_{ij})\} = 0$ and there exist positive constants $\{\delta_j\}_{j=1}^m$ such that $0 < \inf_{j\le m}\delta_j \le \sup_{j\le m}\delta_j < \infty$ and

$$\sup_{1\le j\le m}\left|\mathbb{E}\left\{\psi(\epsilon_{ij}+u)\right\} - \delta_j u\right| = o(u),$$

as $u \to 0$.

(A6) Let $h_i = x_i - x_{i-1}$ with x_i the knots and $h = \max_i h_i$. Then, $\max_i |h_{i+1} - h_i| = o(K^{-1})$ and there exists an $M > 0$ such that $h/\min_i h_i \le M$. In addition, $n^\delta \le K \le c_0 m$ for some small $\delta > 0$ and $c_0 > 0$.

(A7) The discretization points $\{T_j\}_{j=1}^m$ satisfy $T_j = (j-1/2)/m$ with $m \ge n^{\delta'}$ for some $\delta' > 0$.

As discussed in [4], assumptions (A1)–(A5) are broadly applicable and allow one to use even non-smooth loss functions, such as $\rho(x) = |x|$ and its quantile generalization. Assumptions (A6)–(A7) are similar to those in [9] and are similarly very general. Theorem 33.1 presents our main result.

Theorem 33.1. *Assume that assumptions (A1)–(A7) hold and that $\lambda \to 0$ in a such way that $K\min\{\lambda^2K^{2r},\lambda\} \to 0$ as $n\to\infty$ and $n^{\delta-1}K^3 \to 0$[1] as $n \to \infty$ for some $\delta > 0$. Then, if $\mu \in C^j([0,1])$ with $r \le j < p$, it holds that*

$$\|\widehat{\mu}_n - \mu\|^2 = O_{\mathbb{P}}\left(n^{-1}\right) + O_{\mathbb{P}}\left(K^{-2j}\right) + O_{\mathbb{P}}\left(\min(\lambda, \lambda^2K^{2r})\right) \quad \text{as } n \to \infty.$$

Theorem 33.1 is a considerable generalization of [9, Theorem 3.1], which only holds for $\rho(x) = x^2$. The theorem implies that, for densely sampled functional data, $\mu \in C^r([0,1])$, large K and appropriate decay of λ, the rate of convergence is the parametric n^{-1}. For sparsely sampled functional data the rate of convergence is completely determined by K with an upper bound of m^{-2j}, which is in agreement with other findings in the literature (see, e.g., [4]).

33.3 Proof of Theorem 33.1

We begin with two auxiliary lemmas on properties of spline functions and spline approximation of smooth functions. They can be deduced from Lemma 1 in [5] and [2, P. 145–149] respectively.

Lemma 33.1. *Under assumptions (A6)–(A7), there exist constants $0 < c_1 < c_2 < \infty$ independent of n and K such that for any $f \in S_K^p([0,1])$ and $n \ge n_0$ it holds that*

[1] For Lipschitz ψ-functions it suffices that $n^{\delta-1}K^2 \to 0$ as $n \to \infty$ for small $\delta > 0$.

$$c_1 \|f\|^2 \leq m^{-1} \sum_{j=1}^{m} |f(T_j)|^2 \leq c_2 \|f\|^2 .$$

Lemma 33.2. *For every $f \in C^j([0,1])$ there exists a spline function s_f of order p with $p > j$ such that*

$$\sup_{t\in[0,1]} \left|f(t) - s_f(t)\right| \leq c_0 h^j \sup_{|t-y|<\mathbf{x}} \left|f^{(j)}(t) - f^{(j)}(y)\right|,$$

where the constant c_0 depends only on p and j.

Proof (Theorem 33.1).
Let $L_n : S_K^p([0,1]) \to \mathbb{R}_+$ denote the objective function in (33.2), that is,

$$L_n(g) = \frac{1}{nm} \sum_{i=1}^{n} \sum_{j=1}^{m} \rho(Y_{ij} - g(T_j)) + \lambda \int_{[0,1]} |g^{(r)}(t)|^2 \, \mathrm{d}t.$$

Define $C_n = n^{-1} + K^{-2j} + \min(\lambda, \lambda^2 K^{2r})$. On $S_K^p([0,1])$ and for $\lambda \geq 0$ define the norm

$$\|g\|_{r,\lambda}^2 = \|g\|^2 + \lambda \|g^{(r)}\|^2,$$

and observe that $\|g\| \leq \|g\|_{r,\lambda}$. We will show that for every $\epsilon > 0$ there exists a sufficiently large $D \geq 1$ such that

$$\liminf_{n\to\infty} \Pr\left(\inf_{\substack{g\in S_K^p([0,1]) \\ \|g\|_{r,\lambda}=D}} L_n(s_\mu + C_n^{1/2} g) > L_n(s_\mu) \right) \geq 1 - \epsilon, \tag{33.3}$$

where s_μ is the spline approximation to μ constructed with the help of Lemma 33.2. The inequality in (33.3) then implies the existence of a $\widehat{\mu}_n \in S_K^p([0,1])$ such that $\|\widehat{\mu}_n - s_\mu\|^2 \leq \|\widehat{\mu}_n - s_\mu\|_{r,\lambda}^2 \leq DC_n$ and so, by the triangle inequality, Lemma 33.2 and assumption (A6),

$$\|\widehat{\mu}_n - \mu\| \leq \|\widehat{\mu}_n - s_\mu\| + \left\|\mu - s_\mu\right\| = O_{\mathbb{P}}(C_n^{1/2}) + O(K^{-j}) = O_{\mathbb{P}}(C_n^{1/2}),$$

and thus Theorem 33.1 holds.

Hence, it suffices to establish (33.3). To that end we write $R_j = \mu(T_j) - s_\mu(T_j)$ and use the fundamental theorem of calculus along with (A1) to obtain the decomposition

$$L_n(s_\mu + C_n^{1/2} g) - L_n(s_\mu) = I_1(g) + I_2(g) + I_3(g) + I_4(g),$$

with

$$I_1(g) = \frac{1}{nm}\sum_{i=1}^{n}\sum_{j=1}^{m}\int_{R_j}^{R_j+C_n^{1/2}g(T_j)} \mathbb{E}\left\{\psi\left(\epsilon_{ij}+u\right)-\psi(\epsilon_{ij})\right\} \mathrm{d}u + \lambda C_n \left\|g^{(r)}\right\|^2,$$

$$I_2(g) = \frac{1}{nm}\sum_{i=1}^{n}\sum_{j=1}^{m}\int_{R_j}^{R_j+C_n^{1/2}g(T_j)} \left\{\psi\left(\epsilon_{ij}+u\right)-\psi(\epsilon_{ij})-\mathbb{E}\left\{\psi\left(\epsilon_{ij}+u\right)-\psi(\epsilon_{ij})\right\}\right\} \mathrm{d}u,$$

$$I_3(g) = -\frac{C_n^{1/2}}{nm}\sum_{i=1}^{n}\sum_{j=1}^{m}\psi(\epsilon_{ij})g(T_j),$$

$$I_4(g) = 2\lambda C_n^{1/2}\langle g^{(r)}, s_\mu^{(r)}\rangle.$$

The remainder of the proof now consists of showing that

$$\inf_{g\in S_K^p([0,1]):\|g\|_{r,\lambda}=D} I_1(g) \geq c_0 D^2 C_n + O(1)DC_n \tag{33.4}$$

$$\sup_{g\in S_K^p([0,1]):\|g\|_{r,\lambda}\leq D} |I_2(g)| = o_{\mathbb{P}}(1)C_n \tag{33.5}$$

$$\sup_{g\in S_K^p([0,1]):\|g\|_{r,\lambda}\leq D} |I_3(g)| = O_{\mathbb{P}}(1)DC_n \tag{33.6}$$

$$\sup_{g\in S_K^p([0,1]):\|g\|_{r,\lambda}\leq D} |I_4(g)| = O(1)DC_n, \tag{33.7}$$

for a strictly positive constant c_0. From the inequalities (33.4)-(33.7) it then follows that

$$\inf_{g\in S_K^p([0,1]),\|g\|_{r,\lambda}=D} L_n(s_\mu + C_n^{1/2}g) - L_n(s_\mu) \geq c_0 D^2 C_n + O_{\mathbb{P}}(1)C_n + o_{\mathbb{P}}(1)DC_n,$$

from which (33.3) can be established by taking a sufficiently large D and letting $n \to \infty$.

Assertions (33.6) and (33.7) can be established exactly as in the proofs of Theorem 2 in [4] and Theorem 2 in [3], respectively. Hence we focus on proving (33.4) and (33.5) in the two steps below.

Step 1: Proof of (33.4). First, observe that by lemma 33.2,

$$\max_{1\leq j\leq m}\left|R_j\right| = \max_{1\leq j\leq m}\left|\mu(T_j) - s_\mu(T_j)\right| \leq \left\|s_\mu - \mu\right\|_\infty = O(K^{-j}) = o(1).$$

Furthermore, by Proposition 1 in [3], there exists a global $c_0 > 0$ such that for every $g \in S_K^p([0,1])$ it holds that $\|g\|_\infty \leq c_0 K^{1/2}\|g\|$. Therefore, for every $g \in S_K^p([0,1])$ such that $\|g\|_{r,\lambda} \leq D$ we have that $C_n^{1/2}\|g\|_\infty \leq c_0 DKC_n^{1/2} \to 0$, by our limit assumptions. By using (A5) it can now be seen that

$$\frac{1}{m}\sum_{j=1}^{m}\int_{R_j}^{R_j+C_n^{1/2}g(T_j)} \mathbb{E}\left\{\psi\left(\epsilon_{ij}+u\right)-\psi(\epsilon_{ij})\right\} \mathrm{d}u$$
$$= \frac{1}{m}\sum_{j=1}^{m}\int_{R_j}^{R_j+C_n^{1/2}g(T_j)} \mathbb{E}\left\{\psi\left(\epsilon_{ij}+u\right)\right\} \mathrm{d}u$$
$$= \frac{1}{m}\sum_{j=1}^{m}\int_{R_j}^{R_j+C_n^{1/2}g(T_j)} \{\delta_j u + o(u)\}\,\mathrm{d}u$$
$$= \frac{C_n}{m}\sum_{j=1}^{m}\delta_j|g(T_j)|^2(1+o(1))$$
$$+ 2\frac{C_n^{1/2}}{m}\sum_{j=1}^{m}R_j g(T_j)(1+o(1))$$
$$\geq c_0 C_n\|g\|^2 - c_1 C_n\|g\|,$$

for some positive constants c_0 and c_1, where the inequality follows from the positivity and finiteness of δ_j in (A5), Lemma 33.1, the Schwarz inequality and the fact that $\max_j |R_j| \leq c_0 C_n^{1/2}$. From this inequality it follows that

$$I_1(g) \geq c_0 C_n\|g\|^2 + \lambda C_n\|g^{(r)}\|^2 - c_1 C_n\|g\|,$$

so that, by definition of $\|\cdot\|_{r,\lambda}$, we obtain that

$$\inf_{g\in S_K^p([0,1]):\|g\|_{r,\lambda}=D} I_1(g) \geq c_0 C_n D^2 - c_1 D C_n = c_0 C_n D^2 + O(1) D C_n,$$

as asserted.

Step 2: Proof of (33.5). Let $\mathcal{F}$ denote the class of real-valued functions

$$\mathcal{F} = \left\{ f_g(x_1,\ldots,x_m) = \frac{1}{m}\sum_{j=1}^{m}\int_{R_j}^{R_j+C_n^{1/2}g(T_j)}\{\psi(x_j+u)-\psi(x_j)\}\,\mathrm{d}u,\right.$$
$$\left. g\in S_K^p([0,1]), \|g\|_{r,\lambda}\leq D\right\}$$

We aim to apply Lemma 19.36 of [8] to the empirical process based on $\mathcal{F}$. To that end we need (i) a uniform bound for all $f_g \in \mathcal{F}$. (ii) A uniform bound on the squared expectations of $f_g \in \mathcal{F}$ with respect to the distribution $\mathbb{P}$ of $\boldsymbol{\epsilon}_1 = (\epsilon_{11},\ldots,\epsilon_{1m})$ and (iii) an estimate for the bracketing entropy $\log N_{[]}(\epsilon,\mathcal{F},\mathcal{L}^2(\mathbb{P}))$ for all small $\epsilon > 0$.

We begin with the uniform bound. The limits of integration shrink to zero as $n \to \infty$ so, by a change of variables, assumption (A3) and the Schwarz inequality, we obtain

$$
\begin{aligned}
\left|f_g(x_1,\ldots,x_m)\right| &= m^{-1}\sum_{j=1}^{m}\left|\int_{R_j}^{R_j+C_n^{1/2}g(T_j)}\{\psi(x_j+u)-\psi(x_j)\}\,\mathrm{d}u\right| \\
&\le m^{-1}\sum_{j=1}^{m}\left|\int_{0}^{C_n^{1/2}g(T_j)}\left|\psi(x_j+R_j+u)-\psi(x_j)\right|\,\mathrm{d}u\right| \\
&\le M_1 C_n^{1/2}\left\{m^{-1}\sum_{j=1}^{m}|g(T_j)|^2\right\}^{1/2} \\
&\le c_0 C_n^{1/2} D,
\end{aligned}
$$

where the last inequality follows from Lemma 33.1 and the fact that for all $f_g \in \mathcal{F}$ we have $\|g\| \le D$. Thus, $\|f_g\|_\infty \le c_0 C_n^{1/2} D$ uniformly in $f_g \in \mathcal{F}$.

To bound $\mathbb{E}\{|f_g(\boldsymbol{\epsilon}_1)|^2\}$ observe that by applying the Schwarz inequality twice, we get

$$
\begin{aligned}
\mathbb{E}\{|f_g(\boldsymbol{\epsilon}_1)|^2\} &= m^{-2}\mathbb{E}\left\{\left|\sum_{j=1}^{m}\int_{R_j}^{R_j+C_n^{1/2}(g(T_j)}\{\psi(\epsilon_{1j}+u)-\psi(\epsilon_{1j})\}\,\mathrm{d}u\right|^2\right\} \\
&\le m^{-1}\sum_{j=1}^{m}\mathbb{E}\left\{\left|\int_{R_j}^{R_j+C_n^{1/2}(g(T_j)}\{\psi(\epsilon_{1j}+u)-\psi(\epsilon_{1j})\}\,\mathrm{d}u\right|^2\right\} \\
&= m^{-1}\sum_{j=1}^{m}\mathbb{E}\left\{\left|\int_{0}^{C_n^{1/2}g(T_j)}\{\psi(\epsilon_{1j}+R_j+u)-\psi(\epsilon_{1j})\}\,\mathrm{d}u\right|^2\right\} \\
&\le C_n^{1/2}\max_{1\le j\le m}|g(T_j)| \\
&\quad\times m^{-1}\sum_{j=1}^{m}\left|\int_{0}^{C_n^{1/2}g(T_j)}\mathbb{E}\{|\psi(\epsilon_{1j}+R_j+u)-\psi(\epsilon_{1j})|^2\}\,\mathrm{d}u\right|.
\end{aligned}
$$

Using (A4) and the inequality $\|g\|_\infty \le K^{1/2}\|g\|$ for every $g \in S_K^p([0,1])$, we further obtain

$$
\begin{aligned}
\mathbb{E}\{|f_g(\boldsymbol{\epsilon}_1)|^2\} &\le M_2 C_n^{1/2}K^{1/2}m^{-1}\sum_{j=1}^{m}\left|\int_{0}^{C_n^{1/2}g(T_j)}|u+R_j|\,\mathrm{d}u\right| \\
&\le c_0 C_n^{3/2}K^{1/2}m^{-1}\sum_{j=1}^{m}|g(T_j)|^2 + c_0 C_n K^{1/2}\max_{1\le j\le m}|R_j| \le c_0 K^{1/2}C_n^{3/2},
\end{aligned}
$$

where the last inequality follows from Lemma 33.1, i.e., $\max_j |R_j| = O(C_n^{1/2})$. Thus, $\mathbb{E}\{|f_g(\boldsymbol{\epsilon}_1)|^2\} \le c_0 K^{1/2} C_n^{3/2}$ uniformly in $f_g \in \mathcal{F}$.

Finally, to bound the bracketing entropy $\log N_{[]}(\epsilon, \mathcal{F}, \mathcal{L}^2(\mathbb{P}))$, first observe that $\|g\|_{r,\lambda} \leq D$ implies $\|g\| \leq D$ so that it suffices to consider $g \in S_K^p([0,1])$ with $\|g\| \leq D$. Since $S_K^p([0,1])$ is a $(K+p)$-dimensional subspace and $p \leq K$ for all large n, by Theorem 2.6 of [6] we have

$$\log N(\epsilon, \mathcal{F}, \mathcal{L}^2(\mathbb{P})) \leq 2K \log(1 + 4D/\epsilon), \quad \epsilon > 0.$$

Furthermore, note that for any $g_1, g_2 \in S_K^p([0,1])$ such that $\|g_1\| \leq D$ and $\|g_2\| \leq D$, by using (A3) and Lemma 33.1, we obtain that

$$\begin{aligned}\left|f_{g_1}(\mathbf{x}) - f_{g_2}(\mathbf{x})\right| &\leq m^{-1} \sum_{j=1}^{m} \left| \int_{R_j + C_n^{1/2} g_1(T_j)}^{R_j + C_n^{1/2} g_2(T_j)} \{\psi(x_j + u) - \psi(x_j)\} \, \mathrm{d}u \right| \\ &\leq c_0 C_n^{1/2} m^{-1} \sum_{j=1}^{m} |g_1(T_j) - g_2(T_j)| \leq c_0 C_n^{1/2} \|g_1 - g_2\|.\end{aligned}$$

This Lipschitz property combined with Theorem 2.7.11 of [7] now yields

$$\log N_{[]}(\epsilon, \mathcal{F}, \mathcal{L}^2(\mathbb{P})) \leq 2K \log(1 + 4c_0 D C_n^{1/2}/\epsilon), \quad \epsilon > 0.$$

With these bounds we can now apply Lemma 19.36 of [8] from which it can be deduced that under our limit conditions it is sufficient to show that

$$\int_0^{c_0 K^{1/4} C_n^{3/4}} \log^{1/2} N_{[]}(\epsilon, \mathcal{F}, \mathcal{L}^2(\mathbb{P})) \, \mathrm{d}\epsilon = o(n^{1/2} C_n),$$

for (33.5) to hold. By a change of variables, this integral may be bounded by

$$\begin{aligned}\int_0^{c_0 K^{1/4} C_n^{3/4}} \log N_{[]}(\epsilon, \mathcal{F}, \mathcal{L}^2(\mathbb{P})) \mathrm{d}\epsilon &\leq c_0 K^{1/2} \int_0^{c_0 K^{1/4} C_n^{3/4}} \log^{1/2} \left(\frac{4 c_0 D C_n^{1/2}}{\epsilon} + 1 \right) \mathrm{d}\epsilon \\ &\leq c_0 K^{1/2} C_n^{1/2} \int_0^{K^{1/4} C_n^{1/4}} \log^{1/2} \left(\frac{1}{\epsilon} + 1 \right) \mathrm{d}\epsilon \\ &\sim c_0 K^{1/2} C_n^{1/2} K^{1/4} C_n^{1/4} \log^{1/2}(n) \\ &= c_0 K^{3/4} C_n^{3/4} \log^{1/2}(n),\end{aligned}$$

where for two sequences $\{a_n\}_n$ and $\{b_n\}_n$, $a_n \sim b_n$ means $a_n/b_n \to 1$ as $n \to \infty$. Hence,

$$\frac{\int_0^{c_0 K^{1/4} C_n^{3/4}} \log^{1/2} N_{[]}(\epsilon, \mathcal{F}, \mathcal{L}^2(\mathbb{P})) \, \mathrm{d}\epsilon}{n^{1/2} C_n} \sim \frac{\log^{1/2}(n) K^{3/4}}{n^{1/2} C_n^{1/4}} \leq \frac{\log^{1/2}(n) K_n^{3/4}}{n^{1/4}} \to 0,$$

since, by our limit assumptions, $n^{\delta - 1} K^3 \to 0$ for some $\delta > 0$. This completes the proof of (33.5) and thereby establishes the result.

References

1. Cai, T.T., Yuan, M.: Optimal estimation of the mean function based on discretely sampled functional data: phase transition, Ann. Statist. **39**, 2330–2355, (2011)
2. de Boor, C. A Practical Guide to Splines, rev. ed., Springer (2000)
3. Kalogridis, I., Claeskens, G., Van Aelst, S.: Robust and efficient estimation of nonparametric generalized linear models, TEST **32**, 1055–1078, (2023)
4. Kalogridis, I., Van Aelst, S.: Robust optimal estimation of location from discretely sampled functional data, Scand. J. Statist. **50**, 411–451, (2023)
5. Zhou, S., Shen, X., Wolfe, A.: Local asymptotics for regression splines and confidence regions, Ann. Statist. **26**, 1760–1782, (1998)
6. van de Geer, S.: Empirical Processes in M-estimation, Cambridge, (2000)
7. van der Vaart, A.W., Wellner, J.A.: Weak Convergence and Empirical Processes, Springer, (1996)
8. van der Vaart, A.W.: Asymptotic Statistics, Cambridge, (1998)
9. Xiao, L.: Asymptotic properties of penalized splines for functional data, Bernoulli **26**, 2847–2875, (2020)

Chapter 34
Functional Regression with Shape Constraints

Yuwei Jiang and Natalya Pya Arnqvist

Abstract Functional regression is a dynamic research field within functional data analysis, with a wide range of applications across different areas. This paper aims to expand the existing framework of shape-constrained generalized additive models to functional generalized additive models with shape constraints. We introduce an extension of the shape-constrained P-spline (SCOP-spline) approach to a broad class of functional regression models with various shape constraints. Our framework includes parametric and a mixture of constrained and unconstrained smooth effects of functional and scalar covariates. Estimation and inference in this framework build upon the shape-constrained generalized additive models, enabling the use of well-established, robust, and flexible procedures. The described methods are implemented in the user-friendly R package `scam`. Simulation shows performance improvements when modelling with the proposed approach compared to the unconstrained method.

34.1 Introduction

Functional regression models focus on the class of models that involve functional responses and/or functional predictors [1] , and are widely applied in various fields, including, for example, environmental modelling [2], marine ecology [3], and physical activity research [4]. Many approaches to estimating functional regression models have been developed, beginning with the early work of [5] and later improvements by [6]. The models are typically classified into three categories based on the role of functional data: 'scalar-on-function' regression, where the response is a scalar and

Yuwei Jiang (✉)
Department of Mathematics and Mathematical Statistics, Umeå University, 901 87 Umeå, Sweden
e-mail: `yuwei.jiang@umu.se`

Natalya Pya Arnqvist
Department of Mathematics and Mathematical Statistics, Umeå University, 901 87 Umeå, Sweden
e-mail: `natalya.pya@umu.se`

© The Author(s), under exclusive license to Springer Nature Switzerland AG 2025

G. Aneiros et al. (eds.), *New Trends in Functional Statistics and Related Fields*,
Contributions to Statistics, https://doi.org/10.1007/978-3-031-92383-8_34

the predictors are functional; 'function-on-scalar' regression, where the response is functional, and the predictors are scalar; and 'function-on-function' regression, where both the response and predictors are functional. This paper primarily focuses on the first case and demonstrates how the proposed approach can be extended to the other two.

Shape constraints, such as monotonicity and convexity, play a crucial role in many real-world applications. They may reflect prior knowledge about the relationship between the response and predictor or be imposed to ensure the structural consistency of models. Shape-constrained generalized additive models (SCAM) [7] provide a general framework for semi-parametric regression modelling with shape constraints and have been applied across various fields, including genetics [8], sustainability studies [9], biotechnology [10], and ecological and environmental studies [11]. In functional regression, examples of applications that benefit from incorporating shape constraints include modelling growth curves [12] and analysing the effect of the drug on disease severity in longitudinal clinical studies [13].

Research on incorporating shape constraints into functional regression models has gained significant attention, especially in the univariate setting. The theoretical properties of shape-constrained estimators, including isotonic and convex regression, have been studied by [14, 15] among others. Alternative approaches rely on kernel-based monotone regression [16] and parametric monotone regression models [17]. In the context of functional regression, [18] introduced a shape-constrained framework using Bernstein polynomials, enabling the derivation of point-wise asymptotic confidence intervals for constrained estimators. However, this approach is restricted to univariate models and does not cover exponential family responses.

This paper presents a novel framework for shape-constrained functional generalized additive regression, extending the model proposed by [7] to functional regression settings. The framework offers a unified approach that accommodates parametric and a mixture of constrained and unconstrained smooth effects of functional and scalar covariates of univariate and multivariate type, supporting functional and scalar responses that arise from a general exponential family distribution. Incorporating established robust methods of the shape-constrained generalized additive models for model coefficient estimation and smoothness selection enables the proposed modelling approach to be flexible and have wide practical applicability.

34.2 Modelling Framework

To simplify the presentation, the three categories of functional regression models will first be presented in the basic setting with a single functional or scalar predictor, before extending to cases with multiple predictors and beyond continuous, real-valued scalar responses.

Scalar-on-function regression

Let the observed data for the ith subject, $i = 1, \cdots, n$, be $\{Y_i, (z_{ij}, t_{ij})\}$, $j = 1, \cdots, J$, where Y_i is a scalar real-valued response and z_{ij} are discrete realizations of a real-valued, square-integrable smooth function $z_i()$ at observation points t_{ij} over a finite interval $\mathcal{I} \subset \mathbb{R}$. A simple scalar-response functional shape-constrained linear model can be expressed as

$$Y_i = \alpha + \int_{\mathcal{I}} \omega(t) z_i(t) dt + \epsilon_i, \quad \epsilon_i \sim N(0, \sigma^2), \tag{34.1}$$

where $\omega(t)$ is a smooth coefficient function to estimate subject to some shape constraint. To fit the model, we propose to represent the coefficient function in terms of the shape-constrained P-spline (SCOP-spline) basis expansion $\omega(t) = \sum_{j=1}^{q} \gamma_j b_j(t)$, introduced in [7]. Here, b_j are B-spline basis functions of at least second order, γ_j are unknown coefficients, and q is the number of basis functions. Using the first-order derivative feature of the B-splines, shape constraints are imposed by re-parameterizing the spline coefficients $\boldsymbol{\gamma} = \boldsymbol{\Sigma}\tilde{\boldsymbol{\beta}}$, where $\boldsymbol{\gamma} = (\gamma_1, \ldots, \gamma_q)^T$, $\boldsymbol{\Sigma}$ is a matrix of 0's and 1's, $\boldsymbol{\beta} = (\beta_1, \ldots, \beta_q)^T$, and $\tilde{\boldsymbol{\beta}}$ be a vector of $\boldsymbol{\beta}$ elements. Depending on the imposed shape constraints, some of these elements are exponentiated as β_j. For example, in case of a monotonically increasing smooth, a sufficient condition for $\omega_j'(t) \geq 0$ is that the spline coefficients satisfy $\gamma_j > \gamma_{j-1}$ for all j. This condition can be imposed by setting $\tilde{\boldsymbol{\beta}} = \left[\beta_1, \exp(\beta_2), \ldots, \exp(\beta_q)\right]^T$, while $\Sigma_{ij} = 0$ if $i < j$ and $\Sigma_{ij} = 1$ if $i \geq j$. The model matrix elements of the linear functional term become $X_{ij} = \int b_j(t) z_i(t) dt$. In a smoothing context, with each $\omega(x)$ is associated a smoothing penalty, which is quadratic in the β_j and controls the 'wiggliness' of ω. In a matrix form, the penalty can be written as $\boldsymbol{\beta}^T \mathbf{S} \boldsymbol{\beta}$, where $\mathbf{S}$ is a matrix of known coefficients. Given the linearity of the model in $\omega(t)$, and by replacing the integral with a discrete sum approximation (numerical quadrature), the model estimation follows the exact methodology developed for SCAM [7].

While the model described above is a typical case, practical applications often require various extensions, such as incorporating scalar covariates of linear and smooth effect types and multiple functional predictors, where smooth terms can be with or without shape constraints and accommodating response distributions beyond the normal case. So, suppose that for subject i, we observe multiple functional predictors $z_{1i}(t), \ldots, z_{Ki}(t)$, which may be subject to shape constraints and have different domains $\mathcal{I}_1, \ldots, \mathcal{I}_K$, along with scalar covariates $x_{1i}, \ldots, x_{ri}$ with linear or smooth effects. The functional regression model (34.1) naturally extends to a multiple shape-constrained additive model

$$\begin{aligned} Y_i = \mathbf{A}_i \boldsymbol{\alpha} + \sum_l \int_{\mathcal{I}_l} \omega_l(t)\, z_{li}(t)\, \mathrm{d}t + \sum_k \int_{\mathcal{I}_k} \psi_k(t)\, z_{ki}(t)\, \mathrm{d}t + \\ \sum_s m_s(x_{si}) + \sum_r f_r(x_{ri}) + \varepsilon_i, \end{aligned} \tag{34.2}$$

where $\mathbf{A}_i$ is the ith row of a known parametric model matrix with a vector of unknown coefficients $\boldsymbol{\alpha}$; $\omega_l(t)$ and $\psi_k(t)$ are shape-constrained and unconstrained smooth coefficient function correspondingly; $m_s(x_{si})$ and $f_r(x_{ri})$ are smooths of scalar predictors with and without shape constraints.

In case of a non-Gaussian response, for example, a binary outcome, model (34.2) can be generalized to allow responses from the exponential family distributions

$$g(\mu_i) = \mathbf{A}_i\boldsymbol{\alpha} + \sum_l \int_{\mathcal{I}_l} \omega_l(t)\, z_{li}(t)\, \mathrm{d}t + \sum_k \int_{\mathcal{I}_k} \psi_k(t)\, z_{ki}(t)\, \mathrm{d}t + \sum_s m_s(x_{si}) + \sum_r f_r(x_{ri}), \tag{34.3}$$

where $\mu_i = E(Y_i)$ and $g()$ is a known link function. To represent (34.3) for computation, the constrained linear functional terms are represented as described above, while basis expansions of modest ranks are used for constructing unconstrained coefficient functions of the functional predictors, $\psi_k(t)$, resulting in $\sum_l \int_{\mathcal{I}_l} \omega_l(t)\, z_{li}(t)\, \mathrm{d}t + \sum_k \int_{\mathcal{I}_k} \psi_k(t)\, z_{ki}(t)\, \mathrm{d}t = \mathbf{X}_i\tilde{\boldsymbol{\beta}} + \mathbf{Z}_i\boldsymbol{\gamma}$, where $\mathbf{X}$ and $\mathbf{Z}$ are model matrices, and $\tilde{\boldsymbol{\beta}}$, γ are vector of unknown coefficients. The unknown smooth effects of scalar predictors can be represented by analogy via basis expansions with chosen penalties and identifiable constraints. The penalties of the unconstrained smooth terms are quadratic in $\boldsymbol{\gamma}$. By absorbing the corresponding identifiability constraints of scalar covariates, the model matrices for all m_s and f_r can be combined as that $\sum_s m_s(x_{si}) + \sum_r f_r(x_{ri}) = \mathbf{M}_i\tilde{\boldsymbol{\beta}}' + \mathbf{F}_i\boldsymbol{\gamma}'$, $\tilde{\boldsymbol{\beta}}'$ is a vector of a mixture of β_j and exponentianted β'_j reflecting the applied SCOP-splines. The construction of the smooth coefficients of the functional predictors does not require the imposition of the identifiability constraints since smooth terms are fully identifiable in this case. So, (34.3) can be expressed as $g(\mu_i) = \mathbf{A}_i\boldsymbol{\alpha} + \mathbf{X}_i\tilde{\boldsymbol{\beta}} + \mathbf{Z}_i\boldsymbol{\gamma} + \mathbf{M}_i\tilde{\boldsymbol{\beta}}' + \mathbf{F}_i\boldsymbol{\gamma}'$, and by combining the matrices column-wise into one model matrix $\mathbf{X} = [\mathbf{A} : \mathbf{X} : \mathbf{Z} : \mathbf{M} : \mathbf{F}]$, $g(\mu_i) = \mathbf{X}_i\tilde{\boldsymbol{\beta}}$, where $\tilde{\boldsymbol{\beta}}$ now contains all vector of coefficients.

Function-on-scalar regression

Suppose that the response of interest is of functional type, $Y_i(t)$, $i = 1, \ldots, n$, observed over domain $\mathcal{T}$, and for each subject i we observe scalar predictors, $x_{1i}, \ldots, x_{pi}$. The shape-constrained function-on-scalar additive regression model can be written as

$$Y_i(t) = f_0(t) + \sum_j x_{ji} f_j(t) + \sum_k x_{ki} m_k(t) + \epsilon_i(t), \tag{34.4}$$

where x_i are scalar predictors, and the dependence of the functional response on those is captured via coefficient functions f_j and m_k to be estimated correspondingly without or with shape constraints. We assume that errors, $\epsilon_i(t)$, correlated within a subject can be modelled as an AR(1) error process, $\epsilon_i = \rho\epsilon_{i-1} + e_i$, where e_i are independent $N(0, \sigma^2)$. As in the scalar-on-function scenario, the smooth coefficient functions are represented via SCOP-splines in the constrained case or penalized spline basis expansion with the chosen penalties. Multiplication of the smooth terms by the scalar predictors x_{ji} is performed similarly to when implementing 'varying coefficients models' [19] by means of the so-called 'by' variable.

Function-on-function regression

In this case, for the ith subject, we observe the functional response $Y_i(t)$ over domain $\mathcal{T}$, along with multiple functional predictors $z_{1i}(s), \ldots, z_{Ki}(s)$, which may be subject to shape constraints and can have different domains $\mathcal{I}_1, \ldots, \mathcal{I}_K$. For simplicity, we omit the linear and smooth effects of scalar covariates, though the model can be extended to incorporate them. Then, the shape-constrained additive regression model with linear functional effects can be expressed as

$$Y_i(t) = f_0(t) + \sum_l \int_{\mathcal{I}_l} \omega_l(t,s) z_{li}(s) \mathrm{d}s + \sum_k \int_{\mathcal{I}_k} \psi_k(t,s) z_{ki}(s) \mathrm{d}s + \epsilon_i(t), \quad (34.5)$$

where tensor product splines with or without shape constraints are used for representing the bivariate smooth coefficient functions $\omega_l(t, s)$ and $\psi_k(t, s)$ with the associated penalties.

After expressing the model terms using corresponding basis expansions, the proposed shape-constrained functional additive regression models can be formulated in matrix form, $\mu_i = \mathbf{X}_i \tilde{\boldsymbol{\beta}}$. Let $l(\boldsymbol{\beta})$ be the model log likelihood and $\mathbf{S}^k$ is the penalty matrix of known coefficients that correspond to the kth smoothing penalty. The estimated model coefficients are then $\hat{\boldsymbol{\beta}} = \operatorname{argmax}_{\boldsymbol{\beta}} \left\{ l(\boldsymbol{\beta}) - 1/2 \sum_k^M \lambda_k \boldsymbol{\beta}^T \mathbf{S}^k \boldsymbol{\beta} \right\}$, given M values for the smoothing parameter λ_k, controlling the level of penalization. Smoothing parameters are estimated to minimize a prediction error criterion such as GCV or AIC in the outer optimization scheme following the methodology introduced for SCAM [7]. The developed simulation-free approximate Bayesian confidence intervals for the model terms are used for the functional regression models.

34.3 Simulation Study

In this section, we demonstrate the performance of the shape-constrained scalar-on-function regression model under a decreasing constraint in a simple simulated scenario:

$$Y_i = \int \omega(t)\, z_i(t)\, dt + \epsilon_i, \quad \epsilon_i \sim N\left(0, \sigma^2\right),$$

where the residuals $\epsilon_i \sim N\left(0, \sigma^2\right)$ and independent. The true coefficient function is given by $\omega(t) = -4 \exp(4t)/(1+\exp(4t))$, and the values of σ are $\{0.1, 0.5, 1, 2, 3\}$. The observation points t_i were simulated from the uniform distribution over $[-1, 3]$ with sample sizes $n = 100, 300$ and 500. The first six images in Fig. 34.1 provide examples of the generated functional predictors.

The R package `scam` was used to fit the model, and the proposed constrained method was compared to the standard unconstrained GAM implemented in the `mgcv` package [20]. Besides the decreasing constraint, `scam` supports eight different SCOP-spline variants, incorporating monotonicity, concavity, and combinations of both. The two lower right plots in Fig. 34.1 show the estimated coefficient function $\omega(t)$ for the sample size $n = 100$ using both the constrained and unconstrained methods. The results clearly demonstrate that the shape-constrained model performs better than the unconstrained version.

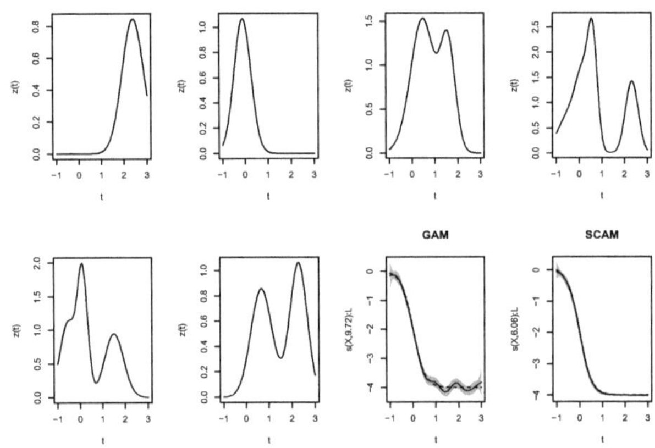

Fig. 34.1: Simulated example of the scalar on function regression with decreasing constraint. The first six plots (from top left to lower right) show six examples of the generated functional predictors. In total, 200 functions were simulated and each at 100 evaluation points. Third lower plot: estimated constrained coefficient function (solid line) and true coefficient function (dashed line). Lower right plot: estimated unconstrained coefficient function (solid line) and true coefficient function (dashed line).

To assess the performance of both methods, 100 replicates were simulated for each combination of five noise levels and three sample sizes. The relative mean square errors (RMSE) are defined as $\left(MSE_{gam} - MSE_{scam}\right)/\overline{MSE_{scam}}$ where $\overline{MSE_{scam}}$ is the average value across replications. These errors quantify the accuracy of the estimated response variable Y_i for both approaches, as shown in Fig.34.2. The results suggest that the SCAM method is more efficient than GAM. Fig. 34.3 further illustrates this by comparing the estimated smooth coefficient functions for both methods at a sample size of $n = 100$ across different noise levels. Notably, as the noise level increases, the advantages of SCAM become more apparent. The same trend is observed for $n = 300$ and $n = 500$, though the performance gap between SCAM and GAM diminishes as the sample size increases (figures not shown).

Simulation studies suggest that SCAM may offer practical advantages over the GAM approach for scalar-on-function regression when modelling a single smooth coefficient variable under the assumption of a normal distribution. Moreover, based

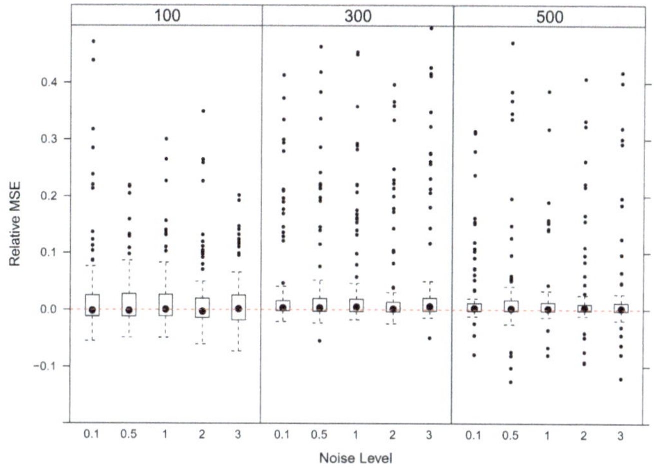

Fig. 34.2: RMSE comparisons between SCAM, GAM for the Gaussian distribution for each of five noise levels. 100 replicate were used.

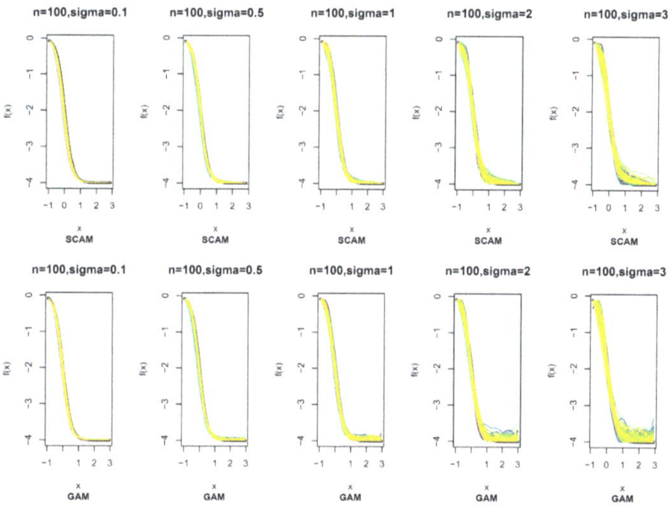

Fig. 34.3: The comparison of the coefficient function graphs for SCAM and GAM estimates is presented for $n = 100$ with 100 repetitions of the sampling.

on the suggested theoretical framework, our method can be extended to include scalar covariates with both linear and smooth effects, as well as multiple functional predictors. The smoothing terms can be either shape-constrained or unconstrained, and the approach is flexible enough to accommodate response distributions beyond the normal case. Additionally, our method extends to function-on-scalar and function-on-function regression, along with their respective generalizations, as outlined in Eqs. (34.4) and (34.5). Further simulation studies will be necessary to assess the performance of these extensions and validate their practical applicability.

References

1. Ramsay, J. O., Silverman, B. W.: Functional Data Analysis Springer-Verlag, New York, (2005)
2. Ghosal, R., Saha, E.: Atmospheric Environment **254**, 118388, (2021)
3. Ghosal, R., Maity, A., Clark, T., Longo, S. B.: Journal of the Royal Statistical Society Series C: Applied Statistics **69**, 565–587, (2020)
4. Goldsmith, J., Liu, X., Jacobson, J. S., Rundle, A. G.: Medicine and science in sports and exercise **48 9**, 1723–9, (2016)
5. Hastie, T., Tibshirani, R.: Journal of the Royal Statistical Society Series B: Statistical Methodology **55**, 757–779, (1993)
6. Reiss, P. T., Huang, L., Mennes, M.: The International Journal of Biostatistics **6**, (2010)
7. Pya, N., Wood, S. N.: Shape constrained additive models. Statistics and computing **25**, 543–559, (2015)
8. Amadeu, R. R., Munoz, P. R., Zheng, C., Endelman, J. B.: Genetics **219**, iyab124, (2021)
9. Grant, D., Zelinka, D., Mitova, S.: Environmental Research Letters **16**, 094022, (2021)
10. Lee, U., Carroll, R. J., Marder, K., Wang, Y., Garcia, T. P.: Biostatistics **22**, 819–835, (2021)
11. McDowell, R., Simpson, Z. P., Ausseil, A., Etheridge, Z., Law, R.: Scientific reports, Nature Publishing Group **11**, 1–14, (2021)
12. Hu, Y., He, X., Tao, J. & Shi, N.: Science in China Series A: Mathematics **52**, 1342–1350, (2009)
13. Ahkim, M., Gijbels, I., Verhasselt, A.: Test **26**, 429–450, (2017)
14. Brunk, H. D.: (University of Missouri-Columbia, 1969)
15. Hanson, D., Pledger, G.: The Annals of Statistics 1038–1050, (1976)
16. Hall, P., Huang, L.-S.: The Annals of Statistics **29**, 624–647, (2001)
17. Banerjee, M.: In Advances in multivariate statistical methods, 249–271 (World Scientific, 2009)
18. Ghosal, R., Ghosh, S., Urbanek, J., Schrack, J. A., Zipunnikov, V.: Computational Statistics & Data Analysis **178**, 107614, (2023)
19. Wood, S. N.: Generalized additive models: an introduction with R (Chapman and Hall/CRC, 2017)
20. Wood, S. N.: Journal of the Royal Statistical Society (B) **73**, 3–36, (2011)

Chapter 35
Enhancing Causal Inference in Functional Data: a Method for Estimating Time-Varying Causal Treatment Effects

Francesca Ieva, Nicole Fontana, Carlo Andrea Pivato, Emanuele Di Angelantonio and Piercesare Secchi

Abstract Functional data analysis has emerged as a powerful statistical tool with applications across various scientific domains, yet its integration with causal inference remains underdeveloped despite the fundamental importance of causation in scientific investigations. A central challenge in causal analyses is the lack of random treatment assignment, which complicates the chance of drawing valid conclusions from observational data. This study focuses on evaluating the causal effect of a binary treatment on functional outcomes, addressing the additional complexity of sparse and irregularly measured data. To overcome non-randomized treatment assignments, we employed a weighting approach to mitigate confounding and constructed a pseudo-population for unbiased causal effect estimation. We effectively test and estimate the causal association between treatment and functional outcomes by applying interval-wise testing procedure and weighted functional-on-scalar modelling. The proposed method is applied for drug repurposing by estimating the causal effect of metformin, a diabetes medication, on kidney function using a real-world biobank.

35.1 Introduction

Functional data analysis (FDA) has gained prominence in modern statistics and has been successfully applied across diverse scientific domains [7]. Despite its growing relevance, the application of FDA methods to causal inference remains underdeveloped, even though causation is a cornerstone of many scientific investigations.

Francesca Ieva (✉), Nicole Fontana, Piercesare Secchi
MOX, Department of Mathematics, Politecnico di Milano, Milan, Italy.
e-mail: francesca.ieva@polimi.it

Francesca Ieva, Nicole Fontana, Emanuele Di Angelantonio
Health Data Science Research Centre, Human Technopole, Milan, Italy.

Carlo Andrea Pivato
Department of Biomedical Sciences, Humanitas University, Pieve Emanuele, Milan, Italy.

© The Author(s), under exclusive license to Springer Nature Switzerland AG 2025
G. Aneiros et al. (eds.), *New Trends in Functional Statistics and Related Fields*,
Contributions to Statistics, https://doi.org/10.1007/978-3-031-92383-8_35

A central aim of many analyses is to determine the causal effect of a treatment or intervention, but these analyses are often complicated by the absence of random assignment to treatments, which undermines the ability to draw valid causal conclusions from observational data. To address this challenge, a variety of methods have been developed in the literature to handle causal inference in observational data. Among these, the propensity score method [9] has played a pivotal role and has seen widespread use, particularly in fields like epidemiology and political science. However, its application to FDA is testified in only a few studies [1, 4]. A promising direction lies in adapting tools from classical causal inference to FDA contexts, tailoring their application to the specific research needs, whether focusing on functional treatments, covariates, or outcomes. This work specifically focuses on a scenario where the treatment is binary and the outcome is functional. In addition, we will deal with a more complicated case where the outcome is sparse and irregularly measured. To handle such irregularly spaced observations, we propose using the conditional expectation (PACE) method for curve reconstruction [11]. While traditional FDA models focus on identifying correlations among scalar/functional predictors and functional/scalar outcomes, they are missing the opportunity to assess causal relationships. Bridging this gap requires the integration of causal inference approaches in the context of functional outcomes. To address the challenge of non-randomized treatment assignment, which is one of the main requirements in causal inference, we employed a weighting approach to construct a pseudo-population free from measurable confounding factors, potentially biasing the association between the treatment and the outcome. By applying an interval-wise testing procedure and a weighted functional-on-scalar model, we effectively test and estimate the causal association between the treatment and the functional outcomes.

This work is motivated by the potential of drug repurposing, a strategy that accelerates therapeutic innovation by offering higher success rates and lower costs compared to traditional drug development. Leveraging large-scale real-world data creates significant opportunities to generate hypotheses and evaluate the effectiveness of repurposed drugs. Our study focuses on assessing the effect of metformin, a type 2 diabetes medication, on kidney function within a causal framework, aiming to provide robust evidence that could guide future clinical trials. Building on the emulated target trial for metformin on chronic kidney disease (CKD) [2], we extend previous approaches by moving beyond single-time-point diagnoses to evaluating kidney function over time. This is achieved using laboratory measurements collected longitudinally in UK Biobank, a large-scale biomedical database, and introducing a novel method for estimating time-varying causal effects.

35.2 Methodology

35.2.1 Elements of Causal Inference in a Functional Setting

Estimating causal effects is essential for evaluating treatment impact, with randomized controlled trials (RCTs) considered as the gold standard. RCTs ensure that all variables, both measured and unmeasured, are equally distributed across treatment and control groups, allowing for the identification of the causal treatment effects. However, they are costly, often impractical, and sometimes unethical, increasing research to rely on observational data. While observational studies offer a practical alternative and address some of the limitations of RCTs, they face challenges in disentangling causal relationships from mere correlations. The main challenge is eliminating confounding bias, which stems from differences in the distribution of confounding variables between the treatment and control groups. Weighting methods are used to achieve similar distributions of these variables between the two groups [8]. The fundamental principle underlying weighting involves assigning appropriate weights to each observation in the dataset to create a pseudo-population in which the only difference between groups is in whether they received treatment. In a functional setting, this concept is extended to ensure that, at the onset of observing a functional outcome, the population is balanced with respect to all potential confounding factors. This requires adapting weighting methods to manage the complexities of functional data. The need for this approach stems from the fact that the likelihood of receiving treatment is influenced by each individual's unique context, which can also affect the baseline functional outcome. Addressing this interplay between treatment allocation and individual-specific factors is crucial, as it helps control for confounding that may distort the interpretation of treatment effects on functional outcomes.

35.2.2 Predictor Trajectories Smoothing

Classical functional principal component analysis (FPCA) methods are designed for densely observed curves and struggle with sparse or irregularly timed longitudinal data. To address these limitations, in this work, we employed the PACE method [11], which provides a robust framework for handling such irregular data. This nonparametric approach models sparse and irregularly timed repeated measurements of the response trajectories by pooling data across subjects and applying smoothing techniques. Under minimal assumptions, it estimates population-level mean and covariance functions. At the individual level, functional principal component scores derived from the PACE method enable the prediction of smooth trajectories, even when only one or very few measurements are available per participant. Let $Y_i(t_{il})$ be the measured outcome at time $t_{il} \in \mathcal{T}$ for the patient i and $\hat{\mu}(t)$ and $\hat{G}(s,t)$ the smooth mean and covariance surface estimate. The estimated trajectory for the ith patient is:

$$\hat{Y}_i(t) = \hat{\mu}(t) + \sum_{k=1}^{K} \hat{\xi}_{ik}\hat{\phi}_k(t), \tag{35.1}$$

where $\hat{\xi}_{ik}$ where estimated using the conditional expectation algorithm and the functional principal components $\hat{\phi}_k(t)$ are the estimated eigenfunctions of the eigenequation:

$$\int_{\mathcal{T}} \hat{G}(s,t)\hat{\phi}_k(s)ds = \hat{\lambda}_k\hat{\phi}_k(t) \tag{35.2}$$

with the constraints $\int_{\mathcal{T}} \hat{\phi}_k(t)^2 dt = 1$ and $\int_{\mathcal{T}} \hat{\phi}_k(t)\hat{\phi}_m(t)dt = 0$ for $m < k$. Functional curves are observed within a time window $T = (a, b) \subset \mathbb{R}$, where the initial point a is defined by applying an exposure density sampling procedure [5]. This approach defines the point in time a at which we start observing an individual, ensuring that the population is synchronized based on whether or not they received the treatment. The slope of the functional curves before the defined start time is used as a covariate during the weighting step to account for prior information.

35.2.3 Weighting and Groups Comparison

A weighting causal inference method was applied to control for baseline confounding variables. Specifically, weights were computed using the entropy balancing method to ensure a balance between the treatment and control groups [3]. Entropy balancing relies on a maximum entropy reweighting scheme that calibrates unit weights so that the reweighted treated and control group satisfy a potentially large set of pre-specified balance conditions that incorporate information about known sample moments. Once the weights were computed, they were truncated at the 1st and 99th percentiles to minimize the influence of extreme values. A covariate is considered balanced between the two groups if the standardized mean difference (SMD) is less than or equal to 0.1. Each individual i in group j is then assigned a weight w_{ij}. Applying these weights to the entire population generates the pseudo-population in which measured confounders are evenly distributed across the groups.

The interval-wise testing procedure described in [6] is used to assess whether there is a difference in the mean between the two groups. Consider two independent samples of random functions in the $L^2(T)$ space $Y_{ij}(t)$ where $j = 0, 1$ is the group index, $i = 1, ..., n_j$ is the unit index, and $T = (a, b) \subset R$. Let $\mu_j(\cdot)$ with $j = 0, 1$ denote the functional means of the two groups. We want to test the following hypotheses:

$$H_0 : \mu_0(\cdot) = \mu_1(\cdot) \text{ against } H_1 : \mu_0(\cdot) \neq \mu_1(\cdot), \tag{35.3}$$

where equality is defined in the L^2 sense. The interval-wise testing procedure performs functional hypothesis testing over every open interval I of the form (t_1, t_2) and its complementary interval $T \setminus (t_1, t_2)$ within the domain. We applied this method to the estimated curves $\hat{Y}(t)$ in our pseudo-population, which was generated using weights w_{ij} derived in the previous step. To do that, we rewrite the test statistics based on the L^2 distance between the two sample means over $\mathcal{I}$ used in the standard interval-wise testing procedure to incorporate weights described in Sec. 35.2.3 appropriately:

$$T^{\mathcal{I}}(Y_{0,1},\ldots,Y_{0,n_1},Y_{1,1},\ldots,Y_{1,n_2}) = \frac{\|\overline{Y}_0 - \overline{Y}_1\|^2_{L^2(\mathcal{I})}}{|\mathcal{I}|} = \frac{1}{|\mathcal{I}|}\int_{\mathcal{I}} (\overline{Y}_0(t) - \overline{Y}_1(t))^2\,dt, \tag{3}$$

where $\overline{Y}_j = \frac{1}{\sum_{i=1}^{n_j} w_{ji}} \sum_{i=1}^{n_j} w_{ji} \cdot Y_{ji}$, $j = 0, 1$ and w_{ji} are the computed weights for the individual i in the group j. An exact permutation test can then be achieved by evaluating the test statistic $T^{\mathcal{I}}$ over all possible permutations of the data across the sample units. In this framework, the p-value of the corresponding test is the proportion of the corresponding test statistics evaluated on permuted data exceeding the statistics evaluated on the original data set.

35.2.4 *Weighted Functional-on-Scalar Model*

To estimate the effect of the binary treatment on the functional outcome, we utilized a functional-on-scalar model:

$$Y_i(t) = \beta_0(t) + \beta_1(t)X_{ij} + \epsilon_i(t), \text{ where } X_{ij} = \begin{cases} 0 & \text{if } j = 0 \ (i \in \text{control group}), \\ 1 & \text{if } j = 1 \ (i \in \text{treatment group}). \end{cases} \tag{35.4}$$

This model was applied to our pseudo-population to determine the difference in the outcome caused by the treatment, controlling for all measurable confounding factors that can influence the baseline outcome and the treatment assignment. To achieve this, we implemented a weighted version of the functional-on-scalar model. Parameter estimation was carried out using weighted least squares estimators by minimizing the following objective function:

$$\sum_{i=1}^{n} w_i \int \left(Y_i(t) - \beta_0(t) - \beta_1(t)X_{i1}\right)^2 dt. \tag{35.5}$$

Thanks to this design, the estimated $\beta_1(t)$ represents the causal time-varying treatment effect on the functional outcomes.

35.3 Data and Setting

This study utilizes data from UK Biobank, a multicenter prospective cohort study where participants aged 40–75 years were recruited across the United Kingdom between 2006 and 2010 [10]. This work expands on a previously designed emulated trial [2], where eligibility criteria, treatment strategies, and confounding factors were defined to mimic a target clinical trial. Building on the well-defined causal methodological framework presented in that paper, where we study the effect of the treatment on a single-time-point diagnosis, we extend the approach here to a functional outcome. The following treatment strategies were compared: metformin initiators (treatment group) versus non-initiators (control group). To estimate the effect of metformin and avoid time-dependent bias, we implemented an exposure density sampling method for matching cohort entry time. This approach allowed us to synchronise the start of follow-up for both groups at the time of exposure. Participants were required to have at least two creatinine measurements during the follow-up period, enabling the data to be modelled as a functional outcome. The outcome of the analysis was the estimated glomerular filtration rate (eGFR), which is a measure measure of kidney function derived from the creatinine blood samples. eGFR was assessed over a four-year period following cohort entry time.

35.4 Results

In the selected cohort, 8116 participants were included in the analysis, of which 6122 (75.4 %) non-initiated metformin and 1994 (24.6 %) initiated metformin. The number of observations ranged from 2 to 29 for each individual, with a median of 7. FPCA with PACE is applied to reconstruct the smoothed eGFR curves. The AIC criterion leads to the choice of three eigenfunctions to reconstruct the curves that describe 99% of the total variability. Weights were computed using entropy balance methods across a set of confounding factors, including three demographic variables, two lifestyle factors, ten comorbidities, ten drugs, and the information on the slope curve prior to matching. Appendix 1 provides an explanation of the covariates considered as confounding factors and the criteria used to define them. The weights were truncated at the 1st and 99th percentiles. After truncation, a strong balance between the groups is obtained, with all covariates achieving a standardized mean difference (SMD) of ≤ 0.1, as shown in Fig. 35.1. The interval-wise testing procedure is applied to our pseudo-population every month, and the results consistently indicate that the difference in mean functions between the two groups is statistically significant across all intervals. The estimated weighted regression coefficient of treatment $\beta_1(t)$ is shown in Fig. 35.2. The shape of this estimate indicates that metformin has a positive effect on kidney function, with the effect size growing throughout the follow-up period. Comparing the treatment effect between the weighted and unweighted populations reveals a difference, as the estimated effect in the unweighted group remains almost constant throughout the period. This suggests that confounding factors in the unweighted analysis may obscure the true causal effect of treatment.

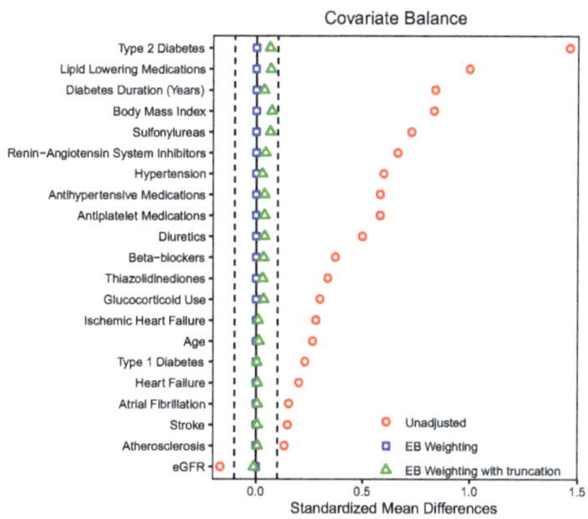

Fig. 35.1: SMD of confounders between treated and control groups unadjusted (circle), with weights (square) and with weights truncated (triangle)

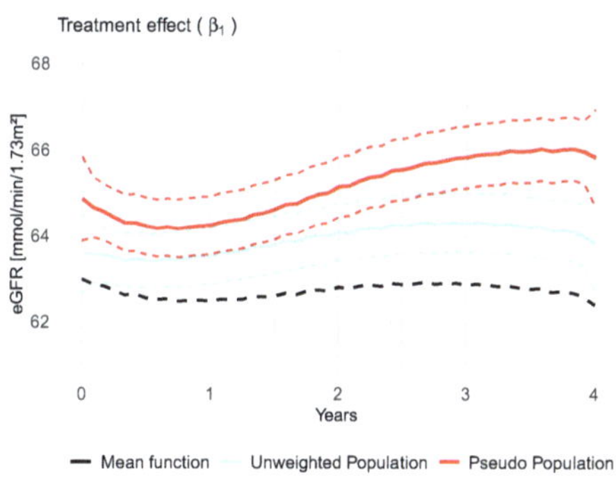

Fig. 35.2: Estimated regression function (β_1) for the unweighted population (light blue) and for the pseudo-population (red)

35.5 Discussion and Conclusion

This study integrates causal inference frameworks with functional outcomes. Assessing outcomes in a functional manner represents a significant advancement, offering a more realistic understanding of how treatments influence outcomes over time and providing more accurate estimates of their true effects. A key strength of this approach lies in its capacity to leverage longitudinal observational data for causal effect estimation, even in scenarios where observations are sparse and irregularly measured. This method enables the estimation of time-varying treatment effects on outcome trajectories, overcoming the inherent limitations of cross-sectional studies. By integrating a widely recognized weighting technique into a functional-on-scalar model, this method effectively addresses bias from measurable baseline confounders enabling causal estimation. The results confirm the importance of including a causal setting for the treatment effect evaluation, as unadjusted models produce biased associations influenced by confounders, potentially leading to inaccurate conclusions. Using the proposed method, this paper enables the study of the time-varying causal effect of metformin on kidney function. The results show an increase in eGFR, indicating better kidney function in the group of metformin initiators. This finding has significant clinical implications, highlighting the need for adequate RCTs to evaluate the potential of repurposing metformin for kidney disease treatment.

The presented approach can be applied to other contexts. One potential future advancement can be expanding the framework to include functional treatments to evaluate the effects of treatment changes over time. Additionally, incorporating functional covariates across the entire follow-up period would enable the model to control for time-varying confounders. These advancements would enhance the ability to estimate the dynamic causal effects of treatments on time-varying outcomes.

Acknowledgements The present research has been supported by MUR, grant Dipartimento di Eccellenza 2023-2027. UKB data are accessible for health-related research in the public interest. This study used data from the UKB Resource under Application Number 102297.

Appendix 1

Confounding variables definition.

A confounder must satisfy the following three requisites:

- being a risk factor for chronic kidney disease progression (among those who are not exposed to metformin),
- being associated with metformin, and
- not being an intermediary between metformin and chronic kidney disease (i.e., it must not lie on the causal pathway).

The following covariates were identified as possible candidates for confounders: demographics (age, sex, ethnicity), comorbidities (type 1 diabetes, type 2 diabetes, diabetes duration, atrial fibrillation, heart failure, ischemic heart disease, atherosclerotic disease, stroke or transient ischemic attack, hypertension), lifestyle factors (smoking and BMI), and concomitant medications (sulfonylureas, thiazolidinedione, insulin, diuretics, lipid-regulating drugs, antiplatelet drugs, antihypertensive, beta-blockers, renin-angiotensin system drugs, glucocorticoid therapy), information on the slope curve prior to matching (considering at most one year before the matching time). Smoking status and information on the slope curve prior to matching are not included in the final set of confounding variables since are already balanced in the two groups.

References

1. Belloni, A., Chernozhukov, V., Fernández-Val, I., Hansen, C.: Program evaluation and causal inference with high-dimensional data. Econometrica **85**(1), 233–298 (2017)
2. Fontana, N., Pivato, C.A., Di Angelantonio, E., Ieva, F.: Effect of metformin on kidney outcomes: a target trial emulation study. Submitted to BMJ (2024)
3. Hainmueller, J.: Entropy balancing for causal effects: a multivariate reweighting method to produce balanced samples in observational studies. Polit. Anal. **20**, 25–46 (2012)
4. Kreske, E., Xavier, L., Lina, S.: Causal inference with a functional outcome. J. R. Stat. Soc. C **73**(1), 221–240 (2024)
5. Ohneberg, K., Beyersmann, J., Schumacher, M.: Exposure density sampling: dynamic matching with respect to a time-dependent exposure. Stat. Med. **38**, 4390–4403 (2019)
6. Pini, A., Vantini, S.: Interval-wise testing for functional data. J. Nonparametr. Stat. **29**(2), 407–424 (2017)
7. Ramsay, J.O., Silverman, B.W.: Functional data analysis (2nd ed.). Springer Ser. Stat. Springer (2005)
8. Robins, J.M., Hernán, M.A., Brumback, B.: Marginal structural models and causal inference in epidemiology. Epidemiology **11**, 550–560 (2000)
9. Rosenbaum, P.R., Rubin, D.B.: The central role of the propensity score in observational studies for causal effects. Biometrika **70**(1), 41–55 (1983)
10. Sudlow, C., Gallacher, J., Allen, N., et al.: UK Biobank: an open access resource for identifying the causes of a wide range of complex diseases of middle and old age. PLoS Med. **12**, e1001779 (2015)
11. Yao, F., Müller, H.G., Wang, J.L.: Functional data analysis for sparse longitudinal data. J. Am. Stat. Assoc. **100**(470), 577–590 (2005)

Chapter 36
Linear and Nonlinear Regression Models for Spatial Downscaling of Particulate Matter

Luigi Ippoliti, Tonio Di Battista, Luigi Di Carlo, Stefania Fensore, Eugenia Nissi, Pasquale Valentini, Carlo Zaccardi

Abstract This paper presents a novel spatial downscaling methodology for PM10 data, offering valuable insights for air quality assessment and public health protection. The proposed framework integrates spectral methods with artificial intelligence techniques to construct both linear and nonlinear spatial regression models. By harmonizing coarse-resolution PM_{10} data obtained from large-scale sources, such as the Copernicus Atmosphere Monitoring Service (CAMS), with point-level measurements from monitoring stations, the methodology enhances spatial resolution and improves the accuracy of population exposure estimates at the municipality level. These fine-scale exposure assessments provide public health authorities with essential tools to identify high-risk areas, design targeted intervention strategies, and evaluate the effectiveness of air quality regulations.

36.1 Introduction to Spatial Downscaling

Spatial downscaling is essential for environmental and public health studies, particularly in estimating particulate matter (PM) concentrations at finer spatial resolutions. It integrates large-scale data sources, such as the Copernicus Atmosphere Monitoring Service (CAMS), with localized monitoring station measurements to address the change of support problem (COSP) [4]. This process enhances spatial precision, improving population exposure assessments at the municipality level and supporting epidemiological research and policy development.

CAMS data offer extensive coverage of atmospheric pollutants but lack the spatial granularity required for local analysis, whereas monitoring stations provide precise

Luigi Ippoliti (✉), Luigi Di Carlo, Eugenia Nissi, Pasquale Valentini , Carlo Zaccardi
Department of Economics, Viale Pindaro 42, 65127 Pescara, Italia
e-mail: luigi.ippoliti@unich.it

Tonio Di Battista, Stefania Fensore
DSEGS, Viale Pindaro 42, 65127 Pescara, Italia

© The Author(s), under exclusive license to Springer Nature Switzerland AG 2025

G. Aneiros et al. (eds.), *New Trends in Functional Statistics and Related Fields*, Contributions to Statistics, https://doi.org/10.1007/978-3-031-92383-8_36

measurements with limited spatial reach. Downscaling harmonizes these sources through statistical techniques such as spatial kriging [4] and Bayesian hierarchical modeling [1].

In this study, we present a comprehensive downscaling framework incorporating both linear and nonlinear models of increasing complexity. Initially, we introduce a spectral method to capture multiscale relationships between CAMS outputs and monitoring data, ensuring computational efficiency and interpretability. We then extend our approach with artificial intelligence techniques, including Multilayer Perceptron (MLP) models [7] and deep-learning-based super-resolution frameworks [8, 9], enabling the modeling of complex, nonlinear relationships between coarse-resolution predictors and fine-resolution targets. This hybrid methodology provides a robust and flexible solution for improving spatial PM predictions and exposure assessments.

36.2 Spatial Downscaling of CAMS Images

Let $X(s)$ denote a proxy measure (e.g., PM10 from the numerical model output CAMS) at grid cell $s \in \mathbb{Z}^2$, and let $Y(v)$ represent the PM10 collected at monitoring site $v \in \mathbb{R}^2$. We assume that $X(s)$ is available throughout the spatial domain, while $Y(v)$ is observed only sparsely. In the following, each point v is associated with the CAMS grid cell s in which it lies. So, all the points v falling in the same square region are assigned the same CAMS output value. This follows the standard interpretation of the CAMS model output, conceptualized as a tiled surface with grid cell resolution, where each tile represents the surface value at that resolution [1].

From a theoretical perspective, assume that both X and Y are weakly stationary, real-valued random fields with zero mean. Let their spatial covariance functions be $R_x(u) = \text{Cov}\big(X(s), X(s+u)\big)$ and $R_y(u) = \text{Cov}\big(Y(s), Y(s+u)\big)$, and their spatial correlation functions be $r_x(u) = R_x(u)/\sigma_x^2$ and $r_y(u) = R_y(u)/\sigma_y^2$, where $\sigma_x^2 = R_x(0)$ and $\sigma_y^2 = R_y(0)$. Additionally, let $\rho_{xy} = \text{Cor}(X, Y)$ denote the simple correlation between the two variables.

The spectral density function of X and Y are defined as follows:

$$\mathcal{S}_x(\omega) = (2\pi)^{-2} \sum_{u \in \mathbb{Z}^2} R_x(u)\, e^{-iu^\top \omega}, \quad \omega \in [-\pi, \pi]^2$$

$$\mathcal{S}_y(\omega) = (2\pi)^{-2} \sum_{u \in \mathbb{Z}^2} R_y(u)\, e^{-iu^\top \omega}, \quad \omega \in [-\pi, \pi]^2.$$

Suppose also that the covariance functions $R_x(u)$ and $R_y(u)$ are absolutely summable:

$$\sum_{u \in \mathbb{Z}^2 2} |R_x(u)| < \infty, \qquad \sum_{u \in \mathbb{Z}^2 2} |R_y(u)| < \infty,$$

and that the spectral density functions are "non-vanishing":

$$\mathcal{S}_x(\omega) \neq 0, \quad \mathcal{S}_y(\omega) \neq 0, \quad \omega \in [-\pi, \pi]^2.$$

36.2.1 Model specifics

The primary objective is to explore the relationship between $X(s)$ and $Y(s)$ across different spatial scales and to utilize $X(s)$ for predicting $Y(s)$. To accomplish this, we introduce here a series of models with progressively increasing complexity, both in terms of parameters and architectural structure. These models are designed to capture both linear and nonlinear relationships between input and output variables, offering significant versatility for diverse applications.

36.2.1.1 Basic Spatial Regression Model

The simplest model which one can use for predicting $Y(s)$ is certainly the basic spatial regression model:

$$y(s) = \beta_1 \, x(s) + \epsilon(s), \tag{36.1}$$

which provides a straightforward and interpretable framework for spatial downscaling. This model captures the primary relationship between coarse-scale predictors and fine-scale responses. Its computational efficiency makes it particularly well-suited for large datasets. However, linear regression models like equation (36.1) have notable limitations when applied to spatial downscaling. The assumption of a globally linear relationship between the predictor $X(s)$ and the response $Y(s)$ may not adequately represent complex spatial processes. Additionally, the fine-scale variability encapsulated in the residual term $\epsilon(s)$ often exhibits spatial dependence, which violates the independence assumption inherent to standard linear regression models. These limitations necessitate extensions or adaptations to better capture the complexities of spatial data.

36.2.1.2 Multi-Resolution Regression Model

To capture the potentially complex relationship between the processes X and Y, we model their dependence in the spectral domain. This allows for different dependence at different spatial scales as each frequency corresponds to a spatial scale, with low frequency corresponding to large spatial scale phenomenon and vise versa.

Their spectral representation is defined as:

$$X(s) = \frac{1}{(2\pi)^2} \int_{\mathbb{R}^2} e^{i\omega^\top s} \mathcal{X}(\omega)\, \mathrm{d}\omega, \quad Y(s) = \frac{1}{(2\pi)^2} \int_{\mathbb{R}^2} e^{i\omega^\top s} \mathcal{Y}(\omega)\, \mathrm{d}\omega,$$

where $\omega \in [-\pi, \pi]^2$ is a frequency. The spectral processes $\mathcal{X}(\omega)$ and $\mathcal{Y}(\omega)$ are defined for all frequencies via the Fourier transform:

$$\mathcal{X}(\omega) = \int_{\mathbb{R}^2} e^{-i\omega^\top s} X(s)\, \mathrm{d}s, \quad \mathcal{Y}(\omega) = \int_{\mathbb{R}^2} e^{-i\omega^\top s} Y(s)\, \mathrm{d}s.$$

The processes $\mathcal{X}(\omega)$ and $\mathcal{Y}(\omega)$ are Gaussian, which implies that X and Y are also Gaussian processes. These processes are characterized by having zero means, *i.e.*, $\mathrm{E}[\mathcal{X}(\omega)] = 0$ and $\mathrm{E}[\mathcal{Y}(\omega)] = 0$. Furthermore, they are mutually independent for $\omega \neq \omega'$. However, at the same frequency, the covariance of the joint spectral process has the form:

$$\mathrm{Cov}\begin{pmatrix} \mathcal{X}(\omega) \\ \mathcal{Y}(\omega) \end{pmatrix} = \begin{pmatrix} \sigma_x^2 \mathcal{S}_x(\omega) & \rho_{xy}\sigma_x\sigma_y\mathcal{S}_{xy}(\omega) \\ \rho_{xy}\sigma_x\sigma_y\mathcal{S}_{xy}(\omega) & \sigma_y^2\mathcal{S}_y(\omega) \end{pmatrix} \tag{36.2}$$

where $\mathcal{S}_x(\omega)$ and $\mathcal{S}_y(\omega)$ are spectral densities that define the marginal spatial dependence in the spectral domain of X and Y, respectively, and $\mathcal{S}_{xy}(\omega)$ is the cross-spectral density determining the dependence between the spectral processes at different frequencies.

It thus follows that, in the spectral domain, the conditional expectation of $\mathcal{Y}(\omega)$ given $\mathcal{X}(\omega)$ can be expressed as:

$$\mathrm{E}[\mathcal{Y}(\omega)|\mathcal{X}(\omega)] = \rho_{xy}\frac{\sigma_y}{\sigma_x}\frac{\mathcal{S}_{xy}(\omega)}{\mathcal{S}_x(\omega)}\mathcal{X}(\omega) = \beta(\omega)\mathcal{X}(\omega),$$

where $\beta(\omega) = \frac{\sigma_y}{\sigma_x}\frac{\sqrt{\mathcal{S}_y(\omega)}}{\sqrt{\mathcal{S}_x(\omega)}}\tau_{xy}(\omega)$, with $\tau_{xy}(\omega) = \rho_{xy}\frac{\mathcal{S}_{xy}(\omega)}{\sqrt{\mathcal{S}_x(\omega)}\sqrt{\mathcal{S}_y(\omega)}}$.

Accordingly, conditionally on X, the regression function can be expressed as:

$$\mathrm{E}[Y(s)|X] = \frac{1}{(2\pi)^2}\int_{\mathbb{R}^2} e^{i\omega^\top s}\beta(\omega)\mathcal{X}(\omega)\, \mathrm{d}\omega. \tag{36.3}$$

From equation (36.3), if we assume that X and Y share the same correlation structure, such that $\mathcal{S}_y(\omega) = \mathcal{S}_x(\omega)$, and the dependence is constant across all frequencies, *i.e.*, $\tau_{xy}(\omega) = \rho_{xy}$, then $\beta(\omega) = \beta = \rho_{xy}\frac{\sigma_y}{\sigma_x}$. In this case, equation (36.3) defines the conditional mean of the simple regression model (36.1). This result can be theoretically justified by assuming separability between the processes X and Y, such that $\mathcal{S}_{xy}(\omega) = \sqrt{\mathcal{S}_x(\omega)\mathcal{S}_y(\omega)}$. Therefore, equation (36.3), can be viewed as a multi-resolution extension of the simple linear model (36.1). In addition, since the function $\beta(\omega)$ is defined for all frequencies $\omega \in [-\pi, \pi]^2$, it can be expressed as a linear combination of K known basis functions:

$$\beta(\omega) = \sum_{k=1}^{K} \phi_k(\omega)\,\beta_k,$$

such that the conditional mean in equation (36.3) results in the following linear model

$$\mathrm{E}[Y(s)|X] = \sum_{k=1}^{K} \beta_k \widetilde{x}_k(s), \tag{36.4}$$

with

$$\widetilde{x}_k(s) = \frac{1}{(2\pi)^2} \int_{\mathbb{R}^2} \phi_k(\omega)\, e^{i\omega^\top s}\, \mathcal{X}(\omega)\, \mathrm{d}\omega. \tag{36.5}$$

For interpretation purposes, it is convenient to select basis functions so that $\sum_{k=1}^{K} \phi_k(\omega) = 1$ for all ω. In this case, if $\beta_k = \beta$ for all k, then $\beta(\omega) = \beta$ for all ω, and the spectral downscaler reduces to the simple linear model.

36.2.1.3 Nonlinear Regression Models

In this section we extend the discussion to the non-linear form of the model, outlining its multilayer architecture using neural network for spatial smoothing and prediction tasks. Specifically, we consider the following general regression model:

$$\bar{x}_l(s) = f_l\Big(\mathbf{W}_l\, \bar{x}_{l-1}(s)\Big),\ l = 1, \ldots, L-1 \tag{36.6}$$

$$y(s) = \sum_{k=1}^{N_L} \beta_k\, \bar{x}_{k,L-1}(s) + \epsilon(s), \tag{36.7}$$

which starts with an input vector of regressors $\bar{x}_{k,0}(s)$ at layer $l = 0$ and processes it through successive layers up to layer $L - 1$. At each layer, the regressors are transformed by computing a weighted sum of the inputs from the previous layer, represented by the weight matrix $\mathbf{W}_l$, and applying a non-linear activation function f_l. The outputs of each layer then serve as inputs to the next layer. This iterative transformation enables the model to capture and learn complex, non-linear relationships in the data. The final activation function at layer L is chosen so as to match the support of $y(s)$ at the output layer (*i.e.*, $f_L(\cdot)$ is the identity function) and one optimizes relative to squared error loss. In this case, the last layer's transformed regressors are combined linearly with coefficients β_k to predict the response variable $y(s)$, with $\epsilon(s)$ representing the error term. Note that, in our case, the inputs $\bar{x}_{k,0}(s)$ are derived from the weighted Fourier basis functions as defined in equation (36.5). This approach aligns with the methodology proposed by [3], where Wendland basis functions are utilized to construct the input vector.

Equations (36.6) and (36.7) provide the foundation of the well-known Multi-Layer Perceptron (MLP) framework [7], renowned for its ability to model intricate nonlinear relationships that standard linear regression methods often fail to capture. Another notable example of a similar structure is the projection pursuit method [6]. While both MLP and projection pursuit share a comparable layered architecture, they differ significantly in how they define their activation functions. In neural networks, the activation functions are predefined and typically involve fixed forms such as rectified linear unit (ReLU) functions. In contrast, projection pursuit employs activation functions that are not fixed but estimated from the data, allowing for greater flexibility in capturing patterns.

A different model architecture emerges if equation (36.6) is interpreted as an Efficient Sub-Pixel Convolutional Neural Network [8]. In this framework, equations (36.6) and (36.7) can be restructured as follows:

$$\begin{aligned} \bar{x}_l^{LR}(s) &= f_l\Big(\mathbf{W}_l\, \bar{x}_{l-1}^{LR}(s)\Big), \quad l = 1, \dots, L-1 \\ x(s)^{SR}(s) &= \mathcal{PS}\Big(\mathbf{W}_L\, \bar{x}_{L-1}(s)\Big), \end{aligned} \tag{36.8}$$

where the architecture begins with several convolutional layers applied directly to a low-resolution image, denoted as X^{LR}. This is followed by a sub-pixel convolution layer that uses a periodic shuffling operator $\mathcal{PS}$ to upscale the low-resolution feature maps and generate a high-resolution image X^{SR}. The periodic shuffling operator transforms a low-resolution feature map with multiple channels, each encoding distinct features, into a higher-resolution spatial feature map. This operation not only increases spatial resolution but also reduces the channel count, making it an effective method in super-resolution tasks. Once the high-resolution image X^{SR} is obtained, it is incorporated into the following regression model:

$$y(s) = \beta_1\, x(s)^{SR}(s) + \epsilon(s)$$

which mirrors the structure of the basic spatial regression model in equation (36.1). However, the key difference here is that Y is now predicted over the entire spatial domain using the high-resolution image X^{SR}. This approach integrates the power of super-resolution techniques with spatial regression, enabling enhanced predictive performance over fine-scale spatial domains.

36.3 Analysis of PM10 in Italy

In this section, we present preliminary results from modeling PM10 concentrations in Italy. The air quality monitoring network provides PM10 measurements from 492 monitoring stations managed by the European Environment Agency across the country. In contrast, PM_{10} concentrations from CAMS provide continuous spatial coverage across the Italian domain through a gridded dataset with a resolution of 11 km × 11 km, encompassing the monitoring locations. For this analysis, we consider daily average data from 2023 and compare predictions from three competing models.

The first model is a simple regression model, as defined in equation (36.1), with residuals modeled using an exponential spatial covariance function. The second model extends this approach to a multivariate regression framework, following equation (36.4), with $K = 6$. Here, Bernstein basis polynomials $\phi_k(\omega)$ are used to construct regressors $\widetilde{x}_k(s)$, capturing large-scale spatial variations at lower values of k and progressively finer-scale features as k increases. The third model is a nonlinear extension of the second, implemented as a multilayer perceptron (MLP). The network's input consists of the regressors $\widetilde{x}_k(s)$ evaluated at the monitoring sites, with one hidden layer containing 50 neurons and a ReLU activation function.

For comparison purposes, we evaluate the predictive performance of the models using leave-one-out cross-validation at the 492 monitoring stations serving as training data. The first model yields a cross-validation error of 4.15, while the second model provides only a slight improvement, with a value of 4.01. This result is not surprising, as the different regressors $\widetilde{x}_k(s)$ capture varying degrees of spatial variability, ultimately conveying the same information as the residual spatial correlation modeled in the basic regression approach. As expected, the first regressor $\widetilde{x}_1(s)$ exhibits a strong correlation with the raw data, suggesting a good agreement between CAMS and the monitoring data for large-scale spatial features. This indicates that the estimated slope coefficient β_1 is primarily influenced by the low-frequency trends shared by both datasets. By incorporating $\widetilde{x}_k(s)$, $k = 1, \ldots, 6$, as an input to the MLP model, the leave-one-out cross-validation error further decreases to 3.4, demonstrating an improvement in prediction accuracy of approximately 18%.

By applying the trained models for spatial prediction over a fine grid covering the Italian municipalities (serving as test data), Fig. 36.1 compares the spatial predictions obtained from the first and the third models, respectively. As expected, the first model produces a much smoother map, consistent with the nature of spatial kriging. In contrast, the predictions from the third model exhibit finer spatial details, better capturing local variations and reflecting the information provided by the monitoring network more accurately.

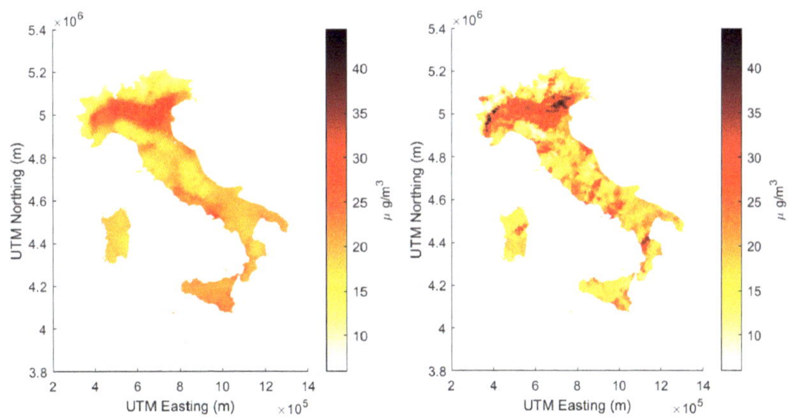

Fig. 36.1: Comparison of spatial predictions across Italian municipalities using the spatial kriging model (*Left*) and the multilayer perceptron model (*Right*).

36.4 Discussion

This study presents a novel spatial downscaling framework for PM10 data, integrating spectral methods with deep learning to develop both linear and nonlinear regression models. Preliminary results indicate that the Multilayer Perceptron (MLP) effectively captures nonlinear spatial patterns. Future work will explore the performance of alternative models based on super-resolution and projection pursuit approaches. Additionally, a key research direction is uncertainty quantification, for which we propose incorporating Extreme Learning Machines (ELMs), a reservoir computing approach that trains only the output layer while maintaining universal approximation properties [10]. By leveraging ensemble-based ELM predictions, we aim to generate uncertainty estimates by computing calibrated prediction intervals as in [5].

Acknowledgements We acknowledge financial support from the National Recovery and Resilience Plan GRINS - PE0000018 - BAC "High-Resolution Data Fusion for Air Quality Mapping - HAIRQ-MAP" , SPOKE 0 E 2-CUP J33C22002910001.

References

1. Berrocal,V.J., Gelfand,A.E., Holland, D.M.: A spatio-temporal downscaler for output from numerical models. Journal ofAgricultural, Biological, and Environmental Statistics. **15**, 176–197 (2010)
2. Bollt, E.: On explaining the surprising success of reservoir computing forecaster of chaos? The universal machine learning dynamical system with contrast to VAR and DMD. Chaos: An Interdisciplinary Journal of Nonlinear Science. **31**, 291—311 (2021)
3. Chen, W. and Li, Y. and Reich, B. J. and Sun, Y.: DeepKriging: Spatially Dependent Deep Neural Networks for Spatial Prediction. Statistica Sinica. **34**, 291—311 (2024)
4. Cressie, N.A.: Statistics for Spatial Data, 2nd edn. Wiley, New York (1993)
5. Daw, R., Wikle, C. K.: Reds: Random ensemble deep spatial prediction. Environmetrics. **34**, 291—311 (2023)
6. Friedman, J. H., Stuetzle, W.: Projection pursuit regression. Journal of the American statistical Association. **76**, 817–823 (1981)
7. Goodfellow, I., Bengio, Y., Courville, A.: Deep Learning. MIT press (2016)
8. Shi, W., Caballero, J. , Huszar, F., Totz, J., Aitken, A. P., Bishop, R., Rueckert, D. and Wang, Z.: Real-Time Single Image and Video Super-Resolution Using an Efficient Sub-Pixel Convolutional Neural Network. In: IEEE Conference on Computer Vision and Pattern Recognition (CVPR), 1874–1883 (2016)
9. Sun, Y., Deng, K., Ren, K., Liu, J., Deng, C., Jin, Y.: Deep learning in statistical downscaling for deriving high spatial resolution gridded meteorological data: A systematic review. ISPRS Journal of Photogrammetry and Remote Sensing. **208**, 14–38 (2024)
10. Zhang, E., Vasconcellos Vargas, D.: A Survey on Reservoir Computing and its Interdisciplinary Applications Beyond Traditional Machine Learning. IEEE ACCESS. **11**, 81033–81070 (2023)

Chapter 37
A Bayesian Non-Parametric Model to Learn Functions with Discontinuities

Alessandro Lanteri, Raffaele Argiento, Silvia Montagna

Abstract Many scientific problems involve dealing with high-dimensional data which can be interpreted as samples drawn from different curves. These so-called functional data might be characterized by the presence of discontinuity points which are shared among different functions. In this work we propose a predictive Bayesian non-parametric model for such data which is capable to cluster similar patterns between different curves. Assuming that each function is composed by the combination of a smooth function and a step function, we use a functional Bayesian latent factor model for the smooth components and a hierarchical normalized generalized Gamma prior for the step functions. This choice allows for flexible but parsimonious estimation and clustering, with a complete sharing of information across curves. To test the performance and the robustness of the proposed model we conduce a simulation study.

37.1 Introduction

In most scientific fields there is a need to study high-dimensional objects which evolve with a behavior that can be well represented by functions. For this reason, in the last decades, functional data analysis (FDA) has become a wide statistical field. There are several monographs regarding FDA, among other [10, 16], while some of

Alessandro Lanteri (✉)
Università degli Studi di Torino, Dipartimento di Economia e Statistica "Cognetti De Martiis"
e-mail: alessandro.lanteri@unito.it

Raffaele Argiento
Università degli Studi di Bergamo, Dipartimento di Scienze Economiche
e-mail: raffaele.argiento@unibg.it

Silvia Montagna
Università degli Studi di Modena e Reggio Emilia, Dipartimento di Economia "Marco Biagi"
e-mail: silvia.montagna@unimore.it

© The Author(s), under exclusive license to Springer Nature Switzerland AG 2025

G. Aneiros et al. (eds.), *New Trends in Functional Statistics and Related Fields*,
Contributions to Statistics, https://doi.org/10.1007/978-3-031-92383-8_37

the recent developments in the field are reviewed for instance in [1, 9]. An important challenge arises when functions present discontinuities. The problem of estimating discontinuity points, sometimes known in the literature as change-point estimation, is itself a prolific branch of statistics [3, 4, 11]. However, in many scientific problems, these discontinuity points are known and the interest of researchers lies rather in learning the functional form of different objects and understanding whether they exhibit similar patterns (e.g., [12, 8, 7]).

The goal of this work is twofold, we propose a flexible model to: 1) learn functions that have known points of discontinuity 2) find patterns of similarity between functions. To achieve this goal we use a FDA Bayesian non-parametric framework for functions estimation and a Bayesian hierarchical model for patterns clustering. Specifically, we assume that each function consists of two components, a smooth function and a step function. The latter can be interpreted as a sequence of random intercepts acting between each contiguous pair of discontinuity points. We model the smooth component with a functional Bayesian latent factor (BLF) [15] while the random intercepts are modeled using normalized generalized Gamma (NGG) processes [13] in a hierarchical fashion in order to exploit their clustering properties [2]. The Bayesian model we propose enables us to estimate each function while fully sharing information across functions in a parsimonious manner, reducing the number of parameters to be estimated. Additionally, the model naturally clusters the random intercepts across different functions, without knowing the number of clusters.

In this work we consider functions that share the same discontinuity points. This is a distinctive feature in a wide variety of scientific problems. For example, in finance analysis some global events (e.g., crises, wars) may simultaneously impact all stock markets; in ecology one may be interested in how different biological species belonging to a small regions react to extreme weather events. Another example of an application that falls into this setting is sports analytics. It is well known that Olympic athletes plan their training regimen to be in their best shape at the time of major competitions such as world championships or Olympics, which calendars are obviously shared among all athletes practicing the same sport. These methods of preparation might lead to large discontinuities in performance, for example, between the pre-Olympic period and the Olympic period. Performance discontinuities are common also in team sports, caused for instance by the player transfer market, which usually occurs between two different seasons. Additionally, many leagues divide seasons into regular season and playoffs, and it is well known that players performance varies greatly between these two sub-seasons. The presence of known change points can occur in contexts beyond functional time series data. For example, one might investigate how different materials react to water as a function of temperature. It is well-known that water undergoes phase transitions at specific temperatures. In each phase, which remains consistent across all functions, materials may exhibit varying levels of reaction. This work is carried out as follows. In Sect. 37.2 we describe the problem and present the model while in Sect. 37.3 we test the properties of the model through a simulation study.

37.2 Model and Priors

Let us assume that we want to learn n functions $y_i(t)$, for $i = 1, \dots, n$ and $t \in [0, 1]$, given data that consists of noisy observation at various points of these functions. The points at which the functions are observed may differ for each function, they are not necessarily equidistant and may vary in number across functions. More formally, each function is observed at T_i points $t_{i1}, ..., t_{i,T_i}$ and we let $y_{ij} = y_i(t_{ij}) + \varepsilon_{ij}$ be a noisy observed measurements of function i evaluated at t_{ij}, for $i = (1, \dots, n)$ and $j = (1, \dots, T_i)$. We consider each $y_i(t)$ to be a non-smooth function and assume that the points where the n curves may present discontinuities are shared by all functions. Let $0 = \tilde{t}_0 < \tilde{t}_1 < \cdots < \tilde{t}_{S-1} < \tilde{t}_S = 1$ be a sequence of points in $[0, 1]$, such that $\Omega_s = [\tilde{t}_{s-1}, \tilde{t}_s)$, for $s = 1, \dots, S - 1$, and $\Omega_S = [\tilde{t}_{S-1}, \tilde{t}_S]$, form a partition of $[0, 1]$. We will call each interval Ω_s a phase. The Bayesian model we propose consists of three additive components:

$$\begin{aligned} y_{ij} &= f_i(t_{ij}) + \mu_i(t_{ij}) + \varepsilon_{ij}, \\ \varepsilon_{ij} &\overset{iid}{\sim} N(0, \xi^2), \quad \text{for } i = 1, \dots, n, \ j = 1, \dots, T_i. \end{aligned} \tag{37.1}$$

In the model above, $f_i(t)$ is a smooth function and $\mu_i(t) = \sum_{s=1}^{S} \mu_{is} \mathbb{1}_{\Omega_s}(t)$ is a step function, where $\mathbb{1}_{\Omega_s}(\cdot)$ is the indicator function of the subset Ω_s and μ_{is} is the mean of $y_i(t)$ conditioned on $t \in \Omega_s$. Each step μ_{is} can be interpreted as a phase-specific random effect for the i-th function. The last component ε_{ij} is a white noise. In the presence of covariates the model can be easily enriched with additional components, to take them into account. We propose the use of a functional BLF model for the smooth components and a hierarchical normalized generalized Gamma (HNGG) prior for the step functions. The model choices are described in details hereafter.

37.2.1 *Priors for the Smooth Component*

We assume that $f_1(t), ..., f_n(t)$ belong to the linear span of a finite set of basis $\{b_1(t), ..., b_p(t)\}$ such that we can write

$$f_i(t) = \sum_{m=1}^{p} \theta_{im} b_m(t)$$

A potential issue with this characterization is that the dimensionality of the problem is order $O(np)$ and a large p might be required to recover a rich enough variety of functions. One way to ease this problem is to use a latent factor model along with a shrinkage prior on the loadings. The idea behind the latent factor model representation is to define $\boldsymbol{\theta}_i = (\theta_{i1}, \dots, \theta_{ip})^T$ as $\boldsymbol{\theta}_i = \boldsymbol{\Lambda}\boldsymbol{\eta}_i + \boldsymbol{\varsigma}_i$, where $\boldsymbol{\Lambda} = (\lambda_{ml})$, $m = 1, \dots, p$ and $l = 1, \dots, k$, is a $p \times k$ factor loading matrix, with $k < p$, while $\boldsymbol{\eta}_i$ is k−dimensional vector of coefficients and $\boldsymbol{\varsigma}_i$ is a residual component. Following

[5] we define the prior distribution on the unknown parameters as follow:

$$
\begin{aligned}
&\lambda_{ml} \overset{iid}{\sim} N(0, \phi_{ml}^{-1}\tau_l^{-1}),\ m = 1, ..., p,\ l = 1, ..., k \\
&\phi_{ml} \sim Ga\left(\frac{\nu_\phi}{2}, \frac{\nu_\phi}{2}\right) \\
&\tau_l = \prod_{h=1}^{l} \delta_h,\ \ \delta_1 \sim Ga(a_1, 1),\ \ \delta_h \sim Ga(a_2, 1) \\
&\boldsymbol{\eta_i} \overset{iid}{\sim} N_k(\mathbf{0}, \boldsymbol{I}_k) \\
&\boldsymbol{\varsigma}_i \overset{iid}{\sim} N_p(\mathbf{0}, \Sigma) \\
&\Sigma_{jj}^{-2} \overset{iid}{\sim} Ga(a_\sigma, b_\sigma)\ \ j = 1, ..., p
\end{aligned}
\qquad (37.2)
$$

where τ_l is a global shrinkage parameter for the l-th column, ϕ_{ml} are local shrinkage parameters and Σ is a diagonal matrix. We refer to [15] for more details on the application of the BLF model to functional data.

37.2.2 Prior for the Step Functions

To estimate the step functions components, we need to estimate all the phase-specific random effects μ_{is}'s. Let n_s, for $s = 1, \ldots, S$, be the number of functions with measurements falling in Ω_s such that $\sum_{s=i}^{S} n_s = n$. The number of μ_{is}'s can be very large, being of order $O(nS)$. To address this issue we assume that, across functions and phases, some μ_{is}'s share the same value. Formally, we assume partial exchangeability on the sequences $(\mu_{1s}, \ldots, \mu_{n_s s})$, for $s = 1, \ldots, S$, - that is to say that the S sequences are exchangeable and all the elements within a sequence are exchangeable, which is a weaker assumption than exchangeability or independence -, and use a hierarchical NGG process prior on the μ_{is}'s. The model we propose for the step functions is

$$
\begin{aligned}
&\mu_{is} | P_s \overset{iid}{\sim} P_s \quad i = 1, ..., n\ s = 1, ..., S \\
&P_s | P \overset{iid}{\sim} NGG(\kappa, \sigma, P)\ \ s = 1, ..., S \\
&P \sim NGG(\kappa_0, \sigma_0, P_0) \\
&P_0 \sim N(\mu_0, \varsigma_0^2)
\end{aligned}
\qquad (37.3)
$$

where $\{\kappa, \sigma, \kappa_0, \sigma_0, \mu_0, \varsigma_0^2\}$ are fixed hyperparameters. We recall that, for a distribution $P \sim NGG(\kappa, \sigma, H)$, H is the base distribution, κ is the scaling parameter which regulates the discretization of the distribution (i.e. if $\kappa \to \infty$ then $P \approx H$, if $\kappa = 0$ then P is a Dirac distribution on a single value) and $0 \le \sigma \le 1$ regulates the "rich gets richer" phenomenon and when $\sigma = 0$ the Dirichlet Process is recovered. Note that each P_s is a discrete distribution with atomic locations drawn from P which is also a discrete distribution. This means that there is a positive probability that there will be ties within samples drawn from the same P_s and ties between samples drawn from different P_s's. This two-layer structure allows a complete sharing of information of the phase-specific random effects, both within and between functions, and allows μ_{is}'s belonging to different function in different phases to assume the same value. We define the set of all the possible values for μ_{is} to be $\boldsymbol{\psi} = \{\psi_1, \ldots, \psi_M\}$, where M is unknown and possibly infinite when nS goes to infinity. We recall that the HNGG includes the well known hierarchical Dirichlet Process (HDP) [17] for $\sigma = \sigma_0 = 0$ and thus provides great flexibility in clustering. Theoretical properties of hierarchical normalized completely random measure can be found in [6]. The computational procedure to implement a MCMC sampler can be derived using the fact that the HNGG can be represented via the well known metaphor of the Chinese Restaurant Franchise process [17].

37.3 Simulation Study

In this section we test the performance and the robustness to prior choice of the proposed modeling method using synthetic data. We generate $n = 50$ functions, each given by the sum of a step function and a smooth component. The support of these functions is $[0, 1]$ and is divided in $S = 3$ phases $\{\Omega_1, \Omega_2, \Omega_3\}$. The levels μ_{is}'s of the step functions $\mu_i(t) = \sum_{s=1}^{S} \mu_{is} \mathbb{1}_{\Omega_s}(t)$ are randomly drawn from $M = 3$ possible values $\boldsymbol{\psi} = \{\psi_1, \psi_2, \psi_3\} = \{-2, 0, 1\}$. Each smooth component is a randomly generated B-spline function. We show two examples of generated functions and the additive components in the left column of Fig. 37.1. For each function, we sample, uniformly in $[0, 1]$, an average of 12 points per phase and add a small Gaussian noise to the corresponding function value.

Here we perform a sensitivity analysis in order to test the robustness of our method to the parameter choices of the HNGG model. We fit the synthetic data described above with our model adopting different choices for $\{\sigma, \sigma_0, \kappa, \kappa_0\}$. To evaluate the goodness-of-fit of the model in each scenario we use the log-pseudo marginal likelihood (LPML), whose values are reported in Table 37.1. Although optimal LPML is obtained in the setting $\sigma = \sigma_0 \approx 0$, which is an approximation of a HDP, the settings with $\sigma > 0$ and $\sigma_0 > 0$ show a lower variability in the LPML for different choices of $\{\kappa, \kappa_0\}$, in particular, the settings with $\{\sigma, \sigma_0\} = \{0.3, 0.1\}$ are better on average, in terms of goodness-of-fit, than the settings with other choices of $\{\sigma, \sigma_0\}$. This suggests that HNGG could be generally preferred over HDP as it seams to be less sensitive to parameter choice and provides more flexibility. To obtain a clustering of the random effects we use the partition which minimizes the loss function induced by the variation of information (VI) and estimate M with the number of obtained clusters, as proposed and discussed in [14, 18]. In Table 37.2 we show the classification rate along with the estimated number of clusters obtained by the VI loss function. In all setting we achieve a classification rate larger than 0.96 and the number of estimated clusters of the random effects is correct (i.e. 3) in

	$\sigma = \sigma_0 \approx 0$			$\sigma = 0.1$ $\sigma_0 = 0.3$			$\sigma = 0.3$ $\sigma_0 = 0.1$		
κ \ κ_0	0.1	1	10	0.1	1	10	0.1	1	10
0.1	-2932	-2934	-2924	-2922	-2914	-2913	-2896	-2918	-2904
1	-2900	-2877	-2911	-2910	-2936	-2921	-2908	-2894	-2901
10	-2924	-2904	-2899	-2913	-2898	-2901	-2915	-2922	-2892

Table 37.1: LPML index for different parameter choices.

almost all settings. Note that the parameter choice when M is overestimated, $\kappa_0 = 10$ and $\sigma_0 = 0.3$, strongly promote the model to adopt more clusters. Anyway, in these cases the extra cluster contains only one element, which is incorrectly classified in all other settings, this is the reason why we achieved the best classification rate even when M is overestimated.

	$\sigma = \sigma_0 \approx 0$			$\sigma = 0.1$ $\sigma_0 = 0.3$			$\sigma = 0.3$ $\sigma_0 = 0.1$		
κ \ κ_0	0.1	1	10	0.1	1	10	0.1	1	10
0.1	0.980 (3)	0.980 (3)	0.980 (3)	0.980 (3)	0.973 (3)	0.980 (4)	0.967 (3)	0.973 (3)	0.967 (3)
1	0.980 (3)	0.980 (3)	0.980 (3)	0.973 (3)	0.973 (3)	0.980 (4)	0.967 (3)	0.967 (3)	0.967 (3)
10	0.980 (3)	0.967 (3)	0.967 (3)	0.967 (3)	0.980 (3)	0.973 (3)	0.967 (3)	0.973 (3)	0.960 (3)

Table 37.2: Classification rate and the number of estimated clusters (in brackets) for different parameter choice.

Now, we investigate how well the proposed model is able to estimate the generated function. In the following we consider only the scenario with parameters which lead to a better performance in the above sensitivity analysis both in terms of LPML and classification index. In the central column of Fig. 37.1 we present the estimated functions along with 95% credibility bands (dashed lines). As we can see, model provides good estimates and succeed in catching both the phase-specific random effects and smooth component. In the right column of Fig. 37.1 we display the box plots of the posterior draws of the μ_{is}'s for the three different phase of the two selected functions. We can appreciate how such box plots are well concentrated on the true values $\boldsymbol{\psi} = \{\psi_1, \psi_2, \psi_3\} = \{-2, 0, 1\}$.

From the above analysis we can say that the model is not too sensible to parameter choice, both in term of classification and goodness of fit. Moreover, it is flexible enough to give accurate estimate of functions which are continuous within phases and might jump between one phase and the next. Additionally, it naturally groups the phase-specific random intercepts in a parsimonious number of clusters. This last property of the model can be used to find similarities in the behavior of

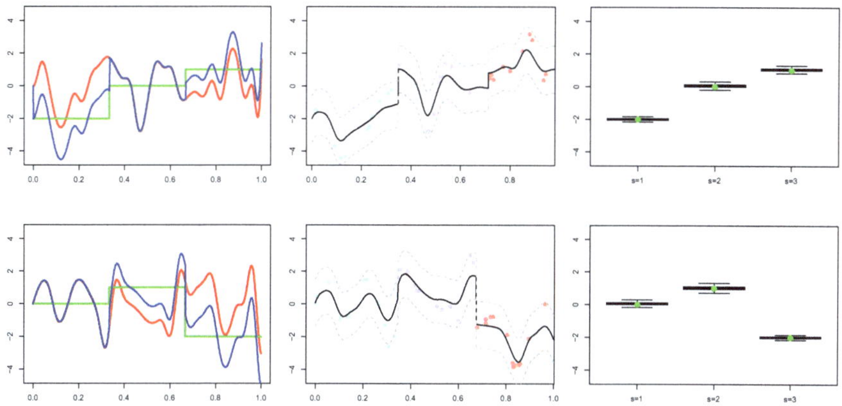

Fig. 37.1: Left column: two randomly selected generated function (blue) along with the two additive components, the smooth function (red) and the step function (green). Central column: estimated function fitting noisy data points on that function, with 95% credibility bands. Points have different color for each phase. Right column: Box plots of the posterior draws of the $\mu_{i,s}$'s, for $s = 1, 2, 3$. The green dots represent the real values of $\mu_{i,s}$.

different curves which might follow patterns of interest for the researcher in real data applications.

References

1. Aneiros G., Horová I., Hušková M., and Vieu P.: On functional data analysis and related topics. Journal of Multivariate Analysis, **189**, 104861, (2022)
2. Argiento R., Cremaschi A., and Vannucci M.: Hierarchical normalized completely random measures to cluster grouped data. Journal of the American Statistical Association, **115**(529), 318–333, (2020)
3. Aue A. and Horváth L.: Structural breaks in time series. Journal of Time Series Analysis, **34**(1), 1–16, (2013)
4. Aue A., Rice G., and Sönmez O.: Detecting and dating structural breaks in functional data without dimension reduction. Journal of the Royal Statistical Society Series B: Statistical Methodology, **80**(3), 509–529, (2018)
5. Bhattacharya A. and Dunson D.B.: Sparse bayesian infinite factor models. Biometrika, 291–306, 2011.
6. Camerlenghi F., Lijoi A., Orbanz P., Prünster I., et al.: Distribution theory for hierarchical processes. The Annals of Statistics, **47**(1), 67–92, (2019)
7. Colombi A., Argiento R., Camerlenghi F., and Paci L.. Hierarchical Mixture of Finite Mixtures (with Discussion). Bayesian Analysis, pages 1 – 29, (2024)
8. Argiento R., Dolmeta, P. and Montagna S.: Bayesian garch modeling of functional sports data. Stat Methods Appl, **32**, 401–423, (2023)

9. Goia A. and Vieu P.: An introduction to recent advances in high/infinite dimensional statistics. Special Issue on Statistical Models and Methods for High or Infinite Dimensional Spaces. Journal of Multivariate Analysis, **146**, 1–6, (2016)
10. Horváth L. and Kokoszka P.: Inference for functional data with applications, volume 200. Springer Science & Business Media, (2012)
11. Horváth L. and Rice G.: Extensions of some classical methods in change point analysis. Test, **23**(2), 219–255, (2014)
12. Liang M., Koslovsky M.D., Hébert E.T., Kendzor D.E., and Vannucci M.. A bayesian non-parametric approach for clustering functional trajectories over time. Statistics and Computing, **34**(6), 215, (2024)
13. Lijoi A., Mena R.H., and Prünster I.: Controlling the reinforcement in bayesian non-parametric mixture models. Journal of the Royal Statistical Society: Series B (Statistical Methodology), **69**(4), 715–740, (2007)
14. Meilă M.: Comparing clusterings - an information based distance. Journal of multivariate analysis, **98**(5), 873–895, (2007)
15. Montagna S., Tokdar S.T., Neelon B., and Dunson D.B.: Bayesian latent factor regression for functional and longitudinal data. Biometrics, **68**(4), 1064–1073, (2012)
16. Ramsay J. and Silverman B.W.: Functional data analysis (springer series in statistics). (1997)
17. Teh Y.W., Jordan M.I., Beal M.J., and Blei D.M.: Hierarchical dirichlet processes. Journal of the American Statistical Association, **101**(476), 1566–1581, (2006)
18. Sara Wade, Zoubin Ghahramani, et al.: Bayesian cluster analysis: Point estimation and credible balls (with discussion). Bayesian Analysis, **13**(2), 559–626, (2018)

Chapter 38
A Novel Approach To Estimate Functional Gaussian Graphical Model Based On Penalized Multivariate Functional Regression Model

Salvatore Latora, Luigi Augugliaro and Gerda Claeskens

Abstract In this article we contribute to the literature on the functional Gaussian graphical model by introducing a new penalty function for the multivariate functional regression model that allows us to have a direct connection between the estimated parameters and conditional dependencies among the functional curves, i.e. the edge set of the graphical model. Introducing the model in infinite dimensional space, we show how to estimate the model in a finite-dimensional subspace. A focus is also given on how the bases and scores for the Karhunen-Loéve expansion of the curves are obtained. The performance of the proposed method is evaluated by a simulation study and compared with that of other better-known models in the literature, showing good performance.

38.1 Introduction

Many authors have contributed to the literature of Gaussian Graphical Models (GGM) for functional data. One of the first works is those of [4] in which, since the covariance operator is compact and therefore non-invertible, the connection between the inverse covariance operator and conditional independence is lost, the authors assume that the functional curves are concentrated to a finite-dimensional subspace by working with an $K-$trunkated version of the Karnhunen-Loève expansion. Therefore,

Salvatore Latora (✉)
Department of Economics, Business and Statistics, University of Palermo, Viale delle Scienze Ed. 13, Palermo - Italy, e-mail: salvatore.latora@unipa.it

Luigi Augugliaro
Department of Economics, Business and Statistics, University of Palermo, Viale delle Scienze Ed. 13, Palermo - Italy, e-mail: luigi.augugliaro@unipa.it

Gerda Claeskens
ORSTAT - Faculty of Economics and Business, KU Leuven, Naamsestraat 69, Leuven - Belgium, e-mail: gerda.claeskens@kuleuven.be

© The Author(s), under exclusive license to Springer Nature Switzerland AG 2025

G. Aneiros et al. (eds.), *New Trends in Functional Statistics and Related Fields*, Contributions to Statistics, https://doi.org/10.1007/978-3-031-92383-8_38

they estimate a finite-dimensional graphical model by a modified lasso criterion. The main problem with this approach is that the authors prove the asymptotic properties in the finite setting, but they are unsatisfied with the infinite-dimensional space of curves. This approach is overcome by [3] which introduces the assumption of *partial separability* of the covariance operator. This allows to have a direct connection between conditional independence and conditional covariance function. Consequently, the functional graphical model is given by a sequence of finite-dimensional Gaussian graphical models. However, the partial separability assumption is a very strong assumption that is not verified in many real-world applications. In [2] the authors provide an edge set learning strategies based on neighborhood selection for functional GGM. After obtaining the neighbourhood set of each node, the final edge set will be obtained by using an **AND/OR** scheme. In this article we contributed to the literature on functional GGM by introducing a novel penalty function that, by working with a neighbourhood selection framework for functional data, allows a direct connection between the matrix of estimated parameters of the regression model and the edge set.
The paper is organized as follows: in the second section the methodological proposal is explained; in the third section a simulation study will be shown; finally, conclusions are provided in the final section.

38.2 Proposal Methodology

Before introducing the methodological innovation brought by this paper, it is appropriate to give some notations and definitions.

38.2.1 Notation and Definitions

Let $f_j \in L^2_{[0,1]}$, for $j = 1, \dots, p$ be a random function, where p denote the number of random functions and $L^2_{[0,1]}$ denote the space of square-integrable functions endowed with the inner product $\langle f_1, f_2 \rangle = \int_0^1 f_1(t) f_2(t) dt$ and associated norm $\|f\| = \sqrt{\langle f, f \rangle}$. We denote with $\boldsymbol{f} = (f_1, \dots, f_p)^T \in \mathcal{F} = \left[L^2_{[0,1]}\right]^p$ the curves in a p-dimensional Hilbert space equipped with inner product $\langle \boldsymbol{f}, \boldsymbol{g} \rangle_p = \sum_{j=1}^p \langle f_j, g_j \rangle$ and associated norm $\|\boldsymbol{f}\|_p = \sqrt{\langle \boldsymbol{f}, \boldsymbol{f} \rangle_p}$. A Functional Gaussian Graphical Model is defined as $\{\boldsymbol{f}, MPG(\mu, \mathcal{G}), G\}$, where the multivariate functional data $\boldsymbol{f}$ follows a *multivariate Gaussian process* $MPG(\mu, \mathcal{G})$ specified by a mean function $\mu(t)$ and a covariance operator $\mathcal{G}(t, s)$, and $G = (V, E)$ is an undirected graph defined by V the node set in which each node is a functional curve f_j, and $E \subset V \times V$ is the edge set. E encodes the presence or absence of conditional independence among the functional curves: $f_j \perp\!\!\!\perp f_h \mid f_l$ if $(h, k) \in E$. The multivariate Gaussian process is an assumption we will carry throughout the paper. We also assume that $\mu_j(t) = E\left[f_j(t)\right] = 0$ and $V\left[f_j(t)\right] < \infty$, for $j = 1, \ldots, p$.

38.2.2 Proposed Methodology

The goal of the proposed model is to recover the edge set E by using a neighbourhood selection approach similar in [2], based on *multivariate functional linear regression* ([6]) by introducing a novel penalty function that allows us to have e direct connection between the parameter estimates and E. In particular, assuming we have n units, for each target node $j \in V$ we have that

$$f_{ij}(t) = \sum_{k \neq j}^{p} \int_0^1 f_{ik}(t)\beta_{jk}(t,t')dt' + \epsilon_{ij}(t), \quad \forall j = 1, \ldots, p, \; i = 1, \ldots, n, \tag{38.1}$$

where $\beta_{jk}(t,t')$ is a functional parameter that captures the effect of the functional curve k to the response j, and $\epsilon_{ij}(t)$ is a functional error term. As in [6] the functional parameter $\beta_{jk}(t,t')$ admits the following expansion

$$\beta_{jk}(t,t') = \sum_{l,l'=1}^{+\infty} b_{jkll'}\phi_l(t)\phi_{l'}(t'), \tag{38.2}$$

where $b_{jkll'} = \int \beta_{jk}(t,t')\phi_l(t)\phi_{l'}(t')$ and $\{\phi_l\}_{l=1}^{\infty}$ is a *complete orthonormal system* (CONS) of $\left[L^2_{[0,1]}\right]^p$, with $\phi_l = (\phi_{1l}, \ldots, \phi_{pl})^T$, such that $\mathcal{G}(t,t') = \sum_{l=1}^{\infty} \gamma_l\phi_l(t)\phi_{l'}(t')$, where $\{\gamma_l\}_{l=1}^{\infty}$ are its non-negative and decreasign eigenvalues. By setting (38.1) in a multivariate penalized functional regression, the problem becomes

$$\hat{\boldsymbol{\beta}}_{.1}, \ldots, \hat{\boldsymbol{\beta}}_{.p} = \underset{\boldsymbol{\beta}_{.1},\ldots,\boldsymbol{\beta}_{.p}}{\arg\min} \sum_{i=1}^{n}\sum_{j}^{p} \|f_{ij}(t) - \sum_{k \neq j}^{p} \langle f_{ik}, \beta_{jk}(t,\cdot)\rangle\|^2 + \lambda_n P(\boldsymbol{\beta}_{.k}), \tag{38.3}$$

where $P(\boldsymbol{\beta}_{.k})$ is a proper penalty function. In particular, we introduce the following penalty function

$$P(\boldsymbol{\beta}_{.k}) = \sum_{\substack{j,k \\ j \neq k}} \sqrt{\|\boldsymbol{\beta}_{jk}\|^2 + \|\boldsymbol{\beta}_{kj}\|^2}, \tag{38.4}$$

which allows us to impose a direct connection between parameter functions and E such that if $\|\beta_{jk}\|^2 = \|\beta_{kj}\|^2 = 0$ then there is not an edge between functional curves j and k. However, the problem in (38.3) is an infinite-dimensional optimization problem, so a way to solve it is to work with an finite-dimensional approximation.

38.2.3 Vector-on-Vector Multivariate Regression

In order to recover the edge set E is useful to work with the Karnhunen-Loève theorem, which says that:

$$f_{ij}(t) = \sum_{l=1}^{\infty} \xi_{ijl}\phi_{jl}(t), \tag{38.5}$$

where $\xi_{ijl} = \langle f_{ij}; \phi_{jl}\rangle$, for $i = 1, \ldots, n$, is an element of $\boldsymbol{\xi}_{jl}$, a r.v. that we assume has a Gaussian distribution with $E\left[\boldsymbol{\xi}_{jl}\right] = 0$ $V\left[\boldsymbol{\xi}_{jl}\right] = \gamma_{jl}$ and $E\left[\boldsymbol{\xi}_{jl}\boldsymbol{\xi}_{jr}\right] = \delta_{lr}$, where δ_{lr} is the Kronecker's delta such that $\delta_{lr} = 1$ if $l = r$. In particular we will work with an L truncated version of (38.5), so for the generic curve j we denote with $\boldsymbol{\xi}_{ij} = (\xi_{ij1}, \ldots, \xi_{ijL})^T$ the vector of scores. As in [2], each vector score can be represented as:

$$\boldsymbol{\xi}_{ij} = \sum_{k\neq j}^{p} \boldsymbol{B}_{kj}\boldsymbol{\xi}_{ik} + \boldsymbol{r}_{ij}, \tag{38.6}$$

where $\boldsymbol{r}_{ij}$ is an error term and $\boldsymbol{B}_{kj}$ is a block of a matrix $\boldsymbol{B}$ of parameters related to the effect each functional curves $k \neq j$ to the response curve j for all the basis L. In particular, the generic term of $\boldsymbol{B}_{kj}$ is given by b_{kjl} scores of the expasion of $\beta_{kj}(t, t')$ in (38.2). The finite version of (38.3) will be

$$\hat{\boldsymbol{B}} = \arg\min_{\boldsymbol{B}} \sum_{i=1}^{n}\sum_{j=1}^{p} ||\boldsymbol{\xi}_{ij} - \boldsymbol{\xi}_{i\backslash j}\boldsymbol{B}_{kj}||_F^2 + \lambda_n \sum_{\substack{j,k\\ j\neq k}} \sqrt{||\boldsymbol{B}_{jk}||_F^2 + ||\boldsymbol{B}_{kj}||_F^2}, \tag{38.7}$$

where $||\cdot||_F^2$ is the Frobenius norm, which can be solved using the ADMM algorithm. Proposed by [7], the ADMM algorithm allows us to write the optimization problem as

$$\hat{\boldsymbol{B}}, \hat{\boldsymbol{Z}} = \arg\min_{\boldsymbol{B},\boldsymbol{Z}} \sum_{i=1}^{n}\sum_{j=1}^{p} ||\boldsymbol{\xi}_{ij} - \boldsymbol{\xi}_{i\backslash j}\boldsymbol{B}_{kj}||_F^2 + \lambda_n \sum_{\substack{j,k\\ j\neq k}} \sqrt{||\boldsymbol{Z}_{jk}||_F^2 + ||\boldsymbol{Z}_{kj}||_F^2} \tag{38.8}$$

$$\text{s.t. } \boldsymbol{B}_{jk} = \boldsymbol{Z}_{jk} \quad \forall\, j, k = 1, \ldots, p,$$

where $\boldsymbol{\xi}_{iæ} = (\xi_{i1}, \ldots, \xi_{ip})\backslash\xi_{ij}$. It is possible to show that the solution of (38.8), at step $t+1$, will be:

$$\begin{aligned}
\hat{\boldsymbol{B}}_{kj}^{(t+1)} &= \left((\xi_{\backslash j})^T \xi_{\backslash j} + \rho I_L\right)^{-1} \left((\xi_{\backslash j})^T \xi_j + \rho\left(\boldsymbol{Z}_{kj}^{(t)} + \boldsymbol{U}_{kj}^{(t)}\right)\right),\\
\hat{\boldsymbol{Z}}_{jk}^{(t+1)} &= \left(1 - \frac{\lambda}{\rho||\boldsymbol{B}_{jk}^{(t+1)} + \boldsymbol{U}_{jk}^{(t)}||_F^2}\right)_+ \left(\boldsymbol{B}_{jk}^{(t+1)} + \boldsymbol{U}_{jk}^{(t)}\right),\\
\hat{\boldsymbol{U}}_j^{(t+1)} &= \boldsymbol{U}_j^{(t)} + \boldsymbol{B}_j^{(t+1)} - \boldsymbol{Z}_j^{(t+1)}.
\end{aligned} \tag{38.9}$$

38.2.4 Choice of the Basis and Scores Estimation

The choice of the basis $\{\phi_{jl}\}_{l=1}^{+\infty}$ used to estimate the vector of scores $\boldsymbol{\xi}_{jl}$ is made by using the PACE method proposed in [5] curve by curve. In particular, for the generic j−curve we assume that each curve is observed with a measurement error on a finite number of random temporal points N_{ij}, i.e., $Y_{ij}(T) = f_{ij}(T_{ij}) + \epsilon_{ij} = \mu_j(T_{ij}) + \sum_{l=1}^{L} \xi_{ijl}\phi_{jl}(T_{ij}) + \epsilon_{ij}, \quad \forall\, T_{ij} \in \mathcal{T}$, with $E\left[\epsilon_{ij}\right] = 0$ and $V\left[\epsilon_{ij}\right] = \sigma_j^2$, where ϵ_{ij} are independent of ξ_{ijl}. First, a local linear smoother is used to obtain a smooth mean function estimate $\hat{\mu}_j$ and a smooth covariance estimate $\hat{\mathcal{G}}_j(f_j(t), f_j(s))$ such that the eigenfunctions and eigenvalues is estimate by the standard eigenequations $\int_{\mathcal{T}} \hat{\mathcal{G}}_j(s,t)\hat{\phi}_{jl}(s)\, ds = \hat{\gamma}_{jl}\hat{\phi}_{jl}(s)$. Assuming that ξ_{ijl} and ϵ_{ij} are jointly Gaussian, and denoted with $\tilde{\boldsymbol{Y}}_{ij} = \left(Y_{ij1}, \ldots Y_{ijN_i}\right)^T$, $\tilde{\boldsymbol{X}}_{ij} = \left(X_{ij}(T_{ij1}), \ldots, X_{ij}(T_{ijN_i})\right)^T$ and $\boldsymbol{\mu}_{ij} = \left(\mu_j(T_{ij1}), \ldots \mu_j(T_{ijN_i})\right)^T$, the best prediction of the scores is given by the conditional expectation

$$\hat{\xi}_{ijl} = E\left[\xi_{ijl} \mid \tilde{\boldsymbol{Y}}_{ij}\right] = \hat{\gamma}_{jl}\hat{\boldsymbol{\phi}}_{jl}^T \hat{\Sigma}_{\boldsymbol{Y}_{ij}}^{-1}\left(\tilde{\boldsymbol{Y}}_{ij} - \hat{\boldsymbol{\mu}}_{ij}\right), \tag{38.10}$$

where $\hat{\Sigma}_{\boldsymbol{Y}_{ij}}$ is the estimate of $\Sigma_{\boldsymbol{Y}_{ij}} = cov\left(\tilde{\boldsymbol{Y}}_{ij}\tilde{\boldsymbol{Y}}_{ij}\right) + \sigma^2 \mathcal{I}_{N_{ij}}$, in which σ^2 is the variance error term, estimated as $\hat{\sigma^2} = \frac{2}{|\mathcal{T}|}\int_{\mathcal{T}_1}\{V\hat{(t)} - G\hat{(t)}\}dt$.

38.3 Simulation Study

The simulation study was carried out by setting the number of curves $p = 17$, the number of equispaced temporal points $d = 60$ in the domain $\mathcal{T} = [0, 1]$ and taking into account five sample sizes: $n = 25, 50, 75, 100, 200$. For the generation of the true scores, a variance-covariance matrix of dimension $pL \times pL$ was generated with the following structure as in [4]

$$\Sigma = \left[\begin{array}{c|c|c|c} \Lambda_1 & \Sigma_{12} & \ldots & \Sigma_{1p} \\ \hline \Sigma_{21} & \Lambda_2 & \ldots & \Sigma_{2p} \\ \hline \vdots & \vdots & \ddots & \vdots \\ \hline \Sigma p1 & \ldots & \ldots & \Lambda_p \end{array}\right], \tag{38.11}$$

where Λ_j is a diagonal matrix of variances of $\left(\xi_{j1}, \ldots, \xi_{jL}\right)^T$, given by $\gamma_{j1} \geq \gamma_{j2} \geq \cdots \geq \gamma_{jL}$. To simulate Σ matrix we follow a similar strategy used in [2]: we generate a first Toepliz matrix $\boldsymbol{T}$ of dimension $L \times L$ such that $T_{jk} = 1$ for $j = k$ and $T_{jk} = 1/2 \mid j - k \mid$ if $j \neq k$. Then a tridiagonal matrix R of dimension $l \times L$ such that $R_{jj} = 1$, $R_{j+1,j} = R_{j,j+1} = 0.5$ and the other entries are set to 0. A first precision matrix $\tilde{\Theta}$ is obtained by setting $\tilde{\Theta}_{jj} = T$, $\tilde{\Theta}_{j,j+1} = \tilde{\Theta}_{j+1,j} = 0.4R$ and $\tilde{\Theta}_{j,j+2} = \tilde{\Theta}_{j+2,j} = 0.2\mathcal{I}_L$. Then a singular value decomposition of each block in the diagonal of $\tilde{\Theta}^{-1}$ was made such that $\Lambda_j^{-1} = U_j M_j V_j$ and a diagonal block matrix $U = diag\left(U_1, \ldots, U_p\right)$ is build. The final Σ is obtained as $\Sigma = U\Theta^{-1}U^T$. We set the number of basis $L^* = 5$, and we generate the true scores ξ^* from a multivariate Gaussian distribution with zero mean and variance-covariance matrix Σ. By using a set of L^* Fourier basis, denoted as ϕ_l^*, the observed curves are obtained as $Y_{ij}(T_{ij}) = \sum_{l=1}^{5} \xi_{il}^{X_1 *} \phi_l^* + \epsilon_{ij}^2$ where $\epsilon_{ij}^2 \sim \mathcal{N}(0, 1)$ is an error term. The performance of our model has been evaluated, on 100 simulations, via the *ROC* curve for 100 different values of λ_n: from λ_{max} such that all parameters will be equal to zero and therefore the edge set E is empty, to λ_{min} such that all parameters will be different from zero and therefore the edge set E is full. These performances have been compared with those of other competitors, in particular, the model proposed by [4] and those proposed by [2], using both **AND/OR** schemes and, as the authors propose, two different methods for the estimation of the scores: a first one by projecting the curves onto their own spaces (denoted by gX), and a second one by projecting the curves onto the response curve's space (denoted by gY). Table 38.1 shows the

Table 38.1: Average AUC for the five sample size.

model	$n = 25$	$n = 50$	$n = 75$	$n = 100$	$n = 200$
proposal	0.8274	0.9224	0.9667	0.9783	0.9976
Zhao.and.gX	0.8222	0.9257	0.9684	0.9848	0.9985
Zhao.and.gY	0.8307	0.9241	0.9675	0.9850	0.9985
Zhao.or.gX	0.7884	0.8964	0.9453	0.9705	0.9961
Zhao.or.gY	0.7924	0.8928	0.9442	0.9705	0.9961
Qiao	0.7396	0.8288	0.8782	0.9091	0.9599

average AUC values, from which it can be noted that our proposal has excellent performances for sample sizes of 50 and above and is similar to the model proposed by [2], in particular using an **AND** scheme, while it has slightly better performances than the same model but using an **OR** scheme, and superior to the model proposed by [4].

38.4 Conclusions

The methodology shown in this article allows to estimate a Gaussian Graphical model for functional data that, remaining within the framework of penalized multivariate functional regression models, represents an excellent alternative to the models already known in the literature: While the advantages over the model proposed by [4] are obvious in that our proposal has superior performance, the advantages over using the methodology proposed by [2] are computational. With our proposal, parameter estimation involves the estimation of the adjacency matrix E, whereas in the method proposed by [2], parameter estimation is followed by the estimation of the neighbourhood set for each functional curve, and then by the estimation of the adjacency matrix E by combining the neighbourhood sets based on an **AND/OR** scheme, the choice of which is subjective and may influence the estimation.

Acknowledgements The authors are grateful to the anonymous reviewers for their valuable comments.
L.A. gratefully acknowledges financial support from the University of Palermo (FFR2022, FFR2023 and FFR2024). L.A. was also financially supported by the European Union - Next Generation EU - Mission 4 Component 2 - CUP: B53D23009480006. G.C. acknowledges the support from project C16/20/002 of the Research Fund KU Leuven, Belgium.

Funding

References

1. Ramsay, J.O., Silverman, B.W.: Functional Data Analysis. Springer Series in Statistics. Springer New York, NY, (1997)
2. Zhao., B., Zhai, P.S., Wang, Y.S., Kolar, M.: High-dimensional functional graphical model structure learning via neighborhood selection approach. Electron. J. Statist. **18**(1), 1042 - 1129, (2024)
3. Zapata, J., Oh, S.Y., Petersen, A.: Partial separability and functional graphical models for multivariate Gaussian processes, Biometrika, **109**(3), 665–681, (2022)
4. Qiao, X., Guo, S., James, G.M.: Functional Graphical Models. Journal of the American Statistical Association, **114**(525), 211–222, (2018)
5. Yao, F., Müller, H. G., Wang, J. L.: Functional Data Analysis for Sparse Longitudinal Data. Journal of the American Statistical Association, **100**(470), 577–590, (2005)
6. Chiou, J.M., Yang, Y.F., Chen, Y.T.: Multivariate functional linear regression and prediction, Journal of Multivariate Analysis, **146**, 301-312, (2016)
7. Boyd, S., Parikh, N., Chu, E., Peleato, B., Eckstein, J.: Distributed Optimization and Statistical Learning via the Alternating Direction Method of Multipliers. Found. Trends Mach. Learn. **3**(1), 1–122, 2011
8. Hsing, T., Eubank, R.: Theoretical foundations of functional data analysis, with an introduction to linear operators. John Wiley & Sons. (2015)

Chapter 39
Local Null Hypothesis Significance Testing on Riemaniann Manifolds

Niels Lundtorp Olsen, Alessia Pini and Simone Vantini

Abstract We discuss the problem of testing hypotheses on functional data defined on manifold domain. Aim of this work is to test hypotheses locally, i.e., to provide a p-value function $\tilde{p}(s)$ that can be used to test a null hypothesis against an alternative one on every point of the manifold. In the following we present three methods to do so: an adjusted p-value function controlling the functional false discovery rate; an adjusted p-value function controlling the ball-wise error rate; an adjusted e-value function that is ball-wise e-valid.

39.1 Introduction

Many recent inferential methods for functional data have started focusing on 'local' techniques: according to such methods, inference is performed locally on the domain, and the identification of the areas of the domain where an effect is present is provided. Local inferential techniques are either based on simultaneous confidence bands, which are provided with a fixed coverage probability of the whole band [2, 3], or on the definition of a p-value function, which provides an adjusted p-value at each point of the domain, guaranteeing a control of a quantity related with the error rate on the whole domain [1, 4]. Focusing on this second line of research, depending on the quantity that is controlled, different methods can be defined. Such methods deal with the extension to functional data of the control of either the family-wise

Niels Lundtorp Olsen
Technical University of Denmark, Copenhagen, Denmark, e-mail: nalo@dtu.dk

Alessia Pini (✉)
Department of Statistical Sciences, Università Cattolica Del Sacro Cuore, Milano, Italy
e-mail: alessia.pini@unicatt.it

Simone Vantini
Department of Mathematics, Politecnico di Milano, Milano, Italy
e-mail: simone.vantini@polimi.it

© The Author(s), under exclusive license to Springer Nature Switzerland AG 2025

G. Aneiros et al. (eds.), *New Trends in Functional Statistics and Related Fields*,
Contributions to Statistics, https://doi.org/10.1007/978-3-031-92383-8_39

error rate (FWER) or of the false discovery rate (FDR) - two well known quantities defined for multivariate data. Most of the existing techniques assume that the domain of functional data is one-dimensional. Given the increasing demand from various applications, we extend this idea to a more general setting, where the domain is a Riemannian manifold. We describe three different approaches, characterized by three different notions of control of the error rate: the false discovery rate in Sect. 39.3; the ball-wise error rate in Sect. 39.4; and the ball-wise e-validity in Sect. 39.5.

39.2 Setting and Pointwise Testing

We assume that the domain of functional data is a *Riemannian manifold*, which is considered fixed, known, and the same for all functions. We start by formally defining the manifold and intrinsic metric, topology, and measure.

Definition 39.1. A Riemaniann manifold M of finite dimension is a smooth manifold together with a smoothly varying 2-tensor field g on M which is an inner product at each point. The inner product g defines a metric d and a measure μ on M, which we refer to as *Riemannian metric* and *Riemannian measure*, respectively.

We assume to observe N continuous functions $\xi_1, \ldots, \xi_N \colon M \to \mathbb{R}$. We aim to test a null hypothesis H_0 against an alternative H_1 locally on the manifold M. In detail, we will denote by H_0^s and H_1^s the pointwise null and alternative hypothesis at point $s \in M$. We denote as $M_0 = \{s \in M : H_0^s \text{is true}\}$ - i.e., the set of all points of M where H_0^s is true - and $M_1 = M \setminus M_0$. We assume that a local test statistic $T : M \to \mathbb{R}$ is available to test point-wise hypotheses. We assume that the map T is continuous for almost all $s \in M$, almost surely. Finally, we assume that a method is available to compute an exact pointwise p-value $p(s)$. In particular, we let point-wise p-values be defined in the usual sense as the probability of observing a more extreme event under H_0:

$$p(s) = P_{H_0}(T(s) > T_{\text{obs}}(s)) \quad s \in M \tag{39.1}$$

where $T(s)$ is the point-wise test statistic and $T_{\text{obs}}(s)$ is the observed point-wise statistic at point $s \in M$. We will refer to the function $p : M \to \mathbb{R}$ defined in (39.1) as the *unadjusted p-value function*. Due to the assumptions of continuity of functional data and of the test statistic on M, we have that the unadjusted p-value function is continuous on the manifold M. Finally, if we assume that the test of H_0^s against H_1^s is exact, we have that the unadjusted p-value function controls the pointwise error rate, namely:

$$\forall s \in M_0,\ P(p(s) \leq \alpha) = \alpha \tag{39.2}$$

Clearly, the control of the pointwise error rate (39.2) is not strong enough since it does not take into account multiplicity. In particular, there is no guarantee about the probability of incorrectly rejecting the null hypothesis on the whole domain. To solve this problem, we propose three different methods, based on different adjustments of the pointwise p-value.

39.3 The Functional False Discovery Rate

The first adjustment that we propose is based on the control of the functional false discovery rate (fFDR). We briefly present the definition of fFDR and of a controlling procedure here. Further details on the methodology, as well as all proofs of the theoretical results and an extensive simulation study, can be found in [4].

Given $M_0, \alpha \in (0, 1)$, and an instance of $p(s)$, let $V_p = \{s \in M_0 : p(s) < \alpha\}$ be the region where the null hypothesis is wrongly rejected, and let $S_p = \{s \in M_1 : p(s) < \alpha\}$ be the region where the null hypothesis is correctly rejected. The set V corresponds to committing type I errors, and in a given research situation, it is desired that V is as small as possible and S is as large as possible.

Definition 39.2. The functional false discovery rate is defined as

$$\text{fFDR} = \mathbb{E}[Q] = \mathbb{E}\left[\frac{\mu(V_p)}{\mu(V_p \cup S_p)} 1_{\mu(V_p \cup S_p)>0}\right]$$

where Q is the *proportion of false discoveries*.

The functional false discovery rate can be seen as the functional version of the classical multivariate false discovery rate: it is the proportion of the measure of the portion of domain that is falsely discovered $\mu(V)$ over the measure of the portion of domain that is discovered $\mu(V \cup S)$. Again analogously to the classical multivariate case, under the positive regression dependence condition on the subset of true null hypotheses (see [4]), it is possible to define the functional analogous of the Benjamini-Hochberg adjusted p-value as follows:

$$\tilde{p}_{\text{fFDR}}(s) = \min_{t \geq p(s)} \left\{1, \frac{\mu(M)t}{\mu(r : p(r) \leq t)}\right\}, \quad s \in M \tag{39.3}$$

As shown in [4], the adjusted p-value function $\tilde{p}_{\text{fFDR}}(s)$ defined in (39.3) controls the fFDR: if $V_{\tilde{p}} = \{s \in M_0 : \tilde{p}_{\text{fFDR}}(s) < \alpha\}$ and $S_{\tilde{p}} = \{s \in M_1 : \tilde{p}_{\text{fFDR}}(s) < \alpha\}$:

$$\mathbb{E}\left[\frac{\mu(V_{\tilde{p}})}{\mu(V_{\tilde{p}} \cup S_{\tilde{p}})} 1_{\mu(V_{\tilde{p}} \cup S_{\tilde{p}})>0}\right] \leq \alpha.$$

39.4 The Functional Ball-Wise Error Rate

The second adjustment that we propose is based on an extension of the family-wise error rate (FWER) to functional data. As discussed in [1] for functional data defined in one-dimensional domains, a good compromise between pointwise control - which is too weak - and FWER control - which is too strong - is the control of the error rate over all intervals of the domain. We briefly summarize the extension of this notion to data defined on manifold domains here. For further details, refer to [5]. We first need to define the notion of ball.

Definition 39.3. For a given manifold M with metric d, define $B(x, \epsilon)$ (the ball of center x and radius ϵ) as

$$B(x, \epsilon) = \{y \in M | d(x, y) < \epsilon\}, \quad x \in M, \epsilon > 0$$

Then, we can define the control of ball-wise error rate (BWER) as follows.

Definition 39.4. A p-value function $\tilde{p} : M \to [0, 1]$ controls the ball-wise error rate if $\forall \alpha \in (0, 1)$:

$$\forall B(x, \epsilon) \subseteq M_0, \; P(\exists s \in B(x, \epsilon) : \tilde{p}(s) \leq \alpha) \leq \alpha$$

As detailed in [5], it is possible to define a BWER-adjusted p-value function as follows:

$$\tilde{p}_{\text{BWER}}(s) = \sup_{B(x,\epsilon) \in M:\, s \in B(x,\epsilon)} p^{B(x,\epsilon)} \quad \text{for } s \in M \tag{39.4}$$

where $p^{B(x,\epsilon)}$ is the p-value for the test of $H_0^{B(x,\epsilon)} = \cap_{s \in B(x,\epsilon)} H_0^s$ against $H_1^{B(x,\epsilon)} = \cup_{s \in B(x,\epsilon)} H_1^s$.

As shown in [5], if the p-values of tests on balls $p^{B(x,\epsilon)}$ are exact, the adjusted p-value function $\tilde{p}_{\text{BWER}}(s)$ controls the BWER:

$$\forall \alpha \in (0, 1) \; \forall B(x, \epsilon) \subseteq M_0, \; P(\exists s \in B(x, \epsilon) : \tilde{p}_{\text{BWER}}(s) \leq \alpha) \leq \alpha$$

39.5 The Functional Ball-Wise e-Validity

A lively area of recent literature about testing hypotheses is focusing on the definition of e-value, as an alternative to p-values (see for instance [7]). Namely, an e-variable is a nonnegative extended random variable $E : \Omega \to [0, +\infty]$ whose expected value under the null hypothesis is at most 1, and an e-value is a value taken by an e-variable. Based on this definition, we can define a pointwise e-value function $e : M \to [0, +\infty]$. In detail, for all $s \in M$, $e(s)$ is the value assumed by a pointwise e-variable $E(s)$ such that $\mathbb{E}_{H_0^s}(E(s)) \leq 1$, and $\mathbb{E}_{H_1^s}(E(s)) > 1$. The fact that $e(s)$ can be used to test the pointwise hypothesis H_0^s against H_1^s follows from Markov's inequality:

$$\mathbb{P}_{H_0^s}\left(E(s) \geq \frac{1}{\alpha}\right) \leq \frac{\mathbb{E}_{H_0^s}(E(s))}{1/\alpha} \leq \alpha.$$

As shown in [8], e-values are often mathematically more tractable, especially in multiple testing. In fact, the e values of single tests can be merged simply by averaging them. This allows us to define an adjusted e-value function in a similar way to the definition of the BWER-adjusted p-value function. In the following, we briefly describe the e-value function and the adjusted e-value function. For a complete discussion of the methodology, the proofs of all theoretical properties, and some examples see [6].

We first define the property of ball-wise e-validity, that is the analogous in terms of e-values of the ball-wise error rate.

Definition 39.5. An e-variable function $\tilde{E}(s)$ is ball-wise e-valid if there exist an e-variable $\overline{E}^{B(x,\epsilon)}$ such that

$$\forall B(x,\epsilon) \in M_0, \tilde{E}(s) \leq \overline{E}^{B(x,\epsilon)} \quad \forall s \in B(x,\epsilon)$$

As shown in [6], it is possible to define the unadjusted e-value function as follows:

$$\tilde{e}(s) = \inf_{B(x,\epsilon):s\in B(x,\epsilon)} e^{B(x,\epsilon)}.$$

If the pointwise e-value function is well defined, it is possible to show that the adjusted e-value function is ball-wise e-valid, according to definition 39.5.

The definition of the adjusted e-value function $\tilde{e}(s)$ is very similar to the definition of the BWER-adjusted p-value function. One important difference between the two methods is how the adjustment is performed in practice: for computing $\tilde{p}_{\text{BWER}}(s)$ one needs to define an exact test on every ball, which could be difficult without making assumptions on the dependence structure of functional data along the domain. Similarly, for computing $\tilde{e}(s)$ one needs to define an e-value for the test on every ball. However, since the average of e-values is still an e-value, $e^{B(x,\epsilon)}$ can be trivially defined as

$$e^{B(x,\epsilon)} = \int_{B(x,\epsilon)} e(s)\mathrm{d}s.$$

Therefore, the computation of $\tilde{e}(s)$ is much easier than the computation of $\tilde{p}_{\text{BWER}}(s)$.

39.6 Numerical Example

We illustrate the different procedures in a numerical example where the manifold M is a sphere. We generate a sample of size $n = 10$ of functional data on the unit sphere S^2. We set the functional mean to be equal to zero, apart from a signal equal to $5(0.7 - d(s_0, s))$ at all points $s \in M$ at distance lower than 0.7 from the point s_0 of coordinates $(0.5, -0.5, 1/\sqrt{(2)})$. We simulated data from a Gaussian random field with a Gaussian kernel. Fig. 39.1 displays - from left to right - the mean of functional data, the sample pointwise mean, and the sample pointwise standard deviation.

We test the pointwise hypothesis $H_0^s : \mathbb{E}[\xi_i(s)] = 0$ against the alternative $H_1^s : \mathbb{E}[\xi_i(s)] \neq 0$. We employed a one-sample permutation test based on the pointwise classical t-test statistic. For computing the p-value $p^{B(x,\epsilon)}$, we employ a permutation test based on the integral test statistic $\int_{B(x,\epsilon)} T(s)\mathrm{d}s$, where $T(s)$ denotes the pointwise t-test statistic at point $s \in M$. Finally, the pointwise e-value $e(s)$ is computed applying the Shafer calibrator [7] to the pointwise p-value: $e(s) = 1/\sqrt{(p(s))} - 1$. The results are displayed in Fig. 39.2.

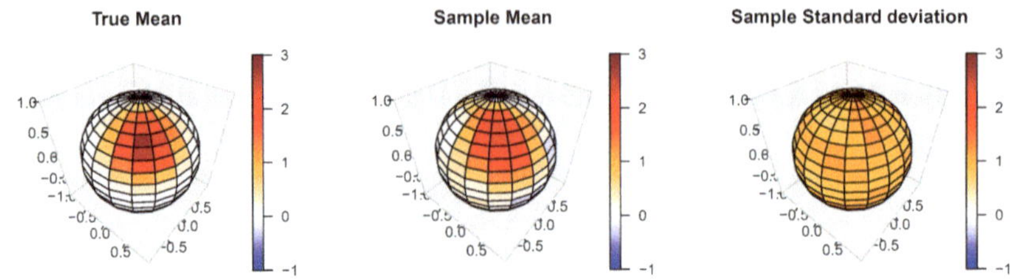

Fig. 39.1: From left to right: true mean of functional data, sample mean and sample standard deviation of a simulated Gaussian sample of size $n = 10$.

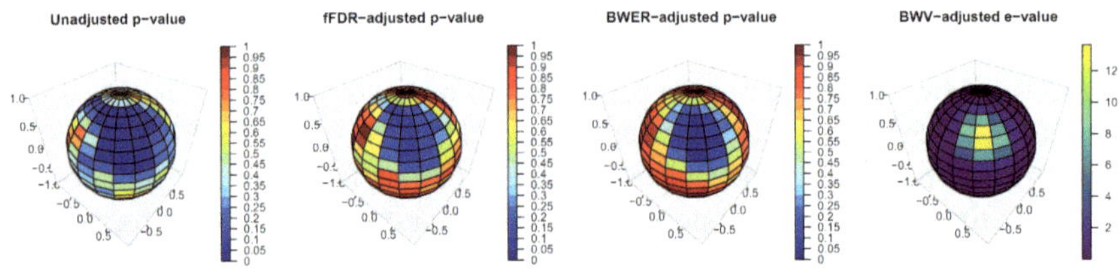

Fig. 39.2: Unadjusted and adjusted p-value and e-value functions obtained from one-sample permutation test on the simulated data.

The unadjusted p-value function is small also in areas of the domain where the null hypothesis is true, while the adjusted p-value and e-value functions detect differences from zero only in the correct area. Interestingly, by exploiting the strong local dependence of data, the BWER-adjusted p-value function $\tilde{p}_{\mathrm{BWER}}(s)$ detects as significant a larger area compared to the fFDR-adjusted p-value function. The ball-wise valid e-value function detects a smaller area of the domain. However, its visualization and interpretation are the clearest, providing evidence to reject the null hypothesis only on a small area close to the point s_0.

39.7 Conclusions

We described three different approaches to test pointwise hypotheses on functional data with manifold domains. The three methods are characterized by three different notions of control of the error rate: the false discovery rate, the ball-wise error rate, and the ball-wise e-validity. Note that all three techniques can be applied to test any functional hypotheses (e.g., testing the mean of a functional sample, testing difference between two samples, testing parameters in functional regression). In order to apply the three approaches, one needs to specify a way to define a pointwise p-value function (for $\tilde{p}_{\text{fFDR}}(s)$ and $\tilde{p}_{\text{BWER}}(s)$), of a pointwise e-value function (for $\tilde{e}(s)$). For defining such a quantity, either parametric or nonparametric techniques can be employed; the only requirement is that pointwise p-values are exact, and pointwise e-values are valid (namely, the pointwise expected value is smaller or equal to one). The only exception is the case of the ball-wise adjusted p-value function $\tilde{p}_{\text{BWER}}(s)$. In such a case, the adjustment procedure also requires to compute exact p-values $p^{B(x,\epsilon)}$ for testing the null hypothesis on every ball of the domain - which could be difficult in practice, especially using a parametric test.

References

1. Abramowicz, K., Pini, A., Schelin, L., Stamm, A., Vantini, S.: Domain selection and familywise error rate for functional data: A unified framework. Biometrics, **79**(2), 1119–1132 (2023)
2. Choi, H., Reimherr, M.: A geometric approach to confidence regions and bands for functional parameters. Journal of the Royal Statistical Society: Series B (Statistical Methodology), **80**(1), 239–260 (2018)
3. Liebl, D., Reimherr, M.: Fast and Fair Simultaneous Confidence Bands for Functional Parameters. Journal of the Royal Statistical Society: Series B (Statistical Methodology), **85**(3) 842—868 (2023)
4. Olsen, N. L., Pini, A., Vantini, S.: False discovery rate for functional data. TEST **30**, 784–809 (2021)
5. Olsen, N. L., Pini, A., Vantini, S.: Local inference for functional data on manifold domains using permutation tests. arXiv.2306.07738 (2023)
6. Pini, A., Vantini, S.: Interval-wise e-value function for testing functional data. MOX technical report (2025)
7. Shafer, G.: Testing by Betting: A Strategy for Statistical and Scientific Communication. Journal of the Royal Statistical Society Series A: Statistics in Society, **184**(2), 407–431 (2021)
8. Vovk., V., Wang, R.: E-values: Calibration, combination and applications. Annals of Statistics, **49**(3) 1736–1754 (2021)

Chapter 40
Efficient Bayesian Linear Models for a Large Number of Observations

Hassan Maatouk, Didier Rullière and Xavier Bay

Abstract Bayesian linear models are widely used as efficient approaches for nonparametric function estimation. In this paper, we present a Bayesian method for generating finite-dimensional linear models that can handle large datasets. This method is based on an efficient Markov chain Monte Carlo algorithm. The advantage of this approach is that sampling is performed before conditioning, rather than after. This enables the use of efficient samplers when the prior covariance matrix exhibits special properties, such as being Toeplitz, block-Toeplitz, or sparse. The performance of the proposed approach is evaluated in the context of nonparametric function estimation with large datasets. A numerical comparison with direct approaches based on Cholesky factorization is provided to illustrate its efficiency in terms of computational runtime.

40.1 Introduction

Gaussian processes (GPs) are commonly used as effective prior distributions over function spaces. They are frequently employed in tasks like regression and classification in machine learning [17, 20]. Over the years, GPs have gained popularity as a Bayesian tool, finding applications across diverse fields such as geosciences, physics, biology, chemistry, engineering, finance, and machine learning [1, 3, 4, 19].

Hassan Maatouk (✉)
LAMPS, Université de Perpignan Via Domitia, 52 av. Paul Alduy, 66860 Cedex 9 Perpignan, France, e-mail: hassan.maatouk@univ-perp.fr

Didier Rullière
Mines Saint-Étienne, Univ Clermont Auvergne, CNRS, UMR 6158 LIMOS, Institut Henri Fayol, Saint-Étienne, F-42023, France, e-mail: drulliere@emse.fr

Xavier Bay
Mines Saint-Étienne, Univ Clermont Auvergne, CNRS, UMR 6158 LIMOS, Institut Henri Fayol, Saint-Étienne, F-42023, France, e-mail: bay@emse.fr

© The Author(s), under exclusive license to Springer Nature Switzerland AG 2025

G. Aneiros et al. (eds.), *New Trends in Functional Statistics and Related Fields*,
Contributions to Statistics, https://doi.org/10.1007/978-3-031-92383-8_40

In this paper, Bayesian linear models with Gaussian random weights for a large number of observations are considered. In this context, *direct* approaches based on Cholesky factorization and eigen-decomposition [11] become computationally prohibitive when handling large datasets, as their computational complexity grows cubically with the number of observations. To address this problem, we propose the use of a Markov Chain Monte Carlo (MCMC) method. This approach builds upon the efficient Elliptical Slice Sampling (ESS) algorithm [16] and requires only the evaluation of the likelihood function at each MCMC iteration. The main advantage of this method is that sampling is performed before conditioning rather than after. Consequently, if the prior covariance (precision) matrix admits special structures, such as Toeplitz or block-Toeplitz (i.e., when evaluating stationary product kernels on a regularly spaced grid) [21, 22], highly efficient samplers can be employed such as the Fast Fourier Transform (FFT) [21] and the fast large-scale approaches developed in [13, 14, 15]. In the numerical examples of this paper, a comparison is conducted between Cholesky factorization and the highly efficient LS.KLE sampler developed in [14].

The paper is organized as follows: in Sect. 40.2, GP regression is briefly reviewed. Sect. 40.3 is divided into two parts. The first part revisits *direct* approaches for generating linear models, while the second part develops a highly efficient MCMC algorithm tailored for large datasets. Sect. 40.4 is dedicated to numerical examples.

40.2 Gaussian Process Regression Review

In this paper, we consider the following regression problem with Gaussian noise, where n vectors $\boldsymbol{x}_i \in \mathbb{R}^d$ and n responses $y_i \in \mathbb{R}$ are observed from the model

$$y_i = f(\boldsymbol{x}_i) + \epsilon_i, \quad i = 1, \dots, n. \tag{40.1}$$

We assume that $\epsilon_i \overset{\text{i.i.d.}}{\sim} \mathcal{N}(0, \sigma^2)$, is an additive i.i.d. zero-mean Gaussian noise with a constant variance of σ^2. Here, $f : \mathcal{X} \subset \mathbb{R}^d \longrightarrow \mathbb{R}$ represents an unknown function that generates the observed target values $\boldsymbol{y} = [y_1, \dots, y_n]^\top \in \mathbb{R}^n$. The sequence $\{(\boldsymbol{x}_i, y_i)\}$ represents the training samples. The set $\mathcal{X}$ is a compact subset of $\mathbb{R}^d$ and without loss of generality, we suppose that $\mathcal{X}$ is the unit hypercube.

A GP is a stochastic process, i.e., a collection of random variables, such that every finite subset of these variables follows a multivariate normal (MVN) distribution. A GP is fully specified by its mean function $\mu(\cdot)$ and covariance function $k(\cdot,\cdot)$, where the covariance function plays a crucial role in controlling the smoothness of the GP sample paths. If we denote this GP by Z, then we can write

$$Z \sim \mathcal{GP}(\mu(\cdot), k(\cdot,\cdot)),$$

where the mean and covariance functions are defined as:

$$\mu(\boldsymbol{x}) = \mathrm{E}[Z(\boldsymbol{x})], \quad \forall \boldsymbol{x} \in \mathcal{X};$$
$$k(\boldsymbol{x}, \boldsymbol{x}') = \mathrm{Cov}\,(Z(\boldsymbol{x}), Z(\boldsymbol{x}')), \quad \forall \boldsymbol{x}, \boldsymbol{x}' \in \mathcal{X}.$$

Additionally, $Z(\cdot) = \mu(\cdot) + Y(\cdot)$, where Y is a zero-mean GP, i.e., $Y(\cdot) \sim \mathcal{GP}(0, k(\cdot,\cdot))$. In the regression framework (40.1), GPs are widely recognized as efficient prior distributions over function spaces [20]. Conditionally on data, we have

$$\{Y(\cdot) | Y(\mathbb{X}) + \boldsymbol{\epsilon} = \boldsymbol{y}\} \sim \mathcal{GP}\left(\widetilde{\mu}(\cdot), \widetilde{k}(\cdot,\cdot)\right),$$

where $\boldsymbol{\epsilon} = [\epsilon_1, \ldots, \epsilon_n]^\top \sim \mathcal{N}(\mathbf{0}, \sigma^2 \mathrm{I}_n)$ and the conditional mean function $\widetilde{\mu}$ and covariance function $\widetilde{k}$ are given by

$$\widetilde{\mu}(\boldsymbol{x}) = \mathrm{E}\,[Y(\boldsymbol{x})|\boldsymbol{y}] = k(\boldsymbol{x}, \mathbb{X})^\top \left(k(\mathbb{X}, \mathbb{X}) + \sigma^2 \mathrm{I}_n\right)^{-1} \boldsymbol{y}; \tag{40.2}$$
$$\widetilde{k}(\boldsymbol{x}, \boldsymbol{x}') = k(\boldsymbol{x}, \boldsymbol{x}') - k(\boldsymbol{x}, \mathbb{X})^\top (k(\mathbb{X}, \mathbb{X}) + \sigma^2 \mathrm{I}_n)^{-1} k(\boldsymbol{x}', \mathbb{X});$$

with I_n the $n \times n$ identity matrix, $\mathbb{X} = [\boldsymbol{x}_1, \ldots, \boldsymbol{x}_n]^\top \in \mathbb{R}^{n \times d}$ the design matrix, $k(\mathbb{X}, \mathbb{X})$ the covariance matrix of the Gaussian vector $Y(\mathbb{X}) \in \mathbb{R}^n$ and $k(\boldsymbol{x}, \mathbb{X})$ the vector of covariance between $Y(\boldsymbol{x})$ and $Y(\mathbb{X})$, i.e., $k(\boldsymbol{x}, \mathbb{X}) = [k(\boldsymbol{x}, \boldsymbol{x}_1), \ldots, k(\boldsymbol{x}, \boldsymbol{x}_n)]^\top$.

40.3 Finite-Dimensional Bayesian Linear Models

In this section, we assume that the parent GP, $\{Y(\boldsymbol{x})\}_{\boldsymbol{x} \in \mathcal{X}}$, is approximated by the following Bayesian linear model with Gaussian random weights, known as the *weight-space* view of GPs [20, Sect. 2.1.1]

$$\check{Y}(\boldsymbol{x}) = \Phi(\boldsymbol{x})^\top \boldsymbol{\eta}, \quad \forall \boldsymbol{x} \in \mathcal{X}, \tag{40.3}$$

where $\boldsymbol{\eta} \in \mathbb{R}^m$ is a zero-mean Gaussian vector with positive-definite covariance matrix $\tau^2 \boldsymbol{K}$, with τ^2 the *signal variance* parameter, and Φ is a sequence of deterministic basis functions such that $\Phi(\boldsymbol{x}) \in \mathbb{R}^m$, for any $\boldsymbol{x} \in \mathcal{X}$. For simplicity of notations, we denote by $\boldsymbol{\Phi} = \Phi(\mathbb{X}) \in \mathbb{R}^{n \times m}$, where $\mathbb{X}$ is the design matrix. Therefore, the set of noisy data $\{\check{Y}(\mathbb{X}) + \boldsymbol{\epsilon} = \boldsymbol{y}\}$ can be written in matrix form as follows:

$$\boldsymbol{\Phi}\boldsymbol{\eta} + \boldsymbol{\epsilon} = \boldsymbol{y},$$

where $\boldsymbol{\epsilon} = [\epsilon_1, \ldots, \epsilon_n]^\top$ is an zero-mean Gaussian noise, i.e., $\boldsymbol{\epsilon} \sim \mathcal{N}(\mathbf{0}, \sigma^2 \mathrm{I}_n)$.

40.3.1 Direct Sampling Approach

In this section, we briefly review the *direct* classical approach for generating the posterior distribution $\{\boldsymbol{\eta} | \boldsymbol{\Phi}, \boldsymbol{y}\}$ and, consequently, Model (40.3) under noisy observations $\{\check{Y}(\mathbb{X}) + \boldsymbol{\epsilon} = \boldsymbol{y}\}$. Using the predictive equations (40.2), we get $\{\boldsymbol{\eta}|\boldsymbol{\Phi}, \boldsymbol{y}\} \sim \mathcal{N}(\widetilde{\boldsymbol{\mu}}, \widetilde{\boldsymbol{K}})$, where [11, Proposition 1]

$$\begin{cases} \widetilde{\boldsymbol{\mu}} = \tau^2(\boldsymbol{\Phi}\boldsymbol{K})^\top(\tau^2\boldsymbol{\Phi}\boldsymbol{K}\boldsymbol{\Phi}^\top + \sigma^2 \mathrm{I}_n)^{-1}\boldsymbol{y}; \\ \widetilde{\boldsymbol{K}} = \tau^2\boldsymbol{K} - \tau^4(\boldsymbol{\Phi}\boldsymbol{K})^\top(\tau^2\boldsymbol{\Phi}\boldsymbol{K}\boldsymbol{\Phi}^\top + \sigma^2 \mathrm{I}_n)^{-1}\boldsymbol{\Phi}\boldsymbol{K}. \end{cases} \tag{40.4}$$

Comments on the *direct* approach (40.4)

The *direct* approach is based on generating the posterior distribution in (40.4) using Cholesky factorization or eigen-decomposition [11]. Its computational complexity grows cubically with respect to the dimension m [8]. Additionally, this approach requires the inversion of an $n \times n$ matrix. It is worth noting that this method is based on sampling after conditioning rather than before. Consequently, if the prior covariance matrix $\boldsymbol{K}$ exhibits a particular structure, such as Toeplitz, or block-Toeplitz [21, 22], banded or sparse [6], this property will be lost in the sampling step.

In the following section, we present an alternative approach for generating samples from the posterior distribution $\{\boldsymbol{\eta}|\boldsymbol{\Phi}, \boldsymbol{y}\}$, where sampling is performed before conditioning rather than after. Consequently, this approach is well-suited for handling high-dimensional spaces (when the prior covariance matrix $\boldsymbol{K}$ exhibits special properties) and large datasets.

40.3.2 Alternative Sampling Approaches

Due to its computational complexity, which grows cubically, *direct* approaches become infeasible for generating the posterior distribution $\{\boldsymbol{\eta}|\boldsymbol{\Phi}, \boldsymbol{y}\}$ when the dimension m and the dataset size n are large. In this section, we present an alternative approach to address this issue. Specifically, we explain how MCMC methods can efficiently generate the posterior distribution when both m and n are large. According to [20, Sect. 2.1.1] and Bayes's rule, we have the following posterior distribution:

$$p(\boldsymbol{\eta}|\boldsymbol{\Phi}, \boldsymbol{y}) := \frac{p(\boldsymbol{y}|\boldsymbol{\Phi}, \boldsymbol{\eta})p(\boldsymbol{\eta})}{p(\boldsymbol{y}|\boldsymbol{\Phi})}, \tag{40.5}$$

where $p(\boldsymbol{y}|\boldsymbol{\Phi})$ is the normalizing constant, also known as the *marginal likelihood.* It is independent of $\boldsymbol{\eta}$ and given by

$$p(\boldsymbol{y}|\boldsymbol{\Phi}) := \int_{\mathbb{R}^m} p(\boldsymbol{y}|\boldsymbol{\Phi}, \boldsymbol{\eta})p(\boldsymbol{\eta})\mathrm{d}\boldsymbol{\eta}.$$

By developing the likelihood and prior in (40.5), we obtain

$$p(\boldsymbol{\eta}|\boldsymbol{\Phi}, \boldsymbol{y}) \propto \exp\left(-\frac{1}{2\sigma^2}[\boldsymbol{y} - \boldsymbol{\Phi}\boldsymbol{\eta}]^\top [\boldsymbol{y} - \boldsymbol{\Phi}\boldsymbol{\eta}]\right) \exp\left(-\frac{1}{2\tau^2}\boldsymbol{\eta}^\top \boldsymbol{K}^{-1}\boldsymbol{\eta}\right). \quad (40.6)$$

The posterior probability density function (pdf) in (40.6) is proportional to the product of a likelihood function and a zero-mean Gaussian prior:

$$p(\boldsymbol{\eta}|\boldsymbol{\Phi}, \boldsymbol{y}) \propto \underbrace{\exp\left(-\frac{1}{2\sigma^2}[\boldsymbol{y} - \boldsymbol{\Phi}\boldsymbol{\eta}]^\top [\boldsymbol{y} - \boldsymbol{\Phi}\boldsymbol{\eta}]\right)}_{\text{likelihood}} \underbrace{\exp\left(-\frac{1}{2\tau^2}\boldsymbol{\eta}^\top \boldsymbol{K}^{-1}\boldsymbol{\eta}\right)}_{\text{Gaussian prior}}$$
$$\propto L(\boldsymbol{\eta})\mathcal{N}(\boldsymbol{\eta}; \boldsymbol{0}, \tau^2\boldsymbol{K}).$$

The logarithm of the likelihood function $L(\boldsymbol{\eta})$ can be expressed as follows:

$$\log[L(\boldsymbol{\eta})] = -\frac{1}{2\sigma^2}||\boldsymbol{y} - \boldsymbol{\Phi}\boldsymbol{\eta}||^2. \quad (40.7)$$

The logarithm function in (40.7), which has a computational complexity of order $O(nN)$ will be evaluated at each MCMC iteration. In this context, sampling from (40.6) can be performed using Metropolis-Hastings (MH) proposals [5, 17]:

$$\boldsymbol{\eta}' = \rho\boldsymbol{\zeta} + \sqrt{1 - \rho^2}\boldsymbol{\eta}, \quad \boldsymbol{\zeta} \sim \mathcal{N}(\boldsymbol{0}, \tau^2\boldsymbol{K}), \quad (40.8)$$

where $\rho \in [-1, 1]$ is a step-size parameter, $\boldsymbol{\eta}$ is the current state, and $\boldsymbol{\eta}'$ is the proposal state. Recall that the MH acceptance ratio, $\alpha = \min\{1, L(\boldsymbol{\eta}')/L(\boldsymbol{\eta})\}$ depends solely on the likelihood ratio and is independent of ρ. Furthermore, this method is straightforward to implement and can be readily applied to a wide range of models with Gaussian priors.

The ESS relies on the parametrization $\rho = \sin(\theta)$ in (40.8), offering an adaptive and automated approach to tuning the step-size parameter ρ. As a result, the MH proposal in (40.8) is reformulated as follows:

$$\boldsymbol{\eta}' = \sin(\theta)\boldsymbol{\zeta} + \cos(\theta)\boldsymbol{\eta}, \quad \boldsymbol{\zeta} \sim \mathcal{N}(\boldsymbol{0}, \tau^2\boldsymbol{K}),$$

where the angle θ is uniformly generated from a $[\theta_{\min}, \theta_{\max}]$ interval which is shrunk exponentially fast until an acceptable state is reached. For a given value of θ, a uniform random number is generated and compared with the likelihood ratio $L(\boldsymbol{\eta}')/L(\boldsymbol{\eta})$. If the proposal $\boldsymbol{\eta}'$ is rejected, one shrinks the bracket of θ, and continues this process until acceptance. Detailed guidelines for shrinking the bracket are provided in [16].

Comments on the MCMC approach

Unlike *direct* methods, the key advantage of the MCMC approach described in this section is its ability to avoid matrix inversion. Moreover, the sampling process is performed before conditioning rather than after. Consequently, highly efficient samplers can be employed when the prior covariance matrix $\boldsymbol{K}$ has a specific structure, such as Toeplitz or block-Toeplitz [21]. In Sect. 40.4, we evaluate the computational performance of the proposed approach, denoted MCMC-ESS, when using an efficient prior sampler. It is worth noting that MCMC-ESS can handle more complex posterior inference problems, including those involving additional shape constraints [12].

40.4 Numerical Performance

In this section, we investigate the performance of the developed approach for large datasets. To this end, we consider the finite-dimensional Bayesian linear model with Gaussian random weights proposed by [10]. This approach has gained significant attention due to its ability to incorporate a wide range of shape constraints across the entire domain, as well as its strong theoretical foundation [2, 9]. We briefly review this approach to ensure that the article is self-contained. Let Y be a zero-mean GP with covariance function k, i.e., $Y \sim \mathcal{GP}(0, k)$. Let $\{t_j\}$ denote a sequence of $m \geq 2$ equally spaced knots on $\mathcal{X}$, i.e., $0 = t_1 < \ldots < t_m = 1$. Then, Y can be approximated by the finite-dimensional Bayesian linear model $\check{Y}$, defined as follows:

$$\check{Y}(x) := \sum_{j=1}^{m} Y(t_j)\phi_j(x) = \sum_{j=1}^{m} \eta_j \phi_j(x) = \Phi(x)^\top \boldsymbol{\eta}, \quad x \in \mathcal{X}, \tag{40.9}$$

where we denote $\eta_j = Y(t_j)$ for $j = 1, \ldots, m$. Consequently, $\boldsymbol{\eta} = [\eta_1, \ldots, \eta_m]^\top$ is a zero-mean Gaussian with positive-definite covariance matrix $\tau^2 \boldsymbol{K}$. Indeed,

$$\mathrm{Cov}(\eta_j, \eta_{j'}) = \mathrm{Cov}(Y(t_j), Y(t_{j'})) = k(t_j, t_{j'}) = \tau^2 \boldsymbol{K}_{j,j'}, \quad \forall j, j' = 1, \ldots, m.$$

Additionally, $\Phi(\cdot) = [\phi_1(\cdot), \ldots, \phi_m(\cdot)]^\top \in \mathbb{R}^m$, with ϕ_j the compactly supported hat basis function associated to the knot t_j, i.e., the B-spline of degree 1 [7, Fig. 1.a]. Precisely, we have for any $j \in \{2, \ldots, m-1\}$

$$\phi_j(x) := \begin{cases} 1 - \frac{|x - t_j|}{t_{j+1} - t_j} & \text{if } x \in [t_{j-1}, t_{j+1}]; \\ 0 & \text{otherwise.} \end{cases}$$

Moreover, we have $\phi_1(x) := 1 - (m-1)|x|$, if $x \in [t_1, t_2]$, and zero otherwise. Similarly, for $j = m$, we have $\phi_m(x) := 1 - (m-1)|x-1|$, if $x \in [t_{m-1}, t_m]$, and zero otherwise. As proved in [10], every sample path of $\check{Y}(\cdot)$ converges uniformly to a sample path of $Y(\cdot)$. It is worth noting that, if a stationary covariance kernel, such as, the Matérn family of covariance functions (40.11), is used, then the prior covariance matrix $\tau^2 \boldsymbol{K}$ exhibits a Toeplitz structure [21, 22]. This property is explored in the numerical examples presented in this section.

Now, we consider the statistical problem of recovering an unknown function $f : \mathcal{X} \longrightarrow \mathbb{R}$ from the training samples $\{(x_i, y_i)\}_{i=1}^n$. The following target function proposed by [18] is considered

$$f(x) = \frac{1}{\left[1 + (10x)^4\right]} + \frac{1}{2}\exp\left\{-100(x-0.5)^2\right\}, \quad x \in \mathcal{X}. \tag{40.10}$$

This function is particularly relevant in our case because it exhibits significant variation. Consequently, a substantial number of observations are required to achieve a *satisfactory* approximation. The training samples are obtained as follows: the covariates $\{x_i\}$ are sampled uniformly over $\mathcal{X}$, and the observed values $\{y_i\}$ were obtained using model (40.1) with the target function f defined in (40.10) and a standard deviation of $\sigma = 0.1$. The positive-definite covariance matrix $\tau^2\boldsymbol{K}$ of the Gaussian vector $\boldsymbol{\eta}$ is obtained using the stationary Matérn family of covariance functions

$$k(x, x') = \tau^2 \frac{2^{1-\nu}}{\Gamma(\nu)} \left(\frac{\sqrt{2\nu}}{\ell}|x - x'|\right)^{\nu} B_\nu\left(\frac{\sqrt{2\nu}}{\ell}|x - x'|\right), \quad \forall x, x' \in \mathcal{X}, \tag{40.11}$$

where $\ell > 0$ is a *length-scale* parameter, $\Gamma(\cdot)$ is the Gamma function and $B_\nu(\cdot)$ denotes the modified Bessel function of the second kind of order ν. It is worth noting that a process with the Matérn kernel of order ν admits sample paths that are $\lceil \nu - 1 \rceil$ times differentiable [20, Sect. 4.2.1].

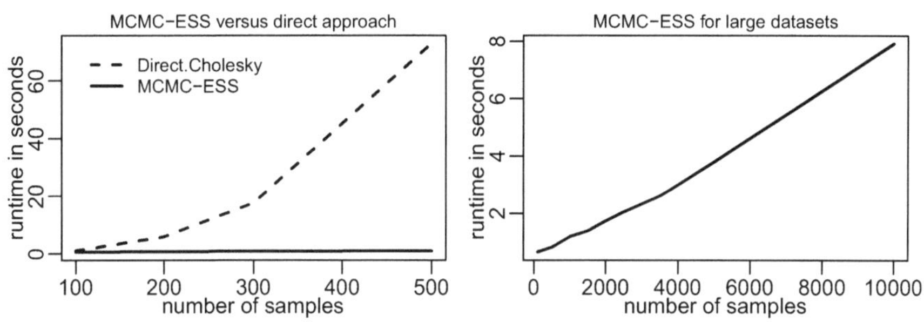

Fig. 40.1: The computational runtime for generating 1,000 posterior sample paths as a function of the number of samples: the proposed MCMC-ESS approach versus the *direct* approach for small datasets (left), and the proposed MCMC-ESS approach for large datasets up to $n = 10,000$ (right).

In Fig. 40.1, Model (40.9) is employed, where the dimension m is fixed at 50. The Matérn covariance function with a smoothness parameter $\nu = 5/2$ is used (40.11) to generate the prior covariance matrix $\boldsymbol{K}$, where the *length-scale* parameter ℓ is chosen such that the correlation at the maximum possible separation between the covariates equals 0.05. The computational runtime (in seconds) required to generate 1,000 posterior sample paths as a function of the number of samples is illustrated. In the left panel, the proposed MCMC-ESS approach, using the efficient LS.KLE sampler [14], is compared to the *direct* approach based on Cholesky factorization, in a setting with a small number of observations. In the right panel, we present the performance of the MCMC-ESS approach for a large number of observations (up to

$n = 10,000$). Unlike the *direct* approach presented in Sect. 40.3.1, the runtime of the MCMC-ESS approach increases linearly with the number of samples. Moreover, generating 1,000 MCMC iterations with $n = 10,000$ observations takes less than 8 seconds, including the parameter selection (noise variance σ^2 and *signal variance* τ^2).

40.5 Conclusion

In this paper, Bayesian linear regression models for a large number of observations are considered. Direct approaches based on Cholesky factorization and eigen-decomposition become computationally prohibitive when applied to large datasets. To address this issue, we present an efficient Markov Chain Monte Carlo (MCMC) algorithm based on Elliptical Slice Sampling (ESS), which requires only the evaluation of the likelihood function at each MCMC iteration. The main advantage of this approach is that sampling is performed before conditioning rather than after. This enables the use of highly efficient methods in the sampling procedure when the prior covariance matrix exhibits special structures, such as Toeplitz, banded and sparse. The effectiveness of this approach is demonstrated using synthetic data in the context of nonparameteric function estimation.

Acknowledgements The authors are grateful to the editor and the reviewer for the time and effort dedicated to providing feedback on our manuscript, which has been of great significance to us.

References

1. Banerjee, S., Carlin, B.P., Gelfand, A.E.: Hierarchical modeling and analysis for spatial data. Chapman and Hall/CRC, (2003)
2. Bay, X., Grammont, L., Maatouk, H.: Generalization of the Kimeldorf-Wahba correspondence for constrained interpolation. Electron. J. Stat. **10**(1), 1580–1595, (2016)
3. Blanchet-Scalliet, C., Demory, B., Gonon, T., Helbert, C.: Gaussian process regression on nested spaces. SIAM/ASA JUQ **11**(2), 426–451, (2023)
4. Chevalier, C., Bect, J., Ginsbourger, D., Vazquez, E., Picheny, V., Richet, Y.: Fast parallel kriging-based stepwise uncertainty reduction with application to the identification of an excursion set. Technometrics **56**(4), 455–465, (2014)
5. Cotter, S.L., Roberts, G.O., Stuart, A.M., White, D.: MCMC methods for functions: modifying old algorithms to make them faster. Stat. Sci. **28**(3), 424–446, (2013)
6. Durrande, N., Adam, V., Bordeaux, L., Eleftheriadis, S., Hensman, J.: Banded matrix operators for Gaussian Markov models in the automatic differentiation era. In: The 22nd International Conference on AISTATS, volume **89**, pp. 2780–2789 PMLR, (2019)
7. Eilers, P.H.C., Marx, B.D.: Flexible smoothing with B-splines and penalties. Stat. Sci. **11**(2), 89–121, (1996)
8. Golub, G., Van Loan, C.F.: Matrix computations. The Johns Hopkins University Press, (1996)
9. Grammont, L., Maatouk, H., Bay, X.: Equivalence between constrained optimal smoothing and Bayesian estimation. J. Nonparametric Stat. **37**(1), 60–81, (2024)

10. Maatouk, H., Bay, X.: Gaussian process emulators for computer experiments with inequality constraints. Math. Geosci. **49**(5), 557–582, (2017)
11. Maatouk, H., Bay, X., Rullière, X.: A note on simulating hyperplane-truncated multivariate normal distributions. Stat. Probabil. Lett. **191**, 109650, (2022)
12. Maatouk, H., Rullière, D., Bay, X.: Efficient constrained Gaussian process approximation using elliptical slice sampling. Working paper or preprint, March (2024)
13. Maatouk, H., Rullière, D., Bay, X.: Large scale Gaussian processes with Matheron's update rule and Karhunen-Loève expansion. In: Hinrichs, A., Kritzer, P., Pillichshammer, F. (eds) Monte Carlo and Quasi-Monte Carlo Methods, pp. 469–487, Cham, (2024). Springer International Publishing
14. Maatouk, H., Rullière, D., Bay, X.: Sampling large hyperplane-truncated multivariate normal distributions. Comput. Stat. **39**(4), 1779–1806, (2024)
15. Maatouk, H., Rullière, D., Bay, X.: Large-scale constrained Gaussian processes for shape-restricted function estimation. Stat. Comput. **35**(7), (2025)
16. Murray, I., Adams, R., MacKay, D.: Elliptical slice sampling. In Proceedings of the thirteenth international conference on AISTATS, pp. 541–548, (2010)
17. Neal, R.M.: Regression and classification using Gaussian process priors. Bayesian Statistics **6**, 475–501, (1999)
18. Pensoneault, A., Yang, X., Zhu, X.: Nonnegativity-enforced Gaussian process regression. TAML **10**(3), 182–187, (2020)
19. Petit, S.: Improved Gaussian process modeling: Application to Bayesian optimization. PhD thesis, Université Paris-Saclay, (2022)
20. Williams, C.K., Rasmussen, C.E.: Gaussian processes for machine learning, volume **2**. MIT press Cambridge, MA, (2006)
21. Wood, A.T.A., Chan, G.: Simulation of stationary Gaussian processes in $[0, 1]^d$. J. Comput. Graph. Stat. **3**(4), 409–432, (1994)
22. Zimmerman, D.L.: Computationally exploitable structure of covariance matrices and generalized convariance matrices in spatial models. J. Stat. Comput. Simul. **32**(1-2), 1–15, (1989)

Chapter 41
Innovative Approach to Wind Direction Data Analyses: A Compositional Periodic Spline Representation in Bayes Spaces

Jitka Machalová and Jana Heckenbergerová

Abstract Analyzing environmental data within their natural sample space is essential for optimizing energy, economics, and transport systems. Wind direction, for example, significantly impacts aircraft landing, power transmission, and wind energy generation. However, its circular nature presents unique challenges. Previous research on wind direction distribution has employed methods such as Newton's optimization of von Mises distribution mixtures, the EM algorithm, and evolutionary algorithms, focusing primarily on the approximation of individual functional observations. In this study, we advanced this field by introducing a compositional periodic spline representation of wind direction data within the Bayes spaces framework. The presented approach is efficient for processing directional data using functional data analysis techniques. Our theoretical framework will be validated on empirical wind direction datasets.

41.1 Introduction

Directional distribution modelling is a sophisticated statistical approach. It is essential for analyses of data with inherent periodicity, such as wind direction, animal movement, and time-of-day patterns. This technique addresses the unique challenges posed by the circular nature of such data, where traditional linear models fail to capture the underlying structure.

Jitka Machalová (✉)
Department of Mathematical Analysis and Applications of Mathematics, Faculty of Science, Palacký University Olomouc, 17. listopadu 12, Olomouc, 77146, Czech Republic,
e-mail: jitka.machalova@upol.cz

Jana Heckenbergerová
University of Pardubice, Faculty of Economics and Administration, Studentská 84, Pardubice, 532 10, Czech Republic e-mail: jana.heckenbergerova@upce.cz

© The Author(s), under exclusive license to Springer Nature Switzerland AG 2025

G. Aneiros et al. (eds.), *New Trends in Functional Statistics and Related Fields*,
Contributions to Statistics, https://doi.org/10.1007/978-3-031-92383-8_41

Similarly, periodical data modelling is crucial for analyzing datasets that exhibit regular and repeating patterns over time. Such data are prevalent in various fields, including meteorology, economics, and biomedical sciences. One outstanding method is restricted cubic splines (RCS), which effectively model non-linear relationships by introducing numerical constraints that account for periodicity. In [8] it was demonstrated that periodic RCS can outperform other methods. They effectively estimate seasonal variations and hormonal fluctuations, reducing variability at the extremes of the period.

Recent advancements have introduced robust statistical models. For instance, a hierarchical von Mises distribution model was developed [1]. This model offers a flexible framework for modelling complex directional datasets, including those with multiple modes and asymmetries. Furthermore, optimized hot spot and directional distribution analyses were employed in [13]. These methods characterize the spatiotemporal variation of large wildfires, demonstrating their applicability in environmental studies [13]. A global sensitivity analysis method for wind farm power output estimation models was proposed, highlighting the use of Shapley effects and vine copulas to account for dependencies among input variables [5]. These methods are used to estimate long-term wind characteristics at a target site, which is crucial for effective wind turbine placement. A modelling framework applicable to the 3D joint distribution of the circular-linear-linear dataset was presented, consisting of a parametric model based on copulas and a nonparametric kernel density estimation model [12]. These improvements highlight the importance of incorporating directionality into statistical models. This progresses the understanding and prediction of phenomena with inherent circular characteristics.

The distributional properties of large-scale database systems can be characterized by a cumulative distribution function or by a probability density function. Densities are non-negative functions integrating to one, which limits their direct analysis using standard functional data analysis (FDA) methods that assume unconstrained functions [11]. This constraint also affects spline-based approximations commonly used in FDA. More importantly, densities exhibit scale invariance and relative scale properties, distinguishing them from standard L^2 functions [7, 2, 3]. Scale invariance implies that multiplying a density by a positive constant does not change the relative information it carries. For bounded domains $I = [a, b] \subset \mathbb{R}$, the Bayes space $\mathcal{B}^2(I)$ models positive densities with square-integrable logarithms [7, 3]. This space, with a separable Hilbert structure, treats densities as scale-invariant, meaning that any positive multiple of a density holds the same relative information [3]. This differs from $L^2(I)$, as $\mathcal{B}^2(I)$ consists of equivalence classes of proportional densities. The vector space structure in $\mathcal{B}^2(I)$ is defined by *perturbation* and *powering* for densities $f, g \in \mathcal{B}^2(I)$ and $\alpha \in \mathbb{R}$

$$(f \oplus g)(x) = \frac{f(x)g(x)}{\int_I f(y)g(y)\,\mathrm{d}y}, \quad (\alpha \odot f)(x) = \frac{f(x)^\alpha}{\int_I f(y)^\alpha\,\mathrm{d}y}, \quad x \in I.$$

The difference between two densities is defined as $f \ominus g = f \oplus [(-1) \odot g]$. The inner product completing the Hilbert space as

$$\langle f, g\rangle_{\mathcal{B}} = \frac{1}{2\eta}\int_I\int_I \ln\frac{f(x)}{f(y)}\ln\frac{g(x)}{g(y)}\,\mathrm{d}x\,\mathrm{d}y,$$

with $\eta = b - a$. The corresponding norm and distance are

$$\|f\|_{\mathcal{B}} = \sqrt{\langle f, f\rangle_{\mathcal{B}}}, \quad d_{\mathcal{B}}(f, g) = \|f \ominus g\|_{\mathcal{B}}.$$

An isometric isomorphism between $\mathcal{B}^2(I)$ and $L^2(I)$ is achieved using the centered log-ratio (clr) transformation [3]

$$\mathrm{clr}(f)(x) = \ln f(x) - \frac{1}{\eta}\int_I \ln f(y)\,\mathrm{d}y \qquad \forall x \in I.$$

This transformation allows for standard operations

$$\mathrm{clr}(f \oplus g)(x) = \mathrm{clr}(f)(x) + \mathrm{clr}(g)(x), \quad \mathrm{clr}(\alpha \odot f)(x) = \alpha \cdot \mathrm{clr}(f)(x),$$

and the inner product becomes

$$\langle f, g\rangle_{\mathcal{B}} = \langle \mathrm{clr}(f), \mathrm{clr}(g)\rangle_{L^2} = \int_I \mathrm{clr}(f)(x)\,\mathrm{clr}(g)(x)\,\mathrm{d}x.$$

The clr transformation imposes the constraint

$$\int_I \mathrm{clr}(f)(x)\,\mathrm{d}x = 0,$$

defining the clr space $L^2_0(I)$. In this contribution, we focus specifically on the construction and application of *periodic splines* within the Bayes space framework. Periodic splines are particularly well-suited for modelling cyclic or repeated patterns in density functions, providing a smooth and continuous approximation that naturally handles boundary conditions. This extension offers methodological advantages for accurately representing periodic behaviors in density data and enhances the interpretability of the resulting models in $\mathcal{B}^2(I)$.

41.2 Periodic Smoothing Splines

First, we recall the basic knowledge about the B-spline representation of splines; for details, see [4], [6]. Let the sequence of knots $\Delta\lambda := \lambda_0 = a < \lambda_1 < \ldots < \lambda_g < b = \lambda_{g+1}$ be given. In the following, $\mathcal{S}_k^{\Delta\lambda}[a, b]$ denotes the vector space of polynomial splines of degree $k > 0$, defined on a finite interval $[a, b]$ with the sequence of knots $\Delta\lambda$. A spline function $s_k(x) \in \mathcal{S}_k^{\Delta\lambda}[a, b]$ is called *periodic* if it satisfies

$$s_k^{(l)}(a) \;=\; s_k^{(l)}(b) \qquad \forall\, l = 0, 1, , \ldots, k-1. \tag{41.1}$$

For a periodic spline $s_k(x) \in \mathcal{S}_k^{\Delta\lambda}[a,b]$ to ensure the fulfillment of the periodic boundary conditions (41.1) it is necessary to add *periodic boundary knots* for $i = 1,\ldots,k$ in the form, [6],

$$\lambda_{-i} = \lambda_{g+1-i} - b + a, \qquad \lambda_{i+g+1} = \lambda_i + b - a. \tag{41.2}$$

This choice of additional knots is suitable for imposing periodicity conditions (41.1). With this choice, it follows that $B_i^{k+1}(x) \equiv B_{g+1-i}^{k+1}(x+b-a)$ for $i = 1,\ldots,k$.Consequently, taking account of properties of B-splines, the condition (41.1) will be satisfied if and only if, [6], $b_{-i} = b_{g+1-i}$ for $i = 1,\ldots,k$. It is obvious that if we denote

$$\mathcal{S}_{k,per}^{\Delta\lambda}[a,b] = \left\{s_k(x) \in \mathcal{S}_k^{\Delta\lambda}[a,b] : \; s_k^{(l)}(a) = s_k^{(l)}(b) \;\forall\, l = 0,1,,\ldots,k-1\right\},$$

then $dim(\mathcal{S}_{k,per}^{\Delta\lambda}[a,b]) = g+1$. Thus every spline $s_k(x) \in \mathcal{S}_{k,per}^{\Delta\lambda}[a,b]$ has a unique representation

$$s_k(x) = \sum_{i=-k}^{g-k} b_i \widetilde{B}_i^{k+1}(x),$$

where

$$\widetilde{B}_i^{k+1}(x) = \begin{cases} B_i^{k+1}(x) & \text{for } i = 0,\ldots,g-k, \\ B_i^{k+1}(x) + B_{g+1+i}^{k+1}(x) & \text{for } i = -k,\ldots,-1. \end{cases} \tag{41.3}$$

In matrix notation, the spline $s_k(x) \in \mathcal{S}_{k,per}^{\Delta\lambda}[a,b]$ can be written as

$$s_k(x) = \widetilde{\mathbf{C}}_{k+1}(x)\widetilde{\mathbf{b}}, \tag{41.4}$$

where $\widetilde{\mathbf{C}}_{k+1}(x) = \left(\widetilde{B}_{-k}^{k+1}(x),\ldots,\widetilde{B}_{g-k}^{k+1}(x)\right)$ is a vector of B-spline functions $\widetilde{B}_i^{k+1}(x)$ defined in (41.3) and vector $\widetilde{\mathbf{b}} = \left(b_{-k},\ldots,b_{g-k}\right)^\top$ is a vector of corresponding B-spline coefficients. If we want to work with periodic splines in $L_0^2[a,b]$ space, it means that we consider subspace

$$\mathcal{Z}_{k,per}^{\Delta\lambda}[a,b] = \{s_k(x) \in \mathcal{S}_{k,per}^{\Delta\lambda}[a,b] : \int_a^b s_k(x)\,\mathrm{d}x = 0\}.$$

It is known, [6], that for $s_k(x) \in \mathcal{S}_{k,per}^{\Delta\lambda}[a,b]$ it holds

$$\int_a^b s_k(x)\,\mathrm{d}x = \frac{1}{k+1}\sum_{i=-k}^{g-k} b_i(\lambda_{i+k+1} - \lambda_i).$$

From this we can write that $s_k(x) \in \mathcal{S}_{k,per}^{\Delta\lambda}[a,b]$ has zero integral if and only if

$$b_{g-k} = -\frac{1}{\lambda_{g+1} - \lambda_{g-k}} \sum_{i=-k}^{g-k-1} b_i(\lambda_{i+k+1} - \lambda_i). \tag{41.5}$$

So one can see that $dim(\mathcal{Z}_{k,per}^{\Delta\lambda}[a,b]) = g$. For representation of spline $s_k(x) \in \mathcal{Z}_{k,per}^{\Delta\lambda}[a,b]$ we define matrix

$$\mathbf{P} = \begin{pmatrix} \mathbf{I}_g \\ \mathbf{a} \end{pmatrix},$$

where

$$\mathbf{a} = -\frac{1}{\lambda_{g+1} - \lambda_{g-k}} \left(\lambda_1 - \lambda_{-k}, \ldots, \lambda_g - \lambda_{g-k-1}\right)$$

and $\mathbf{I}_g$ is an identity matrix of order g. Thus, with respect to (41.4) and (41.5) spline $s_k(x) \in \mathcal{Z}_{k,per}^{\Delta\lambda}[a,b]$ can be written as $s_k(x) = \widetilde{\mathbf{C}}_{k+1}(x)\,\mathbf{P}\,\overline{\mathbf{b}}$, where $\overline{\mathbf{b}} = (b_{-k}, \ldots, b_{g-k-1})^\top$.

41.2.1 Periodic Smoothing Spline in $L_0^2[a,b]$

In [9], [10], the construction of smoothing splines was studied in the space $L_0^2[a,b]$. However, standard, nonperiodic, B-spline basis functions $B_i^{k+1}(x)$ were used. Now we would like to construct smoothing splines from the space $\mathcal{Z}_{k,per}^{\Delta\lambda}[a,b]$. For this purpose, let data (x_i, y_i), $a \leq x_i \leq b$, weights $w_i > 0$, $i = 1, \ldots, n$, sequence of knots $\Delta\lambda$ with periodic boundary knots (41.2) and a parameter $\alpha \in (0,1)$ be given.

Now our task is to find a spline $s_k(x) \in \mathcal{Z}_{k,per}^{\Delta\lambda}[a,b] \subset L_0^2(I)$, which for arbitrary $l \in \{1, \ldots, k-1\}$ minimizes the functional

$$J_l(s_k) = (1-\alpha) \int_a^b \left[s_k^{(l)}(x)\right]^2 \mathrm{d}x + \alpha \sum_{i=1}^{n} w_i \left[y_i - s_k(x_i)\right]^2. \tag{41.6}$$

The choice of parameter α and the derivation l affects the smoothness of the resulting spline. With respect to the notation given above, the solution to this minimization problem is given by formula

$$\overline{\mathbf{b}}^* = \mathbf{G}^{-1}\mathbf{g}, \tag{41.7}$$

where matrix $\mathbf{G}$ and vector $\mathbf{g}$ comes from the given data. The resulting periodic smoothing spline with zero integral is obtained by the formula

$$s_k^*(x) = \sum_{i=-k}^{g-k-1} b_i^* \widetilde{B}_i^{k+1}(x),$$

where $\overline{\mathbf{b}}^* = \left(b_{-k}^*, \ldots, b_{g-k-1}^*\right)^\top$ is given in (41.7).

41.3 Applications

Environmental datasets were collected at Pardubice Airport, located in the center of the Czech Republic (50°00'N, 15°44'E and 230 m above sea level). Data were manually downloaded on a monthly basis from the meteomanz.com website. The measurements span from January 1, 2014, to December 31, 2023, with a temporal resolution of one hour, resulting in approximately 87,500 records. The dataset consists of measurements of temperature, relative humidity, pressure, wind direction, wind speed, and cloudiness.

To obtain pure wind direction datasets for each year, the original measurements required preprocessing. Specifically, instances of calm wind conditions, where wind speed is zero and thus direction is undefined, were excluded from the dataset. The resulting datasets, with an angular resolution of 10 degrees, can be represented using a wind rose or a histogram of absolute frequencies, as illustrated in Fig. 41.1.

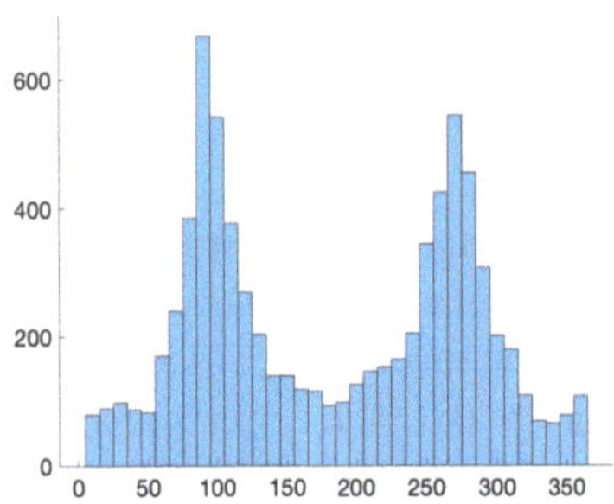

(a) Histogram for the year 2014

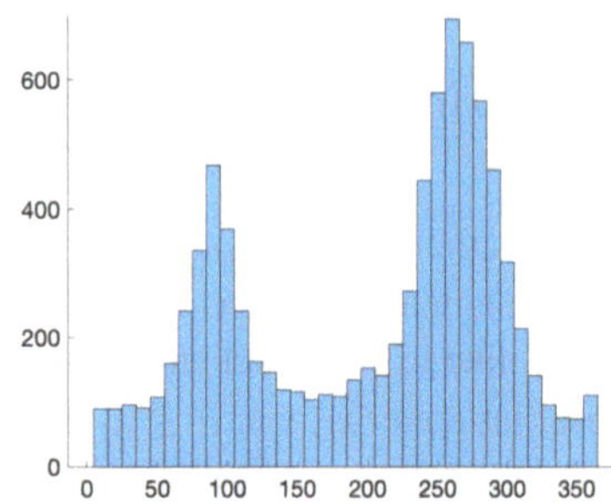

(b) Histogram for the year 2017

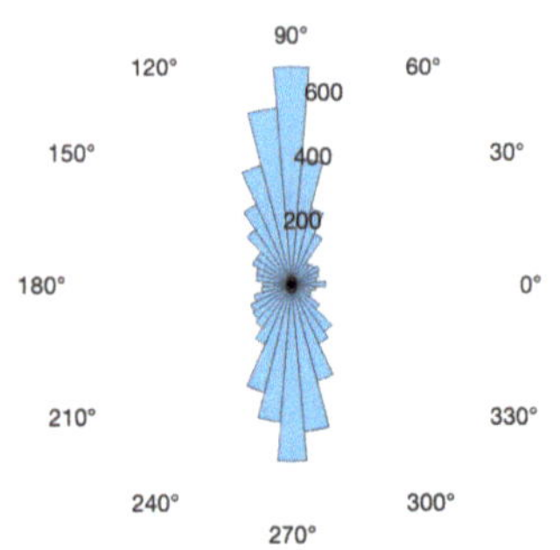

(c) Wind rose for the year 2014

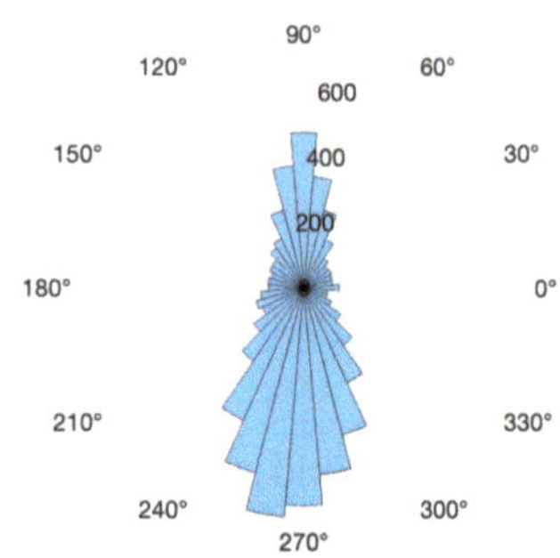

(d) Wind rose for the year 2017

Fig. 41.1: Histogram of absolute frequencies - in cartesian and polar coordinates

To numerically and graphically illustrate the process of finding a periodic spline approximation of wind direction distribution, we utilize the 2014 dataset. The initial step involves converting the wind direction values from degrees to radians. Subsequently, the histogram is transformed into relative frequencies to approximate the probability density function of wind direction, as depicted in Fig. 41.2(a). The next step requires converting the (x,y)-points from the $\mathcal{B}^2(I)$ space to the $L_0^2(I)$ space using the centered log-ratio transformation, as shown in Fig. 41.2(a) by stars.

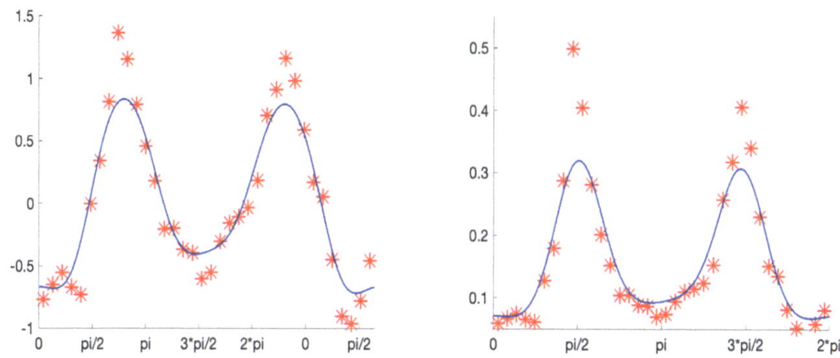

(a) Clr data of the year 2014 (b) Original data of the year 2014

Fig. 41.2: Periodic smoothing spline for clr (left) and original (right) data

For the periodic smoothing splines in Fig. 41.2 we set degree $k = 3$, the derivative in minimized functional (41.6) was $l = 1$, and the parameter $\alpha = 0.8$. We used 9 inner equidistant knots $\Delta\lambda := \{\lambda_i\}_{i=0}^{10}$, where $\lambda_i = \pi i/5$ for $i = 0, 1, \ldots, 10$. Following this, we quantify the vector $\overline{\mathbf{b}}^*$ of smoothing spline parameters according to the methodology specified above in (41.7). The final step involves an inverse transformation of the clr from the $L_0^2(I)$ space to the $\mathcal{B}^2(I)$ space, which yields an approximation of the density of the 2014 directional data, see Fig. 41.2(b).

41.4 Conclusion

In this study, we have introduced a novel approach for the analysis of wind direction data using a compositional periodical spline representation within the Bayes spaces framework. This methodology addresses the inherent circular nature of wind direction data, which poses significant challenges for traditional linear models. By leveraging the properties of Bayes spaces, we have developed a robust statistical framework that allows for the efficient processing and analysis of directional data.

Our approach utilizes periodic smoothing splines to model the wind direction distribution, providing a smooth and continuous approximation that naturally handles boundary conditions. The transformation of directional density from the $\mathcal{B}^2(I)$ space to the $L_0^2(I)$ space using the centered log-ratio transformation, followed by the application of smoothing splines, ensures that the resulting models are both accurate and interpretable.

The theoretical framework presented in this study has been validated through the analysis of empirical datasets, demonstrating its applicability and effectiveness. The results highlight the importance of incorporating directionality in statistical models to improve the understanding and prediction of phenomena with inherent circular characteristics.

Future research could explore the extension of this methodology to other types of practical and actual data with periodic or directional properties. Furthermore, expansion to multivariate cases, such as bivariate joint distribution of directional wind speeds, could enrich the applicability of the Bayes spaces framework in statistical data analysis.

References

1. Benlakhdar, S., Rziza, M., Thami, R.O.H.: Statistical modeling of directional data using a robust hierarchical von mises distribution model: perspectives for wind energy. Computational Statistics, **37**(4), 1599–1619 (2022)
2. van den Boogaart, K. G., Tolosana-Delgado, R.: Compositional data analysis: Theory and applications. John Wiley & Sons, (2010)
3. van den Boogaart, K. G., Egozcue, J. J., Pawlowsky-Glahn, V.: Bayes spaces: Use of Hilbert geometry for the statistical analysis of positive data. Biometrika, **101**(2), 393-412 (2014)
4. de Boor, C.: A Practical Guide to Splines. Springer, New York, (1978)
5. Carta, J. A., Díaz, S., Castañeda, A.: A global sensitivity analysis method applied to wind farm power output estimation models. Applied Energy, **280**, (2020)
6. Dierckx, P.: Curve and Surface Fitting with Splines, Clarendon Press, (1993)
7. Egozcue, J. J., Díaz-Barrero, J. L., Pawlowsky-Glahn, V.: Hilbert space of probability density functions based on Aitchison geometry. Acta Mathematica Sinica, English Series, **22**(4), 1175-1182 (2006)
8. Lusa L., Ahlin Č.: Restricted cubic splines for modelling periodic data. PLoS One. **15**(10), (2020)
9. Machalová, J., Hron, K., Monti, G.S.: Preprocessing of centred logratio transformed density functions using smoothing splines. Journal of Applied Statistics, **43**(8), 1419-1435 (2016)
10. Machalová, J., Talská, R., Hron, K., Gába, A.: Compositional splines for representation of density functions. Computational Statistics, **36**(2), 1031-1064 (2021)
11. Ramsay, J.: Functional data analysis. Encyclopedia of Statistics in Behavioral Science, Wiley Online Library, (2005)
12. Wang, Z., Zhang, W., Zhang, Y., Liu, Z.: Circular-linear-linear probabilistic model based on vine copulas: An application to the joint distribution of wind direction, wind speed, and air temperature. Journal of Wind Engineering and Industrial Aerodynamics, **215**, (2021)
13. Zerbe, K., Polit, Ch., McClain, S., Cook, T.: Optimized Hot Spot and Directional Distribution Analyses Characterize the Spatiotemporal Variation of Large Wildfires in Washington, USA, 1970-2020. International Journal of Disaster Risk Science, **13**(1), 139-150 (2022)

Chapter 42
Bayes Hilbert Space Additive Density-on-Scalar Regression Based on Individual Observations

Eva-Maria Maier, Alexander Fottner, Almond Stöcker and Sonja Greven

Abstract We propose additive regression models for response densities, which may be observed only via a sample from the conditional density given scalar covariates - or even a single observation, nicely linking our functional regression approach to distributional regression models for scalar data. We formulate our models in Bayes Hilbert spaces to incorporate nonnegativity and integration to one constraints for densities. Considering densities and Bayes Hilbert spaces with respect to an arbitrary finite measure allows us to unify models for continuous, discrete (compositional data) and mixed densities. We propose a (penalized) maximum likelihood estimation approach and derive asymptotic results for the estimators, also giving us (asymptotic) inference. We show a computationally useful approximate equivalence of the maximization problem to (penalized) estimation for certain multinomial - or equivalently, Poisson - additive regression models, greatly simplifying our implementation in an R package. We apply our new approach to a motivating question and data set from gender economics, where we model the mixed density of the woman's share in a couple's total labor income - which is continuous on $(0, 1)$ with positive probability mass on 0 and 1 for single-earner couples - in Germany, given age of the youngest child, East vs. West German residence, and year.

Almond Stöcker and Sonja Greven share senior authorship.

Eva-Maria Maier
Humboldt-Universität zu Berlin, Unter den Linden 6, Berlin, Germany,
e-mail: eva-maria.maier@hu-berlin.de

Alexander Fottner
Augsburg University, Augsburg, Germany, e-mail: alexander.fottner@uni-a.de

Almond Stöcker
EPFL Lausanne, Lausanne, Switzerland, e-mail: almond.stoecker@epfl.ch

Sonja Greven (✉)
Humboldt-Universität zu Berlin, Unter den Linden 6, Berlin, Germany,
e-mail: sonja.greven@hu-berlin.de

The original version of this chapter was previously published non-open access. A correction to this chapter is available at https://doi.org/10.1007/978-3-031-92383-8_67

© The Author(s) 2025, corrected publication 2025

G. Aneiros et al. (eds.), *New Trends in Functional Statistics and Related Fields*,
Contributions to Statistics, https://doi.org/10.1007/978-3-031-92383-8_42

42.1 Introduction

Recent developments have led to similar flexibility in modeling functional responses than for modeling scalar data, such as functional additive mixed models [9, 6]. If the response is a density, functional regression models cannot be directly applied - not even for continuous densities - due to the non-negativity and integration to one properties of densities. In [7] we developed additive models for density responses that respect the geometry of densities using so-called Bayes Hilbert spaces. In practice, however, densities are often not directly observed and only samples from densities for each covariate combination are available. We thus here develop an estimation approach that avoids two-step procedures, which would estimate densities in a first step before modeling them in a second step, and instead directly works with the observed data. In the extreme, only a single response observation per covariate combination may be available, nicely linking our density regression model to distributional regression models that model the conditional distribution of scalar observations. In addition, the new likelihood-based estimation approach and our new asymptotic results provide us with statistical inference for the model, in contrast to the approach in [7].

In the following, we first summarize the Bayes Hilbert space framework before introducing our additive regression model for conditional densities. We propose estimation via (penalized) maximum likelihood estimation and derive asymptotic results as well as a computationally useful approximation using Poisson additive models, for which well-developed software exists and can be employed for our implementation in an R package. We close with an application to a gender economic question.

42.2 Bayes Hilbert Spaces

The Bayes Hilbert Space $B^2(\mu)$ [4, 13] generalizes the Aitchison geometry [1] for compositional data, which can be viewed as the discrete special case. Let $(\mathcal{T}, \mathcal{A})$ be a measurable space with finite (reference) measure μ. For example, in our application, $\mathcal{T} = [0, 1]$, $\mathcal{A}$ its Borel σ-algebra, and $\mu = \delta_0 + \lambda + \delta_1$ with λ the Lebesgue measure on $[0, 1]$, and δ_t the Dirac measure at $t \in \{0, 1\}$. Let $\mathcal{M}(\mathcal{T}, \mathcal{A}, \mu)$ be the set of σ-finite measures with the same null sets as μ and identify its elements with the corresponding (μ-a.e.) positive and unique densities with respect to (w.r.t.) μ (Radon-Nikodym derivatives). The set $\mathcal{B}(\mathcal{T}, \mathcal{A}, \mu)$ of equivalence classes w.r.t. proportionality $\propto$ then is called the *Bayes space (with reference measure μ)*. Where possible, we choose the probability density f as a representative of each such equivalence class $[f]$. Functions are considered equivalent if they differ only on a null-set. We first obtain a vector space structure as follows.

Proposition 42.1. *(adapted from [4, 13])*

$\mathcal{B}(\mu) := \mathcal{B}(\mathcal{T}, \mathcal{A}, \mu)$ is a real vector space with addition $\oplus$ and scalar multiplication $\odot$ defined as follows: For $[f], [g] \in \mathcal{B}(\mu)$ and $a \in \mathbb{R}$ define

- *Addition (perturbation) as $[f] \oplus [g] = [f \cdot g]$.*
- *Scalar multiplication (powering) as $a \odot [f] = [f^a]$.*

Now define $B^p(\mu) := \{[f] \in \mathcal{B}(\mu) \mid \int_{\mathcal{T}} |\log f|^p \, \mathrm{d}\mu < \infty\}$ and $L_0^p(\mu) := \{\tilde{f} \in L^p(\mu) | \int_{\mathcal{T}} \tilde{f} \, \mathrm{d}\mu = 0\} \subset L^p(\mu)$. The *centered-log-ratio (clr) transformation* $\mathrm{clr} : B^p(\mu) \to L_0^p(\mu),\ f \mapsto \mathrm{clr}[f]$ for $1 \le p < \infty$ is defined as

$$\mathrm{clr}[f] := \tilde{f} := \log(f) - \frac{1}{\mu(\mathcal{T})} \int_{\mathcal{T}} \log(f) \, \mathrm{d}\mu.$$

Proposition 42.2. *(for $p = 1$ in [13])*
The clr transformation $\mathrm{clr} : B^p(\mu) \to L_0^p(\mu)$ is an isomorphism with inverse $\mathrm{clr}^{-1}(\tilde{f}) = [\exp \tilde{f}]$.

We can finally obtain the following Hilbert space structure for (positive) densities.

Proposition 42.3. *(adapted from [13])*
*$B^2(\mu)$ is a **Hilbert space** with inner product $\langle \cdot, \cdot \rangle_{B^2(\mu)}$,*

$$\begin{aligned}\langle [f], [g] \rangle_{B^2(\mu)} &= \frac{1}{2\mu(\mathcal{T})} \int \int \log \frac{f(s)}{f(t)} \log \frac{g(s)}{g(t)} \, \mathrm{d}\mu(s) \, \mathrm{d}\mu(t) \\ &= \langle \tilde{f}, \tilde{g} \rangle_{L^2(\mu)} = \int_{\mathcal{T}} \tilde{f} \cdot \tilde{g} \, \mathrm{d}\mu, \quad [f], [g] \in B^2(\mu),\end{aligned}$$

the so-called Bayes Hilbert space.

Proposition 42.3 also shows that the clr-transformation is an isometry and we can thus equivalently work with distances and norms in the clr-transformed space.

42.3 Bayes Hilbert Space Additive Regression

Consider pairs of independent observations $(y_i, \mathbf{x}_i) \in \mathcal{Y} \times \mathcal{X}$, $\mathcal{Y} = \mathcal{T}$, $\mathcal{X} = \mathbb{R}^P$, for $i = 1, \dots, N$, $N \in \mathbb{N}$, where the y_i are realizations of the conditional distribution of $Y_i \mid \mathbf{x}_i$, whose density $f_{\mathbf{x}_i} \in B^2(\mu)$ is not observed directly. It is modeled via a structured additive density-on-scalar regression model

$$f_{\mathbf{x}_i} = \bigoplus_{j=1}^{J} h_j(\mathbf{x}_i), \tag{42.1}$$

where $J \in \mathbb{N}$ and $h_j(\mathbf{x}_i) \in B^2(\mu)$ are partial effects depending on no, one, or more covariates in $\mathbf{x} = (x_1, \dots, x_P)$. E.g., $h_j(\mathbf{x})$ can be an intercept β_0, a linear effect $x_k \odot \beta$, a smooth effect $g_1(x_k)$, a group-specific intercept β_{x_m} or smooth effect $g_{x_m}(x_k)$, or a smooth interaction $g_2(x_k, x_l)$, all in $B^2(\mu)$, where x_k, x_l denote continuous covariates and x_m a grouping variable, for $k, l, m \in \{1, \ldots, P\}$ pairwise distinct. For more possible partial effects, see also Table B.1 in Appendix B of [7]. As identifiability constraints, we center main effects around the intercept and interactions around main effects.

Each partial effect is represented using a tensor product basis as

$$h_j(\mathbf{x}_i) = \left(\mathbf{b}_{\mathcal{X},j}(\mathbf{x}_i) \circledast \mathbf{b}_{\mathcal{Y}}\right)^\top \boldsymbol{\theta}_j := \bigoplus_{n=1}^{K_{\mathcal{X},j}} \bigoplus_{m=1}^{K_{\mathcal{Y}}} \theta_{j,n,m} \odot b_{\mathcal{X},j,n}(\mathbf{x}_i) \odot b_{\mathcal{Y},m},$$

where $\mathbf{b}_{\mathcal{X},j} = (b_{\mathcal{X},j,1}, \ldots, b_{\mathcal{X},j,K_{\mathcal{X},j}})^\top : \mathcal{X} \to \mathbb{R}^{K_{\mathcal{X},j}}$ and $\mathbf{b}_{\mathcal{Y}} = (b_{\mathcal{Y},1}, \ldots, b_{\mathcal{Y},K_{\mathcal{Y}}})^\top \in B^2(\mu)^{K_{\mathcal{Y}}}$ are vectors of basis functions over the covariates (chosen depending on the type of effect e.g. as linear, splines etc.) and over $\mathcal{Y}$, respectively, and $\boldsymbol{\theta}_j = (\theta_{j,1,1}, \ldots, \theta_{j,K_{\mathcal{X},j},K_{\mathcal{Y}}})^\top \in \mathbb{R}^{K_{\mathcal{X},j} K_{\mathcal{Y}}}$ is the unknown coefficient vector. To obtain a basis $\mathbf{b}_{\mathcal{Y}} \in B^2(\mu)^{K_{\mathcal{Y}}}$, we transform an appropriate basis in $L^2(\mu)^{K_{\mathcal{Y}}+1}$ (e.g. B-splines, or indicator functions in the discrete case) to $L^2_0(\mu)^{K_{\mathcal{Y}}}$ as in appendix D of [7] and inverse clr-transform component-wise. To obtain a semi-parametric approach, we often choose relatively large $K_{\mathcal{Y}}, K_{\mathcal{X},j}$, increasing with N, and may then employ regularization as discussed below.

To emphasize dependence on $\boldsymbol{\theta}$, we write $f_{\mathbf{x}_i,\boldsymbol{\theta}}$ in the following. Applying the clr transformation to (42.1) and using its linearity yields

$$\begin{aligned}\tilde{f}_{\mathbf{x}_i,\boldsymbol{\theta}} := \mathrm{clr}[f_{\mathbf{x}_i,\boldsymbol{\theta}}] &= \sum_{j=1}^{J} \sum_{n=1}^{K_{\mathcal{X},j}} \sum_{m=1}^{K_{\mathcal{Y}}} \theta_{j,n,m}\, b_{\mathcal{X},j,n}(\mathbf{x}_i)\, \tilde{b}_{\mathcal{Y},m} = \sum_{j=1}^{J} \left(\mathbf{b}_{\mathcal{X},j}(\mathbf{x}_i) \otimes \tilde{\mathbf{b}}_{\mathcal{Y}}\right)^\top \boldsymbol{\theta}_j \\ &= \left(\mathbf{b}_{\mathcal{X}}(\mathbf{x}_i) \otimes \tilde{\mathbf{b}}_{\mathcal{Y}}\right)^\top \boldsymbol{\theta} = \tilde{\mathbf{b}}(\mathbf{x}_i)^\top \boldsymbol{\theta},\end{aligned} \tag{42.2}$$

where $\tilde{\mathbf{b}}_{\mathcal{Y}} := (\tilde{b}_{\mathcal{Y},1}, \ldots, \tilde{b}_{\mathcal{Y},K_{\mathcal{Y}}})^\top$ with $\tilde{b}_{\mathcal{Y},m} := \mathrm{clr}[b_{\mathcal{Y},m}]$ assumed bounded, $\mathbf{b}_{\mathcal{X}} = (\mathbf{b}_{\mathcal{X},1}^\top, \ldots, \mathbf{b}_{\mathcal{X},J}^\top)^\top : \mathcal{X} \to \mathbb{R}^{K_{\mathcal{X}}}$ for $K_{\mathcal{X}} := \sum_{j=1}^{J} K_{\mathcal{X},j}$, $\boldsymbol{\theta} = (\boldsymbol{\theta}_1^\top, \ldots, \boldsymbol{\theta}_J^\top)^\top \in \mathbb{R}^K$ for $K := K_{\mathcal{X}} K_{\mathcal{Y}}$, and $\tilde{\mathbf{b}}(\mathbf{x}) := \mathbf{b}_{\mathcal{X}}(\mathbf{x}) \otimes \tilde{\mathbf{b}}_{\mathcal{Y}} \in L^2_0(\mu)^K$ for $\mathbf{x} \in \mathcal{X}$.

42.4 Estimation and Inference

42.4.1 Penalized Maximum Likelihood Estimation and Asymptotic Results

The density model directly yields the likelihood

$$L(\boldsymbol{\theta}) := \prod_{i=1}^{N} f_{\mathbf{x}_i,\boldsymbol{\theta}}(y_i) = \prod_{i=1}^{N} \frac{\exp(\tilde{f}_{\mathbf{x}_i,\boldsymbol{\theta}}(y_i))}{\int \exp(\tilde{f}_{\mathbf{x}_i,\boldsymbol{\theta}})\, \mathrm{d}\mu},$$

which gives us the log-likelihood

$$\ell(\boldsymbol{\theta}) = \sum_{i=1}^{N} \left(\tilde{\mathbf{b}}(\mathbf{x}_i)(y_i)^\top \boldsymbol{\theta} - \log \int_{\mathcal{Y}} \exp\left[\tilde{\mathbf{b}}(\mathbf{x}_i)^\top \boldsymbol{\theta}\right] \mathrm{d}\mu \right). \tag{42.3}$$

Regularization is obtained if desired by adding a penalty term $\mathrm{pen}(\boldsymbol{\theta}) := -\sum_{j=1}^{J} \boldsymbol{\theta}_j^\top \mathbf{P}_j \boldsymbol{\theta}_j = -\boldsymbol{\theta}^\top \mathbf{P} \boldsymbol{\theta}$, resulting in the penalized log-likelihood $\ell_{\mathrm{pen}}(\boldsymbol{\theta}) := \ell(\boldsymbol{\theta}) + \mathrm{pen}(\boldsymbol{\theta})$. We choose anisotropic penalty matrices $\mathbf{P}_j = \xi_{\mathcal{X},j}(\mathbf{P}_{\mathcal{X},j} \otimes \mathbf{I}_{K_\mathcal{Y}}) + \xi_{\mathcal{Y},j}(\mathbf{I}_{K_{\mathcal{X},j}} \otimes \mathbf{P}_\mathcal{Y})$ with suitable penalty matrices $\mathbf{P}_{\mathcal{X},j} \in \mathbb{R}^{K_{\mathcal{X},j} \times K_{\mathcal{X},j}}$ for $\mathbf{b}_{\mathcal{X},j}$ and $\mathbf{P}_\mathcal{Y} \in \mathbb{R}^{K_\mathcal{Y} \times K_\mathcal{Y}}$ for $\mathbf{b}_\mathcal{Y}$ (e.g. difference or derivative based for splines), and smoothing parameters $\xi_{\mathcal{X},j}, \xi_{\mathcal{Y},j} \geq 0$. $\mathbf{P} = \mathrm{diag}(\mathbf{P}_1, \ldots, \mathbf{P}_J) \in \mathbb{R}^{K \times K}$ is block diagonal.

We have the following result on consistency and asymptotic normality of the estimator.

Theorem 42.1. *Assume that there is a true value* $\boldsymbol{\vartheta} \in \mathbb{R}^K$ *of* $\boldsymbol{\theta}$*, as well as further regularity assumptions on the basis functions and the covariates. Then we have*

1) The probability that a (penalized) MLE $\hat{\boldsymbol{\theta}}$ *maximizing* (42.3) *exists in some arbitrary small neighborhood around* $\boldsymbol{\vartheta}$ *converges to one for* $N \to \infty$, $\hat{\boldsymbol{\theta}}$ *then is unique and a consistent estimator of* $\boldsymbol{\vartheta}$.

2) $\hat{\boldsymbol{\theta}}$ *is asymptotically normally distributed,* $\hat{\boldsymbol{\theta}} \overset{a}{\sim} \mathcal{N}(\boldsymbol{\vartheta}, \mathbf{F}(\boldsymbol{\vartheta})^{-1})$, *or* $\hat{\boldsymbol{\theta}} \overset{a}{\sim} \mathcal{N}(\boldsymbol{\vartheta}, \mathbf{F}(\hat{\boldsymbol{\theta}})^{-1})$, *with (penalized) Fisher information* $\mathbf{F}(\boldsymbol{\theta}) = \mathbb{E}(\mathbf{F}(\boldsymbol{\theta}))$.

2) can be used to e.g. construct confidence regions. The proof generalizes [3], Sect. 33.3, for scalar θ in the unpenalized i.i.d. case to the vector $\boldsymbol{\theta}$ in the penalized and not identically distributed case.

42.4.2 Estimation via Poisson Additive Models

While (42.3) could be optimized directly, the integrals complicate maximization. Instead, we make use of an approximate equivalence to the (penalized) likelihood of a certain Poisson additive model, as detailed in the following.

We first derive equivalence to an approximate multinomial additive model. Let $\mathbf{x}^{(1)}, \ldots, \mathbf{x}^{(S)}$ be the unique combinations of observed covariate values and let $\mathcal{I}^{(s)} := \{i \in \{1, \ldots, N\} \mid \mathbf{x}_i = \mathbf{x}^{(s)}\}$ for $s = 1, \ldots, S \leq N$ denote the corresponding partition of indices. Thus, $\{y_i \mid i \in \mathcal{I}^{(s)}\}$ forms an i.i.d. random sample of $Y^{(s)} \mid \mathbf{x}^{(s)}$ (of size one for continuous covariates). Consider most generally the mixed case with continuous density on I and potentially D values on discrete points $\mathcal{D}$. For each covariate combination s, we construct a histogram on $I \setminus \mathcal{D}$ and counts on $\mathcal{D}$ to approximate the empirical conditional density of $Y^{(s)} \mid \mathbf{x}^{(s)}$. Let $\Gamma^{(s)} := G^{(s)} + D$ be the number $G^{(s)}$ of histogram bins for any continuous part of the density plus the number D of potential discrete points with positive probability mass. Denote by $n_g^{(s)}$ the count in each histogram bin, by $u_g^{(s)}$ the discrete points or the bin mid-points, and by $\Delta_g^{(s)}$ the bin width or weight μ puts on the discrete points.

We view the vectors $(n_1^{(s)}, \dots, n_{\Gamma(s)}^{(s)})$ as realizations of S independent multinomially distributed random variables with probabilities $p_1^{(s)}, \dots, p_{\Gamma(s)}^{(s)}$, conditional on $n^{(s)} := \sum_{g=1}^{\Gamma(s)} n_g^{(s)} = |\mathcal{I}^{(s)}|$. We model them as $p_g^{(s)}(\boldsymbol{\theta}) = \frac{\Delta_g^{(s)} \exp[\tilde{\mathbf{b}}(\mathbf{x}^{(s)})(u_g^{(s)})^\top \boldsymbol{\theta}]}{\sum_{g'=1}^{\Gamma(s)} \Delta_{g'}^{(s)} \exp[\tilde{\mathbf{b}}(\mathbf{x}^{(s)})(u_{g'}^{(s)})^\top \boldsymbol{\theta}]}$. Up to additive constants, the multinomial log-likelihood is

$$\ell^{\mathrm{mn}}(\boldsymbol{\theta}) := \sum_{s=1}^{S} \sum_{g=1}^{\Gamma(s)} n_g^{(s)} \Big(\tilde{\mathbf{b}}(\mathbf{x}^{(s)})(u_g^{(s)})^\top \boldsymbol{\theta} - \log \sum_{g'=1}^{\Gamma(s)} \Delta_{g'}^{(s)} \exp\big[\tilde{\mathbf{b}}(\mathbf{x}^{(s)})(u_{g'}^{(s)})^\top \boldsymbol{\theta}\big] \Big), \tag{42.4}$$

potentially adding the same penalty as before, $\ell_{\mathrm{pen}}^{\mathrm{mn}}(\boldsymbol{\theta}) := \ell^{\mathrm{mn}}(\boldsymbol{\theta}) + \mathrm{pen}(\boldsymbol{\theta})$.

Theorem 42.2. *Set* $\Delta := \max\{\Delta_g^{(s)} \mid s = 1, \dots, S,\ g = 1, \dots, G^{(s)}\}$. *Let* $\hat{\boldsymbol{\theta}}^{\mathrm{mn}}$, $\hat{\boldsymbol{\theta}}$ *be the (penalized) maximum likelihood estimators ((P)MLEs) and* $\mathbf{F}^{\mathrm{mn}}(\boldsymbol{\theta})$, $\mathbf{F}(\boldsymbol{\theta})$ *be the Fisher informations based on* $\ell^{\mathrm{mn}}(\boldsymbol{\theta})$ *respectively* $\ell(\boldsymbol{\theta})$. *Then,*

1. $\lim_{\Delta \to 0} \ell^{\mathrm{mn}}(\boldsymbol{\theta}) = \ell(\boldsymbol{\theta})$.
2. $\lim_{\Delta \to 0} \mathbf{F}^{\mathrm{mn}}(\boldsymbol{\theta}) = \mathbf{F}(\boldsymbol{\theta}) = \lim_{\Delta \to 0} \mathbb{E}^{\mathrm{mn}}(\mathbf{F}^{\mathrm{mn}}(\boldsymbol{\theta})) = \mathbb{E}(\mathbf{F}(\boldsymbol{\theta}))$ *and* $\lim_{\Delta \to 0}(\mathbf{F}^{\mathrm{mn}}(\boldsymbol{\theta}))^{-1} = (\mathbf{F}(\boldsymbol{\theta}))^{-1}$ *if both are invertible.*
3. *Under mild additional assumptions:* $\lim_{\Delta \to 0} \hat{\boldsymbol{\theta}}^{\mathrm{mn}} = \hat{\boldsymbol{\theta}}$.

For discrete densities, we have equality in these statements (without Δ*-limit).*

Instead of a multinomial model, we can make use of a more convenient Poisson model by not conditioning on $n^{(s)}$. View $n_g^{(s)}$ as independent realizations of $Po(\lambda_g^{(s)}(\boldsymbol{\theta}, \tau^{(s)}))$, $s = 1, \dots, S$, $g = 1, \dots, \Gamma^{(s)}$. Model the rates as $\lambda_g^{(s)}(\boldsymbol{\theta}, \tau^{(s)}) = \exp[\log \Delta_g^{(s)} + \tilde{f}_{\mathbf{x}_i, \boldsymbol{\theta}}(u_g^{(s)}) + \tau^{(s)}]$, where the $\tau^{(s)}$ are additional nuisance parameters, more precisely, intercepts capturing the overall number of observations $n^{(s)}$, whereas $\boldsymbol{\theta}$ models their relative distribution across the domain $\mathcal{Y}$. The corresponding (penalized) Poisson log-likelihood then yields (penalized) maximimum likelihood estimators $(\hat{\boldsymbol{\theta}}^{\mathrm{po}}, \hat{\tau}^{(1)}, \dots, \hat{\tau}^{(S)})$.

Theorem 42.3. *1.* $\hat{\boldsymbol{\theta}}^{\mathrm{po}}$ *does not depend on* $(\hat{\tau}^{(1)}, \dots, \hat{\tau}^{(S)})$.
2. $\hat{\boldsymbol{\theta}}^{\mathrm{po}} = \hat{\boldsymbol{\theta}}^{\mathrm{mn}}$ *and the inverse Fisher informations corresponding to* $\boldsymbol{\theta}$ *are identical in both models.*

The proof extends results of [8], who only considers linear predictors without penalization. Putting these results together, we can thus base estimation and inference for the Bayes space additive model (42.1) - instead of on the Bayes space log-likelihood (42.3) - on a(n approximate) Poisson additive model for the counts per histogram bin as detailed above. The quality of the approximation can be controlled by the bin-widths. Note that logspline density estimation, e.g. [11, 10, 5], can be seen as an intercept-only special case of this approach without covariates.

We provide an implementation of our approach in an `R` package, using the `R` package `mgcv` [12] internally for Poisson additive modeling.

Our structured additive regression model has the advantage of (ceteris paribus) interpretability via partial effects of covariates. [7] propose two approaches for effect interpretation in Bayes Hilbert spaces. First, for coefficient estimates on clr-level, vertical differences between values at two different points in the domain can be directly interpreted akin to log-odds-ratios. Second, for any level a, we can interpret the coefficient as corresponding to an increasing [lowering] effect on the probability mass of the density integrated over areas of the domain above [below] a.

42.5 Application to Gender Economics

We analyze the distribution of the woman's share in a couple's total labor income in Germany, based on a data set derived from the German Socio-Economic Panel (SOEP) study. [7] motivated by [2] analyze this data set using a density-on-scalar regression approach requiring preliminary estimation of the response densities from individual observations. Our new method avoids this two-step approach and additionally provides inference for effect functions.

The data contains $N = 154,924$ opposite-sex couples living in one household, with at least one partner reporting positive labor income. We aim to model conditional densities of the woman's share s in a couple's total gross labor income, given three covariates. The binary covariate *West_East* indicates whether the couple lives in *West* or *East* Germany (including Berlin). The categorical covariate *c_age* indicates a youngest child of age *0-6*, *7-18*, or no minor children in the household. The continuous covariate *year* ranges from 1984 (*West*) / 1991 (*East*) to 2016.

We consider the conditional densities $f_{\textit{West\_East, c\_age, year}} : [0,1] \to \mathbb{R}^+$ of the woman's income share s given the three covariates as elements of a mixed Bayes Hilbert space $B^2(\mu) = B^2(\mathcal{Y}, \mathcal{A}, \mu)$ with reference measure $\mu := \delta_0 + \lambda + \delta_1$ (see Sect. 42.2), since boundary values 0 and 1 correspond to single-earner households and have positive probability mass. We estimate the model

$$\begin{aligned} f_{\textit{West\_East, c\_age, year}} = \beta_0 &\oplus \beta_{\textit{West\_East}} \oplus \beta_{c\_age} \oplus \beta_{\textit{c\_age, West\_East}} \\ &\oplus g(year) \oplus g_{\textit{West\_East}}(year) \oplus g_{c\_age}(year) \\ &\oplus g_{\textit{c\_age, West\_East}}(year) \qquad \in B^2(\mu). \end{aligned} \tag{42.5}$$

The model contains main effects of *East* vs. *West* Germany (binary), of the two child groups with children vs. those without (categorical), of each year vs. the reference year 1991 (smooth), as well as several interaction effects.

As an example of our results, we exemplarily discuss the estimated geopolitical effect $\beta_{\textit{West\_East}}$ shown in Fig. 42.1. On the left, we see (in 1991 after re-unification of Germany, for couples without children), that the income share distribution is much closer to symmetric in the former communist *East*, while in the *West*, women more often do not work (horiontal dash at 0) and the income share distribution is shifted to the left, with high probability mass on (very) small income shares. The right plot provides ceteris paribus interpretations of the effect of *East* vs. *West*. As

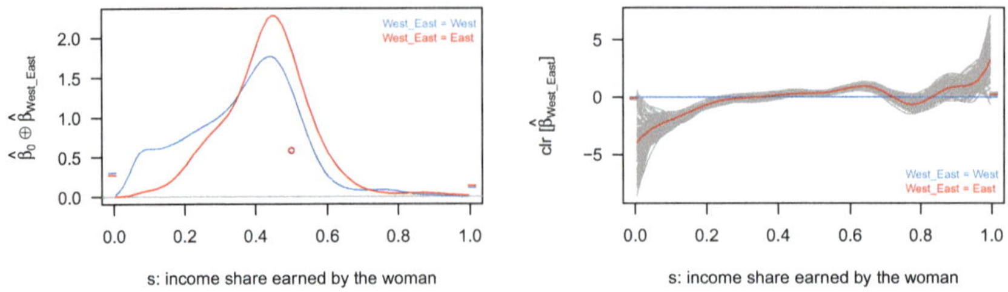

Fig. 42.1: Estimated conditional densities for couples without children in 1991 for *West* vs. *East* Germany [left] and clr transformed estimated effect of *East* relative to the reference *West* Germany ($\text{clr}[\hat{\beta}_{West}] \equiv 0$) with simultaneous confidence bands [right]. Probability masses at 0 and 1 are indicated by horizontal dashes, integrated probability mass on (0, 1) by open circles

$\text{clr}[\hat{\beta}_{East}](s) - \text{clr}[\hat{\beta}_{East}](t) > 0$ for most $s > t$ (and $\text{clr}[\hat{\beta}_{West}] \equiv 0$), the odds for a 'large' income share s vs. a 'small' share t of the woman is larger in *East* than in *West* Germany (positive log-odds) for most values if all other covariates stay the same. Using the second interpretation possibility, we see that the probability mass of densities decreases on $\{\text{clr}(\hat{\beta}_{East}) < a = -0.65\} \approx (0, 0.2)$ in *East* compared to *West* Germany, if all other covariates stay the same. Overall, similarly to [7], in our results we find a more symmetric distribution in *East* than in *West* German couples after reunification and a smaller so-called child penalty comparing couples with and without children. While these East-West differences become smaller, they are persistent over time. With our new approach and inference, we can now also conclude that all main effects in the model are statistically significant.

References

1. Aitchison, J.: The Statistical Analysis of Compositional Data. Chapman & Hall, London - New York (1986)
2. Bertrand, M., Kamenica, E., Pan, J.: Gender Identity and Relative Income within Households. Q. J. Econ. **130**, 571-614 (2015)
3. Cramér, H.: Mathematical Methods of Statistics. Princeton University Press, Princeton (1946)
4. Egozcue, J.J., Díaz-Barrero, J.L., Pawlowsky-Glahn, V.: Hilbert Space of Probability Density Functions Based on Aitchison Geometry. Acta Math. Sin. **22**, 1175–1182 (2006)
5. Eilers, P. H., Marx, B. D.: Flexible smoothing with B-splines and penalties. Stat. Sci. **11**, 89–121 (1996)
6. Greven, S., Scheipl F.: A general framework for functional regression modelling. Stat. Modell. **17**, 1-35 (2017)
7. Maier, E.-M., Stöcker, A., Fitzenberger, B., Greven, S.: Additive Density-on-Scalar Regression in Bayes Hilbert Spaces with an Application to Gender Economics. Ann. Appl. Stat. to appear (2025+)
8. Palmgren, J.: The Fisher information matrix for log linear models arguing conditionally on observed explanatory variable. Biometrika. **68**, 563–566 (1981)

9. Scheipl, F., Staicu, A.-M., Greven, S.: Functional additive mixed models. J. Comput. Graphical Stat. **24**, 477-501 (2015)
10. Stone, C.J.: Asymptotics for doubly flexible logspline response models. Ann. Stat. **19**, 1832–1854 (1991)
11. Stone, C.J., Koo, C.-Y.: Logspline density estimation. Contemp. Math. **59**, 15 (1986)
12. Wood, S. N.: Generalized Additive Models: An Introduction with R. 2nd ed. Chapman and Hall/CRC, New York (2017)
13. van den Boogaart, K.G., Egozcue, J.J., Pawlowsky-Glahn, V.: Bayes Hilbert Spaces. Aust. N.Z. J. Stat. **56**, 171-194 (2014)

Open Access This chapter is licensed under the terms of the Creative Commons Attribution 4.0 International License (https://creativecommons.org/licenses/by/4.0/), which permits use, sharing, adaptation, distribution and reproduction in any medium or format, as long as you give appropriate credit to the original author(s) and the source, provide a link to the Creative Commons license and indicate if changes were made.

The images or other third party material in this chapter are included in the chapter's Creative Commons license, unless indicated otherwise in a credit line to the material. If material is not included in the chapter's Creative Commons license and your intended use is not permitted by statutory regulation or exceeds the permitted use, you will need to obtain permission directly from the copyright holder.

Chapter 43
A Kernel-Based Approach for Testing Mutual Independence of Several Functional Variables

Terence Kevin Manfoumbi Djonguet and Guy Martial Nkiet

Abstract We address the problem of testing for mutual independence of several functional random variables. We first introduce a pairwise independence measure for more than two random elements that aggregates the independence measures between pairs of random elements. From this, we derive an asymptotically normal test statistic, enabling a straightforward testing procedure that avoids the need for permutation or bootstrap resampling techniques. Finally, the proposed testing procedure is extended to handle mutual independence through both asymmetric and symmetric aggregation methods.

43.1 Introduction

For $d \geqslant 2$, we consider random variables $X^{(1)}, \ldots, X^{(d)}$ valued into separable Hilbert spaces $\mathcal{X}^{(1)}, \ldots, \mathcal{X}^{(d)}$, respectively, and we are interested in testing the hypothesis:

$$\mathscr{H}_0 : X^{(1)}, \ldots, X^{(d)} \text{ are mutually independent,}$$

against its negation $\mathscr{H}_1$. The particular case of independence between two random variables ($d = 2$) has been extensively studied in the litterature with several methods based on copulas, distribution or density functions and, more recently, methods relying on Reproducing Kernel Hilbert Space (RKHS) such as the Hilbert-Schmidt Independence Criterion (HSIC) and the Normalized Cross Covariance Operator (NOCCO); see [2, 3, 4]. By opposition, when more than two random variables are considered, the mutual independence testing problem has not been well studied in the literature. [8] and [5] tackled this problem when the $X^{(\ell)}$s are scalar variables, and [13] extended this to the case of multivariate random vectors. In a more general case,

Terence Kevin Manfoumbi Djonguet (✉), Guy Martial Nkiet
Laboratoire de Probabilités, Statistique et Informatique, Unité de Recherche en Mathématiques et Informatique, Université des Sciences et Techniques de Masuku BP 913 Franceville, Gabon,
e-mail: tkmpro95@gmail.com, guymartial.nkiet@univ-masuku.org

© The Author(s), under exclusive license to Springer Nature Switzerland AG 2025

G. Aneiros et al. (eds.), *New Trends in Functional Statistics and Related Fields*, Contributions to Statistics, https://doi.org/10.1007/978-3-031-92383-8_43

[11] introduced the d-variable Hilbert-Schmidt Independence Criterion (dHSIC), a RKHS-based measure of mutual dependence for random variables valued into metric spaces; it is a generalization of the HSIC of [4] to the case where more than two random variables are considered. [6] extended the so-called distance covariance of [12] in order to characterize mutual independence of random vectors via aggregating pairwise dependence. Their approach was latter extended to the context of functional data by [7]. All the methods described above perform well on finite samples but rely on Monte Carlo resampling techniques to conduct independence tests. However, as the sample size increases, the computational cost becomes excessive in practice. In order to overcome such a drawback, [10] proposed a modification of a naive empirical estimator of the dHSIC to obtain an asympotically normal mutual independence test statistic. This advancement enables a simpler testing procedure that rivals the standard dHSIC approach based on sample permutation methods.

In this work, we propose two alternative mutual independence tests based on symmetric and asymmetric aggregating procedures integrating the squared Hilbert-Schmidt norms of the NOCCO of pairs of involved functional variables. More specifically, by aggregating NOCCO two-sample dependence measures, we obtain complete mutual dependence measures that fully characterize $\mathscr{H}_0$. Moreover, following a process of studentization similar to the one used in [9], we show that, under $\mathscr{H}_0$, the obtained test statistics are asymptotically normal. We then propose two testing procedures for mutual independence for several functional variables that have the potential to outperform the standard permutation based testing procedure using dHSIC.

43.2 Preliminary Notions

For $\ell \in \{1, \cdots, d\}$, let us introduce $\mathcal{H}_\ell$ a separable RKHS of functions from $\mathcal{X}_\ell$ to $\mathbb{R}$ with associated kernel $K_\ell : \mathcal{X}_\ell^2 \to \mathbb{R}$. Each K_ℓ is a symmetric function such that, for any $(f, t) \in \mathcal{H}_\ell \times \mathcal{X}_\ell$, one has $K_\ell(t, \cdot) \in \mathcal{X}_\ell$ and $f(t) = \langle K_\ell(t, \cdot), f \rangle_\ell$, where $\langle \cdot, \cdot \rangle_\ell$ denotes the inner product of $\mathcal{X}_\ell$. Throughout this paper, we assume that

$$(\mathscr{A}_1) : \|K_\ell\|_\infty := \sup_{(x,y)\in\mathcal{X}_\ell^2} K_\ell(x, y) < +\infty,\ \ell = 1, \ldots, d.$$

This assumption ensures the existence of the mean elements $m_\ell = \mathbb{E}\big(K_\ell(X^{(\ell)}, \cdot)\big)$, and also that of the cross-covariance operators

$$\Sigma_{\ell j} = \mathbb{E}\big(K_j(X^{(j)}, \cdot) \otimes K_\ell(X^{(\ell)}, \cdot)\big) - m_j \otimes m_\ell,\quad (j, \ell) \in \{1, \ldots, d\}^2,$$

where $\otimes$ denotes the tensor product defined as follows: for all pair $(f_\ell, f_j) \in \mathcal{H}_\ell \times \mathcal{H}_j$, $f_j \otimes f_\ell$ is the linear operator from $\mathcal{H}_j$ to $\mathcal{H}_\ell$ such that $(f_j \otimes f_\ell)(x) = \langle f_j, x \rangle_j f_\ell$. It is known (see [1, 2]) that $\Sigma_{\ell j}$ is an Hilbert-Schmidt operator and that there exists a unique bounded operator $V_{\ell j} : \mathcal{H}_j \to \mathcal{H}_\ell$ such that $\Sigma_{\ell k} = \Sigma_\ell^{1/2} V_{\ell k} \Sigma_k^{1/2}$, where

$\Sigma_\ell := \Sigma_{\ell\ell}$. This operator is the NOCCO of $X^{(\ell)}$ and $X^{(j)}$, and we assume that it is an Hilbert-Schmidt operator. [4] showed that if K_ℓ and K_j are characteristic kernels then the independence of $X^{(\ell)}$ and $X^{(j)}$ is equivalent $\Sigma_{\ell j} = 0$, and also to $V_{\ell j} = 0$. That is why [3] considered as a measure of dependence of these random variables the squared Hilbert-Schmidt norm of their NOCCO, that is

$$N(X^{(\ell)}, X^{(j)}) = \|V_{\ell j}\|_{\mathrm{HS}}^2,$$

where $\|\cdot\|_{\mathrm{HS}}$ denotes the Hilbert-Schmidt norm of operators. This dependence measure will be used here to define measures of mutual independence on which the tests we will propose rely.

43.3 Symmetric and Assymetric Aggregations

Here, we adopt an approach, similar to the one introduced by [6], for defining measures of mutual dependence based on aggregating pairwise dependences. We consider the random variables given, for $\ell = 1, \ldots, d-1$, by

$$X^{(\ell+)} = (X^{(\ell+1)}, \ldots, X^{(d)}),$$

and for $\ell = 2, \ldots, d$, by

$$X^{(-\ell)} = (X^{(1)}, \ldots, X^{(\ell-1)}, X^{(\ell+1)}, \ldots, X^{(d)}).$$

They are valued into the separable Hilbert spaces $\mathcal{X}_{(\ell+)} := \mathcal{X}_{(\ell+1)} \times \cdots \times \mathcal{X}_{(d)}$ and $\mathcal{X}_{(-\ell)} := \mathcal{X}_{(1)} \times \cdots \times \mathcal{X}_{(\ell-1)} \times \mathcal{X}_{(\ell+1)} \times \cdots \times \mathcal{X}_{(d)}$, respectively, and can be used to characterize $\mathscr{H}_0$. Indeed, using similar arguments as in [6], we can show that $\mathscr{H}_0$ is equivalent to either of the following two properties:

$$X^{(\ell)} \perp\!\!\!\perp X^{(\ell+)}, \ \forall \ell \in \{1, \ldots, d-1\}$$

and

$$X^{(\ell)} \perp\!\!\!\perp X^{(-\ell)}, \ \forall \ell \geqslant 1,$$

where $\perp\!\!\!\perp$ denotes stochastic independence. Considering the kernels $K_{(\ell+)}$ and $K_{(-\ell)}$ defined as

$$K_{(\ell+)}(x, y) = \prod_{j=\ell+1}^{d} K_j(x_j, y_j), \quad K_{(-\ell)}(z, t) = \prod_{\substack{j=1 \\ j \neq \ell}}^{d} K_j(z_j, t_j),$$

for any $x = (x_{\ell+1}, \ldots, x_d) \in \mathcal{X}_{(\ell+)}$, $y = (y_{\ell+1}, \ldots, y_d) \in \mathcal{X}_{(\ell+)}$, $z = (z_1, \ldots, z_{\ell-1}, z_{\ell+1}, \ldots, z_d) \in \mathcal{X}_{(-\ell)}$ and $t = (t_1, \ldots, t_{\ell-1}, t_{\ell+1}, \ldots, t_d) \in \mathcal{X}_{(-\ell)}$, we know from [11] that $K_{(\ell+)}$ and $K_{(-\ell)}$ are reproducing kernels on $\mathcal{X}_{(\ell+)}$ and $\mathcal{X}_{(-\ell)}$, respectively, and that when the K_ℓs are continuous and characteristics, it is also the same for

$K_{(\ell+)}$ and $K_{(-\ell)}$. Therefore, the nullity of $N(X^{(\ell)}, X^{(\ell+)})$ (resp. $N(X^{(\ell)}, X^{(-\ell)})$) fully characterizes the independence of $X^{(\ell)}$ and $X^{(\ell+)}$ (resp. $X^{(\ell)}$ and $X^{(-\ell)}$). We then propose the following two measures of mutual dependence of $X^{(1)}, \ldots, X^{(d)}$:

$$R(X) = \sum_{\ell=1}^{d-1} N(X^{(\ell)}, X^{(\ell+)}) \quad \text{and} \quad S(X) = \sum_{\ell=1}^{d} N(X^{(\ell)}, X^{(-\ell)}).$$

Since $\mathscr{H}_0$ is equivalent to having $R(X) = 0$ and, equivalently, $S(X) = 0$, two tests for mutual independence are obtained by taking as test statistics consistent estimators of $R(X)$ and $S(X)$.

43.4 Testing for Mutual Independence

In order to achieve testing for $\mathscr{H}_0$ against $\mathscr{H}_1$, we will introduce estimators of the above measures of mutual dependence and obtain asymptotic normality for centered and rescaled versions of them.

43.4.1 Test Statistics

Let $\{X_i\}_{1 \leqslant i \leqslant n}$, with $X_i = (X_i^{(1)}, \cdots, X_i^{(d)})$, be a i.i.d. sample of $X = (X^{(1)}, \ldots, X^{(d)})$. Putting

$$X_i^{(\ell+)} = (X_i^{(\ell+1)}, \ldots, X_i^{(d)}) \quad \text{and} \quad X_i^{(-\ell)} = (X_i^{(1)}, \ldots, X_i^{(\ell-1)}, X_i^{(\ell+1)}, \ldots, X_i^{(d)}),$$

we consider the empirical means

$$\overline{K}_\ell = \frac{1}{n}\sum_{i=1}^{n} K_\ell(X_i^{(\ell)}, \cdot), \overline{K}_{\ell+} = \frac{1}{n}\sum_{i=1}^{n} K_{(\ell+)}(X_i^{(\ell+)}, \cdot), \overline{K}_{-\ell} = \frac{1}{n}\sum_{i=1}^{n} K_{(-\ell)}(X_i^{(-\ell)}, \cdot),$$

and the empirical covariance and cross-covariance operators

$$\widehat{\Sigma}_\ell = \frac{1}{n}\sum_{i=1}^{n} K_\ell(X_i^{(\ell)}, \cdot) \otimes K_\ell(X_i^{(\ell)}, \cdot) - \overline{K}_\ell \otimes \overline{K}_\ell,$$

$$\widehat{\Sigma}_{\ell+} = \frac{1}{n}\sum_{i=1}^{n} K_{(\ell+)}(X_i^{(\ell+)}, \cdot) \otimes K_{(\ell+)}(X_i^{(\ell+)}, \cdot) - \overline{K}_{\ell+} \otimes \overline{K}_{\ell+},$$

$$\widehat{\Sigma}_{-\ell} = \frac{1}{n}\sum_{i=1}^{n} K_{(-\ell)}(X_i^{(-\ell)}, \cdot) \otimes K_{(-\ell)}(X_i^{(-\ell)}, \cdot) - \overline{K}_{-\ell} \otimes \overline{K}_{-\ell},$$

$$\widehat{\Sigma}_{\ell,\ell+} = \frac{1}{n}\sum_{i=1}^{n} K_{(\ell+)}(X_i^{(\ell+)},\cdot) \otimes K_\ell(X_i^{(\ell)},\cdot) - \overline{K}_{\ell+} \otimes \overline{K}_\ell,$$

$$\widehat{\Sigma}_{\ell,-\ell} = \frac{1}{n}\sum_{i=1}^{n} K_{(-\ell)}(X_i^{(-\ell)},\cdot) \otimes K_\ell(X_i^{(\ell)},\cdot) - \overline{K}_{-\ell} \otimes \overline{K}_\ell.$$

As in [3, 10], we consider a sequence $\{\gamma_n\}_{n\geq 1}$ of strictly positive numbers such that $\lim_{n\to+\infty}\gamma_n = 0$, and the regularized estimators

$$\widehat{V}_{\ell,\ell+} = (\widehat{\Sigma}_\ell + \gamma_n \mathbb{I}_\ell)^{-1/2}\,\widehat{\Sigma}_{\ell,\ell+}(\widehat{\Sigma}_{\ell+} + \gamma_n \mathbb{I}_{\ell+})^{-1/2},$$

$$\widehat{V}_{\ell,-\ell} = (\widehat{\Sigma}_\ell + \gamma_n \mathbb{I}_\ell)^{-1/2}\,\widehat{\Sigma}_{\ell,-\ell}(\widehat{\Sigma}_{-\ell} + \gamma_n \mathbb{I}_{-\ell})^{-1/2},$$

where $\mathbb{I}_\ell$ (resp. $\mathbb{I}_{\ell+}$; resp. $\mathbb{I}_{-\ell}$) denotes the identity operator of the RKHS $\mathcal{H}_\ell$ (resp. $\mathcal{H}_{\ell+}$; resp. $\mathcal{H}_{-\ell}$) to which K_ℓ (resp. $K_{(\ell+)}$; resp. $K_{(-\ell)}$) is associated. We then estimate $N(X^{(\ell)}, X^{(\ell+)})$ and $N(X^{(\ell)}, X^{(-\ell)})$ by

$$\widehat{N}(X^{(\ell)}, X^{(\ell+)}) = \|\widehat{V}_{\ell,\ell+}\|_{\mathrm{HS}}^2 \quad\text{and}\quad \widehat{N}(X^{(\ell)}, X^{(-\ell)}) = \|\widehat{V}_{\ell,-\ell}\|_{\mathrm{HS}}^2,$$

respectively, and take as test statistics the following estimators or $R(X)$ and $S(X)$:

$$\widehat{R}(X) = \sum_{\ell=1}^{d-1} \widehat{N}(X^{(\ell)}, X^{(\ell+)}), \qquad \widehat{S}(X) = \sum_{\ell=1}^{d} \widehat{N}(X^{(\ell)}, X^{(-\ell)}).$$

Remark 1. Expressions allowing to compute the above test statistics in practice can be obtained thanks to the kernel trick. Indeed, from [9] we have

$$\widehat{N}(X^{(\ell)}, X^{(\ell+)}) = \mathrm{Tr}\left(G_\ell(G_\ell + n\gamma_n I_n)^{-1} G_{\ell+}(G_{\ell+} + n\gamma_n I_n)^{-1}\right),$$

$$\widehat{N}(X^{(\ell)}, X^{(-\ell)}) = \mathrm{Tr}\left(G_\ell(G_\ell + n\gamma_n I_n)^{-1} G_{-\ell}(G_{-\ell} + n\gamma_n I_n)^{-1}\right),$$

where G_ℓ, $G_{\ell+}$ and $G_{-\ell}$ are $n \times n$ Gram matrices with terms equal to

$$(G_\ell)_{ij} = \mathcal{K}_{ij}^\ell - \frac{1}{n}\left(\mathcal{K}_{i.}^\ell + \mathcal{K}_{.j}^\ell\right) + \frac{\mathcal{K}_{..}^\ell}{n^2}, \quad (G_{\ell+})_{ij} = \mathcal{K}_{ij}^{\ell+} - \frac{1}{n}\left(\mathcal{K}_{i.}^{\ell+} + \mathcal{K}_{.j}^{\ell+}\right) + \frac{\mathcal{K}_{..}^{\ell+}}{n^2},$$

$$(G_{-\ell})_{ij} = \mathcal{K}_{ij}^{-\ell} - \frac{1}{n}\left(\mathcal{K}_{i.}^{-\ell} + \mathcal{K}_{.j}^{-\ell}\right) + \frac{\mathcal{K}_{..}^{-\ell}}{n^2},$$

where

$$\mathcal{K}_{ij}^\ell = K_\ell(X_i^{(\ell)}, X_j^{(\ell)}), \quad \mathcal{K}_{i.}^\ell = \sum_{j=1}^{n} \mathcal{K}_{ij}^\ell, \quad \mathcal{K}_{.j}^\ell = \sum_{i=1}^{n} \mathcal{K}_{ij}^\ell, \quad \mathcal{K}_{..}^\ell = \sum_{i=1}^{n}\sum_{j=1}^{n} \mathcal{K}_{ij}^\ell,$$

$$\mathscr{K}_{ij}^{\ell+} = \prod_{p=\ell+1}^{d} \mathscr{K}_{ij}^{p},\ \mathscr{K}_{i.}^{\ell+} = \sum_{j=1}^{n} \mathscr{K}_{ij}^{\ell+},\ \mathscr{K}_{.j}^{\ell+} = \sum_{i=1}^{n} \mathscr{K}_{ij}^{\ell+},\ \mathscr{K}_{..}^{\ell+} = \sum_{i=1}^{n}\sum_{j=1}^{n} \mathscr{K}_{ij}^{\ell+},$$

$$\mathscr{K}_{ij}^{-\ell} = \prod_{\substack{p=1\\ p\neq\ell}}^{d} \mathscr{K}_{ij}^{p},\ \mathscr{K}_{i.}^{-\ell} = \sum_{j=1}^{n} \mathscr{K}_{ij}^{-\ell},\ \mathscr{K}_{.j}^{-\ell} = \sum_{i=1}^{n} \mathscr{K}_{ij}^{-\ell},\ \mathscr{K}_{..}^{-\ell} = \sum_{i=1}^{n}\sum_{j=1}^{n} \mathscr{K}_{ij}^{-\ell},$$

and I_n is the $n \times n$ identity matrix.

Remark 2. From what is above, we see that the introduced test statistics can be computed as soon as the $\mathscr{K}_{ij}^{\ell}$s can be computed. In the case where the $X_i^{(\ell)}$s are functional variables belonging in $L^2([0,1])$ and observed on points $t_1, \ldots, t_r$ of a fine grid in $[0,1]$ such that $t_1 = 0$ and $t_r = 1$, the previous terms can be computed or easily approximated, depending on the used kernels. For example, if the Gaussian kernel is used, one has

$$\mathscr{K}_{ij}^{\ell} = \exp\left(-\omega_\ell^2 \int_0^1 \left(X_i^{(\ell)}(t) - X_j^{(\ell)}(t)\right)^2 dt\right),$$

where $\omega_\ell > 0$. These terms can be approximated by using trapezoidal rule, so leading to

$$\mathscr{K}_{ij}^{\ell} \simeq \exp\left(-\omega_\ell^2 \sum_{m=1}^{r-1} \frac{t_{m+1} - t_m}{2}\left(\left(X_i^{(\ell)}(t_m) - X_j^{(\ell)}(t_m)\right)^2 + \left(X_i^{(\ell)}(t_{m+1}) - X_j^{(\ell)}(t_{m+1})\right)^2\right)\right).$$

43.4.2 Studentized Versions of the Test Statistics

For any compact operator A, we denote by $\{\lambda_q(A)\}_{q\geqslant 1}$ the decreasing sequence of its eigenvalues, and for any $\gamma > 0$ and any integer $r \in \{1, 2\}$ we put:

$$d_r(A, \gamma) = \left\{\sum_{q=1}^{+\infty} (\lambda_q(A) + \gamma)^{-r} \lambda_q^r(A)\right\}^{1/r}.$$

Then, for all $\ell \in \{1, \ldots, d\}$, we consider

$$\widehat{D}_{r,n}^{(\ell+)} = d_r(\widehat{\Sigma}_\ell, \gamma_n) d_r(\widehat{\Sigma}_{\ell+}, \gamma_n),\ \ \widehat{D}_{r,n}^{(-\ell)} = d_r(\widehat{\Sigma}_\ell, \gamma_n) d_r(\widehat{\Sigma}_{-\ell}, \gamma_n),$$

$$\widehat{D}_{r,n}^{(+)}(X) = \left\{\sum_{\ell=1}^{d-1} (\widehat{D}_{r,n}^{(\ell+)})^r\right\}^{1/r},\ \ \widehat{D}_{r,n}^{(-)}(X) = \left\{\sum_{\ell=1}^{d} (\widehat{D}_{r,n}^{(-\ell)})^r\right\}^{1/r},$$

and the centered and rescaled versions of the test statistics introduced above, given by:

$$\widehat{T}_R = \frac{n\widehat{R}(X) - \widehat{D}_{1,n}^{(+)}(X)}{\sqrt{2}\widehat{D}_{2,n}^{(+)}(X)}, \qquad \widehat{T}_S = \frac{n\widehat{S}(X) - \widehat{D}_{1,n}^{(-)}(X)}{\sqrt{2}\widehat{D}_{2,n}^{(-)}(X)}. \tag{43.1}$$

Remark 3. The above terms used for centering and rescalling our test statistics can be computed in practice thanks to the kernel trick. Indeed, with similar arguments as in [9], we obtain

$$\widehat{D}_{r,n}^{(\ell+)} = \left\{\mathrm{Tr}\left(G_\ell^r(G_\ell + n\gamma_n I_n)^{-r}\right)\mathrm{Tr}\left(G_{\ell+}^r(G_{\ell+} + n\gamma_n I_n)^{-r}\right)\right\}^{1/r}$$

and

$$\widehat{D}_{r,n}^{(-\ell)} = \left\{\mathrm{Tr}\left(G_\ell^r(G_\ell + n\gamma_n I_n)^{-r}\right)\mathrm{Tr}\left(G_{-\ell}^r(G_{-\ell} + n\gamma_n I_n)^{-r}\right)\right\}^{1/r}.$$

43.4.3 Asymptotic Null Distributions

Now, we will obtain asymptotic normality for the test statistics given in (43.1). We will first make some assumptions that will be needed for obtaining the main result. We say that a compact operator A belongs to the class $\mathscr{C}$ if:

(i) $\lambda_q(A) > 0$ for any $q \geqslant 1$;

(ii) $\sum_{q=1}^{+\infty}(\lambda_q(A))^{1/2} < +\infty$.

Considering the covariance operators given by

$$\Sigma_{\ell+} = \mathbb{E}\big(K_{\ell+}(X^{(\ell+)},\cdot) \otimes K_{\ell+}(X^{(\ell+)},\cdot)\big) - \mathbb{E}\big(K_{\ell+}(X^{(\ell+)},\cdot)\big) \otimes \mathbb{E}\big(K_{\ell+}(X^{(\ell+)},\cdot)\big)$$

and

$$\Sigma_{-\ell} = \mathbb{E}\big(K_{-\ell}(X^{(-\ell)},\cdot) \otimes K_{-\ell}(X^{(-\ell)},\cdot)\big) - \mathbb{E}\big(K_{-\ell}(X^{(-\ell)},\cdot)\big) \otimes \mathbb{E}\big(K_{-\ell}(X^{(-\ell)},\cdot)\big),$$

we make the following assumptions:

$(\mathscr{A}_2)$: For $\ell \in \{1,\ldots,d\}$, the operators Σ_ℓ, $\Sigma_{\ell+}$ and $\Sigma_{-\ell}$ belong to the class $\mathscr{C}$;

$(\mathscr{A}_3)$: the sequence $\{\gamma_n\}_{n\geqslant 1}$ satisfies : $\lim_{n\to+\infty}\left(\gamma_n^{-9}n^{-1}\right) = 0$.

Then, we have :

Theorem 1. Assume $(\mathscr{A}_1)$ to $(\mathscr{A}_3)$. Then under $\mathscr{H}_0$, $\widehat{T}_R$ and $\widehat{T}_S$ converge in distribution, as n goes to $+\infty$, to a standard normal random variable $\mathcal{N}(0,1)$.

Based on these results, we propose the following testing procedures: for a fixed significance level $\alpha \in]0,1[$, one has to reject $\mathscr{H}_0$ when $|\widehat{T}_R| > \Phi^{-1}(1-\alpha/2)$ or

$|\widehat{T}_S| > \Phi^{-1}(1-\alpha/2)$, where Φ is the cumulative distribution function of the standard normal random variable.

43.5 Simulations

In this section, we study the finite sample performances of the tests based on $\widehat{R}(X)$ and $\widehat{S}(X)$ and compare them to that of [11]. We consider the case $d = 3$ and $\mathcal{X}_1 = \mathcal{X}_2 = \mathcal{X}_3 = \mathrm{L}^2([0,1])$, and generate functional data through the following model: $X^{(1)}(t) = \sqrt{2}\sum_{k=1}^{30} \alpha_k \cos(k\pi t)$, $X^{(2)}(t) = \sqrt{2}\sum_{k=1}^{30} \beta_k \cos(k\pi t)$ and $X^{(3)}(t) = \sqrt{2}\sum_{k=1}^{30} \xi_k \cos(k\pi t)$. The coefficients α_k are independent and identically distributed following the Cauchy distribution $C(0, 0.5)$. For a fixed integer $m \in \{1, \dots, 30\}$, and for $1 \leqslant k \leqslant m$, $\beta_k = \xi_k = f(\alpha_k)$, f being a given function. For $m+1 \leqslant k \leqslant d$, the coefficients β_k and ξ_k are independent and identically distributed following the standard normal random variable. In this model, the three processes are mutually independent for $m > 1$, the coefficient m being the parameter controlling the strength of the dependency. For $m = 0$, the processes are independent. The results are presented in the three following tables.

$f(x)$	m	$\widehat{T}_R$		$\widehat{T}_S$		dHSIC	
		$n = 50$	$n = 75$	$n = 50$	$n = 75$	$n = 50$	$n = 75$
x	0	0.032	0.044	0.044	0.038	0.074	0.058
	1	0.712	0.886	0.326	0.464	0.540	0.708
	3	0.998	1.000	0.832	0.954	0.986	1.000
	5	1.000	1.000	0.962	1.000	1.000	1.000
	10	1.000	1.000	1.000	1.000	1.000	1.000
x^2	0	0.032	0.044	0.044	0.038	0.074	0.058
	1	0.958	1.000	0.746	1.000	0.908	1.000
	3	1.000	1.000	0.992	1.000	1.000	1.000
	5	1.000	1.000	1.000	1.000	1.000	1.000
	10	1.000	1.000	1.000	1.000	1.000	1.000
$x^2\sin(x)$	0	0.032	0.044	0.044	0.038	0.074	0.058
	1	0.906	0.972	0.566	0.760	0.806	0.920
	3	1.000	1.000	0.952	0.998	1.000	1.000
	5	1.000	1.000	0.996	1.000	1.000	1.000
	10	1.000	1.000	1.000	1.000	1.000	1.000

Table 43.1: Empirical powers and levels over 500 replications, with nominal significance level fixed as $\alpha = 0.05$

References

1. Baker, C.R.: Joint measures and cross-covariance operators. Trans. Amer. Math. Soc. **186**, 273–289 (1973)
2. Fukumizu, K., Bach, F., Gretton, A: Statistical consistency of kernel canonical correlation analysis. J. Mach. Learn. Res. **8**, 361–383 (2007)
3. Fukumizu, K., Gretton, A., Sun, X., Schölkopf, B: Kernel measures of conditional dependence. Adv. Neural Inf. Process. Syst. (2007)
4. Gretton, A., Bousquet, O., Smola, A., Schölkopf, B: Measuring statistical dependence with Hilbert-Schmidt norms. In: Lecture Notes in Comput. Sci., 63–77 (2005)
5. Han, F., Chen, S., Liu, H.: Distribution-free tests of independence in high dimensions. Biometrika **104**, 813–828 (2017)
6. Jin, Z., Matteson, D.S.: Generalizing distance covariance to measure and test multivariate mutual dependence via complete and incomplete V-statistics. J. Multivariate Anal. **168**, 304–322 (2018)
7. Krzyśko, M., Smaga, L., Kokoszka, P.: Marginal distance and hilbert-schmidt covariances-based independence tests for multivariate functional data. J. Artificial Intelligence Res **73**, 1355–1384 (2022)
8. Leung, D., Drton, M: Testing independence in high dimensions with sums of rank correlations. Ann. Statist. **46**, 280–307 (2018)
9. Manfoumbi Djonguet, T.K., Nkiet, G.M.: An independence test for functional variables based on kernel normalized cross-covariance operator. J. Multivariate Anal. **202**, 105293 (2024)
10. Manfoumbi Djonguet, T.K., Nkiet, G.M.: Kernel-based method for joint independence of functional variables. Comm. Statist. Theory Methods **54**, 921–937 (2025)
11. Pfister, N., Bühlmann, P., Schölkopf, B., Peters, J.: Kernel-based tests for joint independence. J. R. Stat. Soc. Ser. B. Stat. Methodol. **80**, 5–31 (2018)
12. Székely, G.J., Rizzo, M.L., Bakirov, N.K.: Measuring and testing dependence by correlation of distances. Ann. Statist. **35**, 2769–2794 (2007)
13. Yao, S., Zhang, X., Shao, X.: Testing mutual independence in high dimension via distance covariance. J. R. Stat. Soc. Ser. B. Stat. Methodol. **80**, 455–480 (2018)

Chapter 44
Addressing Robustness and Sparsity in Partially Linear Additive Models

Alejandra Mercedes Martínez

Abstract Partially linear additive models combine a linear component for some covariates with an additive nonparametric component for others, offering flexibility, interpretability, and the ability to include categorical variables in the linear part. Variable selection is critical to avoid including irrelevant covariates that reduce predictive performance. To address the impact of outliers and strong assumptions like second moments of the errors, a family of robust estimators is proposed. These estimators perform simultaneous variable selection for both model components. A simulation study demonstrates the advantages of the proposed method over its least-squares counterpart.

44.1 Introduction

Partially linear additive models (PLAM) assume that $(Y_i, \mathbf{Z}_i^{\mathrm{T}}, \mathbf{X}_i^{\mathrm{T}})^{\mathrm{T}} \in \mathbb{R}^{1+q+p}$, $1 \leq i \leq n$, are independent and identically distributed random vectors with the same distribution as $(Y, \mathbf{Z}^{\mathrm{T}}, \mathbf{X}^{\mathrm{T}})^{\mathrm{T}}$ such that

$$Y = m(\mathbf{Z}^{\mathrm{T}}, \mathbf{X}^{\mathrm{T}}) + u = \mu + \boldsymbol{\beta}^{\mathrm{T}}\mathbf{Z} + \sum_{j=1}^{p} \eta_j(X_j) + \sigma\varepsilon\,. \quad (44.1)$$

The constant $\mu \in \mathbb{R}$, vector $\boldsymbol{\beta} \in \mathbb{R}^q$, the univariate functions $\eta_j : \mathbb{R} \to \mathbb{R}$, $1 \leq j \leq p$, and the scale parameter $\sigma > 0$ are the objects to be estimated. In order to ensure identifiability of the additive components η_j, throughout this paper, we assume that $\int \eta_j(x)\, dx = 0$, for $1 \leq j \leq p$. The errors ε are assumed to be independent from the vector of covariates $(\mathbf{Z}^{\mathrm{T}}, \mathbf{X}^{\mathrm{T}})^{\mathrm{T}}$. Furthermore, in the classical setting, the errors ε are assumed to have 0 mean and variance equal to 1, while in the robustness framework

Alejandra Mercedes Martínez (✉)
Departamento de Ciencias Básicas, Universidad Nacional de Luján, and CONICET, Argentina
e-mail: ale_m_martinez@hotmail.com

© The Author(s), under exclusive license to Springer Nature Switzerland AG 2025

G. Aneiros et al. (eds.), *New Trends in Functional Statistics and Related Fields*, Contributions to Statistics, https://doi.org/10.1007/978-3-031-92383-8_44

we will be dealing with, they are considered to have a symmetric distribution $F(\cdot)$ with scale parameter equals 1.

Usually, in a first step of modelling, researchers introduce all the available variables in the model and so those that have a small impact on the response variable will reduce the prediction capability of the estimators. For this reason, variable selection plays an important role. On the other hand, as in linear regression models, estimation and variable selection approaches, under a PLAM, can be seriously affected in the presence of atypical data if non-robust procedures are used.

Even though several methods for variable selection exist, such as stepwise selection or best subset selection, this work considers a regularization procedure. Under PLAM, [4] and [6] proposed different approaches for estimating and selecting variables from both the linear and additive components of the model, while [1] introduced a Bayesian method for generalized partially linear additive models. However, none of these approaches are resistant to atypical data. A more robust alternative was proposed in [7], who developed an estimation and variable selection procedure based on modal regression. Although this approach is more robust than the previouly mentioned methods, it remains highly sensitive to high-leverage points in the linear component of the model.

In this extended abstract, a robust method for estimating and simultaneously selecting variables from both the linear and the additive parts of a sparse partially linear additive model is proposed.

44.2 The Proposed Estimator

To approximate each additive function, we will use a $B-$spline basis. Similar to what is done in [2], η_j is approximated as a $\sum_{s=1}^{k_j-1} c_s^{(j)} B_s^{(j)}(t)$ with $B_s^{(j)}$, for $1 \leq s \leq k_j-1$ and $1 \leq j \leq p$, the sth element of the basis of centered $B-$splines of order ℓ_j, that is, $B_s^{(j)}$ is such that $B_s^{(j)}(t) = \widetilde{B}_s^{(j)}(t) - \int_{\mathcal{I}_j} \widetilde{B}_s^{(j)}(t)\,dt$ with $\{\widetilde{B}_s^{(j)} : 1 \leq s \leq k_j\}$ a spline basis of order ℓ_j. Besides, only $k_j - 1$ elements of the basis are considered instead of the k_j total in order to allow the paramerer μ to be identifiable. Then, the PLAM in (44.1) is approximated by the following linear regression model

$$Y = \mu + \boldsymbol{\beta}^{\mathrm{T}}\mathbf{Z} + \sum_{j=1}^{p}\sum_{s=1}^{k_j-1} c_s^{(j)} B_s^{(j)}(X_j) + \sigma\varepsilon\,. \tag{44.2}$$

Let $\boldsymbol{\beta} = (\beta_1,\ldots,\beta_q)^{\mathrm{T}} \in \mathbb{R}^q$, $\mathbf{c}^{(j)} = (c_1^{(j)},\ldots,c_{k_j-1}^{(j)})^{\mathrm{T}} \in \mathbb{R}^{k_j-1}$, $\mathbf{c} = (\mathbf{c}^{(1)\mathrm{T}},\ldots,\mathbf{c}^{(p)\mathrm{T}}) \in \mathbb{R}^K$, $K = \sum_{j=1}^{p} k_j - p$, and $\mathbf{V} = (\mathbf{V}^{(1)}(X_1)^{\mathrm{T}},\ldots,\mathbf{V}^{(p)}(X_p)^{\mathrm{T}})^{\mathrm{T}} \in \mathbb{R}^K$ where $\mathbf{V}^{(j)}(t) = (B_1^{(j)}(t),\ldots,B_{k_j-1}^{(j)}(t))^{\mathrm{T}} \in \mathbb{R}^{k_j-1}$. Then, (44.2) can be written in a simplified form as

$$Y = \mu + \boldsymbol{\beta}^{\mathrm{T}}\mathbf{Z} + \mathbf{c}^{\mathrm{T}}\mathbf{V} + \sigma\varepsilon.$$

Denote $\mathbf{b} = (b_1, \ldots, b_q)^{\mathrm{T}} \in \mathbb{R}^q$ and $\mathbf{d}^{(j)} = (d_1^{(j)}, \ldots, d_{k_j-1}^{(j)})^{\mathrm{T}} \in \mathbb{R}^{k_j-1}$ two general vectors that belong to the possible set of candidates for $\boldsymbol{\beta}$ and $\mathbf{c}^{(j)}$, for $1 \leq j \leq p$, and $\mathbf{d} = (\mathbf{d}^{(1)\mathrm{T}}, \ldots, \mathbf{d}^{(p)\mathrm{T}}) \in \mathbb{R}^K$. Let $\widehat{\mu}$ be a preliminary robust estimator of μ and $\widehat{\sigma}$ a preliminary robust scale estimator of σ, then we define the penalized robust estimators for the partially linear additive models as

$$(\widehat{\boldsymbol{\beta}}, \widehat{\mathbf{c}}) = \mathrm{argmin}_{\mathbf{b}\in\mathbb{R}^q, \mathbf{d}\in\mathbb{R}^K} PL_{n,\boldsymbol{\lambda},\mathbf{k}}(\mathbf{b}, \mathbf{d}), \tag{44.3}$$

where

$$PL_{n,\boldsymbol{\lambda},\mathbf{k}}(\mathbf{b}, \mathbf{d}) = L_n(\widehat{\mu}, \widehat{\sigma}, \mathbf{b}, \mathbf{d}) + \mathcal{J}_{\boldsymbol{\lambda},\mathbf{k}}(\mathbf{b}, \mathbf{d}), \tag{44.4}$$

$$L_n(a, \varsigma, \mathbf{b}, \mathbf{d}) = \frac{1}{n} \sum_{i=1}^{n} \rho\left(\frac{r_i(a, \mathbf{b}, \mathbf{d})}{\varsigma}\right), \qquad r_i(a, \mathbf{b}, \mathbf{d}) = Y_i - a - \mathbf{b}^{\mathrm{T}}\mathbf{Z}_i - \mathbf{d}^{\mathrm{T}}\mathbf{V}_i,$$

and $\mathcal{J}_{\boldsymbol{\lambda},\mathbf{k}}(\mathbf{b}, \mathbf{d})$ is an arbitrary penalty function (selected by the user) depending on a regularization parameter $\boldsymbol{\lambda} = (\lambda_{1,1}, \ldots, \lambda_{1,q}, \lambda_{2,1}, \ldots, \lambda_{2,p})^{\mathrm{T}} \in \mathbb{R}^{q+p}$ that determines the model complexity, that is, $\lambda_{1,s}$, $1 \leq s \leq q$, and $\lambda_{2,j}$, $1 \leq j \leq p$, are the penalty parameters that control the sparsity of the parametric and nonparametric components, respectively, and $\mathbf{k} = (k_1, \ldots, k_p)^{\mathrm{T}}$ is the parameter vector that controls the $B-$spline approximation of the additive parte of the model. For instance, $\mathcal{J}_{\boldsymbol{\lambda},\mathbf{k}}(\mathbf{b}, \mathbf{d})$ can be defined as

$$\mathcal{J}_{\boldsymbol{\lambda},\mathbf{k}}(\mathbf{b}, \mathbf{d}) = \sum_{s=1}^{q} p_{\lambda_{1,s}}(|b_s|) + \sum_{j=1}^{p} p_{\lambda_{2,j}}(\|\mathbf{d}^{(j)}\|_{H_j}) \tag{44.5}$$

with p_λ a univariate penalty function such as the SCAD or the MCP penalties. In addition, $\|\mathbf{d}^{(j)}\|_{H_j} = (\mathbf{d}^{(j)\mathrm{T}} H_j \mathbf{d}^{(j)})^{1/2}$ with H_j of dimension $(k_j - 1) \times (k_j - 1)$ with its (s, s') element given by $\int_0^1 B_s^{(j)}(t) B_{s'}^{(j)}(t)\, dt$, and ρ is a bounded $\rho-$function as defined in [8].

44.2.1 Selection of the Tuning Parameters

Two vectors of parameters have to be selected in order to compute the proposal: $\mathbf{k}$ which states the number of elements for the $B-$splines approximation and the penality vector $\boldsymbol{\lambda}$ that controls the model complexity. Besides, any robust estimation and variable selection approach also needs robust methods for selecting the regularization and penalty parameters.

For $\mathbf{k}$ fixed, following the ideas in [9], the penalty parameters will be selected by minimizing the robust BIC criterion defined as

$$\mathrm{RBIC}_{\boldsymbol{\lambda}}(\boldsymbol{\lambda}) = \log\left(\widehat{\sigma}^2 \sum_{i=1}^{n} \rho\left(\frac{r_i(\widehat{\beta}_{\boldsymbol{\lambda}}, \widehat{\mathbf{c}}_{\boldsymbol{\lambda}})}{\widehat{\sigma}}\right)\right) + df_c \frac{\log(n)}{n} + df_n \frac{\log(n/K)}{n/K} \tag{44.6}$$

where $(\widehat{\beta}_\lambda, \widehat{\mathbf{c}}_\lambda)$ is the solution of (44.3) with penalty vector $\boldsymbol{\lambda} = (\lambda_{1,1}, \ldots, \lambda_{1,q}, \lambda_{2,1}, \ldots, \lambda_{2,p})^{\mathrm{T}}$, df_c is the number of nonzero parametric components, df_n is the number of nonzero nonparametric components and $\widehat{\sigma}$ is a scale estimator.

On the other hand, any $B-$spline approach requires the selection of a set of knots to define the basis functions. Regarding their locations, equally space knots can be used when the additive functions η_j do not exhibit dramatic changes in their derivatives; otherwise quantile knots are preferred. With respect the number of internal knots, we will consider the robust BIC criterion defined in [2] which is the following:

$$\mathrm{RBIC}_{\mathbf{k}}(\mathbf{k}) = \log\left(\widehat{\sigma}^2 \sum_{i=1}^{n} \rho\left(\frac{r_i(\widehat{\beta}_{\lambda,\mathbf{k}}, \widehat{\mathbf{c}}_{\lambda,\mathbf{k}})}{\widehat{\sigma}}\right)\right) + \frac{\log(n)}{2n} \sum_{j=1}^{p} k_j \tag{44.7}$$

where $\mathbf{k} = (k_1, \ldots, k_p)^{\mathrm{T}}$ as before, $r_i(\widehat{\beta}_{\lambda,\mathbf{k}}, \widehat{\mathbf{c}}_{\lambda,\mathbf{k}})$, for $1 \le i \le n$, are the residuals obtained using a basis of dimension k_j to approximate the additive function η_j when λ is given, ρ is the $\rho-$function used to compute the robust estimator and $\widehat{\sigma}$ is the scale estimator of σ.

44.3 Simulation Study

This section presents the results of a small numerical experiment conducted to compare the performance of the robust proposal against its least-squares counterpart in a high-dimensional setting.

The samples $\{Y_i, \mathbf{Z}_i^{\mathrm{T}}, \mathbf{X}_i^{\mathrm{T}})^{\mathrm{T}}\}_{i=1}^{n}$, $n = 600$, were generated with the same distribution as $(Y, \mathbf{Z}^{\mathrm{T}}, \mathbf{X}^{\mathrm{T}})^{\mathrm{T}}$ satisfying the PLAM in (44.1) with $q = 100$, $p = 50$, $\mu = 0$, $\sigma = 1$, $\boldsymbol{\beta} = (3, \ldots, 3, 1.5, \ldots, 1.5, 2, \ldots, 2, -1.5, \ldots, -1.5, 0, \ldots, 0) \in \mathbb{R}^{100}$ where each nonzero coefficient is repeated five times, yielding to a vector with its first 20 coordinates different from 0, and

$$\eta_1(x) = \eta_6(x) = 5x - 5/2$$

$$\eta_2(x) = \eta_7(x) = 3(2x-1)^2 - 1$$

$$\eta_3(x) = \eta_8(x) = 60x^3 - 90x^2 + 30x$$

$$\eta_4(x) = \eta_9(x) = 2\sin(\pi x) - \frac{4}{\pi}$$

$$\eta_5(x) = \eta_{10}(x) = 2\cos(2\pi x)$$

and $\eta_j \equiv 0$, for $j = 11, \ldots, 50$. Besides, for $i = 1, \ldots, n$, $X_{ij} \sim U(0,1)$ and so $\mathbb{E}\eta_j(X_{ij}) = 0$ for all $1 \le j \le 50$, $\mathbf{Z}_i = (Z_{i1}, \ldots, Z_{iq})^{\mathrm{T}}$ are generated from a multivariate normal distribution with zero mean and a variance-covariance matrix Σ defined as $(\Sigma)_{k,\ell} = 0.5^{|k-\ell|}$, for $1 \le k, \ell \le q$. Besides, X_{ij} are independent from each other and from the vector $\mathbf{Z}_i$. For clean samples, $u \sim N(0,1)$, denoted from now on C_0.

The robust proposal uses the Tukey's bisquare loss which is defined as $\rho_c(t) = \min\{1-(1-(t/c)^2)^3, 1\}$ with $c = 4.685$. For the preliminary robust scale estimator, an S−estimator is used, while for $\widehat{\mu}$ we used the approach in [2] with the Tukey's loss function for the robust proposal and the quadratic function for its least-squares counterpart. From now on, the robust approach is denoted ROB and LS the approach based on least squares which considers $\rho(t) = t^2$. The penalized estimators use the penalty given by (44.5) with p_λ the SCAD function.

Two contaminated scenarios have been considered to study the effect of atypical data:

C_1 : $u \sim 0.85N(0, \sigma^2) + 0.15N(15, \sigma^2)$.
C_2 : A 5% of covariates $\mathbf{Z}_i$ were replaced by $(20, \ldots, 20) \in \mathbb{R}^{100}$.

While contamination C_1 is a bias contamination scheme where 15% of the errors are normally distribution with their location shifted to 15, contamination C_2 is a high-leverage contamination scenario where a 5% of the covariates of the linear part were randomly chosen to take a value that highly affects the least-squares estimators.

For all scenarios, in addition to the penalized estimators, both robust and least-squares oracle estimators were computed to provide an optimal counterpart for comparison. The oracle estimators do not perform variable selection; instead, they use only the significant covariates in the model, that is, the first 20 covariates from the linear part and the first 10 covariates from the additive part.

A total of 100 samples were generated. Two measures were considered to observe the quality of the variable selection procedure. For each sample, we computed

C: the number of zero components correctly estimated to be zero.
IC: the number of zero componentes incorrectly estimated to be zero.

These measures were obtained for the linear component of the model, for the additive component and for the complete model. Since 100 values of each of these measures were obtained for each estimator and each of the three parts of the model (linear, additive and complete), the mean over replications was computed in each case to summarize the results. Recalling that the linear component of the model has 80 zero variables, the C value of the linear part will be better when closer to 80, while for the additive part, C will be better when closer to 40. Finally, for the complete model, we look for estimators that give values closer to 120. With respect to the IC measure, the closer to 0, the better.

To evaluate the performance of the estimator, for each replication, we computed the ASE value defined as

$$\text{ASE} = \frac{1}{n^*}\sum_{i=1}^{n^*} r_{(i)}^2,$$

where $r_i = \widehat{y}_i - m(\mathbf{z}_i^{\mathrm{T}}, \mathbf{x}_i^{\mathrm{T}})$ is the ith error obtained for the observation $(y_i, \mathbf{z}_i^{\mathrm{T}}, \mathbf{x}_i^{\mathrm{T}})^{\mathrm{T}}$, $r_{(1)}^2 \leq \cdots \leq r_{(n)}^2$ are the ordered statistics of the squared residuals, m is the true regression model, that is, $m(\mathbf{z}_i^{\mathrm{T}}, \mathbf{x}_i^{\mathrm{T}}) = \mu + \boldsymbol{\beta}^{\mathrm{T}}\mathbf{z}_i + \sum_{j=1}^{p} \eta_j(x_{ij})$, and $n^* = 90$, which means that the average of the squared residuals is calculated over the 90% of the smallest ones. Then, the mean and the standard deviation of the 100 values of ASE

Table 44.1: Results of the measures for variable selection

		Linear part		Additive part		Complete model	
Cont	Method	C	IC	C	IC	C	IC
C_0	LS	39.44	0	39.82	0.00	79.26	0.00
	ROB	42.85	0	38.36	0.00	81.21	0.00
C_1	LS	0.60	0	15.27	0.13	15.87	0.13
	ROB	72.75	0	37.87	0.07	110.62	0.07
C_2	LS	2.26	0	13.67	0.20	14.5	0.20
	ROB	66.75	0	34.09	0.00	102.5	0.00

obtained by each estimator were calculated to obtain final measures of position and dispersion results. This measure was also calculated for the oracle robust and least-squares estimators.

Table 44.1 presents the results of the variable selection procedure. Under the uncontaminated scenario (C_0), the robust and least-squares approaches yield similar values for the C measure, as well as values at the optimal level for the IC measure, both for the linear and additive parts of the model and for the complete model. However, in the presence of contamination, the robust approach maintains consistently good performance, while the least-squares estimator shows a reduction in its variable selection capability. Specifically, the C measure decreases for all three parts of the model, and the IC measure increases for both the additive part and the complete model.

Table 44.2 summarizes the mean and standard deviations of the ASE values for both robust and least-squares approaches, along with their corresponding oracle estimators. The estimators incorporating a regularization procedure are referred to as Penalized, while those that do not perform variable selection and use only the true significant variables of the model are labelled Oracle. Under C_0, the robust estimator shows slightly higher values than its least-squares counterpart for both oracle and penalized approaches, which is typically attributed to the efficiency loss associated with robust estimators. As expected, both penalized estimators yield higher ASE values compared to their oracle counterparts. In the contaminated scenarios, the robust approach remains relatively stable, producing ASE values close to those observed under the uncontaminated setting. In contrast, the least-squares approach exhibits a dramatic increase in ASE values under C_1, with mean values exceeding 51 times those under C_0 and also larger standard deviations. Notably, under C_1, the mean and standard deviation of the ASE values for the penalized robust estimator are more than four times higher than those in the uncontaminated case.

For a clearer visualization of the ASE values, **Fig. 44.1** displays adjusted boxplots, see [5], for the robust and least-squares penalized and oracle estimators across the three scenarios. Under C_0, both penalized estimators produce boxplots located at higher values and with larger boxes and whiskers compared to their oracle counterparts. Under C_1 and C_2, the least-squares estimators yield significantly higher ASE values, as reflected in higher and wider boxplots. Additionally, the boxplot for the

Table 44.2: Results of the estimation measure for both penalized and oracle, robust and least-squares estimators

Cont	Method	Oracle	Penalized
C_0	LS	0.0642 (0.01)	0.1655 (0.08)
	ROB	0.0688 (0.01)	0.1628 (0.07)
C_1	LS	5.6188 (1.02)	8.5461 (1.40)
	ROB	0.0785 (0.02)	0.6940 (1.57)
C_2	LS	51.3685 (3.24)	9.8175 (0.76)
	ROB	0.0864 (0.01)	0.1470 (0.06)

penalized robust estimator under C_1 reveals the presence of outliers, which have contributed to the higher mean and standard deviation reported in Table 44.2 as mentioned above.

Remark 44.1. In this simulation study, the number of covariates is smaller than the sample size. However, when the number of covariates exceeds the sample size, or even when it becomes excessive due to the B-spline expansion, obtaining initial robust estimators for the additive functions η_j, vector $\boldsymbol{\beta}$, constant μ and standard deviation σ becomes more challenging. As mentioned in [3], a possible approach is to consider a "ridge" $S-$estimator, which does not perform variable selection but enables computation when $q + K$ is large.

In summary, the obtained results show the stability of the robust proposal and the advantages of using a resistant method combined with a penalty for automatic variable selection to provide reliable results when atypical data may arise.

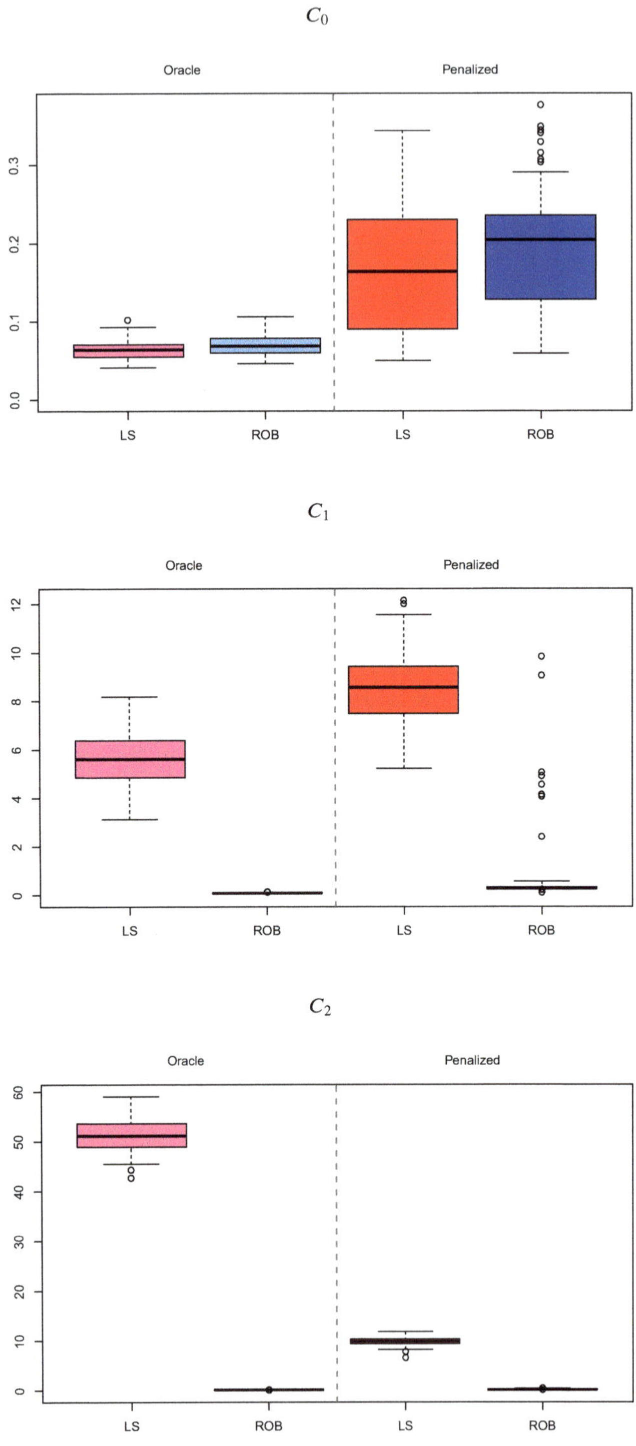

Fig. 44.1: Adjusted boxplot for the ASE values under C_0, C_1 and C_2 for both oracle estimators and both penalized estimators

References

1. Banerjee, S., Ghosal, S.: Bayesian variable selection in generalized additive partial linear models. Stat, **3**, 363-378 (2014)
2. Boente, G., Martínez, A.: A robust spline approach in partially linear additive models. Comput. Stat. Data Anal., **178**, 107611 (2023)
3. Boente, G., Martínez, A.: Robust variable selection for partially linear additive models. Stat. Comput., **34**, 201 (2024)
4. Hu, Y., Zhao, K., Lian, H.: Bayesian quantile regression for partially linear additive models. Stat. Comput., **25**, 651-668 (2015)
5. Hubert, M., Vandervieren, E.: An adjusted boxplot for skewed distributions. Comput. Stat. Data Anal., **52**, 5186-5201 (2008)
6. Kazemi, M., Shahsavani, D., Arashi, M.: Variable selection and structure identification for ultrahigh-dimensional partially linear additive models with application to cardiomyopathy microarray data. Stat., optim. inf. comput., **6**, 373-382 (2018)
7. Lv, J., Yang, H., Guo, C.: Variable selection in partially linear additive models for modal regression. Commun. Stat. Simul. Comput., **46**, 5646-5665 (2017)
8. Maronna, R., Martin, D., Yohai, V., Salibián-Barrera, M.: Robust Statistics: Theory and Methods (with R). John Wiley and Sons (2019)
9. Zhao, Y., Zhang, R., Liu, J., Lv, Y.: Robust and efficient variable selection for semiparametric partially linear varying coefficient model based on modal regression. Ann. Inst. Stat. Math., **66**, 165-191 (2014)

Chapter 45
Covariance Operators for Phonetics: Revisiting Tonal Coarticulation

Valentina Masarotto and Yiya Chen

Abstract We aim to provide tools to tackle a critical question for the linguistic community: what is the most suitable way to study the impact of tonal coarticulation in phonetics? Current methods do not fully capture the intricacies of this phenomenon. We show that we can resolve some controversy on the subject by using what is known in functional data analysis as second-order variation. Our approach treats speech frequency curves as functional observations and leverages a crucial insight: time and frequency covariance functions hold the key to understanding the finer effects of tonal coarticulation. This insight drives our two-step approach. First, we model mean functions using Generalized Additive Models. The residuals of such models are then investigated for any structure nested at covariance level using a state-the-art k-sample test for covariances. If such structure is found, we tap into it using covariance principal component analysis. We apply the method to an articulatory dataset specifically collected to study the cognitive mechanisms of tonal coarticulation.

45.1 Introduction

Most human forms of communication comprise a melodic component. In European languages, this is embodied in the variation of pitch. Volume and intonation are used to capture attention, differentiate affirmations from questions, or serious remarks from sarcastic ones. Pitch variation is an essential part in conveying a message. When moving to tonal languages, pitch variation plays an even more dramatic role: the modulation of the pitch of the sound characterizes the lexical identity of a word.

Valentina Masarotto (✉)
Mathematical Institute, Universiteit Leiden, e-mail: `v.masarotto@math.leidenuniv.nl`

Yiya Chen
Centre for Linguistics and Leiden Institute for Brain and Cognition, Universiteit Leiden, e-mail: `yiya.chen@hum.leidenuniv.nl`

© The Author(s), under exclusive license to Springer Nature Switzerland AG 2025
G. Aneiros et al. (eds.), *New Trends in Functional Statistics and Related Fields*,
Contributions to Statistics, https://doi.org/10.1007/978-3-031-92383-8_45

That is, the meaning of a word can change entirely depending on the tone used to pronounce it. This highlights why the study of pitch variation is central in Phonetics.

An intriguing aspect of tonal languages is tonal coarticulation, which refers to the influence that the pitch or tone of one syllable has on the pitch or tone of adjacent syllables, leading to changes in how tones are realized in context. Tonal coarticulation is as relevant as it is difficult to study it rigorously. The existing body of research primarily focuses on investigating its directionality and magnitude: is the current syllable being modified by the previous one, or will it affect the following one? By how much? Unfortunately studies have revealed diverse and occasionally contradictory patterns ([2] and references thereof). As a whole, research on coarticulation remains scarce, and still lacks a rigorous and general way to be tackled. This makes it a vibrant and significant area of investigation in linguistics.

This paper aims to employ functional data analysis to support existing research and help to adjudicate the current conflicting perspectives on coarticulation and its cognitive mechanisms. At the same time we hope to introduce second order functional inference as a tool to study speech data. We utilize data collected by Ting Zou as part of her Ph.D. thesis [11]. A unique aspect of the data recordings is the introduction of cognitive load, where speakers were asked to perform a memory task while producing speech. The objective of the data collection is to explain which external factors (in this case a stress memory task) have a significant impact on coarticulation. Researchers in linguistics use various ways to transform speech recordings into quantitative data. Most of them are based on the idea of representing sounds through fundamental frequency curves, denoted by f0. An f0 curve is a graphical representation of the fundamental frequency of a sound wave over time. The recent survey [9] gives a wide and well referenced overview of the most common areas of research in phonetics together with the statistical techniques most widely employed. A versatile and popular tool is Generalised Additive Mixed Models (GAMMs). Their drawback is to rely on a "discretized" version of the f0 curves, which instead are most often assumed to be smooth and continuous throughout their trajectories. The intrinsic smoothness of the f0 curves points naturally to the functional data analysis (FDA) domain. Recently there has been a surge of successful application of FDA to linguistic, as it became apparent from the recent works (see references in [9]). What most of these works have in common is that they study f0 curves only with respect to their shape, or their mean structure. However, functional data can exhibit another intriguing kind of variation, statistically known as "second order variation". It refers to the cases where functional covariances display statistical variation. These situations typically arise when several covariance operators can be distinguished from data clusters. Recent literature offers evidence that significant phonetic features are nested in the covariance structure between pitch intensities at different frequencies. In fact, [1] claim that covariances can be considered “a summary of what a language sounds like", while disregarding specific difference at word level.

The novelty of our work consists in proposing a unified approach to study tonal variation of f0 curves both with respect to their shape and their second order covariance structure. Effect of cognitive load on f0 contour is first explored via GAMMs. However, regardless of a good model fit, GAMMs do not seem to provide statistically significant evidence to support the empirical evidence by experts. Thus we conjecture that cognitive load might have an effect at finer resolution, and we offer a way to uncover it, via the functional ANOVA permutation test by [4], which was shown to have state of the art performance. It can be seen that, differently from fitting a regression model, second order analysis yield significant results in all cases considered. Subsequently we tap into this via tangent space covariance PCA, which helps quantify the effect of cognitive load on the spectrum of the covariance operators. To the best of our knowledge, no study considered tonal coarticulation as a second-order effect.

45.2 Statistical methods

As stated in the introduction, we are interested in exploring whether cognitive load has an effect on the oscillation of speech recording *around* the mean, and can thus be identified as a second order effect. This intuition is consistent with [1], where the authors demonstrate that the covariance structure between the intensities at different frequencies captures language-specific variability and does not incorporate differences at the word level. Mathematically it translates into exploring the variation in a finite collection of covariances $\{\Sigma_j\}_{j=1}^N$ on a real separable Hilbert space $\mathcal{H}$, equipped with the inner product $\langle\cdot,\cdot\rangle : \mathcal{H} \times \mathcal{H} \to \mathbb{R}$, and corresponding norm $\|\cdot\| : \mathcal{H} \to [0, \infty)$. Since $\mathcal{H}$ will in principle be infinite-dimensional, we will need to review some basic definitions and notation.

In phonetics, a $f0$ curve is a graphical representation of the fundamental frequency of a sound wave, and it is generally regarded as a smooth function of frequency over time. We will therefore treat $f0$ curves as functional observation a random element X defined over a probability space $(\Omega, \mathcal{F}, \mathbb{P})$ and belonging to $\mathcal{H}$. Unless differently specified, we take $\mathcal{H}$ to be a bounded time domain, such as $X \in L^2[0,T]$ with $\mathbb{E}(||X||^2) < \infty$. The covariance operator is defined as the tensor product

$$\Sigma = \mathbb{E}\left[(X - m) \otimes (X - m)\right] := \int_\Omega (X - m) \otimes (X - m) d\mathbb{P},$$

where m is the mean function $m = \mathbb{E}(X) := \int_\Omega X d\mathbb{P}$.

We work in the scenario of N independent samples $\{X_{1,i}\}_{i=1}^{n_1}, \ldots, \{X_{N,i}\}_{i=1}^{n_N}$ of i.i.d. random elements in $\mathcal{H}$. Each of the N sample is modeled by a prototypical random function X_j, with mean function $\mu_j \in \mathcal{H}$ and covariance operator Σ_j : $\mathcal{H} \times \mathcal{H} \to \mathcal{H}$ and we observe n_j realisations from each population. In applications one inevitably works with finite-dimensional representations of the mean and the covariance functions, which can be seen as functional analogues of the mean vector and variance-covariance matrix in multivariate data analysis. For the rest of the paper, when referring to the mean or the covariance operators, we will actually implicitly refer to their empirical versions, constructed on the basis of the samples of size n from each group. These could simply be the empirical covariances within each group (under a complete observation assumption), or some smoothed estimators. Whichever the case may be, the empirical $\hat{\Sigma}_j$ are finite dimensional, of rank $q \leq n$. In case a smoothing technique is used, we assume that it is such that the $\hat{\Sigma}_j$ share a common range, and can thus be represented as $q \times q$ positive matrices, via a common (tensor product) basis. We refer to [5] for a mathematical justification as to why functional results employed in this paper do hold for the finite dimensional projections.

45.2.1 Functional ANOVA to Compare Unexplained Variation With and Without Cognitive Load and Functional Principal Component Analysis of Covariances

To check the second order effect of coarticulation, we first need a suitable, highly sensitive measure of dissimilarity between second order objects, i.e. covariances. Once we have a suitable metric, we can define a distance-based permutation test, in the fashion of the functional ANOVA procedure of [4]. The main intuition of [4] is summarized hereafter for completeness.

Assume we have N groups of functional data, $\{X_{i,j}\}$, $i = 1, \dots, n$, $j = 1, \dots, N$. Mark as Σ_j the covariances relative to each group, $i = i, \dots, N$. The permutation test of Masarotto et al. exploits elements of optimal transport. Optimal transport is a mathematical field that studies how to optimally deform one distribution of mass to another, while minimizing the cost of such deformation. The optimal deformation is obtained through optimal transport maps. In order to define the testing procedure, we thus need the notion of transport maps. As a measure of dissimilarity we take the Wasserstein-Procrustes distance of optimal transport between covariance operators, whose expression (both in finite and infinite dimension) in the Gaussian case is

$$\Pi^2(\Sigma_1, \Sigma_2) = \inf_{\substack{Z_1 \sim N(0,\Sigma_1) \\ Z_2 \sim N(0,\Sigma_2)}} \mathbb{E}\|Z_1 - Z_2\|^2 = \mathrm{tr}\Sigma_1 + \mathrm{tr}\Sigma_2 - 2\mathrm{tr}\sqrt{\Sigma_1^{1/2}\Sigma_2\Sigma_1^{1/2}}.$$

The test statistics relies on the intuition that under the null, if all covariance operators are the same, then we can build centered Gaussian process with corresponding covariances, and the transport maps to deform one into the other would just be equal to the identity. This intuition gives rise to the following test statistics:

$$T = \sum_{j=1}^{N} \left|\left|\left| \mathbf{t}_j - I \right|\right|\right|_2^2, \tag{45.1}$$

where I is the identity operator, $|||\cdot|||_2$ is the Hilbert-Schmidt norm, and

$$\mathbf{t}_j = \overline{\Sigma}^{-1/2}(\overline{\Sigma}^{1/2}\Sigma_j\overline{\Sigma}^{1/2})^{1/2}\overline{\Sigma}^{-1/2}, \quad j = 1, \dots, N \tag{45.2}$$

are optimal transport maps from the operators to their Fréchet mean. We call $\overline{\Sigma}$ a Fréchet mean of the operators $\{\Sigma_j\}_{j=1}^N$ if $\overline{\Sigma} = \underset{\Gamma \in G}{\arg\min} \sum_{i=1}^K \mathrm{dist}^2(\Sigma_i, \Gamma)$. Fréchet means generalize arithmetic means to general metric spaces. Here G is a suitable parametric function space. We want to stress that the test does not rely on the assumption of Gaussianity. If we have covariance operators, we can build the corresponding centered Gaussian process, and overcome any issue that covariances pose by comparing these Gaussian processes. In fact, we avoid making any parametric assumptions, and calibrate the test statistic by means of permutations.

In practice, we work with the unique empirical Fréchet mean

$$\hat{\Sigma} = \underset{\mathbb{R}^{q \times q} \ni \Gamma \succeq 0}{\arg\min} \sum_{j=1}^{K} \Pi^2(\hat{\Sigma}_j, \Gamma).$$

$\hat{\Sigma}$ can be approximated by the iterative steepest descent procedure [5]. We also work with the empirical versions of the $\mathbf{t}_j$. In [4] it is shown that the transport-maps based test is robust to finite dimensional representation, is state of the art, and extremely sensitive in case of high-frequency functional differences.

If the null hypothesis of equality among covariances is rejected, a second order analysis would entail explaining and understanding the variability of the sample of covariances. As in multivariate analysis in Euclidean spaces, Principal Component Analysis (PCA) is a good candidate for such task. When it comes to covariances, PCA is hindered by their non-linearity. However, [4] solved this obstacle presenting the first instance of tangent space PCA of covariances, whose strength is displayed in Sect. 45.3.

45.3 Results

We present first and second order analysis of the interaction between tonal context and cognitive load. Functional ANOVA is successful in detecting significant differences between covariance operators relative to curves with and without cognitive load.

45.3.1 Data

Twelve native Mandarin speakers participated in the experiment. The study is based on speech recording of balanced non-sensical tonal combination. The stimuli are disyllabic sequences of /ma ma/ syllables with all possible tonal bi-combination of four Mandarin tones: high tone (T1), rising tone (T2), low tone (T3), and falling tone (T4), for a total of 16 possible combinations. We identify them by their tones, thus writing them as $TiTj$, $i, j = 1 \dots, 4$. All tonal combinations were recorded four times each, in two conditions: no-cognitive-load and high-cognitive-load condition.

The cognitive-load conditions were set according to [3] and the reading task was preceded by six one-digit numbers to memorize, and followed by memory testing material. Participants were asked to try their best to remember the digits. In total, there were 128 trials (16 disyllabic combinations, 4 four repetitions, 2 conditions) for each subject (S1-S12). Not all repetitions were usable for all subjects, as some were overloaded with noise. As we would like to filter out subject-specific way of speaking while performing the analysis, and covariances are computed across different speakers, repetitions, and tones, we do not consider the missing recording to be an issue. More details on the data are given on the upcoming publication. A technical report with more detailed analysis is available upon request.

45.3.2 First Order Analysis via Generalized Additive Models

The coarticulatory effects on f0 curves were first examined through generalized additive mixed models via the `mgcv` package [10] in `R` (version 1.8-23).

We worked on the assumption that the bisyllabic $TiTj$ words differ from one another, for every speaker, combination and repetition. Moreover, we accept that each bi-syllabic entry might include speaker- and repetition-specific variational patterns (such as age, gender, and emotional state), which are incorporated as random effects in the model. The statistical significance of random effects is not tested, as it is beyond the scope of the analysis. Fixed effects are compared and analyzed separately for every tonal combination. Tone contour and cognitive load appear as a fixed constant term. Cognitive load and the interaction between cognitive load and following or preceding tone are added as non-linear terms over time. Ordered factors were used to tap into potential cognitive load effect on particular tonal difference. An AR(1) error model is incorporated for residuals, assuming a Gaussian model. Results are given in Table 45.1. Due to space constraint we include only the result relative to the modelling of T1 in syllable 1 as a function of the following tones and cognitive load. We indicate as CL_0 the lack of cognitive load and CL_6 the presence of cognitive load (as to 0 or 6 ciphers to remember). We can see that cognitive load has not at all or very little effect on tones T2 and T4, and only seems slightly detectable on the parametric coefficients, even if the GAMM model provides an acceptable fitting (Fig. 45.1). This contradicts the empirical knowledge and expectations of experts. Similar results can be seen for other tonal combinations, which are available upon request. In the next two Sections, we will investigate if a cognitive load effect exists but it is hidden at a higher resolution, namely into the covariance structure of the residuals.

45.3.3 Functional ANOVA of Covariances

Let X_{i,j,S_n,CL_l}, $i, j = 1, \ldots, 4$, $n = 1, \ldots, 12$, $l \in 0, 6$ be the residuals curves of the GAMM model. We fit a separate model for each TiTx and for each TxTj $i, j = 1, \ldots, 4$. Each gives rise to 384 residual curves (12 speakers, 4 repetitions, 4

Parametric coefficients	p-value	Smooth terms	p-value
(Intercept)	$< 0.01^{***}$	s(time): CL_0 T1.T1	0.45
CL_6 T1.T1	$< 0.01^{***}$	s(time): CL_6 T1.T1	0.69
CL_6 T1.T2	0.20	s(time): CL_6 T1.T2	0.06
CL_6 T1.T3	$< 0.01^{***}$	s(time): CL_6 T1.T3	0.03^{*}
CL_6 T1.T4	0.03^{*}	s(time): CL_6 T1.T4	0.39
		s(time, subject interaction)	$< 0.01^{***}$

Table 45.1: GAMM output to model anticipatory tonal coarticulation effect in T1Tx combination.

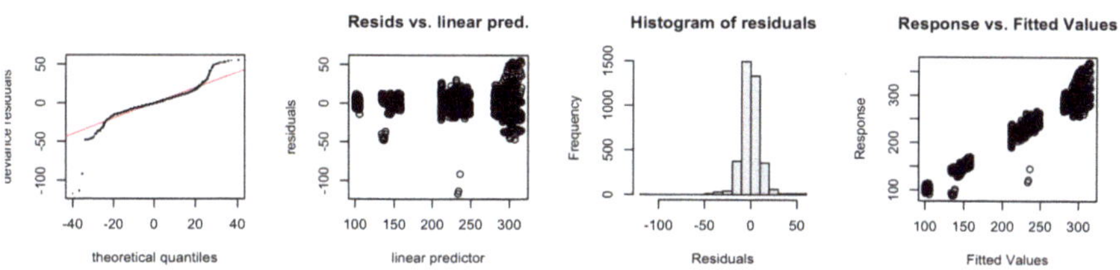

Fig. 45.1: GAMM model check for T1Tx f0 contours

tonal combinations where either the first or the second tone is fixed and two cognitive load conditions). Let $\widehat{\Sigma}^{CL0}_{TiTj}, \widehat{\Sigma}^{CL6}_{TiTj}$ $i, j = 1, \ldots, 4$ be the bi-syllable specific empirical covariances extracted from the residuals of the GAMM model. We employ the permutation-based test defined in Sect. 45.2.1 to verify whether cognitive load has a second order effect. In mathematical terms this translates into testing for every tonal combination $TiTj, i, j = 1, \ldots, 4$ the null hypothesis of equality: $H_0 : \widehat{\Sigma}^{CL0}_{TiTj} = \widehat{\Sigma}^{CL6}_{TiTj}$ versus the alternative hypothesis that the two operators differ. In practice, we generate $2n$ curves (n for each cognitive load scenario). The sample size is $n = 30$. The number of permutations is 500. Table 45.2 collects the p-values for each tonal combination. A significant p-value (less than 0.05) would indicate that there is a significance difference between the covariance structure of the residuals of the GAMMs model, when we compare curves with and without cognitive load. It is evident from Table 45.2 that this is the case, and the covariance structure of the residuals carries a significant difference between the curves with cognitive load, and the curves without. The effect of cognitive load is independent on the tonal combination considered. Given that the cognitive load seems to be a second order effect, it is not surprising that the GAMMs analysis, albeit correct, was not fully successful in modelling it.

After performing the global test, since the null hypothesis H_0 is rejected, one can investigate main directions of variation through a tangent space principal component analysis. We only display results relative to the collection of curves $T1Tj$, $j = 1, \ldots, 4$. More plots and details are to be found in the upcoming publications. Tangent space PCA captures very well the difference between cognitive load and lack thereof, as most scenarios are isolated in at least one plot. More precisely: The first PC captures (part of) the difference between T1T3 (with and without cognitive load) and all the other tonal combinations. This is expected, as moving from any tone to T3 requires a lot of amplitude variation, which is exactly what covariance operators capture. The third PC captures (part of) the difference between T1T3 with or without cognitive load, while the fourth does it for T1T4 (Table 45.3). The first 3 PCs explain almost the full variance of the data, yet marking cognitive load as a subtle effect. Interpretation of the effect of each PCs individually is in general not trivial, making particularly pleasing the fact that we can interpret them in terms of cognitive load in this setting.

	pv		pv		pv		pv		pv		pv		pv		pv
T1T1	0.02	T2T1	0.02	T3T1	0.01	T4T1	0.01	T1T2	0.03	T2T2	0.04	T3T2	0.01	T4T2	0.01
T1T3	0.02	T2T3	0.02	T3T3	0.01	T4T3	0.02	T1T4	0.01	T2T4	0.01	T3T4	0.01	T4T4	0.01

Table 45.2: p-values for the functional ANOVA test applied on the residual covariance operators. A significant p-value indicates a second-order effect of the cognitive load.

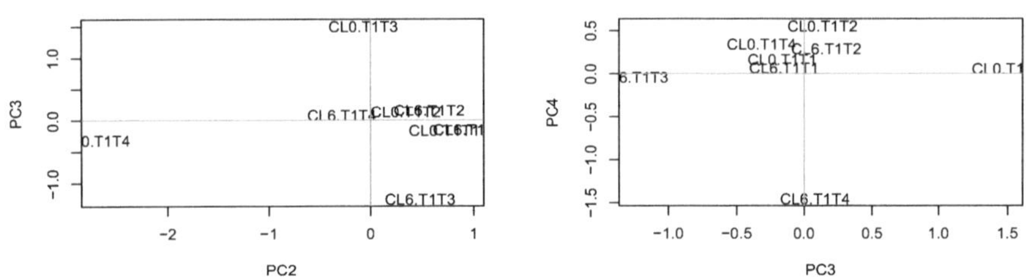

Table 45.3: PCA plot for T1Tj combination

References

1. Aston, J.A., Chiou, J.M., Evans, J.P.: Linguistic pitch analysis using functional principal component mixed effect models. Journal of the Royal Statistical Society Series C: Applied Statistics, **59**(2), 297-317 (2010)
2. Chen, Y. Tonal variation In Cohn, A. C., Fougeron, C. and Huffman, M.K.: The Oxford handbook of laboratory phonology, Oxford University Press, (2012)
3. Lavie, N., Hirst, A., De Fockert, J.W., Viding, E.: Load theory of selective attention and cognitive control. Journal of Experimental Psychology: General, **133**(3), 339 (2004)
4. Masarotto, V., Panaretos, V.M., Zemel, Y.: Transportation-based functional ANOVA and PCA for covariance operators. Electronic Journal of Statistics, **18**(1), 1887-1916 (2024)
5. Masarotto, V., Panaretos, V.M., Zemel, Y.: Procrustes metrics on covariance operators and optimal transportation of Gaussian processes. Sankhya A, **81**, 172-213 (2019)
6. Masarotto, V. Masarotto, G.: Package ‘fdWasserstein’, R (2024)

7. Pigoli, D., Aston, J.A., Dryden, I.L., Secchi, P.: Distances and inference for covariance operators. Biometrika, **101**(2), 409-422 (2014)
8. Pigoli, D., Hadjipantelis, P.Z., Coleman, J.S., Aston, J.A.: The statistical analysis of acoustic phonetic data: exploring differences between spoken Romance languages. Journal of the Royal Statistical Society Series C: Applied Statistics, **67**(5), 1103-1145 (2018)
9. Tavakoli, S., Matteo, B., Pigoli, D., Chodroff, E., Coleman, J., Gubian, M., Renwick, M. EL, Sonderegger, M.: Statistics in phonetics. Annual Review of Statistics and Its Application, 12 (2024)
10. Wood, S.N.: Generalized additive models: an introduction with R. Chapman and Hall/CRC (2006)
11. Zou, T.: Production and perception of tones by Dutch learners of Mandarin. Ph.D. Thesis, Leiden University (2017)

Chapter 46
A-optimal Designs of Experiments in Linear Models with Dynamic Factors and Functional Responses

Caterina May, Theodoros Ladas, Davide Pigoli and Kalliopi Mylona

Abstract Optimal design of experiments with functional data is a novel and highly relevant research topic. We illustrate here a recent work which considers linear models where both the response and one or more factors are continuous functions, for example, of the time. The goal is to find the optimal dynamic experimental conditions to estimate precisely the functional coefficients of the model. After establishing a suitable estimator and obtaining its variance-covariance matrix, the definition of optimal design criteria is extended to the functional data context. In this short paper, some additional examples and numerical results on A-optimal dynamic designs are provided, with the choice of suitable bases of functions to represent the data. The efficiency of estimators is compared under various choices of bases.

46.1 Introduction

Optimal design of experiments (DOE) aims at efficiently planning, conducting, and analyzing experiments to gather the most informative data while minimizing resource expenditure. The theory and methodology of optimal DOE have been widely developed and used in various fields, including engineering, medicine, agriculture, and marketing. Classical references are, for instance, [3], and [4] and [5], among many others. At its core, an optimal design seeks to maximize the precision of pa-

Caterina May (✉)
King's College London, UK and Università degli Studi del Piemonte Orientale, Italy,
e-mail: caterina.may@kcl.ac.uk, caterina.may@uniupo.it

Theodoros Ladas
King's College London, UK, e-mail: theodoros.ladas@kcl.ac.uk

Davide Pigoli
King's College London, UK, e-mail: davide.pigoli@kcl.ac.uk

Kalliopi Mylona
King's College London, UK, e-mail: kalliopi.mylona@kcl.ac.uk

© The Author(s), under exclusive license to Springer Nature Switzerland AG 2025

G. Aneiros et al. (eds.), *New Trends in Functional Statistics and Related Fields*,
Contributions to Statistics, https://doi.org/10.1007/978-3-031-92383-8_46

rameter estimates in a statistical model or to achieve specific experimental objectives. The "optimality" is determined by mathematical criteria tailored to the experiment's goals. By strategically selecting experimental settings (e.g., input variables), optimal DOE minimizes redundancy and focuses resources on the most impactful experiments. This leads to robust and actionable insights while saving time, cost, and effort.

In many modern applications data are collected as curves over time, and it is therefore natural to consider statistical models with curves (or functions) as responses and/or as covariates (see [2] and [6] for a comprehensive discussion on these models). However, the optimal design of experiments in the context of functional data has been explored very little until now. Recent work [1] fills a gap in this field by developing optimal design of experiments for linear models in which both experimental factors and responses have a functional nature. Some more examples and numerical results for A-optimal designs of experiments in linear models with dynamic factors and functional responses are provided in this short paper as an illustration of the methodology proposed.

46.2 A Motivating Example

In the context of forensic entomology, there is an interest in the association between temperature curves and insects' growth curves. This problem is well described and studied in [7]; however, in a laboratory setting, the current practice is to keep the incubator temperature constant for each run, and then having various runs at different temperatures (see Fig. 1). This is a choice based only on practicality, since it doesn't

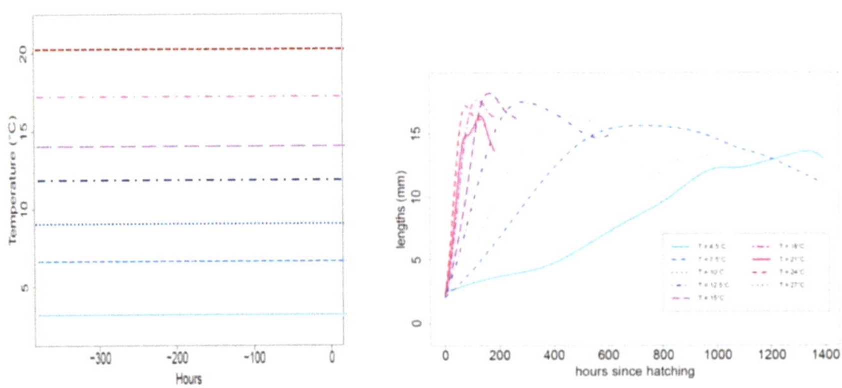

Fig. 46.1: Example from the National History Museum in London: experiments on larval growth curves at constant temperatures. The right side figure is taken from [7]

require the operator to change the incubator temperature during the run. A natural question is if there is a more efficient way to design the experiment in order to explore the relationship between the growth rate curve and the temperature which is in reality a dynamic factor.

46.3 Model, Inferential Results and Optimal Design of Experiments

We consider the following function-on-function linear model:

$$\mathbf{y}(t) = \int_0^{\mathcal{T}} \mathcal{X}(s)\boldsymbol{\beta}(s,t)ds + \boldsymbol{\varepsilon}(t), \tag{46.1}$$

where $\mathbf{y}(t) = (y_1(t), \dots, y_N(t))^T$ is the vector of responses, $\boldsymbol{\varepsilon}(t) = (\varepsilon_1(t), \dots, \varepsilon_N(t))^T$ is the vector of errors, which are i.i.d. zero-mean stochastic processes (not necessarily Gaussian), $\boldsymbol{\beta}(s,t) = (\beta_0(t), \beta_1(s,t), \dots, \beta_p(s,t))^T$ is the functional coefficient to be estimated, $\mathcal{X}(s)$ is the $N \times (p+1)$ model matrix:

$$\mathcal{X}(s) = [\mathbf{1}_N \,|\, \mathrm{X}(s)], \tag{46.2}$$

with $\mathrm{X}(s)$ being the $N \times p$ design matrix of the dynamic predictors:

$$\mathrm{X}(s) = \left[\mathbf{x}_1(s)\; \mathbf{x}_2(s) \dots \mathbf{x}_p(s)\right] = \begin{bmatrix} x_{11}(s) & \dots & x_{1p}(s) \\ x_{21}(s) & \dots & x_{2p}(s) \\ \vdots & & \vdots \\ x_{N1}(s) & \dots & x_{Np}(s) \end{bmatrix} \tag{46.3}$$

In [1], a reduced ranked estimator of the unknown functional coefficient is studied that minimizes the integrated sum of squares. To obtain this estimator, the functional coefficients in model (46.1) are expanded in proper bases of functions (separable representation):

$$\beta_i(s,t) = \sum_{k,l} b^i_{k,l}\, \eta^i_k(s)\, \theta_l(t)$$

and projected in a subspace given by a finite number of elements of the bases. The expression of the variance-covariance of this estimator, which is a matrix B of random coefficients, is then provided (see [1] for details).

46.3.1 Optimal Design of Experiments

In the context of experimental design, the experimenter chooses the predictors for each run $n = 1, ..., N$, according to a specific objective. Our aim here is to construct dynamic A-optimal designs, which minimize the trace of the covariance matrix of the estimator (that is, the total variance). In [1] it is shown that these designs do not depend on the unknown autocovariance structure of the error process and the expression of the criterion to be maximized is provided. The expression derived is still based on the functional expression of the factors. However, in practice it is needed to use some bases of functions to express them:

$$x_{ni}(s) = \sum_{k=1}^{K_x^i} \gamma_{nk}^i \, c_k^i(s),$$

for any $n = 1, ..., N$ and $i = 1, ..., p$, and the optimization will be with respect to the basis coefficients. Crucially, though, no requirements on this basis functions are put, that is, the range of admissible functional shapes can be freely chosen based on the experimental constraints. On the other hand, it is also possible to compare designs based on different basis functions to see which is more efficient.

46.4 Numerical Examples

In this section we illustrate an example of dynamic A-optimal design for model (46.1) and some numerical results. For simplicity, we consider the case of a single predictor. As illustrated in detail in [1], the design depends on the basis $\{c_k(s), 1 \leq k \leq K_x\}$ chosen to expand the predictor, and on the basis $\{\eta_k(s), 1 \leq k \leq K\}$ chosen to expand the functional coefficient in the direction of s.

In [1], examples using B-splines of various degrees are provided and discussed. Here, we complete this exemplification with the case of B-splines of degree 2 for both factor ad coefficient: the exact dynamic A-optimal design obtained with $N = 12$ runs is computed with a coordinate-exchange algorithm (see [8]) implemented in Python 3.10, using 1000 random starts and 100 refinement iterations. In Fig. 46.2, the 12 runs of the A-optimal dynamic factor are displayed, each one represented by a second-degree B-spline basis that yields smooth curves over the interval $[-1, 1]$. Additionally, since the functional coefficient is also expanded using a second-degree B-spline, the factor exhibits multiple fluctuations between its highest and lowest values throughout the duration of the experimental run.

Finally, we provide the numerical results of the relative efficiencies of the functional A-optimal designs obtained for a range of combinations of different B-splines degrees and basis dimensions. Fig. 46.3, 46.4 and 46.5 display these results. The columns indicate the number of breaks in the functional coefficient, while the rows show different parameterizations of the functional factor. The tables provide the full range of efficiencies for zero-, first-, and second-degree B-splines, along with various numbers of equidistant breaks. The relative efficiency, that is, the ratio between the traces of the covariance matrices, is obtained with respect to the last row, that is, the maximum number of breaks of the B-splines expanding the factor. We observe that as the functional coefficient incorporates more breaks, additional breaks in the functional factor are required to achieve efficiency levels closer to 100%.

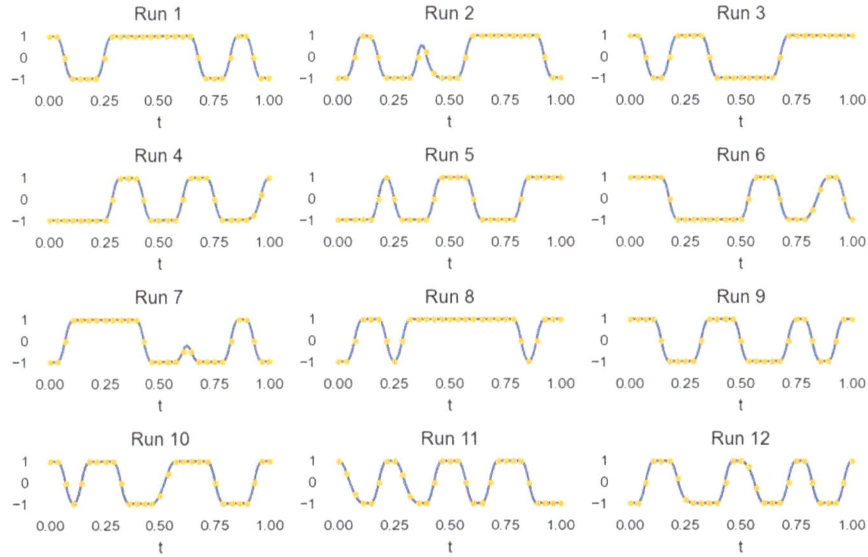

Fig. 46.2: A-optimal design obtained by expanding both the dynamic factor and the functional coefficient with a B-splines basis of degree 2. The number of equidistant breaks is 29 for the dynamic factor and 8 for the coefficient.

46.5 Conclusions

In this paper a strategy proposed in [1] for obtaining A-optimal designs for reduced rank estimators in fully functional linear models is illustrated, and some additional examples and numerical results are provided. A similar method could be used to obtain functional D-optimal designs. Exploring other optimality criteria is planned for future work.

This study fills an important research gap on optimal design of experiments for linear models with functional responses and dynamic factors. A key results is that the optimal designs do not depend on the stochastic distributions of the errors and on the bases of functions used for the responses. Moreover, the basis representation of the factors can be flexibly chosen based on the experimental requirements. On the other hand, the complexity of the coefficients in the direction of the factor is an important modelling choice that impact on the design.

Acknowledgements This work was supported by the major research grant entitled *Multi-objective optimal design of experiments* funded by the EPSRC (Engineering and Physical Sciences Research Council, UK). The first author was supported also by the Italian national project PRIN 2022 [grant number 2022TRB44L].

Fig. 46.3 Relative efficiencies of the A-optimal designs obtained with zero degree B-splines expanding the functional coefficient and different splines expanding the functional factor in the direction of s

Relative Efficiencies		Zero-th Degree							
Degree	Breaks	1	2	3	4	5	6	7	8
Zero-th	1	1.00	-	-	-	-	-	-	-
	2	1.00	1.00	-	-	-	-	-	-
	3	1.00	1.00	1.00	-	-	-	-	-
	5	1.00	1.00	1.00	0.71	1.00	-	-	-
	9	1.00	1.00	1.00	0.83	1.00	0.70	0.77	0.58
	15	1.00	1.00	1.00	0.94	0.86	0.87	0.90	1.02
	19	1.00	1.00	1.00	1.05	0.89	0.93	1.13	0.80
	29	1.00	1.00	1.00	1.00	1.00	1.00	1.00	1.00
First	1	1.00	1.00	-	-	-	-	-	-
	2	1.00	1.00	0.49	-	-	-	-	-
	3	1.00	1.00	0.58	0.44	-	-	-	-
	5	1.00	1.00	0.80	0.67	0.40	0.42	-	-
	9	1.00	1.00	0.91	0.90	0.74	0.71	0.66	0.62
	15	1.00	1.00	0.97	0.96	0.95	0.87	0.85	0.77
	19	1.00	1.00	0.98	0.93	0.99	0.92	0.86	0.94
	29	1.00	1.00	1.00	1.00	1.00	1.00	1.00	1.00
Second	1	1.00	1.00	0.52	-	-	-	-	-
	2	1.00	1.00	0.52	0.18	-	-	-	-
	3	1.00	1.00	0.67	0.37	0.22	-	-	-
	5	1.00	1.00	0.85	0.63	0.48	0.19	0.18	-
	9	1.00	1.00	0.94	0.80	0.77	0.62	0.60	0.45
	15	1.00	1.00	0.98	0.92	0.86	0.82	0.85	0.83
	19	1.00	1.00	0.99	0.99	0.89	0.91	1.02	0.83
	29	1.00	1.00	1.00	1.00	1.00	1.00	1.00	1.00

Bsplines Beta

Fig. 46.4 Relative efficiencies of the A-optimal designs obtained with first degree B-splines expanding the functional coefficient and different splines expanding the functional factor in the direction of s

Relative Efficiencies		First Degree							
Degree	Breaks	1	2	3	4	5	6	7	8
Zero-th	1	-	-	-	-	-	-	-	-
	2	-	-	-	-	-	-	-	-
	3	1.00	1.00	-	-	-	-	-	-
	5	1.00	1.00	0.96	0.96	-	-	-	-
	9	1.00	1.00	0.97	0.95	0.86	0.90	0.94	0.86
	15	1.00	1.00	0.99	0.99	0.96	0.98	0.92	0.93
	19	1.00	1.00	1.00	1.00	1.00	0.96	0.99	0.97
	29	1.00	1.00	1.00	1.00	1.00	1.00	1.00	1.00
First	1	0.57	-	-	-	-	-	-	-
	2	0.57	0.57	-	-	-	-	-	-
	3	0.70	0.70	0.54	-	-	-	-	-
	5	0.92	0.92	0.79	0.55	0.55	-	-	-
	9	0.98	0.98	0.97	0.91	0.85	0.74	0.62	0.52
	15	1.00	1.00	0.99	0.97	0.98	0.93	0.93	0.87
	19	1.00	1.00	1.00	0.99	0.97	0.97	0.97	0.96
	29	1.00	1.00	1.00	1.00	1.00	1.00	1.00	1.00
Second	1	0.63	0.63	-	-	-	-	-	-
	2	0.63	0.63	0.26	-	-	-	-	-
	3	0.79	0.78	0.49	0.33	-	-	-	-
	5	0.95	0.95	0.81	0.60	0.37	0.31	-	-
	9	0.99	0.99	0.93	0.86	0.75	0.71	0.60	0.53
	15	1.00	1.00	0.99	0.96	0.93	0.91	0.84	0.81
	19	1.00	1.00	0.99	0.96	0.97	0.97	0.94	0.89
	29	1.00	1.00	1.00	1.00	1.00	1.00	1.00	1.00

Bsplines Beta

Fig. 46.5 Relative efficiencies of the A-optimal designs obtained with second degree B-splines expanding the functional coefficient and different splines expanding the functional factor in the direction of s.

Relative Efficiencies		**Second Degree**							
Degree	*Breaks*	1	2	3	4	5	6	7	8
Zero-th	1	-	-	-	-	-	-	-	-
	2	-	-	-	-	-	-	-	-
	3	-	-	-	-	-	-	-	-
	5	0.99	0.99	0.79	-	-	-	-	-
	9	0.97	0.99	0.95	0.98	0.91	0.79	0.47	-
	15	1.00	1.00	0.98	0.99	0.94	0.99	1.00	0.98
	19	0.99	1.00	0.99	1.01	0.97	1.01	0.99	1.00
	29	1.00	1.00	1.00	1.00	1.00	1.00	1.00	1.00
First	1	-	-	-	-	-	-	-	-
	2	-	-	-	-	-	-	-	-
	3	0.54	0.54	-	-	-	-	-	-
	5	0.70	0.72	0.60	0.57	-	-	-	-
	9	0.93	0.93	0.92	0.79	0.71	0.62	0.60	0.59
	15	0.98	0.99	0.98	0.96	0.95	0.89	0.79	0.79
	19	0.99	0.99	0.99	1.00	0.96	0.95	0.94	0.89
	29	1.00	1.00	1.00	1.00	1.00	1.00	1.00	1.00
Second	1	0.19	-	-	-	-	-	-	-
	2	0.19	0.19	-	-	-	-	-	-
	3	0.50	0.48	0.31	-	-	-	-	-
	5	0.80	0.78	0.59	0.39	0.33	-	-	-
	9	0.95	0.94	0.84	0.80	0.66	0.60	0.49	0.34
	15	0.99	0.98	0.95	0.93	0.89	0.84	0.82	0.76
	19	1.01	0.99	0.98	0.97	0.95	0.91	0.88	0.86
	29	1.00	1.00	1.00	1.00	1.00	1.00	1.00	1.00

Bsplines Beta

References

1. May, C., Ladas, T., Pigoli, D. and Mylona, K.: Optimal design of experiments for functional linear models with dynamic factors, (2024).
2. Ramsay, J.O., Silverman, B.W.: Functional data analysis. Second edition. Springer, New York, (2005)
3. Silvey, S.D.: Optimal Design: Introduction to the Theory for Parameter Estimation. Monographs on applied probability and statistics. Chapman and Hall. (1980)
4. Pukelsheim, F.: Optimal Design of Experiments. Wiley Series in Probability and Statistics. Wiley, New York, (1993)
5. Atkinson, A.C., Donev, A.N., Tobias, R.D.: Optimum Experimental Designs, With SAS. Oxford Statistical Science Series, vol. 34, p. 511. Oxford University Press, Oxford (2007)
6. Horvath, L., Kokoszka, P.: Inference for Functional Data with Applications. Springer, NY (2012)
7. Pigoli, D., Ferraty, F., Aston, J.A.D., Mazumder, A., Richards, C., Hall, M.J.R.: Estimation of temperature-dependent growth profiles for the assessment of time of hatching in forensic entomology. Journal of the Royal Statistical Society Series C: Applied Statistics **72**(2), 231 253 (2023)
8. Meyer, R.K., Nachtsheim, C.J.: The coordinate-exchange algorithm for constructing exact optimal experimental designs. Technometrics **37**(1), 60–69 (1995)

Chapter 47
Moving Object-Oriented Spatial Statistics Beyond Stationary and Euclidean Paradigms

Alessandra Menafoglio

Abstract This communication explores Object-Oriented Spatial Statistics (O2S2) as a framework for addressing the challenges of analyzing spatially dependent, complex data structures, such as curves, surfaces, or distributions. By highlighting recent advancements in the field, we will demonstrate how geometric principles can be utilized to model spatially distributed functional data, while accounting for non-stationary behaviors and non-Euclidean spatial domains. A particular emphasis will be given to carefully selecting the embedding space for the analysis, particularly when data exhibit specific features and constraints – a key example being the case of probability density functions, which can be effectively modeled within Bayes spaces. Despite notable progress, challenges remain to build effective estimation methods for the covariance structure of functional fields under non-stationary or non-Euclidean settings, with these challenges being only partially alleviated by using integrated notions of spatial dependence, such as the trace-covariogram and trace-variogram.

47.1 Introduction

In recent years, the statistical community has faced significant challenges in analyzing and extracting insights from increasingly complex and varied datasets. Conventional statistical techniques, originally developed for scalar data, often struggle when applied to more complex data structures such as curves, surfaces, or images. To address these limitations, Functional Data Analysis (FDA, [23]) and Object-Oriented Data Analysis (OODA, [12]) have emerged as prominent frameworks, enabling to extend classical statistical methods to accommodate these richer data representations.

Georeferenced data presents an additional layer of complexity, as spatial dependence must be considered alongside the underlying data structure. This has driven a

Alessandra Menafoglio (✉)
MOX, Department of Mathematics, Politecnico di Milano,
e-mail: alessandra.menafoglio@polimi.it

© The Author(s), under exclusive license to Springer Nature Switzerland AG 2025

G. Aneiros et al. (eds.), *New Trends in Functional Statistics and Related Fields*,
Contributions to Statistics, https://doi.org/10.1007/978-3-031-92383-8_47

surge of interest in developing spatial statistical approaches tailored to diverse data types, including functional data, probability distributions, and Riemannian objects. In response to these challenges, Object-Oriented Spatial Statistics (O2S2, [13]) has been introduced as a unifying framework that leverages geometric principles to analyze spatially distributed complex data.

Despite these advancements, several challenges remain in the context of O2S2. Most existing methodologies are tailored for Euclidean spatial domains or for settings that can be locally approximated by the Euclidean geometry [16, 17], and they often rely on the assumption of stationarity (at least in the covariance function). However, many natural processes, especially in the geosciences, evolve in inherently non-Euclidean spaces where distances are more appropriately measured along geodesic paths shaped by physical constraints. These processes also frequently exhibit strong non-stationary behaviors, often driven by the underlying physics. For instance, in hydrological studies, spatial relationships between measurement sites are better represented using water distance along a stream network rather than direct Euclidean distance, which fails to account for the network's topology and flow connectivity. Moreover, such dynamics can significantly influence the local distribution of observed variables, leading to spatially varying moments and covariance structures. In this communication, we will thus present an overview of the object-oriented approach to spatial functional data analysis, with a particular focus on recent advancements in addressing challenges related to non-Euclidean domains and the modeling of spatially non-stationary phenomena.

The remainder of this contributions is structured as follows. Sect. 47.2 introduces the general O2S2 framework for analyzing spatial random fields in Hilbert spaces, outlining key definitions and assumptions. Sect. 47.3 highlights the role of the embedding space, using Bayes Hilbert spaces of probability density functions [7, 24] as a paradigmatic example. Sect. 47.4 explores two major open challenges in the field: the analysis of non-stationary fields and the study of data over non-Euclidean domains. This section also discusses modeling strategies that can be employed to address these issues. Finally, Sect. 47.5 presents the conclusions of this work.

47.2 A General Model for O2S2 in Stationary and Euclidean Settings

Let $(\Omega, \mathfrak{F}, \mathbb{P})$ a probability space, and let $\mathcal{H}$ be a separable Hilbert space equipped with operations $(+, \cdot)$ and an inner product $\langle \cdot, \cdot \rangle$. In particular, the space $\mathcal{H}$ denotes the *feature* space for the geostatistical analysis, as data are regarded as realizations of random points in $\mathcal{H}$. In this communication, $\mathcal{H}$ will always consists of real-valued functions on a compact interval, though the framework also extends to other types of functional observations as well as to manifold data (see, e.g., [22]).

For spatial locations $s_1, ..., s_n$ in a domain $D \subset \mathbb{R}^d$ $(d = 2, 3)$, observations $\mathcal{X}_{s_1}, ..., \mathcal{X}_{s_n}$ form a dataset of spatially dependent objects. Following classical geostatistics [**?**, e.g.,]]menaCressie1993, these are viewed as partial observations of a random field $\{\mathcal{X}_s, s \in D\}$ defined on $(\Omega, \mathfrak{F}, \mathbb{P})$ with values in $\mathcal{H}$. The field is assumed to satisfy $\mathbb{E}[\|\mathcal{X}_s\|^2] < \infty$, allowing the expected value $m_s = \mathbb{E}[\mathcal{X}_s]$ to be equivalently defined via a Bochner integral w.r.t $\mathbb{P}$, or in the Frechét sense.

The spatial covariance function C characterizes the second-order structure of the field, mapping each pair (s_1, s_2) to the cross-covariance operator $C(s_1, s_2) : \mathcal{H} \to \mathcal{H}$ defined, for $x \in \mathcal{H}$ as

$$C(s_1, s_2)x = \mathbb{E}[\langle(\mathcal{X}_{s_1} - m_{s_1}), x\rangle(\mathcal{X}_{s_2} - m_{s_2})]\}. \tag{47.1}$$

Note that $C(s, s)$ is the (auto)covariance operator targeted by several widely used methods in FDA, such as functional principal component analysis [9]). A global measure of spatial dependence is given by the trace-covariogram

$$C(s_1, s_2) = \mathbb{E}[\langle\mathcal{X}_{s_1} - m_{s_1}, \mathcal{X}_{s_2} - m_{s_2}\rangle], \tag{47.2}$$

which represents the trace of the associated cross-covariance operator $C(s_1, s_2)$ (see [14]). A related measure, the trace-variogram, is defined as $2\gamma(s_1, s_2) = \mathbb{E}[\|\mathcal{X}_{s_1} - \mathcal{X}_{s_2}\|^2] - \|m_{s_1} - m_{s_2}\|^2$, for s_1, s_2 in D.

These concepts enable definitions of second-order stationarity. This means requiring that *(i)* the mean is spatially constant (i.e., $\mathbb{E}[\mathcal{X}_s] = m$ for all $s \in D$), and *(ii)* the family of cross-covariance operators can be defined in terms of lag vectors $s_1 - s_2$, instead of pairs (s_1, s_2) (i.e., $\mathbb{E}[\langle\mathcal{X}_{s_1} - m, \cdot\rangle(\mathcal{X}_{s_2} - m)] = C(s_1 - s_2)$, for all s_1, s_2 in D). Formulating a stationary hypothesis based on the trace-covariogram (a.k.a. global second-order stationarity) leads to a weaker requirement *(ii)*, namely to *(ii')* $\mathbb{E}[\langle\mathcal{X}_{s_1} - m, \mathcal{X}_{s_2} - m\rangle] = C(s_1 - s_2)$, for all s_1, s_2 in D. In both cases, the corresponding isotropic cases are obtained by further requiring that the family C or the function C only depend on the distance $\|s_1 - s_2\|_d$. Under these assumptions, which essentially state a homogeneity of the process in space, estimation of the covariance function is possible, as broadly explored in the literature under different settings (e.g., [8, 18, 9, 14, 22]). For instance, estimation of the trace-covariogram can be performed by a two-step procedure, where first an empirical estimator for the trace-variogram is used, and then a parametric model is fitted to guarantee validity (e.g., positive definiteness) of the obtained covariance model.

47.3 On the Importance of the Embedding Space

The choice of feature space is a fundamental aspect of O2S2, as it directly influences the mathematical operations and similarity measures used for analysis. Indeed, it should be noted that all the definitions given in Sect. 47.2 are inherently associated to the chosen geometry of the Hilbert space H, which directly affects how the salient features of the process are extracted. Moreover, operations such as addition and scalar multiplication within the feature space play a crucial role in defining linear predictors like kriging (see [13] for a review). The latter is build as a linear combination of the observed data, given by $\sum_{i=1}^{n} \lambda_i \mathcal{X}_{s_i}$, where the weights $\lambda_1, ..., \lambda_n$ are optimized for prediction [14]. The structure of this linear combination is dictated precisely by the operations defined in $\mathcal{H}$.

Selecting an appropriate feature space is thus critical for ensuring that the mathematical framework accurately captures the characteristics of the data. For instance, a space such as L^2 or a Sobolev space may be suitable for unconstrained functional data but would be inadequate for tensor or distributional data, such as probability density functions (PDFs). These latter types of data are frequently encountered in geosciences. A prime example is particle-size fractions (PSFs) and their continuous counterparts, particle-size densities (PSDs), which are widely used in hydrogeological studies and various applied sciences related to flow and transport modeling. These quantities are closely tied to the porosity and permeability of the medium, and their analysis must account for their compositional nature—i.e., their total sum or integral is constrained to be one. Ignoring these constraints and treating PDFs evaluations on an absolute scale can introduce spurious correlations, leading to biased estimates and misleading results [1, 20]. More in general, embedding continuous PDFs in L^2 and constructing linear approximations or predictors based on a Euclidean geometry can result in negative densities or totals that deviate from one [10, 15]. Such inconsistencies arise because Euclidean spaces are not well-suited for these data types, which naturally belong to a simplex.

From the O2S2 perspective, an analyst must first consider the geometric properties of the data–such as positivity and compositional constraints–before selecting a feature space. For discrete PDFs, a natural choice is the Aitchison geometry for compositional data within the simplex [20], while for continuous PDFs, its continuous counterpart –a.k.a. *Bayes Hilbert space*– offers a more appropriate framework [24, 15, 10]. In Bayes spaces, classical point-wise addition and multiplication by a constant are replaced by perturbation and powering operations, which involve point-wise multiplication or exponentiation by a constant, re-normalized to ensure an integral of one. Moreover, the inner product is defined using log-ratios of point evaluations of the PDFs rather than their absolute values, better reflecting the relative scale of the data imposed by the constraints.

We finally remark that, although the focus of this communication is on Hilbert spaces, the optimal feature space may not be a Hilbert space, but could instead be a Riemannian manifold, or a Banach space, depending on the nature of the data, inducing an additional layer of complexity due to the absence of an inner product. See [21, 17] for recent advances in the area of O2S2 for Riemannian data.

47.4 Beyond Stationary and Euclidean Paradigms

47.4.1 Modelling Non-Stationarity Covariances

While anisotropic extimations are easly envisioned from available methodologies, relaxing the stationary assumption is way more difficult. A natural way out involves decoupling the elements $\mathcal{X}_s$ in a non-stationary mean m_s (typically estimated via a linear model) and a stationary residual δ_s – as done by [14, 11]. Nonetheless, several natural processes display high-degrees of local stationarities and anisotropies, which are hardly captured by available covariates. In these settings, non-stationary spatial covariance functions are better suited for modeling, see, e.g., [19] in the scalar case.

In [6], we recently proposed a mathematical construction yielding a fully non-stationary covariance model for Hilbert space valued random fields. Here, we consider an orthonormal basis $\{e_k\}_{k\in\mathbb{N}}$ for $\mathcal{H}$ and formulate the model as

$$\mathcal{X}_s = m_s + \sum_{u=1}^{r} A^u(s) w_s^u, \tag{47.3}$$

where each $A^u : D \to \mathcal{B}(\mathcal{H})$ are spatially varying bounded linear operators in D, and each w_s^u are zero-mean and square-integrable $\mathcal{H}$-valued spatial random processes. This model is called the *fully non-stationary functional linear model of coregionalization* (FNF-LMC), by analogy with the model considered in classical geostatistical references. By imposing that the scalar projections of the processes w_s^u onto the basis $\{e_k\}_{k\in\mathbb{N}}$ follow the non-stationary anisotropic model of [19], along with a few technical conditions, in [6] we prove that the resulting functional random field is well-defined and fully non-stationary. On this basis, and following a locally stationary paradigm, estimation procedures for this model can be established, offering enhanced flexibility compared to alternative stationary and non-stationary approaches available in the literature.

47.4.2 Non-Euclidean spatial domains

In all the cases explored in Sect. 47.2, definitions of stationarity and isotropy are fundamentally based on Euclidean geometry for the spatial domain—through the Euclidean lag vector $(s_1 - s_2)$ and the Euclidean spatial distance $||s_1 - s_2||$ (see Sect. 47.2). Considering non-Euclidean metrics is far from straightforward under the classical geostatistical paradigm. While spatial stationarity for the marginal distribution of $\mathcal{X}_s$ can be established, defining (and ensuring) spatial homogeneity of the covariance function is less obvious. As a matter of fact, even the concept of stationarity is closely tied to that of spatial metric. Additionally, as discussed extensively in the scalar case (e.g., [5]), popular parametric families may become invalid under non-Euclidean geometry (i.e., they may no longer guarantee positive semi-definiteness of the associated covariances). Although strategies have been developed in the scalar setting (see e.g. [25, 4] for the case of stream networks), developing comprehensive modeling strategies for general non-Euclidean domains remains an open challenge in both scalar and O2S2 settings. However, following a construction similar to that of [6] mentioned in Sect. 47.4.1, the O2S2 challenge can be addressed jointly with the scalar one by leveraging advancements made on scalar models.

For instance, in [2], we extend the framework introduced by [25] to incorporate functional observations into their stream network methodology. Here, using the previous notation, each element of the functional field is expressed as a truncated basis expansion, $\mathcal{X}_s = m_s + \sum_{k=1}^{N} w_s^k e_k$, where $w_s^k = \langle \mathcal{X}_s - m_s, e_k \rangle$ are the projections of the element $\mathcal{X}_s$ onto the truncated orthonormal basis $\{e_k\}_{k=1,...,K}$. Note that the stationarity and covariance structure of the process $\{\mathcal{X}_s, s \in D\}$ are completely governed by these coefficients. To define each scalar field, a moving average (MA) model analogous to that of [25] is considered. This yields a trace-covariogram for the process defined by

$$C(s_i, s_j) = \sum_{k=1}^{N} \mathbb{E}\left[w_{s_i}^k w_{s_j}^k\right] = \sum_{k=1}^{N} C_t^{(k)}(h|\boldsymbol{\theta}), \tag{47.4}$$

where $C_t^{(k)}(h|\boldsymbol{\theta})$ represents the covariance function of the k-th scalar random field $\{w_s^k, s \in D\}$. In practice, the covariance functions $C_t^{(k)}(h|\boldsymbol{\theta})$ can be modeled using any of the valid models proposed by [25], and the same applies to the trace-covariogram $C(s_i, s_j)$.

This paradigmatic example demonstrates that the efforts made to explore the intricate covariance structures that describe natural phenomena in the scalar setting can be leveraged to develop new modeling frameworks valid in the O2S2 context. However, the estimation complexity increases significantly when dealing with Hilbert space data, and this challenge warrants further exploration in future studies. This includes addressing issues related to high dimensionality and the instability inherent in the estimation of covariance operators—problems that are well-known and widely discussed in the literature on functional data. These challenges can be mitigated by considering integrated measures of spatial dependence, such as trace-variograms and trace-covariograms, though this comes at the cost of providing only a global assessment of the spatial dependence in the data.

47.5 Conclusion

In this communication, we explored the emerging field of Object-Oriented Spatial Statistics (O2S2), highlighting its potential to handle spatially dependent complex data, particularly in non-Euclidean settings. By extending classical geostatistical methodologies with the philosophy of Functional Data Analysis (FDA) and Object-Oriented Data Analysis (OODA), we have outlined a mathematical framework that allow for the modeling of spatially varying phenomena with great flexibility. Challenges remain, particularly regarding the modeling of non-stationary spatial processes and the extension of spatial covariance structures to non-Euclidean domains. Nevertheless, recent advancements in non-stationary functional models and in approaches for handling complex spatial domains suggest that these challenges can be addressed by leveraging the results attained in scalar settings through wise model constructions. Finally, although application of O2S2 in the geosciences has been broad and successful, further research in this area will be crucial to refine estimation methods and improve their robustness in the critical situations outlined in this contribution.

Acknowledgements The present research has been supported by MUR, grant Dipartimento di Eccellenza 2023-2027.

References

1. J. Aitchison. The statistical analysis of compositional data. Journal of the Royal Statistical Society. Series B (Methodological), **44**(2), 139–177, (1982)
2. C. Barbi, A. Menafoglio, and P. Secchi. An object-oriented approach to the analysis of spatial complex data over stream-network domains. Spatial Statistics, **58**, 100784, (2023)
3. N. Cressie. Statistics for Spatial data. John Wiley & Sons, New York, (1993)
4. N. Cressie, J. Frey, B. Harch, and M. Smith. Spatial prediction on a river network. Journal of Agricultural, Biological, and Environmental Statistics, **11**, 127–150, (2006)
5. F. Curriero. On the use of non-Euclidean distance measures in geostatistics. Mathematical Geology, **38**, 907–926, 2006.
6. G. De Carlo, R. Scimone, and A. Menafoglio. Fully non stationary functional spatial random fields. Technical report, Politecnico di Milano, (2025)
7. J. J. Egozcue, J. L. Díaz-Barrero, and V. Pawlowsky-Glahn. Hilbert space of probability density functions based on Aitchison geometry. Acta Mathematica Sinica, English Series, **22**(4), 1175–1182, (2006)
8. R. Giraldo, P. Delicado, and J. Mateu. Continuous time-varying kriging for spatial prediction of functional data: An environmental application. Journal of Agricultural, Biological, and Environmental Statistics, **15**(1), 66–82, (2010)
9. L. Horváth and P. Kokoszka. Inference for Functional Data with Applications. Springer Series in Statistics,. New York, NY: Springer, (2012)
10. K. Hron, A. Menafoglio, M. Templ, K. Hrůzová, and P. Filzmoser. Simplicial principal component analysis for density functions in bayes spaces. Computational Statistics & Data Analysis, **94**, 330 – 350, (2016)
11. R. Ignaccolo, J. Mateu, and R. Giraldo. Kriging with external drift for functional data for air quality monitoring. Stochastic Environmental Research and Risk Assessment, **28**(5), 1171–1186, (2014)
12. J. S. Marron and A. M. Alonso. Overview of object oriented data analysis. Biometrical Journal, **56**(5), 732–753, (2014)
13. A. Menafoglio and P. Secchi. Statistical analysis of complex and spatially dependent data: A review of object oriented spatial statistics. European Journal of Operational Research, **258**(2), 401 – 410, (2017)
14. A. Menafoglio, P. Secchi, and M. Dalla Rosa. A Universal Kriging predictor for spatially dependent functional data of a Hilbert Space. Electronic Journal of Statistics, **7**, 2209–2240, (2013)
15. A. Menafoglio, A. Guadagnini, and P. Secchi. A Kriging approach based on Aitchison geometry for the characterization of particle-size curves in heterogeneous aquifers. Stochastic Environmental Research and Risk Assessment, **28**(7), 1835–1851, (2014)
16. A. Menafoglio, G. Gaetani, and P. Secchi. Random domain decompositions for object-oriented kriging over complex domains. Stochastic Environmental Research and Risk Assessment, **32**, 3421–3437, (2018)

17. A. Menafoglio, D. Pigoli, and P. Secchi. Kriging riemannian data via random domain decompositions. Journal of Computational and Graphical Statistics, **30**(3), 709–727, (2021)
18. D. Nerini, P. Monestiez, and C. Manté. Cokriging for spatial functional data. Journal of Multivariate Analysis, **101**(2), 409–418, (2010)
19. C. J. Paciorek and M. J. Schervish. Spatial modelling using a new class of nonstationary covariance functions. Environmetrics, **17**(5), 483–506, (2006)
20. V. Pawlowsky-Glahn, J. J. Egozcue, and R. Tolosana-Delgado. Modelling and analysis of compositional data. John Wiley & Sons, Ltd, (2015)
21. D. Pigoli, A. Menafoglio, and P. Secchi. Kriging prediction for manifold-valued random field. Journal of Multivariate Analysis, **145**, 117–131, (2016)
22. Davide Pigoli, Alessandra Menafoglio, and Piercesare Secchi. Kriging prediction for manifold-valued random fields. Journal of Multivariate Analysis, **145**, 117–131, (2016)
23. J. O. Ramsay and B. W. Silverman. Functional Data Analysis. Springer Series in Statistics. New York, NY: Springer, (2005)
24. K. G. van den Boogaart, J. J. Egozcue, and V. Pawlowsky-Glahn. Bayes Hilbert spaces. Australian & New Zealand Journal of Statistics, **56**, 171–194, (2014)
25. J. M. Ver Hoef and E. E. Peterson. A moving average approach for spatial statistical models of stream networks. Journal of the American Statistical Association, **105**(489), 6–18, (2010)

Chapter 48
The Spherical Depth for Functional Data

Erik Mendroš and Stanislav Nagy

Abstract In the nonparametric analysis of multivariate data, the spherical depth of a point $\boldsymbol{x} \in \mathbb{R}^d$ with respect to a distribution P on $\mathbb{R}^d$ is the probability that a ball, determined by two antipodal points on its boundary that are sampled independently from P, covers $\boldsymbol{x}$. The greatest advantage of the spherical depth is its fast computation, which is, unlike for many other depth functions, not exponential in the dimension d. That makes the spherical depth amenable for the analysis of high-dimensional, and functional data. We explore the theory and practice of spherical depth when applied to data of high dimensionality. In particular, we point to several difficulties with known results in the literature and revise multiple classical propositions on the behavior of spherical depth.

48.1 Introduction: Depth in Multivariate and Function Spaces

Nonparametric statistics of multivariate and functional datasets is difficult because of the absence of a canonical ordering in multivariate spaces. For that reason, one cannot simply define notions as fundamental as quantiles or ranks for multidimensional/functional data. One way of defining those concepts in $\mathbb{R}^d$ is using the statistical depth functions [11, 17, 19], which evaluate the degree of "centrality" of each point $\boldsymbol{x} \in \mathbb{R}^d$ with respect to (w.r.t.) the geometry of the mass of a probability distribution P on $\mathbb{R}^d$. Points that are positioned centrally within the main bulk of P obtain high depth values, and points lying on the periphery of P get low depth. This

Erik Mendroš (✉)
Charles University, Faculty of Mathematics and Physics, Sokolovská 83, 186 75 Prague, Czech Republic, e-mail: erikmendros@gmail.com

Stanislav Nagy
Charles University, Faculty of Mathematics and Physics, Sokolovská 83, 186 75 Prague, Czech Republic, e-mail: nagy@karlin.mff.cuni.cz

© The Author(s), under exclusive license to Springer Nature Switzerland AG 2025
G. Aneiros et al. (eds.), *New Trends in Functional Statistics and Related Fields*, Contributions to Statistics, https://doi.org/10.1007/978-3-031-92383-8_48

way, for example, a depth-based analog of a median of P in $\mathbb{R}^d$ can be defined as the point that attains the highest depth value w.r.t. P.

In multivariate spaces $\mathbb{R}^d$, a great number of depth functions have been proposed. We mention the seminal halfspace (Tukey) depth [17] and the simplicial depth [11]; many other examples can be found in [19]. One of the major problems of these classical depths is their high computational complexity, which typically increases exponentially with the dimension of the data d. Thus, the classical depths cannot be used directly for analyzing high-dimensional (or functional) datasets. In this contribution, we focus on the spherical depth [5], defined for $\boldsymbol{x} \in \mathbb{R}^d$ and P a Borel probability distribution on $\mathbb{R}^d$ as

$$SD(\boldsymbol{x}; P) = \mathsf{P}(\boldsymbol{x} \in S(\boldsymbol{X}_1, \boldsymbol{X}_2)), \tag{48.1}$$

where $\boldsymbol{X}_1, \boldsymbol{X}_2 \sim P$ are independent random vectors sampled from P, and

$$S(\boldsymbol{x}_1, \boldsymbol{x}_2) = \left\{ \boldsymbol{y} \in \mathbb{R}^d : \left\| \boldsymbol{y} - \frac{\boldsymbol{x}_1 + \boldsymbol{x}_2}{2} \right\| \leq \left\| \frac{\boldsymbol{x}_1 - \boldsymbol{x}_2}{2} \right\| \right\} \tag{48.2}$$

is the unique (Euclidean) ball in $\mathbb{R}^d$ that contains both $\boldsymbol{x}_1$ and $\boldsymbol{x}_2$ as antipodal points on its boundary.

The spherical depth (48.1) is an arguably simple depth notion, which does not satisfy, e.g., the important affine invariance property, commonly required from depths in $\mathbb{R}^d$ [19]. Nevertheless, its advantage is its computational simplicity, as for any dimension d, the sample version of (48.1) can be defined as a U-statistic of order 2 given by

$$SD(\boldsymbol{x}; P_n) = \binom{n}{2}^{-1} \sum_{i<j} \mathbb{I}(\boldsymbol{x} \in S(\boldsymbol{X}_i, \boldsymbol{X}_j)). \tag{48.3}$$

Here, P_n is the empirical measure of a random sample $\boldsymbol{X}_1, \ldots, \boldsymbol{X}_n$ from P, and the sum is taken over all $i, j = 1, \ldots, n$, $i < j$. As such, computing (48.3) requires only $O(d\,n^2)$ operations in $\mathbb{R}^d$. This time complexity is an extraordinary improvement over the fastest known algorithms for, e.g., the halfspace or simplicial depth (which are of order at least $O(n^{d-1}\log(n))$, see [14]). Thus, the spherical depth is naturally interesting, especially in the analysis of high-dimensional or even functional (infinite-dimensional) datasets.

In this short paper, we explore the potential of spherical depth when analyzing data of high dimensionality. First, in Sect. 48.2, we formally introduce the spherical depth for functional data and discuss its links to several known depth functions. We emphasize the intimate connection between the spherical depth, the halfspace mass depth from [3], and the classical halfspace depth. Several theoretical properties of SD are outlined in Sect. 48.3. The most notable results include a thorough discussion of the monotonicity properties of SD and of the breakdown point of the associated median. We conclude with a small real data example in Sect. 48.4. Due to space limitations, all our theoretical results are given without proofs; the proofs will appear elsewhere.

Notations. Write $\mathcal{F}$ for a (possibly infinite-dimensional) normed vector space over $\mathbb{R}$; typical examples are the Euclidean space $\mathbb{R}^d$, the space $L_p([0,1])$ of real functions on $[0,1]$ whose p-th power is integrable for $p \in (0,\infty)$, or the space $L_\infty([0,1])$ of (essentially) bounded measurable real functions on $[0,1]$. If $\mathcal{F}$ is a Hilbert space (in particular, for $\mathcal{F} = L_2([0,1])$ or $\mathcal{F} = \mathbb{R}^d$), we denote the standard inner product by $\langle\cdot,\cdot\rangle$ and always consider $\|x\|^2 = \langle x, x\rangle$. The set of all (Borel) probability measures on a measurable space $\mathcal{S}$ is denoted by $\mathcal{P}(\mathcal{S})$. All random variables are defined on a common probability space $(\Omega, \mathcal{A}, \mathsf{P})$. For a random variable $X \sim P \in \mathcal{P}(\mathcal{F})$ and a map $\varphi\colon \mathcal{F} \to \mathcal{S}$, we write also $P_{\varphi(X)} \in \mathcal{P}(\mathcal{S})$ for the distribution of $\varphi(X)$. The symbol $\stackrel{d}{=}$ stands for equality in distribution.

48.2 Spherical Depth for Functional Data

In a normed vector space $\mathcal{F}$, the spherical depth SD from (48.1) generalizes naturally, with the only exception that we take $x \in \mathcal{F}$ and $P \in \mathcal{P}(\mathcal{F})$, and the Euclidean norm in the definition of $S(x_1, x_2)$ in (48.2) is replaced by a general norm $\|\cdot\|$ on $\mathcal{F}$.

Our first observation is that in a Hilbert space $\mathcal{F}$, the depth $SD(x;P)$ has an interesting interpretation, see also Lemma 2.2 in [5]. Starting from the definition of $S(x_1, x_2)$ in (48.2), we have that $y \in S(x_1, x_2)$ if and only if

$$0 \geq \|2\,y - x_1 - x_2\|^2 - \|x_1 - x_2\|^2 = 4\,\langle x_1 - y, x_2 - y\rangle\,,$$

meaning that $SD(x;P)$ can be rewritten into

$$SD(x;P) = \mathsf{P}\left(\langle X_1 - x, X_2 - x\rangle \leq 0\right) = \mathsf{P}\left(X_2 \in H_{x,x-X_1}\right), \qquad (48.4)$$

where $H_{x,u} = \{y \in \mathcal{F}\colon \langle y, u\rangle \geq \langle x, u\rangle\}$ is the closed halfspace in $\mathcal{F}$ whose boundary hyperplane passes through x and its inner normal vector is $u \in \mathcal{F}$, $u \neq 0$. The first expression in (48.4) interprets as the probability that for $X_1, X_2 \sim P$ independent, the angle between vectors $X_1 - x$ and $X_2 - x$ is at least $\pi/2$. the second expression in (48.4) bears resemblance with the halfspace depth in $\mathcal{F}$ defined as

$$HD(x;P) = \inf_{u\in\mathcal{F},\ u\neq 0} \mathsf{P}\left(X \in H_{x,u}\right), \quad \text{where } X \sim P.$$

Indeed, while the halfspace depth evaluates the minimum probability of a halfspace $H_{x,u}$, $u \in \mathcal{F}$, the spherical depth in (48.4) assesses the mean probability of a halfspace $H_{x,x-X_1}$, where the orientation of the halfspace is given by $x - X_1$, $X_1 \sim P$. The latter construction (and the spherical depth itself) is reminiscent of the halfspace mass depth from [3], which is given for $\mathcal{F} = \mathbb{R}^d$ by the right-hand side of (48.4), but with $\boldsymbol{x} - \boldsymbol{X}_1$ uniform on the unit sphere in $\mathbb{R}^d$ and $\boldsymbol{X}_2 \sim P$, $\boldsymbol{X}_1$ and $\boldsymbol{X}_2$ independent. Depths in the same spirit can be found in the literature as the integrated dual depths [4], or integrated rank-weighted depths [16].

A family of depths generalizing the spherical depth are the β-skeleton depth functions from [18], which set on the same groundwork both the spherical depth, and the closely related lens depth [6, 7, 10, 12].

Finally, it is interesting to observe the similarity of the spherical depth and the (2nd order) band depth for functional data from [13], defined for $x \in \mathcal{F} = L_\infty([0,1])$ and $P \in \mathcal{P}(L_\infty([0,1]))$ as $BD(x;P) = \mathsf{P}(x \in B(X_1, X_2))$, where the ball $S(x_1, x_2)$ from (48.2) is replaced by the band $B = B(x_1, x_2)$ given by

$$\begin{aligned} B &= \{y \in \mathcal{F} \colon \min\{x_1(t), x_2(t)\} \le y(t) \le \max\{x_1(t), x_2(t)\} \text{ for all } t \in [0,1]\} \\ &= \{y \in \mathcal{F} \colon y(t) \in S(x_1(t), x_2(t)) \text{ for all } t \in [0,1]\}. \end{aligned}$$

The band $B(x_1, x_2)$ can be seen as a "coordinate-wise" ball (48.2) when applied to the function values $x_1(t)$ and $x_2(t)$ in $\mathbb{R}$. This makes the functional spherical depths share properties similar to those of the well-known band depths for functional data. For further insights and a detailed comparison of the theoretical properties of BD and SD, see [1].

48.3 Some Theory of the Spherical Depth

In this section, we examine the properties of SD, shedding light on when and to what extent is SD meaningful for functional data. For a comprehensive discussion on the properties of statistical depth functions relevant in function spaces, see, e.g., [8, 15].

Theorem 48.1. *The functional spherical depth SD is scalar-affine invariant, that is $SD(ax + b; P_{aX+b}) = SD(x; P_X)$, for any $x, b \in \mathcal{F}$ and $a \in \mathbb{R} \setminus \{0\}$.*

Theorem 48.2. *Let $P_X \in \mathcal{P}(\mathcal{F})$ be centrally symmetric around $\mu \in \mathcal{F}$, meaning that $X - \mu \overset{d}{=} -(X - \mu)$, and let $\mathsf{P}(\|X_1 + X_2 - 2\mu\| = \|X_1 - X_2\|) = 0$, where $X_1, X_2 \sim P_X$ are independent. Then*

$$SD(\mu; P_X) = 1/2. \tag{48.5}$$

Whenever $\mathcal{F}$ is a Hilbert space, the assumption of angular symmetry around μ, defined as $(X - \mu)/\|X - \mu\| \overset{d}{=} -(X - \mu)/\|X - \mu\|$, suffices for (48.5) to hold.

Note that in Theorem 48.2 we do not claim SD to be maximized at μ. In fact, even for distributions in $\mathbb{R}^d$ that are absolutely continuous and (centrally) symmetric around the origin, the maximum SD may exceed the value $1/2$ obtained in (48.5). The following example is motivated by the examples in the Appendix in [7] for the lens depth.

Example 48.1. Let $\mathcal{F} = (\mathbb{R}^d, \langle \cdot, \cdot \rangle)$ and consider the uniform distribution on the set $B = B(-2) \cup B(2)$, where $B(t) = \left\{\boldsymbol{y} \in \mathbb{R}^d \colon \|\boldsymbol{y} - (t, 0, \dots, 0)\| \le 1\right\}$. Then, by Theorem 48.2 we know that $SD(\boldsymbol{0}; P) = 1/2$. Nevertheless, it can be shown that there exists a point $\boldsymbol{x} = (t_0, 0, \dots, 0) \in \mathbb{R}^d$ with $t_0 > 0$ such that $SD(\boldsymbol{x}; P) > 1/2$. A formal proof can be derived based on the alternative expression (48.4) for SD.

From Example 48.1, we see that central symmetry alone does not suffice for SD to be monotone on rays starting from the center of symmetry. Consequently, we adopt the stronger assumption of spherical symmetry. A random vector $\boldsymbol{X}$ is said to be spherically symmetric around $\boldsymbol{\mu} \in \mathbb{R}^d$ if, for every orthogonal matrix $Q \in \mathbb{R}^{d \times d}$, the distributions of $\boldsymbol{X} - \boldsymbol{\mu}$ and $Q(\boldsymbol{X} - \boldsymbol{\mu})$ are the same.

To the best of our knowledge, the only result addressing the monotonicity of SD for unimodal spherically symmetric distributions is from [5].[1] However, as noted in Sect. 5.2 in [10], that proof is incorrect. As our first major contribution, we present a way to fix the proof of Theorem 3.5 from [5], as stated in the following theorem.

Theorem 48.3. *Let $\mathcal{F} = (\mathbb{R}^d, \langle\cdot,\cdot\rangle)$ and let $P \in \mathcal{P}(\mathbb{R}^d)$ be spherically symmetric around $\boldsymbol{\mu} \in \mathbb{R}^d$ and absolutely continuous with density f. Suppose that f is monotone in the sense that $f(\boldsymbol{x}) \geq f(\boldsymbol{y})$ for any $\boldsymbol{x}, \boldsymbol{y} \in \mathbb{R}^d$ such that $\|\boldsymbol{x} - \boldsymbol{\mu}\| \leq \|\boldsymbol{y} - \boldsymbol{\mu}\|$. Then, SD is monotone in the sense that $SD(\boldsymbol{x}; P) \leq SD(\boldsymbol{y}; P)$ for any $\boldsymbol{x}, \boldsymbol{y} \in \mathbb{R}^d$ such that $\|\boldsymbol{x} - \boldsymbol{\mu}\| \geq \|\boldsymbol{y} - \boldsymbol{\mu}\|$.*

Theorem 48.3 is valid in $\mathbb{R}^d$. In function spaces, however, the notion of spherical symmetry fails to be well-defined (Sect. 2 in [2]). A concept of symmetry better suited for function spaces $\mathcal{F}$ would be the elliptical symmetry; a functional equivalent of elliptical symmetry has been established in Definition 2.1 in [2]. Our proof of Theorem 48.3, however, does not extend directly to elliptical symmetry; the question of monotonicity of SD under elliptically symmetric distributions hence remains an interesting open problem.

The next two results state that $SD(x; P)$ decreases uniformly to 0 as x drifts away to infinity, and that the spherical depth map satisfies nice continuity properties in both x and P. Theorem 48.4 for $\mathcal{F} = \mathbb{R}^d$ can be found in Theorem 3.6 in [5].

Theorem 48.4. *For any $P \in \mathcal{P}(\mathcal{F})$ it holds true that $\lim_{\|x\| \to \infty} SD(x; P) = 0$.*

Theorem 48.5. *For any $\mathcal{F}$ separable, the spherical depth map $SD: \mathcal{F} \times \mathcal{P}(\mathcal{F}) \to [0, 1]$ is upper semi-continuous in both of its arguments, meaning that whenever $x_n \to x$ in $\mathcal{F}$ and P_n converges weakly to P in $\mathcal{P}(\mathcal{F})$, then $\limsup_{n\to\infty} SD(x_n; P_n) \leq SD(x; P)$, and also $\limsup_{n\to\infty} \max_{x\in\mathcal{F}} SD(x; P_n) \leq \max_{x\in\mathcal{F}} SD(x; P)$.*

If, in addition, $\mathcal{F}$ is a Hilbert space and the distribution P is smooth in the sense that $P(h) = 0$ for each hyperplane $h \subset \mathcal{F}$, then SD is continuous in the sense that $\lim_{n\to\infty} SD(x_n; P_n) = SD(x; P)$ and $\lim_{n\to\infty} \max_{x\in\mathcal{F}} SD(x; P_n) = \max_{x\in\mathcal{F}} SD(x; P)$.

Theorem 48.5 refines several related results from [5, 6]. From Theorem 48.5, we immediately obtain that the set of spherical medians of any $P \in \mathcal{P}(\mathcal{F})$ given as $\mathfrak{M}(P) = \{y \in \mathcal{F} : SD(y; P) = \max_{x\in\mathcal{F}} SD(x; P)\}$ is always non-empty and closed. Together with Theorem 48.4, this also gives that $\mathfrak{M}(P)$ is closed and bounded. In particular, if the normed vector space $\mathcal{F}$ satisfies the Heine-Borel property (in particular, if $\mathcal{F} = \mathbb{R}^d$ for some finite d), then $\mathfrak{M}(P)$ must be compact.

---

[1] Note that the proof of the monotonicity property in Theorem 4 from [18] is given only for $\beta > 1$, which excludes the case of the spherical depth.

Theorem 48.5 also implies a strong consistency result for SD — if P is smooth as in the claim of the theorem and P_n is the empirical distribution based on a random sample $X_1, \dots, X_n$ from P, we have that $SD(x; P_n)$ converges almost surely to $SD(x; P)$. That follows directly from Theorem 48.5 and the classical Varadarajan theorem about the weak convergence of empirical measures.

Finally, we examine the robustness of the sample spherical median $\theta_n = \arg\max_{x \in \mathcal{F}} SD(x; P_n)$, for P_n empirical measures of random samples of size n from P. As a consequence of Theorem 48.5, the median θ_n is qualitatively robust [9], provided that the population median $\mathfrak{M}(P)$ is unique and $\mathcal{F}$ is a Hilbert space.

The breakdown point (BP) is another notable robustness index for an estimator. For the median θ_n, BP is discussed in Sect. 4.1 in [18], where it is claimed that BP of the finite-dimensional spherical median is at least $(2-\sqrt{2})/2 \approx 0.293$. Unfortunately, that proof seems to be flawed.[2] A similar mistake occurred in the claim about BP of the lens depth median in Sect. 3.1 in [12]. In our efforts to correct that proof, we obtained only a substantially weaker version of the original statement.

Theorem 48.6. *Let* $\mathcal{F} = \left(\mathbb{R}^d, \langle \cdot, \cdot \rangle\right)$. *The BP of* θ_n *is at least* $1 - (1 + 1/(d+1))^{-1/2}$.

While Theorem 48.6 is hardly satisfactory in high dimensions, assuming angular symmetry of P allows us to make a considerably stronger assertion.

Theorem 48.7. *Let* $\mathcal{F}$ *be a Hilbert space and let* $P \in \mathcal{P}(\mathcal{F})$ *be angularly symmetric and smooth in the sense that* $P(h) = 0$ *for each hyperplane* $h \subset \mathcal{F}$. *Then, the breakdown point of* θ_n *is, as* $n \to \infty$, *almost surely at least* $1 - \sqrt{2/3} \approx 0.183$.

In particular, we see that under the assumption of angular symmetry of P, the asymptotic breakdown point of the sample spherical median is positive, and bounded away from zero uniformly in d.

48.4 A Real Data Application

As a simple concluding application, we use the functional spherical depth with a real dataset. Our dataset consists of half-hourly electricity demand measurements on Thursdays in Adelaide, Australia (the data is available as thursdaydemand in R package fds). In Fig. 48.1 we see the results when considering $\mathcal{F} = L_p([0, 24])$ and $p = 2$ (that is, the functions are considered in the Hilbert space of square-integrable functions) and $p = \infty$ (i.e., within the Banach space of bounded functions).

As can be seen from Fig. 48.1, the choice of p does affect the resulting depth. The integral in the computation of the L_p-norm with $p = 2$ (left-hand panel) tends to "smooth out" any local irregularities in the functional data, while the use of the

[2] In the proof of Theorem 12 in [18], it is claimed that the number of contaminating points m must satisfy $\binom{n+m}{2} - \binom{n}{2} > \binom{n}{2}$ to break down θ_n. While this claim is not necessarily wrong, the reasoning given in the proof is insufficient; instead of $\binom{n}{2}$, the right-hand side needs to be a *lower bound* on the SD of θ_n to make sure that θ_n breaks down.

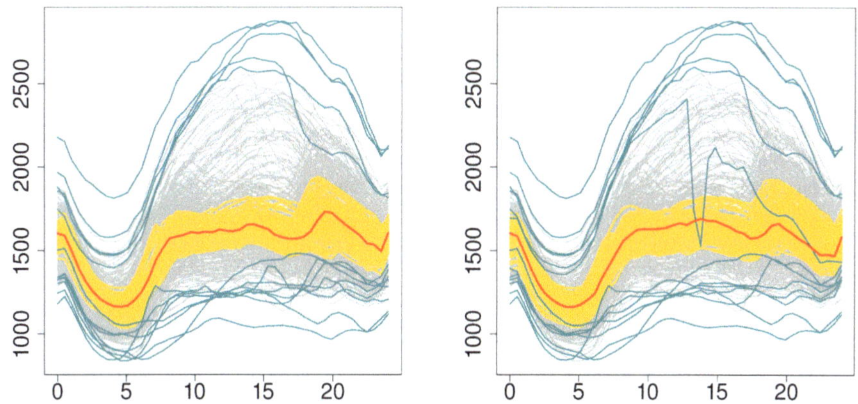

Fig. 48.1: A real data example: SD with $p = 2$ (left-hand panel) and $p = \infty$ (right-hand panel) for a dataset of half-hourly electricity demands on Thursdays in Adelaide between 6/7/1997 and 31/3/2007. The median curve (the deepest function) is plotted in thick red, the 25% of observations attaining the highest values of SD are plotted in thick yellow lines, and the 2.5% of the data curves with the lowest SD are in blue

supremum norm (that is, $p = \infty$) on the right-hand side of Fig. 48.1 also treats local deviations as important. This is particularly well seen in the irregular function with a sharp drop at around noon (12 on the x-axis in Fig. 48.1), which is flagged as an outlier using $p = \infty$, but not using $p = 2$.

Acknowledgements The research of S. Nagy was supported by the Czech Science Foundation (project n. 24-10822S) and the ERC CZ grant LL2407 of the Ministry of Education, Youth and Sport of the Czech Republic.

References

1. Bočinec, F., Mendroš, E., and Nagy, S.: A comparison of band-based approaches to functional depth. In "New Trends in Functional Statistics and Related Fields", (G.Aneiros et al. Ed.s), Springer, (2025)
2. Boente, G., Salibián Barrera, M., and Tyler, D. E.: A characterization of elliptical distributions and some optimality properties of principal components for functional data. J. Multivariate Anal. **131**, 254–264 (2014)
3. Chen, B., Ting, K. M., Washio, T., and Haffari, G.: Half-space mass: a maximally robust and efficient data depth method. Mach. Learn. **100**(2–3), 677–699 (2015)
4. Cuevas, A. and Fraiman, R.: On depth measures and dual statistics. A methodology for dealing with general data. J. Multivariate Anal. **100**(4), 753–766 (2009)
5. Elmore, R. T., Hettmansperger, T. P., Xuan, F.: Spherical data depth and a multivariate median. In: Liu, R. Y., Serfling, R., Souvaine, D. L. (eds.) Data depth: robust multivariate analysis, computational geometry and applications, pp. 87–101. Amer. Math. Soc., Providence, RI (2006)

6. Fraiman, R., Gamboa, F., and Moreno, L.: Connecting pairwise geodesic spheres by depth: DCOPS. J. Multivariate Anal. **169**, 81–94 (2019)
7. Geenens, G., Nieto-Reyes, A., and Francisci, G.: Statistical depth in abstract metric spaces. Stat. Comput. **33**(2), Paper No. 46, 15 (2023)
8. Gijbels, I. and Nagy, S.: On a general definition of depth for functional data. Statist. Sci. **32**(4), 630–639 (2017)
9. Hampel, F. R.: A general qualitative definition of robustness. Ann. Math. Statist. **42**, 1887–1896 (1971)
10. Kleindessner, M. and von Luxburg, U.: Lens depth function and k-relative neighborhood graph: versatile tools for ordinal data analysis. J. Mach. Learn. Res. **18**, Paper No. 58, 52 (2017)
11. Liu, R. Y.: On a notion of data depth based on random simplices. Ann. Statist. **18**(1), 405–414 (1990)
12. Liu, Z. and Modarres, R.: Lens data depth and median. J. Nonparametr. Stat. **23**(4), 1063–1074 (2011)
13. López-Pintado, S. and Romo, J.: On the concept of depth for functional data. J. Amer. Statist. Assoc. **104**(486), 718–734 (2009)
14. Mosler, K. and Mozharovskyi, P.: Choosing among notions of multivariate depth statistics. Statist. Sci. **37**(3), 348–368 (2022)
15. Nieto-Reyes, A. and Battey, H.: A topologically valid definition of depth for functional data. Statist. Sci. **31**(1), 61–79 (2016)
16. Ramsay, K., Durocher, S., and Leblanc, A.: Integrated rank-weighted depth. J. Multivariate Anal. **173**, 51–69 (2019)
17. Tukey, J. W.: Mathematics and the picturing of data. In Proceedings of the International Congress of Mathematicians (Vancouver, B. C., 1974), Vol. 2, pp. 523–531. Canad. Math. Congress, Montreal, Que.
18. Yang, M. and Modarres, R.: β-skeleton depth functions and medians. Comm. Statist. Theory Methods **47**(20), 5127–5143 (2008)
19. Zuo, Y. and Serfling, R.: General notions of statistical depth function. Ann. Statist. **28**(2), 461–482 (2000)

Chapter 49
False Discovery Rate Envelope and its Performance for Local Testing in Functional Data Analysis

Tomáš Mrkvička

Abstract False discovery rate (FDR) is a common way to control the number of false discoveries in multiple testing. There are a number of approaches available for controlling FDR. However, it is of great practical importance to visualize the test statistic together with its rejection or acceptance region. Therefore, the FDR envelope was introduced based on resampling principles. This graphical envelope detects the outcomes of all individual hypotheses by a simple rule: the hypothesis is rejected if and only if the empirical test statistic is outside of the envelope. Such an envelope offers a straightforward interpretation of the test results, similar to the recently developed global envelope testing, which controls the family-wise error rate. Here, I summarise the comparison of the performance of these two envelopes with several procedures designed directly for local testing tasks in functional data analysis. Namely, interval-wise testing, threshold-wise testing, step-down p-min adjustment, and adaptive two-stage FDR procedure. The comparison is done on the base of a simulation study that consists of 1584 scenarios.

49.1 Introduction

Nowadays, as the data becomes more complex, functional test statistics become more popular. The benefit of functional test statistics over univariate test statistics is that they can better reveal characteristics of the studied phenomena. The motivational example studies the functional linear model and shows the importance of the graphical interpretation of the results. It shows that by using richer test statistics than univariate test statistics or functional F-statistics computed for every functional argument, we can obtain results that are richer in interpretation. When the test statistic is complex,

Tomáš Mrkvička (✉)
University of South Bohemia, Faculty of Agriculture and Technology, Studentská 1668, České Budějovice 370 01, Czech Republic, e-mail: mrkvicka.toma@gmail.com

© The Author(s), under exclusive license to Springer Nature Switzerland AG 2025

G. Aneiros et al. (eds.), *New Trends in Functional Statistics and Related Fields*, Contributions to Statistics, https://doi.org/10.1007/978-3-031-92383-8_49

a formal graphical interpretation of the test results helps the user understand the data more deeply.

The global envelope test [10] provides such a formal graphical interpretation for multivariate or functional test statistics under a rigorous family-wise error rate (FWER) control of false positives. It provides the natural graphical interpretation, i.e., the test statistic is outside of the produced envelope if and only if the test is rejected. The global envelope test was applied on functional one-way ANOVA [7], functional two-sample test [4] and functional linear model [8] in functional data analysis. The difference between the first two applications of global envelope test is only in different test statistics. The temporal first application uses means of groups, the second means, medians and trimmed means. The third application uses the linear model parameters as a test statistic and different permutation scheme. The false discovery rate (FDR) envelope [5] provides the same graphical interpretation but under the FDR control. The FWER controls the probability of appearing at least one false positive by the global significance level α, whereas FDR controls the ratio of the number of false positives and the number of all positives. Thus, the FWER is a stronger control, and they are equivalent under the complete null case, where all hypotheses are truly negative.

When the test statistic is functional or multivariate, two kinds of tasks appear. The first task is a global test, i.e., to test if at least one argument of the test statistic can be considered significant. This leads to using multiple testing corrections such as Bonferroni or Holm. In tests where the functional test statistic is resampled under the null hypothesis, the F-max, p-min, or refinements of the p-min procedure [10] can be used. The resampling tests, e.g., permutation or simulation tests, eliminate the problem of homogeneity of the test statistic across the domain, which is an advantage of these tests with respect to parametric ones. Also, the resampling tests can be used for any test statistics, even for those whose limiting distribution is unknown, especially for those test statistics that provide rich information. These two aspects make the resampling tests more general and easier to use than parametric tests, and therefore, we concentrate here on resampling tests only.

The second tool can be called the local test. There is interest in distinguishing between the domain where the null hypothesis should be rejected and where it should not be. Several new methods were recently introduced for solving this problem in the framework of resampling tests: Interval-wise testing [11], threshold-wise testing [1], step-down adjustment [12] and false discovery rate (FDR) envelope [5].

The problem of local testing can also be solved by classical multiple-testing procedures both in the framework of resampling tests and parametrical tests. There are many methods for multiple testing corrections that can be used for local testing, e.g., Bonferroni, Holm, p-min, F-max, linear step-up. The two most important groups of multiple testing methods are those controlling FDR and those controlling family-wise error rate (FWER). The FWER control is stronger than the FDR control; therefore generally, the FDR methods should have a lower type II error rate, whereas FWER methods should have a lower type I error rate. Thus, the choice of control should be made according to the type of errors the user needs to minimize. In the local testing problem, our aim is to treat both types of errors equally. Thus, the choice is not clear. Nevertheless, the methods that control the false discovery rate are, according to our experiences, more suitable for local testing than those controlling family-wise error rate (FWER). Therefore, we have chosen an adaptive linear step-up procedure that controls the false discovery rate [3] as a state-of-the-art method to which we will compare the newly introduced methods in the framework of resampling tests. It is an improvement of the original false discovery rate control method [2] since it incorporates the estimator of the probability that the hypothesis is true null.

Thus, the paper aims to compare the new approaches to local testing with each other and with the state-of-the-art method. Our framework for comparison is the domain selection, i.e., to find the domain where the hypotheses should be rejected and where they should not. Since all the methods have to be applied to the discretized function, the methods are applied to the high-dimensional vector; thus, we also compare them under different discretizations.

The methods use different ways of controlling the type I error rate: Interval-wise error rate - IWER, family-wise error rate - FWER and false discovery rate - FDR. Due to this fact, it is not possible to use FWER or FDR as a measure of precision. It is also not possible to use type I error rate or type II error rate since FWER methods have principally tighter control of type I errors and thus FDR methods should principally control better the type II errors. But it is possible to inspect the proportion of both types of errors among all hypotheses, which we will simply call the error rate. The error rate is chosen as the measure of precision due to the two facts. None of the methods directly controls the error rate and it is a natural measure in the local testing problem, where we aim to distinguish the domains where the tested hypothesis should be rejected and where it should not.

We concentrate only on nonparametrical methods using the resampling strategy because these methods are universal. They have no assumptions about the distribution of the test statistics or about the homogeneity of the test statistics along the tested domain. Therefore, they can also be used for nonstandard test statistics to reach better interpretation [9]. For the comparison of the methods, we chose the functional two-sample problem for simplicity, but all the methods can be used for functional linear models or other kinds of tests.

49.2 Motivational example

Fig. 49.1 (left) shows population growth, defined here as the population at the end of the year divided by the population at the beginning of the year, in years from 1960 to 2014 in Africa, Asia, Europe, and North America, and Latin America, total in 112 countries with more than one million inhabitants in 1950. Fig. 49.1 (right) shows the GDP of every country in the study discounted to the 1960 USD. The discounting was performed according to USD inflation. The data was obtained from the World bank. The missing values of the GDP of a country were extrapolated using the closest known ratio of the GDP of the country and the median GDP in that year.

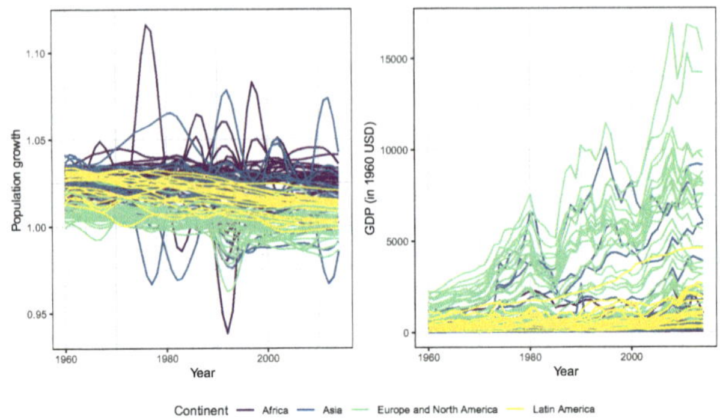

Fig. 49.1: Population growth and GDP (in 1960 USD) curves for 1960–2014 for 112 countries from the four continents.

The missing values of GDP were interpolated using linear interpolation of the two closest ratios.

We explore the functional general linear model (GLM) for the population growth across years 1960-2014 with categorical covariate (continent), continuous covariate (GDP), and interactions (GDP with respect to the continent). For this purpose, we applied three tests, and on each of them, we applied the proposed FDR control to obtain all years which are significant for the studied covariate. In order to motivate the FDR envelope methodology we show only the results for the interactions. More details about the FDR envelope are given in the next Section. Here the permutations using the Freedman-Lane scheme are used to obtain the resamples under the null hypothesis.

To test the effect of the interactions, we assume the factorial model

$$\text{population growth} \sim \text{Continent} * \text{GDP},$$

and we used the test statistic

$$\boldsymbol{\beta}^I = (\underbrace{\beta^I_{1,1960}, \dots \beta^I_{1,2014}}_{\text{1: Africa}}, \underbrace{\beta^I_{2,1960}, \dots \beta^I_{2,2014}}_{\text{2: Asia}}, \underbrace{\beta^I_{3,1960}, \dots \beta^I_{3,2014}}_{\text{3: Europe and North America}}, \underbrace{\beta^I_{4,1960}, \dots \beta^I_{4,2014}}_{\text{4: Latin America}}), \tag{49.1}$$

where $\beta^I_{1,i}, \dots \beta^I_{4,i}$ are interaction parameters of the univariate GLM model for year i, computed with condition $\sum_{j=1}^{4} \beta^I_{j,i} = 0$. This test statistics, that contains every element of the model in the non-cumulative nature allows us to reveal rich information from the data. The permutations of residuals from the null model (population growth $\sim$ Continent + GDP) was used. The graphical output of the FDR envelope algorithm is shown in Fig. 49.2. It provides a rich interpretation of the interaction component. The population growth of countries in Africa is significantly positively influenced by GDP up to 1974, whereas it is significantly less influenced than the whole world from 1998 onwards. The population growth of Asian countries is more positively influenced by GDP than the whole world from 1996 onwards. The same holds for Europe and North America. On the other hand, the population growth in Latin American countries is negatively influenced by GDP up to 1974. We also

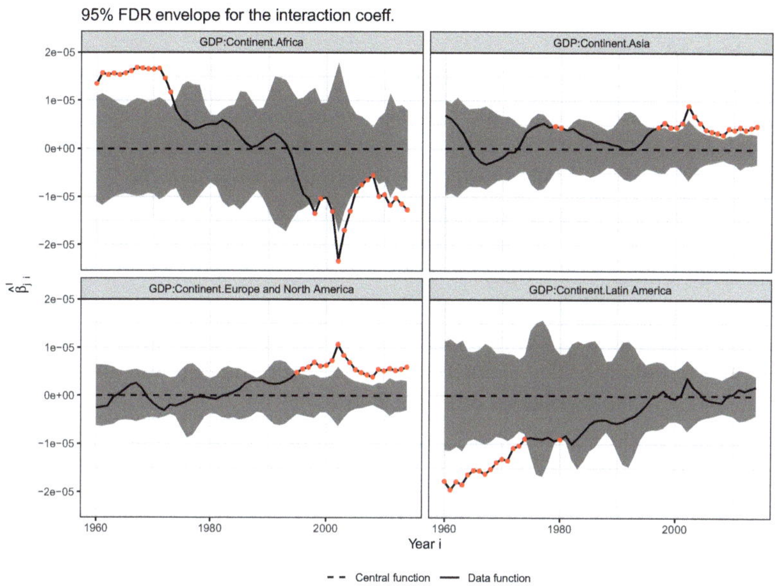

Fig. 49.2: Test for the interaction effect of continent and GDP (in 1960 USD) using the IATSE algorithm for FDR control with 5000 permutations. The null model includes the main effects.

applied Adaptive two-stage (ATS) FDR algorithm to the test vectors (49.1). ATS rejected one year less than the FDR envelope in total, and it gave no information about the reasons (direction) of rejection and variability of the test statistic under the null. Definitely, the results of FDR envelope presented in Fig. 49.2 are much richer and allow for interpreting the results in much deeper detail than the results provided by ATS or other non-graphical FDR algorithm which gives only q-values or the rejections as an output. For comparison, if the F-statistic is used instead of the test vector (49.1), we obtain even less information from the test. This F-test of the interaction effect only gives information about whether or not to reject the interaction effect in a year. Specifically, it rejected the hypotheses for the years 1960-1982, 1988, and 1995-2014. If the univariate test statistic is used, e.g., integral of F statistic, then only information about rejecting the effect of interaction is obtained.

49.3 FDR Envelope

The FDR envelope is described in detail in [5]. Here, we just roughly explain the procedure in words.

1. The multivariate/functional test statistic T_0 is chosen.
2. This test statistic is simulated under the null hypothesis, so $T_1^1, \ldots, T_n^1$ resamples are obtained.
3. The set of continuous min-max envelopes that are driven by $T_1^1, \ldots, T_n^1$ and whose width is determined by a single parameter γ is constructed.
4. Another set of test statistics $T_1^2, \ldots, T_n^2$ under the null hypothesis is produced.
5. The width $\hat{\gamma}$ of the FDR envelope is determined such that the estimated FDR of this envelope is equal to the predetermined FDR level (0.05). A set of test statistics independent of the min-max envelope is necessary here to obtain an unbiased estimate of the FDR.
6. The estimation of FDR is done according to the Algorithm 4.3 defined in [5]. The estimation is adaptive, i.e., it adjusts the procedure by estimating the amount of true null hypothesis. The estimation is based on estimating the number of false positives from the second set of test statistics.

49.4 Comparison of the Performance in the Local Testing Task

The simulation study was reported in detail in [6]. It consisted of 1584 cases, where 11 models were compared with 6 varieties of the strength of functional autocorrelation structure, with 3 resolutions, with 4 different numbers of permutations used and 2 types of errors. We performed a simulation study to investigate the performance of the methods collected in Table 49.1 for local testing.

Table 49.1: Multiple testing methods, that were investigated in the simulation study for local testing.

Abbr.	Control	Description	Timing (seconds)		
			50	200	800
ERL	FWER (weak)	Global envelope test based on the extreme rank length (ERL) measure [10]	0.07	0.13	0.36
Area	FWER (weak)	Global envelope test based on the area measure [10]	0.06	0.11	0.34
SDA	FWER (strong)	Step-down adjustment of p-min [12]	3.47	6.33	15.50
IWT	FWER (weak)	Integral tests with p-value adjusted over all subintervals [11]	2.49	139	8896
TWT	FWER (weak)	Integral tests with p-value adjusted over all sublevel and superlevel sets [1]	0.06	0.50	5.69
ATS	FDR	ATS [3] with non-parametric p-values	0.01	0.03	0.18
IATSE	FDR	FDR envelope using the IATSE algorithm [5]	0.02	0.06	0.25

Apart from the obvious observation that the FDR methods have larger type I errors than the FWER method and vice versa for type II errors, we observed the following general behavior of the different methods.

The SDA method has a high total error proportion that increases with decreasing correlation. This is due to the use of ranking without breaking the ties. (ERL and area methods use tie-breaking, but they only provide weak FWER control.) A greater number of permutations can eliminate this problem, but enormous numbers of permutations may be required when handling a very large number of hypotheses. For example, for a brain study, ten million permutations would be required [8].

The TWT method was not robust with respect to heavy tails and also non-constant variance. Further, IWT is rather computational. In our simulation study setup, a test with a (rather low) resolution $m = 800$ took 2,5 hours, which was extremely long in comparison to the other methods, which took some seconds (see Table 49.1).

Interestingly, the TWT and IWT methods tend to have problems detecting true positives for more autocorrelated functions, whereas the ERL and Area methods have problems detecting true positives for less autocorrelated functions.

Both studied FDR methods had no weak situation for any model, any autocorrelation, any level of discretization, nor any number of permutations. Therefore, it seems that FDR methods are the most suitable for the problem of local testing when the same weight is given to both types of errors. We observed a bit lower total error proportions for the ERL and Area methods for highly autocorrelated functions, but the difference between them and the FDR methods does not seem to be large.

The difference in performance between the two FDR methods was quite minor. Still, the FDR envelope method performs better for lower numbers of simulations, which can be an advantage for highly dimensional problems.

When one gives more weight to the type I error, the FWER method should be preferred. From the studied FWER methods, the TWT and ERL or Area seem to be the competitors in the Gaussian case. In our study, the TWT method performed better for wiggly data, whereas ERL and Area performed better for smoother data. However, since the TWT method does not show robustness concerning heavy tails and non-constant variance, it seems that ERL or Area method should be preferred. On the other hand, the IWT method gives better results than ERL or Area for most of the cases with more or less independent data, i.e., for highly wiggly functions. Thus, our conclusion is that for smoother data ERL or Area should be preferred, while for very wiggly data, there is not consistently the best method. It should be noted that all FWER methods of this study except SDA control FWER only weakly. If strong control of FWER is required, from the tested methods only the SDA method can be used.

49.5 Conclusions and Discussion

The false discovery rate methods consistently showed the best performance of all methods. When higher importance is given to type I error rate, the family-wise error rate methods should be preferred. From these methods, the global envelope test ERL, Area performed the best for smoother functions. None of the tested methods was consistently the best for very wiggly functions.

Thus, both envelope methods (FDR and FWER) performed very well with respect to other methods in local testing tasks. On top of that, they provide a formal graphical interpretation that is very useful for practitioners to understand the problem under the study in more detail. Also, they allow for employing any test statistic in the testing procedure, which can help in obtaining more rich information from the study, as was demonstrated by the motivational example.

Finally, the ERL, Area, and FDR envelope methods are provided in the R package **GET** [10] for general functional linear model, and also for specific statistical problems where functional test statistics appear. It is also provided for any other test with functional or multivariate test statistics under the assumption that the simulations under the null hypothesis and the computations of the test statistics are provided by the user. Thus, users can use the **GET** package and decide if FDR or FWER control is preferred.

References

1. Abramowicz, K., Pini, A., Schelin, L., Sjöstedt, S., Stamm, A., Vantini, S.: Domain selection and familywise error rate for functional data: A unified framework. Biometrics (2023)
2. Benjamini, Y., Hochberg, Y.: Controlling the false discovery rate: A practical and powerful approach to multiple testing. Journal of the Royal Statistical Society Series B (1995)
3. Benjamini, Y., Krieger, A.M., Yekutieli, D.: Adaptive linear step-up procedures that control the false discovery rate. Biometrika (2006)
4. Lopez-Pintado, S., Qian K.: A depth-based global envelope test for comparing two groups of functions with applications to biomedical data. Statistics in Medicine (2021)
5. Mrkvička, T., Myllymäki, M.: False discovery rate envelopes. Stat Comput (1995)
6. Mrkvička, T., Myllymäki, M.: Comparison of approaches for local testing with functional test statistics. Journal of Statistical Computation and Simulation
7. Mrkvička T., Myllymäki M., Jílek M., Hahn U.: A one-way ANOVA test for functional data with graphical interpretation, Kybernetika
8. Mrkvička, T., Myllymäki, M., Kuronen, M., Narisetty, N.: New methods for multiple testing in permu-tation inference for the general linear model. Statistics in Medicine. (2022)
9. Mrkvička, T., Roskovec, T., Rost, M.: A nonparametric graphical tests of significance in functional GLM. Methodology and Computing in Applied Probability. (2021)
10. Myllymäki, M., Mrkvička, T.: GET: Global Envelopes in R. Journal of Statistical Software (2024)
11. Pini, A., Vantini, S.: Interval-wise testing for functional data. Journal of Nonparametric Statistics. (2017)
12. Xu, M., Reiss, P.: Distribution-free pointwise adjusted p-values for functional hypotheses. In: Aneiros G, Horová I, Huškováa M, et al., editors. Handbook of spatial statistics. Springer, Cham; 2020. Functional and High-Dimensional Statistics and Related Fields. IWFOS 2020. Contributions to Statistics.

Chapter 50
Interpretable Functional Boxplots

Stanislav Nagy

Abstract The functional boxplot is perhaps the best-known visualization tool for functional data. We revisit the definition of the functional boxplot and demonstrate that its classical construction, when used in conjunction with the modified band depth for functional data, does not result in a well-defined boxplot. As a remedy to this problem, we argue that the only functional depths that induce well-behaved functional boxplots are those of infimal type. Several practical issues connected with infimal depth-based boxplots are discussed, such as the problem of possible degeneracy and difficulties with identifying shape outlying functions.

50.1 Introduction: Functional Boxplots

The functional boxplot is a celebrated visualization tool applicable in functional data analysis. Proposed in [17], the boxplot combines functional depth with a visualization based on the graphs (and bands) of the functional data; for an example, see Fig. 50.1. In this contribution, we review the definition of the functional boxplot and show that its best-known construction based on a depth of integrated type (left-hand panel in Fig. 50.1) is ill-defined and fails to possess a graphical interpretation. Instead, we argue that the often neglected infimal depths are the only functional depths that result in well-defined functional boxplots (right-hand panel of Fig. 50.1).

First, in Sect. 50.2, we introduce the background necessary for the construction of the functional boxplot: (i) the functional depths and (ii) the bands of functions, which constitute the two building blocks of functional boxplots. Then, in Sect. 50.3, we give examples of why the classical boxplots fail to be interpretable. We identify that the crucial property that makes the boxplots ill-defined is the lack of band convexity of the underlying functional depth. In our main result, we show that the only functional

Stanislav Nagy (✉)
Charles University, Faculty of Mathematics and Physics, Sokolovská 83, 186 75 Prague, Czech Republic, e-mail: nagy@karlin.mff.cuni.cz

© The Author(s), under exclusive license to Springer Nature Switzerland AG 2025

G. Aneiros et al. (eds.), *New Trends in Functional Statistics and Related Fields*, Contributions to Statistics, https://doi.org/10.1007/978-3-031-92383-8_50

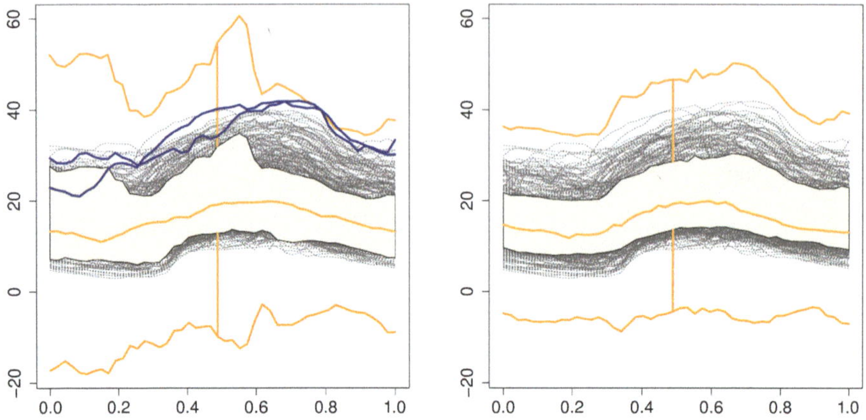

Fig. 50.1: Two versions of functional boxplots for the real dataset `fridaytempkent` from R package **fds**. A boxplot based on MBD (an integrated depth, left-hand panel) and a boxplot based on an infimal depth (right-hand panel). The median function (thick orange line) lies in the middle of the 50%-central region (beige-colored band of functions in the middle of the data cloud). The whiskers (orange curves at the top and bottom of the plots) delimit the region containing the non-outlying functions (outliers are plotted in thick blue in the left-hand panel). The two displayed types of boxplot give quite different results; the infimal depth-based boxplot is less erratic and more stable

depths that are band convex are depths of infimal type. Consequently, only infimal depth-based boxplots have a natural interpretability property. Finally, in Sect. 50.4, we discuss several apparent shortcomings of infimal depth-based boxplots, such as (i) possible inconsistency and degeneracy of infimal depths, (ii) the problem with ties in depth values, and (iii) difficulties with identifying functions outlying in shape and/or phase.

This contribution is partially based on [13], where detailed proofs of all the theoretical results claimed in this paper can be found.

Notations. We write $\mathcal{F}$ for the function space of continuous real-valued functions on $[0, 1]$. The set of all (Borel) probability measures on a measurable space $\mathcal{S}$ is denoted by $\mathcal{P}(\mathcal{S})$. All random variables are defined on a common probability space $(\Omega, \mathcal{A}, \mathsf{P})$. For $X \sim P \in \mathcal{P}(\mathcal{F})$ in a function space $\mathcal{F}$ and $t \in [0, 1]$, the marginal distribution of the uni-variate random variable $X(t)$ is denoted by $P_t \in \mathcal{P}(\mathbb{R})$.

50.2 Prerequisites: Functional Depths and Bands

(i) Depth for Functional Data. The classical concept of a boxplot for one-dimensional ($\mathbb{R}$-valued) data uses quantiles—the median, the quartiles, and the inter-quartile range. For functional data, no canonical concept of quantiles exists. A well-studied construction used as a surrogate for quantiles in multidimensional and function spaces is the statistical depth [7, 8, 18]. A depth D orders points in the (multivariate or functional) sample space $\mathcal{F}$ according to their "centrality" when considered within the mass of a given probability distribution $P \in \mathcal{P}(\mathcal{F})$—points $x \in \mathcal{F}$ that are located centrally within P are assigned higher depth values $D(x; P) \in [0, 1]$. This way, the deepest (data) point is a natural analog of the median for $\mathcal{F}$-valued data, and sets of points where the function $D(\cdot; P)$ exceeds given thresholds $\alpha > 0$ are the so-called α-central regions of the distribution P. Based on depth functions, boxplots for multivariate data have been proposed in [16]. For univariate data, a classical depth is the halfspace (or Tukey) depth D_1, which is for F the distribution function of $Q \in \mathcal{P}(\mathbb{R})$ given by

$$D_1(u; Q) = \min\left\{F(u); 1 - \lim_{v \to u-} F(v)\right\} \qquad \text{for } u \in \mathbb{R}. \tag{50.1}$$

Depths have also been successfully introduced to functional data analysis. Many functional depths exist, notably the modified band depth [8] (MBD, belonging to the larger family of integrated depth functions [1, 12]), or various depths of infimal type [9, 14]. The original construction of the functional boxplot from [17] uses MBD for functional data. The MBD is also currently set as the default choice in virtually all implementations of functional boxplots in statistical software (function `fbplot` in package **fda**, `fncBoxPlot` in **depthProc**, and `fBoxplot` in **GET** in R; function `fboxplot` in module **statsmodels** and `Boxplot` in **scikit-fda** in Python, etc.).

The *modified band depth* [8] of a function $x \in \mathcal{F}$ and $P \in \mathcal{P}(\mathcal{F})$ is

$$MBD(x; P) = \mathsf{E}\,\lambda\left(\{t\colon \min\{X_1(t), X_2(t)\} \le x(t) \le \max\{X_1(t), X_2(t)\}\}\right), \tag{50.2}$$

where λ is the Lebesgue measure on $[0, 1]$, and $X_1, X_2 \sim P$ are independent. The depth (50.2) interprets as the mean proportion of time points $t \in [0, 1]$ such that x is contained in between the graphs of the (random) functions X_1 and X_2.

The MBD is a member of a more general class of depths called the *integrated depths*. For a one-dimensional depth function D_1 such as (50.1), a general integrated depth is defined for $x \in \mathcal{F}$ and $P \in \mathcal{P}(\mathcal{F})$ as

$$D(x; P) = \int_0^1 D_1(x(t); P_t)\,\mathrm{d}t. \tag{50.3}$$

It is simple to observe ([12], Sect. A.3) that MBD from (50.2) can be written as the integrated depth (50.3) based on the one-dimensional simplicial depth [7].

Another important family of functional depths is that of *infimal depths* given by

$$D(x; P) = \inf_{t \in [0,1]} D_1(x(t); P_t). \tag{50.4}$$

Infimal depths were first considered in [9]. Under different names, they have been used and amended in [10, 11, 14]. Besides integrated and infimal depths, other functional depths exist in the literature, see, e.g., [5].

(ii) Bands of Functions as a Visualization Tool. An essential element of a functional boxplot is the choice of visualization based on bands. Loosely speaking, a band of a collection of functions $B \subseteq \mathcal{F}$ is the set of functions whose graphs, when plotted in $\mathbb{R}^2$, lie "in between" the graphs of the functions in B. Formally, a *band* of $B \subseteq \mathcal{F}$ is

$$\mathcal{B}(B) = \left\{ x \in \mathcal{F} \colon \inf_{y \in B} y(t) \leq x(t) \leq \sup_{y \in B} y(t) \text{ for all } t \in [0,1] \right\} \subseteq \mathcal{F}. \tag{50.5}$$

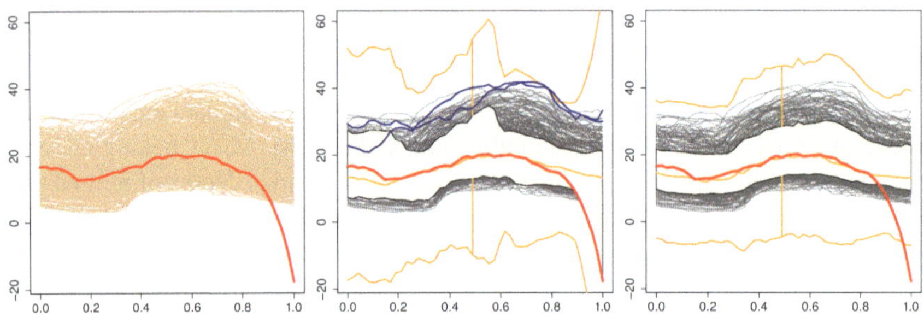

Fig. 50.2: The functional sample from Fig. 50.1, with a single outlier y (thick red curve). The function y is, in most of the domain, located quite deeply in the sample, meaning that the integral in MBD in (50.2) and (50.3) attaches to y a high depth value. On the other hand, the extreme values of $y(t)$ for $t \approx 1$ make the infimum in (50.4) quite low. While for the integrated depth, y is claimed to be so deep that it forms a part of the central band of the boxplot (central panel), the infimal depth correctly identifies y as a deviating function, and the boxplot remains unperturbed (right-hand panel)

The bands fit a visualization paradigm neatly. In contrast to the convex hulls, which are for n-element sets B always at-most-n-dimensional subsets of $\mathcal{F}$, the band of two functions $x_1 \neq x_2 \in \mathcal{F}$ is always an infinite-dimensional object. Thus, using bands for constructing the central region of a boxplot, one cleverly exploits the natural geometry of $\mathcal{F}$ to obtain full-dimensional sets capturing the modes of variability of $P \in \mathcal{P}(\mathcal{F})$.

50.3 Band Convexity and Functional Boxplots

Just as the one-dimensional boxplot, also the functional boxplot from [17] has three main components: (i) the median function (as an analog of the median from $\mathbb{R}$), (ii) the central region (the region delimited by the first and the third quartile in $\mathbb{R}$), and (iii) an outlier-detection rule (the construction of whiskers in $\mathbb{R}$).

The functional boxplot in $\mathcal{F}$ is built using a functional depth D. The functional version of the median function in (i) is clear—the median is replaced by the D-based deepest (sample) function. The central region in (ii) is constructed as the band of an appropriate central region using D, that is, the band of 50% of the D-deepest functional data. Finally, similarly as for $\mathbb{R}$-valued data, the decision rule (iii) is based on inflating the central region from (ii) by a certain (constant and pre-defined) factor around the median curve from (i).

The difficulty with the original definition of the MBD-based boxplot from [17] lies in step (ii) and the construction of the band of the central region. Let $X_1, \ldots, X_n \in \mathcal{F}$ be the sample of observed functional data, and write $P \in \mathcal{P}(\mathcal{F})$ for the corresponding empirical measure. Let $B = \{X_{(1)}, \ldots, X_{(\lceil n/2 \rceil)}\}$ be the $\lceil n/2 \rceil$ deepest functions from the sample, and let $\mathcal{B}(B)$ be the central band of those 50% deepest sample functions B. For $\mathcal{B}(B)$ to be interpretable, we must have for any function $x \in \mathcal{F}$ that

$$x \in \mathcal{B}(B) \quad \text{if and only if} \quad D(x; P) \geq \inf_{y \in \mathcal{B}(B)} D(y; P).$$

In other words, x must be claimed to be "non-central" if and only if its depth is smaller than the depth of any "central curve." This property of a functional depth D is new in the literature, but is crucial to ensure that the bands are well-defined. We call it the *band convexity* of functional depths:

(C) For any $P \in \mathcal{P}(\mathcal{F})$ and any $\alpha \in [0, 1]$, the central region $D_\alpha(P) = \{x \in \mathcal{F} : D(x; P) \geq \alpha\}$ is a band, i.e. $D_\alpha(P) = \mathcal{B}(D_\alpha(P))$.

The MBD from (50.2), as well as all other integrated depths (50.3), do not satisfy (C). This can be seen by considering the dataset in the left-hand panel of Fig. 50.2—the thick outlying function y has a very high MBD, but does not belong to the band of all the other (non-outlying) functional data, even though some sample functions have a much lower MBD than y. In Fig. 50.2 we also see what can go wrong with an MBD-based boxplot. An array of additional examples of ill-conceived boxplots using integrated depths is in [13].

The main result from [13] is Theorem 2, which states that the only functional depths D that satisfy (C) are depths of infimal type (50.4), with an appropriately chosen one-dimensional depth D_1. The statement of the following theorem is admittedly imprecise; for a rigorous argumentation, see the original paper [13].

Theorem 1. *A functional depth* $D : \mathcal{F} \times \mathcal{P}(\mathcal{F}) \to [0, 1]$ *satisfies condition (C) if and only if D can be written as an infimal depth* (50.4) *for properly selected one-dimensional depth functions* $D_1 : \mathbb{R} \times \mathcal{P}(\mathbb{R}) \to [0, 1]$.

Theorem 1 states that the only functional depths that give a natural graphical interpretation to a functional boxplot are depths of infimal type. To compare an integrated depth-based boxplot with an interpretable boxplot based on an infimal depth, see Fig. 50.2. In addition to this result, in [13], it is also proved that the properties of a functional boxplot are completely driven by the properties of the one-dimensional depth D_1, and for reasonable choices of D_1 (such as D_1 the one-dimensional half-space depth from (50.1)), the functional boxplots have good properties: they are always well-defined, their central bands are interpretable, and any function flagged as an outlier can be identified as deviating in a specific part of the domain $[0, 1]$.

50.4 The Practice of Functional Boxplots

We saw that infimal depth-based boxplots are the only ones with a natural graphical interpretation. There are, however, also several problematic points associated with the use of infimal depths and bands. Notably, three problems make a straightforward use of functional boxplots challenging:

(a) possible degeneracy and inconsistency of the functional infimal depth [4],
(b) the problem with ties when ranking functions using infimal depths, and
(c) a difficulty with identifying functions outlying in shape.

(a) Degeneracy/Inconsistency. It is known that in certain "high-dimensional" functional models $X \sim P \in \mathcal{P}(\mathcal{F})$, it may happen that for the infimal depth we have $D(X; P) = 0$ almost surely [4]. In that case, no sensible ordering of functional data arises. This problem occurs, for example, for P the distribution of a Brownian motion on $[0, 1]$, and is caused by the extremely erratic nature of $X \sim P$ in the neighborhood of $t = 0$. The same phenomenon then causes that the infimal depth cannot be estimated consistently from its sample version.

The degeneracy problem causes serious trouble from the theoretical point of view. In practice, it is, however, irrelevant since the functional data are always represented by finite-dimensional vectors of their discretized function values. For finite-dimensional data, such a problem does not arise [3]. At the same time, under mild assumptions such as the continuity of the functional data and an appropriate assumption of absolute continuity of the marginal distributions P_t of P, it can be shown that the infimal depths do possess uniformly consistent sample versions [4]. In that situation, the functional boxplots are also naturally well-defined and consistent.

(b) Ties in Depth-Ranking. The discrete nature of the infimal depth implies that when ranking functional data, some observations may attain the same tied depths. In such situations, finite sample adjustments to the resulting boxplots must be made. Several elaborate tie-breaking procedures have been proposed in the literature to tackle this problem. Starting with the tie-breaking procedure behind the extremal depth [14] that breaks ties in depth D according to the length of the domain where each function x attains its smallest point-wise depth $D_1(x(t); P_t)$, a wealth of well-performing tie-breaking procedures have been developed based on various indices of outlyingness. For an overview of some of these, we refer to [10] and the references therein. While inherently, all these methods are heuristic, incorporating them into the construction of functional boxplots improves the performance of the boxplot considerably.

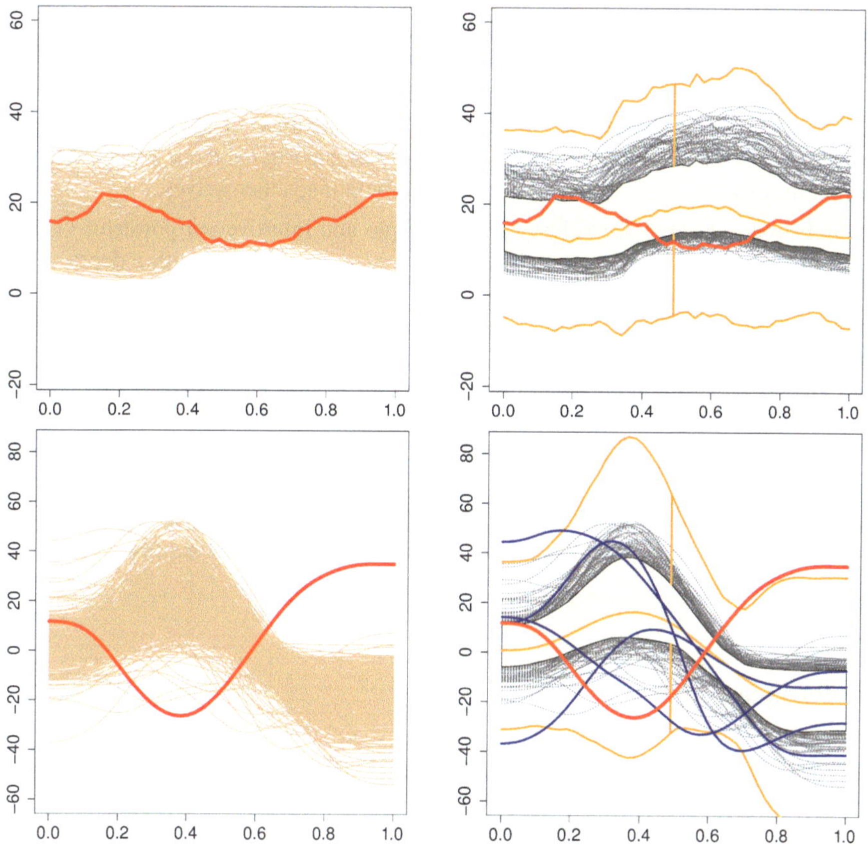

Fig. 50.3: The same dataset as in Fig. 50.1 (top panels) with a shape outlying function (thick red). This function cannot be identified using a basic infimal depth-based boxplot (top right panel). Upon differentiating the appropriately smoothed sample functions (bottom panels), the red curve stands out as a magnitude outlier (bottom left panel). Applying infimal depth-based boxplot to the derivatives, the outlier is correctly flagged as deviating (bottom right panel)

(c) Boxplots and Shape Outliers. Using the bands (50.5) for visualizing functional data has a natural downside when functions outlying in shape may contaminate the data, see the top panels of Fig. 50.3. This property should be considered a feature rather than a downside of a functional boxplot. The outliers detected using (an infimal depth-based) functional boxplot must necessarily stand out of the central band of functions, thanks to the band convexity (C) itself.

If detecting functions outlying in shape is a concern, one must take a different approach and first pre-process (transform) the functional data so that after the transformation, the desired type of outliers turn into magnitude-outlying functions. A simple yet well-performing transform of such a type is a derivative of (properly smoothed) trajectories of random functions, as done in [6]; see the bottom panels of Fig. 50.3. A wider collection of sequential transformations that allow detection of more general shape-outlying functions has been proposed in [2]; besides derivatives, those transforms include the integrals (primitive functions) of the original data, or applying the functional boxplots simultaneously to the original data, warped (that is, time-aligned) data [15], and the estimated warping maps themselves.

Conclusion. The functional boxplot is a simple and intuitive yet powerful visualization tool. To be interpretable, it must be used in conjunction with a depth of an infimal type; integrated depths do not result in well-defined boxplots. While depths of infimal type are known to have several weaknesses, and the resulting boxplots inherit some of them, methods for dealing with these problems have been proposed in the literature. A comprehensive study of the more advanced properties of functional boxplots, such as establishing well-justified rules for the construction of whiskers, is an exciting avenue for future research in functional data analysis.

Acknowledgements This work was supported by the Czech Science Foundation (project n. 24-10822S) and the ERC CZ grant LL2407 of the Ministry of Education, Youth and Sport of the Czech Republic.

References

1. Cuevas, A. and Fraiman, R.: On depth measures and dual statistics. A methodology for dealing with general data. J. Multivariate Anal. **100**(4), 753–766 (2009)
2. Dai, W., Mrkvčka, T., Sun, Y., and Genton, M. G.: Functional outlier detection and taxonomy by sequential transformations. Comput. Statist. Data Anal. **149**, 106960, 17 (2020)
3. Donoho, D. L. and Gasko, M.: Breakdown properties of location estimates based on halfspace depth and projected outlyingness. Ann. Statist. **20**(4), 1803–1827 (1992)
4. Gijbels, I. and Nagy, S.: Consistency of non-integrated depths for functional data. J. Multivariate Anal. **140**, 259–282 (2015)
5. Gijbels, I. and Nagy, S.: On a general definition of depth for functional data. Statist. Sci. **32**(4), 630–639 (2017)
6. Hlubinka, D., Gijbels, I., Omelka, M., and Nagy, S.: Integrated data depth for smooth functions and its application in supervised classification. Comput. Statist. **30**(4), 1011–1031 (2015)
7. Liu, R. Y.: On a notion of data depth based on random simplices. Ann. Statist. **18**(1), 405–414 (1990)
8. López-Pintado, S. and Romo, J.: On the concept of depth for functional data. J. Amer. Statist. Assoc. **104**(486), 718–734 (2009)
9. Mosler, K.: Depth statistics. In: Becker, C., Fried, R., and Kuhnt, S. (eds.) Robustness and complex data structures, pp. 17–34. Springer, Heidelberg (2013)
10. Myllymäki, M. and Mrkvička, T.: GET: Global envelopes in R. J. Stat. Softw. **111**(3), 1–40 (2024)
11. Myllymäki, M., Mrkvička, T., Grabarnik, P., Seijo, H., and Hahn, U.: Global envelope tests for spatial processes. J. R. Stat. Soc. Ser. B. Stat. Methodol. **79**(2), 381–404 (2017)
12. Nagy, S., Gijbels, I., Omelka, M., and Hlubinka, D.: Integrated depth for functional data: Statistical properties and consistency. ESAIM Probab. Stat. **20**, 95–130 (2016)
13. Nagy, S., Mrkvička, T., and Elías, A.: Which depth to use to construct functional boxplots? arXiv:2409.18603 (2024)

14. Narisetty, N. N. and Nair, V. N.: Extremal depth for functional data and applications. J. Amer. Statist. Assoc. **111**(516), 1705–1714 (2016)
15. Ramsay, J.: Curve registration. In: The Oxford handbook of functional data analysis, pp. 235–258. Oxford Univ. Press, Oxford (2011)
16. Rousseeuw, P. J., Ruts, I., and Tukey, J. W.: The bagplot: A bivariate boxplot. Am. Stat. **53**(4), 382–387 (1999)
17. Sun, Y. and Genton, M. G.: Functional boxplots. J. Comput. Graph. Statist. **20**(2), 316–334 (2011)
18. Zuo, Y. and Serfling, R.: General notions of statistical depth function. Ann. Statist. **28**(2), 461–482 (2000)

Chapter 51
Scalar-on-Function Regression with Partially Observed Covariate

Silvia Novo, Alessandro Palummo and Laura M. Sangalli

Abstract In the context of functional regression with scalar response, we investigate the case of partially observed functional covariate. Both functional linear model and functional single-index model are considered, based on functional principal component analysis for partially observed functional data.

51.1 Introduction

Many recent advances in functional data analysis concern functional regression, particularly models where the response is scalar, commonly referred to as "scalar-on-function" regression (see, e.g., [21]). Designed to capture the (linear) relationship between a real-valued response Y and a functional covariate $\mathcal{X} = \{\mathcal{X}(t),\ t \in [a, b] \subset \mathbb{R}\}$ that takes values in a separable Hilbert space $\mathcal{H}$, the functional linear model (FLM) remains the most widely used and studied method (see, e.g., [3], [4], [2], [9], [12]). Assuming that both $\mathcal{X}$ and Y are centered, and given a sample of pairs $\{(\mathcal{X}_i, Y_i)\}_{i=1}^n$ independent and identically distributed (i.i.d.) as $(\mathcal{X}, Y)$, the FLM is defined as:

$$Y_i = \langle \theta_0, \mathcal{X}_i \rangle + \varepsilon_i, \tag{51.1}$$

where $\langle \cdot, \cdot \rangle$ denotes the inner product in $\mathcal{H}$, $\theta_0 \in \mathcal{H}$ is the functional parameter to be estimated, and ε_i represents the random error term. While the FLM provides a straightforward and interpretable framework, its limited flexibility has motivated statisticians to explore alternative models that relax the linearity assumption while

Silvia Novo (✉)
Department of Statistics, Instituto Flores de Lemus, UC3M-Santander Big-Data Institute, Universidad Carlos III de Madrid, Spain, e-mail: `snovo@est-econ.uc3m.es`

Alessandro Palummo
MOX - Dipartimento di Matematica, Politecnico di Milano, Italy
e-mail: `alessandro.palummo@polimi.it`

Laura M. Sangalli
MOX - Dipartimento di Matematica, Politecnico di Milano, Italy
e-mail: `laura.sangalli@polimi.it`

© The Author(s), under exclusive license to Springer Nature Switzerland AG 2025
G. Aneiros et al. (eds.), *New Trends in Functional Statistics and Related Fields*, Contributions to Statistics, https://doi.org/10.1007/978-3-031-92383-8_51

still preserving the interpretability of the outputs. This led to significant advancements in scalar-on-function semiparametric regression (see, e.g., [18]). One notable example are the several papers dealing with the so-called functional single-index model (FSIM) (see, e.g., [11], [1], [19]), where a smooth link function $r(\cdot)$ transforms the result of the inner product in (51.1), giving rise to the model:

$$Y_i = r\left(\langle\theta_0, \mathcal{X}_i\rangle\right) + \varepsilon_i. \tag{51.2}$$

Both the FLM and FSIM have been extensively studied in the case where the covariate is fully observed, as evidenced by the cited references. These models have also been investigated in scenarios where some values of the response variable are missing (see, e.g., [10], [6] for the FLM, and [17] for the FSIM). However, in real applications it often happens that the functional covariate is observed solely over a subset of the domain $[a, b]$, i.e., fragments or large portions of the functional covariate are missing. This setting hinders the applicability of classical as well as recent approaches to scalar-on-function regression. To best of our knowledge, [5] is the first study to address scalar-on-function regression, and specifically the FLM, when the covariate is partially observed.

Partially observed functional variables are frequently encountered across diverse applied domains, such as for instance in environmental and geospatial sciences (see, e.g., [20]), in industry (see, e.g., [15], [8]), or in medical studies (see [22], [16]). As an illustration, Fig. 51.1 shows the curvature (left) and radius (right) profiles of the Internal Carotid Artery (ICA) for the 65 patients included in the Aneurisk dataset (see [22]). The data are derived from the reconstruction of three-dimensional angiographic images obtained from individuals suspected of having cerebral aneurysms. The region of the observed ICA varies across subjects (as can be observed by looking at the starting and ending points of these curves in Fig. 51.1), with some patients having longer sections recorded, while others have shorter

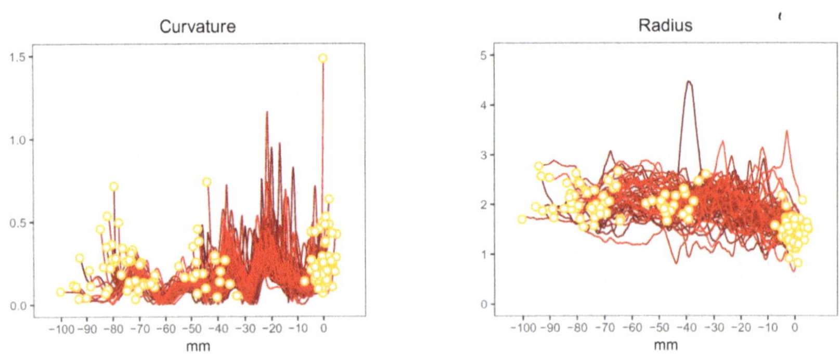

Fig. 51.1: Curvature (left) and radius (right) profiles of the 65 patients. Yellow markers indicate the beginning and end of each subject's curve, highlighting the partial observability of these functional data

sections, depending on the focal point of the medical scan. Thus, the ICA radius and curvature are partially observed functional variables.

One of the primary objectives of the Aneurisk Project [22] was to investigate the relationship between the morphology of the ICA (its radius and curvature) and the location of the aneurysm, i.e., a scalar variable. Focusing on one of the functional variables, we could attempt to model this problem using the FLM or FSIM, considering only the common portion of the domain of the curves. However, discarding portions of the domain where not all subjects are observed could lead to erroneous conclusions.

This highlights the necessity of developing more sophisticated techniques to handle partially observed covariates in regression analysis. A first attempt in this direction is offered by [5], who proposed a FLM for scalar-on-function regression with partially observed functional covariates, based on a two-step approach, where they first use a reconstruction method to estimate the missing portion of the covariates, using the method proposed in [15], and then apply a standard FLM to the reconstructed functional data. Reconstruction methods for partially observed functional data have indeed been the object of an extensive literature; see, e.g., the proposals by [13], [16], [7], [23], [15], or [20].

In this work we propose innovative methods for scalar-on-function regression with partially observed functional covariates, both FLM and FSIM. Differently from [5], the proposed approach does not involve reconstruction of the missing portion of the functional covariates. Instead, as detailed in Sec. 2, it leverages regularized functional principal component analysis (fPCA) for partially observed functional data, developed by [20].

The remainder of the paper is structured as follows. Sect. 2 briefly outlines the proposed estimation methods for the FLM and FSIM. Sect. 3 provides a brief discussion on additional challenges associated with the Aneurisk data.

51.2 Methodology

This section aims to demonstrate how functional principal component analysis (fPCA) for partially observed data can be leveraged to estimate the FLM and FSIM when the functional covariate is incomplete.

51.2.1 Functional Principal Component Analysis of the Partially Observed Covariate

We consider the fPCA method described in [20], which in turn relies on the proposal by [14], improving upon the approach presented by [23]. Below, we provide a summary of the procedure.

Let $\mathcal{H}$ be $L^2[a,b]$. Let Σ denote the covariance operator of $\mathcal{X}_i$, and assume that $\mathbb{E}[\mathcal{X}_i] = 0$. The functional PC expansion of $\mathcal{X}_i$ is given by $\mathcal{X}_i(t) = \sum_{k=1}^{\infty} s_{ik} f_k(t)$ where $f_k(t)$ is the k-th eigenfunction of Σ, and $s_{ik} = \langle \mathcal{X}_i(t), f_k(t)\rangle$ is the corresponding score for the subject i and component k. The functions $f_k(t)$ are referred to as principal components (PCs).

These PCs can also be characterized through the "best M-basis approximation" property, which ensures that the first M PCs provide the best reconstruction of $\mathcal{X}_i(t)$ in the L^2 sense, among all orthonormal bases of dimension M. Specifically, for any M, the first M PCs satisfy:

$$\{f_k\}_{k=1}^{M} = \underset{\{\phi_k\}_{k=1}^{M}:\langle\phi_k,\phi_h\rangle=\delta_{kh}}{\arg\min} \mathbb{E}\left[\int_a^b \left\{\mathcal{X}_i(t) - \sum_{k=1}^{M}\langle \mathcal{X}_i(t), \phi_k(t)\rangle \phi_k(t)\right\}^2 dt\right], \tag{51.3}$$

where δ_{kh} denotes the Kronecker delta.

But in real-world applications we usually don't observe continuous realizations of $\mathcal{X}_i(t)$, but only (noisy) measurements at some points of the domain, which may vary with the realization. Assume that $\mathcal{X}_i$ is partially observed for $i = 1, \ldots, n$, and let O_i be the set of all indexes ℓ where $\mathcal{X}_i(t_\ell)$ is not missing. Then the sample version of expression (51.3) for $M = 1$ becomes:

$$\sum_{i=1}^{n} \sum_{\ell \in O_i} \{\mathcal{X}_i(t_\ell) - s_{i1} f_1(t_\ell)\}^2 .$$

However, the estimation of the infinite-dimensional PC function f_1 from the above expression, using discrete observations, does not incorporate any information on the smoothness of the functional data. To address this, proper regularization must be introduced. Denoting $\boldsymbol{s}_1 = (s_{11}, \ldots, s_{n1})^\top$, the first principal component is estimated by solving the following optimization problem, subject to $\|\boldsymbol{s}_1\|_2 = 1$:

$$\underset{\boldsymbol{s}_1, f_1}{\arg\min} \sum_{i=1}^{n} \sum_{\ell \in O_i} \{\mathcal{X}_i(t_\ell) - s_{i1} f_1(t_\ell)\}^2 + \lambda \boldsymbol{s}_1^\top \boldsymbol{s}_1 \int_a^b \{f_1(t)\}^2 \, dt, \tag{51.4}$$

where $\lambda > 0$ is a smoothing parameter. This functional can be minimized using a majorize–minimization (MM) algorithm, as detailed in [20]. This reasoning can be applied to each of the M principal components. The PC functions thus computed provide the basis for the functional regression models discussed in Sect. 51.2.2.

51.2.2 Model Estimation

In the case of FLM, we can estimate the model using a fPCA approach (see, e.g. [3])), using PCs obtained as in [20], as reviewed in Sect. 51.2.1.

Regarding FSIM, we propose to estimate the function-parameter θ_0 in (51.2), that summarizes the information the covariate carries about the response, by expanding it on the PC basis, obtained as described in Sect. 51.2.1; we moreover expand on the PC basis also the functional covariate. We thus follow a similar approach as in [1], but where the B-spline basis is replaced by the PC basis.

To estimate the function $r(\cdot)$, we can use kernel smoothing techniques, particularly, kernel-based or k-nearest-neighbors-based estimators (see [19]). Specifically, setting $r_{\theta_0}(\chi) = r(\langle\theta_0, \chi\rangle)$ for $\chi \in \mathcal{H}$, we can define kernel-based and k-nearest-neighbour-based (kNN-based) estimators for each $\theta, \chi \in \mathcal{H}$, respectively, as

$$\widehat{r}_{h,\theta}(\chi) = \frac{\sum_{i=1}^{n} Y_i K\left(h^{-1} d_\theta(\mathcal{X}_i, \chi)\right)}{\sum_{i=1}^{n} K\left(h^{-1} d_\theta(\mathcal{X}_i, \chi)\right)}, \quad \widehat{r}^{*}_{k,\theta}(\chi) = \frac{\sum_{i=1}^{n} Y_i K\left(H_{k,\chi,\theta}^{-1} d_\theta(\mathcal{X}_i, \chi)\right)}{\sum_{i=1}^{n} K\left(H_{k,\chi,\theta}^{-1} d_\theta(\mathcal{X}_i, \chi)\right)}. \tag{51.5}$$

Here, K is a kernel function, h and k are the tuning parameters and $d_{\theta_0}(\chi_1, \chi_2) = |\langle\theta_0, \chi_1 - \chi_2\rangle|$, for $\chi_1, \chi_2 \in \mathcal{H}$, is the projection semimetric and $H_{k,\chi,\theta} = \min\left\{h \in \mathbb{R}^+ \text{ such that } \sum_{i=1}^{n} 1_{B_\theta(\chi,h)}(\mathcal{X}_i) = k\right\}$ with $B_\theta(\chi, h) = \{z \in \mathcal{H} : d_\theta(\chi, z) \le h\}$.

To address tuning parameters selection and estimation of θ_0, we can use leave-one-out cross-validation (LOOCV). That is, considering the following objective functions

$$CV(h,\theta) = n^{-1} \sum_{j=1}^{n} \left(Y_j - \widehat{r}_{h,\theta}^{(-j)}(\mathcal{X}_j)\right)^2 \text{ and } CV^*(k,\theta) = n^{-1} \sum_{j=1}^{n} \left(Y_j - \widehat{r}_{k,\theta}^{*(-j)}(\mathcal{X}_j)\right)^2,$$

respectively, and minimising them respect to both parameters.

51.3 Discussion

In this work, we briefly outlined an innovative proposal to estimate the FLM and FSIM when the covariate is partially observed, which avoids the reconstruction of the missing portion of the functional covariates. This study was motivated by the Aneurisk dataset, where both the radius and curvature of the ICA are partially observed. However, this is not the only challenge presented by the Aneurisk dataset. Another critical issue concerns the response variable, that is the location of the aneurysm: indeed, for some subjects in the dataset, this variable is right-censored, as discussed, e.g., in [22]. This characteristic must be accounted for, to accurately model the data, underscoring the need for even more general models that can better capture the complexities of such data.

Acknowledgements S. Novo acknowledges funding from Universidad Carlos III de Madrid through Ayudas para la Movilidad del Programa Propio de Investigación 2024. A. Palummo acknowledges funding by European Union-NextGenerationEU. We also acknowledge the research project Dipartimento di Eccellenza 2023-2027, Dipartimento di Matematica, Politecnico di Milano, funded by the Italian Ministry of University and Research.

References

1. Ait-Saïdi, A., Ferraty, F., Kassa, R., Vieu, P.: Cross-validated estimations in the single-functional index model. Stat. **42**(6), 475–494 (2008)
2. Cai, T. T., Hall, P.: Prediction in functional linear regression. Ann. Stat. **34**(5), 2159–2179 (2006)
3. Cardot, H., Ferraty, F., Sarda, P.: Functional linear model. Stat. Probab. Lett. **345**, 11–22 (1999)
4. Cardot, H., Ferraty, F., Sarda, P.: Spline estimators for the functional linear model. Stat. Sin. **13**, 571–591 (2003)
5. Crambes, C., Daayeb, C., Gannoun, A., Henchiri, Y.: Functional linear model with partially observed covariate and missing values in the response. J. Nonparametr. Stat. **35**(1), 172–197 (2023)
6. Crambes, C., Henchiri, Y.: Regression imputation in the functional linear model with missing values in the response. J. Stat. Plan. Inference **201**, 103–119 (2019)
7. Delaigle, A., Hall, P.: Approximating fragmented functional data by segments of Markov chains. Biometrika **103**(4), 779–799 (2016)
8. Elías, A., Jiménez, R., Paganoni, A. M., Sangalli, L. M.: Integrated depths for partially observed functional data. J. Comput. Graph. Stat. **32**(2), 341–352 (2023)
9. Febrero-Bande, M., Galeano, P., González-Manteiga, W.: Measures of influence for the functional linear model with scalar response. J. Multivar. Anal. **101**(2), 327–339 (2010)
10. Febrero-Bande, M., Galeano, P., González-Manteiga, W.: Estimation, imputation and prediction for the functional linear model with scalar response with responses missing at random. Comput. Stat. Data Anal. **131**, 91–103 (2019)
11. Ferraty, F., Peuch, A., Vieu, P.: Modèle à indice fonctionnel simple. C. R. Math. **336**(12), 1025–1028 (2003)
12. García-Portugués, E., González-Manteiga, W., Febrero-Bande, M.: A goodness-of-fit test for the functional linear model with scalar response. J. Comput. Graph. Stat. **23**(3), 761–778 (2014)
13. Goldberg, Y., Ritov, Y. A., Mandelbaum, A.: Predicting the continuation of a function with applications to call center data. J. Stat. Plan. Inference **147**, 53–65 (2014)
14. Huang, J. Z., Shen, H., Buja, A.: Functional principal components analysis via penalized rank one approximation. Electron. J. Stat. **2**, 678–695 (2008)
15. Kneip, A., Liebl, D.: On the optimal reconstruction of partially observed functional data. Ann. Stat. **48**(3), 1692–1717 (2020)
16. Kraus, D.: Components and completion of partially observed functional data. J. R. Stat. Soc. Ser. B Stat. Methodol. **77**(4), 777–801 (2015)
17. Ling, N., Cheng, L., Vieu, P., Ding, H.: Missing responses at random in functional single index model for time series data. Stat. Pap. **63**, 1–28 (2022)
18. Ling, N., Vieu, P.: On semiparametric regression in functional data analysis. Wiley Interdiscip. Rev. Comput. Stat. **13**(6), e1538 (2021)
19. Novo, S., Aneiros, G., Vieu, P.: Automatic and location-adaptive estimation in functional single-index regression. J. Nonparametr. Stat. **31**(2), 364–392 (2019)
20. Palummo, A., Arnone, E., Formaggia, L., Sangalli, L. M.: Functional principal component analysis for incomplete space–time data. Environ. Ecol. Stat. **31**(2), 555–582 (2024)
21. Reiss, P. T., Goldsmith, J., Shang, H. L., Ogden, R. T.: Methods for scalar-on-function regression. Int. Stat. Rev. **85**(2), 228–249 (2017)

22. Sangalli, L. M., Secchi, P., Vantini, S., Veneziani, A.: A case study in exploratory functional data analysis: geometrical features of the internal carotid artery. J. Am. Stat. Assoc. **104**(485), 37–48 (2009)
23. Stefanucci, M., Sangalli, L. M., Brutti, P.: PCA-based discrimination of partially observed functional data, with an application to AneuRisk65 data set. Stat. Neerl. **72**(3), 246–264 (2018)

Chapter 52
Efficient Physics-Informed Smoothing of Space-time Functional Data

Alessandro Palummo, Eleonora Arnone, Letizia Clementi and Laura M. Sangalli

Abstract In this work, we introduce an efficient estimation procedure for smoothing space-time functional data observed over multidimensional irregular domains. We discuss the limitations of existing approaches when applied to large datasets of spatio-temporal functional data, observed over complex supports, such as those encountered in neuroimaging applications, and propose a novel iterative procedure for the solution of physics-informed nonparametric regression problems. The proposed method combines computational efficiency with high accuracy. Moreover, it provides the basis to address more complex and large-scale functional data analysis problems, at the population level, concerning for instance functional principal component analysis and functional clustering.

52.1 Introduction

The availability of increasingly large datasets, combined with growing data complexity, poses several challenges to statistical methods designed to analyze them. This work focuses on functional data observed over multidimensional irregular domains and simultaneously varying along an additional dimension, such as time or frequency. Data of this form are increasingly prevalent across diverse fields. In environmental

Alessandro Palummo (✉), Laura M. Sangalli
MOX, Department of Mathematics, Politecnico di Milano, Italy,
e-mail: alessandro.palummo@polimi.it, laura.sangalli@polimi.it

Eleonora Arnone
Department of Management, Univesità degli Studi di Torino, Italy,
e-mail: eleonora.arnone@unito.it

Letizia Clementi
Fondazione IRCCS Istituto Neurologico Carlo Besta, SC Neurologia, Salute Pubblica, Disabilità, Italy
e-mail: letizia.clementi@istituto-besta.it

© The Author(s), under exclusive license to Springer Nature Switzerland AG 2025

G. Aneiros et al. (eds.), *New Trends in Functional Statistics and Related Fields*,
Contributions to Statistics, https://doi.org/10.1007/978-3-031-92383-8_52

sciences, for instance, it is of interest to analyze satellite data observed over time across the globe, or over water basins with complex shorelines, or over land regions with complex conformation. In life sciences, it is crucial to analyze biological signals measured within an organ's anatomical structure and recorded by medical scanners. In neuroscience, for instance, functional Magnetic Resonance Imaging (fMRI) returns a four-dimensional biological signal, the Blood Oxygenation Level Dependent (BOLD) signal, which is used to investigate functional activity and connectivity in the brain, as well as their alterations caused by trauma, diseases, or disorders. BOLD signals are functional data measuring neural cell activation over time and across a non-convex three-dimensional domain, the grey matter, with its formidably complex anatomy, characterized by numerous sulci and gyri. As a reference example, Fig. 52.1 shows the Power Spectrum Density (PSD) of the BOLD signal recorded at 3 spatial locations inside the brain volume. The analysis of these functional signals poses several challenges, primarily due to the high dimensionality of the data and the complex geometry of their support. Classical spatio-temporal techniques that rely on the Euclidean distance are unsuitable for complex geometries like the brain's anatomy. Such methods neglect the domain's folded structure, leading to artifacts in the estimates. In this work, we propose a computationally efficient procedure for smoothing functional data observed over space and time, while accounting for the complex geometry of the spatial domain.

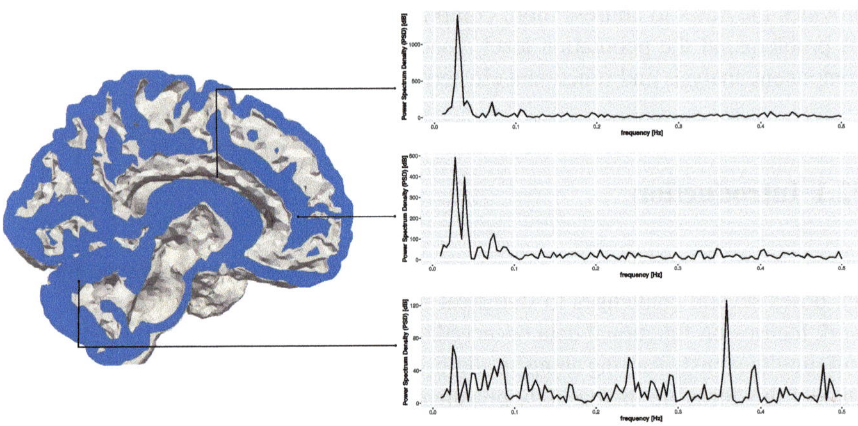

Fig. 52.1: Left: slice of brain volume highlighting the complicated geometry of the problem under analysis. Right: BOLD Power Spectrum Density recorded at 3 spatial locations located inside the brain volume

It should be noted that, over the past decades, several techniques have been developed to analyze spatially dependent functional data, such as methods based on kriging for functional data (see, e.g., the reviews in [11, 22, 21]) or smoothers like thin-plate splines or soap film smoothing [31, 5, 20]. Some of these methods can handle spatially dependent functional data that vary across non-convex two-dimensional domains [23, 5, 20]. However, none of the aforementioned techniques can address functional data that vary spatially over three-dimensional non-convex domains, such as non-trivial two-dimensional Riemannian manifolds or non-convex volumes. Furthermore, all these methods face significant challenges when applied to large datasets and become computationally unsustainable for massive ones.

The method proposed in this work finds its foundations in the physics-informed nonparametric regression framework [7], which has emerged as a powerful tool for analyzing spatial and functional data, borrowing physical knowledge in the statistical modeling via Partial Differential Equations (PDEs). Starting from the work presented in [8] for space-time data smoothing, we propose an efficient iterative approach to its resolution, that departs from the monolithic solution in [8]. Unlike approaches in [1, 25], which assume prior physical knowledge encoded as a single parabolic PDE, our method employs a more flexible separable penalty, featuring two smoothness terms that regularize the functional data over the spatial and temporal dimensions. The presented procedure is shown to reach comparable accuracy to the monolithic approach, while significantly reducing computational cost, thereby enabling the analysis of massive datasets.

52.2 Preliminaries: the Smoothing Problem

Let $\mathcal{D} \subset \mathbb{R}^d$, with $d = 1, 2, 3$, be a multidimensional bounded domain of interest, and let $[0, T] \subset \mathbb{R}$ be a time interval. We consider functional data defined over the support $\mathcal{D} \times [0, T]$. For each functional realization f, we only observe a noisy and discrete realization $\{y_{ij}\}_{ij}$, with $i = 1, \ldots, n$ and $j = 1, \ldots, m$, over the observational grid $\{(\boldsymbol{p}_i, t_j)\}_{ij}$, for $i = 1, \ldots, n$ and $j = 1, \ldots, m$. In particular, we assume the following nonparametric model

$$y_{ij} = f(\boldsymbol{p}_i, t_j) + \epsilon_{ij} \qquad i = 1, \ldots, n, j = 1, \ldots, m,$$

where $f : \mathcal{D} \times T \to \mathbb{R}$ is the unknown, continuous and exact realization of the functional datum, and the ϵ_{ij} are independent errors with zero mean and constant variance $\sigma^2 < \infty$. Likewise in [8, 4], we estimate f by minimizing the following penalized sum of squared error functional

$$J(f) = \sum_{i=1}^{n} \sum_{j=1}^{m} (y_{i,j} - f(\boldsymbol{p}_i, t_j))^2 + \lambda_{\mathcal{D}} \int_T \int_{\mathcal{D}} (Lf - u)^2 + \lambda_T \int_{\mathcal{D}} \int_T \left(\frac{\partial^2 f}{\partial t^2} \right)^2, \quad (52.1)$$

where $\lambda_{\mathcal{D}} > 0$ and $\lambda_T > 0$ are two smoothing parameters that weight the penalizations respectively in space and time. The first regularizing term involves the Partial Differential Equation $Lf = u$, where L is a linear, second order elliptic operator [12]; this equation encodes the available prior physical knowledge, allowing for non-isotropic and non-stationary smoothing over $\mathcal{D}$. The second regularizing term involves instead the second derivative in time, a classical roughness penalty extensively used in functional data analysis; see, e.g., [27].

The minimizer of the functional in Equation (52.1) cannot be determined analytically, necessitating the use of numerical discretization techniques to obtain an approximate solution. Let $\mathcal{T}$ be a suitable triangulation of the domain $\mathcal{D}$. Following the finite element method [26], we define on $\mathcal{T}$ a set of N Lagrangian basis functions, $\psi_1, \ldots, \psi_N$, which in this work we set to be linear polynomials once restricted to any element of $\mathcal{T}$. For the temporal dimension, inspired by [8], one possible approach is to introduce a set of M cubic B-spline basis functions [30], denoted as $\phi_1, \ldots, \phi_M$, defined over the time domain $[0, T]$. The unknown spatio-temporal field $f(\boldsymbol{p}, t)$ can then be expressed using a separable tensor-product basis expansion as $\sum_{i=1}^{N} \sum_{j=1}^{M} c_{ij} \psi_i(\boldsymbol{p}) \phi_j(t)$, where c_{ij} are the coefficients of the expansion and form the primary variables of the discretized estimation problem. This choice results in a monolithic approach in which the estimation of the expansion coefficient vector requires solving an $NM \times NM$ sparse linear system. However, as N or M increase, the computational cost of this approach grows rapidly, rendering it impractical for large-scale problems.

Moreover, a significant computational bottleneck of this smoothing problem is the accurate selection of the smoothing parameters $\lambda_{\mathcal{D}}$ and λ_T, which is crucial to achieving optimal results. This selection can be performed by minimizing the classical Generalized Cross-Validation (GCV) [14] index, which here takes the form

$$\mathrm{GCV}(\lambda_{\mathcal{D}}, \lambda_T) = \frac{nm}{(nm - \mathrm{Tr}[S])^2} \|\boldsymbol{y} - \hat{\boldsymbol{y}}\|_2^2,$$

where S denotes the $NM \times NM$ smoothing matrix. However, the computation of the trace of S quickly becomes computationally prohibitive, even for moderately sized problems, which can be considered orders of magnitude smaller than those encountered for the neuroimaging problem discussed above.

52.3 Iterative Estimation Procedure

The $NM \times NM$ linear system that leads to the solution of the smoothing problem can be approximated by the iterative solution of M coupled linear systems of dimension $3N \times 3N$. In the contest of large discretizations, such as those necessary to accurately smooth complex functional data as those represented in Fig. 52.1, such iterative procedure dramatically reduces the computational cost, enabling the analysis of massive data analysis problems.

To do so, instead of employing B-spline basis functions in time, we here consider a discretization of the time dimension using finite differences. Specifically, we employ a uniformly spaced grid in the time domain $[0, T]$, having nodes $\tau^0, \ldots \tau^M$ such that $\tau^0 = 0$, $\tau^k = k\Delta t$ and $\tau^M = T$, being Δt a fixed time step. As classically done in treatment of parabolic problems [29], we proceed by first obtaining a semi-discrete problem discretizing the spatial dimension via finite elements [9].

Subsequently we recover a fully discrete formulation using central finite differences for the discretization of the second order time derivatives. This enables us to develop an iterative procedure where we have to solve, at each step of the algorithm, M linear systems of dimension $3N \times 3N$, providing a competitive advantage with respect to the monolithic approach. Fig. 52.2 gives an idea of the different dimensionality involved in the monolithic and the proposed iterative algorithm.

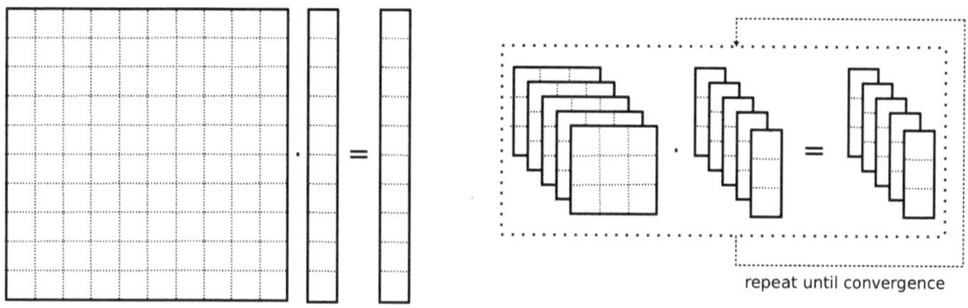

Fig. 52.2: Sketch of the dimensionality involved in the linear system associated with the monolithic (left) and iterative (right) approach

To accurately select the smoothing parameters $\lambda_{\mathcal{D}}$ and λ_T, we propose a stochastic approximated method following [2], that avoids the exact computation of $\mathrm{Tr}[S]$. Specifically, let $\boldsymbol{u}$ be a vector of samples from a Rademacher distributed random variable. Using Hutchinson's stochastic estimator [17] the trace $\mathrm{Tr}[S]$ can be approximated as $\mathbb{E}[\boldsymbol{u}^\top S\boldsymbol{u}]$. Notably, $S\boldsymbol{u}$ can be obtained by solving a penalized least square problem as in Equation (52.1) once we treat $\boldsymbol{u}$ as observation vector. This allows us to compute $S\boldsymbol{u}$ directly within the iterative estimation procedure, enabling the evaluation of the GCV index for any pair of smoothing parameters without incurring additional computational overhead beyond that of the iterative procedure itself.

52.4 Future Extensions towards Population Studies

The proposed approach offers the grounds from where we can aim addressing more complex and massive functional data analysis problems, at population level, based on a sample of functional observations over space and time. One major example in this sense is functional Principal Component Analysis (fPCA); see, e.g., the textbooks by [27, 13, 15, 18]. In particular, the seminal work of [16] introduced an fPCA approach where smoothing is enforced directly on the principal components, rather than on the single functional data. Such approach has later been extended to functional data observed over non-convex supports by [19, 3, 10], and further generalized to space-time functional data by [24], considering differential roughness penalties, capable to deal with the brain's intricate anatomical structure. However, these methods require iterative solutions of smoothing problems (see, e.g. [6, 8]), which become prohibitive in the context of massive dataset. The efficient approach here presented could thus be profitably adapted to the solution of these space-time fPCA problems. In the study of fMRI data, this would for instance enable the identification of the main variability patterns in the space-frequency domain of BOLD signals, a problem currently beyond reach of existing methods.

Besides fPCA, the proposed procedure could be integrated within other important functional data analysis problems, such as clustering. In particular, a fascinating direction for future research consists in developing appropriated regularized functional k-mean algorithms, based on the regularized approaches presented in [28]. For the considered application, this would enable the exploration of different subgroups of subjects displaying different neuronal activity patterns.

Acknowledgements A. Palummo acknowledges funding by European Union - NextGenerationEU. We also acknowledge the research project Dipartimento di Eccellenza 2023-2027, Dipartimento di Matematica, Politenico di Milano, funded by the Italian Ministry of University and Research.

References

1. Arnone, E., Azzimonti, L., Nobile, F., Sangalli, L. M.: Modeling spatially dependent functional data via regression with differential regularization. Journal of Multivariate Analysis. **170**, 275-295 (2019)
2. Arnone, E., De Falco, C., Formaggia, L., Meretti, G., Sangalli, L. M: Computationally efficient techniques for spatial regression with differential regularization. International Journal of Computer Mathematics. **100**(10), 1971-1991 (2023)
3. Arnone, E., Negri, L., Panzica, F., Sangalli, L. M.: Analyzing data in complicated 3D domains: Smoothing, semiparametric regression, and functional principal component analysis. Biometrics. **79**(4), 3510-3521 (2023)
4. Arnone, E., Sangalli, L. M., Vicini, A.: Smoothing spatio-temporal data with complex missing data patterns. Statistical Modelling. **23**(4), 327-356 (2023)
5. Augustin, N. H., Trenkel, V. M., Wood, S. N., Lorance, P.: Space-time modelling of blue ling for fisheries stock management. Environmetrics. **24**(2), 109-119 (2013)
6. Azzimonti, L., Nobile, F., Sangalli, L. M., Secchi, P.: Mixed finite elements for spatial regression with PDE penalization. SIAM/ASA Journal on Uncertainty Quantification. **2**(1), 305-335 (2014)
7. Azzimonti, L., Sangalli, L. M., Secchi, P., Domanin, M., Nobile, F.: Blood flow velocity field estimation via spatial regression with PDE penalization. Journal of the American Statistical Association. **110**(511), 1057-1071 (2015)
8. Bernardi, M. S., Sangalli, L. M., Mazza, G., Ramsay, J. O.: A penalized regression model for spatial functional data with application to the analysis of the production of waste in Venice province. Stochastic environmental research and risk assessment. **31**, 23-38 (2017)
9. Ciarlet, P. G.: The finite element method for elliptic problems. Society for Industrial and Applied Mathematics, Philadelphia (2002)
10. Clementi, L., Arnone, E., Santambrogio, M. D., Franceschetti, S., Panzica, F., Sangalli, L. M.: Anatomically compliant modes of variations: New tools for brain connectivity. Plos one. **18**(11), e0292450 (2023)
11. Delicado, P., Giraldo, R., Comas, C., Mateu, J.: Statistics for spatial functional data: some recent contributions. Environmetrics. **21**(3-4), 224-239 (2010)
12. Evans, L. C.: Partial differential equations. American Mathematical Society, Rhode Island (2022)

13. Ferraty, F., Vieu, P.: Nonparametric Functional Data Analysis. Springer, Berlin (2006)
14. Golub, G. H., Heath, M., Wahba, G.: Generalized cross-validation as a method for choosing a good ridge parameter. Technometrics. **21**(2), 215-223 (1979)
15. Hsing, T., Eubank, R.: Theoretical foundations of functional data analysis, with an introduction to linear operators. John Wiley & Sons, Ltd., Chichester (2015)
16. Huang, J., Shen, H., Buja, A.: Functional principal components analysis via penalized rank one approximation. Electronic Journal Of Statistics. **2**, 678-695 (2008)
17. Hutchinson, M. F.: A stochastic estimator of the trace of the influence matrix for Laplacian smoothing splines. Communications in Statistics-Simulation and Computation, **18**(3), 1059-1076 (1989)
18. Kokoszka, P., Reimherr, M.: Introduction to Functional Data Analysis. CRC Press, Boca Raton (2017)
19. Lila, E., Aston, J. A., Sangalli, L. M.: Smooth principal component analysis over two-dimensional manifolds with an application to neuroimaging. Ann. Appl. Stat. **10**(4), 1854-1879 (2016)
20. Marra, G., Miller, D. L., Zanin, L.: Modelling the spatiotemporal distribution of the incidence of resident foreign population. Statistica Neerlandica. **66**(2), 133-160 (2012)
21. Martínez-Hernández, I., Genton, M. G.: Recent developments in complex and spatially correlated functional data. Braz. J. Probab. Stat. **34**(2), 204-229 (2020)
22. Menafoglio, A., Secchi, P.: Statistical analysis of complex and spatially dependent data: a review of object oriented spatial statistics. European J. Oper. Res. **258**(2), 401-410 (2017)
23. Menafoglio, A., Gaetani, G., Secchi, P.: Random domain decompositions for object-oriented kriging over complex domains. Stochast. Environ. Res. Risk Assessment, **32**(12), 3421–3437 (2018)
24. Palummo, A., Arnone, E., Formaggia, L., Sangalli, L. M.: Functional principal component analysis for incomplete space–time data. Environmental and Ecological Statistics. **31**(2), 555-582. (2024)
25. Ponti, L., Perotto, S., Sangalli, L. M.: A PDE-regularized smoothing method for space–time data over manifolds with application to medical data. International Journal for Numerical Methods in Biomedical Engineering **38**(12), e3650 (2022)
26. Quarteroni, A., Quarteroni, S.: Numerical models for differential problems (Vol. 2). Springer, Milan (2009)
27. Ramsay, J. O., Silverman, B. W.: Functional data analysis. Springer, New York (2005)
28. Sun, W., Wang, J., Fang, Y.: Regularized k-means clustering of high-dimensional data and its asymptotic consistency. Electron. J. Stat. **6**, 148-167 (2012)
29. Thomée, V.: Galerkin finite element methods for parabolic problems (Vol. 25). Springer Science & Business Media, Berlin (2007)
30. Wahba, G.: Spline models for observational data. Society for industrial and applied mathematics, Philadelphia (1990)
31. Wood, S. N., Bravington, M. V., Hedley, S. L.: Soap film smoothing. Journal of the Royal Statistical Society Series B: Statistical Methodology, **70**(5), 931-955 (2008)

Chapter 53
Ordinal-on-Function Dimensionality Reduction

Giulia Patanè, Federica Nicolussi, Alexander Krauth, Günter Gauglitz, Bianca Maria Colosimo, Luca Dede' and Alessandra Menafoglio

Abstract In this communication, we tackle the challenge of dimensionality reduction of high dimensional objects, such as functional profiles, while maximizing the predictive power of an ordinal target variable. This issue is particularly important in sensor-driven decision-making processes, where the observable signal displays complexity and the aim is to evaluate a multi-level (ordinal) variable such as the degree of deterioration, quality, risk or stress of an object of interest. We describe a recent method called functional-ordinal Canonical Correlation Analysis (foCCA, [5]), which allows to perform ordinal-on-function dimensionality reduction. FoCCA reduces the dimensionality of observable features while optimizing their ability to

Giulia Patanè (✉)
Politecnico di Milano, Piazza Leonardo da Vinci 32, 20133 Milano, Italy,
e-mail: giulia.patane@polimi.it

Federica Nicolussi
Politecnico di Milano, Piazza Leonardo da Vinci 32, 20133 Milano, Italy,
e-mail: federica.nicolussi@polimi.it

Alexander Krauth
Eberhard Karls Universität Tübingen, Auf der Morgenstelle 18, 72076 Tübingen, Germany,
e-mail: alexander.krauth@uni-tuebingen.de

Günter Gauglitz
Eberhard Karls Universität Tübingen, Auf der Morgenstelle 18, 72076 Tübingen, Germany,
e-mail: guenter.gauglitz@uni-tuebingen.de

Bianca Maria Colosimo
Politecnico di Milano, Piazza Leonardo da Vinci 32, 20133 Milano, Italy,
e-mail: biancamaria.colosimo@polimi.it

Luca Dede'
Politecnico di Milano, Piazza Leonardo da Vinci 32, 20133 Milano, Italy,
e-mail: luca.dede@polimi.it

Alessandra Menafoglio
Politecnico di Milano, Piazza Leonardo da Vinci 32, 20133 Milano, Italy,
e-mail: alessandra.menafoglio@polimi.it

© The Author(s), under exclusive license to Springer Nature Switzerland AG 2025
G. Aneiros et al. (eds.), *New Trends in Functional Statistics and Related Fields*,
Contributions to Statistics, https://doi.org/10.1007/978-3-031-92383-8_53

distinguish between consecutive levels of an ordinal target variable. Differently from existing dimensionality reduction tools for functional data, foCCA fully integrates the ordinal nature of the target variable: in this way, foCCA captures and represents the relative differences between consecutive levels of the ordinal target, while also explaining these differences through the functional features. Extensive simulations show that foCCA performs better than current state-of-the-art dimensionality reduction tools in prediction accuracy within the reduced feature space. In particular, foCCA both improved predictive power and enhanced interpretability compared to other approaches.

53.1 Introduction

In industrial settings, statistical process control monitors key targets like quality, stress, risk, and degradation using observable features. Automated decision-making typically requires encoding these targets as ordinal variables with discrete levels [12], which are then the object of monitoring and predictive statistical procedures. This work focuses on predicting an ordinal variable that describes a multi-level stage of a process using high-dimensional or functional features, while jointly reducing their dimensionality. When employing functional data to predict an ordinal target, two approaches can be adopted: either using directly the high dimensional object for ordinal-on-function modeling, e.g. through Generalized Linear Models (GLM), or performing dimensionality reduction to project the object in a lower dimensional space. The first approach allows for a better interpretation of the dependence between the high dimensional object and the ordinal target. An example is provided by [3], which introduces an ordinal logit model to predict the probability that a unit i, based on a functional signals, falls within a level $l_i \leq l$ for any $l \in \{1, ..., \mathfrak{L}\}$. Direct approaches, such as the one presented in [3], offer the advantage of easy interpretability. However, ordinal-on-function GLM models may be more susceptible to the curse of dimensionality, especially when the model assumptions oversimplify the relation between the functional object and the ordinal target. For example, in the study by [3], the odds ratio is assumed to remain constant across levels, which may not accurately capture the true underlying dynamics of the data.

In this communication, we explore the second approach. Several existing methods for functional data analysis, indeed, include a dimensionality reduction step, such as Functional Principal Component Analysis (fPCA) [6] and Fisher's Discriminant Analysis [7], but fail to account for the ordinal nature of the target variable. FPCA reduces data dimensionality by capturing feature variability, while Fisher's Discriminant Analysis aims to enhance discriminative power. However, these methods struggle to effectively predict ordinal variables, limiting their use in this context. To address this, in [5] we proposed a novel methodology, named functional-ordinal Canonical Correlation Analysis (foCCA). FoCCA reduces the dimensionality of functional data while maximizing correlation with the ordinal variable and establishing interpretable differences between levels. It effectively reduces data complexity, performs supervised classification of ordinal levels, and enhances model interpretation. In this communication, we describe the methodology and computational challenges for its implementation, and show via simulations its performances in comparison with competitors.

53.2 State of the Art based on Dimensionality Reduction

FPCA is a cornerstone in functional dimensionality reduction [6]. It finds the functional directions of a dataset, which maximize the variance in the reduced space. When the direction of maximum variance and the one of the variance explained by the ordinal variable is the same, fPC's scores can be employed to build a predictor for the ordinal target. However, sometimes the variance explained by the ordinal variable is maximized along a different direction with respect to the fPCs. In this cases, fPCA may fail to effectively support prediction of the target ordinal variable.

Functional Fisher's Discriminant Analysis [7] detects the functional directions along which the ratio between variance between classes and variance within classes is maximized. This method is suitable when all the classes are separated along the same direction. However, it can fail when the consecutive levels are highly non-equidistant. Indeed, if a direction separates only a subset of all the pairs of consecutive levels, the variance between groups along that direction can be relatively small.

For these reasons, a method that is able to separate levels while taking into account non-equidistance is needed.

53.3 Functional-Ordinal Canonical Correlation Analysis

Our objective is to reduce the dimensionality of functional data while enhancing the prediction accuracy of ordinal levels in the reduced space. Building on the methodology in [5], we utilize Canonical Correlation Analysis (CCA, [2]) to identify directions that maximize the correlation between two datasets. This approach can be adapted to functional datasets by incorporating smoothness constraints through penalization [4]. We extend this concept to find functional canonical directions that can distinguish an ordinal variable based on the reduced functional signal. An ordinal variable is a categorical variable were the categories have a natural order and the distances between them is not known a priori [1]. Proper encoding of the ordinal variable is crucial to reflect the order and potential non-equidistance of its levels. We introduce an encoding criterion for the ordinal variable that is compatible with CCA in a ordinal-on-function framework and provide an interpretation of the canonical directions. On this basis, foCCA identifies the smoothed directions of maximum association between the functional and the (encoded) ordinal.

More precisely, foCCA receives as input a functional-ordinal data set $\{X_i, L_i\}_{i=1}^N$, where $X_i \in L^2(\Omega)$ are functional data and L_i's are ordinal labels which take value in an ordered set $\{1, ..., \mathfrak{L}\}$. For each $i \in \{1, ..., N\}$, we encode the label L_i through the variable Y_i, which is a binary vector of length $\mathfrak{L} - 1$, where $\mathfrak{L}$ is the number of distinct levels. For any $l \in \{1, ..., \mathfrak{L}\}$, $Y_i[l] = \mathbb{1}_{L_i > l}$. The output of foCCA, fixing equal to K the number of desired components, consists of K pairs $\{\beta_k, \theta_k\}_{k=1}^K$, where $\beta_k \in L^2(\Omega)$ and $\theta_k \in \mathbb{R}^{\mathfrak{L}-1}$, which maximize the following object:

$$ccorsq_{\lambda_1,\lambda_2}(\beta, \theta) = \frac{\widehat{\text{Cov}}(x_\beta, y_\theta)^2}{\left(\widehat{\text{Var}}(x_\beta) + \lambda_1 Pen(\beta)\right)\left(\widehat{\text{Var}}(y_\theta) + \lambda_2 Pen(\theta)\right)} \quad (53.1)$$

i.e. the sample correlation between the functional scores, i.e. the projection x_β of X's on β, and the ordinal scores, i.e. the projection y_θ of Y's on θ. In [5], we provide an extensive guide to interpret the results carried out through foCCA. Indeed, the K pairs of functional and ordinal scores can be employed to analyze the relation between the functional and the ordinal variable. In particular, for any $k \in \{1, ..., K\}$, we show that the coefficients of the vector θ_k represent the separation of each pair of consecutive levels, along the respective direction β_k. This allows us to highlight the interpretability gained through our method. Moreover, the functional scores can be used to classify the data into the ordinal categories.

In this communication we provide an overview of the discretization of the foCCA maximization problem. In particular, functional data are discretized via B-spline basis expansion: this allows to obtain a matrix form of the optimization problem, which is proved to be equivalent to a generalized eigenvalue problem, whose solution provides the coefficients of the B-spline basis for the functional component β and of the ordinal component (vector) θ. In [5], we discuss details on the penalization and a method to tune the penalization parameter. The same parameter selection is used for the competitors, in order to fairly compare them in the ordinal prediction framework.

53.4 Simulation Study

We here present the main results of an extensive simulation study, consisting of two scenarios with varying degree of severity. They allow us to systematically analyze and compare how each method handles the specific challenges of the underlying data structures. This simulation study provides valuable insights into their respective advantages and drawbacks in real-world applications.

In **scenario A**, the functional dataset exhibits, along a specific functional direction, a portion of variance explained by the ordinal variable that is less than the global variance of the dataset. In particular, this condition can challenge fPCA. The degree of severity of this scenario is regulated by varying the expected ratio, called q_a, between the global variance and the variance explained by the ordinal variable. The lower the expected ratio q_a, the higher the severity of the scenario, i.e. the difficulty to capture the ordinal separability. In Fig. 53.1 we report an example of simulated scenarios of type A.

In **scenario B**, the successive levels of the ordinal variable exhibit non-equidistance. In this scenario, the functional dataset exhibits, along a first functional direction, a portion of variance explained only by the first half of the levels (lowest), and along a second functional direction, another portion of variance explained by the second half of levels (highest). The ratio between the variance explained by the first levels (along the first direction) and the second ones (along the second direction) is quantified through $q_b \in [0, 1]$. Therefore, the degree of severity of this scenario is regulated by varying the ratio q_b. In principle, scenario B can challenge all the methods. In Fig. 53.2 we report an example of simulated scenarios of type B.

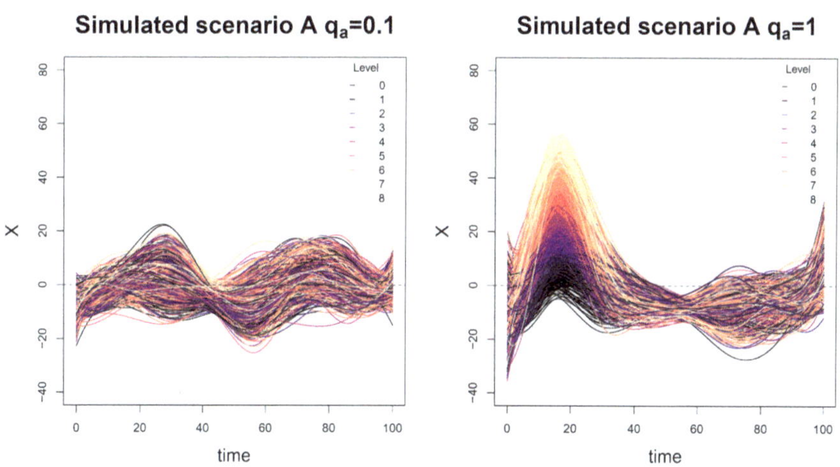

Fig. 53.1: Example of scenario A with different degrees of severity

53.4.1 *Results*

In Fig. 53.3 we report the results of a Monte Carlo simulation with 100 samples to evaluate the performance of classifiers in a 2D reduced space under both scenario A and scenario B, with maximum ($q_a = 0.1$, $q_b = 0.1$) and minimim ($q_a = 1$, $q_b = 1$)

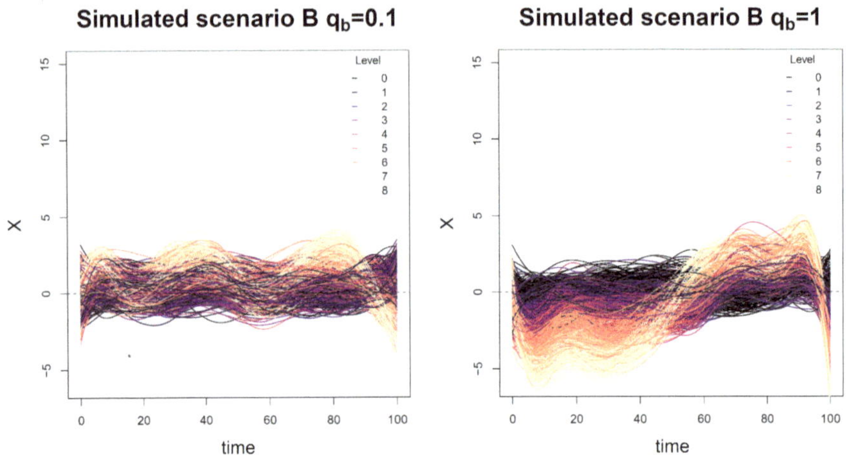

Fig. 53.2: Example of scenario B with different degrees of severity

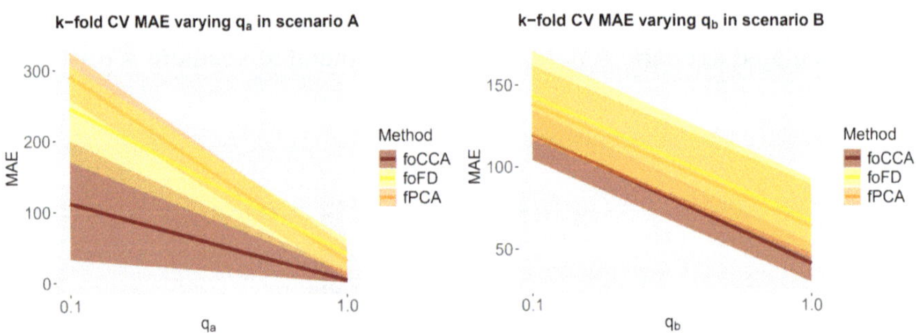

Fig. 53.3: Monte Carlo simulation with 100 samples to evaluate and compare the Mean Absolute Error of the ordinal predictor, in the the reduced spaces obtained through the three methods: foCCA, fPCA and foFD. The line represents the mean values, while the bands' boundaries are the pointwise quantiles 5% and 95%

degree of severity. We here compare three dimensionality reduction techniques: fPCA, Fisher's discriminant analysis adapted for ordinal data (foFD), and foCCA. As one can observe in Fig. 53.3, foCCA leads to lower error (MAE) than fPCA and foFD in both the scenarios and extremes of severity. A further analysis can be found in [5], where the same simulation is computed with 500 samples, varying the ratios q_a and q_b within the set $\{0.1, 0.2, \ldots 1\}$. The results demonstrate that foCCA consistently outperforms, in terms of ordinal predictive ability, the competitors across all levels of severity.

53.5 Conclusions

This work addresses the challenge of reducing the dimensionality of functional data while optimizing the prediction of an ordinal target. To this end, we introduced foCCA, detailed in [5]. Our extensive simulation study revealed that this method significantly outperforms existing techniques, such as fPCA and ordinal Fisher's Discriminant Analysis, in two realistic scenarios across all levels of severity. Additionally, it provides the distances between consecutive levels, which can be intrinsically non-homogeneous. This approach is particularly well-suited for ordinal prediction based on high-dimensional data or when the differences between consecutive levels are of particular interest. For instance, in [5] we present a case study in which light intensity signals evolving in time have to be used to predict the levels of antigen concentration in a region of interest. In this study, the levels are nominally assigned, but the intrinsic difference among consecutive levels are not known a priori. FoCCA allows to understand in which functional features they differ and to quantify this difference. When the intrinsic difference between two levels, based on the functional data, turns out to be near to 0, the expert can decide to merge the two levels for the next analyses.

Acknowledgements This research has received funding by the European Commission under the "HORIZON-CL4-2021-DIGITALEMERGING-01 project BioProS - Biointelligent Production Sensor to Measure Viral Activity" (grant agreement no. 101070120), 2022-2026". GP, FN, LD, and AM acknowledge the initiative "Dipartimento di Eccellenza 2023–2027", MUR, Italy, Dipartimento di Matematica, Politecnico di Milano. LD acknowledges his membership to INdAM GNCS - Gruppo Nazionale per il Calcolo Scientifico (National Group for Scientific Computing), Italy.

References

1. Agresti A.: Categorical Data Analysis (3 ed.). Hoboken, New Jersey: John Wiley Sons (2013)
2. Härdle W., Simar L.: Canonical Correlation Analysis, Applied Multivariate Statistical Analysis, Springer (2007)
3. Jacques J., Samardžić S.: Analysing Cycling Sensors Data Through Ordinal Logistic Regression with Functional Covariates. Journal of the Royal Statistical Society Series C : Applied Statistics, **71**(4), 969–986 (2022)
4. Leurgans S. E., Moyeed R. A., Silverman B. W.: Canonical correlation analysis when the data are curves. Journal of the Royal Statistical Society, **55**, 725–740 (1993)
5. Patanè G., Nicolussi F., Krauth A., Colosimo B. M., Dede' L., Menafoglio A.: Functional-Ordinal Canonical Correlation Analysis With Application to Data from Bioprinting Sensors, arXiv 2501.18317, (2025)
6. Ramsay, J.O., Silverman, B.W.: Functional Data Analy-sis. Springer Series in Statistics. New York, NY: Springer NewYork. (2005)
7. Hastie T., Buja A., Tibshirani R.: Penalized discriminant analysis. Annals of Statistics, **23**, 73–102 (1995)
8. Kundu P., Darpe A. K., Kulkarni M. S., Zuo M.: Online damage severity level classification in gears under natural damage progression. The International Journal of Advanced Manufacturing Technology (2022)

9. Li Y., Li B., Ding J., Feng Y., Ma M., Han Z., Xu Y., and Xia L.: A Novel Framework for Forecasting Mental Stress Levels Based on Physiological Signals. Neural Information Processing, 123–138 (2023)
10. Movahed R. A., Jahromi G. P., Shahyad S., Meftahi G. H.: A Novel Framework for Classifying Major Depressive Disorder Using EEG Signals. Journal of Neuroscience Methods (2021)
11. Ruan H., Dai X., Chen S., Qiu X.: Arrhythmia Classification and Diagnosis Based on ECG Signal: A Multi-Domain Collaborative Analysis and Decision Approach. Electronics, **11**(19), 3251 (2022)
12. Zhang G., Lu J., Gao Y.: Multi-level Decision Making: Models, Methods and Applications. Springer (2015)

Chapter 54
Varying Coefficient Regression Models on Fluvial Networks

Nicola Pronello, Rosaria Ignaccolo and Luigi Ippoliti

Abstract Regression modeling on fluvial networks focuses on capturing covariate effects on a target variable while considering the complex spatial heterogeneity of river systems. Traditional methods assume covariate effects are spatially constant, a limitation in understanding fluvial processes. To address this, we propose a spatially varying coefficient linear regression model that integrates concepts from Graph Signal Processing for nonparametric estimation. Our proposal introduces an estimator of a smooth covariance matrix-valued function over the network, allowing for local coefficient estimation. This framework is well-suited for environmental applications where local variability is critical, offering robust tools for analyzing spatial ecological data across fluvial systems. The proposed methodology is illustrated using examples from a river network in Piemonte, Italy.

54.1 Introduction

Regression on a fluvial network is a well-studied problem in spatial statistics, beginning with the seminal work [6], and also extended for Hilbert data in [1]. In this framework, covariates linearly affect the target variable Y, while spatial heterogeneity is accounted for using a correlation function applied to the error term. This function incorporates both the water's flow direction and stream distance. However,

Nicola Pronello (✉)
Dipartimento di Economia e Statistica "Cognetti de Martiis", Università degli Studi di Torino, Italy
e-mail: nicola.pronello@unito.it

Rosaria Ignaccolo
Dipartimento di Economia e Statistica "Cognetti de Martiis", Università degli Studi di Torino, Italy
e-mail: rosaria.ignaccolo@unito.it

Luigi Ippoliti
Dipartimento di Economia, Università degli Studi "Gabriele d'Annunzio", Italy
e-mail: luigi.ippoliti@unich.it

© The Author(s), under exclusive license to Springer Nature Switzerland AG 2025

G. Aneiros et al. (eds.), *New Trends in Functional Statistics and Related Fields*,
Contributions to Statistics, https://doi.org/10.1007/978-3-031-92383-8_54

a significant limitation of this approach is that the effects of covariates are assumed to be constant across the entire fluvial network. This assumption may be problematic as riverine systems are inherently complex, and there is no guarantee that a covariate's influence remains uniform throughout the domain. To overcome this limitation, we propose a varying coefficient linear regression model on a fluvial network. Our approach bridges the gap between linear regression models on networks (see [8] for functional response) and spatially varying coefficient models (see [2] for a comprehensive discussion of the topic). Moreover, our proposal includes a novel estimation procedure. By extending concepts from *Graph Signal Processing* [5, 3], we estimate a smooth covariance matrix-valued function over the fluvial network in a non-parametric fashion. This enables straightforward estimation of the model parameters.

Fig. 54.1: Left: Fluvial network in Piemonte region with superimposed locations (red dots) of water bodies with available measurements in Summer 2020. Right: Network representation, where each water body is shown as a red vertex, and connections represent morphological contiguity

54.2 The Model

Assume that the p-dimensional process $\mathbf{Z}(\mathbf{c}) = \{Y(\mathbf{c}), \mathbf{X}(\mathbf{c})\}$, where $Y(\mathbf{c})$ represents a target variable and $\mathbf{X}(\mathbf{c})$ denotes $p-1$ covariates. This process is observed on a network domain $\mathcal{L}$ with n (ordered) sites, each having coordinates $\mathbf{c} \in \mathbb{R}^k$. Such a domain can be obtained, for example, by considering the morphological contiguity of a fluvial network to translate this last one in a simplified network where vertices correspond to monitored water bodies, as shown in Fig. 54.1 for the fluvial network of Piemonte, a region in the North-Western part of Italy. Note that, in this context, no repeated measurements are commonly available at each site.

We consider the case in which the variables follow a Gaussian distribution such that

$$\mathbf{Z}(\mathbf{c}) \sim \mathcal{N}(\mu(\mathbf{c}), \mathbf{\Sigma}(\mathbf{c})).$$

Our main goal is to study the evolution of $Y(\mathbf{c})$ by means of spatially varying coefficient regression models, i.e.

$$Y(\mathbf{c}) = \mathbf{X}(\mathbf{c})\boldsymbol{\beta}(\mathbf{c}) + \epsilon(\mathbf{c}) \text{ where } \epsilon(\mathbf{c}) \perp \mathbf{X}(\mathbf{c}) \text{ and } \epsilon(\mathbf{c}) \sim \mathcal{N}(0, \sigma^2).$$

Jointly, $Y(\mathbf{c})$ and $X(\mathbf{c})$ can be modeled as

$$\begin{bmatrix} Y(\mathbf{c}) \\ X(\mathbf{c}) \end{bmatrix} \sim \mathcal{N}\left(\mu(\mathbf{c}) = \begin{bmatrix} \mu_y(\mathbf{c}) \\ \mu_x(\mathbf{c}) \end{bmatrix}, \mathbf{\Sigma}(\mathbf{c}) = \begin{bmatrix} \sigma_y^2(\mathbf{c}) & \sigma_{xy}'(\mathbf{c}) \\ \sigma_{xy}(\mathbf{c}) & \mathbf{\Sigma}_x(\mathbf{c}) \end{bmatrix}\right)$$

and, by using the properties of the multivariate normal distribution, the conditional distribution of $Y(\mathbf{c})$ given $X(\mathbf{c})$ can be expressed as

$$Y(\mathbf{c})|X(\mathbf{c}) \sim \mathcal{N}\left(\mu_{y|x}(\mathbf{c}), \sigma_{y|x}^2(\mathbf{c})\right)$$

with

$$\mu_{y|x}(\mathbf{c}) = \mu_y(\mathbf{c}) + \mathbf{\Sigma}_x(\mathbf{c})^{-1}\boldsymbol{\sigma}_{xy}(\mathbf{c})\,(\mathbf{x}(\mathbf{c}) - \boldsymbol{\mu}_x(\mathbf{c})) \quad \text{and}$$

$$\sigma_{y|x}^2(\mathbf{c}) = \sigma_y^2(\mathbf{c}) - \boldsymbol{\sigma}_{xy}(\mathbf{c})'\mathbf{\Sigma}_x(\mathbf{c})^{-1}\boldsymbol{\sigma}_{xy}(\mathbf{c}).$$

Here, when $\boldsymbol{\mu}(\mathbf{c}) = \mathbf{0}$, $\mathbf{\Sigma}_x(\mathbf{c})^{-1}\boldsymbol{\sigma}_{yx}(\mathbf{c})$ represents the coefficient vector $\boldsymbol{\beta}(\mathbf{c})$ in a linear regression model. To overcome possible issues due to limited sample size in each location $\mathbf{c}$, our proposal relies on a plug-in estimator of $\boldsymbol{\beta}(\mathbf{c})$ by means of nonparametric smoothing of the covariance matrix $\mathbf{\Sigma}(\mathbf{c})$; then the coefficient estimates, when data are centered, are given by

$$\widehat{\boldsymbol{\beta}}(\mathbf{c}) = \widehat{\mathbf{\Sigma}}_x(\mathbf{c})^{-1}\widehat{\boldsymbol{\sigma}}_{yx}(\mathbf{c}). \tag{54.1}$$

This approach allows for incorporating spatial heterogeneity, making it well-suited for analyzing environmental processes that vary across space.

54.3 Nonparametric Estimator of $\mathbf{\Sigma}(\mathbf{c})$

The estimate of $\boldsymbol{\beta}(\mathbf{c})$ by means of Equation (54.1) is possible only after defining a suitable estimator of $\mathbf{\Sigma}(\mathbf{c})$ varying on the network $\mathcal{L}$, that is one for each coordinate $\mathbf{c}$. However, constructing an estimator of $\mathbf{\Sigma}(\mathbf{c})$ in a nonstationarity framework represents a statistical challenge. By considering mean corrected data for the sake of simplicity ($\mathbf{z}(\mathbf{c})$ denotes these ones in the following), a naive estimate of $\mathbf{\Sigma}(\mathbf{c})$ is given by the raw covariance

$$\mathbf{\Sigma}^*(\mathbf{c}) = \left\{\mathbf{z}(\mathbf{c})_j \mathbf{z}(\mathbf{c})_r^T\right\}_{j,r=1,\ldots,p},$$

that exploits only the information in one datum $\mathbf{z}(\mathbf{c})$ and is not full rank since by construction $\text{rank}(\mathbf{\Sigma}^*(\mathbf{c})) = 1$. Then to obtain a reliable estimator of $\mathbf{\Sigma}(\mathbf{c})$ we propose to exploit information from the neighbours of the unit $\mathbf{c}$ in the network by means of linear smoothing.

Given n units in the network $\mathcal{L}$ with coordinates $\mathbf{c}_i$, with $i = 1, \ldots, n$, following [4] and [7] we consider the class of linear smoother estimators defined by

$$\widehat{\mathbf{\Sigma}}(\mathbf{c}) = \sum_{i=1}^{n} \omega_i(\mathbf{c}) \mathbf{\Sigma}^*(\mathbf{c}_i),$$

where the coefficients $\omega_i(\mathbf{c})$ of the linear combination need to be determined. This class includes a variety of estimators, such as the Gaussian process regression estimator, the Kernel smoother, and the local polynomial estimator.

It can be shown that finding $\omega(\mathbf{c}) = (\omega_1(\mathbf{c}), ..., \omega_n(\mathbf{c}))$ is equivalent to solve an optimization problem, that is:

$$\widehat{\mathbf{\Sigma}}(\mathbf{c}) = \text{argmin}_{\mathbf{\Sigma}(\mathbf{c})} \sum_{i=1}^{n} K(\mathbf{c}_i, \mathbf{c}) d(\mathbf{\Sigma}(\mathbf{c}), \mathbf{\Sigma}^*(\mathbf{c}_i))^2,$$

where and $K(\cdot,\cdot) : \mathcal{L} \times \mathcal{L} \to \mathbb{R}$ is a kernel function and $d(\cdot,\cdot)$ is a suitable distance between covariance matrices. This leads to obtain a Nadaraya-Watson (Kernel smoother) solution with

$$\omega_i(\mathbf{c}) = \frac{K(\mathbf{c}_i, \mathbf{c})}{\sum_{i=1}^{n} K(\mathbf{c}_i, \mathbf{c})}.$$

54.3.1 Kernel Definition

We propose constructing kernel weights $\omega_i(\mathbf{c})$ suitable for a network domain, which was possible by extending some common theory on Graph Signal Processing [3]. Usually, a kernel function is defined as a function of distances so that, in a spatial fluvial network, distances among water bodies would be among their barycenters. To properly consider the nature domain where covariance matrices live, we define the kernel from a Laplacian operator $\mathbf{L}$ resulting from the neighbourhood structure of $\mathcal{L}$. Let $\mathbf{W}$ be an adjacency matrix such as such that entries are equal to 1 if two units are neighbours, and 0 otherwise, the resulting Laplacian $\mathbf{L}$ is then defined as:

$$\mathbf{L} = \mathbf{D} - \mathbf{W}$$

with $\mathbf{D} = diag\{\mathbf{W}\mathbf{1}_n\}$. Finally, the kernel values $K(\mathbf{c}_i, \mathbf{c})$ for the linear smoother are obtained from the elements of the matrix

$$\mathbf{K} = (\mathbf{I} + \gamma \mathbf{L})^{-1},$$

where γ is a smoothing parameter (to be fixed) that controls the Kernel width. In fact, γ tunes the values of the weights and the extension of the neighborhood of each unit in $\mathcal{L}$, such as for $\gamma \to \infty$ one has $\omega_i \to 1/n$. As shown in [5] this characterization of the Kernel comes by applying a real valued function on the eigenvalues of the Laplacian operator $\mathbf{L}$ corresponding to a regularization functional. Fig. 54.2 shows two examples of kernel weights for the city of Turin, highlighting the effect on $K(\mathbf{c}_i, \mathbf{c})$ of an increase of the γ value.

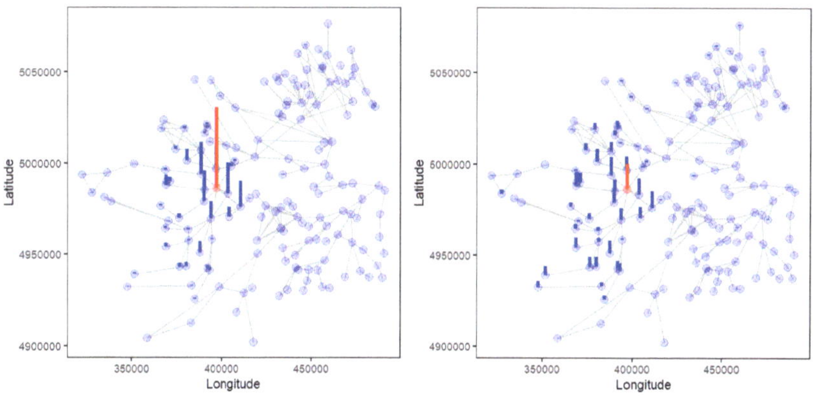

Fig. 54.2: Kernel weights $K(\mathbf{c}_i, \mathbf{c})$ as vertical blue bars for i = Turin (red bar) with $\gamma = 1.2$ (left) and $\gamma = 12$ (right)

54.4 Implementation

To apply the proposed methodology, firstly, we need to obtain mean-corrected data: to this goal, we utilize a nonparametric estimator for $\boldsymbol{\mu}(\mathbf{c})$, as described in [7].

As highlighted in the previous section, it is necessary to choose a value for the smoothing parameter γ and, among the various methodologies proposed in the literature, we focus on the cross-validation approach. In particular, we adopt the leave-one-out criterion (*loocv*), based on a log-likelihood formulation as proposed in [7]:

$$\mathrm{CV}(\gamma) = \frac{1}{n}\sum_{c=1}^{n} \boldsymbol{z}(\boldsymbol{c})^{T}\hat{\boldsymbol{\Sigma}}_{-\boldsymbol{c}}^{-1}\boldsymbol{z}(\boldsymbol{c}) - \log\left(\left|\hat{\boldsymbol{\Sigma}}_{-\boldsymbol{c}}^{-1}\right|\right),$$

where $\hat{\boldsymbol{\Sigma}}_{-\boldsymbol{c}}$ represents the Nadaraya-Watson estimator evaluated at $\mathbf{c}$ while excluding the observation at that location, and $\boldsymbol{z}(\boldsymbol{c})$ refers to the mean-corrected data observed at $\mathbf{c}$. The optimal bandwidth γ is determined by minimizing $\mathrm{CV}(\gamma)$. The selection process examines the variation in CV(γ) across different values of γ to identify an optimal balance and avoid excessive smoothing.

As an example, Fig. 54.3 shows the estimated varying - along the fluvial network - coefficients $\hat{\beta}_{HNO_3}(\mathbf{c})$ related to nitrate nitrogen concentration (HNO_3) when the response variable is mercury concentration (Hg) and other 30 pollutant and chemical features are considered as regressors. Indeed, public decision-makers are particularly interested in studying local correlations between a measure of water bodies' ecological status and other variables that are possibly easily accessible (like temperature, morphological features, anthropogenic pressures, etc.). This will be further investigated in an extended version of this work.

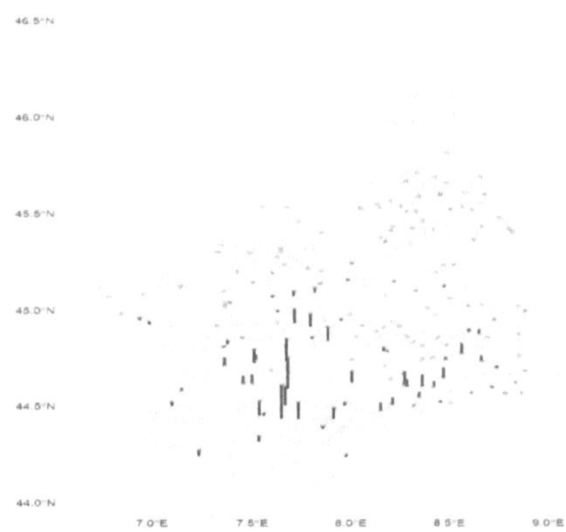

Fig. 54.3: $\hat{\beta}_{HNO_3}(\mathbf{c})$, in blue negative estimated values and in red positive values, as varying marginal effects of HNO_3 on Hg

Acknowledgements All the authors acknowledge financial support under the National Recovery and Resilience Plan (NRRP), Mission 4, Component 2, Investment 1.1, Call for tender No. 104 published on 2.2.2022 by the Italian Ministry of University and Research (MUR), funded by the European Union – NextGenerationEU– *Complex Environmental Data and Modeling (Co-Env)* - 2022E3RY23 – CUP D53D23011080006- Grant Assignment Decree No. 1109 adopted on 20.07.2023 by the Italian Ministry of Ministry of University and Research (MUR).

References

1. Barbi, C., Menafoglio, A., Secchi, P.: An object-oriented approach to the analysis of spatial complex data over stream-network domains. Spatial Statistics **58**, 100784 (2023)
2. Dambon, J. A., Sigrist, F., Furrer, R.: Maximum likelihood estimation of spatially varying coefficient models for large data with an application to real estate price prediction. Spatial Statistics **41**, 100470 (2021)

3. Hammond, D., Vandergheynst, P., Gribonval, R.: Wavelets on graphs via spectral graph theory. Applied and Computational Harmonic Analysis **30**(2), 129–150 (2011)
4. Petersen, A., Deoni, S., Müller H.G.: Fréchet estimation of time-varying covariance matrices from sparse data, with application to the regional co-evolution of myelination in the developing brain. The Annals of Applied Statistics **13**(1), 393-419 (2019)
5. Smola, A.J., Kondor, R.: Kernels and Regularization on Graphs. In: Schölkopf, B., Warmuth, M.K. (eds.) Learning Theory and Kernel Machines. pp. 144–158. Springer, Berlin, Heidelberg (2003)
6. Ver Hoef, J. M., Peterson, E.: A Moving Average Approach for Spatial Statistical Models of Stream Networks. J. Am. Stat. Assoc. **105**, 6–18 (2010)
7. Yin, J., Geng, Z., Li, R., Wang, H.: Nonparametric covariance model. Statistica Sinica **20**(1), 469-479 (2010)
8. Zhu, X., Cai, Z.: Network Functional Varying Coefficient Model. J. Am. Stat. Assoc. **117**, 540, (2022)

Chapter 55
Non-Parametric Testing of Time Reversibility in Functional Data

Hedvika Ranošová and Daniel Hlubinka

Abstract Three different notions of time reversibility of stochastic processes are introduced. Based on a random sample, we construct a statistical test for each version of time reversibility without any distributional or moment assumptions. The characteristic functional is utilized in the Cramér-von Mises test statistic with a pre-chosen weighting measure. Several simulations illustrate the performance of the proposed test.

55.1 Introduction

Time reversibility as a property of stochastic processes has been extensively discussed, namely for Markov chains with discrete and continuous time, where reversibility can be described through transition probabilities ([6]) or for linear processes ([11]) in relation to Gaussianity.

Previously, [3] showed that the third-order moment $\mathsf{E}\left[X(t+r)X(t+s)X(t)\right]$ of a time-reversible stationary time series $\{X(t)\}$ is centrally symmetric and its spectrum must be real. On the other hand, [8] used the difference between two bicovariances $\mathsf{E}\left[X(t)^i X(t-k)^j\right] - \mathsf{E}\left[X(t)^j X(t-k)^i\right]$ to test the time reversibility of order m, for which the difference of bicovariances is zero for any k and $i+j \leq m$. A different path is examined in this chapter, as we are interested in testing the property of time reversibility given a random sample of observed continuous-time stochastic processes so that we may employ methods from functional data analysis. Thus, we are interested in inference about the whole underlying distribution of

Hedvika Ranošová (✉)
Department of Probability and Mathematical Statistics, Charles University, Sokolovská 83, 186 75 Praha 8, Czech Republic, e-mail: ranosova@karlin.mff.cuni.cz

Daniel Hlubinka
Department of Probability and Mathematical Statistics, Charles University, Sokolovská 83, 186 75 Praha 8, Czech Republic, e-mail: hlubinka@karlin.mff.cuni.cz

© The Author(s), under exclusive license to Springer Nature Switzerland AG 2025

G. Aneiros et al. (eds.), *New Trends in Functional Statistics and Related Fields*, Contributions to Statistics, https://doi.org/10.1007/978-3-031-92383-8_55

the process. A characteristic functional is chosen for the construction of the test statistic. Reversibility in the context of FDA can be seen as symmetry of the sample with respect to the vertical axis (in further settings placed at the midpoint of the observation interval) and as a complement to horizontal symmetry, the test of which was introduced in [9].

In what follows, we consider a stochastic process $X = \{X(t) : t \in [0,1]\}$ defined on a probability space $(\Omega, \mathcal{A}, \mathsf{P})$ and $X_1, \ldots, X_n$ its independent and identically distributed copies. To construct test statistic, we employ the characteristic functional that is nowadays popular tool in functional data analysis, especially for non-parametric and semi-parametric problems (see [1], [2], [4], [9]). For that X is usually assumed to be a random element of the real separable Hilbert space of square-integrable functions $L^2[0,1]$ or the Banach space of continuous functions $C[0,1]$ to ensure that the characteristic functional characterizes the distribution. The characteristic functional is then a function defined on the dual space. Based on the discussion in [5], Chapter 7.4, if we additionally assume that X is jointly measurable with respect to σ-algebra $\mathcal{B}[0,1] \times \mathcal{A}$ and the trajectories are square-integrable, then X is a random element of $L^2[0,1]$.

Alternatively, assuming that X is a random element of $C[0,1]$, the domain is its dual space (the space of all signed measures), which is a superset of $L^2[0,1]$. An extension to càdlàg functions (to include counting and Lévy processes) may also be possible as the Skorokhod space is a Banach space (and its dual is described e.g. in [7]). As in [4], it is sufficient to work with characteristic functions defined on $L^2[0,1]$ in all three cases.

The chapter is organized as follows. Sect. 55.2 introduces three types of reversibility with examples. The characteristic functional is defined in Sect. 55.3, and we construct a test statistic based on the characteristic functional in Sect. 55.4. Sect. 55.5 contains a small simulation study to illustrate the testing procedure.

55.2 Types of Reversibility

We introduce three types of reversibility that may be of interest:

- *Raw reversibility.* We say that the process X is *(raw) reversible* if

$$\{X(t) : t \in [0,1]\} \stackrel{d}{=} \{X(1-t) : t \in [0,1]\}. \tag{55.1}$$

- *Anchored reversibility.* Let X be a stochastic process such that $X(0) = 0$. Denote $\tilde{X}(t) = X(1-t) - X(1)$, $t \in [0,1]$. Then X is *anchored reversible* if

$$\{X(t) : t \in [0,1]\} \stackrel{d}{=} \{\tilde{X}(t) : t \in [0,1]\}. \tag{55.2}$$

- *Bridged reversibility.* Let X be a stochastic process and denote $\tilde{X}(t) = X(t) - t \cdot X(1) - (1-t) \cdot X(0)$, $t \in [0,1]$. Then X is *brigded reversible* if

$$\{\tilde{X}(t) : t \in [0,1]\} \overset{d}{=} \{\tilde{X}(1-t) : t \in [0,1]\}. \tag{55.3}$$

Let us discuss some examples of reversible processes in light of the stated definition. The stationary Ornstein-Uhlenbeck process U (a centered Gaussian process with $\mathsf{E}[U(t)U(s)] = e^{-|t-s|}$, $s, t \in [0, 1]$) is raw reversible (55.1), as can be seen from the Gaussianity of the process and its covariance function $\mathsf{E}[U(1-t)U(1-s)] = \mathsf{E}[U(t)U(s)]$. Similar calculations show that Brownian bridge $\{B(t) : t \in [0, 1]\}$ is raw reversible as for $s, t \in [0, 1]$

$$\begin{aligned}\mathsf{E}[B(1-s)B(1-t)] &= \min\{1-s, 1-t\} - (1-s)(1-t)\\ &= 1 - \max\{s,t\} - 1 + s + t - st = \min\{s,t\} - st.\end{aligned}$$

The second definition is tailored to fit the Wiener process $\{W(t) : t \in [0, 1]\}$ as for $s, t \in [0, 1]$ we have

$$\begin{aligned}\mathsf{E}[(W(1-t) - W(1))(W(1-s) - W(1))] &= \min\{1-s, 1-t\} - 1 + s - 1 + t + 1\\ &= 1 - \max\{s,t\} + s + t - 1 = \min\{s,t\}.\end{aligned}$$

The last version of reversibility (55.3) is directly inspired by the construction of the Brownian bridge from the Wiener process and deals with an additional linear trend. The Brownian bridge $\{B(t) : t \in [0, 1]\}$ is clearly bridged reversible as $B(0) = B(1) = 0$ a.s., and the raw reversibility of the Brownian bridge. The Wiener process is bridged reversible because the modified process (which is the Brownian bridge) is raw reversible.

As anchored and bridged reversibilities are simply a raw reversibility of a modified process we may define the testing procedure as testing the null hypothesis of

$$H_0 : \{X(t) : t \in [0,1]\} \overset{d}{=} \{\tilde{X}(t) : t \in [0,1]\} \tag{55.4}$$

against an alternative of inequality where $\tilde{X}$ is a modification of X which differs based on a chosen reversibility. Raw reversibility can also be extended to other transformations in the time domain. Set $v : [0, 1] \to [0, 1]$ and we may be interested in testing

$$\{X(t) : t \in [0,1]\} \overset{d}{=} \{X(v(t)) : t \in [0,1]\}. \tag{55.5}$$

55.3 Empirical Characteristic Functionals

Let X be a stochastic process (satisfying the assumptions in Sect. 55.1). The *characteristic functional* (CF) of X is defined as $\varphi_X(u) = \mathsf{E}\left[e^{i\langle u, X\rangle}\right]$, $u \in L^2[0, 1]$, where $\langle u, X\rangle = \int_0^1 u(t)X(t)\,\mathrm{d}t$ and $i = \sqrt{-1}$.

The empirical CF based on a random sample $X_1, \ldots, X_n$ is defined as

$$\hat{\varphi}_X(u) = \frac{1}{n}\sum_{j=1}^{n} e^{i\langle u, X_j\rangle}, \quad u \in L^2[0,1].$$

For a fixed $u \in L^2[0,1]$, the empirical CF is a consistent and unbiased estimator of the CF (based on the LLN in $L^2[0,1]$) and the asymptotic distribution based on the CLT.

55.4 Test Statistic

The test statistic for (55.4) is constructed with the difference between the empirical CF $\hat{\varphi}_X$ based on $X_1,\dots,X_n$ and the empirical CF $\hat{\varphi}_{\tilde{X}}$ of the modified sample $\tilde{X}_1,\dots,\tilde{X}_n$ which are both random samples but not mutually independent. Let Q be a probability measure on $L^2[0,1]$, then we define the test statistic of the Cramér-von Mises type

$$T_n(Q) = \int_{L^2[0,1]} \left|\hat{\varphi}_X(w) - \hat{\varphi}_{\tilde{X}}(w)\right|^2 \mathrm{d}Q(w). \tag{55.6}$$

Let us also define the theoretical value of the test statistic based on the true CFs as

$$T_X(Q) = \int_{L^2[0,1]} \left|\varphi_X(w) - \varphi_{\tilde{X}}(w)\right|^2 \mathrm{d}Q(w). \tag{55.7}$$

The test statistic $T_n(Q)$ and its theoretical counterpart $T_X(Q)$ are both non-negative. If the distributions of X and $\tilde{X}$ coincide, then $\varphi_X(w) = \varphi_{\tilde{X}}(w)$ for each $w \in L^2[0,1]$ and $T_n(Q) = 0$. Therefore, the test should reject for large values of the test statistic.

The critical values for the test statistic T_n are obtained using Efron bootstrap with a procedure similar to [9] using the *pooled* sample. In each bootstrap repetition, we sample with replacement n observations $X_1^*,\dots,X_n^*$ from the pooled sample $(X_1,\dots,X_n,\tilde{X}_1,\dots,\tilde{X}_n)'$. The bootstrap test statistic is then computed from $X_1^*,\dots,X_n^*$ and $\tilde{X}_1^*,\dots,\tilde{X}_n^*$.

55.4.1 Choice of Measure Q

We adapt the discussion of the choice of Q for the reversibility test from [4] and [9]. If we denote φ_Q the (real part of the) CF corresponding to Q then $T_n(Q)$ can be rewritten similarly to [4] and [9] with contrasts $X_i - X_j$, $X_i - \tilde{X}_j$ and $\tilde{X}_i - \tilde{X}_j$ as

$$T_n(Q) = \frac{1}{n^2} \sum_{i,j=1}^{n} \Big[\varphi_Q(X_i - X_j) + \varphi_Q(\tilde{X}_i - \tilde{X}_j) - 2 \cdot \varphi_Q(X_i - \tilde{X}_j) \Big].$$

A common choice is a centered Gaussian measure Q with

$$\varphi_Q(w) = e^{-\frac{1}{2}\langle w, C_Q w \rangle}$$

and is used in the simulation study with different covariance operators. If $\tilde{X}(t) = X(1-t)$ and Q is by itself a raw reversible process, then additionally $\varphi_Q(X_i - X_j) = \varphi_Q(\tilde{X}_i - \tilde{X}_j)$ since

$$\begin{aligned}
\varphi_Q(\tilde{X}_i - \tilde{X}_j) &= \mathsf{E}_{Y\sim Q} e^{i\langle Y, \tilde{X}_i - \tilde{X}_j \rangle} \\
&= \mathsf{E}_{Y\sim Q} \exp\left\{ i \int_0^1 Y(t)(X_i(1-t) - X_j(1-t))\,\mathrm{d}t \right\} \\
&= \mathsf{E}_{Y\sim Q} \exp\left\{ i \int_0^1 Y(1-t)(X_i(t) - X_j(t))\,\mathrm{d}t \right\} \\
&= \mathsf{E}_{Y\sim Q} e^{i\langle \tilde{Y}, X_i - X_j \rangle} \\
&= \varphi_Q(X_i - X_j).
\end{aligned}$$

As functional observations cannot be observed directly and are stored as discrete observations, we assume that all inner products are sufficiently well approximated by the Riemann sums. In functional data analysis, data can sometimes be smoothed first using a fixed basis (e.g. Fourier or spline), the inner product is then reduced to the standard dot product of coefficients.

55.5 Simulation

The proposed test is investigated in a small simulation study. The raw reversibility is tested with the Brownian bridge in the null hypothesis (as an example of a raw reversible process) and the Wiener process as an example of a non-reversible process. The sample size is set as $n = 20$ and the covariances chosen for the measure Q from Sect. 55.4.1 are similar to [4] and [9] – the covariances of the Brownian bridge (BB), Wiener process (BM), Ornstein-Uhlenbeck process (OU), the Gaussian kernel covariance[1] (GK), and identity. Additionally, two data-driven covariances are included (sample covariance and its inverse).

The results are summarized in Table 55.1 for two different nominal levels based on 500 simulations with 200 bootstrap repetitions. The number of points where

[1] By that we mean a covariance function $c_{GK}(s,t) = \exp\{-(s-t)^2\}$, $s, t \in [0, 1]$.

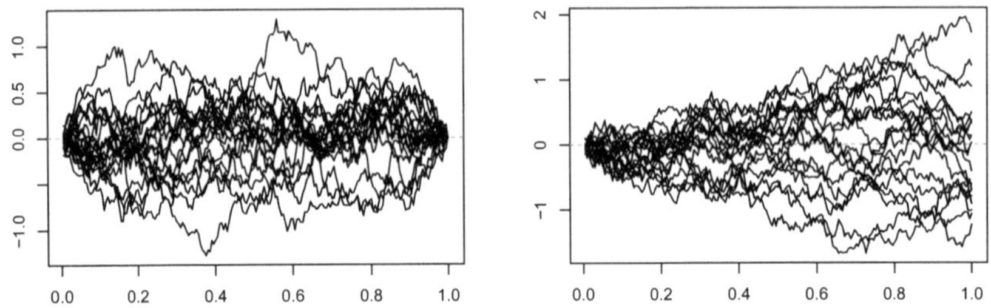

Fig. 55.1: Random samples of 20 functions from a time (raw) reversible distribution (Brownian bridge) and non-reversible distribution (Wiener process).

processes were observed was set to 30, which is larger than $n = 20$. The first two columns present the behavior of the test under H_0 and in all cases the level holds well for all covariances. Under the alternative (for a random sample of Wiener processes), the best behavior (the largest power) occurs for covariances of the Wiener process, Ornstein-Uhlenbeck, and the Gaussian kernel. In our case, the sample covariance may be close to the true covariance of the Wiener process, and similar results were expected. The test failed to reject with the Brownian bridge covariance and with identity. Note that both OU and BB covariances correspond to the time-reversible choice of Q.

	Brownian Bridge		Wiener Process	
	$\alpha = 5\%$	$\alpha = 10\%$	$\alpha = 5\%$	$\alpha = 10\%$
BB	4.8	10.4	5.4	11.8
WP	6.0	9.4	58.2	83.4
OU	5.4	10.2	40.4	72.8
GK	5.4	9.2	51.0	80.6
$\hat{C}$	3.0	9.0	49.4	76.0
Id	4.4	10.2	4.8	10.4
$\hat{C}^{-1}$	5.0	9.4	31.8	45.0

Table 55.1: Empirical size for a random sample of $n = 20$ Brownian bridges and empirical size for $n = 20$ Wiener processes for nominal levels 5 and 10 %.

Acknowledgments. The research of HR was supported by the Charles University Grant Agency, grant GAUK 70324. The research of DH was supported by the Czech Grant Agency project no. 23-05737S.

References

1. Čoupek, Petr, Viktor Dolník, Zdeněk Hlávka,and Daniel Hlubinka. "Fourier approach to goodness-of-fit tests for Gaussian random processes." Statistical Papers **65.5** (2024): 2937-2972.
2. Henze, Norbert, and María Dolores Jiménez-Gamero. "A test for Gaussianity in Hilbert spaces via the empirical characteristic functional." Scandinavian Journal of Statistics **48.2** (2021): 406-428.
3. Hinich, Melvin J., and Philip Rothman. "Frequency-domain test of time reversibility." Macroeconomic Dynamics **2.1** (1998): 72-88.
4. Hlávka, Zdeněk, Daniel Hlubinka, and Kateřina Koňasová. "Functional ANOVA based on empirical characteristic functionals." Journal of Multivariate Analysis **189** (2022): 104878.
5. Hsing, Tailen, and Randall Eubank. Theoretical foundations of functional data analysis, with an introduction to linear operators. Vol. **997**. John Wiley & Sons, 2015.
6. Löpker, Andreas, and Zbigniew Palmowski. "On time reversal of piecewise deterministic Markov processes." (2013): 1-29.
7. Pestman, Wiebe R. "Measurability of linear operators in the Skorokhod topology." Bulletin of the Belgian Mathematical Society-Simon Stevin **2.4** (1995): 381-388.
8. Ramsey, James B., and Philip Rothman. "Time irreversibility and business cycle asymmetry." Journal of Money, Credit and Banking **28.1** (1996): 1-21.
9. Ranošová, Hedvika, and Daniel Hlubinka. "A Symmetry Test for Functional Data via the Empirical Characteristic Functional." (2024+) *Submitted.*
10. Tanaka, Hiroshi. "Time reversal of random walks in one-dimension." Tokyo Journal of Mathematics **12.1** (1989): 159-174.
11. Weiss, Gideon. "Time-reversibility of linear stochastic processes." Journal of Applied Probability **12.4** (1975): 831-836.

Chapter 56
An LRD Spectral Test for Irregularly Discretely Observed Functional Time Series in Manifolds

María D. Ruiz–Medina and Rosa M. Crujeiras

Abstract Recently, in [18], a statistical hypothesis test for Long Range Dependence (LRD) in manifold–supported functional time series has been formulated, in the spectral domain, for fully observed functional data. The asymptotic Gaussian distribution of the proposed test statistic, based on the weighted periodogram operator, under the null hypothesis, and the consistency of the test have been derived. In this paper, under a stochastic spatial manifold sampling design, we analyze the asymptotic properties of this statistical test when discretized and irregularly–distributed, as well as contaminated versions of the functional values of the correlated data are available.

56.1 Introduction

Hypothesis testing in the spectral domain for stationary functional time series has been mainly developed in the weak–dependent case. Under a Short Range Dependence (SRD) scenario, assuming suitable summability conditions of the L^2 norms of the cumulant operators of a stationary functional time series, the asymptotic properties of the weighted periodogran operator are derived in [12]. Specifically, bias L^2 asymptotics analysis of the empirical cumulant spectral operators is achieved, obtaining, in particular, the consistency of the weighted periodogram operator in the integrated mean square error sense, and the asymptotic Gaussian distribution of the functional discrete Fourier transform (fDFT) and the weighted periodogram operator. A spectral version of covariance tests (see, e.g., [2]; [13]; [14]), based on the Hilbert–Schmidt operator norm, to compare the second–order structure of

María D. Ruiz–Medina (✉)
Universidad de Granada, Spain e-mail: mruiz@ugr.es

Rosa M. Crujeiras
CITMAga, Universidade de Santiago de Compostela, Spain e-mail: rosa.crujeiras@usc.es

© The Author(s), under exclusive license to Springer Nature Switzerland AG 2025
G. Aneiros et al. (eds.), *New Trends in Functional Statistics and Related Fields*,
Contributions to Statistics, https://doi.org/10.1007/978-3-031-92383-8_56

two functional time series in terms of their spectral density operator families, is formulated in [19].

Beyond the weak–dependent and stationary assumptions, LRD functional time series analysis is developed in [8] via the pure point spectral decomposition of the long-run covariance operator. This paper opens new research lines regarding the consistent estimation of the dimension and the orthonormal functions spanning the dominant subspace, where the projected curve process displays the largest dependence range. The results are applied to the statistical analysis of fractionally integrated functional autoregressive moving average processes.

In the stationary case, a characterization of LRD in the spectral domain for functional time series is derived in [17]. The asymptotic unbiasedness of the integrated periodogram operator in the Hilbert–Schmidt operator norm is obtained. The spectral density operator family is assumed to be diagonalized in terms of a common resolution of the identity (spectral kernel). The LRD operator also admits a weak–sense diagonalization in terms of this spectral kernel, and the corresponding pure point spectrum is parameterized. Under a Gaussian scenario, the weak–consistent minimum contrast estimation of the LRD operator is derived in a semiparametric spectral framework. The results in [17] can be applied to stationary functional time series in connected and compact two points homogeneous spaces, under the invariance assumption with respect to the group of isometries of the manifold (see, e.g., [11]).

Asymptotics for bias analysis of the integrated empirical second and fourth order cumulant spectral density operators is addressed in compact–manifold–supported LRD functional time series in [18]. In this context, a statistical hypothesis test, based on the weighted periodogram operator, for detecting LRD is also formulated in the double spectral domain. Note that in this work the eigenvalues, in the weak–sense pure point spectral representation of the LRD operator, are not parameterized as in [17]. The asymptotic normality of the proposed test statistic under the null hypothesis follows from an extended version of the results derived in [12], on the asymptotic Gaussian distribution of the weighted periodogram operator in SRD stationary functional time series. Based on the proved L^1 consistency of the integrated weighted periodogram operator under LRD, the almost surely divergence of the Hilbert–Schmidt operator norm of the proposed test statistic is obtained, yielding the consistency of the test.

The results derived in all the above cited references on spectral analysis of functional time series are obtained under the assumption of fully observed functional data. This paper addresses, in the spectral domain, the problem of testing LRD in compact–manifold–supported functional time series from irregularly spatially discretely observed, and contaminated correlated functional data. Spatial nonparametric series regression is applied to approximate the underlying functional values of the data, under a stochastic spatial manifold sampling design. Uniform and L^2 convergence rates are derived, under suitable interaction of the restrictions on the time–varying increasing rates of the volume of the spatial sampling region, the spatial sampling frequency, the dimension of the sieve basis, and the dimension, norm and variance of the associated spatial weak–dependent random matrix sequences, to ensure that the spectral asymptotic results obtained in [18] hold for SRD and LRD.

56.2 Formulation of the Hypothesis Testing Problem

The spectral preliminary elements involved in our testing problem formulation are introduced in the next two sections.

56.2.1 Notation and Background Elements and Assumptions

In what follows, $L^2(\mathbb{M}_d, d\nu, \mathbb{R})$ and $L^2(\mathbb{M}_d, d\nu; \mathbb{C})$ respectively denote the space of real– and complex– valued square integrable functions on a Riemannian manifold $\mathbb{M}_d$, given by a connected and compact two–points homogeneous space embedded into $\mathbb{R}^{d+1}$. The topological dimension of $\mathbb{M}_d$ is d, and $d\nu$ denotes the normalized Riemannian measure on $\mathbb{M}_d$ (see, e.g., [9]). Let $X = \{X_t,\ t \in \mathbb{Z}\}$ be an $L^2(\mathbb{M}_d, d\nu, \mathbb{R})$–valued correlated sequence, defined on the basic probability space $(\Omega, \mathcal{Q}, \mathcal{P})$, which is assumed to be strictly stationary (in time and space) with zero–mean. In particular, X satisfies $\mathcal{P}\left(X_t \in L^2(\mathbb{M}_d, d\nu, \mathbb{R})\right) = 1$, for every $t \in \mathbb{Z}$. Suitable Hölder regularity conditions will also be assumed in space in Sect. 56.3. Consider the family of nuclear covariance operators $\{\mathcal{R}_\tau,\ \tau \in \mathbb{Z}\}$, with $\mathcal{R}_\tau = E[X_s \otimes X_{s+\tau}] = E[X_{s+\tau} \otimes X_s]$, for every $s, \tau \in \mathbb{Z}$. Their functional Fourier transforms $\{\mathcal{F}_\omega,\ \omega \in [-\pi, \pi]\}$ are defined here in the norm of the space $\mathcal{S}(L^2(\mathbb{M}_d, d\nu; \mathbb{C}))$ of Hilbert–Schmidt operators on $L^2(\mathbb{M}_d, d\nu; \mathbb{C})$ as follows:

$$\mathcal{F}_\omega \underset{\mathcal{S}(L^2(\mathbb{M}_d,d\nu;\mathbb{C}))}{=} \frac{1}{2\pi}\sum_{\tau\in\mathbb{Z}} \exp(-i\omega\tau)\,\mathcal{R}_\tau$$

$$\underset{\mathcal{S}(L^2(\mathbb{M}_d,d\nu;\mathbb{C}))}{=} \sum_{n\in\mathbb{N}_0} f_n(\omega) \sum_{j=1}^{\Gamma(n,d)} S^d_{n,j} \otimes \overline{S^d_{n,j}}, \quad \omega \in [-\pi, \pi]. \tag{56.1}$$

The last equality in (56.1) holds under the assumption of invariance of the covariance and spectral density operators of X with respect to the group of isometries of $\mathbb{M}_d$. Here, $\{S^d_{n,j} \otimes \overline{S^d_{n,j}},\ j = 1, \ldots, \Gamma(n,d),\ n \in \mathbb{N}_0\}$ is given in terms of the orthonormal basis $\{S^d_{n,j},\ j = 1, \ldots, \Gamma(n,d),\ n \in \mathbb{N}_0\}$ of eigenfunctions of the Laplace–Beltrami operator Δ_d on $L^2(\mathbb{M}_d, d\nu, \mathbb{C})$ (see, e.g., [6]); [7]). For every $n \in \mathbb{N}_0$, $\Gamma(n,d)$ represents the dimension of the eigenspace $\mathcal{H}_n$ associated with the eigenvalue $\lambda_n(\Delta_d)$ of the Laplace Beltrami operator Δ_d (see, e.g., Sect. 2.1 in [9]). Along the paper the operator integrals are understood as improper operator Stieltjes integrals which converge strongly (see, e.g., Sect. 8.2.1 in [16]).

The fDFT $\widetilde{X}_{\omega}^{(T)}$ of X, based on a functional sample $\{X_t,\ t = 0, \dots, T-1\}$ of size $T \geq 2$, is given by

$$\widetilde{X}_{\omega}^{(T)}(x) = \frac{1}{\sqrt{2\pi T}} \sum_{t=0}^{T-1} X_t(x) \exp(-i\omega t), \quad x \in \mathbb{M}_d, \quad \omega \in [-\pi, \pi]. \quad (56.2)$$

The kernel $p_{\omega}^{(T)}$ of the periodogram operator $\mathcal{P}_{\omega}^{(T)} = \widetilde{X}_{\omega}^{(T)} \otimes \widetilde{X}_{-\omega}^{(T)}$ satisfies, for every $\omega \in [-\pi, \pi]$,

$$p_{\omega}^{(T)}(x, y) = \frac{1}{2\pi T} \sum_{t=0}^{T-1} \sum_{s=0}^{T-1} X_t(x) X_s(y) \exp(-i\omega[t-s]),\ \forall x, y \in \mathbb{M}_d. \quad (56.3)$$

Denote by $\mathcal{F}_{\omega}^{(T)}$ the spectral cumulant operator of order 2 of the fDFT $\widetilde{X}_{\omega}^{(T)}$ in (56.2) over the diagonal $\omega \in [-\pi, \pi]$. Its kernel is then given by

$$f_{\omega}^{(T)}(x, y) = \mathrm{cum}\left(\widetilde{X}_{\omega}^{(T)}(x), \widetilde{X}_{-\omega}^{(T)}(y)\right) = E\left[p_{\omega}^{(T)}(x, y)\right],\ x, y \in \mathbb{M}_d,\ \omega \in [-\pi, \pi].$$

This cumulant operator involves the Féjer kernel

$$F_T(\omega) = \frac{1}{T} \sum_{t=0}^{T-1} \sum_{s=0}^{T-1} \exp\left(-i(t-s)\omega\right) = \frac{1}{T} \left[\frac{\sin\left(T\omega/2\right)}{\sin(\omega/2)}\right]^2$$

in its definition, as given in the following identity:

$$\begin{aligned} \mathcal{F}_{\omega}^{(T)} &= \left[F_T * \mathcal{F}_{\bullet}\right](\omega) \\ &= \int_{-\pi}^{\pi} F_T(\omega - \xi) \mathcal{F}_{\xi} d\xi, \quad T \geq 2,\ \omega \in [-\pi, \pi]. \end{aligned}$$

The weighted periodogram operator $\widehat{\mathcal{F}}_{\omega}^{(T)}$ has kernel $\widehat{f}_{\omega}^{(T)}(x, y)$ satisfying, for every $\omega \in [-\pi, \pi]$,

$$\widehat{f}_{\omega}^{(T)}(x, y) = \left[\frac{2\pi}{T}\right] \sum_{s=0}^{T-1} W^{(T)}\left(\omega - \frac{2\pi s}{T}\right) p_{\frac{2\pi s}{T}}^{(T)}(x, y),\ x, y \in \mathbb{M}_d, \quad (56.4)$$

where the weight function $W^{(T)}$ is such that $W^{(T)}(x) = \sum_{j \in \mathbb{Z}} \frac{1}{B_T} W\left(\frac{x + 2\pi j}{B_T}\right)$, with B_T being the positive bandwidth parameter. Note that function W is assumed to be a real–valued, positive, even, and bounded in variation, function defined on $\mathbb{R}$, with $W(x) = 0$, if $|x| \geq 1$, $\int_{\mathbb{R}} |W(x)|^2\, dx < \infty$, and $\int_{\mathbb{R}} W(x) dx = 1$.

56.2.2 SRD and LRD Scenarios

We are interested on testing SRD against LRD. It is well-known that SRD in real–valued stationary time series is characterized by the absolutely integrability of the covariance function. This notion is extended to the infinite–dimensional framework in terms of the summability of the sequence of nuclear norms of the covariance operator family. That is, SRD is understood in the following sense:

$$\sum_{\tau\in\mathbb{Z}} \|\mathcal{R}_\tau\|_{L^1(L^2(\mathbb{M}_d,d\nu,\mathbb{R}))} = \sum_{\tau\in\mathbb{Z}}\sum_{n\in\mathbb{N}_0} \Gamma(n,d)\left|\int_{-\pi}^{\pi} \exp(i\omega\tau)\, f_n(\omega)d\omega\right| < \infty.$$

Note that this equality follows from (56.1) under the invariance property assumed. Here, $L^1(L^2(\mathbb{M}_d, d\nu, \mathbb{R}))$ denotes the space of trace or nuclear operators on $L^2(\mathbb{M}_d, d\nu, \mathbb{R})$. Series X is said to display LRD when this summability condition fails. The LRD scenario tested has been introduced in [17] in the spectral domain, under the following functional semiparametric modelling

$$\mathcal{F}_\omega = \mathcal{M}_\omega|\omega|^{-\mathcal{A}}, \quad \omega\in[-\pi,\pi], \tag{56.5}$$

where the invariant positive self–adjoint operators $\mathcal{M}_\omega$ and $|\omega|^{-\mathcal{A}}$ are composed to define $\mathcal{F}_\omega$. Note that, here, $\mathcal{A}$ denotes the LRD operator on $L^2(\mathbb{M}_d, d\nu; \mathbb{C})$, defining the functional parameter in the spectral semiparametric framework adopted. Operator $|\omega|^{-\mathcal{A}}$ in (56.5) is interpreted as in the framework of operator self-similar processes (see, e.g., [3] and [15]). In this framework, $\mathcal{A}$ would play the role of operator–valued Hurst coefficient in the setting of fractional Brownian motion. In (56.5), as $\omega\to 0$, since $\sin(\omega)\sim\omega$, $|1-\exp(-i\omega)|^{-\mathcal{A}} = [4\sin^2(\omega/2)]^{-\mathcal{A}/2}\sim|\omega|^{-\mathcal{A}}$.

The regular spectral operator $\mathcal{M}_\omega$ is such that X is Markovian when the null space of $\mathcal{A}$ coincides with $L^2(\mathbb{M}_d, d\nu; \mathbb{C})$. Assume that $\mathcal{A}$ satisfies

$$\mathcal{A}(f)(g) = \int_{\mathbb{M}_d\times\mathbb{M}_d} f(x)g(y)\sum_{n\in\mathbb{N}_0}\alpha(n)\sum_{j=1}^{\Gamma(n,d)} S^d_{n,j}(x)\overline{S^d_{n,j}}(y)d\nu(x)d\nu(y), \tag{56.6}$$

for $f, g\in C^\infty(\mathbb{M}_d)$, the space of infinitely differentiable functions with compact support in $\mathbb{M}_d$, where sequence $\{\alpha(n),\ n\in\mathbb{N}_0\}$ is such that $l_\alpha\leq\alpha(n)\leq L_\alpha$, for every $n\in\mathbb{N}_0$, and $l_\alpha, L_\alpha\in(0,1/2)$. Hence, $\mathcal{A}$ and $\mathcal{A}^{-1}$ are in the space $\mathcal{L}(L^2(\mathbb{M}_d, d\nu, \mathbb{C}))$ of bounded linear operators on $L^2(\mathbb{M}_d, d\nu, \mathbb{C})$, with the norm $\|\cdot\|_{\mathcal{L}(L^2(\mathbb{M}_d,d\nu,\mathbb{C}))}$. In particular, $\|\mathcal{A}\|_{\mathcal{L}(L^2(\mathbb{M}_d,d\nu,\mathbb{C}))} < 1/2$. Note that operator $|\omega|^{-\mathcal{A}}$ is also interpreted in the weak sense as in (56.6). Operator $\mathcal{M}_\omega$ in (56.5) is a trace operator with kernel $\mathcal{K}_{\mathcal{M}_\omega}(x,y)$ satisfying

$$\mathcal{K}_{\mathcal{M}_\omega}(x,y) = \sum_{n\in\mathbb{N}_0} M_n(\omega)\sum_{j=1}^{\Gamma(n,d)} S^d_{n,j}\otimes\overline{S^d_{n,j}}(x,y), \quad x,y\in\mathbb{M}_d, \tag{56.7}$$

in terms of the sequence of positive eigenvalues $\{M_n(\omega),\ n \in \mathbb{N}_0\}$, $\omega \in [-\pi, \pi]$. For each $n \in \mathbb{N}_0$, $M_n(\omega)$, $\omega \in [-\pi, \pi]$, is a continuous positive slowly varying function at $\omega = 0$ in the Zygmund's sense (see Definition 6.6 in [1], and Assumption IV in [17]). As commented, X displays SRD, when $\alpha(n) = 0$, for every $n \in \mathbb{N}_0$. Note that, under the above introduced LRD scenario,

$$\int_{-\pi}^{\pi} \|\mathcal{F}_\omega\|^2_{\mathcal{S}(L^2(\mathbb{M}_d, d\nu, \mathbb{C}))}\, d\omega < \infty, \tag{56.8}$$

and the positive function sequence $\{f_n(\omega),\ \omega \in [-\pi, \pi],\ n \in \mathbb{N}_0\}$ in equation (56.1) satisfies:

$$f_n(\omega) = \frac{M_n(\omega)}{|\omega|^{\alpha(n)}}, \quad \omega \in [-\pi, \pi],\ n \in \mathbb{N}_0. \tag{56.9}$$

The following testing problem is considered:

$$H_0 :\ f_n(\omega) = M_n(\omega),\ \omega \in [-\pi, \pi],\ \forall n \in \mathbb{N}_0 \tag{56.10}$$
$$H_1 :\ f_n(\omega) = M_n(\omega)\, |\omega|^{-\alpha(n)},\ \omega \in [-\pi, \pi],\ \forall n \in \mathbb{N}_0. \tag{56.11}$$

The test statistic $\mathcal{S}_{B_T}$ proposed in [18] is given by:

$$\mathcal{S}_{B_T} = \sqrt{B_T T} \int_{[-\sqrt{B_T}/2, \sqrt{B_T}/2]} \widehat{\mathcal{F}}^{(T)}_\omega \frac{d\omega}{\sqrt{B_T}} = \int_{-\pi}^{\pi} \frac{\sqrt{B_T T}\widehat{\mathcal{F}}^{(T)}_\omega \mathbb{I}_{[-\sqrt{B_T}/2, \sqrt{B_T}/2]}}{\sqrt{B_T}} d\omega, \tag{56.12}$$

where $\mathbb{I}_{[-\sqrt{B_T}/2, \sqrt{B_T}/2]}$ denotes the indicator function on the interval $[-\sqrt{B_T}/2, \sqrt{B_T}/2]$. Note that $\frac{\mathbb{I}_{[-\sqrt{B_T}/2, \sqrt{B_T}/2]}}{\sqrt{B_T}}$ converges, as $T \to \infty$, in the weak sense, i.e., in the sense of generalized functions (see [5]), to a Dirac Delta distribution at zero frequency.

56.3 The Methodological Approach

This section describes the strategy adopted to discriminate between SRD and LRD in stationary functional time series, under the scenario of irregularly spatially discretely observed contaminated functional data, extending the asymptotic spectral results derived in [18] for fully observed data. Our proposal is based on nonparametric series least–squares estimation of the underlying functional values of the data, when only a discretized and contaminated version of these data is observed. To this aim, under a stochastic spatial sampling design, a spatial functional formulation, in a connected and compact two–points homogeneous space $\mathbb{M}_d$, of the nonparametric series regression results obtained in [4] for weak–dependent data is derived.

Let $\{X_t,\ t = 0, \ldots, T-1\}$ be, as before, a functional sample of size T of the strictly stationary sequence X introduced in Sect. 56.2. We consider the following observation model: for $t = 0, \ldots, T-1$,

$$Y_t(x_i) = X_t(x_i) + \varepsilon_t(x_i), \quad i = 1, \ldots, M(T),$$

where $\varepsilon_t(x_i)$, $i = 1, \ldots, M(T)$, are independent and identically distributed random variables satisfying $E[\varepsilon_t(x_i)/X_t(x_i)] = 0$, for $i = 1, \ldots, M(T)$. Here, $M(T)$ denotes the time–varying spatial sampling frequency. As before, $X_t(x_i)$ is the pointwise observed value of X_t at spatial location $x_i \in \mathbb{M}_d$, for $i = 1, \ldots, M(T)$. We also assume that x_i, $i = 1, \ldots, M(T)$, are the observed values of $\mathcal{X}_1, \ldots, \mathcal{X}_{M(T)}$, which are independent and identically distributed manifold–valued random vectors, with probability distribution given by the uniform probability measure on $\mathbb{M}_d$. In particular, since in Theorem 4.5 in [18], $B_T = T^{-\beta}$, $\beta \in (0, 1)$, we will consider $M(T) = T^{\eta}$, $\eta \in (0, 1)$, such that $\sqrt{M(T)/K(T)}B_T^{-1}T^{-1} \to 0$, as $T \to \infty$, for preserving consistency of the LRD test (see also Assumption 5 in [4]). Here, $K(T) = \widetilde{K}(M(T))$ denotes the time–varying dimension of the tensor product sieve basis. In particular, when β– mixing conditions are assumed to model the spatial weak–dependence structure, Lemma 2.2 in [4] provides some specific restrictions on $M(T)$, $K(T)$ and β–mixing coefficients in order to ensure Assumption 5 in [4] holds. The main aim is to controlling the dimension, norm and variance measure of the involved spatially weak–dependent stochastic matrices, constructed from sieve bases (see, e.g., Theorem 4.2 and Corollary 4.2 in [4]).

Consistent estimation requires the spatial sampling region $\mathcal{S}_d(T) \subset \mathbb{M}_d$ expands, when the spatial sampling frequency $M(T)$ increases. Both divergence rates, respectively associated with the volume $|\mathcal{S}_d(T)| = \int_{\mathcal{S}_d(T)} d\nu(x)$ of the spatial sampling region $\mathcal{S}_d(T)$, and with $M(T)$, are considered to be asymptotically equivalent, i.e., $|\mathcal{S}_d(T)|/M(T) = O(1)$, $T \to \infty$. The case where $M(T)$ asymptotically dominates the volume $|\mathcal{S}_d(T)|$ in a slowly way, i.e., $|\mathcal{S}_d(T)|/M(T) = o(1)$, with slow convergence to zero, is also analyzed here, and constitutes a more favorable case. Similar assertions hold for the sieve basis dimension $K(T)$, satisfying $K(T)/M(T) \to 0$, $T \to \infty$, in relation to the time–varying spatial sampling frequency $M(T)$ and the functional sample size T. These interacting divergence rates, respectively characterizing the increasing of $|\mathcal{S}_d(T)|$, $K(T)$, and $M(T)$ also lead to some restrictions to be imposed on the sieve basis functions. Thus, under suitable regularity and spatial weak–dependent conditions, spatial uniform and L^2 convergence rates are derived for the functional data series regression estimators, preserving the spectral asymptotic analysis achieved in [18].

Summarizing, the main goal of our paper is to obtain a suitable fitting of the increasing rates, when $T \to \infty$, of the volume $|\mathcal{S}_d(T)|$ of the spatial sampling region, the spatial sampling frequency $M(T)$, the dimension $K(T)$ of the sieve basis, and the dimension, norm and variance of the associated spatial weak–dependent random sieve basis matrix sequences, as well as of B_T^{-1}, to ensure the spectral asymptotic results obtained under SRD (the null hypothesis H_0), and LRD (the alternative hypothesis H_1) in [18] hold. In particular, a suitable interaction between these quantities allows us to prove the asymptotic Gaussian distribution of the test statistic under H_0, and the consistency under LRD, when irregularly spatially distributed, discretely observed and contaminated functional data are available.

A simulation study illustrates the performance of several sieve bases, including random spherical harmonics (see, e.g., [10]). We also analyze the case of compactly supported spherical wavelet or spline bases. Note that, the results derived also hold under an arbitrary stochastic manifold sampling design, beyond the uniform sampling distribution on the manifold. This numerical analysis will also be carried out to illustrate, that, under the above commented restrictions on $|\mathcal{S}_d(T)|$, $K(T)$, B_T^{-1}, and $M(T)$, leading to suitable slow rates of convergence to zero of $|\mathcal{S}_d(T)|/M(T)$, $K(T)/M(T)$, and $B_T^{-1}T^{-1}$, the analyzed sparsely and irregularly-observed settings lead to an equivalent asymptotic behavior (as $T \longrightarrow \infty$) to the fully observed functional data case of our testing procedure.

References

1. Beran, J.: Mathematical foundations of time series analysis. Springer, Switzerland (2017)
2. Boente, G., Rodriguez, D., Sued, M.: Testing equality between several populations covariance operators. Annals of the Institute of Statistical Mathematics. **70**, 919–950 (2018)
3. Characiejus, V., Räckauskas, A.: Operator self-similar processes and functional central limit theorems. Stochastic Process Appl. **124**, 2605–2627 (2014)
4. Chen, X., Christensen, T. M.: Optimal uniform convergence rates and asymptotic normality for series estimators under weak dependence and weak conditions. J. Econometrics. **188**, 447–465 (2015)
5. Gelfand, I. M., Vilenkin, N. Y.: Generalized functions, Vol. 4. Academic Press, New York (1964)
6. Giné, M.: The addition formula for the eigenfunctions of the Laplacian. Advances in Mathematics. **18**, 102–107 (1975)
7. Helgason, S.: Differential operators on homogeneous spaces. Acta Math. **102**, 239–299 (1959)
8. Li, D., Robinson, P.M., Shang, H.L.: Long-range dependent curve time series. Journal of the American Statistical Association. **115**, 957–971 (2019)
9. Ma, C., Malyarenko, A.: Time varying isotropic vector random fields on compact two points homogeneous spaces. J Theor Probab. **33**, 319–339 (2020)
10. Marinucci, D., Rossi, M., Wigman, I.: The asymptotic equivalence of the sample trispectrum and the nodal length for random spherical harmonics. Ann Inst Henri Poincaré Probab Stat. **56**, 374–390 (2020)
11. Ovalle–Muñoz, D. P., Ruiz–Medina, M. D.: LRD spectral analysis of multifractional functional time series on manifolds. *TEST*. **33**, 564–588 (2024)
12. Panaretos, V. M., Tavakoli, S.: Fourier analysis of stationary time series in function space. Ann. Statist. **41**, 568–603 (2013a)
13. Paparoditis, E., Sapatinas, T.: Bootstrap-based testing of equality of mean functions or equality of covariance operators for functional data. Biometrika. **103**, 727–733 (2016)
14. Pigoli, D., Aston, J. A., Dryden, I. L., Secchi, P.: Distances and inference for covariance operators. Biometrika. **101**, 409–422 (2014)
15. Rackauskas, A., Suquet, Ch.: Operator fractional brownian motion as limit of polygonal lines processes in Hilbert space. Stochastics and Dynamics. **11**, 49–70 (2011)
16. Ramm, A.G.: Random fields estimation. Longman Scientific & Technical, England (2005)
17. Ruiz–Medina, M. D.: Spectral analysis of long range dependence functional time series. Fractional Calculus and Applied Analysis. **25**, 1426–1458 (2022)
18. Ruiz–Medina, M.D., Crujeiras, R.M.: Testing LRD in the spectral domain for functional time series in manifolds. arXiv:2411.07731 (2024)
19. Tavakoli, S., Panaretos, V. M.: Detecting and localizing differences in functional time series dynamics: a case study in molecular biophysics. Journal of the American Statistical Association. **111**, 1020–1035 (2016)

Chapter 57
Prediction Regions for Functional-Valued Random Forests

Diego Serrano and Eduardo García-Portugués

Abstract We propose prediction regions for Random Forests (RFs) with functional output. Our approach is based on a metric specification and builds on the notion of Fréchet regression. It leverages the Out-Of-Bag (OOB) observations naturally generated during the training of RFs to estimate the uncertainty in the prediction, using the complete dataset. We outline the assumptions underpinning the construction of the prediction regions through OOB errors. A numerical experiment with quantile curves on the response and scalar predictor illustrates the prediction regions and shows that four types of nominal coverages are honored.

57.1 Introduction

Random Forests (RFs) [1] are a popular nonparametric regression method renowned for its adaptability and strong predictive power. Although the original formulation of RFs was designed exclusively for Euclidean data, several authors have generalized this algorithm for more complex data types. With this goal, Fréchet Random Forests (FRF) were developed in [5, 4] to extend RFs for metric data by incorporating the Fréchet mean in the splitting criterion and the aggregation of trees. This method can be applied to obtain predictions when the response and the predictors are functional variables, whenever a suitable distance is defined in the corresponding space of functions $\mathcal{F}$. A common example are Lebesgue spaces over an interval $T \subset \mathbb{R}$, defined as $L^q(T) := \{f : T \to \mathbb{R} : \int_T |f(t)|^q \,\mathrm{d}t < \infty\}$, with the L^q distance $d_q(f,g)^q := \int_T |f(t) - g(t)|^q \,\mathrm{d}t$. Building on the work in [5], a nonparametric

Diego Serrano (✉)
Department of Statistics, Universidad Carlos III de Madrid, Spain,
e-mail: dieserra@est-econ.uc3m.es

Eduardo García-Portugués
Department of Statistics, Universidad Carlos III de Madrid, Spain,
e-mail: edgarcia@est-econ.uc3m.es

© The Author(s), under exclusive license to Springer Nature Switzerland AG 2025

G. Aneiros et al. (eds.), *New Trends in Functional Statistics and Related Fields*,
Contributions to Statistics, https://doi.org/10.1007/978-3-031-92383-8_57

locally adaptive kernel generated by random forests was developed in [8], along with theoretical guarantees for the consistency of the estimator.

Once a RF prediction is obtained, it is important to quantify its uncertainty to weight prediction statements. However, despite their predictive power, RFs have historically lacked strong statistical inference tools. This issue was addressed in [9] by leveraging the Out-Of-Bag (OOB) observations to develop confidence intervals for Euclidean data. The OOB observations arise naturally in the training of a RF and are a trustworthy estimate of its predictive performance, bypassing the need of sample splitting.

The goal of this work is to generalize the ideas from [9] to FDA in order to develop prediction regions for a RF with functional response. Our prediction regions are balls centered on the RF predicted function, with a radius given by the $1-\alpha$ quantile of the empirical distribution of the OOB errors. Both the prediction of the functional variable and the estimation of the uncertainty are obtained using the complete dataset, at the cost of fitting a single forest. We show through a numerical experiment in the Wasserstein space that these prediction regions respect the nominal coverage rates for four different types of coverage.

The rest of the work is organized as follows: Sect. 57.2 introduces basic concepts regarding RFs and their extension for FDA; Sect. 57.3 defines OOB prediction regions; and Sect. 57.4 gives a numerical experiment in the Wasserstein space.

57.2 Random forests and FDA

RFs are an ensemble method based on combining the predictions from multiple decision trees via bootstrap aggregation. We first introduce decision trees.

57.2.1 *Fréchet decision trees with functional response*

Let Y denote a response functional random variable defined in a functional space $(\mathcal{F}, d_{\mathcal{F}})$ where $d_{\mathcal{F}}$ denotes a distance function in $\mathcal{F}$. Let $\boldsymbol{X}$ be a p-dimensional random vector in $[0,1]^p$ acting as the predictor (see the end of this subsection for a note on the extension to metric-valued predictor). Consider a sample $\mathcal{L}_n := \{(\boldsymbol{X}_i, Y_i)\}_{i=1}^n \subset [0,1]^p \times \mathcal{F}$ with multivariate predictors and functional response. The application of RFs to functional data is achieved by adapting the prediction of results after the tree is grown. This requires the generalization of the mean to spaces where no inner product structure is defined (for example, L^q spaces with $q \neq 2$). Fréchet [6] defined the Fréchet mean $y_{\oplus}$ as the following extension of the mean to metric spaces:

$$y_{\oplus} := \arg\min_{y\in\mathcal{F}} \mathsf{E}\big(d_{\mathcal{F}}(Y,y)^2\big).$$

The Fréchet mean is based on the mean-squared error minimization property of the mean in Euclidean spaces. The Fréchet mean may not necessarily exist nor be unique. Similarly, the classical regression model for Euclidean data is generalized in [7] to Fréchet regression, through the conditional Fréchet mean

$$m(\boldsymbol{x}) := \arg\min_{y \in \mathcal{F}} \mathsf{E}\big(d_{\mathcal{F}}(Y, y)^2 \mid \boldsymbol{X} = \boldsymbol{x}\big). \tag{57.1}$$

For Euclidean responses, $m(\boldsymbol{x}) = \mathsf{E}(Y \mid \boldsymbol{X} = \boldsymbol{x})$.

For multivariate predictors, given a node $A \subset \mathcal{L}_n$, a split variable index $j \in \{1, \ldots, p\}$, and a threshold value c_j, the left and right child nodes are defined, respectively, as

$$A_{j,\ell} := \left\{(\boldsymbol{X}_i, Y_i) \in A : X_{ij} \leq c_j\right\} \quad \text{and} \quad A_{j,r} := \left\{(\boldsymbol{X}_i, Y_i) \in A : X_{ij} > c_j\right\}.$$

The standard way of measuring the quality of a split is the Classification And Regression Trees (CART) criterion [2]. For each split variable X_j, the best threshold c_j is selected over the observed values of X_{ij}, $i = 1, \ldots, n$, as the maximizer of the variance decrease in the response due to the split. The optimal split for node A is the value that maximizes the variance decrease among the p selected thresholds. Once the tree is built, the prediction of a new observation $\boldsymbol{x}$ is calculated as the Fréchet mean of the responses from the training set contained in the terminal node of $\boldsymbol{x}$.

The regression setting can be extended to cases in which the predictors lie on a metric space (in particular, functional spaces). For that, the RF splitting criterion can be generalized using the 2-means algorithm as a split function, and assigning each datum to the closest of the two obtained centroids. The CART criterion is adapted for metric predictors by replacing the sample variance with an empirical version of the Fréchet variance, which is defined as the expected squared distance from the Fréchet mean (see [4]). For easiness in the exposition, we keep restricting to the case where the predictor is multivariate and bounded (as in [8]).

57.2.2 Random Forests with Functional Response

The aggregation of predictions of the training responses through the Fréchet mean used in Fréchet trees can be used to build FRFs through the aggregation of multiple tree predictions [5]. Every forest is comprised of a fixed number $B \in \mathbb{N}$ of trees, each trained with a random resample $\mathcal{L}_n^{*b}$ of $\mathcal{L}_n$, $b = 1, \ldots, B$. The absence of a theoretical analysis of the consistency of FRFs in [5, 4] motivated the development of a locally adaptive kernel generated by random forests with theoretical guarantees in [8]. This estimator of (57.1) is

$$m_n(\boldsymbol{x}) := \arg\min_{y \in \mathcal{F}} \sum_{i=1}^{n} \omega_i(\boldsymbol{x}) d_{\mathcal{F}}(Y_i, y)^2, \tag{57.2}$$

where the weights $\omega_i(\boldsymbol{x})$ are generated during the Fréchet tree-building procedure. Specifically, for a given value of the predictors $\boldsymbol{x} \in [0, 1]^p$, the prediction of the bth tree provides a sequence of weights $\omega_i^b(\boldsymbol{x}, \boldsymbol{\theta}_b) := |\tau_{\boldsymbol{x}}^b|^{-1} 1_{\{(x_i, y_i) \in \tau_{\boldsymbol{x}}^b\}}$, where $|\tau_{\boldsymbol{x}}^b|$

denotes the number of elements in the corresponding terminal node, $b = 1, \ldots, B$. Here, $\boldsymbol{\theta}_b$ represents the randomization parameter vector that governs the growth of the bth tree, determining which variables are considered for each split. The weights computed with the individual trees are then averaged to obtain a single weight $\omega_i(\boldsymbol{x})$ for the ith sample observation, $\omega_i(\boldsymbol{x}) := B^{-1} \sum_{b=1}^{B} \omega_i^b(\boldsymbol{x}, \boldsymbol{\theta}_b)$.

57.2.3 *Out-of-Bag Errors*

For a given tree built with a resample $\mathcal{L}_n^{*b}$, we say that $(\boldsymbol{X}_i, Y_i)$ is OOB if $(\boldsymbol{X}_i, Y_i) \in \mathcal{L}_n \setminus \mathcal{L}_n^{*b}$. We denote by $\widehat{Y}_{(i)}$ the OOB prediction of Y_i, which is based only on trees for which $(\boldsymbol{X}_i, Y_i)$ is OOB. Our goal is to estimate the uncertainty in the prediction $\widehat{Y}$ of a functional response Y using the OOB errors

$$\widehat{R}_i^{\mathrm{oob}} := d_{\mathcal{F}}(Y_i, \widehat{Y}_{(i)}), \quad i = 1, \ldots, n, \tag{57.3}$$

as an estimate of the actual prediction errors $d_{\mathcal{F}}(Y_i, m(\boldsymbol{X}_i)), i = 1, \ldots, n$. Notice that $\widehat{Y}_{(i)}$ is quite different from the leave-one-out cross-validation prediction, denoted by $\widehat{Y}_{-i}$, which is formed by the prediction on $\boldsymbol{X}_i$ of the forest trained with $\mathcal{L}_n \backslash \{(\boldsymbol{X}_i, Y_i)\}$.

Once a RF is trained using [8], we outline the procedure to obtain a specific OOB prediction $\widehat{Y}_{(i_0)}$. By construction, certain observations are left out of each random resample $\mathcal{L}_n^{*b}$, and hence are OOB for the bth tree. Let $\mathcal{B}_{i_0}$ denote the indices of the trees for which Y_{i_0} is OOB. If $b \in \mathcal{B}_{i_0}$, calculate $\omega_i^b(\boldsymbol{x}_{i_0}, \boldsymbol{\theta}_b)$ as explained in Sect. 57.2.2. For each $b \notin \mathcal{B}_{i_0}$, set $\omega_i^b(\boldsymbol{x}_{i_0}, \boldsymbol{\theta}_b) = 0$ for every $i = 1, \ldots, n$. Finally, aggregate the weights as $\omega_i(\boldsymbol{x}_{i_0}) = |\mathcal{B}_{i_0}|^{-1} \sum_{b=1}^{B} \omega_i^b(\boldsymbol{x}_{i_0}, \boldsymbol{\theta}_b)$, and plug them in (57.2) to obtain $\widehat{Y}_{(i_0)}$. The procedure described shows that OOB predictions estimate the predictive power of the RF at no additional training cost, since it is only required to select among the already grown trees.

57.3 Out-of-Bag Prediction Regions for FDA

We develop confidence regions $\mathcal{P}_{1-\alpha}$ that contain the true functional variable Y within a given probability $1 - \alpha$, as a generalization for functional response and multivariate predictors of the confidence intervals in [9]. Since the OOB errors (57.3) provide a reliable estimate of the RF errors, it is natural to measure the uncertainty in a RF prediction through the quantile of the empirical distribution function of the OOB errors. This idea can be applied to FDA in the following way.

Definition 57.1 (Prediction region). The prediction region for functional predictors $\boldsymbol{x} \in [0, 1]^p$, with significance level $\alpha \in (0, 1)$ is defined as

$$\mathcal{P}_{1-\alpha}(\boldsymbol{x}, \mathcal{L}_n) = \left\{ y \in \mathcal{F} : d_{\mathcal{F}}(\widehat{m}(\boldsymbol{x}), y) < \widehat{R}_{[1-\alpha, n]} \right\}, \tag{57.4}$$

where $\widehat{m}(\boldsymbol{x})$ is a RF estimation of the conditional Fréchet mean (57.1), and $\widehat{R}_{[1-\alpha,n]}$ denotes the $(1-\alpha)$-quantile of the empirical distribution based on $\widehat{R}_1^{\text{oob}}, \ldots, \widehat{R}_n^{\text{oob}}$.

We consider four probability coverage types, each suited to different statistical contexts. For a significance level $\alpha \in (0,1)$:

- Type I: $\mathsf{P}\{Y \in \mathcal{P}_{1-\alpha}(\boldsymbol{X}, \mathcal{L}_n)\}$.
- Type II: $\mathsf{P}\{Y \in \mathcal{P}_{1-\alpha}(\boldsymbol{X}, \mathcal{L}_n) \mid \mathcal{L}_n\}$.
- Type III: $\mathsf{P}\{Y \in \mathcal{P}_{1-\alpha}(\boldsymbol{X}, \mathcal{L}_n) \mid \boldsymbol{X} = \boldsymbol{x}\}$.
- Type IV: $\mathsf{P}\{Y \in \mathcal{P}_{1-\alpha}(\boldsymbol{X}, \mathcal{L}_n) \mid \mathcal{L}_n,\ \boldsymbol{X} = \boldsymbol{x}\}$.

In the Euclidean case, guarantees of asymptotic convergence of these probabilities to $1-\alpha$ as n diverges to infinity are provided in [9].

The following conditions are assumed for the application of the prediction regions:

(*c.1*) $(\boldsymbol{X}, Y), (\boldsymbol{X}_1, Y_1), \ldots, (\boldsymbol{X}_n, Y_n) \overset{\text{iid}}{\sim} \mathbb{G}$.
(*c.2*) The data generation process $\mathbb{G}$ is such that:

 (*c.2.1*) There exists a unique conditional Fréchet mean (57.1) for every $\boldsymbol{x} \in [0,1]^p$ $\mathbb{G}$-a.s.
 (*c.2.2*) The radial error $d_{\mathcal{F}}(Y, m(\boldsymbol{X}))$ is independent from the predictor $\boldsymbol{X}$.

These assumptions generalize some of the fundamental conditions defined in [9]. Assumption (*c.1*) ensures the identical distribution of the OOB radial errors (57.3). Existence and uniqueness of the conditional Fréchet mean from assumption (*c.2.1*) is required for the regression model to be well defined, and is a common assumption that ensures the consistency of the RF estimators, see [8] or [3]. We conjecture that, similarly to [9], the prediction regions (57.4) have correct asymptotic coverage under conditions (*c.1*)–(*c.2*), also under metric-valued predictor. We leave this proof for a future work, and assess the empirical coverage of the prediction regions through simulations in the following section.

57.4 A Numerical Experiment

We consider a simple simulation scenario to illustrate the prediction regions and test their empirical coverage for Types I–IV. The functional space $\mathcal{F}$ considered is the 2-Wasserstein space $\mathcal{W}_2(\mathbb{R})$ of probability distributions on the real line with finite second moment and endowed with the 2-Wasserstein metric $d_{\mathcal{W}_2}$. This space is isomorphic to the subset of $L^2[0,1]$ of quantile functions, so the distance $d_{\mathcal{W}_2}$ between two probability distributions P and Q of $\mathcal{W}_2(\mathbb{R})$ can be expressed through the respective quantile functions F_{P}^{-1} and F_{Q}^{-1}: $d_{\mathcal{W}_2}(\text{P},\text{Q})^2 = \int_0^1 \left|F_{\text{P}}^{-1}(u) - F_{\text{Q}}^{-1}(u)\right|^2 du$. This isomorphism motivates the definition of regression models in $\mathcal{W}_2$ in terms of the corresponding quantile functions. Following [7, Section 6.1], we consider the regression function

$$m(x)(\cdot) = \mathsf{E}(Y(\cdot) \mid X = x) = \frac{1}{4} - \log(1+x) + \left(\frac{1}{2} + x^2\right)\Phi^{-1}(\cdot),$$

for $x \in [0,1]$, where Φ^{-1} denotes the quantile function of a $\mathcal{N}(0,1)$ distribution. To generate the response, consider

$$Y(\cdot) = C - \log(1+X) + (S + X^2)\Phi^{-1}(\cdot).$$

with $C \sim \Gamma(1/2, 1/2)$ and $S \sim \mathrm{Exp}(2)$ independent of X, so that the model satisfies (*c.2.2*) and the conditional Fréchet mean of $Y(\cdot) \mid X = x$ is $m(x)(\cdot)$ (*c.2.1*). We consider that the predictor X follows a $U(0,1)$ distribution. The data generation process of $(X, Y(\cdot))$, which we denote by $\mathbb{G}$, is illustrated in Fig. 57.1.

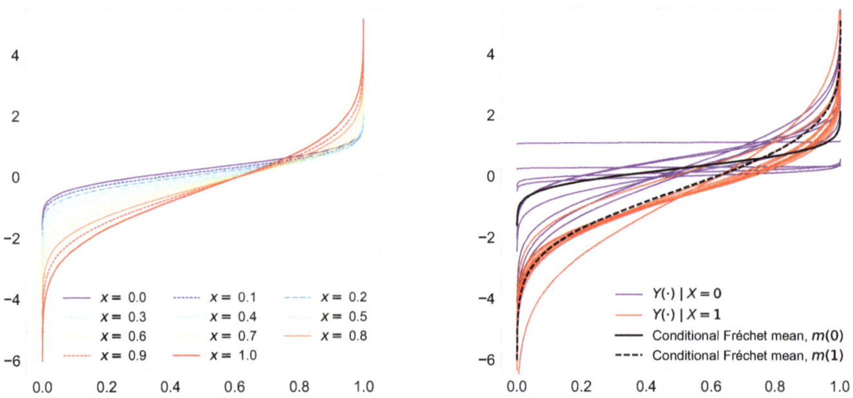

(a) Conditional Fréchet mean $m(x)$ (b) Realizations of $Y(\cdot) \mid X$

Fig. 57.1: Description of the data generation process $\mathbb{G}$. In Panel (a), the regression function (conditional Fréchet mean) is illustrated for different fixed values of the predictor. Panel (b) contains ten realizations of the functional response $Y(\cdot) \mid X = 0$ and ten of $Y(\cdot) \mid X = 1$. In each case, the realizations of $Y(\cdot) \mid X = x$ are centered around the regression function $m(x)$.

We generated $N = 500$ samples of size $n = 400$ according to the aforementioned data generation process, and used the estimator (57.2) in [8] for constructing the RF. Every forest is formed by $B = 200$ trees, and each tree is trained with a subsample of size n generated with nonparametric bootstrap with replacement, using the implementation from [3] in the `pyfréchet` package for Python. Three different values are considered for the significance level α: $0.01, 0.05$, and 0.10.

For Types I–IV, each probability is estimated using $M = 500$ Monte Carlo samples. In the case of Type I, the probability is estimated as follows:

$$\mathsf{P}\left\{Y \in \mathcal{P}_{1-\alpha}(X, \mathcal{L}_n)\right\} \approx \frac{1}{M}\sum_{j=1}^{M} 1_{\left\{Y_j \in \mathcal{P}_{1-\alpha}\left(X_j, \mathcal{L}_n^{(j)}\right)\right\}}$$

for $((X_j, Y_j), \mathcal{L}_n^{(j)})$ such that, for every $j = 1, \ldots, M$, (X_j, Y_j) is independent of $\mathcal{L}_n^{(j)}$ and $(X_j, Y_j) \sim \mathbb{G}$, $\mathcal{L}_n^{(j)} \sim \mathbb{G}^n$. For Type IV, the probabilities are estimated as

$$\mathsf{P}\{Y \in \mathcal{P}_{1-\alpha}(X, \mathcal{L}_n) \mid X = x, \mathcal{L}_n\} \approx \frac{1}{M} \sum_{j=1}^{M} 1_{\{Y_j \in \mathcal{P}_{1-\alpha}(x, \mathcal{L}_n)\}}, \tag{57.5}$$

for $Y_j \mid X = x$ induced by $\mathbb{G}$, and $x \in \{0, 0.5, 1\}$. The estimation of the probabilities in Types II and III is analogous.

For Type I, coverages of 0.992, 0.946, and 0.91 were obtained, respectively, for $\alpha = 0.01, 0.05, 0.10$. For Type III with $X = 0$ the simulations showed coverages of 0.998, 0.956, and 0.928. With $X = 0.5$, the coverages were 0.992, 0.958, and 0.916. Finally, with $X = 1$, the coverages were 0.988, 0.916, and 0.864. The reported coverages of Types II and IV are shown in Fig. 57.2. The results show that our prediction regions achieve overall correct coverage rates across the three significance levels considered.

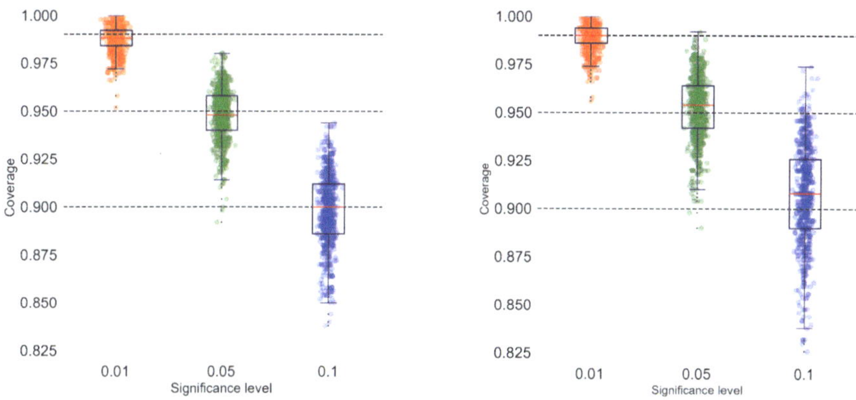

(a) Type II (b) Type IV ($X = 0.5$)

Fig. 57.2: Reported coverage (Types II and IV) across the $N = 500$ simulated datasets. For Type IV, the value of the predictor $X = 0.5$ was considered. Each dot in the boxplots is associated to a dataset $\mathcal{L}_n^{(k)}$, $k = 1, \ldots, N$, and its vertical position measures the estimated coverage probability (as in (57.5)) conditioned to that dataset.

Fig. 57.3 shows the shape of the prediction regions for different values of the predictor X and the significance level α. The prediction regions adapt to the shape of the predicted curve as X varies between 0 and 1.

Acknowledgements The first author acknowledges support from the INFLUENTIA-CM-UC3M project. The second author acknowledges support from "Convocatoria de la Universidad Carlos III de Madrid de Ayudas para la recualificación del sistema universitario español para 2021–2023", funded by Spain's Ministerio de Ciencia, Innovación y Universidades.

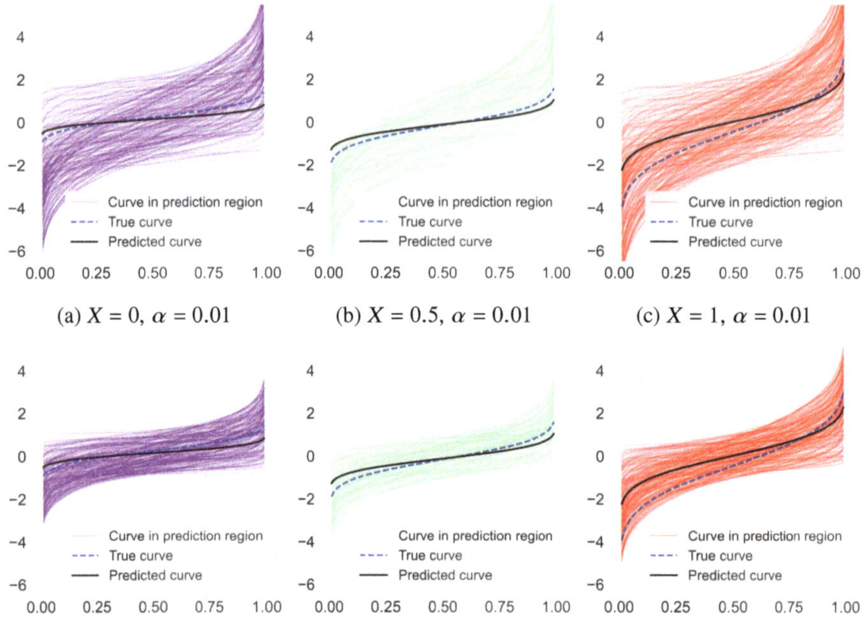

(a) $X = 0$, $\alpha = 0.01$ (b) $X = 0.5$, $\alpha = 0.01$ (c) $X = 1$, $\alpha = 0.01$

(d) $X = 0$, $\alpha = 0.10$ (e) $X = 0.5$, $\alpha = 0.10$ (f) $X = 1$, $\alpha = 0.10$

Fig. 57.3: Examples of prediction regions for different values of X and significance levels α. To represent graphically the shape of each prediction region, 100 Monte Carlo samples of $Y \mid X = x$ inside the prediction region are shown.

References

1. Breiman, L.: Random forests. Machine Learning **45**, 5–32 (2001)
2. Breiman, L., Friedman, J., Stone, C., Olshen, R.: Classification and Regression Trees. Taylor & Francis (1984)
3. Bulté, M., Sørensen, H.: Medoid splits for efficient random forests in metric spaces. Computational Statistics & Data Analysis (2024)
4. Capitaine, L., Bigot, J., Thiébaut, R., Genuer, R.: Fréchet random forests for metric space valued regression with non Euclidean predictors. Journal of Machine Learning Research **25**(355), 1–41 (2024)
5. Capitaine, L., Genuer, R., Thiébaut, R.: Fréchet random forests (2019)
6. Fréchet, M.: Les éléments aléatoires de nature quelconque dans un espace distancié. Annales de l'Institut Henri Poincaré **10**(4), 215–310 (1948)
7. Petersen, A., Müller, H.G.: Fréchet regression for random objects with Euclidean predictors. The Annals of Statistics **47**(2), 691–719 (2019)
8. Qiu, R., Yu, Z., Zhu, R.: Random forest weighted local Fréchet regression with random objects. Journal of Machine Learning Research **25**(107), 1–69 (2024)
9. Zhang, H., Zimmerman, J., Nettleton, D., Nordman, D.J.: Random forest prediction intervals. The American Statistician **74**(4), 392–406 (2020)

Chapter 58
Sample Size Estimation for Two-Sample Functional Hypothesis Test

Mohammad Reza Seydi, Johan Strandberg, Todd C. Pataky, and Lina Schelin

Abstract This study provides guidance for researchers who work with functional (curve) data and aim to perform a prior sample size estimation. Focusing on the two-population framework, we test mean differences between two populations — a scenario common in fields such as human movement science. Through simulations, we examine how standard deviation and smoothness influence sample size requirements to achieve 0.80 statistical power, using four methods with control the family-wise error rate: interval-wise testing (IWT), threshold-wise testing (TWT), F-max, and Extreme Rank Length (ERL) global envelope. For instance, increasing the standard deviation from 5 to 10 can raise the sample size from approximately 10 to over 30. Adjusting the smoothness parameter from 5 to 45 can lead to varied outcomes: the required sample size may increase to over 50, remain near 10, or even decrease, depending on the method and data characteristics. Three key findings are: (1) higher noise levels require larger sample sizes, (2) smoother data necessitate more samples when mean differences span larger domains, and (3) TWT and IWT are more efficient for large-domain differences, while ERL and F-max perform better for differences on narrower domains.

58.1 Introduction

Analysing functional data has become increasingly important in biomechanics and human movement sciences. However, the corresponding data collection methods,

Mohammad Reza Seydi (✉), Johan Strandberg, Lina Schelin
Department of Statistics, Umeå School of Business and Economics, Umeå University, Umeå, Sweden, e-mail: mohammad.seydi@umu.se, johan.strandberg@umu.se
e-mail: lina.schelin@umu.se

Todd C. Pataky
Department of Human Health Sciences, Kyoto University Graduate School of Medicine, Kyoto, Japan, e-mail: pataky.todd.2m@kyoto-u.ac.jp

© The Author(s), under exclusive license to Springer Nature Switzerland AG 2025

G. Aneiros et al. (eds.), *New Trends in Functional Statistics and Related Fields*, Contributions to Statistics, https://doi.org/10.1007/978-3-031-92383-8_58

especially a priori power calculations for sample size estimation, have been less explored. Such calculations are crucial for ensuring that studies are adequately powered to detect meaningful effects, avoiding underpowered results that could mislead conclusions or waste resources. Despite their importance, fewer than 23% of experimental studies in the *Journal of Biomechanics*, published in 2018 and 2019, applied a priori power analysis before data collection [13].

Although there have been some studies on statistical power analysis and sample size estimation based on random field theory (RFT), i.e., the statistical parametric mapping (SPM) method [5, 7, 11], functional data analysis (FDA) literature lacks sufficient focus on this topic. Furthermore, studies on sample size estimation using SPM have examined cases with fixed noise smoothness, varying either the amplitude and full-width-at-half-maximum (FWHM) of Gaussian signals [5] or the signal types and noise standard deviation [13]. In FDA literature, popular nonparametric methods, such as IWT [6], threshold-wise testing (TWT) [1], and global envelope tests [8], have been compared in [10]. Similarly, a comparison of IWT, SPM, and Benjamini-Hochberg (BH) procedures is provided in [12], but neither study addresses sample size estimation.

This study provides guidelines for researchers who perform a priori sample size estimation in two-sample functional hypothesis tests, a common scenario in human movement studies. In a simulation study, we explore how the standard deviation of noise functions (curves) and their smoothness (FWHM), given a fixed difference curve between two groups' mean functions (signal), influences sample size requirements to achieve 0.80 statistical power, using methods that control the family-wise error rate (FWER).

58.2 Methods

Functional hypothesis tests can be expressed both locally and globally. In a two-sample functional test, the global test examines the equality of mean functions of two populations over the domain, $\mathcal{D}$. However, it does not indicate at what interval in the domain these differences are significant, whereas the local test does.

The local functional hypothesis test can be formulated as

$$\begin{cases} H_0^{\mathcal{T}} : \forall t \in \mathcal{T} \subseteq \mathcal{D}, & M_1^{\mathcal{T}}(t) = M_2^{\mathcal{T}}(t) \\ H_a^{\mathcal{T}} : \exists t \in \mathcal{T} \subseteq \mathcal{D} \quad \text{s.t.} & M_1^{\mathcal{T}}(t) \neq M_2^{\mathcal{T}}(t). \end{cases}$$

Here, $\mathcal{T}$ represents a subset (or subsets) of $\mathcal{D}$, while $M_1^{\mathcal{T}}(t)$ and $M_2^{\mathcal{T}}(t)$ refer to the restricted mean functions over $\mathcal{T}$. The sample functional (curve) data, y_{iq}, can be modeled as

$$y_{iq}(t) = \mu_i(t) + \epsilon_{iq}(t), \quad t \in \mathcal{D},$$

where $i = 1, 2$ denotes the group and $q = 1, ..., SS$ represents the individuals and SS is the sample size in each group. Also, μ_i indicates the mean function for group i and also ϵ_{iq} denotes the noise function q for group i.

Statistical power is estimated as the probability of rejecting the null hypothesis at least once within the domain $\mathcal{D}$, given generated data under the alternative hypothesis H_a, i.e. $\mu_1(t) \neq \mu_2(t)$. Consequently, the sample size, SS, is determined as the minimum sample size per group for which the confidence interval for the power includes 0.80. This study explores four methods with control over FWER, and estimates the required sample size to obtain the targeted power. Each of the four methods is briefly explained in the following subsections.

58.2.1 Interval-Wise Testing (IWT)

The IWT performs local testing for functional hypotheses [6]. This method provides interval-wise control of the FWER. For this method, the adjusted p-value function is defined using a unified family that includes all intervals in the domain. So, the family of all intervals is defined as $[d_1, d_2]$ for all $d_1, d_2 \in \mathcal{D}$ s.t. $d_1 < d_2$. For each interval in the defined family, approximate a test statistic $T^{\mathcal{I}}$ based on the $\mathcal{L}^2$ distance

$$T^{\mathcal{I}} = \frac{||\mu_1^{\mathcal{I}}(t) - \mu_2^{\mathcal{I}}(t)||^2_{\mathcal{L}^2}}{|\mathcal{I}|}, \tag{58.1}$$

where $|\mathcal{I}|$ indicates the Lebesgue measure of the interval $\mathcal{I}$. Next, estimate p-value, $p^{\mathcal{I}}$, using B permutations for each interval $\mathcal{I}$ in the defined family. Estimate the adjusted p-value function as the supremum over all p-values of the intervals in the defined family containing point t,

$$\hat{p}_{\text{IWT}}(t) = \sup_{I \ni t} p^{\mathcal{I}}.$$

58.2.2 Threshold-Wise Testing (TWT)

The TWT method is a data-dependent approach for functional hypotheses [1]. Unlike IWT, TWT corrects unadjusted p-values without using a predefined family. To calculate the adjusted p-value function, partition the domain $\mathcal{D}$ into J equal-sized intervals. For each interval $\mathcal{Z}_j$, $j = 1, \ldots, J$, compute the test statistic $T^{\mathcal{Z}_j}$ as in Eq. (58.1), and estimate unadjusted p-values $p^{\mathcal{Z}_j}$ using B permutations. Next, define a finite set of thresholds $\Theta = \theta$ within $[0, 1]$. For each $\theta \in \Theta$, classify intervals as two sets S^+ and S^- which represent intervals with unadjusted p-value larger or equal and smaller than threshold θ, respectively. Then, compute corresponding p-values p_+ and p_-. Define the threshold-adjusted p-value p^θ as

$$p^{\theta} = \begin{cases} p_{+} & \text{if} \quad \mathcal{Z}_j \in \mathcal{S}^{+} \\ p_{-} & \text{if} \quad \mathcal{Z}_j \in \mathcal{S}^{-}, \end{cases}$$

based on the combined test statistic. Finally, estimate the adjusted p-value function as the maximum over Θ,

$$\hat{p}_{\text{TWT}}(t) = \max_{\theta \in \Theta} p^{\theta}, \quad t \in \mathcal{Z}_j$$

TWT provides weak FWER control for large B.

58.2.3 F-max

This method proposed in [4] is based on not only the F-statistic, the t-statistic is also very common to use specifically in biomechanics literature.

For each of the B permutations, the maximum value of the test statistic over all points in the domain is determined. This gives the empirical distribution of the maximum test statistic, denoted by T_M. Then, the adjusted p-value function is calculated as

$$\forall t \in \mathcal{D}, \quad \hat{p}_{\text{F-max}}(t) = P\Big(T_M \geq T(t)\Big),$$

where $T(t)$ represents the test statistic based on the original samples. This method provides strong control of the FWER.

58.2.4 Extreme Rank Length (ERL)

ERL envelope, a global envelope method introduced in [8], provides a technique to test functional hypotheses using simultaneous confidence bands [9]. The $100(1-\alpha)\%$ global envelope is defined by $C^{\alpha}(t)$ set of lower and upper confidence bands, satisfying

$$P(\eta_b^*(t) \notin C^{\alpha}(t) \quad \text{for any} \quad t \in \mathcal{D}) = \alpha,$$

where $\eta_b^*(t)$ is the difference between two groups' mean function, and $b = 1, \ldots, B$ denotes the random permutations. The procedure starts with computing $\eta(t)$, the pointwise difference between mean curves, over all $t \in \mathcal{D}$. Using B permutations $\eta_b^*(t)$ is calculated. Using the raw rank $r_b(t)$, which represents the rank of $\eta_b^*(t)$ for fixed t, the pointwise rank is determined as $R_b(t) = \min(r_b(t), B + 1 - r_b(t))$. Sorting $R_b(t)$ from smallest to largest yields R_b. $R^{(\alpha)}$ can now be defined as the largest R_b such that the number of b satisfies $(R_b < R^{(\alpha)}) \leq \alpha B$. The index set $I_{\alpha} = \{b \mid E_b \geq R^{(\alpha)}\}$ for $b = 1, \ldots, B$ is the base for constructing the simultaneous confidence bands. Here,

$$E_b = \frac{1}{B} \sum_{b'=1}^{B} I(R_{b'} \prec R_b),$$

where

$$R_{b'} \prec R_b \Longleftrightarrow \exists n^* \in \mathcal{D} \quad R_{b'}(t) = R_b(t) \quad \forall t < n^*, \quad R_{b'}(n^*) < R_b(n^*).$$

Finally, $C^\alpha(t)$ is defined as

$$C^\alpha(t) = (\min_{b \in I_\alpha} \eta^{*b}(t), \max_{b \in I_\alpha} \eta^{*b}(t)) \quad \text{for} \quad t \in \mathcal{D}.$$

The ERL method offers weak FWER control.

58.3 Simulation Study

This section begins with a brief explanation of the simulation setup, followed by reporting the simulation outcomes.

58.3.1 Setup

The hip flexion angle data displayed in the left panel of Fig. 58.1, shows the degree of hip flexion for the left limb during the gait cycle [2]. The muscle force data illustrated in the right panel of Fig. 58.1 represents the average force time series of the ankle extensor muscle (Soleus) [3]. Each curve in Fig. 58.1 shows the mean function, denoted as $\mu_i(t)$ for $i = 1, 2$ and $t \in [0, 100]$. The degree of smoothness, Noise FWHM, is defined as FWHM $= 2\sigma\sqrt{2\ln(2)}$, where σ is the standard deviation of the Gaussian kernel. The uncorrelated Gaussian noise function, $\epsilon_{iq}(t)$, with five different values of Noise FWHM, is added to mean functions. We also evaluated two different values for the standard deviation of the noise functions (Noise SD) chosen corresponding to the data characteristics. Note that a larger Noise FWHM value indicates smoother curves. For each dataset, 2500 simulations are used to determine the required sample size to achieve a power of 0.80.

58.3.2 Results

When two mean curves differ over a large portion of the domain, such as the Hip Flexion Angle data (left panel in Fig.58.1), increasing smoothness results in larger sample sizes being required. This growth in required sample size is slight for small

SD values, where it increases from around 5 to 11, but it becomes substantial for larger SD values, rising from approximately 10 to 40 (left panel in Fig. 58.2).

When the difference between two mean curves is limited to a smaller portion of the domain, such as in the Muscle Force data (right panel in Fig. 58.1), the sample size for the F-max and ERL methods remains almost constant as noise curves become smoother (right panel in Fig. 58.2). For larger SD values, the sample size for the F-max and ERL methods stabilizes around 18 as smoothness increases, whereas the TWT and IWT methods require 41 and 55 samples, respectively, for smoothed data with Noise FWHM equal to 45.

There is a general tendency for the TWT and IWT methods to require smaller sample sizes than the F-max and ERL methods when the portion of the domain over which the two mean curves differ is large. Conversely, when this portion is narrow, the F-max and ERL methods tend to require significantly smaller sample size per group (Fig. 58.2). Finally, with all other parameters held constant, increasing Noise SD results in a larger required sample size to achieve a statistical power of at least 0.80 (Fig. 58.2).

58.4 Discussion

This study compares the sample size requirement for four different methods which control FWER. The purpose of this study is to provide guidelines to researchers who work with functional (curve) data. Paying attention to the statistical power and consequently the sample size requirements is necessary not only prior to collecting data, but also during the analysis of the data.

The datasets utilised in this study cover two effect types, one involving the difference between two mean curves on a large part of the domain and the other involving a smaller part. Although several methods address functional hypothesis tests, the four methods mentioned here are widely recognized for their application to the local functional two-sample hypothesis test defined in Sect. 58.2. In this study, the noise curves represent the combination of curve variability and noise measurement. Additionally, the type of noise and parameters can be adjusted based on the characteristics of the data that are intended to be addressed. In summary, our findings highlight the importance of conducting a priori sample size estimation and selecting methods that align with the characteristics of the collected data to ensure valid analysis. We identified three key patterns of the sample size requirement behaviour when different methods are applied to the data:

1. Increasing the standard deviation of noise functions leads to a larger required sample size, regardless of the data characteristics and methods.
2. When the domain region where the two mean curves differ is narrow, increasing the smoothness does not increase the required sample size for F-max and ERL. Smoother data generally requires a larger sample size for all methods if this difference is wide.

3. For wide regions of difference, the TWT and IWT methods tend to require a lower sample size. Conversely, when the region of difference is narrow, the ERL and F-max methods need a lower sample size.

If you have prior knowledge of the expected curve structure or access to pilot data, it should be used to identify the relevant curve characteristics when conducting a prior sample size estimation.

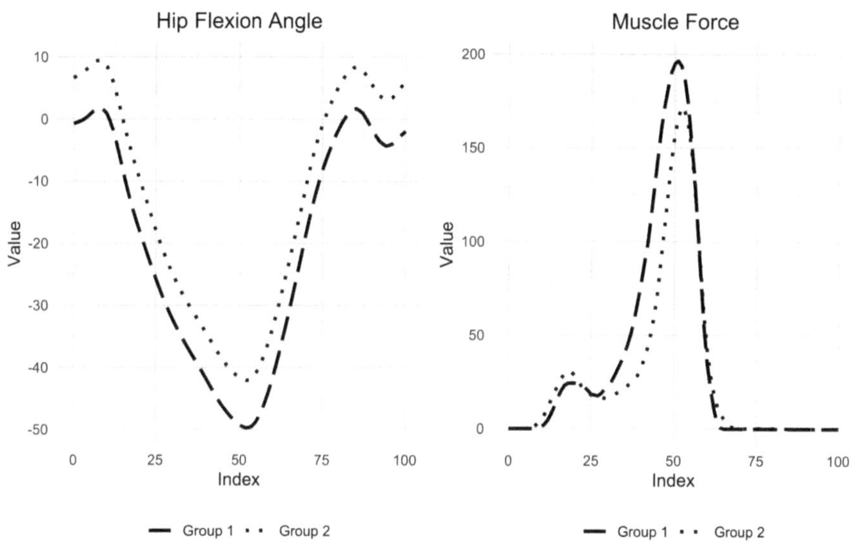

Fig. 58.1: An illustration of two datasets and their mean functions applied as the baselines for simulations. The line types represent the two groups mean functions

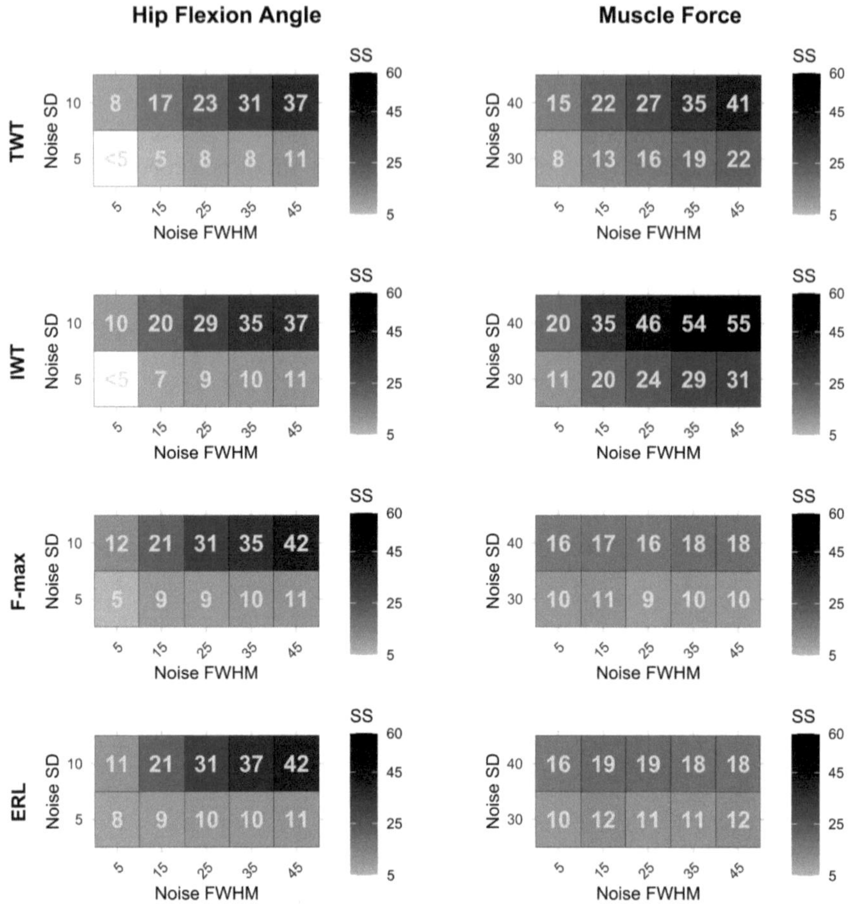

Fig. 58.2: The estimated sample size (SS) for two datasets with different Noise FWHM (smoothness parameter) and SD (standard deviation). The reported SS is the minimum sample size in each group needed to attain a statistical power of 0.80

References

1. Abramowicz, K., et al.: Domain selection and familywise error rate for functional data: A unified framework. Biometrics **79**(2), 1119–1132 (2023)
2. Bakke, D., Besier, T.: Shape model constrained scaling improves repeatability of gait data. J. Biomech. **107**, 109838 (2020)
3. Gomes, A.A., et al.: Muscle force distribution of the lower limbs during walking in diabetic individuals with and without polyneuropathy. J. Neuroeng. Rehabil. **14**, 1–13 (2017)
4. Holmes, A.P., et al.: Nonparametric analysis of statistic images from functional mapping experiments. J. Cereb. Blood Flow Metab. **16**(1), 7–22 (1996)
5. Luciano, F., Ruggiero, L., Pavei, G.: Sample size estimation in locomotion kinematics and electromyography for statistical parametric mapping. J. Biomech. **122**, 110481 (2021)
6. Pini, A., Vantini, S.: Interval-wise testing for functional data. J. Nonparametr. Stat. **29**(2), 407–424 (2017)
7. Pataky, T.C.: Power1D: a Python toolbox for numerical power estimates in experiments involving one-dimensional continua. PeerJ Comput. Sci. **3**, e125 (2017)
8. Myllymäki, M., et al.: Global envelope tests for spatial processes. J. R. Stat. Soc. B **79**(2), 381–404 (2017)
9. Mrkvička, T., et al.: A one-way ANOVA test for functional data with graphical interpretation. Kybernetika **56**(3), 432–458 (2020)
10. Mrkvička, T., Myllymäki, M.: Comparison of approaches for local testing with functional test statistics. J. Stat. Comput. Simul. **94**(16), 3555–3572 (2024)
11. Pataky, T.C., Robinson, M.A., Vanrenterghem, J.: A computational framework for estimating statistical power and planning hypothesis-driven experiments involving one-dimensional biomechanical continua. J. Biomech. **66**, 159–164 (2018)
12. Pataky, T.C., et al.: Simultaneous inference for functional data in sports biomechanics: Comparing statistical parametric mapping with interval-wise testing. AStA Adv. Stat. Anal. **105**, 1–24 (2021)
13. Robinson, M.A., Vanrenterghem, J., Pataky, T.C.: Sample size estimation for biomechanical waveforms: Current practice, recommendations and a comparison to discrete power analysis. J. Biomech. **122**, 110451 (2021)

Chapter 59
Forecasting Age Distribution of Deaths at Subnational Level

Han Lin Shang

Abstract This paper presents several forecasting methods to model and forecast subnational age distribution of death counts. The age distribution of death counts resembles strong similarities with probability density functions, which are nonnegative and have a constrained integral and thus live in a constrained nonlinear space. To address the nonlinear nature of objects, we implement a compositional data analytic approach known as centered log-ratio (clr) transformation to transform multiple densities into unconstrained functions. However, the clr transformation cannot handle the presence of zero values. This motivates us to consider a new cumulative distribution function (CDF) transformation with additional monotonicity. Using the Japanese subnational life-table death counts obtained from the [4], we evaluate the forecast accuracy of the transformations and forecasting methods. The improved forecast accuracy of life-table death counts is of great interest to demographers for estimating regional age-specific survival probabilities and life expectancy and actuaries for the pricing of annuities and setting of reserves.

59.1 Introduction

We present some forecasting methods to model and forecast subnational age distribution of deaths, which also find its usefulness in annuity pricing in actuarial science. For modeling the age distribution of deaths, [8] demonstrated that using compositional data analysis (CoDa) to forecast the age distribution of death counts does not lead to more pessimistic results than forecasting age-specific mortality rates. Within the CoDa framework, applying the Lee-Carter model to model and forecast unconstrained data is common. [10] extended this approach by considering a functional data analytic approach, including more than one principal component. To address the presence of zero values, [11] introduced a cumulative distribution

Han Lin Shang (✉)
Department of Actuarial Studies and Business Analytics, Level 7, 4 Eastern Rd, Macquarie University e-mail: hanlin.shang@mq.edu.au

© The Author(s), under exclusive license to Springer Nature Switzerland AG 2025
G. Aneiros et al. (eds.), *New Trends in Functional Statistics and Related Fields*, Contributions to Statistics, https://doi.org/10.1007/978-3-031-92383-8_59

function (CDF) transformation by converting an age distribution of deaths for each year to a probability bounded within the unit interval. Via the cumulative sum, one can transform a normalized life-table death count, i.e., probability density function, into a CDF. In doing so, it has the additional benefit of monotonicity (see also [7]). With a time series of CDFs, it is common to model its pattern via a logistic transformation.

When multiple density-valued objects are observed over time, it relates to high-dimensional functional time series. We aim to contribute to this growing literature by introducing some forecasting methods and comparing their forecast accuracy with applications to subnational life-table death counts.

In Sect. 59.2, we describe our motivating data set. In Sect. 59.3, we present centered log-ratio and cumulative distribution function transformations. Within each transformation, we consider several joint forecasting methods in Sect. 59.4 to model and forecast unconstrained functional time series. In Sect. 59.5, we evaluate and compare point forecast accuracy using several distance metrics. Sect. 59.6 concludes with some ideas on how the methodology presented can be extended further.

59.2 Age Distribution of Deaths in 47 Japanese Prefectures

Obtained from the [4], we study Japanese life-table death counts from 1975 to 2022, where the life-table radix is fixed at 10^5 at age 0 for each year. We consider ages from 0 to 110 in single years of age. Fig. 59.1 shows rainbow plots of the female and male age-specific life-table death counts in Japan from 1975 to 2022. The time ordering of the curves follows the color order of a rainbow, where curves from the distant past are shown in red, and the more recent curves are shown in purple.

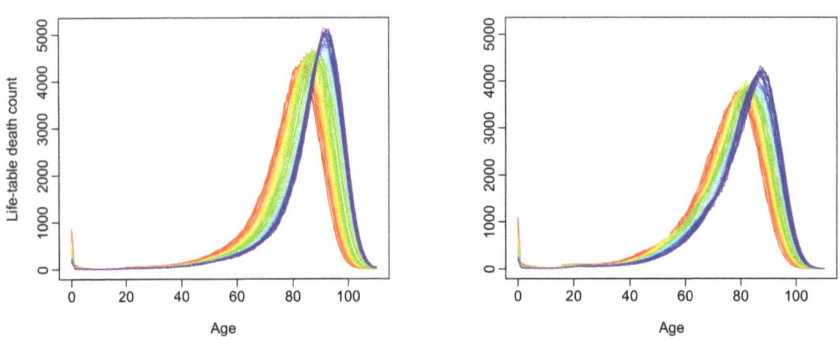

(a) Japanese female data (b) Japanese male data

Fig. 59.1: Functional time series graphical displays of age-specific life-table death counts from 1975 to 2022 in a single-year group. The oldest years are shown in red, with the most recent in violet. Curves are ordered chronologically according to the colors of the rainbow

Both figures show typical mortality curves for a developed country, with a decreasing trend in infant death counts. A typical negatively skewed distribution for the life-table death counts is apparent where the peaks shift to higher ages for both females and males. This shift is a leading source of longevity risk, a significant issue for insurers and pension funds, especially in the selling and risk management of annuity products (see [2] for a discussion). Moreover, the spread of the distribution indicates lifespan variability. A decrease in variability over time can be observed directly and measured by the Gini coefficient ([1, 14]).

59.3 Transformations

59.3.1 Compositional Data Analysis

Let $d^g_{t,s}(u)$ be the age distribution of deaths for age u, state s, gender g at year t, respectively. There are 111 ages, $S = 47$ states, and 48 years in our Japanese data set. The life-table death counts are non-negative and sum to 10^5 each year. Because of the two constraints, we consider extrinsic transformations from the so-called Aitchison's simplex to real-valued space.

When age u is treated as a continuum, the clr transformation can be expressed as

$$\text{clr}[d^g_{t,s}(u)] := \mathcal{Y}^g_{t,s}(u) = \ln d^g_{t,s}(u) - \frac{1}{\eta}\int_u \ln d^g_{t,s}(u)du, \qquad (59.1)$$

where $t = 1, 2, \ldots, 48$, $s = 1, 2, \ldots, 47$, $u \in [0, 110+]$, $\ln(\cdot)$ denotes the natural logarithm, η denotes the length of the age interval, and $\frac{1}{\eta}\int_u \ln d^g_{t,s}(u)du$ represents the geometric mean. Via the inverse clr transformation, the life-table death counts can be reconstructed as

$$\widehat{d}^g_{t,s}(u) = \frac{\exp[\widehat{\mathcal{Y}}^g_{t,s}(u)]}{\int_u \exp[\widehat{\mathcal{Y}}^g_{t,s}(u)]du} \times 10^5.$$

The clr transformation is an example of logarithmic and exponential maps, which are powerful notions in Riemannian manifolds (see also [13], Chapter 7). It tends to magnify variation in variables with low average proportions. Because of the logarithmic transformation in (59.1), it cannot handle the presence of zero death count.

59.3.2 Cumulative Distribution Function Transformation

In the CDF transformation, we first convert an age distribution of death for each year to a probability by dividing its radix 10^5. Via the cumulative sum, we transform a probability density function into a CDF, $D^g_{t,s}(x) = \sum_{u=1}^{x} d^g_{t,s}(u), \quad x = 1, \dots, 111,$ where the cumulative sum at the last age $D^g_{t,s}(111) = 1$. In doing so, it enjoys the additional benefit of monotonicity (see also [7]).

Then, we perform a logistic transformation,

$$Z^g_{t,s}(x) = \text{logit}[D^g_{t,s}(x)] = \ln\left[\frac{D^g_{t,s}(x)}{1 - D^g_{t,s}(x)}\right].$$

Since $D^g_{t,s}(111) = 1$, the last column is removed to avoid the undefineness of the logistic transformation. Considering age as a continuum, $Z^g_{t,s}(x)$ is an example of a high-dimensional functional time series.

To transform $Z^g_{t,s}(x)$ back to the original scale, we first perform an inverse logit transformation, where we obtain

$$D^g_{t,s}(x) = \frac{\exp^{Z^g_{t,s}(x)}}{1 + \exp^{Z^g_{t,s}(x)}},$$

where we add back the last column of ones. Then, we take the first-order differencing to obtain

$$d^g_{t,s}(u) = [\Delta^u_{x=1} D^g_{t,s}(x)] \times 10^5, \quad u = 2, \dots, 111,$$

where Δ represents the first-order differencing, where $\Delta^1_{x=1} D^g_{t,s}(1) = D^g_{t,s}(1)$.

59.4 Functional Time-Series Forecasting Models

We obtain a set of unconstrained functional time series via the clr or CDF transformation. We generally denote it as $\{\mathcal{X}^g_{t,s}(u)\}$, which is an arbitrary functional time series defined on a standard probability space. It is assumed that the observations $\mathcal{X}^g_{t,s}$ are elements of the Hilbert space $\mathcal{H} = \mathcal{L}^2(\mathcal{I})$ equipped with the inner product.

59.4.1 Univariate Functional Time-Series Forecasting Method

The Karhunen-Loève expansion of a stochastic process $\mathcal{X}^g_{t,s}(u)$ can be expressed as

$$\mathcal{X}^g_{t,s}(u) = \mu^g_s(u) + \sum_{k=1}^{K} \beta^g_{t,k,s}\phi^g_{k,s}(u) + e^g_{t,s}(u),$$

where principal component score $\beta^g_{t,k,s}$ is given by the projection of $[\mathcal{X}^g_{t,s}(u) - \mu^g_s(u)]$ in the direction of the kth eigenfunction $\phi^g_{k,s}(u)$, i.e., $\beta^g_{t,k,s} = \langle \mathcal{X}^g_{t,s}(u) - \mu^g_s(u), \phi^g_{k,s}(u)\rangle$; and $e^g_{t,s}(u)$ denotes the model error function with a mean of zero and a finite variance, and $K < n$ is the number of retained principal components.

We consider the eigenvalue ratio criterion of [6], defined as

$$K_s^g = \underset{k=1,\dots,k_{\max}}{\arg\min} \left\{ \frac{\widehat{\lambda}_{k+1,s}^g}{\widehat{\lambda}_{k,s}^g} \times \mathbb{1}\left\{ \frac{\widehat{\lambda}_{k,s}^g}{\widehat{\lambda}_{1,s}^g} \geq \eta \right\} + \mathbb{1}\left\{ \frac{\widehat{\lambda}_{k,s}^g}{\widehat{\lambda}_{1,s}^g} < \eta \right\} \right\}, \tag{59.2}$$

where $\mathbb{1}\{\cdot\}$ denotes the binary indicator function, $\eta = \frac{1}{\ln[\max(\widehat{\lambda}_{1,s}^g, n)]}$ is a small positive value. Instead of searching through n, we restrict our searching range by setting $k_{\max} = \#\{k | \widehat{\lambda}_{k,s}^g \geq \frac{1}{n} \sum_{k=1}^n \widehat{\lambda}_{k,s}^g, k \geq 1\}$. For comparison, we also consider $K_s^g = 6$ as suggested in [3].

59.4.2 Multivariate Functional Time-Series Forecasting Method

The univariate functional time-series method does not incorporate a correlation between female and male data within the same region. To potentially improve forecast accuracy, we consider a multivariate functional time-series method to jointly model and forecast multiple series that could be correlated with each other.

Let $\mathcal{X}_{t,s}^F(u)$ and $\mathcal{X}_{t,s}^M(u)$ represent female and male subnational age distribution of death counts. As our multiple functional time series have the same function support, we consider data where each observation consists of $\omega \geq 2$ functions, i.e., $[\mathcal{X}_{t,s}^F(u), \mathcal{X}_{t,s}^M(u)]$, where $u \in \mathcal{I}$.

The multivariate functional time series are stacked in a vector. Let $\nu_s^F(u)$ and $\nu_s^M(u)$ represent the mean function. For $u, v \in \mathcal{I}$, the theoretical cross-covariance function can be defined with elements

$$c_s(u, v) := \text{Cov}[\mathcal{X}_s^F(u), \mathcal{X}_s^M(v)] = E\left\{[\mathcal{X}_s^F(u) - \nu_s^F(u)][\mathcal{X}_s^M(u) - \nu_s^M(u)]\right\},$$

where $\nu_s^F(u)$ and $\nu_s^M(v)$ denote the mean functions for the female and male series. Based on the sample covariance (222×222) matrix, the Karhunen-Loève expansion of a stochastic process can be expressed as

$$\boldsymbol{\mathcal{X}}_s(u) = \boldsymbol{\nu}_s(u) + \boldsymbol{\Phi}_s(u)\boldsymbol{\beta}_s^\top,$$

where $\boldsymbol{\mathcal{X}}_s(u) = [\boldsymbol{\mathcal{X}}_s^F(u), \boldsymbol{\mathcal{X}}_s^M(u)]$ and $\boldsymbol{\mathcal{X}}_s^F(u) = [\mathcal{X}_{1,s}^F(u), \dots, \mathcal{X}_{n,s}^F(u)]$ denote stacked historical functions. Moreover, $\boldsymbol{\beta}_s = (\boldsymbol{\beta}_{1,s}, \dots, \boldsymbol{\beta}_{K,s})$ and $\boldsymbol{\beta}_{1,s} = (\beta_{1,1,s}, \dots, \beta_{n,1,s})$ is the vector of the basis function coefficients, $\boldsymbol{\nu}_s(u) = [\nu_s^F(u), \nu_s^M(u)]$, and

$$\boldsymbol{\Phi}_s(u) = \begin{bmatrix} \phi_{1,s}^F(u) \cdots \phi_{K,s}^F(u) & \cdots \\ \cdots & \phi_{1,s}^M(u) \cdots \phi_{K,s}^M(u) \end{bmatrix}_{2 \times (K \times 2)},$$

where the same value of K for the female and male data is determined via the eigenvalue ratio criterion, and the off-diagonal captures cross-correlation.

Conditioning on the past functions $\mathcal{X}_s(u) = [\mathcal{X}_s^F(u), \mathcal{X}_s^M(u)]$ and the estimated functional principal components $\mathbf{\Phi}_s(u)$, the h-step-ahead point forecast of $\mathcal{X}_{n+h,s}(u)$ can be expressed as

$$\widehat{\mathcal{X}}_{n+h|n,s}(u) = E\left[\mathcal{X}_{n+h,s}(u)|\mathcal{X}_s(u), \mathbf{\Phi}_s(u)\right] = \widehat{\nu}_s(u) + \widehat{\mathbf{\Phi}}_s(u)\widehat{\boldsymbol{\beta}}_{n+h|n,s}^{\top},$$

where $\widehat{\boldsymbol{\beta}}_{n+h|n,s}$ denotes the time-series forecasts of the principal component scores corresponding to the female and male series.

59.5 Results

59.5.1 Expanding Window Scheme

An expanding window analysis of a time-series model is used to assess model and parameter stability and prediction accuracy over time. The expanding window analysis determines the constancy of a model's parameter by computing parameter estimates and their forecasts over an expanding window through the sample. Using the first 32 years from 1975 to 2006 in the Japanese subnational age- and sex-specific life-table death counts, we produce one- to 16-step-ahead forecasts. Through an expanding-window approach, we re-estimate the parameters in the time-series forecasting models using the first 33 years from 1975 to 2007. Forecasts from the estimated models are then produced for one- to 15-step-ahead forecasts. We iterate this process by increasing the sample size by one year until reaching the end of the data period in 2022. This process produces 16 one-step-ahead forecasts, 15 two-step-ahead forecasts, ..., and one 16-step-ahead forecast. We compare these forecasts with the holdout samples to determine the accuracy of the out-of-sample forecast.

59.5.2 Forecast Accuracy Criteria

Since life-table death counts can be considered a probability density function, we utilize some density evaluation measures. The measures include the discrete version of the Kullback-Leibler divergence ([5]) and the square root of the Jensen-Shannon divergence ([12]). For two probability density functions, denoted by $d_{m+\xi}$ and $\widehat{d}_{m+\xi}$, the symmetric version of the Kullback-Leibler divergence is defined as

$$\begin{aligned}\text{KLD}(h) &= \text{D}_{\text{KL}}(d_{m+\xi}||\widehat{d}_{m+\xi}) + \text{D}_{\text{KL}}(\widehat{d}_{m+\xi}||d_{m+\xi}) \\ &= \frac{1}{111 \times (H+1-h)} \sum_{\xi=h}^{H} \sum_{u=1}^{111} d_{m+\xi}(u) \cdot \left[\ln d_{m+\xi}(u) - \ln \widehat{d}_{m+\xi}(u)\right] + \\ &\quad \frac{1}{111 \times (H+1-h)} \sum_{\xi=h}^{H} \sum_{u=1}^{111} \widehat{d}_{m+\xi}(u) \cdot \left[\ln \widehat{d}_{m+\xi}(u) - \ln d_{m+\xi}(u)\right],\end{aligned}$$

which is non-negative, for $h = 1, 2, \dots, H$ and $H = 16$ in our forecast evaluation.

An alternative metric is the Jensen-Shannon divergence, defined by

$$\text{JSD}(h) = \frac{1}{2} D_{\text{KL}}(d_{m+\xi}||\delta_{m+\xi}) + \frac{1}{2} D_{\text{KL}}(\widehat{d}_{m+\xi}||\delta_{m+\xi}),$$

where $\delta_{m+\xi}$ measures a common quantity between $d_{m+\xi}$ and $\widehat{d}_{m+\xi}$. Following the early work by [10], we consider the geometric mean $\delta_{m+\xi} = \sqrt{d_{m+\xi}\widehat{d}_{m+\xi}}$.

59.5.3 Comparison of Point Forecast Accuracy

With an average of 16 forecast horizons, we evaluate and compare overall forecast accuracy between the CDF and clr transformations using the univariate or multivariate functional time-series method. As measured by the KLD and JSD, Fig. 59.2 displays violin plots based on the averaged errors across 47 prefectures. The clr transformation, in combination with the multivariate functional time-series method, provides the most accurate forecasts.

59.6 Conclusion

We present two transformations for modeling and forecasting the age distribution of death counts at the Japanese prefectures. The CDF and clr transformations are studied to transform objects from simplex to real-valued space. Using the Japanese subnational data, the clr transformation is advantageous, especially with the multivariate functional time-series method, which can model the correlation between the female and male series within a prefecture. Both transformations belong to the extrinsic approach to model and forecast density-valued objects. Another research venue is the intrinsic approach studied in [9]. It would be interesting to compare the extrinsic and intrinsic approaches from a viewpoint of forecast accuracy.

Acknowledgements The author is grateful for the financial support from the Australian Research Council Discovery Project DP230102250 and Future Fellowship FT240100338.

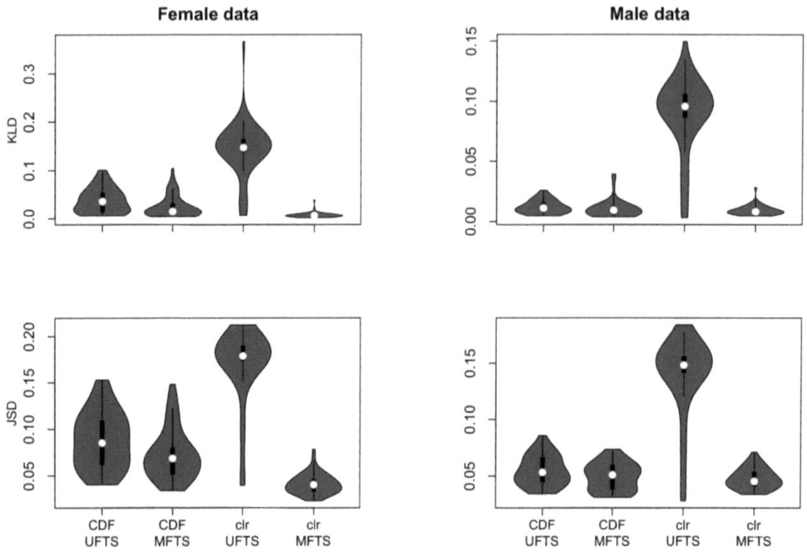

Fig. 59.2: A comparison of forecast accuracy between the CDF and clr transformation using the univariate or multivariate functional time-series method

References

1. Debón, A., Chaves, L., Haberman, S., Villa, F: Characterization of between-group inequality of longevity in EU countries. Insurance: Mathematics and Economics. **75**, 151-165 (2017)
2. Denuit, M., Devolder, P., Goderniaux, A.-C: Securitization of longevity risk: Pricing survivor bonds with Wang transform in the Lee-Carter framework. The Journal of Risk and Insurance. **74**, 87-113 (2007)
3. Hyndman, R. J., Booth, H., Yasmeen, F: Coherent mortality forecasting: the product-ratio method with functional time series models. Demography, **50**, 261-283 (2013)
4. Japanese Mortality Database: National Institute of Population and Social Security Research. Available via `https://www.ipss.go.jp/p-toukei/JMD/index-en.aspCited18September2024`
5. Kullback, S., Leibler, R. A: On information and sufficiency. The Annals of Mathematical Statistics, **22**(1), 79-86 (1951)
6. Li, D., Robinson, P. M., Shang, H. L: Long-range dependent curve time series. Journal of the American Statistical Association: Theory and Methods. **115**, 957-971 (2020)
7. Mayhew, L., Smith, D: A new method of projecting populations based on trends in life expectancy and survival. Population Studies. **67**, 157-170 (2013)
8. Oeppen, J.: Coherent forecasting of multiple-decrement life tables: A test using Japanese cause of death data. In: European Population Conference. Barcelona, Spain. (2008)
9. Petersen, A., Müller, H-G: Functional data analysis for density functions by transformation to a Hilbert space. Annals of Statistics. **44**(1), 183-218 (2016)
10. Shang, H. L., Haberman, S: Forecasting age distribution of death counts: An application to annuity pricing. Annals of Actuarial Science. **14**, 150-169 (2020)

11. Shang, H. L., Haberman, S: Forecasting age distribution of deaths: Cumulative distribution function transformation. Insurance: Mathematics and Economics, in press (2025)
12. Shannon, C. E.: A mathematical theory of communication. Bell System Technical Journal, **27**(3), 379-423 (1948)
13. Shi, J. Q., Choi, T: Gaussian Process Regression Analysis for Functional Data. CRC Press, Boca Raton, FL (2011)
14. Wilmoth, J. R., Horiuchi, S: Rectangularization revisited: Variability of age at death within human populations. Demography. **36**, 475-495 (1999)

Chapter 60
Statistical Analysis of Bivariate Densities with Compositional Splines

Stanislav Škorňa and Jitka Machalová

Abstract In general, densities occur naturally as data as a result of massive data aggregation. Thus, their suitable approximation for reliable (functional) data analysis is crucial not only for the univariate case but also for multivariate in general. However, their specific features, i.e. scale invariance and relative scale, have to be taken into account to build a suitable spline basis. This can be achieved using Bayes space methodology, which also allows to decompose densities into their interactive and independent part, which offers a powerful tool for the study of the dependence structure between random variables. Therefore, it is desirable for the spline approximation to admit the same decomposition. In addition, Bayes spaces methodology enables to convert densities to a standard Lebesgue space of square integrable functions using centered log-ratio transformation in order to use standard operations of the Lebesgue space, where, as a consequence, densities fulfill zero integral constraint. We propose a new bivariate spline basis for clr-transformed densities respecting the zero integral constraint. The practical impact of the proposed basis will be demonstrated on a regression model with a functional response for real geochemical data.

Stanilav Škorňa (✉)
Department of Mathematical Analysis and Applications of Mathematics, Faculty of Science, Palacký University Olomouc, 17. listopadu 12, Olomouc, 77146, Czech Republic,
e-mail: `stanislav.skorna01@upol.cz`

Jitka Machalová
Department of Mathematical Analysis and Applications of Mathematics, Faculty of Science, Palacký University Olomouc, 17. listopadu 12, Olomouc, 77146, Czech Republic,
e-mail: `jitka.machalova@upol.cz`

© The Author(s), under exclusive license to Springer Nature Switzerland AG 2025

G. Aneiros et al. (eds.), *New Trends in Functional Statistics and Related Fields*, Contributions to Statistics, https://doi.org/10.1007/978-3-031-92383-8_60

60.1 Theoretical background

60.1.1 Bayes spaces

Bayes Hilbert spaces, originally developed for σ-finite measures on compact interval $I \subset \mathbb{R}$ [1], can be extended to bivariate case considering a domain $\Omega = \Omega_X \times \Omega_Y = [a, b] \times [c, d]$, see [2]. However, Bayes spaces can be represented equivalently with densities and together with operations *perturbation, powering* and *inner product*, we obtain a complete Hilbert structure as described in [1, 3].

Let $\mathcal{B}(\Omega)$ be Bayes space defined on Ω and $f \in \mathcal{B}(\Omega)$ be bivariate density. Bayes space methodology enables to orthogonally decompose bivariate density into its *interactive part* f_{int} and *independent part* f_{ind}, where the latter is formed by univariate *geometrical marginals* f_1 and f_2, as derived in [3], where the decomposition parts are defined directly in $\mathcal{B}(\Omega)$. However, for convenient use of standard operations and methods of L^2 space, we can use *clr transformation* defined as

$$f^c = \operatorname{clr}(f) = \ln f - \frac{1}{(d-c)(b-a)} \iint_\Omega \ln f(x, y) \,\mathrm{d}x \,\mathrm{d}y,$$

which represents one-to-one transformation between $\mathcal{B}(\Omega)$ and $L^2(\Omega)$. Moreover, the resulting clr transformed function fulfills zero integral constraint, which ensures that every clr transformed density is an element of $L_0^2(\Omega)$. For decomposition, it holds that for any f^c one has corresponding clr transformed marginals defined as

$$f_1^c(x, y) = \frac{1}{d-c} \int_c^d f^c(x, t) \,\mathrm{d}t, \quad f_2^c(x, y) = \frac{1}{b-a} \int_a^b f^c(t, y) \,\mathrm{d}t. \tag{60.1}$$

Consequently, it holds that

$$f^c = f_{int}^c + f_{ind}^c = f_{int}^c + f_1^c + f_2^c. \tag{60.2}$$

See [3] for more details.

60.1.2 Spline Representation on $L_0^2(\Omega)$

Let

$$\Delta\lambda = \{\lambda_i\}_{i=0}^{g+1}, \quad a = \lambda_0 < \lambda_1 < \ldots < \lambda_g < \lambda_{g+1} = b, \tag{60.3}$$

$$\Delta\mu = \{\mu_j\}_{j=0}^{h+1}, \quad c = \mu_0 < \mu_1 < \ldots < \mu_h < \mu_{h+1} = d \tag{60.4}$$

be sequences of knots and let $\mathcal{S}_{kl}^{\Delta\lambda,\Delta\mu}(\Omega)$ be the vector space of tensor product splines $s_{kl}(x, y)$ defined on Ω of degree $k \in \mathbb{N}_0$ in x with knots $\Delta\lambda$ and of degree

$l \in \mathbb{N}_0$ in y with knots $\Delta\mu$. Since the standard B-spline representation, see [4], does not guarantee integration to zero, we designed a new spline basis as follows.

We define subspace $\mathcal{Z}_{kl}^{\Delta\lambda,\Delta\mu} \subset \mathcal{S}_{kl}^{\Delta\lambda,\Delta\mu}$ of a splines $s_{kl}(x, y)$ having zero integral. Considering the coincident boundary knots, we define *ZB*-splines

$$Z_i^{k+1}(x) = \frac{\mathrm{d}}{\mathrm{d}x} B_i^{k+2}(x), \quad Z_j^{l+1}(y) = \frac{\mathrm{d}}{\mathrm{d}y} B_j^{l+2}(y)$$

for $i = -k, \ldots, g-1,\ j = -l, \ldots, h-1$ in variable x and y, respectively, which have zero integral by construction, see [5]. It can be proven that

$$\dim\left(\mathcal{Z}_{kl}^{\Delta\lambda,\Delta\mu}(\Omega)\right) = (g + k + 1)(h + l + 1) - 1$$

and every spline $s_{kl}(x, y) \in \mathcal{Z}_{kl}^{\Delta\lambda,\Delta\mu}(\Omega)$ has a unique representation

$$s_{kl}(x, y) = \sum_{i=-k}^{g-1} \sum_{j=-l}^{h-1} z_{ij}\, Z_{ij}^{k+1,l+1}(x, y) + \sum_{i=-k}^{g-1} v_i\, Z_i^{k+1}(x, y) + \sum_{j=-l}^{h-1} u_j\, Z_j^{l+1}(x, y), \tag{60.5}$$

where $z_{ij}, v_i, u_j,\ i = -k, \ldots, g-1,\ j = -l, \ldots, h-1$ are corresponding spline coefficients and

$$Z_{ij}^{k+1,l+1}(x, y) := Z_i^{k+1}(x) Z_j^{l+1}(y),\ Z_i^{k+1}(x, y) := Z_i^{k+1}(x),\ Z_j^{l+1}(x, y) := Z_j^{l+1}(y).$$

In accordance with (60.1) and (60.2), every spline $s_{kl}(x, y) \in \mathcal{Z}_{kl}^{\Delta\lambda,\Delta\mu}(\Omega)$ can be decomposed into its interactive part $s_{kl}^{int}(x, y)$ and independent part $s_{kl}^{ind}(x, y)$, where

$$s_{kl}^{int}(x, y) := \sum_{i=-k}^{g-1} \sum_{j=-l}^{h-1} z_{ij}\, Z_{ij}^{k+1,l+1}(x, y),$$

$$s_{kl}^{ind}(x, y) := \sum_{i=-k}^{g-1} v_i\, Z_i^{k+1}(x, y) + \sum_{j=-l}^{h-1} u_j\, Z_j^{l+1}(x, y),$$

with clr marginals

$$s_k^1(x, y) := \sum_{i=-k}^{g-1} v_i\, Z_i^{k+1}(x, y), \quad s_l^2(x, y) := \sum_{j=-l}^{h-1} u_j\, Z_j^{l+1}(x, y).$$

Let f_{ij} be observations at points $(x_i, y_j) \in \Omega,\ i = 1, \ldots, m,\ j = 1, \ldots, n$ representing measures of an $L_0^2(\Omega)$ function. For fixed degrees $k, l \in \mathbb{N}_0$ and sequences of knots (60.3), (60.4) for variable x and y, respectively, the spline approximation of the density can be found as a minimizer of functional

$$J_{p,q}(s_{kl}) = \sum_{i=1}^{m}\sum_{j=1}^{n}\left[f_{ij} - s_{kl}(x_i, y_j)\right]^2 + \rho \iint_{\Omega} \left[s_{kl}^{(p,q)}(x, y)\right]^2 \mathrm{d}x\,\mathrm{d}y, \qquad (60.6)$$

where $p \in \{0, 1, \ldots, k-1\}$ and $q \in \{0, 1, \ldots, l-1\}$ denote degrees of derivatives of spline $s_{kl}(x, y)$ in variables x and y and $\rho > 0$ represents *smoothing parameter*. The spline $s_{kl} \in \mathcal{Z}_{kl}^{\Delta\lambda,\Delta\mu}(\Omega)$, which minimizes functional $J_{p,q}(s_{kl})$ is called the *tensor smoothing spline*. Proofs and details on bivariate *ZB*-splines can be found in [6].

60.2 Application

We applied the proposed methodology on a real dataset from the Register of Contaminated Areas of the Czech Republic, where we considered the soil concentration of copper (Cu) and zinc (Zn) in 74 Czech administrative districts with common range of $[0.588, 4.58]$ for Cu and $[1.459, 5.663]$ for Zn measured in $mg \cdot kg^{-1}$ [9]. Our focus was to find a relation between soil contamination (of copper and zinc) and agriculture and industry. For that purpose, we used a ratio between the agricultural land and the total area in each district as an agriculture indicator, and for the industry one, we took the CZ-NACE classification (category B-E, which covers the whole industrial sphere), which gives the information about the number of economic subjects registered in each district [8]. We divided it by the population of the corresponding region for comparability since each subject can be registered in an arbitrary district. Finally, the data were standardized to zero mean and unit variance to eliminate the effect of different scales.

The first crucial step is the preprocessing of the densities. In each district, the number of observations varies. Therefore, we aggregated the data to histograms to obtain samples of the density functions and used the clr transformed histogram values, which represent samples from $L_0^2(\Omega)$, where $\Omega = [0.588, 4.58]\times[1.459, 5.663]$. Consequently, we employed tensor smoothing splines as defined in Sect. 60.1.2, resulting in the estimation of the bivariate density of Cu and Zn soil contamination for every district. The density approximation strongly depends on the choice of degrees k, l of spline, the number and position of knots $\Delta\lambda$, $\Delta\mu$, the number of histogram classes m, n, degrees of derivatives p, q and the penalized parameter ρ. Typically, degrees k, l with p, q are set by the user. However, to ensure a good quality of the estimate of the density function, various statistical methods are used to determine the optimal values of the parameters. For instance, if the true density function is known, one can use (mean) integrated square error [6]. Generally, to obtain the optimal value of smoothing parameter ρ, (generalized) cross validation [7] is used frequently defined as

$$\mathrm{GCV} = \frac{\frac{1}{nm}\sum_{i=1}^{n}\sum_{j=1}^{m}\left(f_{ij} - s_{kl}^{*}(x_i, y_j)\right)^2}{(1 - \mathrm{trace}(\mathbf{H}/nm)^2},$$

where n, m stands for the number of histogram classes in direction of Cu and Zn, respectively, f_{ij} represents clr transformed histogram frequencies, $s^*_{kl}(x, y)$ is a tensor smoothing spline and $\mathbf{H}$ stands for so-called *hat matrix* [7]. The reliability and efficiency of the approximation can be further improved by performing a simulation study as conducted in [6].

After preprocessing the data and estimating the density function for each district, we set up a linear model with a functional response and scalar regressors as

$$\mathrm{clr}(\varphi_i) = \mathrm{clr}(\beta_0) + \mathrm{clr}(\beta_1)\, w_{1,i} + \mathrm{clr}(\beta_2)\, w_{2,i} + \mathrm{clr}(\epsilon_i), \tag{60.7}$$

for $i = 1, \ldots, 74$, where $\mathrm{clr}(\varphi_i)$ represents estimated clr transformed density, $\mathrm{clr}(\beta_0)$, $\mathrm{clr}(\beta_1)$, $\mathrm{clr}(\beta_2)$ indicates unknown functional coefficients for intercept, agriculture and industry, respectively, $w_{1,i}$, $w_{2,i}$ stands for agriculture and industry data and, finally, $\mathrm{clr}(\epsilon_i)$ means random error. Unknown parameters can be commonly estimated using the least square criterion. However, since every density is determined uniquely by its *ZB*-spline coefficients, we assumed a multivariate regression model

$$\mathrm{clr}(\boldsymbol{\varphi}_i) = \mathrm{clr}(\boldsymbol{\beta}_0) + \mathrm{clr}(\boldsymbol{\beta}_1)\, w_{1,i} + \mathrm{clr}(\boldsymbol{\beta}_2)\, w_{2,i} + \mathrm{clr}(\boldsymbol{\epsilon}_i),$$

where $\mathrm{clr}(\boldsymbol{\varphi}_i)$, $\mathrm{clr}(\boldsymbol{\beta}_0)$, $\mathrm{clr}(\boldsymbol{\beta}_1)$, $\mathrm{clr}(\boldsymbol{\beta}_2)$, $\mathrm{clr}(\boldsymbol{\epsilon}_i)$ represent matrices of *ZB*-spline coefficients of corresponding functional counterparts from (60.7). Consequently, for the estimation of *ZB*-spline coefficients of unknown parameters, we used the least square criterion in the form

$$\delta(\boldsymbol{\beta}_0, \boldsymbol{\beta}_1, \boldsymbol{\beta}_2) = \sum_{i=1}^{74} \left\| \mathrm{clr}(\boldsymbol{\varphi}_i) - \left(\mathrm{clr}(\boldsymbol{\beta}_0) + \mathrm{clr}(\boldsymbol{\beta}_1)\, w_{1,i} + \mathrm{clr}(\boldsymbol{\beta}_2)\, w_{2,i}\right) \right\|^2_{L^2}$$

with vectorized form of matrices $\mathrm{clr}(\boldsymbol{\varphi}_i)$, $\mathrm{clr}(\boldsymbol{\beta}_0)$, $\mathrm{clr}(\boldsymbol{\beta}_1)$, $\mathrm{clr}(\boldsymbol{\beta}_2)$. After the estimation of the *ZB*-spline coefficients of the regression parameters, each matrix can be expressed in the form

$$\mathbf{R} = \begin{pmatrix} \mathbf{Z} & \mathbf{v} \\ \mathbf{u}^\top & 0 \end{pmatrix},$$

with the given matrix $\mathbf{Z} = (z_{ij})_{i=-k,j=-l}^{g-1,h-1}$ and vectors $\mathbf{v} = (v_{-k}, \ldots, v_{g-1})^\top$, $\mathbf{u} = (u_{-l}, \ldots, u_{h-1})^\top$ of *ZB*-spline coefficients defined in (60.5), see [6] for details. It implies that the estimated regression coefficients inherit the same structure as clr transformed densities $\mathrm{clr}(\varphi_i)$ and as a consequence of (60.1) and (60.2), estimated regression coefficients can be further decomposed to their interactive part and independent parts with corresponding clr marginals. This is highly beneficial to study not only the dependency structure but also the local regression effect when we filter out the interactive information from the sample. One can consequently construct various regression models employing the decomposition part in order to discover local marginal effects, which could be normally overshadowed by the simultaneous information. However, due to the *ZB*-spline representation of the regression parameters and orthogonality of the decomposition of the Bayes space, the parameters also follow the same decomposition as the densities themselves, which allows to equivalently fit the separate regression models for decomposition parts.

The resulting functional regression parameters constructed from their *ZB*-spline coefficients and expressed in $\mathcal{B}^2(\Omega)$ are pictured in Fig. 60.1. Overall, the shape of β_1 (parameter for agriculture) indicates that an increase in the agriculture index tends to lower values of soil concentration of both elements. The estimate of β_2 (i.e. parameter for industry), on the other hand, indicates that increasing the industry index also leads to higher concentrations of Cu compared to β_1. Interestingly, the shapes of $\beta_{int,1}$ (i.e. interactive part of β_1), $\beta_{1,1}$ (i.e. Cu marginal of β_1), and $\beta_{2,1}$ (i.e. Zn marginal of β_1) imply that the agriculture index has a strong lowering effect on the concentration of Zn. The decomposition of the regression coefficients thus allows us interesting insights, distinguishing effects on the (geometric) marginals for the distributions of Cu and Zn from those on their dependence structure.

60.3 Conclusion

Bayes spaces methodology represents a powerful tool for processing and analysis of density probability functions preserving their properties. Moreover, clr transformation enables one-to-one transformation to L_0^2 space, where the standard operations can be used without the loss of the information and further methods of functional data analysis can be employed efficiently. Proposed bivariate *ZB*-spline representation in L_0^2 respects the zero integral constraint naturally without any additional conditions on the spline coefficients, which is highly beneficial and density data analysis (as demonstrated on the regression model) can be performed effectively.

Furthermore, Bayes spaces methodology provides a framework for the decomposition of densities into their interactive and independent part. Our novel *ZB*-spline representation allows a corresponding decomposition, which is highly beneficial and could serve as a keystone for the study of the dependency structure.

The proposed *ZB*-spline basis might serve as a foundation for the extension to the general multivariate case, which is highly desirable for handling the approximation and analysis of multivariate densities in general as well as for the study of dependency structure between multiple random variables.

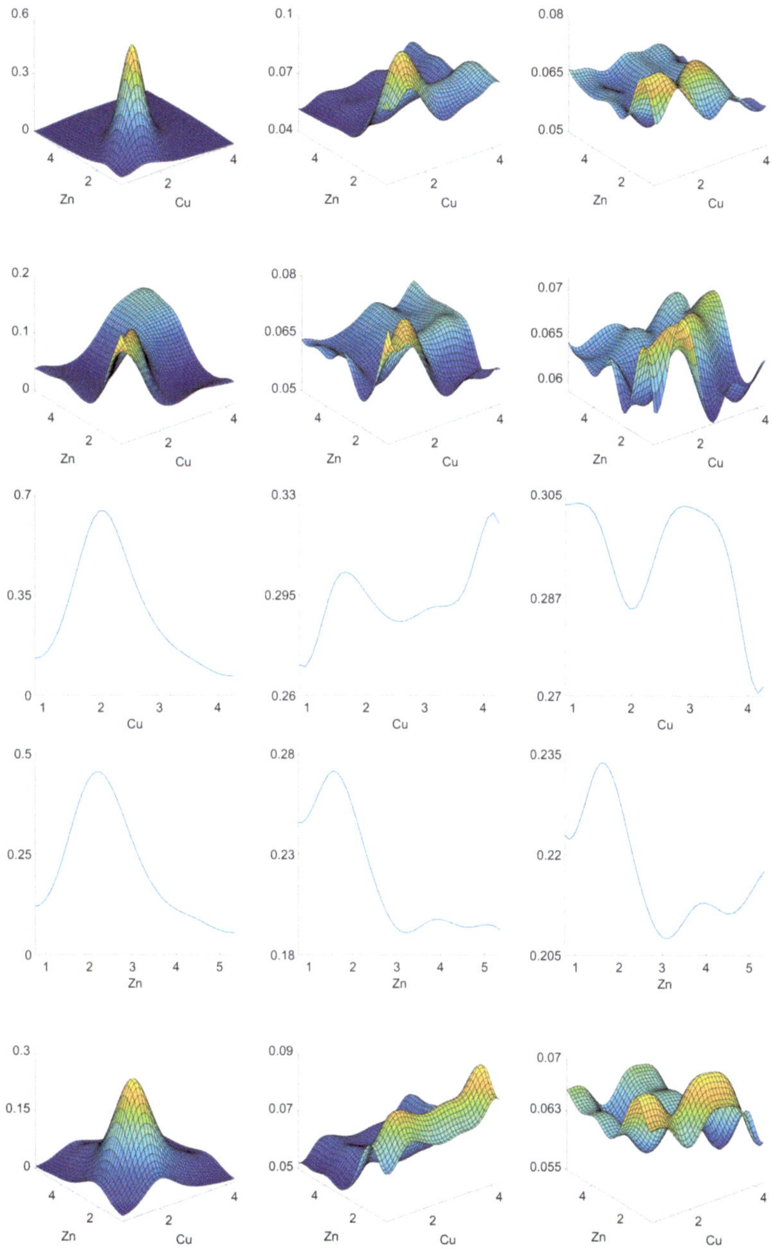

Fig. 60.1: Estimated regression parameters $\widehat{\beta}_0, \widehat{\beta}_1, \widehat{\beta}_2$ (from left to right) for the overall density (first row) and its decomposition parts $\widehat{\beta}_{int,0}, \widehat{\beta}_{int,1}, \widehat{\beta}_{int,2}$ (from left to right) for the interactive part (second row), $\widehat{\beta}_{1,0}, \widehat{\beta}_{1,1}, \widehat{\beta}_{1,2}$ (from left to right) for the x-marginal (Cu) (third row), $\widehat{\beta}_{2,0}, \widehat{\beta}_{2,1}, \widehat{\beta}_{2,2}$ (from left to right) for the y-marginal (Zn) (fourth row) and $\widehat{\beta}_{ind,0}, \widehat{\beta}_{ind,1}, \widehat{\beta}_{ind,2}$ (from left to right) for the independent part (fifth row)

References

1. van den Boogaart, K.G., Egozcue J., Pawlowsky-Glahn, V.: Hilbert Bayes spaces, Australian & New Zealand Journal of Statistics **54**, 171-194, (2014)
2. Hron, K., Machalová, J., Menafoglio, A.: Bivariate densities in Bayes spaces: orthogonal decomposition and spline representation, Statistical Papers **64**, 1629-1667, (2023)
3. Genest, C., Hron, K., Nešlehová, J.: Orthogonal decomposition of multivariate densities in Bayes spaces and relation with their copula-based representation, Journal of Multivariate Analysis **198**, 105228, (2023)
4. Dierckx, P.: Curve and Surface Fitting with Splines, Oxford University Press, New York, (1993)
5. Machalová, J., Talská, R., Hron, K., Gába, A.: Compositional splines for representation of density functions, Computational Statistics **36**, 1031-1064, (2021)
6. Škorňa, S., Machalová, J., Burkotová, J., Hron, K., Greven, S.: Approximation of bivariate densities with compositional splines ArXiv 2405.11615
7. Ramsay, J., Silverman, B. W.: Functional Data Analysis, Springer (Springer), New York, (2005)
8. Český statistický úřad, Public database, `https://vdb.czso.cz/vdbvo2/faces/en/index.jsf?page=home`
9. Matys Grygar, T., Elznicová, J., Tůmová, Š., Kylich, T., Skála, J., Hron, K., Álvarez Vázques, M. A.: Moving from geochemical to contamination maps using incomplete chemical information from long-term high-density monitoring of Czech agricultural soils, Environmental Earth Sciences, **82** (2022)

Chapter 61
Prediction with Mixed Effects Smooth Models by using P-Splines

Veronika Šmajserová and Jitka Machalová

Abstract This contribution explores the application of mixed effects smooth models, as introduced in [1], to the analysis of real-world data with missing observations. These models combine P-splines with linear mixed models, allowing robust and interpretable predictions while providing control over the smoothness of the resulting curves through penalization. We first validate the functionality of the proposed approach on simulated data and subsequently apply the theoretical framework described in [1] to a dataset describing the relationship between light reflectance and wavelength in trees. One of the main challenges in this dataset was the presence of missing values, which we addressed using a one-stage prediction approach that incorporates both observed and unobserved data. The results include smooth and accurate predictions, together with confidence intervals that reflect the variability of the predictions and provide additional information on the behaviour of the model and the uncertainty of the predictions.

61.1 Theoretical Framework

61.1.1 Prediction with P-Splines

P-splines are a smoothing technique that combines B-splines as basis functions with a penalty on the coefficients to achieve a balance between data fidelity and

Veronika Šmajserová (✉)
Department of Mathematical Analysis and Applications of Mathematics, Faculty of Science, Palacký University Olomouc, 17. listopadu 12, Olomouc, 77146, Czech Republic,
e-mail: veronika.smajserova@upol.cz

Jitka Machalová
Department of Mathematical Analysis and Applications of Mathematics, Faculty of Science, Palacký University Olomouc, 17. listopadu 12, Olomouc, 77146, Czech Republic,
e-mail: jitka.machalova@upol.cz

© The Author(s), under exclusive license to Springer Nature Switzerland AG 2025
G. Aneiros et al. (eds.), *New Trends in Functional Statistics and Related Fields*,
Contributions to Statistics, https://doi.org/10.1007/978-3-031-92383-8_61

smoothness. This approach has been widely described in [4]. Mathematically, for given data $(\mathbf{x}, \mathbf{y})$, the P-spline optimization problem minimizes the penalized least squares criteria

$$\delta(\mathbf{b}) = \|\mathbf{y} - \mathbf{C}_{k+1}(\mathbf{x})\mathbf{b}\|^2 + \rho \mathbf{b}^\top \mathbf{P} \mathbf{b}, \tag{61.1}$$

where $\mathbf{C}_{k+1}(\mathbf{x})$ is the collocation matrix derived from B-splines, $\mathbf{b}$ is the vector of spline coefficients, $\mathbf{P}$ is the penalty matrix, and $\rho > 0$ is the smoothing (penalized) parameter. The penalty matrix $\mathbf{P}$ is defined in terms of a difference matrix $\mathbf{D}_d$, which is constructed to enforce smoothness by penalizing higher-order differences of the spline coefficients. Specifically, $\mathbf{P} = \mathbf{D}_d^\top \mathbf{D}_d$, where $\mathbf{D}_d$ represents the difference operator of order d. The first term of (61.1) measures the fit of the model to the data, while the second term enforces smoothness by penalizing rapid changes in the spline coefficients. This combination allows P-splines to be computationally efficient, flexible, and suitable for a wide range of regression problems, including missing data imputation and smoothing in high-dimensional settings. To obtain the vector of spline coefficients $\widehat{\mathbf{b}}$, we solve a penalized least squares problem, i.e. we find a minimum of (61.1) which is given by $\widehat{\mathbf{b}} = \left(\mathbf{C}_{k+1}^\top(\mathbf{x})\mathbf{C}_{k+1}(\mathbf{x}) + \rho \mathbf{D}_d^\top \mathbf{D}_d\right)^{-1} \mathbf{C}_{k+1}^\top(\mathbf{x})\mathbf{y}$.

Suppose that we have an extended set of inputs and outputs $(\mathbf{x}_+, \mathbf{y}_+)$, where $\mathbf{x}_+$ combines the original inputs $\mathbf{x}$ and additional inputs $\mathbf{x}_p$, and $\mathbf{y}_+$ includes both observed $\mathbf{y}$ and unknown values $\mathbf{y}_p$. The extended model, as described in [1], is given by $\mathbf{y}_+ = \mathbf{C}_{k+1}^+(\mathbf{x}_+)\mathbf{b}_+ + \boldsymbol{\epsilon}$, where $\mathbf{C}_{k+1}^+(\mathbf{x}_+)$ is the extended collocation matrix, $\mathbf{b}_+$ is the vector of spline coefficients for the extended dataset, and $\boldsymbol{\epsilon}$ represents the noise vector. The coefficients $\mathbf{b}_+$ can then be estimated as

$$\widehat{\mathbf{b}}_+ = \left((\mathbf{C}_{k+1}^+(\mathbf{x}_+))^\top \mathbf{M} \mathbf{C}_{k+1}^+(\mathbf{x}_+) + \rho \mathbf{P}_+\right)^{-1} (\mathbf{C}_{k+1}^+(\mathbf{x}_+))^\top \mathbf{M} \mathbf{y}_+, \tag{61.2}$$

where $\mathbf{M} = (m_{ij})$ is a diagonal matrix, s.t. $m_{ii} = 1$ if the i-th position corresponds to observed data, and $m_{ii} = 0$ otherwise. The matrix $\mathbf{P}_+$ is the extended penalty matrix that incorporates information on the observed and unobserved parts of the data.

61.1.2 Linear Mixed Models

In this section, we follow the framework described in [2]. Let there be m subjects, each with an observation vector $\mathbf{y}_i$ of length n_i, for $i = 1, \ldots, m$. First, consider the fixed part of the model, represented by r explanatory variables whose values for the i-th subject are contained in the matrix $\mathbf{X}_i \in \mathbb{R}^{n_i \times r}$, $i = 1, \ldots, m$, of full column rank. Now, for the random part of the model, let $\mathbf{Z}_i \in \mathbb{R}^{n_i \times s}$ represent the matrix of explanatory variables for the random effects for the i-th subject, $i = 1, \ldots, m$. The linear mixed model for the i-th subject is given by $\mathbf{y}_i = \mathbf{X}_i \boldsymbol{\beta} + \mathbf{Z}_i \boldsymbol{\alpha}_i + \boldsymbol{\epsilon}_i$, where $\boldsymbol{\beta}$ is the vector of fixed effects of length r, $\boldsymbol{\alpha}_i$ is the vector of random effects of length s specific to the i-th subject, $\boldsymbol{\epsilon}_i$ is the residual vector of length n_i. It is assumed that $\boldsymbol{\alpha}_i \sim \mathcal{N}(\mathbf{0}, \mathbf{D})$, and $\boldsymbol{\epsilon}_i \sim \mathcal{N}(\mathbf{0}, \mathbf{R}_i)$, where $\mathbf{D} \in \mathbb{R}^{s \times s}$ and $\mathbf{R}_i \in \mathbb{R}^{n_i \times n_i}$ are symmetric and positive definite variance matrices. Furthermore, $\boldsymbol{\alpha}_i$ and $\boldsymbol{\epsilon}_i$ are assumed to be independent for all $i = 1, \ldots, m$. The linear mixed model for all subjects can then be expressed in matrix form as $\mathbf{y} = \mathbf{X}\boldsymbol{\beta} + \mathbf{Z}\boldsymbol{\alpha} + \boldsymbol{\epsilon}$, where $\mathbf{y} = (\mathbf{y}_1^\top, \ldots, \mathbf{y}_m^\top)^\top$ is the response vector of length $n = \sum_{i=1}^m n_i$, $\mathbf{X} = [\mathbf{X}_1^\top, \ldots, \mathbf{X}_m^\top]^\top$ is the design matrix for fixed effects $\boldsymbol{\beta}$ and $\mathbf{Z}$ is the design matrix for random effects $\boldsymbol{\alpha}$ with $\mathbf{Z}_i$ on the diagonal and $\boldsymbol{\epsilon} = (\boldsymbol{\epsilon}_1^\top, \ldots, \boldsymbol{\epsilon}_m^\top)^\top$ is the residual error. Random effects and errors are assumed to follow multivariate normal distributions, specifically $\boldsymbol{\alpha} \sim \mathcal{N}(0, \mathbf{G})$ and $\boldsymbol{\epsilon} \sim \mathcal{N}(0, \mathbf{R})$, where $\mathbf{G}$ and $\mathbf{R}$ are block diagonal matrices whose diagonals are composed of m matrices $\mathbf{D}$ and matrices $\mathbf{R}_i$, $i = 1, \ldots, m$, respectively.

61.1.3 Mixed Effects Smooth Models

Now, we focus on linking P-splines with linear mixed models, resulting in so-called mixed effects smooth models, where a key aspect of this connection is setting the smoothing parameter to the ratio of the variances of random error and random effects, i.e. $\rho = \sigma_\epsilon^2/\sigma_\alpha^2$. Unless stated otherwise, the following is based on [1]. Consider the linear mixed model

$$\mathbf{y} = \mathbf{X}\boldsymbol{\beta} + \mathbf{Z}\boldsymbol{\alpha} + \boldsymbol{\epsilon}, \tag{61.3}$$

where $\boldsymbol{\alpha} \sim N(\mathbf{0}, \sigma_\alpha^2\mathbf{G})$ and $\boldsymbol{\epsilon} \sim N(\mathbf{0}, \sigma_\epsilon^2\mathbf{I}_n)$. There are numerous ways to transform the original model $\mathbf{y} = \mathbf{C}_{k+1}(\mathbf{x})\mathbf{b} + \boldsymbol{\epsilon}$ into this mixed model. The aim is to find a transformation Ω such that $\mathbf{C}_{k+1}(\mathbf{x})\boldsymbol{\Omega} = [\mathbf{X}|\mathbf{Z}]$ and $\boldsymbol{\Omega}^\top\mathbf{b} = \left[\boldsymbol{\beta}^\top \mid \boldsymbol{\alpha}^\top\right]^\top$, where $\boldsymbol{\Omega}$ is an orthogonal matrix. The matrix $\boldsymbol{\Omega}$ is divided into two submatrices corresponding to fixed and random components, i.e. $\boldsymbol{\Omega} = [\boldsymbol{\Omega}_f|\boldsymbol{\Omega}_r]$. Thus, we have $\mathbf{X} = \mathbf{C}_{k+1}(\mathbf{x})\boldsymbol{\Omega}_f$ and $\mathbf{Z} = \mathbf{C}_{k+1}(\mathbf{x})\boldsymbol{\Omega}_r$. There are multiple ways to choose the submatrix Ω_r. In this paper, we proceed based on the spectral decomposition of the penalty matrix $\mathbf{P} \in \mathbb{R}^{c\times c}$, which is known to be positive semi-definite. Specifically, we consider the decomposition $\mathbf{P} = \mathbf{U}\widetilde{\boldsymbol{\Sigma}}\mathbf{U}^\top$, where $\widetilde{\boldsymbol{\Sigma}} \in \mathbb{R}^{c\times c}$ is a diagonal matrix whose diagonal entries are the eigenvalues of $\mathbf{P}$, arranged in ascending order, and $\mathbf{U} \in \mathbb{R}^{c\times c}$ contains the corresponding eigenvectors. Since matrix $\mathbf{P}$ is semidefinite, its eigenvalues are nonnegative and it means that the number of positive eigenvalues of $\mathbf{P}$ equals the rank of $\mathbf{P}$. In [5] it was shown that the rank of $\mathbf{P}$ is equal to the rank of $\mathbf{D}_d$, as $\mathbf{P} = \mathbf{D}_d^\top\mathbf{D}_d$. Since $\mathbf{D}_d \in \mathbb{R}^{(c-d)\times c}$ is of full row rank, the number of positive eigenvalues of $\mathbf{P} \in \mathbb{R}^{c\times c}$ is $c - d$, and the number of zero eigenvalues of $\mathbf{P}$ is d. The matrix $\widetilde{\boldsymbol{\Sigma}}$ can thus be written as

$$\widetilde{\boldsymbol{\Sigma}} = \begin{bmatrix} \mathbf{0}_{d\times d} & \mathbf{0}_{d\times(c-d)} \\ \mathbf{0}_{(c-d)\times d} & \boldsymbol{\Sigma} \end{bmatrix},$$

where $\boldsymbol{\Sigma} \in \mathbb{R}^{(c-d)\times(c-d)}$ is a diagonal matrix with positive eigenvalues of $\mathbf{P}$, and the matrix $\mathbf{U}$ can be expressed as $\mathbf{U} = [\mathbf{U}_f|\mathbf{U}_r]$, where $\mathbf{U}_f$ contains eigenvectors corresponding to the zero eigenvalues of $\mathbf{P}$, and $\mathbf{U}_r$ contains eigenvectors corresponding to the positive eigenvalues of $\mathbf{P}$. The matrix $\boldsymbol{\Omega}_r$ can then be chosen as $\boldsymbol{\Omega}_r = \mathbf{U}_r\boldsymbol{\Sigma}^{-1/2}$. The penalized least squares criteria (61.1) can then be rewritten in the context of the mixed model (61.3) as $\delta(\mathbf{b}) = (\mathbf{y} - \mathbf{X}\boldsymbol{\beta} - \mathbf{Z}\boldsymbol{\alpha})^\top(\mathbf{y} - \mathbf{X}\boldsymbol{\beta} - \mathbf{Z}\boldsymbol{\alpha}) + \rho\mathbf{b}^\top\mathbf{P}\mathbf{b}$.

Prediction in the context of mixed-effects models can be performed using a two-stage approach, where the model is first constructed and then predictions are performed. However, in this paper, we focus on a one-stage approach, where model construction and predictions occur simultaneously, offering the flexibility to choose any transformation.

61.2 Prediction with Mixed Effects Smooth Models

The one-stage approach can be utilized with any choice of reparameterization, and for appropriately selected transformations, it yields the same results as the two-stage approach. This method is also known as the extended mixed model approach, as it directly incorporates $\mathbf{y}_p$ into the model, albeit with infinite variance (that is, zero weight). The model is given by $\mathbf{y}_+ = \mathbf{X}_+\boldsymbol{\beta} + \mathbf{Z}_+\boldsymbol{\alpha}_+ + \boldsymbol{\epsilon}_+$, where $\boldsymbol{\alpha}_+ \sim N(0, \sigma_\alpha^2 \mathbf{G}_+)$ and $\boldsymbol{\epsilon}_+ \sim N(0, \sigma_\epsilon^2 \mathbf{R}_+)$. Here, $\mathbf{R}_+ \in \mathbb{R}^{n_+, n_+}$ is a diagonal weighting matrix whose diagonal elements are 1 for observed data $\mathbf{y}$ and infinity for predicted data $\mathbf{y}_p$. The infinite value in the matrix $\mathbf{R}_+$ corresponds to the fact that we have no information about the predicted data. According to [6], the estimates of fixed and random effects are derived from the equations of the extended mixed model. The resulting estimates are given by $\widehat{\widehat{\boldsymbol{\beta}}} = \left(\mathbf{X}_+^\top \widetilde{\mathbf{V}}_+^{-1} \mathbf{X}_+\right)^{-1} \mathbf{X}_+^\top \widetilde{\mathbf{V}}_+^{-1} \mathbf{y}_+$, $\widehat{\widehat{\boldsymbol{\alpha}}} = \sigma_\alpha^2 \mathbf{G}_+ \mathbf{Z}_+^\top \widetilde{\mathbf{V}}_+^{-1} \left(\mathbf{y}_+ - \mathbf{X}_+ \widehat{\widehat{\boldsymbol{\beta}}}\right)$, where $\widetilde{\mathbf{V}}_+ = \sigma_\alpha^2 \mathbf{Z}_+ \mathbf{G}_+ \mathbf{Z}_+^\top + \sigma_\epsilon^2 \mathbf{R}_+$ and therefore

$$\widetilde{\mathbf{V}}_+^{-1} = \frac{1}{\sigma_\epsilon^2} \mathbf{R}_+^{-1} - \frac{1}{\sigma_\alpha^2} \mathbf{R}_+^{-1} \mathbf{Z}_+ \left(\frac{1}{\sigma_\alpha^2} \mathbf{G}_+^{-1} + \mathbf{Z}_+^\top \frac{1}{\sigma_\epsilon^2} \mathbf{R}_+^{-1} \mathbf{Z}_+ \right)^{-1} \mathbf{Z}_+^\top \mathbf{R}_+^{-1}. \tag{61.4}$$

Furthermore, it holds that $\mathbf{Z}_+ = \mathbf{C}_{k+1}^+(\mathbf{x}_+) \boldsymbol{\Omega}_r^+$, where $\boldsymbol{\Omega}_r^+$ is an orthogonal transformation matrix satisfying $\mathbf{C}_{k+1}^+(\mathbf{x}_+) \left[\boldsymbol{\Omega}_f^+ \mid \boldsymbol{\Omega}_r^+\right] = [\mathbf{X}_+ \mid \mathbf{Z}_+]$, and $\left[\boldsymbol{\Omega}_f^+ \mid \boldsymbol{\Omega}_r^+\right]^\top \mathbf{b}_+ = \left[\boldsymbol{\beta}^\top \mid \boldsymbol{\alpha}_+^\top\right]^\top$. For penalties based on differences, we choose the matrix $\boldsymbol{\Omega}_r^+$ based on the spectral decomposition of the penalty matrix $\mathbf{D}_+^\top \mathbf{D}_+$. This ensures the desired equality $\mathbf{C}_{k+1}^+(\mathbf{x}_+) \mathbf{b}_+ = \mathbf{X}_+ \boldsymbol{\beta} + \mathbf{Z}_+ \boldsymbol{\alpha}_+$. From (61.4), it follows that to determine fixed and random effects, we do not need the matrix $\mathbf{R}_+$, but rather its inverse $\mathbf{R}_+^{-1}$. According to [5], $\mathbf{R}_+^{-1}$ is also diagonal, with 1 on the diagonal where $\mathbf{R}_+$ has 1, and 0 where $\mathbf{R}_+$ has infinity.

61.2.1 Application to Simulated Data

We apply the one-stage approach to the simulated data shown in Fig. 61.1. Two parts of the data are treated as missing, while the rest are considered to be observed. Our goal is to reconstruct the missing values and then focus on predicting future observations. In general, we proceed as follows. Consider the model described in [3]

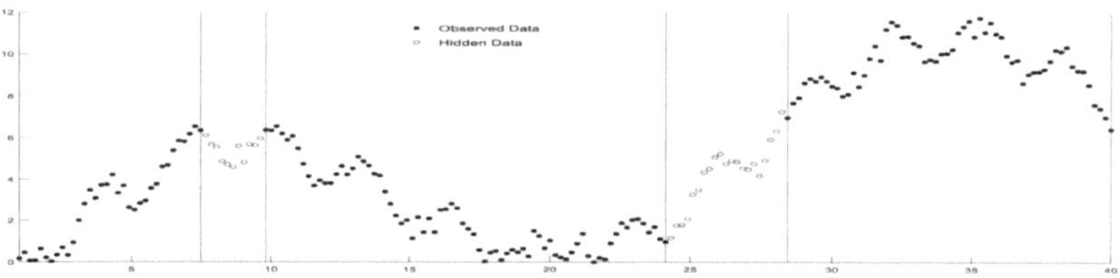

Fig. 61.1: Observed and hidden data points.

$$E(y_i \mid x_i) = f(x_i) + \sum_{j=1}^{J} \left(g_j(x_i)\cos(j\omega x_i) + h_j(x_i)\sin(j\omega x_i)\right), \tag{61.5}$$

where f is a smooth function representing the trend, g and h are smooth functions capturing the amplitudes of cosine and sine waves, respectively, and $\omega = \frac{2\pi}{p}$, p is the period. We set $J = 1$. The model (61.5) can be written as

$$\mathbf{y} = \check{\mathbf{C}}\check{\mathbf{b}} + \boldsymbol{\epsilon}, \tag{61.6}$$

where $\boldsymbol{\epsilon} \sim N(\mathbf{0}, \sigma_\epsilon^2 \mathbf{I}_n)$, $\check{\mathbf{C}} = [\mathbf{C}_{k+1}(\mathbf{x}) \mid \mathbf{K}\mathbf{C}_{k+1}(\mathbf{x}) \mid \mathbf{S}\mathbf{C}_{k+1}(\mathbf{x})]$, and $\check{\mathbf{b}} = (\mathbf{b}, \mathbf{b}_k, \mathbf{b}_s)^\top$. Here, $\mathbf{K}$ and $\mathbf{S}$ are diagonal matrices with $\cos(\omega\mathbf{x})$ and $\sin(\omega\mathbf{x})$ as their diagonals, respectively. To estimate the coefficient vector $\check{\mathbf{b}}$, we minimize the function in the form $\delta(\check{\mathbf{b}}) = (\mathbf{y} - \check{\mathbf{C}}\check{\mathbf{b}})^\top(\mathbf{y} - \check{\mathbf{C}}\check{\mathbf{b}}) + (\check{\mathbf{b}})^\top \check{\mathbf{P}}\check{\mathbf{b}}$, where $\mathbf{P} = \check{\boldsymbol{\rho}} \otimes (\check{\mathbf{D}})^\top \check{\mathbf{D}}$ and $\otimes$ denotes the tensor product. Here, $\check{\boldsymbol{\rho}}$ is a diagonal matrix with diagonal entries (ρ, ρ_k, ρ_s), $\check{\mathbf{D}}$ is a block diagonal matrix with matrices $\mathbf{D}_d^\top \mathbf{D}_d$ and $\mathbf{I}_2 \otimes \mathbf{D}_{d_{ks}}^\top \mathbf{D}_{d_{ks}}$ on its diagonal. We set $d = 2$ and $d_{ks} = 1$, which are standard choices; see [1]. Since we consider a one-stage approach, extending the model (61.6) is necessary. We define the matrix $\check{\mathbf{C}}_+ = \left[\mathbf{C}_{k+1}^+(\mathbf{x}) \mid \mathbf{K}_+\mathbf{C}_{k+1}^+(\mathbf{x}_+) \mid \mathbf{S}_+\mathbf{C}_{k+1}^+(\mathbf{x}_+)\right]$, where $\mathbf{K}_+$ and $\mathbf{S}_+$ are diagonal matrices with their diagonals composed of the vectors $\cos(\omega\mathbf{x}_+)$ and $\sin(\omega\mathbf{x}_+)$, respectively. The extended penalty matrix $\check{\mathbf{P}}_+$ is a block-diagonal with the matrices $\rho(\mathbf{D}_d^+)^\top\mathbf{D}_d^+$, $\rho_k(\mathbf{D}_{d_k}^+)^\top\mathbf{D}_{d_k}^+$ and $\rho_s(\mathbf{D}_{d_s}^+)^\top\mathbf{D}_{d_s}^+$ on its diagonal. Then, we can use the formula (61.2) where we replace the matrix $\mathbf{C}_{k+1}^+(\mathbf{x}_+)$ with $\check{\mathbf{C}}_+$ and the matrix $\rho\mathbf{P}_+$ with $\check{\mathbf{P}}_+$. The matrices $\check{\mathbf{X}}$ and $\check{\mathbf{Z}}$ are given as

$$\check{\mathbf{X}} = \left[\mathbf{1}_n \mid \mathbf{x} \mid \mathbf{x}^2 \mid \cdots \mid \mathbf{x}^{d-1} \mid \cos(\omega\mathbf{x}) \mid \sin(\omega\mathbf{x})\right], \quad \check{\mathbf{Z}} = [\mathbf{Z} \mid \mathbf{K}\mathbf{Z}_{ks} \mid \mathbf{S}\mathbf{Z}_{ks}], \tag{61.7}$$

where $\cos(\omega\mathbf{x})$ and $\sin(\omega\mathbf{x})$ are column vectors of the same length as the vector $\mathbf{x}$, $\mathbf{Z} = \mathbf{C}_{k+1}(\mathbf{x})\boldsymbol{\Omega}_r$ and $\mathbf{Z}_{ks} = \mathbf{C}_{k+1}(\mathbf{x})\check{\boldsymbol{\Omega}}_r$, where $\check{\boldsymbol{\Omega}}_r = \check{\mathbf{U}}_r(\check{\boldsymbol{\Sigma}})^{-\frac{1}{2}}$, and $\check{\mathbf{U}}_r$ and $\check{\boldsymbol{\Sigma}}$ are derived from the spectral decomposition of the matrix $\mathbf{D}_{d_{ks}}^\top \mathbf{D}_{d_{ks}}$. The smoothing parameters ρ, ρ_k, and ρ_s are determined as the ratios of the variance parameter of the random error to the variance parameters of the random effects, σ^2, σ_k^2, and σ_s^2. The variance matrix $\check{\mathbf{G}}$ is a block-diagonal with the matrices $\mathbf{G} = \sigma^2\mathbf{I}_{c-d}$, $\mathbf{G}_k = \sigma_k^2\mathbf{I}_{c-1}$ and $\mathbf{G}_s = \sigma_s^2\mathbf{I}_{c-1}$ on the diagonal.

Now, we move on to the specific procedure. We consider a spline of degree 3, that is, $k = 3$ we set $d = 2$ and $d_{ks} = 1$. Since we consider a one-stage approach, we define the vector $\mathbf{x}_+$ of length 200, which contains 168 components of the vector $\mathbf{x}$ and 32 components of the vector $\mathbf{x}_p$, for which we aim to predict unknown values $\mathbf{y}_p$. Next, using the extended vector of knots of length 27 and the vector $\mathbf{x}_+$, we can construct the collocation matrix for the spline of degree 3 as $\mathbf{C}_4^+(\mathbf{x}_+) \in \mathbb{R}^{200,23}$. Subsequently, we create the matrix $\mathbf{D}_2^+$ and determine the spectral decomposition of the matrix $(\mathbf{D}_2^+)^\top\mathbf{D}_2^+$, which results in the matrices $\mathbf{U}_+$ and $\widetilde{\boldsymbol{\Sigma}}_+$, and thus also $\mathbf{U}_r^+$ and $\boldsymbol{\Sigma}_+$. These matrices are then used to determine the transformation matrix $\boldsymbol{\Omega}_r^+$, which is used to determine the matrix $\mathbf{Z}_+$. We then assemble the matrix $\check{\mathbf{X}}_+$ according to the modified relation (61.7), choosing $p = \pi$, that is, $\omega = 2$, and the matrix $\mathbf{D}_{d_{ks}}^+$, where $d_{ks} = 1$, for which we again find the matrices that form the spectral decomposition of $(\mathbf{D}_1^+)^\top\mathbf{D}_1^+$. Using these, we again construct the transformation matrix denoted as $\check{\boldsymbol{\Omega}}_r^+$, and subsequently determine the matrix $\mathbf{Z}_{ks}^+$. After creating the matrices $\mathbf{K}_+$ and $\mathbf{S}_+$, we determine the matrix $\check{\mathbf{Z}}_+$ according to (61.7). Now we can use the matrices $\check{\mathbf{X}}_+$ and $\check{\mathbf{Z}}_+$ to estimate the variance parameters using the REML method, applying the so-called SOP algorithm described in [7]. Using the code provided in the appendix of [1], we obtain the values $\sigma_\epsilon = 0.1008$, $\sigma^2 = 0.8316$, $\sigma_k^2 = 0.0000$, $\sigma_s^2 = 0.0458$. With these values, we define the matrix $\check{\boldsymbol{\rho}}$ as a diagonal matrix with diagonal elements $\rho = \sigma_\epsilon/\sigma^2$, $\rho_k = \sigma_\epsilon/\sigma_k^2$, $\rho_s = \sigma_\epsilon/\sigma_s^2$. Subsequently, we determine the matrices $\mathbf{G}_+$, $\mathbf{G}_k^+$, and $\mathbf{G}_s^+$ construct the block-diagonal matrix $\check{\mathbf{G}}_+$. Next, we create the matrix $\mathbf{R}_+^{-1}$, so that its diagonal elements are $\frac{1}{\sigma_\epsilon^2}$ at the positions corresponding to the components of the vector $\mathbf{x}$ and 0 at positions corresponding to the components of the vector $\mathbf{x}_p$. From the mixed model equations, [2], we compute the estimates of fixed and random effects, which are then used to estimate the vector $\mathbf{y}_+$ and compute the confidence interval. We can determine the curves for the trend and seasonal components, along with their confidence intervals. The resulting curves, including the original input points, are shown in Fig. 61.2. By comparing the spline fitted to all data with the spline obtained using only the observed data, see Fig. 61.3, we confirmed the validity of the one-stage approach. Most hidden data points fall within the confidence interval, and almost 95% are captured by the prediction intervals. In the next part of this contribution, we apply this approach to real-world data.

61.2.2 Application to Real-World Data

We applied the mixed effects smooth model to real-world data describing the relationship between reflectance and the wavelength of light for 120 trees. Data range from 440.7 to 2337.5 nanometers on the x-axis, but certain intervals have missing

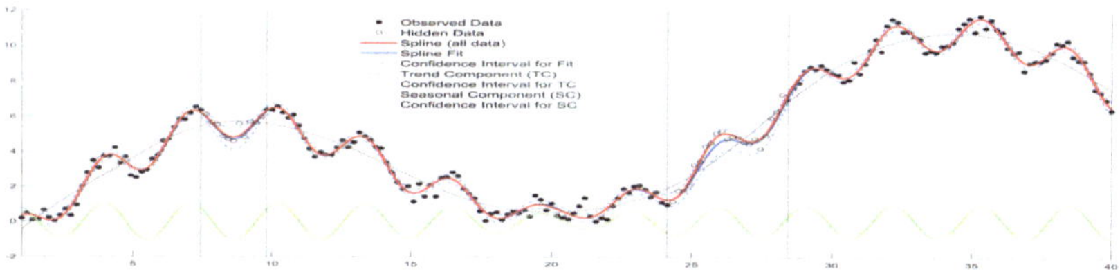

Fig. 61.2: The resulting graph with confidence intervals for the spline, trend and seasonal components.

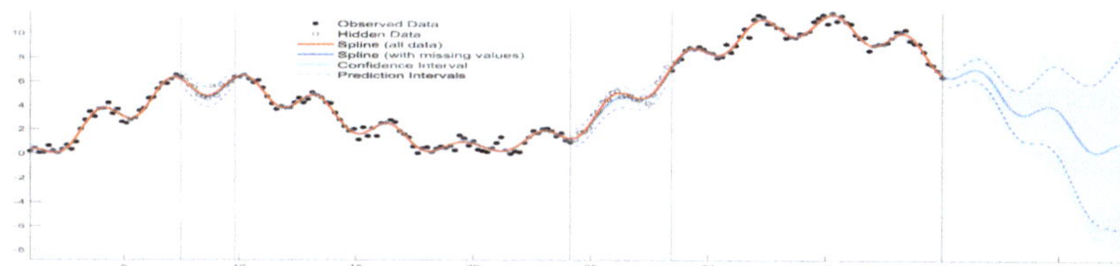

Fig. 61.3: The resulting graph with confidence and prediction intervals for the final spline.

values, as shown in Fig. 61.4. Our goal was to impute missing values and predict future observations using the one-stage approach described in [1].

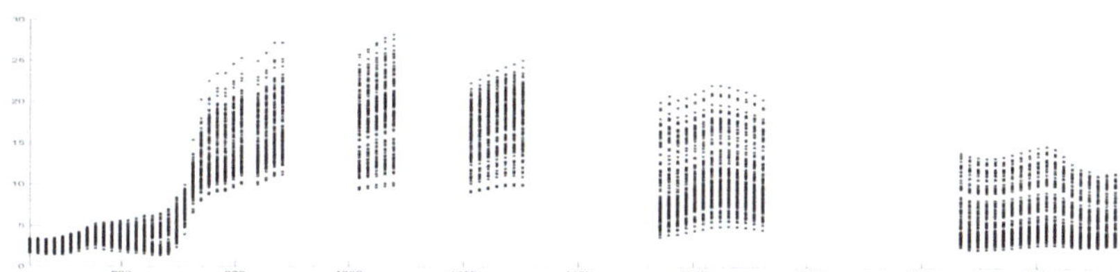

Fig. 61.4: Observed data points.

For demonstration, we first analyze the data for a single object. The dataset consists of vectors $\mathbf{x}$ and $\mathbf{y}$ of length 75. The same model structure as in Sect. 61.2.1 was used, with a trend function combined with periodic components represented by cosine and sine functions (61.5), where $J = 1$ and $\omega = \frac{2\pi}{p}$. We extended the model to include missing values $\mathbf{y}_p$ using the one-stage prediction framework. Testing different values of the period p proved essential, as the true period was not directly observable in the data. After evaluating several values, we selected $p = 200$ for analysis. The estimated variance parameters were obtained using the REML method and the SOP algorithm, yielding the values $\sigma_\epsilon = 0.0700$, $\sigma^2 = 18.7797$, $\sigma_k^2 = 3.3348$, and $\sigma_s^2 = 16.9877$. These parameters were used to calculate the smoothing parameters ρ, ρ_k, and ρ_s. The curves for the trend and seasonal components, along with their confidence intervals for the first object, are shown in Fig. 61.5. This graph can be generated for

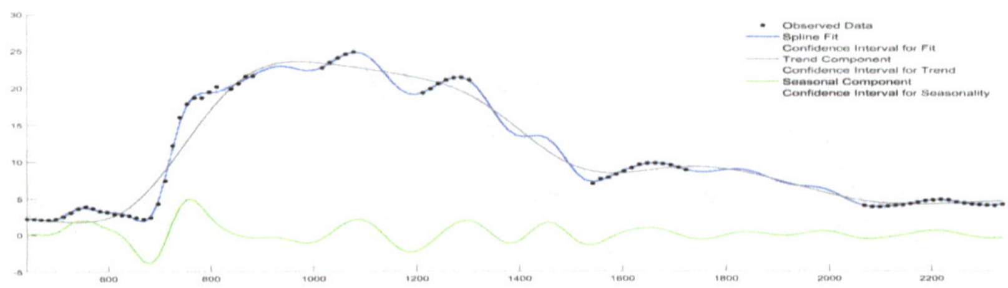

Fig. 61.5: The spline, trend and seasonal components, and confidence intervals for the first object.

each of the 120 objects; however, displaying all results in a single graph would lack clarity. Therefore, we focus only on the resulting splines for each object, which are of primary interest to us, as shown in Fig. 61.6.

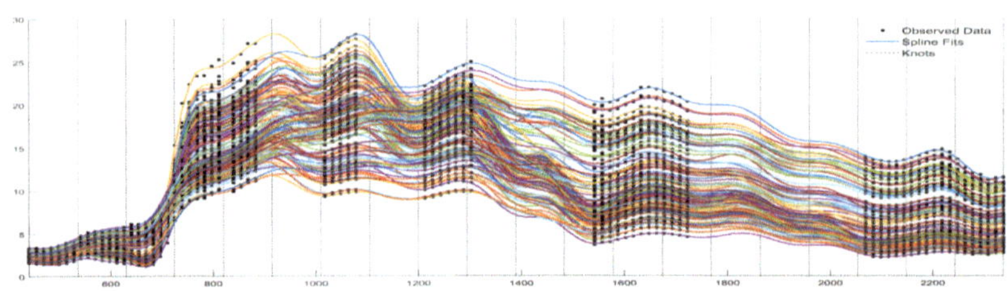

Fig. 61.6: The resulting splines for all objects.

61.3 Conclusion

Mixed effects smooth models provide a comprehensive framework for prediction tasks, particularly in scenarios involving complex dependencies or missing observations. By combining the strengths of P-splines and linear mixed models, they achieve robust, smooth, and interpretable predictions. This contribution demonstrates their theoretical foundation and application to real-world data.

References

1. Carballo, A., Durban, M., Kauermann, G., Lee, D.-J.: A general framework for prediction in penalized regression. Statistical Modelling, **21**(4), 293–312 (2021)
2. Collins, D.: The performance of estimation methods for generalized linear mixed models. University of Wollongong Thesis Collection, (2009)
3. Eilers, P., Gampe, J., Marx, B., Rau, R.: Modulation models for seasonal time series and incidence tables. Statistics in Medicine, **27**, 3430–3441 (2008)
4. Eilers, P., Marx, B.: Practical Smoothing: The Joys of P-splines. Cambridge University Press, Cambridge (2021)
5. Harville, D. A.: Matrix Algebra From a Statistician's Perspective. Springer, New York (1997)
6. Henderson, C. R.: Best Linear Unbiased Estimation and Prediction under a Selection Model. Biometrics, **31**, 423–447 (1975)
7. Rodríguez-Álvarez, M., Durban, M., Dae-Jin, L., Eilers, P.: On the estimation of variance parameters in non-standard generalised linear mixed models: Application to penalised smoothing. Statistics and Computing, **29**, 483–500 (2019)

Chapter 62
Locally Sparse Estimation for Functional Linear Models with Scalar Response

Marco Stefanucci, Mauro Bernardi and Antonio Canale

Abstract In regression models with functional covariates, it is of paramount importance to estimate coefficient functions in an interpretable way. We develop a method able to produce estimated functions which are exactly zero in a subset of positive Lebesgue measure of their domain. We achieve this by combining the B-splines basis with the Overlap Group Lasso penalty, a generalization of the ℓ_2 penalty which allows groups to have common elements. The resulting optimization problem is efficiently solved using the Alternating Direction Method of Multipliers (ADMM) algorithm. The advantage of the proposed method over classical approaches is supported by simulations and a real data analysis.

62.1 Introduction

This work deals with locally sparse estimation in functional linear regression with scalar response. In Functional Data Analysis, a *locally sparse* estimator is a function $\widehat{\psi} \in L_2(\mathcal{T})$, with $\mathcal{T} \subset \mathbb{R}$, that is equal to zero for a subset $\mathcal{T}_0 \subset \mathcal{T}$ with positive Lebesgue measure. The main advantage of such estimators is that null values indicate the absence of relation between input and output variable, leading to a straightforward interpretation of $\widehat{\psi}$.

Marco Stefanucci (✉)
University of Rome Tor Vergata, Via Columbia 2, Rome, Italy,
e-mail: marco.stefanucci@uniroma2.it

Mauro Bernardi
University of Padova, Via Cesare Battisti 241, Padova, Italy,
e-mail: mauro.bernardi@unipd.it

Antonio Canale
University of Padova, Via Cesare Battisti 241, Padova, Italy,
e-mail: antonio.canale@unipd.it

© The Author(s), under exclusive license to Springer Nature Switzerland AG 2025
G. Aneiros et al. (eds.), *New Trends in Functional Statistics and Related Fields*,
Contributions to Statistics, https://doi.org/10.1007/978-3-031-92383-8_62

There exist some attempts in the literature addressing this problem. [6] proposed a flexible estimator, able to produce functions which exhibit zero in one of their derivatives. Their work is based on a generalization of the ℓ_1 (or Lasso) penalty [11]. Along the same line, [8] employ the Lasso penalty on the coefficients of a basis expansion of the regression function. However, this approach does not guarantee that the resulting function has the local sparsity property. Connections between the B-spline basis and local sparsity has been studied by [13] who proposed a two-stage procedure to achieve the desired solution. Finally, [9] introduced an extension of the SCAD penalty [3] to the functional setting.

Here we propose a different approach based on an Overlap Group Lasso penalty [7, 1] and its combination with a B-splines representation [2]. By grouping together adjacent spline coefficients, we encourage them to be all zero or all different from zero, achieving sparsity patterns that are guaranteed to be of positive Lebesgue measure, and cannot be achieved with plain Lasso or Group-Lasso [12] regularization.

62.2 Model

In our modeling formulation, we assume to observe a scalar response variable Y, a functional covariate $X(t)$ and r scalar covariates $Z_1, \dots, Z_r$ for a sample of size n and a regular grid $\mathbf{t} = (t_1, \dots, t_p)$ of size p, resulting in $\mathbf{y} = (y_1, \dots, y_n)^\top$, $\mathbf{X} = (x_1(\mathbf{t}), \dots, x_n(\mathbf{t}))^\top$ and $\mathbf{Z} = (\mathbf{z}_1, \dots, \mathbf{z}_n)^\top$ where $\mathbf{z}_i = (z_{i1}, \dots, z_{ir})^\top$ for all $i = 1, \dots, n$. We further assume a linear model of the form:

$$y_i = \int_{\mathcal{T}} x_i(t)\psi(t)dt + \mathbf{z}_i^\top \boldsymbol{\gamma} + \epsilon_i \quad i = 1, \dots, n, \tag{62.1}$$

where $\psi(t)$ and $\boldsymbol{\gamma}$ are unknown regression coefficients and ϵ_i are independent error terms. An intercept can be incorporated in model (62.1) by including a constant covariate.

In order to define the optimization problem related to our locally sparse estimator, let $\{\varphi_l(t)\}_{l=1}^L$ be a B-spline basis of order d defined without loss of generality in $\mathcal{T} = [0, 1]$ by the knot sequence $\{\tau_l\}_{l=1}^{L-d+2}$ with $\tau_1 = 0$ and $\tau_{L-d+2} = 1$ and $\boldsymbol{\Phi}$ be the $L \times p$ matrix of basis functions evaluations, i.e. $\boldsymbol{\Phi}[lj] = \varphi_l(t_j)$. We approximate $\psi(t)$ with the B-spline basis as $\sum_{l=1}^L \psi_l \varphi_l(t) = \boldsymbol{\psi}^\top \boldsymbol{\varphi}(t)$, its evaluation being $\boldsymbol{\Phi}^\top \boldsymbol{\psi}$. We observe that a single zero in the vector of basis coefficients $\boldsymbol{\psi}$ is not sufficient to induce a function with the local sparsity property. Indeed, there should be exactly d consecutive null coefficients in $\boldsymbol{\psi}$ in order to have $\psi(t) = 0$ in the interval (τ_l, τ_{l+1}).

Motivated by this, and noting that $\int X(t)\psi(t)dt \approx \mathbf{X}\boldsymbol{\Phi}^\top \boldsymbol{\psi} = \mathbf{W}\boldsymbol{\psi}$, we estimate the unknown coefficient function $\psi(t)$ and coefficient vector $\boldsymbol{\gamma}$ based on observed data as

$$(\widehat{\boldsymbol{\psi}}, \widehat{\boldsymbol{\gamma}}) = \arg\min_{\boldsymbol{\psi}, \boldsymbol{\gamma}} \frac{1}{2} \|\mathbf{y} - \mathbf{W}\boldsymbol{\psi} - \mathbf{Z}\boldsymbol{\gamma}\|_2^2 + \lambda \sum_{b=1}^B \|\mathbf{F}_b \boldsymbol{\psi}\|_2, \tag{62.2}$$

where the first part of the objective is the usual sum-of-squares loss and the second part is the Overlap Group-Lasso penalty term. Although this term looks the same as the Group-Lasso regularization, there are important differences between our proposal and the Group-Lasso penalty. First, the B groups involved in equation (62.2) are *overlapping*, as introduced in [7] and [1], meaning that one coefficient can belong to more than one group. In this setting, in order to set a coefficient equal to zero, all groups containing such coefficient should be nullified. Second, traditional algorithm for solving the optimization problem cannot be used in this context due to the non-separability of such penalty. This problem is addressed in the next section.

The construction of the groups follows the nature of the issue we mentioned: in order represent a locally sparse function using B-splines, it is not sufficient to set to zero isolated coefficients (like a Lasso penalty would do) but at least d consecutive coefficients should be nullified. For this reason we group together d consecutive basis coefficients, the generic group being $\boldsymbol{\psi}_b = (\psi_b, \dots, \psi_{b+d-1})$. The $(d \times L)$ matrices $\mathbf{F}_b = \upsilon_b \mathbf{H}_b \mathbf{T}$ include a positive group-specific weight υ_b, a normalized group-specific selection matrix $\mathbf{H}_b$ of dimension $(d \times L)$, and a $(L \times L)$ diagonal weighting matrix $\mathbf{T}$ that allows for non-uniform weights for the variables. In particular, the matrices $\mathbf{H}_b$ can be constructed starting from the not normalized selection matrices having unitary entries corresponding to coefficients included in the b-th group, and zero entries elsewhere. Let

$$\mathbf{S}_1 = \begin{bmatrix}\mathbf{I}_d \ \mathbf{0}_{d,L-d}\end{bmatrix}, \ \dots, \ \mathbf{S}_b = \begin{bmatrix}\mathbf{0}_{d,b-1} \ \mathbf{I}_d \ \mathbf{0}_{d,L-d-b+1}\end{bmatrix}, \ \dots, \ \mathbf{S}_B = \begin{bmatrix}\mathbf{0}_{d,L-d} \ \mathbf{I}_d\end{bmatrix}$$

be such not normalized selection matrices and set $\mathbf{S}^\star = (\sum_{b=1}^{B} \mathbf{S}_b^\top \mathbf{S}_b)^{-1/2}$. $\mathbf{S}^\star$ is a diagonal matrix with entries equal to $\mathbf{S}^\star[ll] = 1/\sqrt{\sum_{b=1}^{B} I_{\{\boldsymbol{\psi}_b \ni \psi_l\}}}$. Due to our peculiar group construction, the first $d-1$ coefficients are under-represented, belonging respectively to only $1, 2, \dots, d-1$ groups while all other coefficient are in d groups (same applies for the last $d-1$ coefficients). Thus, we use this matrix do upweight boundary coefficients and downweight the central ones by computing each $\mathbf{H}_b$ as $\mathbf{H}_b = \mathbf{S}_b \mathbf{S}^\star$. Note that by construction $\sum_{b=1}^{B} \mathbf{H}_b^\top \mathbf{H}_b = \mathbf{I}$.

62.3 Computation

Because of the overlapping nature of the groups, the objective function cannot be separated with respect to the variables and standard algorithms like coordinate descend cannot be applied. For this reason, here we illustrate an optimization procedure based on the ADMM algorithm. ADMM algorithms make use of an artificial variable duplication in order to employ the method of multipliers. It follows that the optimization can be done alternating between the original and duplicated variable in a straightforward manner. This strategy has been successfully employed for a large number of penalized problems. Consider rewriting equation (62.2) as

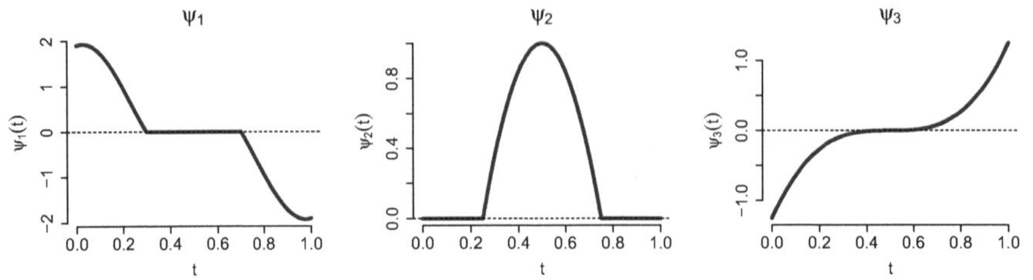

Fig. 62.1: Synthetic regression functions considered in the simulation study

$$(\widehat{\psi},\widehat{\gamma},\widehat{\zeta}) = \arg\min_{\psi,\gamma,\zeta} \frac{1}{2}\|\mathbf{y} - \mathbf{W}\psi - \mathbf{Z}\gamma\|_2^2 + \lambda\sum_{g=1}^{G}\|\zeta_b\|_2 \quad (62.3)$$
$$\text{subject to} \qquad \mathbf{F}\psi = \zeta,$$

where $\mathbf{F} = (\mathbf{F}_1^\top, \dots, \mathbf{F}_B^\top)^\top$ and $\zeta = (\zeta_1^\top, \dots, \zeta_B^\top)^\top$. The matrix $\mathbf{F}$ can be alternatively thought as $\mathbf{F} = \boldsymbol{\Upsilon}\mathbf{HT}$, where $\boldsymbol{\Upsilon} = (\upsilon_1\mathbf{I}_d, \dots, \upsilon_B\mathbf{I}_d)^\top$ and $\mathbf{H} = (\mathbf{H}_1^\top, \dots, \mathbf{H}_B^\top)^\top$. Now the objective appears as a separable problem with respect to the variables, and parameter updates can be done in the following way:

$$\begin{aligned}
\psi^{(t+1)} &= (\mathbf{W}^\top\mathbf{W} + \rho\mathbf{F}^\top\mathbf{F})^{-1}[\mathbf{W}^\top(\mathbf{y} - \mathbf{Z}\gamma^{(t)}) + \rho\mathbf{F}^\top(\zeta^{(t)} - \mathbf{u}^{(t)})],\\
\gamma^{(t+1)} &= (\mathbf{Z}^\top\mathbf{Z})^{-1}[\mathbf{Z}^\top(\mathbf{y} - \mathbf{W}\psi^{(t+1)})],\\
\zeta_b^{(t+1)} &= \mathcal{S}_{\lambda/\rho}(\mathbf{F}_b\psi^{(t+1)} + \mathbf{u}^{(t)}) \qquad\qquad \text{for } b = 1, \dots, B, \qquad (62.4)\\
\mathbf{u}^{(t+1)} &= \mathbf{F}\psi^{(t+1)} - \zeta^{(t+1)} + \mathbf{u}^{(t)},
\end{aligned}$$

where $\mathcal{S}_{\lambda/\rho}(\cdot)$ is the group (or block) soft thresholding operator with parameter λ/ρ, i.e. $\mathcal{S}_\kappa(\mathbf{x}) = (1 - \kappa/\|\mathbf{x}\|_2)_+\mathbf{x}$. Most of the quantities involved in the updates of equation (62.4) can be computed once, notably the terms $(\mathbf{W}^\top\mathbf{W} + \rho\mathbf{F}^\top\mathbf{F})^{-1}$ and $(\mathbf{Z}^\top\mathbf{Z})^{-1}$. The number of iterations of the algorithm can be reduced by allowing ρ to vary at each iteration but this possibility should be balanced with the fact that in this case the quantity $(\mathbf{W}^\top\mathbf{W} + \rho\mathbf{F}^\top\mathbf{F})^{-1}$ should be updated as well, slowing the algorithm. Note that if the group weights and variable weights are constant, a further simplification is $\mathbf{F}^\top\mathbf{F} = \mathbf{I}$. The common case of absence of scalar covariates can be derived from updates in equation (62.4) by setting $\mathbf{Z} = \mathbf{0}$.

62.4 Simulation

In this section we show the empirical performance of the proposed method. We consider three possible regression functions: the first one is

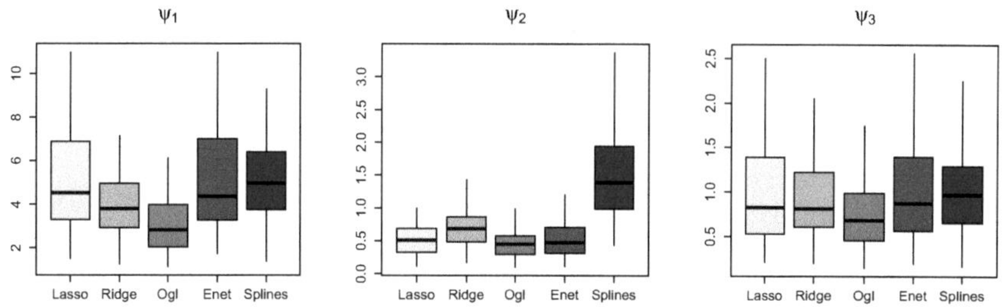

Fig. 62.2: Integrated Squared Error (ISE) over 100 repetitions for the estimation of $\psi_k(t), k = 1, 2, 3$. Ogl = Overlap Group Lasso, Enet = Elastic Net, Splines = Regression Splines

$$\psi_1(t) = \begin{cases} 2(1-t)\sin(2\pi(t+0.2)) & \text{if } 0 \le t \le 0.3, \\ 0 & \text{if } 0.3 \le t \le 0.7, \\ 2t\sin(2\pi(t-0.2)) & \text{if } 0.7 \le t \le 1, \end{cases} \tag{62.5}$$

introduced in [9], which vanishes in the interval $[0.3, 0.7]$; the second one is

$$\psi_2(t) = \begin{cases} 1-(4(t-0.5))^2 & \text{if } 0.25 \le t \le 0.75, \\ 0 & \text{elsewhere,} \end{cases} \tag{62.6}$$

which is a rescaled version of the Epanechnikov kernel, vanishing outside the interval $[0.25, 0.75]$; and the third one is

$$\psi_3(t) = 10(t-0.5)^3, \tag{62.7}$$

which is a cubic function crossing zero at $t = 0.5$. The three functions are presented in Fig.62.1. We set $n = 50$ and generate data $x_1(t), \ldots, x_n(t)$ from a B-spline basis with 25 internal knots, where each basis coefficient is drawn from an independent standard Gaussian random variable, and evaluate such functions on a regular grid in the interval $\mathcal{T} = [0, 1]$. We then generate the response as in model (62.1), with $\gamma = 0$, $\varepsilon_i \sim \mathrm{N}(0, \sigma^2)$, and σ^2 chosen to have a signal-to-noise ratio equal to 4. Unfortunately, none of the competitor models mentioned in the introduction have an available source code, thus we compare our method with alternative penalization schemes such as the ℓ_1 (Lasso), ℓ_2 (Ridge) and mixed ℓ_1/ℓ_2 (Elastic Net, see [14]) employed on basis coefficients, implemented using the `R` package `glmnet`. We also add a classical regression splines estimator [10]. Our locally sparse estimator can be implemented through the function `f2sSP` of the `R` package `fdaSP`, available on `CRAN`. We set the number of basis functions equal to 15 and choose the optimal value of the tuning parameter by minimizing a cross-validated prediction error.

We repeat this procedure 100 times and compute the Integrated Squared Error (ISE) in estimating the true function $\psi_k(t), k = 1, 2, 3$. Fig.62.2 summarizes the

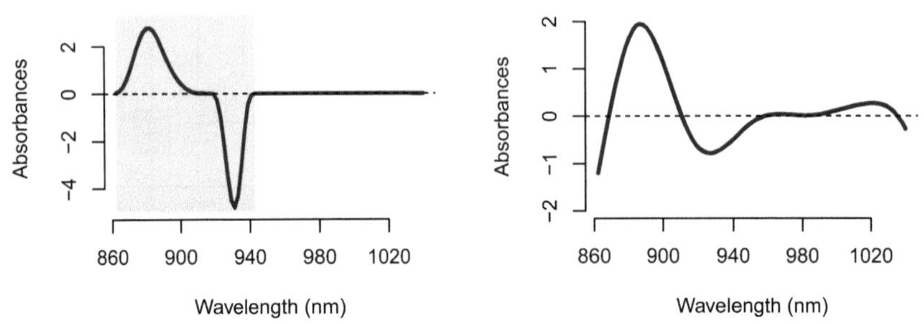

Fig. 62.3: Results on the Tecator dataset. Left: locally sparse regression. Right: spline regression

results. For all regression functions, our procedure attains the lowest ISE on average, resulting also in the least variable procedure among repetitions. For $\psi_1(t)$ and $\psi_2(t)$ we compute also the ISE related to the true null ($\mathcal{T}_0$) and non-null ($\mathcal{T}_1$) region (not reported here) and observe that the ℓ_1 penalty compares similarly to us in $\mathcal{T}_0$ but fails to recover the true function in $\mathcal{T}_1$, while the converse applies to regression splines and ridge approaches.

62.5 Application

The proposed estimator has been applied to the well-known Tecator dataset [4]. This dataset contains the spectral measurements of light absorbance at different wavelengths ω in the interval $\Omega = [850, 1050]$ nm for each of $n = 215$ meat samples. Additional information regarding the protein content is available through chemical analyses of the meat. Here, for each subject, we treat the spectral information as the independent variable $x_i(\omega)$ and the protein content as the dependent variable y_i, postulating a model of the kind

$$y_i = \gamma_0 + \int_\Omega x_i(\omega)\psi(\omega)d\omega + \varepsilon_i \quad i = 1, \ldots, n, \tag{62.8}$$

where γ_0 and $\psi(\omega)$ should be estimated from data and ε_i are independent errors with zero mean. The optimal value of the tuning parameter has been chosen by minimizing the cross-validated prediction error and then applying the 1-standard error rule [5]. The resulting estimate $\widehat{\psi}(\omega)$ is displayed in Fig.62.3, together with a regression spline estimator computed using the `refund` package. Compared to the spline estimate, the proposed estimator shows a similar behaviour in the first part of the domain, where the two most relevant peaks are recognized, but a different behaviour in the second part. Specifically, for wavelengths larger than 940 nm the estimate is equal to zero, reflecting absence of relevant association between the covariate and the response. This result highlights the prominent role that wavelengths in the interval $\Omega_1 = [850, 940]$ have in the determination of the protein content of the meat. Additionally, a practical consequence would be the utilization of instruments capable to track in the signal in Ω_1 only, since the protein content can be properly predicted by using this subset of the domain.

Acknowledgements The first author is supported by the MUR–PRIN 2022 project "CARONTE" (2022KBTEBN, funded by the European Union, NextGeneration EU).

References

1. Bach, F., Jenatton, R., Mairal, J. and Obozinski G.: Optimization with Sparsity-Inducing Penalties. Foundations and Trends® in Machine Learning, **4**, 1, 1–106 (2012)
2. Bernardi, M., Canale, A. and Stefanucci, M.: Locally Sparse Function-on-Function Regression. Journal of Computational and Graphical Statistics, **32**, 3, 985–999 (2023)
3. Fan, J. and Li, R.: Variable Selection via Nonconcave Penalized Likelihood and Its Oracle Properties. Journal of the American Statistical Association, **96**, 1348–1360 (2001)
4. Ferraty, F. and Vieu, P.: Nonparametric Functional Data Analysis: Theory and Practice. Springer-Verlag, Berlin. (2006)
5. Hastie, T., Tibshirani, R. and Friedman, J.: The Elements of Statistical Learning, 2nd edition. Springer, New York. (2009)
6. James, G. M., Wang, J. and Zhu, J.: Functional linear regression that's interpretable. The Annals of Statistics, **37**, 2083–2108. (2009)
7. Jenatton, R., Audibert, J. and Bach, F.: Structured variable selection with sparsity-inducing norms. The Journal of Machine Learning Research, **12**, 2777-2824. (2011)
8. Lee, E. R. and Park, B. U.: Sparse estimation in functional linear regression. Journal of Multivariate Analysis, **105**, 1, 1–17. (2012).
9. Lin, Z., Cao, J., Wang, L. and Wang, H.: Locally sparse estimator for functional linear regression models. Journal of Computational and Graphical Statistics, **26**, 2, 306–318. (2017)
10. Ramsay, J.O. and Silverman, B.W.; Functional data analysis, Springer, New York. (2005)
11. Tibshirani, R.: Regression shrinkage and selection via the lasso. Journal of the Royal Statistical Society: Series B (Statistical Methodology), **58**, 1, 267–288. (1996)
12. Yuan, M. and Lin, Y.: Model selection and estimation in regression with grouped variables. Journal of the Royal Statistical Society: Series B (Statistical Methodology), **68**, 1, 49–67. (2006)
13. Zhou, J., Wang, N.Y. and Wang, N.: Functional linear model with zero-value coefficient function at sub-regions. Statistica Sinica, **23**, 1, 25–50. (2013)
14. Zou, H. and Hastie, T.: Regularization and Variable Selection Via the Elastic Net. Journal of the Royal Statistical Society: Series B (Statistical Methodology), **67**, 2, 301–320. (2005)

Chapter 63
Factor Models for High-Dimensional Functional Time Series

Shahin Tavakoli, Gilles Nisol, and Marc Hallin

Abstract We set up the theoretical foundations for a high-dimensional functional factor model approach in the analysis of large cross-sections (panels) of functional time series (FTS). We first establish a representation result stating that, under mild assumptions on the covariance operator of the cross-section, we can represent each FTS as the sum of a common component driven by scalar factors loaded via functional loadings, and a mildly cross-correlated idiosyncratic component. Our model and theory are developed in a general Hilbert space setting that allows for mixed panels of functional and scalar time series. We then turn to the identification of the number of factors, and the estimation of the factors, their loadings, and the common components. We provide a family of information criteria for identifying the number of factors, and prove their consistency. We provide average error bounds for the estimators of the factors, loadings, and common component; our results encompass the scalar case, for which they reproduce and extend, under weaker conditions, well-established similar results. Under slightly stronger assumptions, we also provide uniform bounds for the estimators of factors, loadings, and common component, thus extending existing scalar results. Our consistency results in the asymptotic regime where the number N of series and the number T of time observations diverge thus extend to the functional context the "blessing of dimensionality" that explains the success of factor models in the analysis of high-dimensional (scalar) time series.

Shahin Tavakoli (✉)
Research Institute for Statistics and Information Science
University of Geneva, Switzerland, e-mail: `shahin.tavakoli@unige.ch`

Gilles Nisol
ECARES and Département de Mathématique
Université libre de Bruxelles, Belgium, e-mail: `gnisol@gmail.com`

Marc Hallin
ECARES and Département de Mathématique
Université libre de Bruxelles, Belgium, e-mail: `mhallin@ulb.ac.be`

© The Author(s), under exclusive license to Springer Nature Switzerland AG 2025

G. Aneiros et al. (eds.), *New Trends in Functional Statistics and Related Fields*,
Contributions to Statistics, https://doi.org/10.1007/978-3-031-92383-8_63

63.1 Introduction

Throughout the last decades, researchers have been dealing with datasets of increasing size and complexity. In particular, Functional Data Analysis (see, e.g. [15, 4]) has received much interest. In Functional Data Analysis, the observations are taking values in some functional space, usually some Hilbert space H—often, in practice, the space $L^2([0,1],\mathbb{R})$ of real-valued squared-integrable functions. When an ordered sequence of functional observations exhibits serial dependence, we enter the realm of *Functional Time Series* (FTS) [9]. Many standard univariate and low-dimensional multivariate time-series methods have been adapted to this functional setting, either using a time-domain approach [10], a frequency domain approach under stationarity assumptions [14, 13, 11, 12] or under local stationarity assumptions [20].

Parallel to this development of functional time series analysis, data in high dimensions have become pervasive in data sciences and related disciplines where, under the name of Big Data, they constitute one of the most active subjects of contemporary statistical research.

This contribution, which is an extended abstract of Hallin et al. [8], Tavakoli et al. [19], stands at the intersection of these two strands of literature, cumulating the challenges of function-valued observations and those of high dimension. Datasets, in this context, consist of large collections of N scalar or functional time series—equivalently, scalar or functional time series in high dimension (from fifty, say, to several hundreds)—observed over a period of time T. Typical examples are continuous-time series of concentrations for a large number of pollutants, or/and collected daily over a large number of sites, daily series of returns observed at high intraday frequency for a large collection of stocks (see, e.g., Fig. 63.1), or intraday energy consumption curves[1], to name only a few. Not all component series in the dataset are required to be function-valued, though, and it is very important for practice that mixed panels of scalar and functional series can be considered as well, such as the following:

$$
\begin{array}{llllllll}
\texttt{(AAPL intraday)} & x_{11} & x_{12} & \cdots & x_{1t} & \cdots & x_{1T} \\
\texttt{(Yield Curve)} & x_{21} & x_{22} & \cdots & x_{2t} & \cdots & x_{2T} \\
\texttt{(daily returns)} & x_{31} & x_{32} & \cdots & x_{3t} & \cdots & x_{3T} \\
\vdots & \vdots & \vdots & \ddots & \vdots & \ddots & \vdots \\
\texttt{(Overnight return)} & x_{i1} & x_{i2} & \cdots & x_{it} & \cdots & x_{iT} \\
\vdots & \vdots & \vdots & \ddots & \vdots & \ddots & \vdots \\
 & x_{N1} & x_{N2} & \cdots & x_{Nt} & \cdots & x_{NT}
\end{array}
$$

In order to model such datasets, we develop a class of *high-dimensional functional factor models* inspired by the factor model approaches developed, mostly, in time series econometrics [16, 17, 1], which have proven effective, flexible, and quite

[1] available, for instance, at `data.london.gov.uk/dataset/smartmeter-energy-use-data-in-london-households`

efficient in the scalar case. We start (Sect. 63.2.3) with a stochastic representation result establishing the existence, under a functional version of the traditional assumptions on dynamic panels, of a factor model representation of the observations with scalar factors and functional loadings. We then develop an estimation theory for the unobserved factors, loadings, and common components (Sect. 63.3), followed (in Sect. 63.4) by theoretical results. In Sect. 63.5 we provide numerical experiments and an application to mortality curve data. Since our methods can deal with mixed panels of scalar and functional time series, our results encompass and extend, sometimes under weaker assumptions, the classical ones [2, 1, 16, 3].

63.2 A Representation Theorem

63.2.1 Notation

Throughout, we denote by

$$\mathcal{X}_{N,T} := \{x_{it},\ i = 1, \dots, N,\ t = 1, \dots, T\} \quad N \in \mathbb{N},\ T \in \mathbb{N} \qquad (63.1)$$

an observed $N \times T$ panel of time series, where the random elements $\{x_{it} : t \geq 1\}$ take values in a real separable Hilbert space H_i equipped with the inner product $\langle \cdot, \cdot \rangle_{H_i}$ and norm $\|\cdot\|_{H_i} := \langle \cdot, \cdot \rangle_{H_i}^{1/2}$, $i = 1, \dots, N$. These series can be of different types. A case of interest is the one for which some series are scalar ($H_i = \mathbb{R}$) and some others are square-integrable functions from some closed real interval to $\mathbb{R}$, that is, $H_i = L^2([a_i, b_i], \mathbb{R})$, where $a_i, b_i \in \mathbb{R}$ with $a_i < b_i$. The typical inner-product on $L^2([a_i, b_i], \mathbb{R})$ is $\langle f, g \rangle_{H_i} := \int_{a_i}^{b_i} f(\tau) g(\tau) d\tau$ for $f, g \in L^2([a_i, b_i], \mathbb{R})$. While the results could be derived for these special cases only, the generality of our approach allows the practitioner to choose the Hilbert spaces $H_1, \dots, H_N$ to suit their specific application[2]. It is also important to note that our abstract Hilbert space notation—as opposed to assuming them to be the standard $L^2([0, 1], \mathbb{R})$ space—emphasizes that one cannot take sums of the data across the index i since the random elements are of distinct nature across i (e.g., they could be intraday stock prices for $i = 1$ but yield curves for $i = 2$, making their sum meaningless).

We tacitly assume that all x_{it}'s are random elements with mean zero and finite second-order moments defined on some common probability space $(\Omega, \mathcal{F}, \mathbb{P})$; we also assume that $\mathcal{X}_{N,T}$ constitutes the finite realization of some second-order t-stationary double-indexed process $\mathcal{X} := \{x_{it},\ i \in \mathbb{N},\ t \in \mathbb{Z}\}$.

Define $\boldsymbol{H}_N := H_1 \oplus H_2 \oplus \cdots \oplus H_N$ as the (orthogonal) direct sum of the Hilbert spaces $H_1, \dots, H_N$. The elements of $\boldsymbol{H}_N$ are of the form $\boldsymbol{v} := (v_1, v_2, \dots, v_N)^\top$ or $\boldsymbol{w} := (w_1, w_2, \dots, w_N)^\top$, where $v_i, w_i \in H_i, i = 1, \dots, N$, and $\cdot^\top$ denotes transposition. The i-th component of an element of $\boldsymbol{H}_N$ is denoted by $(\cdot)_i$; for instance, $(\boldsymbol{v})_i$ stands for v_i. The space $\boldsymbol{H}_N$, naturally equipped with the inner product

[2] For instance, a different inner product could be chosen for the spaces $L^2([a_i, b_i], \mathbb{R})$, or H_i could be chosen to be a reproducing kernel Hilbert space.

$$\langle \boldsymbol{v}, \boldsymbol{w} \rangle_{\boldsymbol{H}_N} := \sum_{i=1}^{N} \langle v_i, w_i \rangle_{H_i},$$

is a real separable Hilbert space. When no confusion is possible, we write $\langle \cdot, \cdot \rangle$ for $\langle \cdot, \cdot \rangle_{\boldsymbol{H}_N}$ and $\|\cdot\| := \langle \cdot, \cdot \rangle^{1/2}$ for the resulting norm. Let $\mathcal{L}(H_1, H_2)$ denote the space of bounded (linear) operators from H_1 to H_2, and use the shorthand notation $\mathcal{L}(H)$ for $\mathcal{L}(H, H)$. For $u_1 \in H_1, u_2 \in H_2$, define $u_1 \otimes u_2 \in \mathcal{L}(H_2, H_1)$ by $(u_1 \otimes u_2)(v_2) = \langle u_2, v_2 \rangle_{H_2} u_1$ for all $v_2 \in H_2$. If $\alpha \in \mathbb{R}, u \in H$, we define $u\alpha := \alpha u$. We shall denote by $\mathbb{R}^{p \times q}$ the space of $p \times q$ real matrices, and will identify this space with the space of mappings $\mathcal{L}(\mathbb{R}^q, \mathbb{R}^p)$. We write $\mathbb{N}$ for the set of positive integers, that is, $\mathbb{N} := \{1, 2, \ldots\}$, and denote the Euclidean norm in $\mathbb{R}^p$ by $|\cdot|$.

When working with the panel data, let $\boldsymbol{x}_t := (x_{1t}, x_{2t}, \ldots, x_{Nt})^\top \in \boldsymbol{H}_N$ where, in order to keep the presentation simple, the dependence on N does not explicitly appear. We denote by $\lambda^x_{N,1}, \lambda^x_{N,2}, \ldots$ the eigenvalues, in decreasing order of magnitude, of $\boldsymbol{x}_t$'s covariance operator $\mathbb{E}\left[(\boldsymbol{x}_t - \mathbb{E}\boldsymbol{x}_t) \otimes (\boldsymbol{x}_t - \mathbb{E}\boldsymbol{x}_t)\right] \in \mathcal{L}(\boldsymbol{H}_N)$, which maps $\boldsymbol{y} \in \boldsymbol{H}_N$ to

$$\mathbb{E}\left[\langle \boldsymbol{x}_t - \mathbb{E}\boldsymbol{x}_t, \boldsymbol{y} \rangle (\boldsymbol{x}_t - \mathbb{E}\boldsymbol{x}_t)\right] \in \boldsymbol{H}_N.$$

In view of stationarity, these eigenvalues do not depend on t. Unless otherwise mentioned, convergence of sequences of random variables is in mean square.

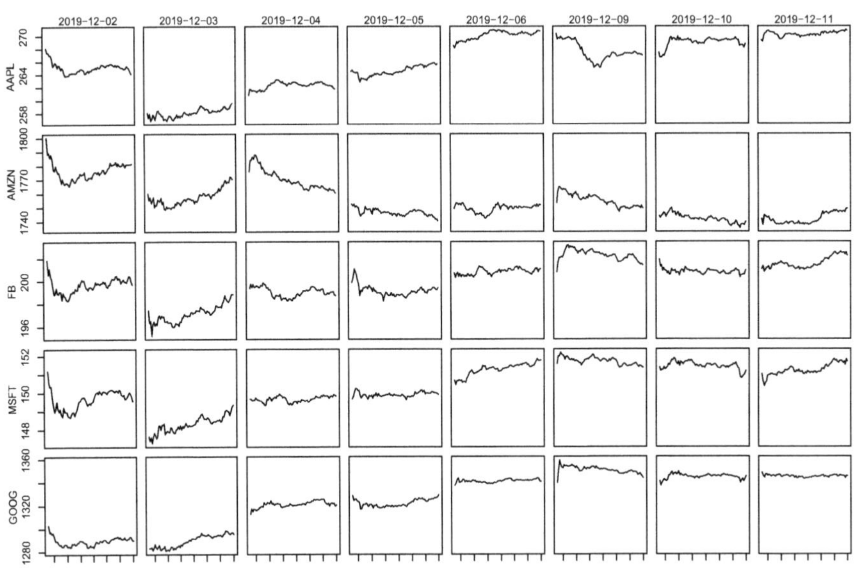

Fig. 63.1: Functional time series of intra-day values for 5 stocks over several consecutive opening days. The data was obtained at https://www.alphavantage.co/.

63.2.2 A Functional Factor Model

The basic idea in all factor-model approaches to the analysis of high-dimensional time series consists in decomposing the observation x_{it} into the sum $\chi_{it} + \xi_{it}$ of two unobservable and mutually orthogonal components, the *common* component χ_{it} and the *idiosyncratic* one ξ_{it}.

Definition 63.2.1 *The functional zero-mean second-order stationary stochastic process*

$$\mathcal{X} := \{x_{it} : i \in \mathbb{N}, \ t \in \mathbb{Z}\}$$

admits a (high-dimensional) functional factor model representation *with* $r \in \mathbb{N}$ factors *if there exist* loadings $b_{il} \in H_i$, $i \in \mathbb{N}$, $l = 1, \ldots, r$, *H_i-valued processes* $\{\xi_{it} : t \in \mathbb{Z}\}$, $i \in \mathbb{N}$, *and a real r-dimensional second-order stationary process* $\{\boldsymbol{u}_t = (u_{1t}, \ldots, u_{rt})^\top : t \in \mathbb{Z}\}$, *co-stationary with $\mathcal{X}$, such that*

$$x_{it} = \chi_{it} + \xi_{it} = b_{i1}u_{1t} + b_{i2}u_{2t} + \cdots + b_{ir}u_{rt} + \xi_{it}, \qquad i \in \mathbb{N}, \ t \in \mathbb{Z} \qquad (63.2)$$

(χ_{it} and ξ_{it} unobservable) holds with

(i) $\mathbb{E}\boldsymbol{u}_t = \boldsymbol{0}$ *and* $\mathbb{E}\left[\boldsymbol{u}_t\boldsymbol{u}_t^\top\right] = \boldsymbol{\Sigma}_{\boldsymbol{u}} \in \mathbb{R}^{r\times r}$ *is positive definite;*
(ii) $\mathbb{E}\left[u_{lt}\xi_{it}\right] = 0$ *and* $\mathbb{E}\left[\xi_{it}\right] = 0$ *for all* $t \in \mathbb{Z}$, $l = 1, \ldots, r$, *and* $i \in \mathbb{N}$*;*
(iii) denoting by $\lambda^{\xi}_{N,l}$ *the l-th (in decreasing order of magnitude) eigenvalue of the covariance operator* $\mathbb{E}\left[\boldsymbol{\xi}_t \otimes \boldsymbol{\xi}_t\right]$ *of* $\boldsymbol{\xi}_t := (\xi_{1t}, \ldots, \xi_{Nt})^\top$, $\lambda^{\xi}_1 := \sup_N \lambda^{\xi}_{N,1} < \infty$*;*
(iv) denoting by $\lambda^{\chi}_{N,l}$ *the l-th (in decreasing order of magnitude) eigenvalue of the covariance operator* $\mathbb{E}\left[\boldsymbol{\chi}_t \otimes \boldsymbol{\chi}_t\right]$ *of* $\boldsymbol{\chi}_t := (\chi_{1t}, \ldots, \chi_{Nt})^\top$, $\lambda^{\chi}_r := \sup_N \lambda^{\chi}_{N,r} = \infty$.

The r scalar random variables u_{lt} are called factors*; the b_{il}'s are the* (functional) factor loadings*; χ_{it} and ξ_{it} are called x_{it}'s* common *and* idiosyncratic component, *respectively.*

The functional factor representation (63.2) also can be written, with obvious notation $\boldsymbol{\chi}_t$ and $\boldsymbol{\xi}_t$, in vector form

$$\boldsymbol{x}_t = \boldsymbol{\chi}_t + \boldsymbol{\xi}_t = \boldsymbol{B}_N\boldsymbol{u}_t + \boldsymbol{\xi}_t,$$

where $\boldsymbol{B}_N \in \mathcal{L}(\mathbb{R}^r, \boldsymbol{H}_N)$ is the linear operator mapping the l-th canonical basis vector of $\mathbb{R}^r$ to $(b_{1l}, b_{2l}, \ldots, b_{Nl})^\top \in \boldsymbol{H}_N$.

It is important to note that our (high-dimensional) functional factor model differs from other functional factor models [6, 7, 18] because ours allows different type of (functional) data across the index i, whereas the competitor models assume all the data (across i) fall into the same Hilbert space.

63.2.3 Representation Theorem

The following results shows that, under adequate condiions on its covariance operator, a process $\mathcal{X}$ admits a functional factor model representation (in the sense of Definition 63.2.1) with $r \in \mathbb{N}$ factors. These conditions are functional versions of the conditions considered in the scalar case, and involve the eigenvalues $\lambda^x_{N,j}$ of the covariance operator of the observations $\boldsymbol{x}_t$—while Definition 63.2.1 involves the eigenvalues $\lambda^\chi_{N,j}$ and $\lambda^\xi_{N,j}$ of the covariance operators of unobserved common and idiosyncratic components. Moreover, when $\mathcal{X}$ admits a functional factor model representation of the form (63.2), its decomposition into a common and an idiosyncratic component is unique.

Let $\lambda^x_j := \lim_{N\to\infty} \lambda^x_{N,j} = \sup_N \lambda^x_{N,j}$. This limit (which is possibly infinite) exists, as $\lambda^x_{N,j}$ is monotone increasing with N.

Theorem 63.2.2 *The process $\mathcal{X}$ admits a (high-dimensional) functional factor model representation with r factors if and only if $\lambda^x_r = \infty$ and $\lambda^x_{r+1} < \infty$.*

The common component χ_{it} is asymptotically identifiable, see Hallin et al. [8] for details. The proof of the Theorem is inspired from [5]. Notice, however, that, unlike this reference, our results do not require the minimal eigenvalue of the covariance of $\boldsymbol{x}_t$ to be bounded from below—an unreasonable assumption in the functional setting.

63.3 Estimation

Assuming that a functional factor model with $r \in \mathbb{N}$ factors holds for $\mathcal{X}$, we shall estimate the factors $\boldsymbol{u}_t$ and loading operator $\boldsymbol{B}_N$ via a least squares criterion.

Since the true number r of factors is unknown, we present estimators of the loadings, factors and common component for various $k \geq 1$, and then provide information criteria for estimating r. For given k, our estimators $\boldsymbol{B}^{(k)}_N$ of the factor loadings and $\boldsymbol{U}^{(k)}_T = (\boldsymbol{u}^{(k)}_1, \ldots, \boldsymbol{u}^{(k)}_T)$ of the factor values are the minimizers, over $\mathcal{L}(\mathbb{R}^k, \boldsymbol{H}_N)$ and $\mathbb{R}^{k\times T}$, respectively, of

$$P\left(\boldsymbol{B}^{(k)}_N, \boldsymbol{U}^{(k)}_T\right) := \sum_t \left\|\boldsymbol{x}_t - \boldsymbol{B}^{(k)}_N \boldsymbol{u}^{(k)}_t\right\|^2. \tag{63.3}$$

Our method for solving this minimization problem is given in Tavakoli et al. [19, Proposition 2.1], and results in the estimators $\widetilde{\boldsymbol{B}}^{(k)}_N$ and $\widetilde{\boldsymbol{U}}^{(k)}_T$.

The information criteria that we will use for identifying the number of factors are of the form

$$\mathrm{IC}(k) := V(k, \widetilde{\boldsymbol{U}}^{(k)}_T) + k\, g(N,T), \tag{63.4}$$

which is the sum of a goodness-of-fit measure

$$V(k,\widetilde{\boldsymbol{U}}_T^{(k)}) := \min_{\boldsymbol{B}_N^{(k)} \in \mathcal{L}(\mathbb{R}^k, \boldsymbol{H}_N)} \frac{1}{NT} \sum_{t=1}^{T} \left\| \boldsymbol{x}_t - \boldsymbol{B}_N^{(k)} \tilde{\boldsymbol{u}}_t^{(k)} \right\|^2 \tag{63.5}$$

($\tilde{\boldsymbol{u}}_t^{(k)} \in \mathbb{R}^k$ is the t-th column of $\widetilde{\boldsymbol{U}}_T^{(k)}$) and a penalty term $k\,g(N,T)$. For a given set $\{x_{it} : i = 1, \ldots, N; t = 1, \ldots, T\}$ of observations, we estimate the number of factors by

$$\hat{r} := \arg \min_{k=1,\ldots,k_{\max}} \mathrm{IC}(k), \tag{63.6}$$

where $k_{\max} < \infty$ is some pre-specified maximal number of factors. The next section provides consistency results for $\hat{r}$.

63.4 Theoretical Results

Consistency results for $\hat{r}$ and the estimators described in the previous section require regularity assumptions, which are functional versions of standard assumptions in scalar factor models (see, e.g., [1]). These are available as Assumptions A, B, C, and D in Tavakoli et al. [19].

The following result gives sufficient conditions on the penalty term $g(N,T)$ in (63.4) for the estimated number of factors (63.6) to be consistent as both N and T go to infinity.

Theorem 63.4.1 *Let* $C_{N,T} := \min\{\sqrt{N}, \sqrt{T}\}$ *and consider* $\hat{r}$ *as defined in* (63.6). *Under Assumptions A, B, C, and D, if* $g(N,T)$ *is such that*

$$g(N,T) \to 0 \quad \text{and} \quad C_{N,T}\,g(N,T) \to \infty \quad \text{as} \quad C_{N,T} \to \infty,$$

then $\mathbb{P}(\hat{r} = r) \longrightarrow 1$ *as N and T go to infinity.*

In addition to these theoretical results, Tavakoli et al. [19] presents average and uniform error bounds for the loadings, factors, and common component.

63.5 Numerical experiments and application

Some numerical experiments and an application to forecasting Japanese mortality curves are provided in Tavakoli et al. [19].

References

1. J. Bai and S. Ng: Determining the number of factors in approximate factor models. *Econometrica*, **70**(1), 191–221 (2002)
2. G. Chamberlain and M. Rothschild: Arbitrage, factor structure, and mean-variance analysis on large asset markets. *Econometrica*, **51**, 1305–1324, (1983)
3. J. Fan, Y. Liao, and M. Mincheva: Large covariance estimation by thresholding principal orthogonal complements. *J. Roy. Stat. Soc. B*, **75**(4), 603–680, (2013)

4. F. Ferraty and P. Vieu: Nonparametric Functional Data Analysis: Theory and Practice. Springer, 2006.
5. M. Forni and M. Lippi: The generalized dynamic factor model: Representation theory. *Economet. Theor.*, **17**(06), 1113–1141, (2001)
6. Y. Gao, H. L. Shang, and Y. Yang: High-dimensional functional time series forecasting: An application to age-specific mortality rates. *J. Multivariate Anal.*, **170**, 232–243, (2018)
7. S. Guo, X. Qiao, Q. Wang, and Z. Wang: Factor modelling for high-dimensional functional time series, ArXiv: 2112.13651 (2024)
8. M. Hallin, G. Nisol, and S. Tavakoli: Factor models for high-dimensional functional time series i: Representation results. *J. Time Ser. Anal.*, **44**(5-6), 578–600, (2023)
9. S. Hörmann and P. Kokoszka: Weakly dependent functional data. *Ann. Stat.*, **38**, 1845–1884, (2010)
10. S. Hörmann, L. Horváth, and R. Reeder: A functional version of the ARCH model. *Economet. Theor.*, **29**, 267–288, (2013)
11. S. Hörmann, L. Kidziński, and M. Hallin: Dynamic functional principal components. *J. Roy. Stat. Soc. B*, **77**, 319–348, (2015)
12. S. Hörmann, P. Kokoszka, and G. Nisol: Testing for periodicity in functional time series. *Ann. Stat.*, **46**(6A), 2960–2984, (2018)
13. V. M. Panaretos and S. Tavakoli: Fourier analysis of stationary time series in function space. *Ann. Stat.*, **41**(2), 568–603, (2013)a
14. V. M. Panaretos and S. Tavakoli: Cramér–Karhunen–Loève representation and harmonic principal component analysis of functional time series. *Stoch. Proc. Appl.*, **123**(7), 2779–2807, (2013)b
15. J. O. Ramsay and B. W. Silverman: Functional Data Analysis. Springer Series in Statistics. Springer, New York, second edition, 2005.
16. J. H. Stock and M. W. Watson: Forecasting using principal components from a large number of predictors. *J. Am. Stat. Assoc.*, **97**(460), 1167–1179, (2002)a
17. J. H. Stock and M. W. Watson. Macroeconomic forecasting using diffusion indexes. *J. Bus. Econ. Stat.*, **20**(2), 147–162, (2002)b
18. C. Tang, H. L. Shang, Y. Yang, and Y. Yang: Forecasting high-dimensional functional time series with dual-factor structures, ArXiv: 2109.04146 (2024)
19. S. Tavakoli, G. Nisol, and M. Hallin: Factor models for high-dimensional functional time series ii: Estimation and forecasting. *J. Time Ser. Anal.*, **44**(5-6), 601–621, (2023)
20. A. van Delft and H. Dette: A similarity measure for second order properties of non-stationary functional time series with applications to clustering and testing. *Bernoulli*, **27**(1), 469–501, (2021)

Chapter 64
Anomaly-Driven Visualization of Functional Data

Romain Valla, Pavlo Mozharovskyi, and Florence d'Alché-Buc

Abstract Functional data analysis recently became a widely acceptable remedy for handling massive amounts of data measurable as one or several variables dependent on a certain argument, with insightful low-dimensional representations of this intrinsically infinite-dimensional phenomenon increasingly gaining popularity. While most of these are based on classical dimension reduction methods, like, *e.g.*, functional principle component analysis, they often lack robustness and thus experience difficulties in correctly visualizing abnormal observations. We introduce a novel method for visualization of functional data, focused on anomalies: functional abnormal component analysis (fACA). Extending already existing—for the Euclidean space—counterpart, we need to overcome difficulties of the functional setting. We showcase the applicability of the developed method within an existing taxonomy of anomalies, on real-world data sets.

64.1 Introduction

Due to rapid development and ubiquitous deployment of means for acquisition of high-frequency data, functional data analysis (FDA) [18] (see also [10]) gained popularity in recent decades. These volumes of data and the development of novel standards emphasize the importance of detecting anomalies [2] in functional data.

Romain Valla
LTCI, Telecom Paris, Institut Polytechnique de Paris, France
e-mail: romain.valla@telecom-paris.fr

Pavlo Mozharovskyi (✉)
LTCI, Telecom Paris, Institut Polytechnique de Paris, France
e-mail: pavlo.mozharovskyi@telecom-paris.fr

Florence d'Alché-Buc
LTCI, Telecom Paris, Institut Polytechnique de Paris, France
e-mail: florence.dalche@telecom-paris.fr

© The Author(s), under exclusive license to Springer Nature Switzerland AG 2025
G. Aneiros et al. (eds.), *New Trends in Functional Statistics and Related Fields*, Contributions to Statistics, https://doi.org/10.1007/978-3-031-92383-8_64

Meanwhile, the simple strategy of first projecting the data onto a finite-dimensional basis and then employing there a multivariate anomaly detection tool faced a major problem of identifying and tuning the correct basis: a task particularly impeded by the presence of anomalies.

As a remedy, together with not that numerous machine learning methods for handling data directly in the functional space (like, *e.g.*, functional isolation forest [21]), depth-based (see [15] for a recent survey, but also [23]) methods witnessed high performance at anomaly detection tasks [12, 20] (see also [17, 11] for a review on functional data depth).

Among the existing variety, meaningful visualization and explanation of anomalies remain a challenging task, requiring case-wise solutions. In the present work, we introduce the *functional abnormal component analysis* (shortly fACA), which constitutes a non-trivial extension of its multivariate counterpart [22], ACA.

64.2 Taxonomy of Functional Anomalies

Probably the main distinguishing feature of the functional setting, compared to the d-variate Euclidean space $\mathbb{R}^d$, is its richness as a natural consequence of the infinite dimensionality. This aspect also makes (basically) impossible the creation of a tool able to universally detect anomalies whatever the data, and gave rise to numerous directions of research focused on analysis and detection of functional anomalies (and outliers).

Thus, to ensure performance of single methods, systematize approaches to anomaly detection, but also develop reliable benchmarks, the scientific community witnessed multiple attempts to split the vast variety of functional anomalies into several types. The work by [12] (much adopted among contemporary FDA literature) suggests systematizing anomalies according to the following taxonomy (see also [20]). Here, we illustrate this on "Wine" data set [13] (see Fig. 64.1), which consists of the proton nuclear magnetic resonance (NMR) analysis of 40 different wine samples (in grey) contaminated with different types of anomalies (in red). First, anomalies can be classified according to their duration: *isolated anomalies* appear over a short period of time and often constitute peak values that are (noticeably) higher than normal values, while *persistent* anomalies cover a larger part of the studied interval. These persistent anomalies can be (approximately) divided into: *amplitude* (shift and scale) *anomalies* which roughly repeat the shape of normal data but differ in absolute values due to an additive or multiplicative shift (*e.g.*, a differing scale range) and *shape anomalies* which exhibit a distinct (from normal data) behavioural pattern.

Two more aspects are necessarily to be taken into account in practice. First, it can sometimes be difficult to assign an anomaly to a single category because it may indeed be a mixture of two or more of them. Second, while *global anomalies* reflect abnormal behaviour w.r.t. the majority of data, *local* (or contextual) *anomalies* (see,

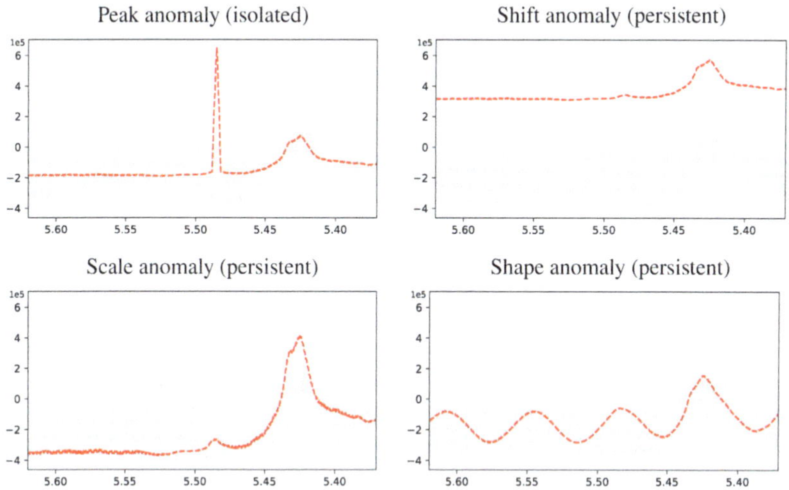

Fig. 64.1: Diverse anomaly types added to normal observations from "Wine" data set [13]: Isolated peak and persistent amplitude (shift and scale) and shape anomalies

e.g., [5, 19]) reveal abnormal behaviour only if focused on a subset of data (*e.g.*, an observation abnormal for a single cluster but located between several clusters).

64.3 Projection-Based Visualization of Functional Anomalies

64.3.1 Functional Setting for Real-World Data

Our object of interest is the random function $\boldsymbol{F}$ that takes its values in a Hilbert space, say the space $L^2([0,1])$ of square integrable functions on the unit interval. In the unsupervised anomaly detection framework, mathematical developments are usually based on possessing a finite sample of n such unlabelled functions: $\widetilde{\mathcal{F}} = \{\widetilde{f}_1, ..., \widetilde{f}_n\} \subset L^2([0,1])$. In practice, one never observes the elements of $\widetilde{\mathcal{F}}$ directly, but measurements are rather given in the form of ensembles of evaluations of $\widetilde{f}_i$ at some discretization points $t_{i1} \leq t_{i2} \leq ... \leq t_{il_i}$

$$[\widetilde{f}_i(t_{i1}), \widetilde{f}_i(t_{i2}), ..., \widetilde{f}_i(t_{il_i})]\,,$$

in general neither equidistant nor common ones, with $\min_i t_{i1} \geq 0$ and $\max_i t_{il_i} \leq 1$. To obtain the working data set of functions $\mathcal{F} = \{f_1, ..., f_n\}$ on [0,1], which are normally continuous and smooth at all points besides a finite set, interpolation techniques are typically employed such as, *e.g.*, splines [4], or piecewise linear interpolation (reconstructing f as a sequence of line segments connecting point $(t_{ij}, \widetilde{f}_i(t_{ij}))$ for $j = 1, ..., l_i$ and, *e.g.*, setting $f_i(t) = \widetilde{f}_i(t_{i1})$ when $0 \le t \le t_{i1}$ and $f_i(t) = \widetilde{f}_i(t_{il_i})$ when $t_{il_i} \le t \le 1$; the reader is also referred to [14] for a recent reference).

64.3.2 The Φ-Depth for Anomalies

Following the original ACA algorithm by [22] and inspired by [16], the following developments shall be based on the (slightly modified and directly adopting our particular case) notion of the Φ-depth (of a function f w.r.t. the data set $\mathcal{F}$):

$$D_\Phi(f|\mathcal{F}) = \inf_{d \in \mathcal{D}} D^1(\langle f, d\rangle | \langle f_1, d\rangle, ..., \langle f_n, d\rangle), \tag{64.1}$$

where $D^1 : \mathbb{R} \times \mathbb{R}^n \to [0, 1]$, $(x, x_1, ..., x_n) \mapsto D^1(x|x_1, ..., x_n)$ is a notion of univariate data depth and $\mathcal{D}$ is a dictionary of orthogonal(izable) (given a scalar product) features $d_1, d_2, ... \in L^2([0, 1])$. In words, it basically corresponds to the weak projection property introduced by [7] for the multivariate space $\mathbb{R}^d$, here in the functional setting. This would result from Definition 1 by [16] when taking (in Banach space) $\varphi : E \to \mathbb{R}$, $\varphi(f) \mapsto \langle d, f\rangle$ (for some $d \in \mathcal{D} \subset E$ uniquely corresponding to φ and a chosen scalar product) and focusing on output dimension $d = 1$; for practical reasons, we ignore the conditions on φ here.

Due to the limited space, we skip here the recall of ACA [22] as well as of the weak projection property [7], and instead directly introduce the methodology for anomaly-driven visualization of functional data: the functional abnormal component analysis.

For the univariate depth, we stick to the notion of *projection depth* [23], which—in the univariate case—narrows down to (for some $x \in \mathbb{R}$ and a sample $\{x_1, ..., x_n\} \subset \mathbb{R}$):

$$D^1(x|x_1, ..., x_n) = 1\Big/\left(1 + \frac{|x - \text{med}(x_1, ..., x_n)|}{\text{MAD}(x_1, ..., x_n)}\right),$$

where med(·) and MAD(·) correspond to (the usual univariate) median and median absolute deviation from the median, respectively.

Further, it is the dictionary $\mathcal{D}$ that determines which aspects of the anomalies are most likely to be highlighted when projected on the features contained there (see Sect. 64.4 for an example). With the functional space being vast, a particular dictionary can focus on only some aspect(s) of the data, *e.g.*, on one (or a few) of the categories describable by the taxonomy of functional anomalies (*cf.*, Sect. 64.2).

64.3.3 Functional Abnormal Component Analysis

We are now ready to state the main algorithm:

Input: Data set $\mathcal{F}$, scalar product $\langle\cdot,\cdot\rangle$, dictionary $\mathcal{D}$, number of components k.

1. *Initialization:* Abnormal components' storage: $\mathcal{B} = \emptyset$.
   For i from 1 to k :

   2. Find the ith abnormal component:

$$\boldsymbol{b}_i = \underset{\boldsymbol{d}\in\mathcal{D}}{\text{argmin}} \min_{j=1,\ldots,n} D_\Phi(\boldsymbol{f}_j|\mathcal{F})$$
$$\text{s.t.:} \quad \langle \boldsymbol{d}, \boldsymbol{b}\rangle = 0 \quad \forall \quad \boldsymbol{b}\in\mathcal{B}\,.$$

   3. Store the ith abnormal component: $\mathcal{B} = \mathcal{B}\cup\boldsymbol{b}_i$.

Output: The set of k orthogonal abnormal components: $\mathcal{B}$.

64.4 Parameters and Implementation Details

When employing the functional abnormal component analysis, several settings should be selected (*cf.*, Sect. 64.3.3). We choose the Sobolev-type scalar product, accounting for both the function and its derivative (*i.e.*, the level and the slope):

$$\langle \boldsymbol{f}, \boldsymbol{d}\rangle = \lambda\frac{\langle \boldsymbol{f}, \boldsymbol{d}\rangle_{L^2}}{\|\boldsymbol{f}\|\,\|\boldsymbol{d}\|} + (1-\lambda)\frac{\langle \boldsymbol{f}', \boldsymbol{d}'\rangle_{L^2}}{\|\boldsymbol{f}'\|\,\|\boldsymbol{d}'\|}, \quad \lambda\in[0,1]\,. \tag{64.2}$$

With the dictionary tasked with underlining a type of anomalies in question, it is expected to substantially impact the performance of the method; throughout the following Sect. 64.5 we experiment with the most prominent choices. An example of such dictionary (with parameter $I\in\mathbb{N}_+$) would be the *dyadic indicator dictionary* with a single feature being (*cf.* also Fig. 64.2); the reader is further referred to [21], where a number of dictionaries for functional data is presented:

$$\boldsymbol{d}_{dind}(t) = 2^i\cdot\mathbb{1}_{\{t\in[\frac{j}{2^i},\frac{j+1}{2^i}]\}} \text{ with } j=0,1,\ldots,2^i-1 \text{ and } i=0,1,\ldots,I\,. \tag{64.3}$$

The number of components k can be chosen intuitively, using *a priori* information.

Several practical shortcuts can be made when implementing the algorithm in Sect. 64.3.3. Thus, we first generate a number of dictionary's features $\boldsymbol{d}_1,\boldsymbol{d}_2,\ldots,\boldsymbol{d}_l$ and save them; this number l can be considered sufficient if adding more features does not improve the algorithm's stability. Instead of running an optimization routine (as it is done, *e.g.*, in ACA [22]), here we simply brute-force all the l stored features. Another detail to keep in mind is the numerical precision of features (sometimes impairing exact orthogonality); numerical errors can accumulate in particular when employing an orthogonalization procedure, *e.g.*, the Gram-Schmidt process.

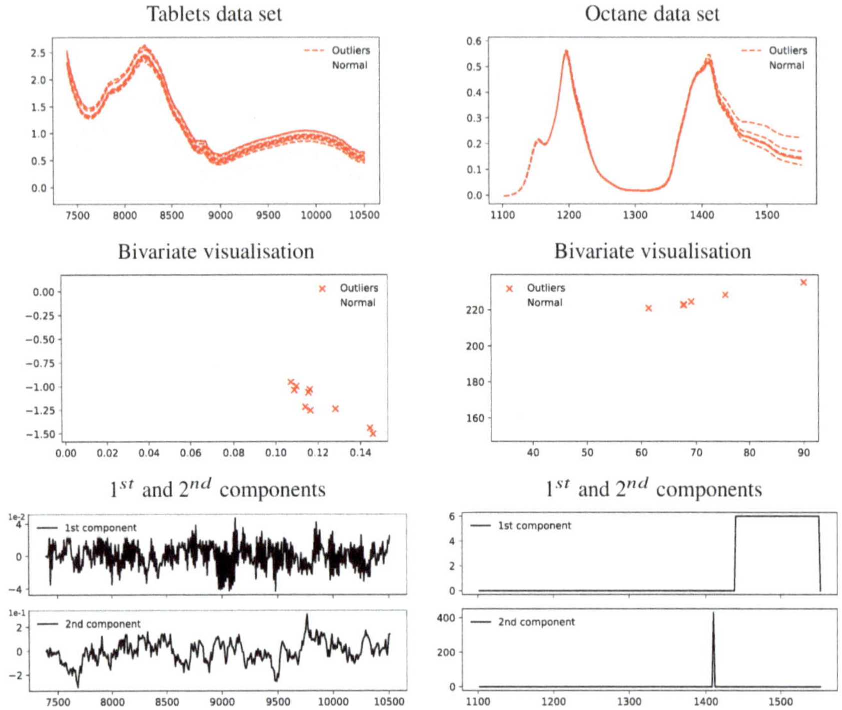

Fig. 64.2: fACA applied to "Tablets" (left) and "Octane" (right) data sets

64.5 Real-Data Illustrations

Table 64.1: Data sets description

Name	% of anomalies	Length	# of observations	# of variables
Wine [13]	2.5	396	40	1
Tablets [8]	12.5	404	80	1
Octane [9]	15.4	226	39	1
Character trajectories [1]	5.4	100	184	2

In this section, we illustrate the application of fACA on three real-world data sets (for classification problems, we add anomalies from another class): two containing univariate and one containing bivariate functional data, see Table 64.1 (referring to

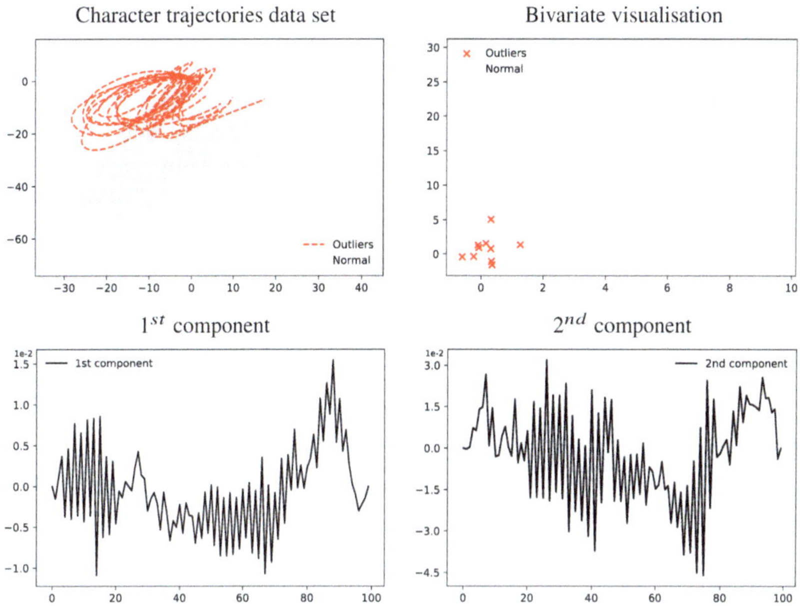

Fig. 64.3: fACA applied to a bivariate data set "Character trajectories" contaminated with persistent anomalies.

the corresponding sources for data description) that also contains "Wine" data set used in Sect. 64.2.

First, we treat "Tablets" data with the Brownian motion dictionary, and observe that trajectories (more or less) wiggling around a constant are sufficient to well separate in visualization 10 persistent shit anomalies; see Fig. 64.2 (left). Second, "Octane" data witness amplitude anomalies over shorter argument interval, distinguishable using dyadic indicator dictionary (64.3), where the 1st component underlines the duration of the 6 abnormal observations and the 2nd one focuses on the peak; see Fig. 64.2 (right).

Third, we use the Brownian bridge dictionary on multivariate functional data: "Character trajectories" data set consisting of multiple-output square integrable functions $\in L^2([0,1],\mathbb{R}^2)$ with the scalar product being the sum of (64.2) over the components. This time, certain (random) trajectories are necessary to separate the 10 persistent anomalies, which indicate the intervals that explain abnormality (when substantially differing from zero).

64.6 Conclusion

fACA presents an *ad hoc* ready-to-use method for an explainable visualization of functional anomalies. It is flexible in parametrization allowing one to adjust the number of components k (*cf.* [3] arguing that important information can be 'hidden' in higher-order components, see also [6]), the dictionary, and the scalar product. While the choice of scalar product for multivariate functional data here seeming arbitrary, it is a simple applicable practice (see, *e.g.*, [21]). Further options include separate handling of functional components or use of a multivarite depth in (64.1).

While fACA provides rapidly a first glance at the nature of anomalies contained in the data set, several limitations should be mentioned. The main limitation of the method is narrowing down the information content of the data to a single score, and thus it cannot be as efficient as a careful in-detail multiple-aspect study of the data set, in particular when using robust methodology. In implementation, the procedure for obtaining an orthogonal dictionary can sometimes be cumbersome and even cause tractability problems. Needless to say, optimization over the dictionary should be properly done depending on its richness.

Acknowledgements The authors gratefully acknowledge the Young Researcher grant of the French National Agency for Research (ANR JCJC 2021) in category Artificial Intelligence number ANR-21-CE23-0029-01 and the support of the CIFRE grant number 2021/1739.

References

1. Bache K., Lichman M.: UCI machine learning repository. (2013)
2. Chandola, V., Banerjee, A., Kumar, V.: Anomaly detection: A survey. ACM Comput. Surv. **41**, 1–58 (2009)
3. Cook, R.D.: Fisher Lecture: Dimension reduction in regression. Statist. Sci. **22**, 1–26, (2007)
4. De Boor, C.: A practical guide to splines. Springer-Verlag, (1978)
5. De Vries, T., Chawla, S., Houle, M. E.: Finding Local Anomalies in Very High Dimensional Space. 10th IEEE ICDM. 128–137 (2010)
6. Delaigle, A., Hall, P.: Achieving near-perfect classification for functional data. J. R. Stat. Soc. Ser. B Methodol. **74**, 267–286 (2012)
7. Dyckerhoff, R.: Data depths satisfying the projection property. Allgemeines Statistisches Arch. **88**, 163–190 (2004)
8. Dyrby, M., Engelsen, S.B., Nørgaard L., Bruhn M., Lundsberg-Nielsen L.: Chemometric quantitation of the active substance in a pharmaceutical tablet using near-infrared (NIR) transmittance and NIR FT-Raman spectra. Appl. Spectrosc. **56**, 579–585 (2002)
9. Esbensen K.: Multivariate data analysis in practice, 5th edn. Camo Software (2001)
10. Ferraty, F., Vieu, P.: Nonparametric Functional Data Analysis. Springer, New York (2006)
11. Gijbels, I., Nagy, S. On a general definition of depth for functional data. Statist. Sci. **32**, 630–639 (2017)
12. Hubert, M., Rousseeuw, P.J., Segaert, P.: Multivariate functional outlier detection. Stat. Methods. Appl. **24**, 177–202 (2015)
13. Larsen, F.H., van den Berg, F., Engelsen, S.B.: An exploratory chemometric study of ^{1}H NMR spectra of table wines. J. Chemom. **20**, 198–208 (2006)

14. Mosler, K., Mozharovskyi, P.: Fast DD-classification of functional data. Stat. Papers **58**, 1055–1089 (2017)
15. Mosler, K., Mozharovskyi, P.: Choosing among notions of multivariate depth statistics. Statist. Sci. **37**, 348–368 (2022)
16. Mosler, K., Polyakova, Y.: General notions of depth for functional data. arXiv:1208.1981 (2012)
17. Nieto-Reyes, A., Battey, H.: A topologically valid definition of depth for functional data. Statist. Sci. **31**, 61–79 (2016)
18. Ramsay, J.O., Silverman, B.W.: Functional Data Analysis. Springer, New York (2005)
19. Saligrama, V. Zhao, M.: Local anomaly detection. In: Lawrence, N.D., Girolami, M. (eds.) 15th AISTAT PMLR, 969–983 (2012)
20. Staerman, G., Adjakossa, E., Mozharovskyi, P., Hofer, V., Sen Gupta, J., Clémençon, S.: Functional anomaly detection: a benchmark study. Int. J. Data Sci. Anal. **16**, 101–117 (2023)
21. Staerman, G., Mozharovskyi, P., Clémençon, S., d'Alché-Buc, F.: Functional isolation forest. In: Lee, W.S., Suzuki, T. (eds.) 11th ACML PMLR. 332–347 (2019)
22. Valla, R., Mozharovskyi, P., d'Alché-Buc, F.: Abnormal component analysis. arXiv:2312.16139 (2023)
23. Zuo, Y., Serfling, R.: General notions of statistical depth function. Ann. Stat. **28**, 461–482 (2000)

Chapter 65
Leveraging Data Exchangeability for a More Reliable and Interpretable Functional Data Analysis

Simone Vantini

Abstract Most statistical methods heavily rely on assumptions about the specific data distribution family (i.e., parametric Statistics). In this work, I highlight that while these assumptions can enhance interpretability and strengthen statistical analysis from a mathematical, computational, and purely statistical perspective, they may also lead to unreliable statistical conclusions in modern contexts such as functional data analysis. As an alternative, I propose an approach that, rather than relying on the specific distribution of the data, leverages the exchangeability property induced by the i.i.d. assumption. Permutation tests are an example of this solution in the framework of functional hypothesis tests and conformal prediction an example in the context of probabilistic prediction of functional data.

65.1 Introduction

From its very beginning, Statistics has been driven by the need to solve real-world problems using data. In 1662, John Graunt [15] utilized publicly available records and manually collected data to estimate the population of London. Later, in 1710, John Arbuthnot [1] analyzed birth records to test the assumption that newborns had an equal probability of being male or female, thus using statistical reasoning to test hypotheses. Even today, the fundamental goal of statistics remains the same: to develop data-driven models and algorithms driven by the need of solving real-life problems. I refer to this foundamental goal of Statistics as statistical engineering emphasizing the similarity with other branches of engineering where other scientific or technological tools are used to solve problems. In this sense, statistical engineering can be viewed as a specialized branch of mathematical engineering, where statistical thinking plays a central role in development of the solution. This perspective does not

Simone Vantini (✉)
MOX - Department of Mathematics, Politecnico di Milano, Piazza Leonardo da Vinci 32, 20133, Milan, Italy e-mail: `simone.vantini@polimi.it`

© The Author(s), under exclusive license to Springer Nature Switzerland AG 2025

G. Aneiros et al. (eds.), *New Trends in Functional Statistics and Related Fields*, Contributions to Statistics, https://doi.org/10.1007/978-3-031-92383-8_65

want to diminish the importance of theoretical research in statistics. On the contrary, a strong mathematical foundation is essential to ensure that developed solutions are not only effective in addressing a problem but also readily adaptable to other problems of different applicative nature but similar from a modeling point of view.

65.2 Cornerstones of Statistical Engineering

An engineering solution, to be truly valuable in solving real-world problems, must satisfy some key properties. In my view, the most critical are effectiveness, efficiency, reliability, and explainability. These attributes define the quality of statistical engineering solutions, with the first two often well accounted for, while the latter two frequently overlooked.

Effectiveness: Solving the Right Problem. Effectiveness refers to the ability of a solution to address the actual problem under investigation (this characteristic is in some contexts unfortunately less common than one might expect). A tool must be designed for the right purpose: using a hammer to drive a screw into wood is ineffective, no matter how well the hammer itself functions. In statistical engineering, effectiveness can be intended as the ability of constructing models and algorithms based on assumptions that are aligned with the data generative process of the phenomenon under study. A model that is mathematically elegant but disconnected from the real-world data it aims to describe will ultimately fail in its practical utility.

Efficiency: Solving the Problem Optimally. With respect to an engineering solution, efficiency ensures that a problem is not just solved but solved in an optimal way, according to some well-defined shared criteria. Efficiency is a well-known concept in statistics and is often understood in two ways: statistical efficiency (for example, estimators should have a small mean squared error, hypothesis tests should be powerful, and confidence or prediction intervals should be narrow) and/or computational efficiency (the developed solutions should be executable within a reasonable time and with reasonable computational resources.

Reliability: Ensuring Effectiveness Beyond Ideal Conditions. Reliability is paramount in engineering and statistical applications alike. A solution should remain effective even when applied in conditions that deviate from those it was originally designed for. In statistical engineering, reliability can be understood as the ability of a model, an algorithm, or an approach in general to retain its effectiveness (and possibly its efficiency) even in application contexts where the assumptions underlying its development do not hold. This property is often referred to as robustness with respect to assumptions.

While statisticians prioritize effectiveness and efficiency when developing solutions, reliability is often overlooked. This may stem from the strong mathematical perspective that dominated the 20th-century Statistics. In mathematics, the concept of reliability is indeed basically irrelevant, as assumptions are treated as axioms (either true or false by definition) making solutions inherently correct or incorrect. However, in Statistics, discussing reliability should be essential, as assumptions can be violated whether mildly or strongly. The use or misuse of the Central Limit Theorem to develop asymptotic solutions that are only approximately valid in the statistical practice, is a well-known example of this type of reasoning. When dealing with complex problems, such as inference in high-dimensional settings, where well-established best practices are either unavailable or not widely accepted within the statistical community, asymptotic arguments are often applied indiscriminately. This has led to a vast body of literature that is largely worthless due to its lack of reliability in real-world applications. As a result, Statistics has even recently faced criticism of being a discipline disconnected from practical concerns, with some even provocatively questioning its future [9].

Explainability: Understanding and Advancing Knowledge. Explainability (or interpretability) describes an engineering solution that provides insights that experts can interpret and validate. Unlike effectiveness, efficiency, and reliability (which determine the practical utility of a solution) explainability contributes to the broader advancement of science. Understanding why one solution works while another does not enables a self-reinforcing cycle of innovation, leading to the development of improved methods and deeper domain understanding. Statistical thinking has traditionally placed strong emphasis on explainability. Statistical inference, which applies inductive reasoning to estimate and validate probabilistic models from data, is a prime example of this emphasis. However, even though there are different ways in which interpretability can be interpreted, in Statistics, interpretability is often intended as a property of a solution to be mathematically explicit and concise.

In complex data problems, however, this standard notion of interpretability can become problematic. The search for explicit and simple probabilistic models to describe complicate random phenomena may lead to a severe risk of oversimplification, potentially undermining both the effectiveness and the reliability of a solution. This can be summarized by Einstein's 1933 famous quote at Princeton University Club, "Everything should be made as simple as possible, but not simpler." An example of oversimplification in the context of functional data analysis is the use of Gaussian models. While Gaussian processes provide mathematically elegant and easily interpretable solutions, real-world functional data often exhibit clear departures from Gaussianity. Unfortunately, these deviations are frequently ignored and the assumption of Gaussianity not even questioned in favor of mathematical convenience.

This is why an alternative yet complementary notion of interpretability for functional data analysis will hereby be proposed in the context of hypothesis testing and probabilistic prediction. This alternative notion of interpretability will be based on the idea of viewing random functions as continuously infinite sets of ordered one-dimensional random variables, each representing the function's evaluation at a specific domain point and will be thus identifiable as a property of a solution of being explainable in terms of these random variables. This perspective partially contrasts with the mainstream approach to functional data analysis, which tends to treat functions as indivisible units in the statistical analysis. However, it aligns with the original perspective taken in functional data analysis introduced by Ramsay [29].

65.3 Leveraging Exchangeability in FDA

Since its early times, functional data have proven themselves to be a relevant laboratory of experimentation for statistical engineering. Due to their unique nature, analyzing functional data has necessitated the development of new statistical tools, even for problems that were long solved in the context of standard multivariate data [30]. Some solutions have been relatively straightforward extensions of multivariate methods based on the Hilbert structure of the functional spaces in which functional data are typically embedded. Others, like the ones described below, are instead original proposals specifically designed for functional data, taking advantage of their inherent smoothness and continuity.

In the following sections, we will show how exchangeability can be leveraged to develop reliable and interpretable solutions for two key problems in functional data analysis: hypothesis testing and probabilistic prediction of functional data. For simplicity: in the hypothesis testing setting, we will focus on the standard two-sample test, and in the probabilistic prediction setting, we will consider the case of i.i.d. functional data in the absence of covariates. While we focus on these specific cases for clarity, both approaches can be extended (though with varying levels of complexity) to more general and challenging settings.

65.3.1 Local Permutation Tests for Functional Data

Consider a scenario where we need to perform a two-sample test for functional data. Suppose we have two independent samples of i.i.d. functional data, possibly drawn from two different functional distributions, and we seek to test whether these distributions are identical. This is a typical statistical problem where standard parametric multivariate approaches cannot be directly extended due to the infinite-dimensional nature of functional data.

Permutation tests provide a viable alternative, both in the multivariate and functional settings, as they do not rely on assumptions about the dimensionality or distributional family of the data. Although permutation tests were first introduced exactly a century ago by Fisher in 1925 [10], they have struggled to find their place in mainstream Statistics. Often overlooked in textbooks, misclassified as a variant of bootstrap methods, and criticized for lacking theoretical rigor, permutation tests have long been regarded as an emergency fallback, used only when no other option is available. However, these criticisms are largely unfounded. The works of Pitman [26, 27, 28] in the 30s and 40s and Lehmann [18, 19] in the 50s and 60s established indeed rigorous theoretical foundations to the approach.

Permutation tests were first proposed by Fisher in his 1925 paper on the statistical comparison of different ethnic groups [10]. In that work, Fisher introduced permutation tests as a straightforward method to explain the rationale behind hypothesis testing to scientists with little mathematical background. He introduced what is now known as a permutational two-sample test. The problem he addressed was comparing the height distributions of English and French men using independent samples of 100 individuals from each group, randomly selected. To convey that observing a positive difference between the sample means of the English and French men was insufficient to conclude that the English were stochastically taller, Fisher devised a virtual experiment. He pooled all 200 observations into a single dataset and then divided them into two groups of 100 in all $\binom{200}{100}$ possible ways . For each possible grouping, he computed the corresponding difference in sample means and then calculated the proportion of groupings that produced a difference greater than or equal to the original one. In essence, Fisher supported then the correct intuition that, if the two distributions were identical, this proportion would follow a (discrete) uniform distribution over the discrete interval $\{1/\binom{200}{100}, 2/\binom{200}{100}, \ldots, 1\}$. Instead, if the English were indeed stochastically taller than the French, the distribution of this proportion (still preserving the same support) would concentrate towards zero, making the original grouping more probable to be found among those with larger differences. In simple terms, Fisher introduced the notion of permutational p-value. For many years, Fisher's virtual experiment remained computationally infeasible and was regarded as a purely theoretical exercise of little practical interest. However, in later years, with the development of computing technologies, permutation tests gained renewed attention.

Indeed, ten years later, just before the outbreak of the Second World War, Pitman was the one who first proved (through randomization arguments) the finite-sample exactness of permutation tests [26], their asymptotic consistency [27], and their asymptotic equivalence to parametric tests [28]. Regarding the theoretical formalization of permutation tests, the works of Lehmann in the 50s and 60s [18, 19] represent a turning point. He reinterpreted the entire theory of permutation tests within the framework of conditional inference concerning a sufficient statistic. Simply put, he emphasized that, in Fisher's procedure, it was the 200 observed heights that could be considered permuted, rather than the 200 individuals themselves. Essentially, he demonstrated that computing the p-value, as Fisher suggested, was equivalent to computing a classical p-value after conditioning the distribution of the test statistic under the null hypothesis on a (always!) sufficient statistic (namely, the unordered set of observed values). This insight directly allows to establish the general validity of the approach, independent of the specific data distribution.

Extending this reasoning to functional data simply requires defining appropriate test statistics in a functional space. One of the first examples in this direction is found in the 2002 paper by Hall and Tajvidi [16], where a permutation test is applied using the same permutational strategy proposed by Fisher in his 1925 work and using the squared L2 distance between the two functional sample means as test statistic. This approach yields a valid p-value regardless of the functional data distribution and sample sizes.

While the aforementioned approach ensures reliability, it presents limitations regarding interpretability (in the sense introduced above). Indeed, if the procedure leads to rejection, one might ask which probabilistic features make the two distributions significantly different. From the perspective of interpretability, a possible answer is to identify specific regions of the domain where the function evaluations show significant differences between the two groups. This approach allows to localize over the domain the rejection of the null hypothesis, providing a richer and more informative output. Our simple proposal is replacing the traditional scalar p-value with a p-value function defined at each point in the functional domain [22]. A simple trivial yet limited solution is defining the p-value function as the pointwise p-values of function evaluations. However, this method, while computationally efficient, has some strong limitations from a statistical point of view: performing an infinite number of tests separately ensures marginal validity but not joint validity, meaning that joint control of Type I error is not guaranteed.

To address this issue, we propose two adjustmnent methods: the former one is the so-called Interval-wise Testing [22]. This method adjusts pointwise p-values locally to ensure simultaneous control of Type I error across each subinterval of the domain. Based on a closed testing approach [20], this method involves performing a test (such as that proposed by Hall and Tajvidi) for each subinterval. The adjusted p-value at each point is then defined as the supremum of all p-values corresponding to intervals that include that point of the domain.

The second solution instead locally adjust the p-values such to control what we define as the functional false discovery rate (i.e., the expected ratio of the measure of the subset of the domain in which the null hypothesis is wrongly rejected, and the measure of the subset of the domain in which the null hypothesis is rejected no matter if wrongly or correctly) [21]. This solution is built on a continuous extension of the classical Benjamin Hockberg procedure.

These proposals have been applied to various testing problems within the frameworks of functional analysis of variance [23], functional linear regression [2, 3, 12, 32], functional linear mixed models [25], and multi-aspect testing of functional data [24].

65.3.2 Conformal Prediction Bands of Functional Data

Exchangeability can also be leveraged to address another problem closely related to null hypothesis testing: constructing prediction sets for new functional observations. In the Gaussian setting, this problem is fully solved for univariate data (Student's prediction intervals) building on the 1908 paper by Student [31] and for the multivariate case (Fisher's prediction ellipsoids) in the 1930-31 papers by Fisher [11] and Hotelling [17]. Similar to their corresponding hypothesis tests, these methods rely on the assumption of normality. However, unlike the tests, these solutions lack asymptotic validity, meaning they cannot be used to predict new observations if the data are non-Gaussian, regardless of sample size.

In 1999, Vovk, Gammerman, and Shafer [33] introduced a fully nonparametric approach to probabiltic prediction, known as conformal prediction [13, 14], which utilizes exchangeability in a manner similar to permutation tests. The key idea is to frame the probabilistic prediction problem as a two-sample test, where the first sample consists of observed data (the training set) and the second consists of the new observation to be predicted. Also in this case, a test statistic measuring the discrepancy between observed data and the new unobserved data is introduced (i.e. the non-conformity score of the new observation in this literature). A permutation scheme basically identical to the one propose in the 1925 Fihser's paper is applied.

Two are the main differences with respect to standard permutation tests: (i) the number of possible permutations is small, limited to $n + 1$ (the sample size of the pooled sample), eliminating computational issues related to the permutation strategy; (ii) the second sample, containing the unique unknown observation, is inherently unobservable, meaning the actual p-value cannot be computed. However, also in this case it can be proven that due to exchangeability, the unobserved p-value under the null hypothesis follows a discrete uniform distribution over the discrete interval $\{\frac{1}{n+1}, \frac{2}{n+1}, \ldots, 1\}$. Exploiting this property, conformal prediction operates in a scenario-analysis by testing all possible values of the new observation against the training set and computing the corresponding p-value for each. The final prediction set consists of all possible values of the new observation whose p-value is at least α ensuring (by construction) that the new observation will belong to this set with probability $1 - \alpha$.

Similarly to permutation tests, also conformal prediction could be extended naturally to the functional settings by introducing appropriate nonconformity scores measuring the dissimilarity of the new functional observation to the observed functional data set. It is indeed very simple to assess if a possible realization of the new observation does or does not belong to the prediction set. However, this approach may in general face two limitations: the obtained prediction sets lack explicit formulations, making interpretation and visualization difficult, and their numerical approximation is infeasible due to the infinite dimensionality of functional data. These would limit the prediction set to a tool more suited for process control rather than actual prediction.

To overcome these issues, exploiting the flexibility with respect to the choice of the non-conformity score, we propose using a nonconformity score that automatically generate prediction regions in the functional space having the geometrical shape of functional bands (making interpretation intuitive) and whose identification is computationally feasible [8]. Our proposal is to defines the nonconformity score as the supremum over the functional domain of the absolute difference between the function to be predicted and the functional sample mean of the training sample, rescaled by the pointwise sample standard deviation. This idea can be simply extended also to the case in which some covariates either scalar or functional are available just by replacing the functional sample mean with a functional prediction of the new observation (for example, based on a functional regression model) and the standard deviation function with a function estimating point-wise, the magnitude of the typical prediction error in that point of the domain [8]. This approach was also extended to the case of multivariate functional data [7] and to the case of multidimensional domains [4, 5].

65.4 Conclusions

Most statistical methods for analyzing independent and identically distributed data heavily rely on additional assumptions about the specific distribution family (i.e., parametric Statistics). In this work, I highlighted that while these assumptions can enhance interpretability and strengthen statistical analysis from a mathematical, computational, and purely statistical perspective, they may also lead to unreliable statistical conclusions in modern contexts such as functional data analysis. As an alternative, I proposed an approach that, rather than relying on the specific distribution of the data, leverages the exchangeability property induced by the i.i.d. assumption. Permutation tests are an example of these solutions in the framework of hypothesis tests and conformal prediction an example in the context of probabilistic prediction.

Additionally, I argued that in functional data analysis, it may sometimes be useful to go back to the original foundations of functional data analysis, where random functions are viewed not merely as abstract mathematical entities in a functional space but also as an infinite continuum of smoothly varying scalar random variables. In certain cases, this old-fashion perspective can inspire the development of tools that are much more interpretable and impactful in real-world applications. The p-value function is an example in the context of hypothesis testing, while conformal prediction bands an example in probabilistic prediction.

Finally, I want to clarify that my argument is not a conviction of the use of models but only to an indiscriminate and dangerous misuse of models. I indeed fully agree with Box's vision on modeling [6]. However, I think that his well-known quote "All models are wrong, but some are useful" has been often used to justify the use of clearly wrong models. This statement does not sufficiently emphasize indeed a crucial aspect of Box's vision: models must be chosen carefully (which ties back to my notion of effectiveness). A completely wrong model can lead to entirely unreliable statistical conclusions, which may be deceptively and effectively masked under a veil of mathematical formalism or computational complexity. For this reason, I believe that, as statisticians, we should always be aware that all models are wrong, some are of course useful, but some can also be harmful.

Acknowledgements These reflections and proposals stem from the many enriching and long-standing interactions with all my coauthors, as well with my colleagues of the MOXstat group at the MOX Laboratory of the Department of Mathematics of Politecnico di Milano. This work is supported by the grant "Excellent Department 2023-27" of MIUR (Italian Ministry of the University and Research). ChatGPT was used for language improvement.

References

1. Arbuthnot, J.: An argument for Divine Providence, taken from the constant regularity observed in the births of both sexes. Philosophical Transactions of the Royal Society of London, **27**, 186–190, (1710)

2. Abramowicz, K., Häger, C.K., Pini, A., Schelin, L., Sjöstedt de Luna, S., Vantini, S.: Non-parametric inference for functional-on-scalar linear models applied to knee kinematic hop data after injury of the anterior cruciate ligament. Scandinavian Journal of Statistics, **45**(4), 1036–1061, (2018)
3. Abramowicz, K., Pini, A., Schelin, L., Sjösted, S., Stamm, A., Vantini, S.: Domain selection and familywise error rate for functional data: A unified framework. Biometrics, **79**(2), 1119–1132, (2023)
4. Ajroldi, N., Diquigiovanni, J., Fontana, M., Vantini, S.: Conformal prediction bands for two-dimensional functional time series. Computational Statistics and Data Analysis, **187**, 107821, (2023)
5. Bortolotti, T., Casu, F., Virelli, M., Tapete, D., De Cumis, M. S., Menafoglio, A., Vantini, S.: Remote Monitoring of Ground Motion Exploiting Functional Data Analysis and Nonparametric Prediction on Multi-Temporal DInSAR Data. Proceedings of the 75th International Astronautical Congress (IAC 2024), Milan, Italy, (2024)
6. Box, G. E. P.: Robustness in the strategy of scientific model building. In Robustness in Statistics (pp. 201–236). Academic Press, (1979)
7. Diquigiovanni, J., Fontana, M., Vantini, S.: Conformal prediction bands for multivariate functional data. Journal of Multivariate Analysis, **189**, 104879, (2022)
8. Diquigiovanni, J., Fontana, M., Vantini, S.: The Importance of Being a Band: Finite-Sample Exact Distribution-Free Prediction Sets for Functional Data. Statistica Sinica, **35**(2), (accepted), (2024)
9. Efron, B.: Prediction, estimation, and attribution. Journal of the American Statistical Association, **115**(530), 636–655, (2020)
10. Fisher, R. A.: The coefficient of racial likeness and the future of craniometry. Journal of the Royal Anthropological Institute of Great Britain and Ireland, **55**, 57–63, (1925)
11. Fisher, R. A.: The moments of the distribution for normal samples of an indefinitely large population. Proceedings of the Cambridge Philosophical Society, **26**, 605–618, (1930)
12. Fontana, M., Tavoni, M., Vantini, S.: Functional Data Analysis of high-frequency load curves reveals drivers of residential electricity consumption. PLOS ONE, **14**(6), e0218702, (2019)
13. Fontana, M., Zeni, G., Vantini, S.: Conformal prediction: A unified review of theory and new challenges. Bernoulli, **29**(1), 1–23, (2023)
14. Gammerman, A., Vovk, V.: Algorithmic Learning in a Random World. Springer, (2005)
15. Graunt, J.: Natural and Political Observations Made upon the Bills of Mortality. London: John Martyn and James Allestry, (1662)
16. Hall, P., Tajvidi, N.: Permutation tests for equality of distributions in high-dimensional settings. Biometrika, **89**(2), 359–374, (2002)
17. Hotelling, H.: The generalization of Student's ratio. The Annals of Mathematical Statistics, **2**(3), 360–378, (1931)
18. Lehmann, E. L.: Permutation tests for two-sample problems. The Annals of Mathematical Statistics, **30**(1), 220–236, (1959)
19. Lehmann, E. L.: Testing Statistical Hypotheses. John Wiley & Sons, (1959)
20. Marcus, R., Peritz, E., Gabriel, K. R.: On closed testing procedures with special reference to ordered analysis of variance. Biometrika, **63**(3), 655–660, (1976)
21. Olsen, N. L., Pini, A., Vantini, S.: False discovery rate for functional data. Test, **30**(3), 784-809, (2021)
22. Pini, A., Vantini, S.: Interval-wise Testing for Functional Data. Journal of Non-parametric Statistics, **29**(2), 407-424, (2017)
23. Pini, A., Vantini, S., Colosimo, B. M., Grasso, M.: Domain-Selective Functional Analysis of Variance for Supervised Statistical Profile Monitoring of Signal Data. Journal of the Royal Statistical Society – Series C, **67**(1), 55-81, (2018)
24. Pini, A., Spreafico, L., Vantini, S., Vietti, A.: Multi-aspect local inference for functional data: Analysis of ultrasound tongue profiles. Journal of Multivariate Analysis, **170**, 162-185, (2019)

25. Pini, A., Sørensen, H., Tolver, A., Vantini, S.: Local inference for functional linear mixed models. Computational Statistics and Data Analysis, **181**, 107688, (2023)
26. Pitman, E. J. G.: Significance tests which may be applied to samples from any population. Journal of the Royal Statistical Society. Supplement, **4**(1), 119–130, (1937)
27. Pitman, E. J. G.: Significance tests which may be applied to samples from any population. II. The correlation coefficient test. Journal of the Royal Statistical Society. Supplement, **4**(2), 225–232, (1937)
28. Pitman, E. J. G.: Significance tests which may be applied to samples from any population. III. The analysis of variance test. Biometrika, **29**(3/4), 322–335, (1938)
29. Ramsay, J. O.: When the data are functions. Psychometrika, **47**(4), 379–396, (1982)
30. Ramsay, J. O., Silverman, B. W.: Functional Data Analysis. Springer, (1997)
31. Student (Gosset, W. S.): The probable error of a mean. Biometrika, **6**(1), 1–25, (1908)
32. Torti, A., Pini, A., Vantini, S.: Modelling time-varying mobility flows using function-on-function regression: Analysis of a bike sharing system in the city of Milan. Journal of the Royal Statistical Society. Series C, **70**(1), 226–247, (2021)
33. Vovk, V., Gammerman, A., Shafer, G.: Machine-learning applications of algorithmic randomness. Proceedings of the Sixteenth International Conference on Machine Learning (ICML'99), 444–453, (1999)

Chapter 66
A Family of Moment Operators for Functional Data and Its Discriminative Properties

Marc Vidal

Abstract We define the p-th moment operator of a Hilbert-valued random element and study its discriminative properties for $p > 2$ under Gaussian mixtures of homoscedastic random variables. We show that the optimal discriminant functional is associated with a spectral subspace of these operators, initially acting as a perturbation, while residing in a joint eigenspace of structurally identical kernels that support independent realizations within a shared functional space. In the empirical case, we investigate how increasing the sample size perturbs the moment operator, leading it to converge toward the optimal discriminant solution in the tail (lower-order eigenelements).

66.1 Introduction

The notion of higher moments for Hilbert-valued random elements remains poorly understood and has yet to be systematically formalized. Their definition requires rigorous handling of tensor norms, which determine how higher-order interactions are measured, as well as careful consideration of the interplay between weak and Bochner forms of measurability and integrability [3]. While covariance operators play a central role in functional data analysis, higher-order moments have received comparatively less attention. Only a few studies address their applicability in the context of functional independent component analysis techniques [7, 8, 10], leaving a rich area for further investigation. Of particular interest are the discriminating properties of these operators, which have been sparsely studied in their multivariate counterparts [4, 6]. A first attempt in this direction was made by Peña et al. [7], who associated the eigendecomposition of a kurtosis operator with a generalized Fisher's discriminant function under a homoscedastic Gaussian model with differing means.

Marc Vidal (✉)
Department of Data Analysis, Ghent University; Max-Planck-Instituts für Kognitions- und Neurowissenschaften. e-mail: marc.vidalbadia@ugent.be

© The Author(s), under exclusive license to Springer Nature Switzerland AG 2025
G. Aneiros et al. (eds.), *New Trends in Functional Statistics and Related Fields*, Contributions to Statistics, https://doi.org/10.1007/978-3-031-92383-8_66

More recently, Vidal et al. [10] extended these results to functional independent component analysis, and interpreted them through the Feldman-Hájek dichotomy, which has also been linked to the phenomenon of near-perfect asymptotic classification [2].

The aim of this paper is to revisit Theorem 1 in [7], which relates the eigendecomposition of the kurtosis operator to Fisher's subspace, and to extend it to a family of moment operators as defined in Sect. 66.2. The main result of the current contribution is Theorem 1 (Sect. 66.3), which, while not technically demanding, presents an interpretation that is far from trivial, with implications extending well beyond the original proof in [7]. To provide empirical evidence of the result, in Sect. 66.3.1 we analyze the effect of perturbing higher-moment operators by simultaneously increasing the sample size and mean differences, using simulated data as in [7, 10].

66.2 Preliminaries and Definition

Let X be a real-valued random element taking values on a separable Hilbert space H with inner product $\langle \cdot, \cdot \rangle$ and induced norm $\| \cdot \|$. The triplet $(\Omega, \mathcal{A}, \mathbb{P})$ is the underlying probability space where the random element is defined, which is assumed to be complete. The H-valued random element is then a measurable function $X : (\Omega, \mathcal{A}, \mathbb{P}) \rightarrow (H, \mathbb{B}_H)$, where $\mathbb{B}_H$ is the Borel σ-algebra generated by the open sets in H. This means that for every measurable set $B \in \mathbb{B}_H$, the preimage $X^{-1}(B) = \{\omega \in \Omega : X(\omega) \in B\} \in \mathcal{A}$ defines an event in the probability space. Since X is Borel measurable and separably valued (i.e., almost surely takes values in a separable subset of H), X is said to be Bochner measurable [3]. Moreover, if $\mathsf{E}\|X\| \equiv \int_\Omega \|X(\omega)\| \mathrm{d}\mathbb{P}(\omega) < \infty$, then X is Bochner integrable. As a consequence, the mean (or first moment) $\mathsf{E}X$ is well-defined as an element of H. Throughout this paper, we assume that $\mathsf{E}\|X\|^p < \infty$, for all $p \in \mathbb{N}$, and that $\mathsf{E}X = 0$, unless stated otherwise.

In the class of bounded operators, we consider compact and self-adjoint operators that admit a spectral decomposition consisting solely of eigenvalues with finite multiplicity, which accumulate at zero. A compact operator A is said to be trace-class (or nuclear) if its trace, defined as $\mathrm{tr}(\mathsf{A}) = \sum_j \left|\langle \mathsf{A}\varphi_j, \varphi_j \rangle\right|$, converges for any orthonormal basis $(\varphi_j)_{j\in\mathbb{N}}$ of H. A trace-class operator A belongs to the Hilbert-Schmidt class if $\mathrm{tr}(\mathsf{A}^*\mathsf{A}) < \infty$, where A^* denotes the adjoint of A.

As a classical example of a Hilbert-Schmidt operator, consider the rank-1 bilinear form $x \otimes y : H \rightarrow H$, called the tensor product operator, and defined by $(x \otimes y)(\cdot) = \langle \cdot, x \rangle y$ for all $x, y \in H$. It seems intuitive to define the p-th moment of X using tensor products. However, the bilinear nature of the standard definition limits its applicability to higher-order tensors. To address this shortcoming, we propose constructing higher-moment operators ($p \geq 2$) as p-fold compositions of $X \otimes X$ in the Hilbert space H.

Definition 66.1. For all $p > 1$, the p-th moment operator of a random element X taking values in H is defined as

$$\mathsf{K}_X^{\otimes p} = \mathsf{E}\left\{(X \otimes X)^{p/2}\right\}.$$

According to Definition 66.1, odd moments naturally relate to fractional powers of the tensor product, specifically square roots of higher powers of $X \otimes X$. This leads to the following sequence of moment operators: $\mathsf{K}_X^{\otimes 2} = \mathsf{E}(X \otimes X)$, $\mathsf{K}_X^{\otimes 3} = \mathsf{E}\{\sqrt{(X \otimes X)^3}\}$, $\mathsf{K}_X^{\otimes 4} = \mathsf{E}\{(X \otimes X)^2\}, \ldots$. All moment operators are self-adjoint, positive definite, and satisfy the trace-class property if and only if $\mathsf{E}\|X\|^{p/2} < \infty$. To compute them, it follows from the tensor product properties that, for a p-fold composition, the general result would be $(X \otimes X)^p = \|X\|^{2(p-1)}(X \otimes X)$, with $\sqrt{\cdot}$ applied for odd p.

As previously indicated, we work with compositions of the kind $\mathcal{X} \circ \mathcal{X} \circ \ldots$, where $\mathcal{X} = X \otimes X$, and $\circ$ is the operator composition. To better capture the full dimensionality of the moments, one could replace $\circ$ with $\tilde{\otimes}$, which here denotes the tensor product of operators [5]. This adjustment allows moment operators with $p > 2$ to acquire a new interpretation within the tensor product space. Then, it follows that $\mathcal{X}\tilde{\otimes}\mathcal{X}$ admits the representation

$$\mathcal{X}\tilde{\otimes}\mathcal{X} = \sum_{j,k}\sum_{l,m} \langle \mathcal{X}e_j, e_k\rangle \langle \mathcal{X}\varphi_l, \varphi_m\rangle (e_k \otimes \varphi_m) = \sum_{j,k}\sum_{l,m} x_{j,k}x_{l,m}(e_k \otimes \varphi_m),$$

where $(e_j)_{j\in\mathbb{N}}, (\varphi_l)_{l\in\mathbb{N}}$ are two arbitrary orthonormal basis, and the coefficients $x_{j,k}x_{l,m}$ form a 4-dimensional tensor (or order-4 matrix). Moment operators will be denoted as the mapping $\mathsf{K}^{\tilde{\otimes} p}$ when using this tensor algebra.

Since the third and fourth moments aim at reflecting both the tail behavior and peakedness of the marginal distributions, it is of interest to remove second-moment dependencies and scale effects from X. In that respect, [9] defined the set of operators

$$\mathcal{W} = \left\{\Psi : \Psi^*\Psi = \mathsf{K}_X^{\otimes 2\dagger} \wedge \mathsf{K}_{\mathbb{X}}^{\otimes 2} = \mathsf{P}_{\overline{\mathsf{R}}(\mathsf{K}_X^{\otimes 2})}\right\},$$

where $\mathsf{K}_X^{\otimes 2\dagger}$ refers to the pseudoinverse of $\mathsf{K}_X^{\otimes 2}$, P is the projection operator and $\mathbb{X} \equiv \Psi(X)$, with X the Karhunen–Loève (KL) expansion of X. Note that $\mathcal{W}$ has effective domain on $\overline{\mathsf{R}}(\mathsf{K}_X^{\otimes 2})$, the closure of the range space of $\mathsf{K}_X^{\otimes 2}$. Due to the unbounded nature of the inverse covariance operator, certain assumptions on its injectivity and eigenvalues decay are often required for $\mathbb{X}$ to be defined, whereas regularization is usually expedient when such conditions are not met [10].

The main issue in defining higher-moment operators through $\mathbb{X}$ is that $\operatorname{tr}(\mathsf{K}_{\mathbb{X}}^{\otimes 2}) = \infty$, which implies Bochner integrability no longer holds. Additionally, $\mathsf{K}_{\mathbb{X}}^{\otimes 2}$ does not satisfy the compactness property necessary for a convergent spectral decomposition, nor do any of its compositions. It has been shown, however, that the spectral decomposition and the existence of a spectral family hold for general self-adjoint operators, even in the absence of compactness or trace-class properties [1]. For convenience in our developments, we consider the spectral decomposition

$$\mathsf{A} = \lambda_0 E_0 + \int_{\sigma(\mathsf{A})\setminus(\lambda_0)} \lambda E(\mathrm{d}\lambda), \tag{66.1}$$

where $\{E(\lambda) : -\infty < \lambda < \infty\}$ is a spectral family of the operator A, and $\lambda_0 E_0$ is a perturbation with eigenvalue λ_0. Unless regularization or finite-dimensional dependency is considered to define $\mathbb{X}$, the operator $\mathsf{K}_{\mathbb{X}}^{\otimes p}$ does not satisfy strong integrability conditions, but its structure can still be discussed, say, in a weak sense, primarily through the above spectral representation. In the case of $\mathsf{K}_{\mathbb{X}}^{\otimes 2}$, the Stieltjes integral in (66.1) collapses to a point mass at $\lambda = 1$, reflecting the operator's invariance across the entire RKHS. No perturbation component is yet considered; its role will be clarified further in Sect. 66.3.

66.3 Identifying Fisher's Discriminant Function as a Spectral Subspace of Higher-Moment Operators

Let us now take H to be L^2, the space of square-integrable functions defined on a compact Euclidean set to the real line. Without essential loss of generality, we will assume that the compact set is the interval $\mathcal{I} = [0, 1]$ and $L^2 \equiv L^2(\mathcal{I})$. Consider an L^2-valued random element $(\mathsf{G}, X) : \Omega \to \{1, 2\} \times L^2$ to be a mixture of two Gaussian functional random variables with continuous trajectories. The mean of $X = X(t)$ is denoted by $\mathsf{E}X(t) = m(t)$, and the class-means as $m_\ell(t) = \mathsf{E}[X(t) \mid \mathsf{G} = \ell]$. Let $\alpha = \mathbb{P}(\mathsf{G} = \ell)$. The decomposition $m(t) = \alpha m_1(t) + (1 - \alpha) m_2(t)$ then follows from the linearity of expectations. Furthermore, the marginal class-covariance functions are given by

$$K_\ell(s, t) = \mathsf{E}\{(X(s) - m_\ell(s))\,(X(t) - m_\ell(t)) \mid \mathsf{G} = \ell\} \quad s, t \in \mathcal{I}, \tag{66.2}$$

which are assumed to be equal, therefore have common form denoted by K_W (within class covariance) with Mercer expansion $K_W(s, t) = \sum_j \lambda_j \gamma_j(s) \gamma_j(t)$. These kernels are related to the covariance function of $X(t)$ as follows:

$$K_X(s, t) = \alpha(1 - \alpha)\mathsf{E}\Big\{\sum_\ell (X_\ell(s) - m(s))\,(X_\ell(t) - m(t))\Big\} = K_W(s, t) + K_B(s, t), \tag{66.3}$$

where $K_B(s, t) = \alpha\,(1 - \alpha)\,\{m_2(s) - m_1(s)\}\,\{m_2(t) - m_1(t)\}$ (between class covariance). The associated induced operators are denoted by $\mathsf{K}_X^{\otimes 2}, \mathsf{K}_W^{\otimes 2}, \mathsf{K}_B^{\otimes 2}, \ldots$, and defined by the relation $(\mathsf{K}^{\otimes 2}\gamma_j)(t) = \int_{\mathcal{I}} K(s, t)\gamma(t)\mathrm{d}t$, for $\gamma(t) \in L^2$.

Now consider $X(t)$ is already in $H_0(K_X) \equiv \mathsf{R}(K_X(\cdot, t))$, so that

$$X(t) = \begin{cases} m_1 + \sum_j \xi_{1,j}\gamma_j & \text{w.p. } \alpha \\ m_2 + \sum_j \xi_{2,j}\gamma_j & \text{w.p. } 1 - \alpha, \end{cases}$$

where $\xi_{\ell,j} = \int_{\mathcal{I}} (X_\ell(t) - m_\ell(t))\gamma_j(t)\mathrm{d}t$ are independent standard normal random variables. We aim to find a random function $\beta \setminus \{0\}$ on $\mathcal{I}$ that optimally maximizes the separation between the groups of $X(t)$. Specifically, $\mathrm{span}(K_B)$ contains the information to maximize the distances between the projected class centers, while

$\text{span}(K_W)$ encodes the information needed to minimize within-class variation. The problem is typically formulated as

$$\mathsf{F}(\beta) \equiv \langle \beta, \mathsf{K}_W^{\otimes 2}\beta\rangle_{L^2}^{-1} \langle \beta, \mathsf{K}_B^{\otimes 2}\beta\rangle_{L^2},$$

where $\mathsf{K}_W^{\otimes 2}$ generally has an unbounded inverse. To address this, let us consider the space $H(K_W) \equiv \overline{\mathsf{R}}(K_W(\cdot, t))$ as the closure of $H_0(K_W)$, with inner product $\langle f, g\rangle_{H(K_W)} = \langle f, \mathsf{K}_W^{\otimes 2\dagger} g\rangle_{L^2}$ and norm $\|\cdot\|_{H(K_X)}$. For $f \in H_0(K_W)$ to lie in $H(K_W)$ a critical requirement referred to as Picard's criterion is often assumed: $\sum_j \langle f, \gamma_j\rangle_{L^2}^2/\lambda_j < \infty$, which means that $\|\cdot\|_{H(K_X)}$ is only finite if the λ_j's decay is slow enough in relation to the decay of the squared KL coefficients (always assuming injectivity on $\mathsf{K}_W^{\otimes 2}$). Under these conditions, it is well-known that the function that maximizes F is given by $\beta(t) = c \cdot \mathsf{K}_W^{\otimes 2\dagger}(m_\Delta)(t)$, for some constant c and $m_\Delta(t) \equiv m_2(t) - m_1(t)$. See [7] for alternative regularized solutions.
With the above-outlined setting, we now present the main result of the section.

Theorem 1. *For $p > 2$, $\mathsf{K}_{\mathbb{X}}^{\tilde{\otimes} p}(\tilde{\tilde{m}}_\Delta) = \kappa\tilde{\tilde{m}}_\Delta$, where $\tilde{\tilde{m}}_\Delta = \mathsf{K}_W^{\otimes 2\dagger}(m_\Delta)$.*

Proof. Let $\Psi_W = \sum_j \pi_j(\gamma_j \otimes \gamma_j)$[1] and $m_\Delta = \sum_j \nu_j\gamma_j$. In Sect. 66.4, we derive the expressions for the moment operators used below. The conditions for a function to be an eigenfunction of a self-adjoint operator are also discussed there. As an illustrative example, we start with the case $p = 2$ (see also [7]).

Operator $\mathsf{K}_{\mathbb{X}}^{\otimes 2}$. Let us consider $\pi_j = \lambda_j^{-1/2}$, and define the function $\tilde{m}_\Delta = \Psi_W(m_\Delta)$. Based on (66.4), the combined form of the mixture under the metric $\langle\cdot,\cdot\rangle_{H(K_W)}$ acting on $\tilde{m}_\Delta$ is

$$\begin{aligned}\mathsf{K}_{\mathbb{X}}^{\otimes 2}(\tilde{m}_\Delta) &= \sum_j \pi_j^3\lambda_j\nu_j\gamma_j + \alpha(1-\alpha)\sum_j\Big(\sum_k \pi_k^2\nu_k^2\Big)\pi_j\nu_j\gamma_j = \sum_j 1\cdot\pi_j\nu_j\gamma_j + \\ &\quad + \alpha(1-\alpha)\sum_k \pi_k^2\nu_k^2\sum_j \pi_k\nu_j\gamma_j = \Big(1 + \alpha(1-\alpha)\sum_j \pi_j^2\nu_j^2\Big)\tilde{m}_\Delta.\end{aligned}$$

Thus, we see that $\tilde{m}_\Delta$ is indeed a spectral subspace (direction) of $\mathsf{K}_{\mathbb{X}}^{\otimes 2}$, with eigenvalue $\kappa = 1 + \alpha(1-\alpha)\sum_j \pi_j^2\nu_j^2$, where $\sum_j \pi_j^2\nu_j^2$ corresponds to the norm $\|m_\Delta\|_{H(K_X)}$ induced by the metric $\langle\Psi_W(\cdot), \Psi_W(\cdot)\rangle_{L^2}$. Therefore, $\tilde{m}_\Delta$ aligns properly with the inner product structure of the operator (i.e., it belongs to its domain)–this is, however, not the case for $\tilde{\tilde{m}}_\Delta$. It follows that for any function ψ_k satisfying $\psi_k \perp \tilde{m}_\Delta$, ψ_k is mapped to itself by $\mathsf{K}_{\mathbb{X}}^{\otimes 2}$, so the corresponding eigenvalue is 1.

From Proposition 1 in [9], we know that $\mathsf{K}_{\mathbb{X}}^{\otimes 2}$ has a constant spectrum of 1. Consequently, the additional contribution must necessarily vanish, $\alpha(1-\alpha)\sum_i \pi_j^2\nu_j^2 = 0$, since the associated eigenvalue cannot be higher than one as $\alpha(1-\alpha) \geq 0$ for any $\alpha \in [0, 1]$, $\pi_j^2 \geq 0$ and $\nu_j^2 \geq 0$, making their product non-negative. In other words,

[1] One could interpret Ψ_W as an approximation to a whitening operator, although for simplicity in our developments, we later assume $\pi_j = \lambda_j^{-1/2}$.

the function $\tilde{m}_\Delta$ initially plays a perturbative role within the kernel but is eventually shifted into a higher spectral subspace as the difference in means increases. Further intuition can be gained from the next case.

Operator $\mathsf{K}_{\mathbb{X}}^{\tilde{\otimes}4}$. Similarly, using (66.5), when $\mathsf{K}_{\mathbb{X}}^{\tilde{\otimes}4}$ maps $\tilde{\tilde{m}}_\Delta$ we obtain

$$\mathsf{K}_{\mathbb{X}}^{\tilde{\otimes}4}(\tilde{\tilde{m}}_\Delta) = \sum_j 3\pi_j^2\nu_j\gamma_j + \alpha(1-\alpha)\Big\{2\sum_j\Big(\sum_k \pi_k^2\nu_k^2\Big)\pi_j^2\nu_j\gamma_j + \\ + \sum_j\Big(\sum_k \pi_k^4\nu_k^4\Big)\pi_j^2\nu_j\gamma_j\Big\} = \Big\{3 + \alpha(1-\alpha)\Big(2\sum_j \pi_j^2\nu_j^2 + \sum_j \pi_j^4\nu_j^4\Big)\Big\}\tilde{\tilde{m}}_\Delta.$$

By the same reasoning discussed earlier, $\sum_j \pi_j^4\nu_j^4$ corresponds to a norm induced by the metric $\langle \mathsf{K}_W^\dagger(\cdot), \mathsf{K}_W^\dagger(\cdot)\rangle_{L^2}$, or equivalently, $\langle \Psi_W(\cdot), \Psi_W(\cdot)\rangle_{L^2}^2$, so $\tilde{\tilde{m}}_\Delta$ and $\tilde{m}_\Delta$ belong to the operator's domain. Again, it follows that for any function ψ_k orthogonal to $\tilde{\tilde{m}}_\Delta$ or $\tilde{m}_\Delta$, we have $\mathsf{K}_{\mathbb{X}}^{\tilde{\otimes}4}(\psi_k) = 3\psi_k$. However, we encounter the same issue: as the difference in means increases, $\tilde{\tilde{m}}_\Delta$ perturbs or is effectively shifted out from $\mathsf{K}_{\mathbb{X}}^{\tilde{\otimes}4}$. Instead, $\tilde{\tilde{m}}_\Delta$ can be thought as residing in the direct sum of two complementary subspaces, i.e., $\tilde{\tilde{m}}_\Delta \in H_1(K_X) \oplus H_2(K_X)$ induced by the direction of $\mathsf{K}_{\mathbb{X}}^{\otimes 2}$ complementary to the one spanned by $\tilde{m}_\Delta$, say, when $\mathsf{K}_{\mathbb{X}}^{\otimes 2}$ corresponds to the identity. Here, $H_1(K_X)$ and $H_2(K_X)$ represent structurally equal but functionally complementary kernels that serve as supports for independent realizations within the same shared space. With $\tilde{\tilde{m}}_\Delta \perp \tilde{m}_\Delta$, we have that $(\mathsf{K}_{\mathbb{X}}^{\otimes 2} \circ \mathsf{K}_{\mathbb{X}}^{\otimes 2})(\tilde{\tilde{m}}_\Delta) = \tilde{\tilde{m}}_\Delta$, so that the corresponding eigenvalue is 1, which in terms of kurtosis indicates maximum bimodality [10].

Operator $\mathsf{K}_{\mathbb{X}}^{\tilde{\otimes}3}$. To proceed, we will assume that $\mathsf{E}(\sqrt{\cdot}) \approx \sqrt{\mathsf{E}(\cdot)}$. This is a common approximation in expectation calculations. Then, using (66.6) one arrives to

$$\mathsf{K}_{\mathbb{X}}^{\tilde{\otimes}3}(\tilde{\tilde{m}}_\Delta) = \Big\{\sum_j 15\cdot\pi_j^4\nu_j^2 + \alpha(1-\alpha)\Big(9\sum_j\Big(\sum_k \pi_k^2\nu_k^2\Big)\pi_j^4\nu_j^2 + \\ +3\sum_j\Big(\sum_k \pi_k^4\nu_k^4\Big)\pi_j^4\nu_j^2 + \sum_j\Big(\sum_k \pi_k^6\nu_k^6\Big)\pi_j^4\nu_j^2\Big)\Big\}^{1/2}\gamma_j = \Big\{15+\alpha(1-\alpha)\Big(9\sum_j \pi_j^2\nu_j^2 + \\ +3\sum_j \pi_j^4\nu_j^4 + \sum_j \pi_j^6\nu_j^6\Big)\Big\}^{1/2}\tilde{\tilde{m}}_\Delta.$$

Incorporating the coefficients into the square root ensures the operator scales $\tilde{\tilde{m}}_\Delta$ uniformly, preserving its structure and confirming that it remains a spectral component with a proportional scalar factor. That is, $\mathsf{K}_{\mathbb{X}}^{\tilde{\otimes}6}(\tilde{\tilde{m}}_\Delta) = \mathsf{K}_{\mathbb{X}}^{\tilde{\otimes}3}(\mathsf{K}_{\mathbb{X}}^{\tilde{\otimes}3}(\tilde{\tilde{m}}_\Delta)) = \kappa_0\mathsf{K}_{\mathbb{X}}^{\tilde{\otimes}3}(\tilde{\tilde{m}}_\Delta) = \kappa_0^2\tilde{\tilde{m}}$, thus we see that $\kappa_0 = \sqrt{\kappa}$ and $\mathsf{K}_{\mathbb{X}}^{\tilde{\otimes}3}(\tilde{\tilde{m}}_\Delta) = \sqrt{\kappa}\tilde{\tilde{m}}_\Delta$.

By analogous reasoning, the result extends to all higher-order even and odd moment operators. Thus, the proof is complete.

66.3.1 Empirical Validation of the Result

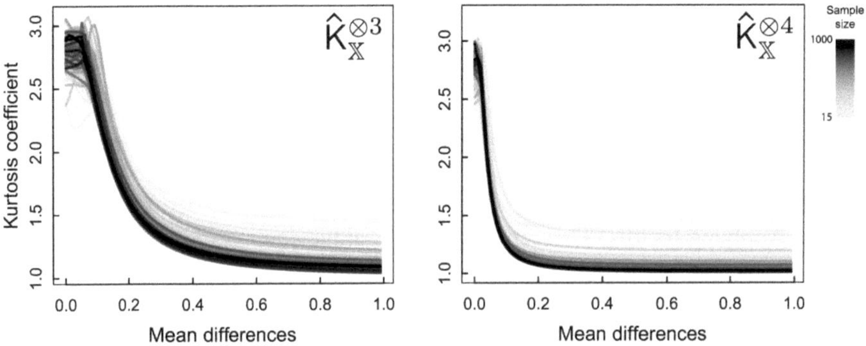

Fig. 66.1: Perturbation of the tail projectors evaluated using the kurtosis coefficient.

We use Simulation 1 in [10] as an illustration of our theoretical results (see case 1; bivariate homoscedastic Gaussian, refer to details therein). Vidal et al. [10] demonstrated that if the kurtosis coefficient of rank-1 score-projections of the data (or any transformations of it) approaches 1 as $n \to \infty$, the classification error tends to 0 (near-perfect asymptotic accuracy). We assess the sample operators $\hat{\mathsf{K}}_{\mathbb{X}}^{\otimes 3}$ and $\hat{\mathsf{K}}_{\mathbb{X}}^{\otimes 4}$ by examining the kurtosis of the projections of the whitened functional data $\mathbb{X}$ onto their respective last eigenfunction, using sample sizes $n = 15, 25, 35, \ldots, 1000$. We know that the last eigenfunction of these operators represents the boundary of the RKHS, as determined by Picard's criterion [10], making it the most susceptible to perturbations. Results in Fig. 66.1 show that as n increases and the mean differences become larger, the kurtosis coefficient for $\hat{\mathsf{K}}_{\mathbb{X}}^{\otimes 4}$ converges more rapidly to 1 compared to $\hat{\mathsf{K}}_{\mathbb{X}}^{\otimes 3}$, with both ultimately converging to the optimal discriminant solution.

66.4 Supplementary Definitions and Proofs

Definition 66.2. A function $f \in H$ is an eigenfunction of a self-adjoint operator A if $f \in \mathsf{Dom}(\mathsf{A})$, $\mathsf{A}(f) = \lambda f$ with $\lambda \in \mathbb{R}$, and $\langle f_j, f_k \rangle = 0$ for $\lambda_j \neq \lambda_k$.

Proof (Theorem 1). From [7], consider $\mathbb{X}_1(t) = \sum_j \left(\xi_{1,j} + (1-\alpha)\nu_j\right)\pi_j\gamma_j(t)$ and $\mathbb{X}_2(t) = \sum_j \left(\xi_{2,j} - \alpha\nu_j\right)\pi_j\gamma_j(t)$. Since X is Bochner-integrable, the expectation and the operator Ψ_W commute. This allows us to operate in the usual metric. Now, recall (66.3). To derive the moment-operator representation, first consider

$$\mathsf{K}_X^{\otimes 2}(u) = \alpha\mathsf{K}_{X_1}^{\otimes 2} + (\alpha-1)\mathsf{K}_{X_2}^{\otimes 2} = \alpha\mathsf{E}(\langle X_1, u\rangle_{L^2}X_1) + (\alpha-1)\mathsf{E}(\langle X_2, u\rangle_{L^2}X_2),$$

where $u = \sum_j u_j \gamma_j$. Disregarding the probabilities temporarily, one obtains:

$$\begin{aligned} \mathsf{K}^{\otimes 2}_{X_\ell}(u) &= \mathsf{E} \sum_j \Big(\xi_{\ell,j} + (-1)^{\ell+1} \nu_j\Big) u_j \sum_k \Big(\xi_{\ell,k} + (-1)^{\ell+1} \nu_k\Big) \gamma_k \\ &= \mathsf{E} \sum_{j,k} \left(\xi_{\ell j} u_j \xi_{\ell k} \gamma_k + \nu_j u_j \nu_k \gamma_k\right), \end{aligned} \tag{66.4}$$

for $\ell = 1, 2$. Using (66.4), we can construct the expression for $\mathsf{K}^{\tilde{\otimes} 4}_{X_\ell}(u)$ by computing the cross-products in $\{\mathsf{E} \sum_{j,k} (\xi_{\ell,j}\xi_{\ell,k} + \nu_j \nu_k) \sum_{m,n} (\xi_{\ell,m}\xi_{\ell,n} + \nu_m \nu_n)\} u$, and similarly for the sixth-order moment, whose expression is not included for the sake of brevity. Applying Isserlis' Theorem, one arrives at the common form operators

$$\mathsf{K}^{\tilde{\otimes} 4}_X(u) = \sum_j 3\lambda_j^2 u_j \gamma_j + 2 \sum_j \Big(\sum_k \lambda_k^{1/2} \nu_k u_k\Big) \lambda_j^{1/2} \nu_j \gamma_j + \sum_j \Big(\sum_k \nu_k^2 u_k\Big) \nu_j^2 \gamma_j; \tag{66.5}$$

$$\begin{aligned} \mathsf{K}^{\tilde{\otimes} 6}_X(u) = \sum_j 15\lambda_j^3 u_j \gamma_j + 9 \sum_j \Big(\sum_k \nu_k u_k\Big) \lambda_j^2 \nu_j \gamma_j + \\ + 3 \sum_j \Big(\sum_k \nu_k^2 u_k\Big) \lambda_j \nu_j^2 \gamma_j + \sum_j \Big(\sum_k \nu_k^3 u_k\Big) \nu_j^3 \gamma_j. \end{aligned} \tag{66.6}$$

Acknowledgements The author acknowledges the support from the FWO project G013024N and the Max-Planck-Gesellschaft.

References

1. Bernau, S.J.: The square root of a positive self-adjoint operator. J. Aust. Math., **8**(1), 17-36 (1968)
2. Berrendero, J.R., Cuevas, A., Torrecilla, J.L.: On the use of reproducing kernel Hilbert spaces in functional classification. J. Am. Stat. Assoc., **113**(523), 1210–1218 (2018).
3. Janson, S., Kaijser, S.: Higher moments of Banach space valued random variables. Memoirs of the American Mathematical Society, Providence, Rhode Island (2015)
4. Loperfido, N.: Skewness and the linear discriminant function. Stat. Probab. Lett., **83**(1), 93–99 (2013).
5. Oertel, F.: Beyond trace class – Tensor products of Hilbert spaces and operator ideals in quantum physics. arXiv preprint 2308.04627 (2024).
6. Peña, D., Prieto, F.J., Viladomat, J.: Eigenvectors of a kurtosis matrix as interesting directions to reveal cluster structure. J. Multivar. Anal., **101**(9), 1995–2007 (2010).
7. Peña, D., Prieto, F.J., Rendón, C.: Independent components techniques based on kurtosis for functional data analysis. In: DES - Working Papers. Statistics and Econometrics. WS 141006, Universidad Carlos III de Madrid (2014) Available via DIALOG.
8. Vidal, M., Rosso, M., Aguilera, A.M.: Bi-smoothed functional independent component analysis for EEG artifact removal. Mathematics, **9**(11), 1243 (2021).
9. Vidal, M., Aguilera, A.M.: Novel whitening approaches in functional settings. Stat, **12**(1), e516 (2023)
10. Vidal, M., Leman, M., Aguilera, A.M.: Functional independent component analysis by choice of norm: a framework for near-perfect classification. Adv. Data Anal. Classif. (2025).

Correction to: Bayes Hilbert space additive density-on-scalar regression based on individual observations

Eva-Maria Maier, Alexander Fottner, Almond Stöcker and Sonja Greven

Correction to:
Chapter 42 in G. Aneiros et al. (eds.), *New Trends in Functional Statistics and Related Fields*, Contributions to Statistics, https://doi.org/10.1007/978-3-031-92383-8_42

The chapter "Bayes Hilbert space additive density-on-scalar regression based on individual observations", written by Eva-Maria Maier, Alexander Fottner, Almond Stöcker and Sonja Greven, was inadvertently published without open access due to an oversight from publisher.

Due to this error, the copyright of the chapter has been changed to © The Author(s) 2025 and the chapter is now distributed under the terms of the Creative Commons Attribution 4.0 International License.

Open Access This chapter is licensed under the terms of the Creative Commons Attribution 4.0 International License (https://creativecommons.org/licenses/by/4.0/), which permits use, sharing, adaptation, distribution and reproduction in any medium or format, as long as you give appropriate credit to the original author(s) and the source, provide a link to the Creative Commons license and indicate if changes were made.

The images or other third party material in this chapter are included in the chapter's Creative Commons license, unless indicated otherwise in a credit line to the material. If material is not included in the chapter's Creative Commons license and your intended use is not permitted by statutory regulation or exceeds the permitted use, you will need to obtain permission directly from the copyright holder.

The updated version of this chapter can be found at
https://doi.org/10.1007/978-3-031-92383-8_42

© The Author(s) 2025

G. Aneiros et al. (eds.), *New Trends in Functional Statistics and Related Fields*,
Contributions to Statistics, https://doi.org/10.1007/978-3-031-92383-8_67

Authors Index

© The Author(s) 2025

G. Aneiros et al. (eds.), *New Trends in Functional Statistics and Related Fields*,
Contributions to Statistics, https://doi.org/10.1007/978-3-031-92383-8

MIX
Papier aus verantwortungsvollen Quellen
Paper from responsible sources
FSC® C105338
FSC
www.fsc.org

If you have any concerns about our products,
you can contact us on
ProductSafety@springernature.com

In case Publisher is established outside the EU,
the EU authorized representative is:
Springer Nature Customer Service Center GmbH
Europaplatz 3, 69115 Heidelberg, Germany

Printed by Libri Plureos GmbH
in Hamburg, Germany